Teacher, Student, and Parent
One-Stop Internet Resources

Log on to BSCSblue.com

Online Study Tools

- Check and Challenge Quizzes
- Interactive Tutor
- Chapter Test Practice
- Vocabulary PuzzleMaker
- Vocabulary e-Flashcards
- Multilingual Science Glossary

Online Research

- Pre-AP Resources
- WebQuest Projects
- Prescreened Web Links
- Career Links
- In the News

Interactive Online Student Edition

- Complete Interactive Student Edition
- Textbook Updates

For Teachers

- Teacher Bulletin Board
- Teaching Today—Professional Development

BLUE VERSION **ninth edition**

BSCS Biology

A Molecular Approach

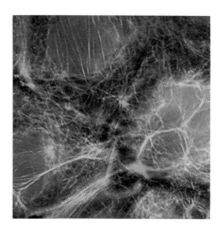

BSCS
5415 Mark Dabling Blvd.
Colorado Springs, CO 80918-3842

 **Glencoe**

New York, New York   Columbus, Ohio   Chicago, Illinois   Peoria, Illinois   Woodland Hills, California

BSCS Biology: A Molecular Approach, 9th Edition

BSCS Administrative Staff
Carlo Parravano, Chair, Board of Directors
Rodger W. Bybee, Executive Director
Janet Carlson Powell, Director and Chief Science Education Officer
Pamela Van Scotter, Director, The BSCS Center for Curriculum Development
Marcia Mitchell, Director of Finance

BSCS Project Staff
Project Director: Rita Stephens
Revision Coordinator: Dottie Watkins
Production Manager: Barbara Perrin
Production Staff: Stacey Luce, Lisa Rasmussen
Reviewers: Mark Bloom, Steve Getty, Jerry Phillips

Cover Image The cover image is a color-enhanced micrograph of pancreatic epithelial cells showing the distribution of DNA (blue), microtubules (green), and actin (purple) ($\times 2520$ at 70 mm).

 Glencoe

The *McGraw-Hill* Companies

Glencoe/McGraw-Hill
8787 Orion Place
Columbus, Ohio 43240-4027

ISBN 0-07-866427-6

5 6 7 8 9   058/111   12 11 10 09 08

FOREWORD

All around us, life on Earth is engaged in an ancient and delicate dance of interdependence. Although we humans are only recent partners in this performance, we are deeply involved because all life is related. We are also involved mentally because evolution has given us mental capacities that allow us to probe the complexity of life in ever-greater detail and—today—to manipulate living systems in ways that were science fiction a generation ago. And we are involved emotionally because biology engages the human spirit at very basic levels. We not only *know* that we are part of the living world, we *feel it*—in our love of animals and wilderness, in the pleasure we feel while in our gardens, and in our fascination with the primates that are our closest cousins.

These days it is virtually impossible to avoid exposure to the implications of modern biology. Television news programs announce the discovery of a gene that is associated with breast cancer; newspapers report on the use of DNA analysis in criminal cases; popular magazines discuss the most recent discoveries about human evolution; and celebrities warn about over-population, starvation, and threats to tropical rain forests. Although such issues may appear unrelated at first, they illustrate a rather small number of major principles that are common to all living systems. These principles help shape our study of biology, and they provide the foundation for this book:

- Evolution: patterns and products of change
- Genetic continuity and reproduction
- Energy, matter, and organization
- Science, technology, and society
- Interaction and interdependence
- Growth, development, and differentiation
- Maintenance of a dynamic equilibrium

BSCS Blue Version approaches these seven principles largely from the perspective of molecular biology and focuses on minute structures such as cells and genes, as well as on the processes related to them. It would be a mistake, however, to assume that one can understand life on Earth only by studying its smallest parts because all of these parts and their processes ultimately exert their effects in whole organisms, which in turn interact with their external environments. This book will introduce you to some of those interactions. Equally important, *BSCS Blue Version* will introduce you to the nature of science. It will require you to use the intellectual tools of inquiry that are common to all of science, which is a unique and powerful system for asking questions about the natural world.

As you work with your fellow students and your teacher to improve your understanding of biology, you will encounter a variety of intellectual challenges (and maybe even some frustration) about the intricacies of life on Earth. But we hope you also will find great rewards. For as you uncover some of the mysteries of living systems, you likely will develop an increased sense of wonder and respect for both life's stunning complexity and its elegant simplicity.

If we have done our job well, this book will leave you with an improved understanding of the myriad steps in life's dance of interdependence. In addition, you will acquire the insights necessary to make informed decisions about personal and social issues that have their roots in biology, and you will acquire the skills necessary to debate those issues.

We hope you, your fellow students, and your teacher will let us know whether we have accomplished our goals. We welcome your feedback, and we welcome you to the ninth edition of *BSCS Blue Version*.

BSCS Blue Version Revision Team
Jon Greenberg, Project Director

CONTENTS

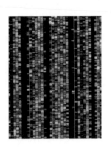

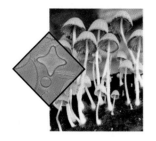
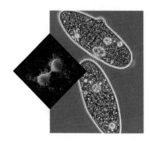

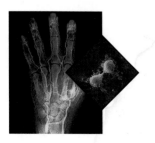

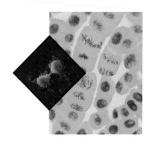
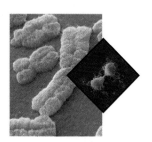

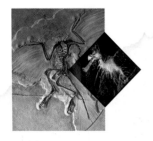

Chapter 24   Ecosystem Structure and Function 632

The Structure of Ecosystems—24.1 Abiotic Factors, 24.2 Energy in Food Webs, 24.3 Relationships in Ecosystems

Ecosystem Dynamics—24.4 Nutrient Cycles, 24.5 Limiting Factors, 24.6 Population Dynamics

Chapter 25   Change in Ecosystems 650

Ecosystems in Space—25.1 Terrestrial Biomes, 25.2 Aquatic Systems

Ecosystems in Time—25.3 Species on the Move, 25.4 Succession

Human Interactions with Ecosystems—25.5 Dependence, 25.6 Dominance, 25.7 Sustainability

Biological Challenges

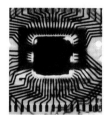

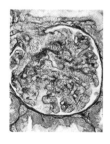

Appendices

xiii

ACKNOWLEDGMENTS

CONTRIBUTING WRITERS AND EDITORS

Robert Bouchard, Ph.D., College of Wooster, Wooster, OH (Chapter 7)

James Curtsinger, Ph.D., University of Minnesota, St. Paul, MN (Chapter 16)

Michael Dougherty, Ph.D., BSCS and Hampden-Sydney College (Prologue, Chapters 1–3 and 7)

Charles Ellis, Ph.D., Northeastern University, Boston, MA (Chapter 10)

Ellen Friedman, San Diego, CA (Chapters 3, 9, 12, 14, 19, and 20)

Francine Galko, Austin, TX (Chapter 22, Glossary)

Judy Hammett, Ph.D., Indiana University, Bloomington, IN (Theory Boxes)

S. B. Kater, Ph.D., University of Utah School of Medicine, Salt Lake City, UT (Chapter 21)

Joseph D. McInerney, BSCS and Johns Hopkins University (Chapter 17)

Jeffrey C. Murray, M.D., University of Iowa Hospital, Iowa City, IA (Chapter 15)

Jerry L. Phillips, Ph.D., BSCS, Colorado Springs, CO (Chapter 23, Investigations)

Mahendra Rao, MBBS, Ph.D., University of Utah School of Medicine, Salt Lake City, UT (Chapter 21)

Kodi Ravichandran, Ph.D., University of Virginia, Charlottesville, VA (Chapter 23)

Sam Stoler, Ph.D., University of Massachusetts, Amherst, MA (Chapter 8)

Richard Wright, Ph.D., South Hamilton, MA (Chapter 25)

REVIEWERS

Nathan Dubowsky, Ph.D., Westchester Community College, State University of New York, Valhalla, NY (Chapter 19)

Dorothy B. Engle, Ph.D., Xavier University, Cincinnati, OH (Chapters 6, 8, 10, 11, 13, 14, 15, 18, 21, 23, and 24)

Erica Goldman, University of Washington, Seattle, WA (Chapter 22)

Andre Jagendorf, Ph.D., Cornell University, Ithaca, NY (Chapter 4)

Kenneth E. Kinman, Hays, KS (Chapter 18)

Martin K. Nickels, Ph.D., Illinois State University, Normal, IL (Chapter 20)

Patsye Peebles, University Lab School, Louisiana State University, Baton Rouge, LA (Investigations)

William B. Provine, Ph.D., Cornell University, Ithaca, NY (Chapters 16, 19, and 20)

Tim L. Setter, Ph.D., Cornell University, Ithaca, NY (Chapter 11)

Randy Wayne, Ph.D., Cornell University, Ithaca, NY (Chapter 6)

Michael P. Yaffe, Ph.D., University of California, San Diego, La Jolla, CA (Chapter 5)

Julie Yetman, Midland, MI (Chapter 24)

Miriam Zolan, Ph.D., Indiana University, Bloomington, IN (Chapter 13)

ADDITIONAL CONTRIBUTORS AND CONSULTANTS

Ken Andrews, Ph.D., Colorado College, Colorado Springs, CO (Microbiology)

John Bannister-Marx, Camp Verde High School, Camp Verde, AZ (Investigations)

Lisa Chilberg, Graphic Consultant, BSCS (Brief Survey of Organisms)

David Desmarais, Ph.D., Ames Research Center, Moffett Field, CA (Chapter 17)

Ann Lanari, Research Assistant, BSCS, Colorado Springs, CO

REVIEWERS OF THE EIGHTH AND EARLIER EDITIONS

Steve Almond, Halliburton Services, Oxnard, CA; Aimee Bakken, University of Washington, Seattle, WA; Bob Barela, Bishop Manogue High School, Reno, NV; Warren Bennett, Hampton High School, Hampton, VA; Richard Benz, Wickliffe High School, Wickliffe, OH; Mary Ann Braus, Chadwick School, Palos Verdes Peninsula, CA; Lornie Bullerwell, Dedham High School, Dedham, MA; Jack Carter, Colorado College, Colorado Springs, CO; Ron Clarno, Sunset High School, Beaverton, OR; Donna Coffman, Colorado Springs, CO; Steve DeGusta, J. F. Kennedy High School, Sacramento, CA; Wilma M. Giol de Rivera, Colegio San Ignacio de Loyola, Puerto Rico; Kevin de Queiroz, University of California, Berkeley, CA; Raymond G. Edwards, Tamaqua Area High School, Tamaqua, PA; James Enderson, Colorado College, Colorado Springs, CO; Frank Fitch, Ben May Institute, Chicago, IL; Al Fruscione, Lexington High School, Lexington, MA; Suzanne S. Galando, Mt. Pleasant Area Senior High School, Mt. Pleasant, PA; Lotte R. Geller, The Roeper School for Gifted Children, Bloomfield Hills, MI; JoAnne Gray, Corliss High School, Chicago, IL; Barbara Grosz, Pine Crest School, Ft. Lauderdale, FL; Edward A. J. Hall, Catonsville High School, Catonsville, MD; Robert C. Heck, Moniteau High School, West Sunbury, PA; Bobbie S. Hinson, Providence Day School, Charlotte, NC; Michael Hoffman, Colorado College, Colorado Springs, CO; Dick Howick, Belmont High School, Belmont, MA; Marlene H. Jacoby, Charlotte Country Day School, Charlotte, NC; Duane Jeffery, Brigham Young University, Provo, UT; Marian S. Johnson, River Dell Regional Schools, Oradell, NJ; Eric E. Julien, Turlock High School, Turlock, CA; Robert Jurmain, San Jose State University, San Jose, CA; John C. Kay, Iolani School, Honolulu, HI; Donald W. Lamb, Manning High School, Manning, IA; Barbara Lester, Ransom Everglades School, Coconut Grove, FL; Stephen R. Lilley, J. E. B. Stuart High School, Falls Church, VA; Daniel R. Lipinski, Bishop Eustace Preparatory School, Pensauken, NJ; Lynn Margulis, University of Massachusetts, Amherst, MA; Robert Martin, Colorado State University, Fort Collins, CO; Robert T. Mills, P.S.B.G.M., Montreal, Quebec; Janice Morrison, Hathaway Brown School, Shaker Heights, OH; Jeff Murray, University of Iowa Hospitals, Iowa City, IA; Joseph D. Novak, Cornell University, Ithaca, NY; Larry Ochs, Norwich Free Academy, Norwich, CT; Gordon Peterson, San Marino High School, San Marino, CA; Annette Prioli-Lee, St. Bernard High School, Playa Del Ray, CA; Robert L. Ragley, Beachwood High School, Beachwood, OH; Ken Rainis, Ward's Natural Science Establishment, Rochester, NY; R. Ward Rhees, Brigham Young University, Provo, UT; John A. Rhodes, Avery Coonley School, Downers Grove, IL; Ailene Rogers, National Cathedral School, Mt. St. Albans, Washington, DC; Parker Small, University of Florida, Gainesville, FL; William G. Smith, Moorestown Friends School, Moorestown, NJ; Richard D. Storey, Colorado College, Colorado Springs, CO; Douglas Swartzendruber, University of Colorado, Colorado Springs, CO; Jean Paul Thibault, Cape Elizabeth High School, Cape Elizabeth, ME; Kent VanDeGraaff, Brigham Young University, Provo, UT; Alex Vargo, Colorado College, Colorado Springs, CO; William H. Wagstaff, Mead Senior High School, Spokane, WA; Bruce Wallace, Virginia Tech, Blacksburg, VA

CONTENTS

LEARNING OUTCOMES

By the end of this chapter you will be able to:

A   Discuss the importance of biology and biotechnology in everyday life.

B   Relate the problem-solving methods of science to the development of a theory of
evolution.

C   Summarize Darwin's theory of evolution by natural selection.

D   Recognize the differences between pseudoscience and true science.

Biology and the Molecular Perspective

■ *What could a sample of your DNA reveal about you?*

■ *How might the use of this technology affect your life?*

A thousand years ago, ideas about health and nature were characterized by tradition and magic, not by skepticism and experimentation. For example, many people believed that disease was the result of evil forces. Illness often was attributed to "bad blood," not germs, and bleeding a person was a common therapy. The cells and molecules that cause disease were unknown. Similarly, the effects of environmental changes on the living world were not well understood.

Science provides a structure for studying the world in a way that explains natural phenomena. Unlike beliefs about evil forces, scientific explanations can be tested. The answers to complex questions are not obvious, but we have made dramatic progress. When your grandparents were in high school, the structure of DNA was unknown. Your parents went to school when animal cloning was still science fiction. Now, genetically modified organisms manufacture drugs, and the complete genetic sequences of many organisms are known. What will the science of the third millennium be like?

This chapter introduces biology, the study of life. Biology unites scientific methods with technology to search for answers to fundamental questions about the living world.

The New Biology

P.1 Biology in Your World

Many scholars believe that biology will be the most influential science of the 21st century. When you consider the prevalence of biology in the news, that prediction seems reasonable. Nearly every day there is a story about medicine or disease, genetic engineering, nutrition, or environmental pollution (Figure P.1). There also are many topics that affect your life every day. How do antibiotics work? Does a family history of cancer guarantee that you will get cancer? How much do genes affect your behavior? What is cloning?

Answers to these and similar questions lie at the levels of cells and molecules. Understanding the molecular basis of life is the focus of this book. Instead of simply describing whole organisms (living things), you will learn how genes, molecules, and cells make organisms function. For example, scientists now understand the functions of many of the genes in viruses. That knowledge is helping medical researchers design new vaccines.

The biology of the new millennium will confront you with questions and choices that your parents never imagined. Should a person's genetic sequence be available to insurance companies? Is it ethical to engineer cells genetically that can be passed to future generations? Who will resolve the social issues that stem from new technologies? Without an understanding of how living systems function and how they are interrelated, it is impossible to make intelligent decisions about issues that affect your life and the lives of those around you. This book will help you acquire a basic knowledge of biology, which, in turn, will enable you to understand biological issues and to make informed judgments in the future.

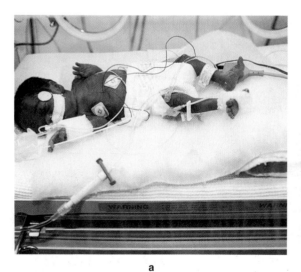
a

b

c

FIGURE P.1

Biology in daily life. Examples include new techniques for the care of premature infants **(a),** the benefits of staying in shape **(b),** and the hazards of water pollution **(c).**

P.2 A Biological View of AIDS

AIDS (acquired immune deficiency syndrome) is an example of how biology can shed light on problems of global importance. AIDS is caused by a virus that infects a type of cell in the immune system, which normally helps fight infection. As the virus destroys these cells, the immune response gradually is destroyed as well. Infections that would be minor in a healthy person can be fatal in a person with AIDS. Because an immune system weakened by the virus provides the opportunity for these infections, they are called "opportunistic infections."

The virus responsible for AIDS was identified in 1983 and later named human immunodeficiency virus, or HIV (Figure P.2). Although HIV is a major concern of health officials in the United States, more than 95% of all HIV-infected people live in developing countries.

The genetic material of the virus is copied inside infected cells and then becomes part of the cells' own genetic material. As a result, infected cells lose their ability to function normally and become "factories," producing more HIV. Each infected cell releases many new viruses, each capable of infecting other cells. Infected cells also may merge with nearby uninfected cells to form a large mass. This mode of spreading between cells worries researchers, because it means that drugs or vaccines that cannot enter the cells themselves will not prevent the spread of infection.

Because HIV affects the body at the cellular and molecular levels, biologists use their knowledge of the functions of molecules and cells within the immune system to develop treatments for AIDS. One of the first drugs used to treat people with AIDS was AZT. AZT interferes with the reproduction of HIV. Today, physicians prescribe a combination of drugs that interrupt different processes involved in HIV infection. Combination therapy is more effective at keeping infection under control than AZT alone. Interestingly, some people who have been repeatedly exposed to HIV never become infected. Scientists are studying a gene that appears to protect those individuals. Despite these advances, many HIV-infected individuals eventually develop an infection that physicians cannot treat, and they die. To date, an estimated 25 million people worldwide have died from the AIDS epidemic and another 40 million people are infected with the virus.

Because treating infection is so difficult, many biologists are focusing, instead, on developing a vaccine that would prevent infection. One approach to vaccine development uses genetic engineering to insert parts of the genetic material from HIV into the genetic material of harmless viruses. The modified virus might then be injected into the body, triggering an immune reaction against the HIV proteins. Then, if the vaccinated individual is exposed to HIV, the immune system should recognize the virus and destroy it before it can cause harm.

The genetic material of HIV mutates, or changes, rapidly. Therefore, a vaccine that is effective against one strain, or form, of HIV may not be effective against another strain. The virus evolves so rapidly that different strains develop within the course of infection in a person. Today, after testing more than 15 different vaccines, scientists have not yet discovered one that is safe and effective. Scientists hope, however, that a better

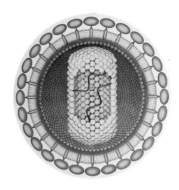

FIGURE P.2

A drawing of HIV. About 0.0001 mm in diameter, it is covered by a two-layer membrane (blue) studded with proteins (green). Other proteins, together with the genetic material (brown and ivory), compose the core.

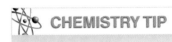

Proteins are some of the fundamental molecules of life.

CONNECTIONS

Science and society are inseparable. AIDS is a social problem that science can help solve. Society must decide how to educate people about AIDS.

understanding of the molecular basis of infection will lead to a protective vaccine. Until then, avoiding exposure to the virus through protective behavior is the most effective way to avoid becoming infected.

P.3 Growth Hormone: New Solution, New Problem

What do you consider "short" (Figure P.3)? Is a person short if *he* is 5 feet 2 inches (about 157 cm) tall? If *she* is 5 feet 2 inches tall? Ideas about "normal" height vary widely. If there were a way to make each of us as tall as we wanted to be, would that be a wise or foolish use of biological knowledge?

Treatment is now available for a disorder called growth hormone deficiency. Human growth hormone (GH), a protein that helps to control growth, is made by a small gland in the brain called the pituitary. Most people produce enough growth hormone to attain a "normal" height. Children with growth hormone deficiency, however, do not produce enough GH. They cannot reach a height in the normal range unless they are given extra hormone. However, GH treatment does not increase the final adult height of all children who are short as a result of heredity. The search for a solution to growth hormone deficiency must begin at the level of molecules because the deficiency involves the interaction of molecules in cells.

In the past, children with the deficiency were treated with GH isolates from the brains of people who had died. Many pituitary glands were needed to obtain even small amounts of GH; therefore, the procedure was expensive. More recently, this procedure was found to be dangerous. Some of the pituitary glands contained a deadly pathogen that was transmitted to the children along with the hormone.

Genetic engineering now enables biologists to use bacteria to produce GH. Human DNA, the genetic material, is inserted into bacteria. The bacteria then can manufacture human GH. GH produced by bacteria is good news for children with growth hormone deficiency, but it has raised some troubling questions. For example, should GH be prescribed for normal teenagers who merely wish to add a few inches of height to gain an advantage in athletics? GH can have some serious side effects. If doctors do not make the hormone available, will aspiring athletes obtain it illegally and use it without supervision?

The dilemma of how to use GH is an example of the issues raised by new technologies. The fact that all organisms share the same genetic material—DNA—has led to startling breakthroughs for health and medicine. New technologies may improve our ability to fight disease, solve environmental problems, and increase life expectancy; but they also create new, often difficult choices for society.

FIGURE P.3

Athletes of average and above average height. Ideas of what constitutes "normal" height can vary for many reasons. A professional basketball player 6 feet 11 inches tall is considered normal, while a player 5 feet 7 inches tall is considered short.

CHEMISTRY TIP

DNA (deoxyribonucleic acid) is a nucleic acid, a molecule necessary for life.

Try Investigation PA   **Analyzing Ethical Issues.**

Biological Challenges

Biology, Ethics, and Public Policy

An improved understanding of biology and an ability to manipulate some basic life processes raise many new and troubling issues. For example, medical technology can save many severely handicapped infants who might have died had they been born only 15 years ago. Should attempts be made to save such babies? Should fetuses found to have a genetic disorder or a birth defect be aborted? On whom shall we test new vaccines against AIDS, especially if some small risk exists that the vaccine will cause AIDS? Technology now allows us to identify and test for certain genes that cause disease. Is it beneficial to patients to diagnose a condition for which there is no known treatment? (See Figure P.4.)

These issues involve a mixture of science and ethics—the study of right and wrong actions. Many universities now train people in bioethics, the application of ethics to biological issues. Bioethicists work in hospitals and research institutions.

Science can be very helpful in telling us what we *can* do and in predicting the effects of each alternative. Science, however, cannot tell us what we *should* do. Although evidence might point toward one alternative, the final decision often is a matter of values. Those values are represented in public policy—laws and regulations that govern how science is applied.

Public policy should be based on sound ethical judgments, but that is not always enough to make policies workable. A policy may be ethically sound, but if most people do not agree with it or if it cannot be enforced, it probably will not work. Many of the issues in this book are so new that there are no

national policies to govern them, and there still is sharp ethical disagreement about many of them. Each person has a responsibility to understand and contribute to debates about the ethical and policy aspects of advances in science and technology. The first step is understanding the science involved. This book will help you do that. The second step is to analyze bioethical issues. Investigation PA is a model that will help with the second step.

FIGURE P.4

Young people with unknown risks of developing cancer.
Several genes appear to be involved in inherited forms of breast cancer. Once a test to identify carriers is available, should these young women be tested, even though there is no cure for breast cancer yet?

Check and Challenge

1. Why is knowledge of biology important today?
2. How do technology and biology affect each other?
3. Why is knowledge of cells and molecules important in understanding problems like AIDS and growth hormone deficiency?
4. How are moral and ethical issues involved in modern biology?

The Methods of Science

P.4 Solving Problems

Everyone makes decisions every day. Most of the decisions people make are based on what they *feel* will be the best solution. Feelings and judgments of how others feel play a major role in how you choose to resolve your day-to-day problems. Effective problem solving, however, goes beyond feelings. To solve a problem, you must also combine what you already know with new observations. Then, based on your knowledge of the situation, you can evaluate the problem and formulate the best plan to solve it.

Imagine that it's the weekend and you are to meet your friends at the movies at noon. Suddenly you notice it's 11:45. You race to get ready and then jump in the car. What route is the fastest? You choose the best route based on what you know about the distance, the amount of traffic on the roads at that time, and the number of stoplights you are likely to encounter. If a lot of other drivers are using the same route, it might be faster to use an alternative route. In essence, you make observations and combine those observations with your best guesses to arrive at a possible solution. If you are late, next time you may try another route. Eventually, through trial and error, you find the fastest route.

In science, problem solving is based on the interpretation of **data,** which is information gained through observation. For example, you could time every possible route to the theater twice—once in heavy traffic, and once with light traffic. After several more trials, you could then interpret the results. If one route was consistently faster than others, you would have solved the problem by using scientific methods.

Solving problems with scientific methods can help you determine what is reasonable and what is not. It can help you evaluate claims made by others about products or events. It can help explain how organisms function and how they interact in the environment.

The easiest way for you to become familiar with these methods is to use them yourself (as you frequently will in this course) and to see how other people use them. One good example of scientific problem solving involves the topic of evolution. Evolution is one of the major unifying themes of

Try Investigation PB ▽ Scientific Observation.

CONNECTIONS

Science as inquiry—the understanding that all is not known and that concepts must be revised and restructured as new data become available—pervades all scientific methods.

Biology Online BSCSblue.com/check_challenge

a

b

c

d

biology. How widely does evolution apply, and how does it occur? What data provide evidence of evolution? In the following sections, you will see how scientific methods were used to develop a theory that explains how organisms change through time.

P.5 A Mechanism for Evolution: Science at Work

The world is filled with beautiful and intriguing organisms. It also contains a fossil record of equally amazing extinct forms of life. Observe the pictures of organisms in Figure P.5. Which organisms are similar and in what ways? Which organisms are different and in what ways? More important, how did these diverse forms arise?

By the beginning of the nineteenth century, biologists had observed that living organisms were different from the fossil organisms found in rocks. These observers developed a theory—evolution—that the organisms of the past had given rise to the organisms of the present, and that organisms had changed through time. A **theory** explains current observations and predicts new observations—in this case, observations about how organisms change through time. The theory of evolution, for example, predicts that there should be observable differences between modern organisms and fossils found in rocks, as well as observable similarities. Evidence from thousands

FIGURE P.5

Similarities and differences between living animals and their extinct relatives. What are the similarities between Canada geese, *Branta canadensis* **(a)**, and the fossil *Archaeopteryx* **(b)**? How are they different? Some animals look alike yet are dissimilar, such as the mountain lion, *Felis concolor* **(c)**, and the extinct saber-toothed cat *Smilodon* **(d)**.

of observations of modern and fossilized organisms from all over the world supports the theory of evolution. (For more on theories, see "Theory in Science" in Section P.6.)

How does evolution occur? A French biologist, Jean Baptiste Lamarck (1744–1829), was one of the first to propose that organisms change through time. He proposed that a change in the environment produces a *need* for change in animals. Thus, if an animal needs to use one part of its body frequently, that part will become stronger and more well-developed. Conversely, if an animal uses some part of its body infrequently, that part will slowly weaken, become smaller, and may disappear. Lamarck assumed that these acquired characteristics would be passed on to the offspring in that changed form.

As with other scientific ideas, Lamarck's ideas could be evaluated through a series of **hypotheses**—explanations that are testable through experimentation or observation. One hypothesis based on the theory of evolution could be stated as follows: Ancient organisms have given rise to modern organisms by evolution.

Predictions that follow from hypotheses can be stated in an *If . . . then* format. One prediction based on Lamarck's explanation of evolution through inheritance of acquired characteristics could be stated as follows: *If* a male and a female increase the size of their muscles through weight training, *then* their children will be born with large muscles. We know from observations that this is not the case. Therefore, the hypothesis must be rejected.

Lamarck's explanation does not do a very good job of predicting observations about the changes of organisms through time. Extensive experimentation for more than 100 years has failed to show that acquired characteristics are inherited.

The British naturalist Charles Darwin, who lived from 1809 to 1882, developed a theory of evolution that scientists still use. Many things influenced Darwin during his life. One was his experience as a naturalist on the five-year voyage of the *Beagle* (Figure P.6) during which he observed

a

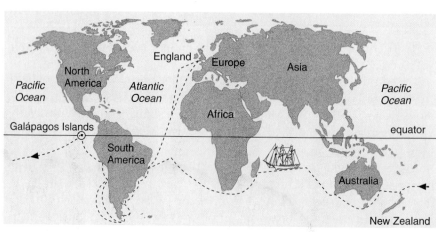

b

FIGURE P.6

Darwin and the route of his exploration. Young Charles Darwin (a) sailed around the world on the *Beagle*. The route taken (b) included stops in South America and the Galápagos Islands. During the trip, Darwin found evidence that would later support the theory of evolution by natural selection.

a   **b**

FIGURE P.7
Two very different examples of protective adaptations. How are the red and blue arrow-poison frog, *Dendrobates pumilio* **(a)**, and the winter flounder, *Pseudopleuronectes americanus* **(b)**, well adapted?

unique and diverse organisms from around the world. Another influence was a book written by Darwin's friend, the geologist Charles Lyell (1797–1875). Lyell promoted a hypothesis, first developed by a Scot, James Hutton, that natural forces existing in the past were the same as those that exist today. This view, called **uniformitarianism,** maintains that geological forces produced changes on Earth in the past, and it predicts that those same forces will continue to produce changes in the future.

Darwin raised the following questions: If Earth has had a long history of change, what was it like before now? Could it have supported the diversity of life it has now? What other forms of life might have lived before now? Darwin reviewed existing data and made numerous observations around the world as he formulated his answers to these questions.

E T Y M O L O G Y

uni- = one (Latin)
formis = form (Latin)
Forces in the past and present were **uniform.**

P.6 The Theory of Natural Selection

On the basis of his extensive research, Darwin published a book in 1859 titled *The Origin of Species.* Darwin's theory stated that new forms of life are produced by means of **natural selection,** the survival and reproduction of organisms that are best suited to their environment (Figure P.7). Natural selection occurs because some members of a population or species have physical or behavioral characteristics that enable them to survive and to produce more offspring than others. (A **species** is a group of similar organisms that naturally reproduce with one another—see Figure P.8.)

a   **b**

FIGURE P.8
Different species. Even though two animals may look similar, they are not necessarily closely related. The lion, *Panthera leo* **(a)**, and the leopard, *Panthera pardus* **(b)**, are classified as different species, even though they closely resemble one another.

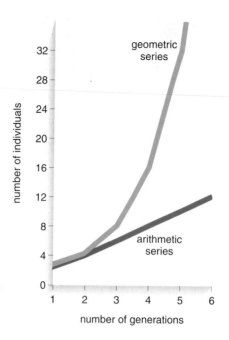

FIGURE P.9

Arithmetic and geometric growth. In a geometric series, each number is the product of a constant factor and the preceding number, for example, $2 \times 4 = 8$, $2 \times 8 = 16$, and so forth. This causes rapid growth. Arithmetic growth is linear ($2 + 2 = 4$, $2 + 4 = 6$, and so forth).

Because offspring tend to inherit the characteristics of their parents, certain characteristics become more common in succeeding generations. Characteristics that increase the chances that an organism will survive and reproduce in its particular environment are called **adaptations.**

Darwin based his theory on several major observations, including the work of other scholars such as the economist Thomas Malthus. Malthus proposed that the number of organisms in any species or population tends to increase from generation to generation in a geometric progression, whereas the food supply increases in an arithmetic progression (Figure P.9). If humans, for example, continued to reproduce at the same rate, they would eventually outstrip the food supply.

Malthus's writings provided Darwin with an important idea: Not all organisms live long enough to reproduce because of limited resources, such as food, water, and other environmental factors. This difference in survival among members of populations has kept those populations in check. Darwin realized that competition for survival could be a powerful force in the evolution of species. Some organisms survive this competition; others do not.

Darwin also observed that naturally occurring **variations,** or small differences (Figure P.10), occur within populations. On the Galápagos Islands, which Darwin visited during his travels, some members of the same finch species have short, thick beaks while others have longer, thinner beaks. On the basis of these observations, Darwin concluded that some variations would help members of a species survive in a particular

FIGURE P.10

A population of ladybird beetles, ×2. How many differences can you see within this population? What is the range in number of spots present? What important characteristic of populations does this illustrate? What is responsible for these differences?

environment (Figure P.11), whereas other variations would not be helpful. His theory proposes that natural selection tends to eliminate those organisms with variations that are not advantageous. For example, scientists have observed that during drought Galápagos finches with long, thin beaks tend to be at a disadvantage because they cannot crack the tough seeds that are plentiful under those conditions.

FIGURE P.11

A variation that aids adaptation. Notice the shape of the beak on this finch. Under what environmental conditions might a long, thin beak be advantageous?

b

a

FIGURE P.12

Evolution of a new species. (a) Cotton bollworms (*Helicoverpa zea,* ×2) and **(b)** tobacco budworms (*Helicoverpa armigera,* ×2) are destructive agricultural pests that once were the same species. Differences evolved in the structure of their sex organs, however, that now prevent mating between the two groups.

Try Investigation PC ◇ The Compound Microscope.

Darwin's theory suggests a number of predictions. For example, *if* organisms with favorable variations are most likely to survive and reproduce, *then* those organisms with unfavorable variations would be less successful at reproduction and would, therefore, die out. Because variations can be inherited, the favorable variations—adaptations—would accumulate through time. Further, *if* organisms with those favorable variations become so different from members of the original species that they can no longer reproduce together, *then* a new species may have evolved (Figure P.12).

Extensive fieldwork and experimentation have supported Darwin's theory of natural selection as a mechanism for evolution. For example, scientists studying fossils predicted that they would find fossils that represented intermediate forms between living organisms and fossils discovered in the older layers of Earth. This prediction has been confirmed in an overwhelming number of cases by scientists all over the world who have discovered fossils of organisms such as birdlike reptiles and dog-sized horses.

Lamarck and Darwin proposed different theories to explain the same observation: Organisms change through time. Darwin's theory of evolution by natural selection has prevailed because it explains existing observations and predicts future observations.

The theory also explains the similarities between humans and other animals, which has some practical consequences. For example, animals such as rats and monkeys share certain features with other animals, including humans. Thus, medical scientists can often apply what they learn from lab animals to humans.

THEORY

Theory in Science

What do most people mean when they say they have a theory about something? Perhaps you have a theory about why the football team lost its first game. When scientists use the word *theory*, they do not mean guess or idea. Instead, a theory is a comprehensive explanation of facts, laws, and reasoning that is supported by many observations and multiple lines of evidence. Theories are accepted by the scientific community.

Theories are important in science because they help organize vast numbers of facts and observations generated by experimentation. In the case of evolution, Darwin built a framework for explaining the relationships that he, and many naturalists before him, observed among living and extinct organisms. Descent with modification and natural selection are two theories encompassed by the broader theory of evolution, a phrase that Darwin did not use.

Because theories lead to logical hypotheses, they allow scientists to make and test predictions. This is an important feature of science. Darwin's theory of **descent with modification,** which means that related organisms share a common ancestor, allows predictions that can be tested. For example, chimpanzees and gorillas share a great number of physical characteristics, but they also show some differences. Descent with modification predicts that the greater the similarity between two groups of organisms, the closer is their relationship. An examination of the DNA of chimps and gorillas suggests that both evolved from a common ancestor that lived about 12 million years ago. Chimps and baboons, which share fewer characteristics, are less closely related. DNA evidence suggests a common ancestor about 25 million years ago (see figure to the right).

The theory of natural selection leads to the prediction that changes in a species' environment will tend to favor organisms with certain characteristics or adaptations. For example, around 1900, a biologist named H. L. Bolley used natural selection to predict changes in the characteristics of flax plants. He reasoned that, if he collected seeds from the occasional flax plants that did not die in the presence of a certain fungus, then he could mate those plants and eventually produce a population of flax plants resistant to the fungus. Bolley was correct, and the flax farmers benefited from his application of the theory of evolution.

Theories such as evolution theory, atomic theory, and the germ theory of disease have been supported repeatedly by data resulting from hypotheses. In addition, virtually no opposing data have yet been found, further strengthening these theories. Any theory, however, can be changed or even discarded if new experiments and observations do not support the model. All theories mentioned here have been modified as new research has been done. And undoubtedly these theories will be modified in the future. This continuous modification is another example of the way science works.

A chimpanzee, a gorilla, and a baboon. Chimpanzees **(a)** and gorillas **(b)** are more closely related to each other than to baboons **(c),** but all three organisms share many characteristics. What physical characteristics support the existence in the past of a common ancestor for all three?

Check and Challenge

1. Describe how evidence and hypotheses are characteristics of science.
2. How does a hypothesis differ from a theory? Give a concrete example of each.
3. Describe the theory of evolution by means of natural selection.
4. Describe how natural selection operates on species.
5. How do Lamarck's theory and Darwin's theory differ?

Science as a Way of Knowing

P.7 Scientific Perspectives

There are many ways to explain natural phenomena. For example, history, religion, art, philosophy, and sociology all provide ways for people to examine the world. In addition, they all contribute to today's body of knowledge.

Science is one way of explaining the natural world. Several characteristics generally define science.

1. Science is based on the assumption that the natural world can be investigated and explained in terms we can understand.
2. Science is based on the results of observations and controlled experiments.
3. The results of these observations and experiments must be (at least in principle) repeatable and verifiable by other scientists.
4. The findings of science must be refutable. In other words, if a hypothesis is not supported by evidence and observations, then the hypothesis must be rejected or modified. The same principle holds true for theories.

Methods of explaining natural phenomena that do not share these characteristics do not qualify as science.

Nonscientific ways of explaining natural phenomena are valuable as well, but they should not be confused with science. For example, poetry conveys metaphorical messages that can be quite powerful, and the exact meaning of a poet's words must be interpreted by each individual. Science, on the other hand, is more literal. In science, words are chosen to convey as precise and clear a meaning as possible.

Science is a human enterprise, so it can be influenced in some ways by personal biases and by politics. Consider the great astronomers Copernicus and Galileo. Copernicus (Figure P.13a) developed a theory that Earth is one

 BSCSblue.com/check_challenge

| a | b |

FIGURE P.13

Two great astronomers. (a) Mikolaj Kopernik (1473–1543), a Polish astronomer, like other European scholars of his time, wrote under the Latin form of his name, Nicolaus Copernicus. Copernicus claimed that the Sun was at the center of the universe—not Earth, as was believed. Earth revolves around the Sun, not vice versa. Although he "proved" this was so, Copernicus died without his ideas being accepted. **(b)** Galileo Galilei (1564–1642), an Italian mathematician and astronomer, made observations with an improved telescope that supported Copernicus's hypothesis.

of several planets orbiting around the Sun. Galileo (Figure P.13b) observed moons orbiting around the planet Jupiter. Their conclusions challenged the authority of the church, which interpreted the Scriptures at that time to say that the Sun moved around Earth, which was the center of the universe. Both scientists were chastised by the church. Galileo was forced to sign and publish a statement saying that his work was incorrect, and he was then sentenced to permanent house arrest.

No matter what Galileo signed, it did not change the fact that Earth is not the center of the solar system. Rejecting science does not change the science—it merely prevents people from learning and understanding what it says.

P.8 Your Role as a Biologist

Science is a method of answering questions and explaining natural phenomena. You have already encountered a few of the many questions biologists ask when they study living things. Throughout this course, you will be asked to answer such questions as, To what environment is this organism adapted? What is being released into my environment that could be hazardous to my health? How does a change in DNA affect the proteins of an organism?

Focus On

Pseudoscience

Pseudoscience means, literally, false science. Some investigators claim their research is scientific, but it does not meet the definition of science. Such work can be classified as pseudoscience. Some examples of pseudoscience include astrology, "miracle cures" for diseases such as cancer and arthritis, homeopathy, and some dieting programs and health practices. A great deal of time, effort, and money can be saved by learning to evaluate pseudoscientific claims carefully (Figure P.14). The following example shows the consequences of confusing science and pseudoscience.

In the 1980s, the states of Arkansas and Louisiana enacted laws requiring science teachers to devote equal time in class to teaching evolution and creationism, or "creation science." The writers of the Arkansas law defined creationism as involving the creation of all living things in six 24-hour days by a supreme being, as in the literal interpretation of the first chapter of Genesis in the Bible. The law excluded other creation accounts. Both states' laws were struck down by the courts because requiring the teaching of creationism in public schools would establish the teaching of a particular type of religion, thus violating the U.S. Constitution's first amendment, which requires separation of church and state.

"Creation science," also called intelligent design, is not science because it does not follow scientific methods. The claim that a deity created the world cannot be tested. Thus, it is not a hypothesis. Furthermore, some creationists resist modifying their model even when observations fail to support it. The idea of creation by a supreme being is a matter of faith, not of science. This does not mean that creationism is wrong, only that it is not science.

The theory of evolution by natural selection says nothing about the existence of a supreme being. Some people believe that Charles Darwin must have been an atheist to propose his theory. Darwin's book, *The Origin of Species* (second and subsequent editions), refutes this charge. Darwin recognized that the question of a deity is a religious, and not a scientific one. Deities cannot be investigated scientifically, and, therefore, are outside the realm of science.

FIGURE P.14

Alternative medicine. Unproven "natural medicinal herbs," "healing" crystals, and "aromatherapy" oils. Extraordinary claims should be viewed with skepticism. Is it likely that the makers of these products are adhering to the principles of science?

ETYMOLOGY

pseudo- = false or sham (Greek)

As you deal with biological issues in this course, and throughout your life, ask at least five basic questions.

1. **What is the question?** Identify the critical question and find out what it really means before you attempt to answer it.

2. **What are the data and how were they obtained?** Determine what the facts are before you attempt to make a decision. Learn to separate data from opinions. Be sure that the data were gathered in a scientific manner and can be validated.

3. **What do the data mean?** Determine if the data support the arguments that are being offered on several sides of an issue. Again, it is important to separate facts from opinions.

4. **Who is reporting the data?** Not all people are reliable sources for data. Determine the credentials of persons who claim to be experts on an

issue. Is this person reporting as a scientist or as someone with personal interests? Learn to question everyone but especially those who do not have the background to know the relevant data and understand the scientific aspects of problems you will be investigating. Most important, remember that even experts can make mistakes. To reduce mistakes, scientists routinely participate in teams to evaluate and eliminate possible sources of bias in the design of investigations and data analysis. Their findings are often subjected to peer-reviewed scientific journals so that others may evaluate and analyze the findings.

5. How complete is the present state of knowledge? Is our knowledge about a subject sufficient to answer the question?

These five questions also can help you connect new information to what you already know. Such connections can increase your understanding of basic concepts in biology and other areas. Establishing these connections can help you detect errors in logic or fact—a crucial ability when you must make important decisions.

Try Investigation PD ◇ Developing Concept Maps.

For example, discoveries about the chemical substances that make up living things help scientists connect theories about evolution and heredity with their understanding of atoms and molecules. Experiments have revealed that inherited traits are encoded in the chemical structure of DNA molecules. Scientists and engineers have used their knowledge of DNA to make GeneChip® DNA microarrays, such as the one in the photo that opened this chapter. The color of each square in the microarray indicates whether the DNA attached there matches part of the DNA being tested. This medically useful technology depends on a chemical theory of heredity based on the structure of DNA. A continuous web of theories and experimental results that support them connects this theory with scientific explanations of heredity and with Darwin's evolving theory of evolution.

In addition to the science you will learn in this course, your own ethics, feelings, and values also will affect every decision you make. As you study biology this year and learn about the molecular basis of life, you will be able to examine closely your positions on some controversial issues. As you do so, you will be developing skills that will serve you for the rest of your life—and in many areas, not just in the study of biology.

Check and Challenge

1. Describe the characteristics of science.
2. What is pseudoscience, and how does it differ from science?
3. Why is creationism considered pseudoscience?

Chapter
HIGHLIGHTS

Summary

Knowledge of biology increases every day and presents everyone with new choices. An understanding of biology at the cellular and molecular level is crucial if we are to make educated and informed decisions about biological issues that affect us individually and as a society. These issues include AIDS and the appropriate uses of human growth hormone, now easily available through the techniques of genetic engineering.

By using scientific methods, you can discover new ways to examine life around you. Darwin used scientific methods to develop the theory of evolution by means of natural selection. He made observations and assumptions and developed a theory to explain them. His theory has withstood years of testing and serves as a model for explaining and predicting observations in biology.

Science is one of many different and valid ways to look at the world. The characteristics of science determine what qualifies as science. Science is based on the assumption that the natural world can be investigated and explained. Pseudoscience is false science. One example of pseudoscience is creationism; it is largely a matter of faith, and its assumptions cannot be investigated or rejected by the methods of science.

Science is a means of answering questions. Learning to analyze the nature of a question, to interpret data, and to assess the reliability of a source will help you evaluate new information. Developing connections between new knowledge and what you already know can help you detect errors in logic and the content of new material. With increased knowledge, you are more prepared to cope with the moral and ethical issues raised as a result of the growth of scientific knowledge.

Key Concepts

Make a list of the major concepts you have learned in this chapter. Group the terms so related ideas are together.

Reviewing Ideas

1. Explain the relationships among technology, biology, and ethics.
2. Give some examples of how biology affects your life.
3. How does an understanding of cells and molecules relate to treating or preventing AIDS?
4. How are hypotheses, observations, and experiments used in science?
5. What are the principles behind the theory of evolution by means of natural selection?
6. How did Darwin use the methods of science to develop his theory?
7. Why were Lamarck's ideas rejected? On what assumptions were his ideas based?
8. Compare and contrast a hypothesis and a theory.
9. What is the relationship between natural selection and the environment?
10. What are the characteristics of science, and how does it differ from pseudoscience?
11. What type of questions should you ask when considering biological issues, and why?

Using Concepts

1. Imagine you are going to investigate the origin of Earth. How would you go about it? What would have to be done? What problems would you encounter?

 BSCSblue.com/vocabulary_puzzlemaker

that indicated every element is made of minute particles. Dalton believed these particles could not be broken into smaller particles, so he named them atoms. **Atoms** are the smallest unit of an element that still has the chemical properties of that element. Dalton stated several principles to describe an atom's chemical behavior. Those principles form the basis of the most important of all chemical theories, the atomic theory. New data have changed the atomic theory since Dalton's time, but this theory is still basic to an understanding of chemistry and biology.

Molecules are made of atoms (the smallest units of elements) that have combined chemically. Molecules may be made from more than one type of atom (in which case the molecule is called a compound, such as water), or from atoms of the same type. For example, hydrogen gas consists of hydrogen molecules, each of which consists of two hydrogen atoms linked together (see Figure 1.2). Similarly, an oxygen molecule consists of two oxygen atoms. (Oxygen also forms molecules of ozone, which contain three atoms of oxygen.) Elements can combine chemically in many ways to form the millions of compounds that give Earth its variety of materials.

Chemists have given each element a symbol of letters from the element's name. H stands for hydrogen, O for oxygen, C for carbon, and N for nitrogen. Iron, however, has two letters. Its symbol is Fe, derived from the word *ferrum*, reflecting that some symbols come from an element's Latin or Greek name. (See Appendix 1A, "The Periodic Table of the Elements.")

Despite all the variety of materials, organisms are made of a limited number of compounds, of which water is the most abundant. About 97% of the compounds present in organisms contain only six elements—carbon (C), hydrogen (H), oxygen (O), nitrogen (N), phosphorus (P), and sulfur (S). The remaining 3% contain small amounts of other elements. The basic six elements are essential to every organism. Some 20 others are essential, too, but only in smaller amounts (Figure 1.3).

The number of atoms of each element in a molecule is shown by the number, called a subscript, following the symbol for the element (the number 1 is always understood and not written). For example, the formula for carbon dioxide, CO_2, means that a molecule of this gas contains one carbon atom and two oxygen atoms. A molecule of ammonia, written as NH_3, contains one nitrogen atom and three hydrogen atoms. The formula tells what elements are in each molecule and how many atoms of each element the molecule contains.

water molecule	hydrogen molecule	oxygen molecule
oxygen atom	hydrogen atom	oxygen atom
2 hydrogen atoms	hydrogen atom	oxygen atom

FIGURE 1.2
Structures of some simple molecules. Molecules of water, hydrogen, and oxygen are made from combinations of atoms, as shown in these models.

ETYMOLOGY

a- = not (Greek)
tomo- = cut or divide (Greek)

An **atom** is a particle that cannot be divided into smaller particles of the same substance.

CONNECTIONS

The presence of the same six elements in all living organisms is an example of patterns in nature.

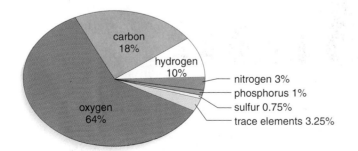

FIGURE 1.3
Elements present in the human body. The large proportions of hydrogen and oxygen reflect the fact that living organisms consist mostly of water. The dry matter in organisms consists mostly of various compounds of carbon with other elements, especially, hydrogen, oxygen, and nitrogen.

carbon 18%
hydrogen 10%
nitrogen 3%
phosphorus 1%
sulfur 0.75%
trace elements 3.25%
oxygen 64%

Molecular Models

What is a model of a molecule? In a sense, a molecular model has some things in common with models of planes, ships, and automobiles that you can purchase in a hobby shop. A model of a ship, for example, is an approximation of the real object, except that the model is much smaller and probably does not function the way the ship does. The model makes it easier to understand how a ship is built and how it operates, but it is not the real thing.

In a similar way, a molecular model gives an idea of a molecular structure. Unlike the ship, however, we do not know what the molecule actually looks like. It is impossible to observe the structure of a molecule directly because a molecule is too small for the eye to see, even with a microscope. Molecular models are constructed from **inferences**—conclusions that follow logically from laboratory data such as X-ray diffraction. As more is learned about a molecule, the model is changed and improved to be consistent with the new information. In this way, a model can serve as a learning tool and be used in developing new hypotheses. The symbol H_2O represents a molecule of water. H and O are printed symbols used to represent the elements hydrogen and oxygen in water, and H_2O is a printed model of a molecule of water. As you will see, the 2 indicates that there are two atoms of hydrogen in one water molecule. (In one drop of water there are about a billion trillion molecules of H_2O.)

The figures and symbols used in this chapter are models that can help you visualize and understand the structure and the function of elements and molecules.

1.2 The Structure of Atoms

Although Dalton didn't know it, atoms themselves are built of many smaller **subatomic** particles. The subatomic particles of atoms that are basic to an understanding of biology are electrons, protons, and neutrons. The **electron** carries a negative electric charge. A **proton** has a positive charge, and a **neutron** has no charge (it is neutral).

Protons and neutrons remain in the center, or nucleus, of the atom. The electrons, however, seem to be everywhere at once except in the nucleus. The rapidly moving electrons form a negatively charged "cloud" around the nucleus. Electrons tend to stay in this cloud because their negative charges are attracted to the positive charges of the protons in the nucleus. Electrons are distributed throughout the cloud based on differing levels of energy (or attraction) called **electron shells.** Electrons in shells near the nucleus are held more tightly than those in shells farther from the nucleus.

The simplest of all atoms is hydrogen. A hydrogen nucleus has a single proton, no neutron, and a single electron that orbits in the energy shell closest to the nucleus (Figure 1.4a). Because two electrons can fit in that shell, hydrogen has room for another electron. That vacancy makes hydrogen very reactive.

Atoms of other elements are more complex than the hydrogen atom. For example, an atom of carbon (Figure 1.4b) has six protons and six neutrons in its nucleus and six electrons orbiting the nucleus (two in the innermost shell, four in an outer shell). The number of neutrons may vary, but carbon atoms always contain six protons. Similarly, nitrogen has seven protons and seven electrons, and oxygen has eight protons and eight electrons (Figure 1.4c, d). In both nitrogen and oxygen, there are two electrons in the innermost shell. Nitrogen has five electrons in its outer

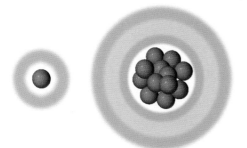
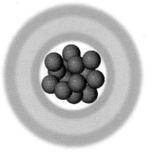
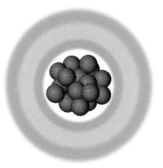

a **hydrogen atom**
(1 proton, 1 electron)

b **carbon atom**
(6 protons, 6 neutrons,
6 electrons)

c **nitrogen atom**
(7 protons, 7 neutrons,
7 electrons)

d **oxygen atom**
(8 protons, 8 neutrons,
8 electrons)

FIGURE 1.4

Some simplified models of atoms. Simplified models of hydrogen **(a)**, carbon **(b)**, nitrogen **(c)**, and oxygen **(d)**, are shown here. The models show electrons as a cloud of negative charge around the nucleus. The number of electrons in the outer level determines the chemical activity. Protons are shown in blue, and neutrons are shown in brown.

shell, and oxygen has six in its outer shell. That second shell can hold eight electrons, as can the third shell. Atoms with filled electron shells, such as helium and neon, are more stable than atoms with unfilled shells, such as nitrogen and oxygen.

Every atom has an equal number of protons and electrons. Thus, the charges are balanced, and the atom itself has no overall electric charge. However, the atoms of most elements can undergo chemical change by gaining, losing, or sharing one or more electrons with other atoms. Atoms with unfilled shells have a strong tendency to lose or gain electrons to complete their outer shells. As you will see in the next section, this is the basis of chemical reactions.

Atoms of different elements differ in their number of protons, neutrons, and electrons. Atoms of the same element always have the same number of protons and electrons, but they may differ in their number of neutrons. Atoms of the same element that differ in their number of neutrons are called **isotopes.** For example, 99% of oxygen atoms are like the one in Figure 1.4d. This atom is called the oxygen-16 isotope, named for the sum of its 8 protons and 8 neutrons. Oxygen-17 and oxygen-18 also exist, with 9 and 10 neutrons, respectively. Some isotopes have unstable atomic nuclei that break down, releasing radiation energy. These isotopes are called radioisotopes. They are useful in biological research because they can help to determine some of the chemical reactions organisms carry out. (See Appendix 1B, "Radioisotopes and Research in Biology.")

E T Y M O L O G Y

iso = equal (Greek)
topos = place (Greek)

Isotopes of a chemical element occupy the same position in the periodic table of the elements.

Check and Challenge

1. What is the relationship between atoms and elements?
2. What are the particles of an atom and how do they interact?
3. What six elements are most important in organisms?

Reactions in Living Cells

1.3 Chemical Reactions

Chemical bonds are the attraction, sharing, or transfer of outer shell electrons from one atom to another. Those bonds between atoms can be broken, the atoms rearranged, and new bonds formed. A **chemical reaction** involves the making and breaking of chemical bonds. During a chemical reaction, substances interact and form new bonds and new substances.

Only electrons in the outer shells of the atoms pictured in Figure 1.4 are involved when the atoms react during a chemical change. For atoms with electrons in more than one energy shell, only the outermost electrons normally interact during chemical changes. The outer electrons are a reliable indicator of the reactivity of an atom, but the structure of the entire atom also influences chemical reactions.

Chemical reactions occur in the cells of all living organisms. **Cells** are the basic units of life, much as atoms are the basic units of matter (Figure 1.5). Chemical reactions are important to a cell for two reasons. First, they are the only way to form new molecules that the cell requires for such things as growth and maintenance. Second, the making and breaking of bonds involves changes in energy. As a result of chemical reactions in a cell, energy may be stored, used to do work, or released.

Chemical reactions can be represented as short statements called chemical equations. For example, the breakdown of water can be represented by the following equation:

$$2\,H_2O \quad \xrightarrow[\text{energy}]{\text{electric}} \quad 2\,H_2 \quad + \quad O_2$$

| 2 molecules | 2 molecules | 1 molecule |
| of water | of hydrogen | of oxygen |

Why is it necessary to use two molecules of water in the equation? Remember that hydrogen and oxygen molecules each consist of two atoms. A single molecule of water does not yield enough oxygen atoms to make an oxygen molecule, but two molecules of water do. The equation is written to

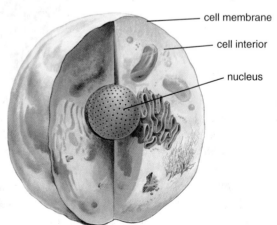

— cell membrane

— cell interior

— nucleus

FIGURE 1.5

A generalized animal cell. A membrane encloses all cells. In animals, the genetic material is enclosed in a structure called the nucleus.

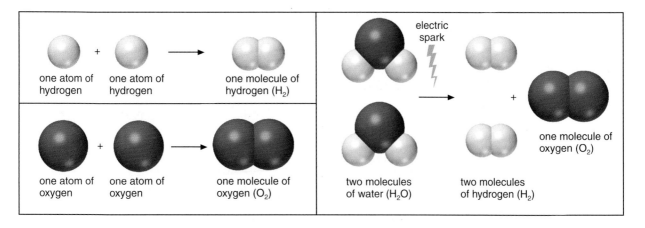

FIGURE 1.6

Models of chemical reactions. In the decomposition of water, twice as many hydrogen molecules as oxygen molecules are produced.

balance the number of atoms on both sides of the arrow. Balancing chemical equations illustrates one of the most basic laws of chemistry: the **law of conservation of matter,** which states that matter is neither created nor destroyed in chemical reactions.

The direction of the arrow in a chemical reaction points from reactants to products. Notice that the number of molecules is shown by a numeral *preceding* the formula for the molecule. Figure 1.6 shows models that represent chemical reactions.

When molecules collide, they may or may not react, depending on the energy and orientation of the molecules. **Activation energy** is the energy needed to get a chemical reaction started. Sometimes an outside source of extra energy is necessary to initiate a reaction. For example, hydrogen (H_2) and oxygen (O_2) gases can be mixed without reacting until a lighted match, a spark, or ultraviolet light adds energy. This relatively small amount of energy causes some hydrogen molecules and oxygen molecules to react to form water. The reaction releases energy that heats the remaining molecules of hydrogen and oxygen, causing them to react to form more water. The reaction is often explosive. The product, water, has less energy than the hydrogen and oxygen gases had separately in the mixture. The difference in energy is accounted for mainly in the form of light and heat produced by the reaction.

1.4 Chemical Bonds

When atoms interact, they can form several types of chemical bonds. One type forms when electrons move from one atom to another atom. This type of chemical bond occurs in many substances, including table salt, also known as sodium chloride (NaCl). As the latter name suggests, table salt is made of two elements, sodium (Na) and chlorine (Cl). When atoms of these two elements react, an electron passes from a sodium atom to a chlorine atom (Figure 1.7). The resulting sodium atom is positively charged, for it has one less electron than protons. It becomes a sodium ion, written as Na^+.

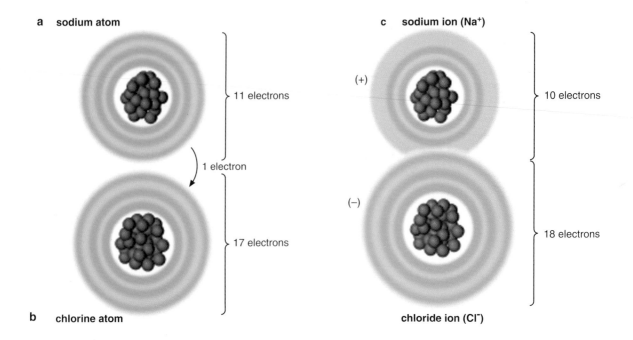

a sodium atom

11 electrons

1 electron

c sodium ion (Na⁺)

(+)

10 electrons

(−)

18 electrons

b chlorine atom

17 electrons

chloride ion (Cl⁻)

sodium + chlorine ⟶ sodium chloride

FIGURE 1.7

An example of ionic bonding. (a) Sodium and **(b)** chlorine can react to form **(c)** the salt sodium chloride (NaCl). By losing one electron, sodium achieves two filled shells (2 + 8) and becomes a stable positive ion. By gaining one electron, chlorine achieves three filled shells (2 + 8 + 8) and becomes a stable negative ion, chloride.

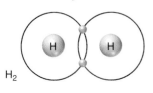

H_2

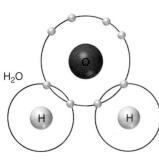

H_2O

FIGURE 1.8

An example of covalent bonding. Covalent bonds form when two atoms share a pair of electrons. Only electrons in the outer energy levels are shown here.

An **ion** is an atom or a molecule that has acquired a positive or negative charge as a result of gaining or losing electrons. The chlorine atom is negatively charged, for it has one more electron than protons. It becomes a chloride ion, written as Cl⁻. Note the change in name from chlorine to chloride. An **ionic bond** is the attraction between oppositely charged ions, such as the sodium chloride bond (Figure 1.7).

In a second type of chemical bond, called a **covalent bond,** two atoms share one or more pairs of electrons (Figure 1.8). Two atoms of hydrogen join to form a molecule of hydrogen gas (H_2) by sharing a pair of electrons. This sharing gives each hydrogen atom a filled electron shell. In a molecule of water, each of the two hydrogen atoms shares a pair of electrons with the same oxygen atom. This gives each hydrogen atom two electrons and fills the outer shell of oxygen with eight electrons.

The chemical behavior of water indicates that the atoms do not share the electrons equally. The larger oxygen atom attracts the electrons more strongly than the smaller hydrogen atoms do. If the electrons of a bond are not shared equally, the bond is called a polar covalent bond. In contrast, the electrons in a molecule of hydrogen gas are shared equally, and the resulting covalent bond is said to be nonpolar.

The unequal sharing of electrons in a water molecule gives the oxygen atom a slight negative charge and each hydrogen atom a slight positive charge (Figure 1.9). Such a molecule is known as a polar molecule. The polar nature of water is biologically significant. Most cells and tissues contain large amounts of water—up to 95% in some, with an average of 70–80% water throughout all organisms. Molecules must dissolve in water in order to move easily in and between living cells. Polar molecules, such as sugar, and ions, such as Na⁺, dissolve in water because of the electric attraction between them and the water molecules. Nonpolar molecules, such as fats and oils, do not dissolve in water.

Polar molecules may form still another type of chemical bond. A weak attraction can occur between a slightly positive hydrogen atom in a molecule and a nearby slightly negative atom of another molecule (or of the same molecule if it is large enough). This type of attraction is called a **hydrogen bond** (Figure 1.10). In compounds found in organisms, hydrogen bonds usually involve hydrogen atoms that are bonded to oxygen or nitrogen. Hydrogen bonds provide an attractive force between water molecules, which explains why water is a liquid at room temperature and not a gas. A large number of hydrogen bonds can be quite strong, but single hydrogen bonds are much weaker than covalent bonds.

1.5 Ions and Living Cells

When table salt dissolves in water, the ionic bonds are broken. Na⁺ and Cl⁻ ions separate, or dissociate, but remain as ions in solution. The positive sodium ion is attracted to the slightly negative end of water, and the negative chloride ion is attracted to the slightly positive end. Sodium ions are important in regulating water balance in organisms. Other ions,

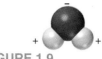

FIGURE 1.9

The polar nature of water. An oxygen atom attracts electrons more strongly than hydrogen, giving the oxygen a slight negative charge and each hydrogen a slight positive charge. The net charge of the molecule is zero.

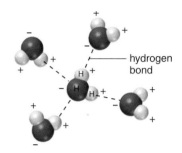

hydrogen bond

FIGURE 1.10

Hydrogen bonding. A hydrogen bond is a weak attraction between a slightly positive hydrogen atom in one polar molecule and a slightly negative atom in another. This water molecule is forming hydrogen bonds with four other water molecules.

Focus On

Structural Formulas

Structural formulas are models that show both the number and arrangement of atoms in molecules. In writing structural formulas of compounds, a covalent bond is indicated by a line. For example, the structural formula for hydrogen gas is written H—H. Sometimes two atoms share two pairs of electrons between them, forming a double bond that is shown by a double line (=), as in O=C=O, carbon dioxide. When three pairs of electrons are shared by two atoms, they form a triple bond that is represented by three lines (≡). For example, the nitrogen gas (N_2) that makes up most of the atmosphere consists of pairs of nitrogen atoms joined by a triple bond. Figure 1.11 shows several methods of representing molecules. Note that the molecular formula tells you only the number of atoms of each type present in the molecule, whereas the structural formula indicates how the atoms are bonded. The space-filling and ball-and-stick models provide pictures of the spatial arrangements of the atoms.

name/molecular formula	structural formula	models	
		space-filling	ball-and-stick
hydrogen/H_2	H—H		
water/H_2O	H—O—H		

ammonia/NH_3	H—N—H
	    |
	    H		

methane/CH_4	H
	    |
	H—C—H
	    |
	    H		

FIGURE 1.11

Several ways of representing the structures of molecules. A variety of symbols and models are used to represent chemical formulas.

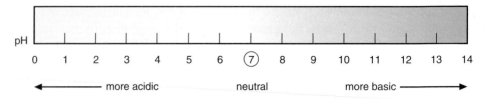

FIGURE 1.12

The pH scale. This scale is used to denote the acidity and alkalinity of solutions.

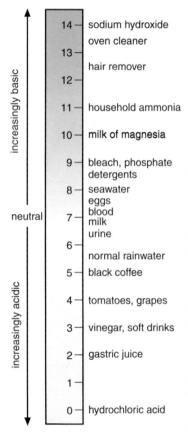

FIGURE 1.13

The pH of some common substances. Note that most biological substances are slightly acidic (pH 6–7). The greater acidity of soft drinks (pH 3) is partly responsible for contributing to tooth decay.

including potassium ions (K^+) and calcium ions (Ca^{+2}), are involved in many reactions inside cells.

When a nonionic compound, such as water, is converted to ions, the process is called ionization. The ionization of water is a vital reaction in living cells. When a water molecule separates, one of its H—O bonds breaks. The result is a positively charged hydrogen ion (H^+) and a negatively charged hydroxide ion (OH^-). The hydrogen ion—a proton—quickly combines with a water molecule to form a hydronium ion (H_3O^+). It still is convenient, however, to refer to the number of hydrogen ions (H^+) in a solution, even though that really means the concentration of hydronium ions (H_3O^+).

Only about one in every 500 million water molecules ionizes in living cells, yet all life processes depend on this tiny amount of ionization. Indeed, living cells must maintain their internal levels of H^+ and OH^- ions within narrow limits, because even small changes greatly influence important reactions.

The level of H^+ and OH^- ions in solution is described by a range of numbers known as the **pH scale** (Figure 1.12). The scale runs from 0 to 14. A solution (a mixture in water) that has the same number of H^+ and OH^- ions is neutral and has a pH of 7. Pure water has a pH of 7. A solution having more H^+ than OH^- is **acidic** and has a pH less than 7 (low pH). A solution that has more OH^- than H^+ ions is **basic** (or alkaline) and has a pH greater than 7 (high pH). Thus a solution with a pH of 2 is highly acidic, and a solution with a pH of 10 is highly basic.

The pH scale is a logarithmic scale. This means that a change of one pH unit is equal to a tenfold change in the level of H^+ ions. For example, a solution with a pH of 3 is 10 times more acidic than a solution with a pH of 4. Figure 1.13 shows the pH of several common substances.

pH is important in biology because the pH of a cell's interior helps regulate the cell's chemical reactions. As discussed earlier, all cells rely on chemical reactions for growth and survival. Thus, it is not surprising that organisms have ways to control pH and to respond to changes in the pH of their environment (Figure 1.14). For example, certain fungi and bacteria can grow in acidic solutions but not in basic ones. Some marine organisms have become so adapted to the slightly basic pH (7.8–8.6) of seawater that they cannot live in less basic solutions. The pH of normal urine is about 6, but human blood must remain near pH 7.4. Death nearly always results if the blood pH falls to and stays at 6.8 or rises to and stays at 8.0. Failed kidney function is most often the reason a person cannot maintain normal blood pH.

a   b

FIGURE 1.14

Adaptations of organisms to differences in soil pH. (a), Conifers (needle-leaf trees) such as these Douglas firs thrive in the acidic soil of Olympic National Forest in Washington state. **(b),** Deciduous hardwood trees such as oaks, maples, and beeches compete more successfully in the less acidic soil of this forest.

Check and Challenge

1. How are chemical reactions important in cells?
2. Distinguish among the various types of chemical bonds.
3. Describe the law of conservation of matter.
4. Why is pH important to living organisms?
5. How does the polarity of water assist the movement of molecules?

> **Try Investigation 1A** ⬩ **Organisms and pH.**

Biochemistry

1.6 Organic Compounds and Life

Many chemical compounds besides water are needed for life to exist. The most important are **organic** compounds, in which carbon atoms are combined with hydrogen and usually oxygen. Organic compounds frequently also contain nitrogen, sulfur, or phosphorus. A few carbon compounds are not included with organic compounds: carbon dioxide (CO_2), carbon monoxide (CO), and carbonic acid (H_2CO_3).

The word *organic* was coined long ago when it was thought these compounds could be formed only by living cells. Since then, millions of different organic compounds have been synthesized in the laboratory. There is no longer any reason to call these compounds organic, but the name is so well established that it is still widely used to describe nearly all carbon compounds, essential for life or not.

> **Try Investigation 1B** ⬩ **Compounds of Living Things.**

Carbon atoms can combine in long chains that form the backbone of large complex molecules, or **macromolecules.** The backbone of carbon atoms is called the carbon skeleton (Figure 1.15). Other atoms and molecules can attach to the carbon skeleton, giving each macromolecule a particular structure and, therefore, a particular function. The following sections of the chapter discuss the characteristics of the four most important classes of molecules in living cells—carbohydrates, lipids, proteins, and nucleic acids. Figure 1.16 shows the relationship between the cell's building-block molecules and their macromolecular forms.

1.7 Carbohydrates

All known types of living cells contain **carbohydrates.** In addition to carbon atoms, carbohydrates contain hydrogen and oxygen atoms in the same two-to-one ratio as water. The simplest carbohydrates are single sugars called **monosaccharides,** which may contain three to seven carbon atoms in their carbon skeletons. Figure 1.17a shows the structural formulas of two forms of the common monosaccharide glucose. Most organisms use glucose (which is also referred to as blood sugar) as a source of energy. The energy in glucose, and in all molecules, is contained in the atoms and bonds of the molecule itself.

Biologically important sugars often have a phosphate group attached to the carbon skeleton and are called sugar-phosphates. The phosphate

ETYMOLOGY

carbo- = carbon (French, Latin)
hydro- = water (Greek)
mono- = single (Latin)
di- = two (Greek)
poly- = many (Greek)
saccharo- = sweet or sugar (Latin, Greek)

Carbohydrates are compounds of carbon, hydrogen, and oxygen in the same 2:1 proportion in which they occur in water (CH_2O). A **monosaccharide** is a single sugar molecule. A **disaccharide** consists of two joined sugar molecules, and a **polysaccharide** is composed of many sugar subunits.

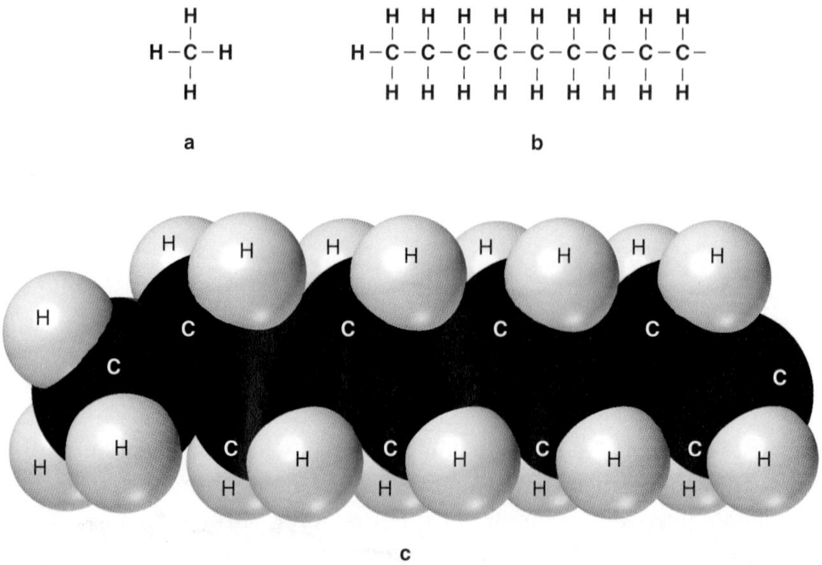

FIGURE 1.15

Structural formulas of some carbon compounds. Carbon atoms (black) make up a carbon skeleton and link with other atoms, such as hydrogen (white). In each formula, a short line indicates a covalent bond: **(a),** a molecule of methane; **(b),** part of the carbon skeleton of a larger molecule; **(c),** a view of **(b)** that indicates the three-dimensional structure of the molecule. Other atoms, such as oxygen (usually shown in red) or nitrogen (usually shown in blue), can be attached to the carbon skeleton instead of hydrogen.

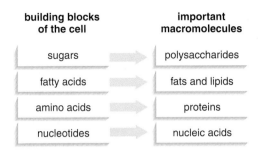

building blocks of the cell		important macromolecules
sugars	⇒	polysaccharides
fatty acids	⇒	fats and lipids
amino acids	⇒	proteins
nucleotides	⇒	nucleic acids

FIGURE 1.16

The building blocks of the cell. Most macromolecules are polymers—long chains of similar subunits (monomers). For example, proteins are polymers made from many amino-acid monomers. Sugars and polysaccharides are carbohydrates.

group, which is composed of an atom of phosphorus and four atoms of oxygen (see Figure 1.17b).

Two simple sugar molecules, or **monosaccharides,** (Figure 1.18a) may bond to form a double sugar, or **disaccharide,** represented in Figure 1.18b. The most familiar of all disaccharides is sucrose, commonly called table sugar. Sucrose contains glucose and another monosaccharide, fructose. Lactose, or milk sugar, is a disaccharide formed of glucose and the monosaccharide galactose. Maltose, or malt sugar, is a common disaccharide made of two glucose molecules.

Several glucose molecules may bond to form complex carbohydrates called **polysaccharides** (Figure 1.18c). Starch and cellulose are the complex carbohydrates commonly formed by plants. Starch is an energy-storage and carbon-reserve compound in many plants and is an important food source for humans. Cellulose is a structural molecule found in the rigid walls surrounding plant cells, and it is an important part of wood and cotton fibers. The human liver and muscles store carbohydrates in the form of glycogen, also called animal starch. Molecules of starch, cellulose, and glycogen consist of thousands of glucose units and have no fixed size.

straight chain form	ring form
a glucose	

straight chain form	ring form
b glucose-6-phosphate	

FIGURE 1.17

The structures of a sugar and a sugar-phosphate. (a), In solution, glucose, a 6-carbon sugar, can exist in two forms: a straight chain and a ring form. The ring form, in which five carbon atoms and an oxygen atom form a closed ring, is by far the most abundant. Although not shown, carbon atoms exist at the points of the hexagon. Also, actual glucose molecules are not solid, as depicted; their centers are mostly open space. **(b),** The two forms of glucose-6-phosphate are shown. Cells often add a phosphate group to glucose after glucose enters the cell.

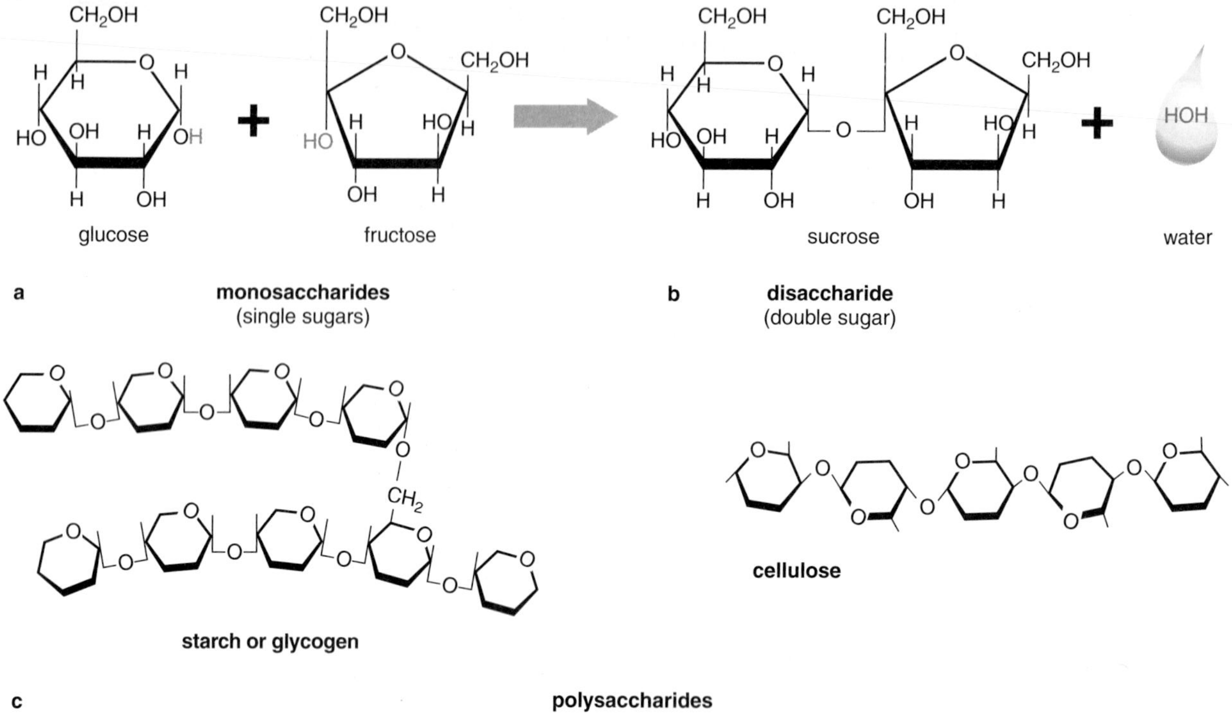

a **monosaccharides**
(single sugars)

b **disaccharide**
(double sugar)

c **polysaccharides**

FIGURE 1.18

Formation of complex carbohydrates. Monosaccharides **(a)**, can combine to form disaccharides **(b)**. Starch and cellulose **(c)**, are polysaccharides formed by linking together many monosaccharide subunits. Glycogen has a structure very similar to starch. Note the glucose subunits are linked differently in cellulose and starch. Human digestive enzymes cannot break down the linkages in cellulose; hence we cannot digest it.

1.8 Lipids

Lipids, or fats and oils, are macromolecules that have two primary functions: long-term storage of energy and carbon and building of structural parts of cell membranes. Lipids generally do not dissolve in water, because they are nonpolar. Like carbohydrates, lipids contain carbon, hydrogen, and oxygen, but not in a fixed ratio. Building blocks of lipids, called fatty acids and glycerol, make up the simple fats most common in our diets and bodies. Three fatty-acid molecules and one glycerol molecule join to form a simple fat, or triglyceride, as shown in Figure 1.19.

The biologically important properties of simple fats depend on their fatty acids. For example, the fats in meat are different from the oils in vegetables because the fatty acids are different. The properties of fatty acids, in turn, depend on the length of the carbon chains and the type of bonds between the carbons. Common fatty acids have a total of 16 or 18 carbon atoms. Fatty acids in which single bonds join the carbon atoms are saturated fatty acids (one way to remember this: the carbon atoms are saturated with hydrogens). Unsaturated fatty acids are fatty acids in which double bonds join some of the carbon atoms. Figure 1.19 shows both types of fatty acids.

Unsaturated fats (fats containing unsaturated fatty acids) tend to be oily liquids at room temperature. Olive oil, corn oil, and sunflower oil consist

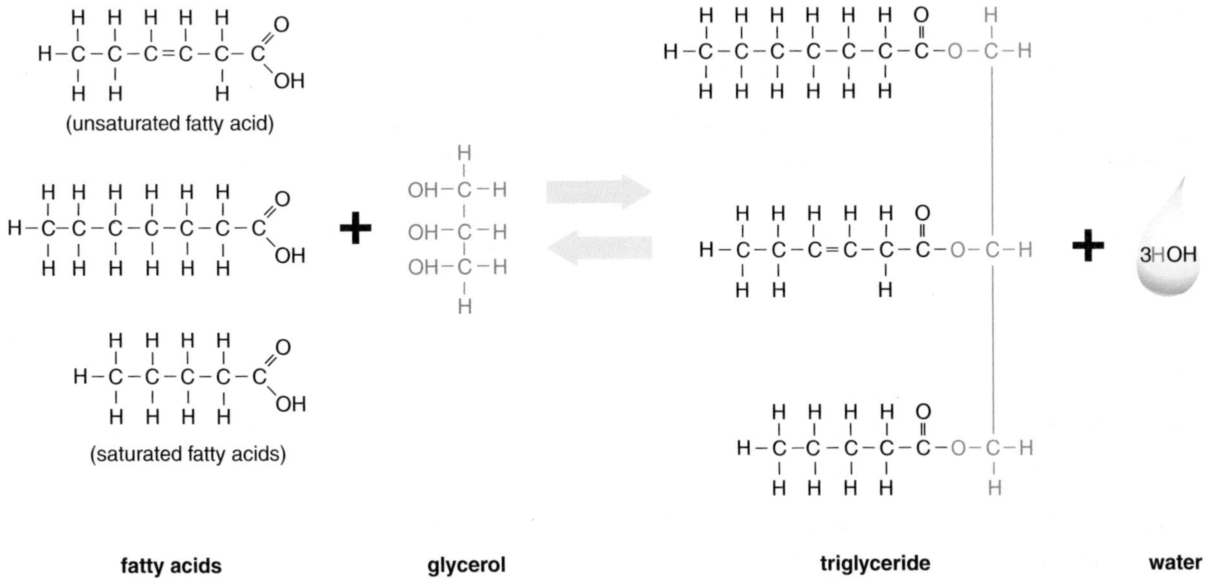

FIGURE 1.19

Formation of a fat molecule. To form a fat, one molecule of glycerol combines with three molecules of fatty acids. The fatty acids in one fat may be alike or different. Examine the diagram. What is a by-product of this reaction?

mostly of unsaturated fats. Saturated fats tend to be solids at room temperature. Butter and lard consist mostly of saturated fats. Fats are a more efficient form of energy storage than are carbohydrates because fats contain a larger number of hydrogen atoms and less oxygen. You will learn later how the amount of hydrogen and oxygen in fats and carbohydrates relates to energy, as well as the important roles that saturated and unsaturated fats play in the diet.

Two other types of lipids important in cells are phospholipids and cholesterol. Phospholipids form when a molecule of glycerol combines with two fatty acids and a phosphate group (Figure 1.20a and b). The polar phosphate group allows one end of the lipid molecule to associate with water. Together with proteins, phospholipids form cellular membranes (Figure 1.20c). Membranes are critical to cell survival because they separate a cell's internal chemical reactions from the outside environment. They also help control which chemicals enter and leave the cell. Cholesterol (Figure 1.20d) is part of the membrane structure of animal cells and is important in nutrition. Cells in the human body manufacture many essential substances, such as sex hormones, from cholesterol.

1.9 Proteins

Every living cell contains from several hundred to several thousand different macromolecules known as **proteins.** Proteins are structural components of cells as well as messengers and receivers of messages (also called receptors) between cells. They play an important role in defense

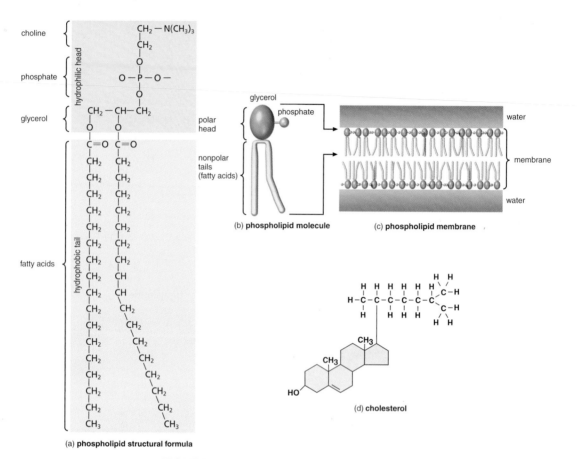

(a) phospholipid structural formula

(b) phospholipid molecule

(c) phospholipid membrane

(d) cholesterol

FIGURE 1.20

Structures of two types of complex lipids. Glycerol joins with two fatty acids and a polar phosphate group to form a phospholipid **(a)** and **(b).** When phospholipids form membranes **(c),** the polar head associates with water on the membrane surface, and the nonpolar tail faces the interior of the membrane, away from water. Membranes prevent the cell contents from mixing with the external environment. Cholesterol **(d),** has a fused four-ring structure with additional side groups. Cholesterol is important in membrane structure, and the sex hormones are derived from it.

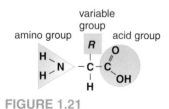

FIGURE 1.21

The structure of an amino acid. Amino acids (except for proline) have a central carbon atom bonded to a hydrogen atom, an amino group, an acid group, and an *R*-group. The *R*-group is any of 20 arrangements of C, H, O, N, and S atoms, depending on, and giving unique structure to, each amino acid.

against disease. Skin, hair, muscles, and parts of the skeleton are made of proteins. Their most essential role, however, is as enzymes, specialized molecules that assist the many reactions occurring in cells.

Cells make proteins by linking building blocks called amino acids. **Amino acids** are small molecules that contain carbon, hydrogen, oxygen, and nitrogen atoms; two also contain sulfur atoms. Figure 1.21 shows the general structural formula for an amino acid. Amino acids have a central carbon atom to which is attached a hydrogen atom, an amino group ($-NH_2$), an acid group ($-COOH$), and a variable group, symbolized by R, which may be one of 20 different atoms or groups of atoms. Observe in Figure 1.22a that in the amino acid glycine, R is an H atom, and in alanine, R is a $-CH_3$ group. Some R groups are polar, such as the amino group in lysine; others are nonpolar, such as the carbon chain in leucine. The polarity of amino acids affects their ability to dissolve in water.

Any two amino-acid molecules may combine when a chemical bond forms between the acid group of one molecule and the amino group of the other (Figure 1.22a). Covalent bonds of this sort, formed between amino acids, are called **peptide bonds.** Additional peptide bonds may form, resulting in a long chain of amino acids, or **polypeptide** (Figure 1.22b). Longer polypeptide chains form proteins; protein molecules range from about 50 to 3,000 amino-acid units. The type, number, and sequence of its amino acids distinguishes a protein from all others.

The sequence of amino acids in a polypeptide chain forms the **primary structure** (Figure 1.23a) of a protein. In most proteins, the chain folds or twists to form local structures known as **secondary structures** (Figure 1.23b). The most common secondary structures, alpha helices and beta sheets, are stabilized by hydrogen bonds. More complex folding creates a **tertiary structure** (Figure 1.23c), which usually is globular, or spherical. One of the major forces controlling how a protein folds is **hydrophobicity,** or the tendency for nonpolar amino acids to avoid water. As a result, nonpolar amino acids tend to be buried inside folded proteins, where they are shielded from contact with water. Some proteins have nonpolar amino acids on the outside, where they can interact with lipid membranes.

The sequence of amino acids in a protein (primary structure) determines its three-dimensional shape (secondary and tertiary structure). The tertiary structure, in turn, determines a protein's function. Each individual protein has a unique shape and, therefore, a specific function. A few proteins

CONNECTIONS

The structure-function concept is critical in understanding the relationship between chemistry and biology. Biological function depends on chemical structure. Biological molecules do what they do because of their unique chemical structures.

ETYMOLOGY

hydro- = water (Greek)
-phobic = fear or dislike (Greek)
tert- = third (Latin)
quaterni- = four (Latin)

Hydrophobic molecules or parts of molecules do not mix easily with water. **Tertiary** structure is the third level of protein structure; **quaternary** structure is the fourth level.

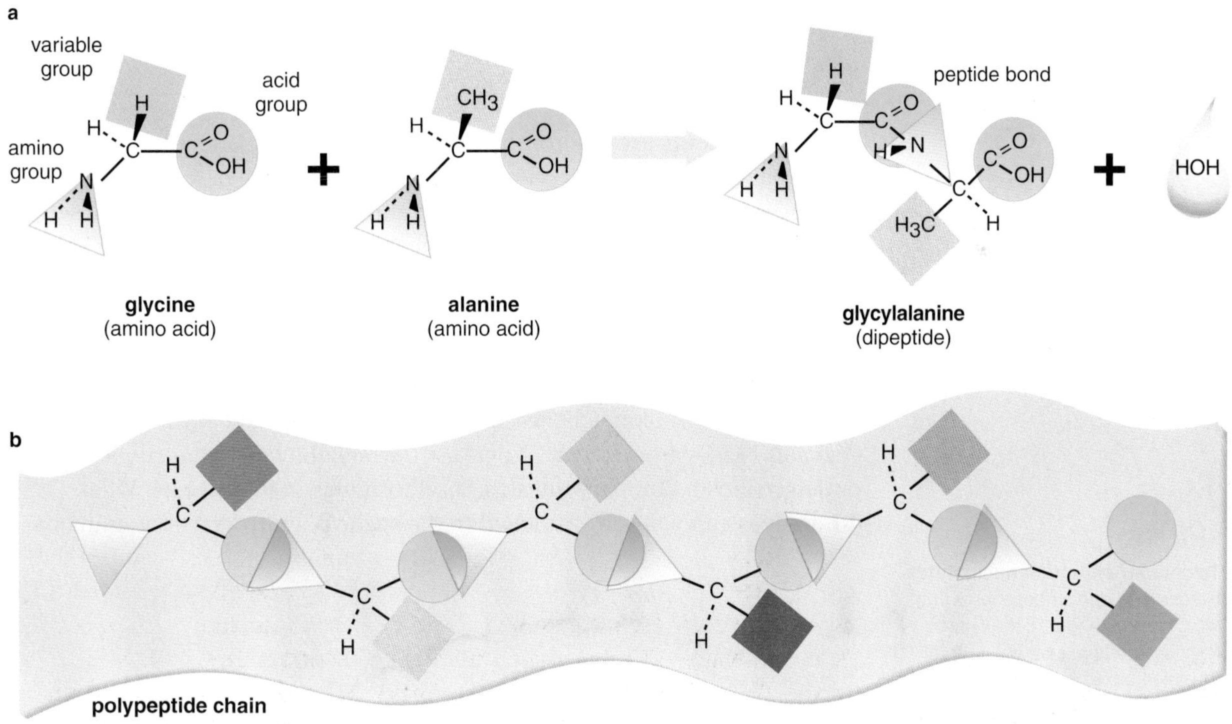

FIGURE 1.22

Formation of a polypeptide. A peptide bond forms when the acid group (—COOH) of one amino acid joins to the amino group (—NH_2) of another amino acid **(a).** The acid group loses an atom of hydrogen and an atom of oxygen. The amino group loses an atom of hydrogen. These atoms combine to form a water molecule. More peptide bonds can form in the same way, joining a chain of amino acids to form a polypeptide **(b).**

a primary structure

amino acid sequence

b secondary structure

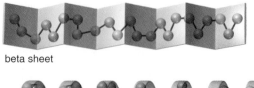

beta sheet

alpha helix

c tertiary structure

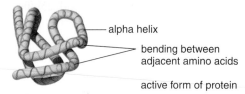

alpha helix

bending between
adjacent amino acids

active form of protein

d quaternary structure

several polypeptides
combine to form
quaternary structure

FIGURE 1.23

Stages in the development of a functional protein. Only the backbone atoms of the secondary structures are shown, not the entire amino acid.

become active only when two or more tertiary forms combine to form a complex **quaternary structure** (Figure 1.23d). Specific geometric and chemical interactions determine which surfaces of the proteins combine to create the quaternary structure.

1.10 Nucleic Acids

Nucleic acids are macromolecules that dictate the amino-acid sequence of proteins, which in turn control the basic life processes. Nucleic acids also are the source of genetic information in chromosomes, which are passed from parent to offspring during reproduction. Nucleic acids are thus the chemical link between generations, dating back to the beginning of life on Earth. How can one type of molecule play such a dominant role in all living things? The answer lies in the structure of the molecules.

Nucleic acids are made of relatively simple units called **nucleotides** connected to form long chains. Each nucleotide consists of three parts. One part is a 5-carbon sugar (a pentose), which may be either ribose or deoxyribose. Compare the structural formulas in Figure 1.24. What differences can you see? Attached to the sugar is a nitrogen-containing base,

FIGURE 1.24

Two sugars found in nucleotides. Ribose and deoxyribose form ring structures with four of their carbon atoms joined by an oxygen atom. The blue boxes highlight the difference between the two sugars. Ribose has a hydroxyl group (—OH), whereas deoxyribose has only a hydrogen atom at the same carbon. Compared to ribose, deoxyribose is missing one oxygen atom; *deoxy* in the name means "minus oxygen."

R   ribose

D   deoxyribose

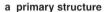

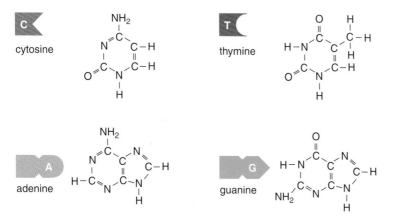

FIGURE 1.25

Four nitrogen bases that occur in nucleotides. In this book, symbols are used to represent the nucleotides adenine, guanine, cytosine, and thymine. Note that cytosine and thymine have a single-ring structure; adenine and guanine have a double ring.

which is a single or double ringlike structure of carbon, hydrogen, and nitrogen (Figure 1.25). The third part of a nucleotide is a phosphate group (P) (see Figure 1.26).

Nucleic acids that contain ribose in their nucleotides are called ribonucleic acids, or **RNA.** Nucleotides containing deoxyribose form deoxyribonucleic acids, or **DNA.** In DNA, each of the four different nucleotides contains a deoxyribose, a phosphate group, and one of the four bases—adenine, thymine, guanine, or cytosine. The possible DNA nucleotides that can form from these bases are illustrated in Figure 1.26. Figure 1.27 shows a chain of nucleotides connected to form a nucleic-acid molecule.

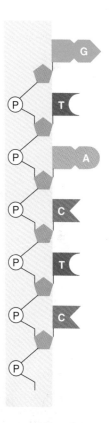

FIGURE 1.27

A short section of a model of nucleic acid. The bases adenine, guanine, thymine, and cytosine are attached to a backbone of alternating deoxyribose sugar and phosphate groups.

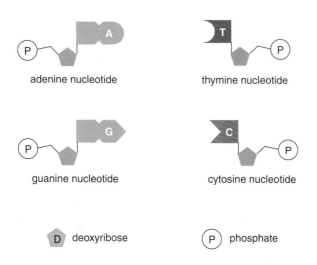

FIGURE 1.26

The components of DNA. Each nucleotide contains one of the bases, one sugar, and one phosphate group. The sugar is deoxyribose, making these the four nucleotides of DNA.

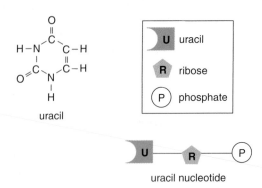

uracil

U uracil

R ribose

P phosphate

uracil nucleotide

FIGURE 1.28

Another difference between DNA and RNA. In addition to containing ribose in place of deoxyribose, RNA contains uracil nucleotides in place of the thymine nucleotides found in DNA.

RNA is a nucleic acid much like DNA, except that it contains the sugar ribose instead of deoxyribose. Also, in RNA the nitrogen base uracil, shown in Figure 1.28, replaces the base thymine. The other three bases, adenine, guanine, and cytosine, are the same in RNA and DNA.

Structurally, there is another difference between DNA and RNA. DNA always occurs in cells as a double-stranded helix (Section 1.11); RNA is single stranded, although it can fold into complex shapes. There are three types of RNA in cells, each of which performs a different role in the synthesis of proteins.

The number and arrangement of the nucleotides vary in RNA as they do in DNA. The different arrangements of nucleotides in a nucleic acid provide the key to how these molecules contribute to the diversity of organisms living on Earth.

Check and Challenge

1. Describe how building-block molecules combine to form each of the four types of biological macromolecules.

2. What is the relationship between the primary and tertiary structure of a protein?

3. Describe the three-part structure of a nucleotide.

4. How do the chemical structures of ribose and deoxyribose differ?

Biology Online BSCSblue.com/check_challenge

Genetic Coding in Cells

1.11 The Double Helix

In a short scientific paper published in 1953, James Watson and Francis Crick (Figure 1.29a and b) proposed a model for the structure of the DNA molecule that is still accepted today (although with some modification). They founded their model on the principle of specific pairing of the nucleotides, shown in Figure 1.30. Hydrogen bonds form only between the nucleotide bases of adenine (A) and thymine (T) or cytosine (C) and guanine (G). In other words, there is a specific interaction between the surfaces of the nucleotide molecules. Experiments done by Rosalind Franklin (Figure 1.29c) in the laboratory of Maurice Wilkins (Figure 1.29d) suggested to Watson and Crick that DNA molecules have a double-helix structure.

The DNA double helix is composed of two long chains of nucleotides. The nucleotides of each chain are connected between their deoxyribose sugars by phosphate groups. (This forms a so-called sugar-phosphate backbone.) The two chains run next to each other, but, chemically speaking, the sugar-phosphate backbones run in opposite directions. The two chains are connected by hydrogen bonding between nitrogen bases to form a long double-stranded molecule. Because of the specific pairing between bases,

a, b   **c**   **d**

FIGURE 1.29

The discoverers of the DNA structure. In 1953, James Watson **(a),** and Francis Crick **(b),** proposed a model for the DNA molecule based partly on the X-ray diffraction studies of Rosalind Franklin **(c),** and Maurice Wilkins **(d).** Watson, Crick, and Wilkins shared the Nobel Prize in 1962. Franklin died in 1958 and was therefore not eligible for the Nobel Prize.

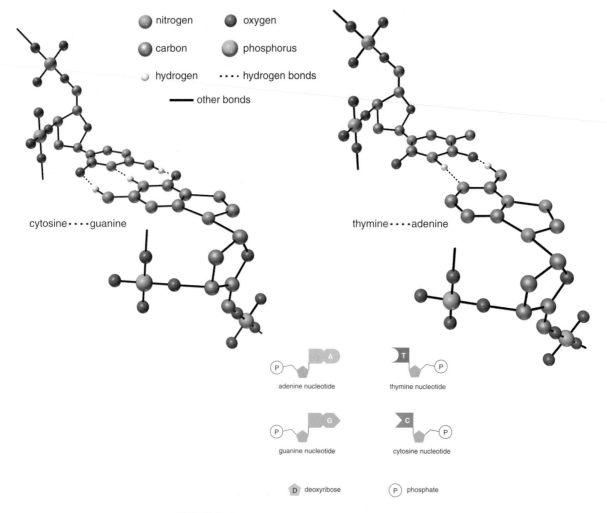

nitrogen   oxygen
carbon   phosphorus
hydrogen   ···· hydrogen bonds
──── other bonds

cytosine····guanine

thymine····adenine

P — A
adenine nucleotide

T — P
thymine nucleotide

P — G
guanine nucleotide

C — P
cytosine nucleotide

D deoxyribose   P phosphate

FIGURE 1.30

A diagram showing the pairing of two nitrogen bases. Hydrogen bonds connect cytosine to guanine and thymine to adenine.

each strand is the complement of the other. For example, a guanine on one strand indicates the presence of cytosine on the other strand, and vice versa. (As you will see later, complementarity is the basis for copying DNA.) The two strands intertwine, forming a double helix that winds around a central axis like a spiral staircase (Figure 1.31). As mentioned earlier, RNA also can fold into complex shapes, although those shapes result from interactions within a single-stranded molecule (or with proteins). Nonetheless, the folds in RNA are held together by the same type of base pairing as in DNA (Figure 1.32).

CONNECTIONS

The observation that DNA is the genetic material in all organisms is evidence of a common evolutionary origin for life.

1.12 The Functions of DNA

DNA forms the **genes,** units of genetic information, that pass from parent to offspring. The structure of DNA explains how DNA functions as the molecule of genetic information. In brief, DNA stores information

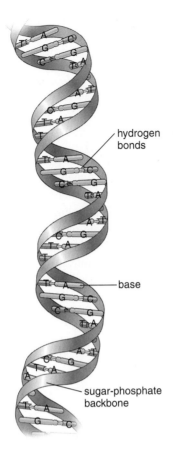

hydrogen
bonds

base

sugar-phosphate
backbone

FIGURE 1.31

The structure of DNA. The backbone of each chain (shown as ribbons) is composed of alternating sugar and phosphate molecules that connect the nucleotides. The two chains are held together by specific patterns of hydrogen bonding between pairs of bases—A to T and G to C—and the chains coil around a central axis to form a double helix. The diameter of the double helix is equal along its length because a double-ring A or G base (called purines) always pairs with a single-ring T or C base (called pyrimidines). There are ten nucleotides per turn of the helix, and only a small portion of a model of a DNA molecule is shown.

in a code consisting of units that are three nucleotides long; these short sequences are called triplet codons. Certain codons are translated by the cell to mean certain amino acids. For example, the codon GGA means glycine when translated by the cell. Thus the sequence of nucleotides in DNA can indicate a sequence of amino acids in protein. As you have seen, the sequence of amino acids in a protein determines its shape, which then

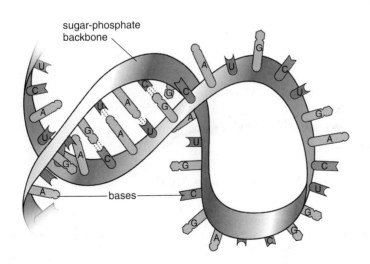

sugar-phosphate
backbone

bases

FIGURE 1.32

A model of a folded RNA molecule, showing single- and double-stranded regions. Most of the molecule is single stranded, but the shaded region shows hydrogen bonding between base pairs, which forms a double-stranded region. Note the four bases in RNA.

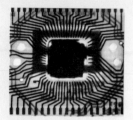

The Scanning Tunneling Microscope

Electron microscopes are powerful tools for exploring the structure of microscopic objects such as cells. Unlike light microscopes, which use light and glass lenses to make details visible, electron microscopes use electron beams and magnetic coils. Electron microscopes permit much greater magnification of cellular structures than do light microscopes. The scanning tunneling microscope uses electrons already in the sample, rather than an external electron beam. This allows scientists to see surface structures clearly. A characteristic of all materials is that some electrons form an electron cloud around the material. The scanning tunneling microscope probes this electron cloud to reveal structure.

To scan the surface of a sample, the tip of the microscope (a needle) is pushed toward the sample until the electron clouds of the tip and the sample touch. Then an electric voltage is applied between the tip and the sample, causing electrons to flow through a narrow channel in the electron clouds. This electron flow is called the tunneling current. The tunneling current is extremely sensitive to the distance between the microscope tip and the surface of the sample and allows precise measurements of the vertical surface features of the sample. A feedback mechanism maintains the tip at a constant height as it sweeps across the sample surface, following the contours of the surface atoms. The motion of the tip is read and processed by a computer and displayed on a screen. A three-dimensional image of the surface is obtained by sweeping the tip in parallel lines (see Figure 1.33). To achieve high-resolution images of surface structures, the microscope must be shielded from even the smallest of vibrations, such as a footstep or sound.

Investigators probing the structure of DNA have generated images depicting the helical twists and turns of a strand of DNA (Figure 1.34). Because scanning tunneling microscopes can operate equally well with samples in air or liquid, the DNA can be observed under normal atmospheric conditions. (In an electron microscope, samples must be observed in a vacuum.)

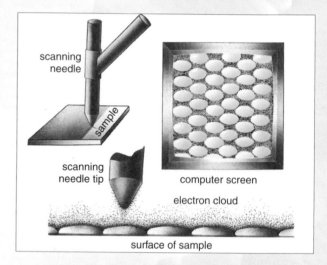

FIGURE 1.33

Mechanism of the scanning tunneling microscope. The shape of the electron cloud above a surface reveals the molecular structure of the surface.

FIGURE 1.34

A DNA molecule seen with the scanning tunneling microscope, (×2,000,000). The microscope reveals the shape of the DNA molecule.

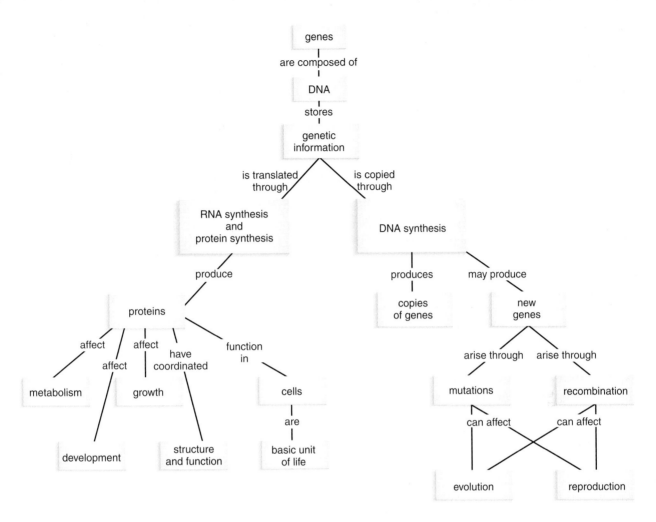

FIGURE 1.35

Functions of DNA. This concept map summarizes the major cellular processes that involve DNA. Note that both the information stored in the structure of DNA and the copying of that information during DNA synthesis have important effects on the life of a cell.

determines function. Proteins perform most of the chemical reactions that keep cells alive. The structure of DNA also accounts for its ability to be copied and passed through inheritance from one generation to the next. You will learn more about these processes in later chapters, but for now they are summarized in the concept map in Figure 1.35.

Check and Challenge

1. What determines which nitrogen bases form pairs in DNA?
2. If one DNA strand has the sequence AGTTC, what is the sequence of the opposite, or complementary, strand?
3. List four life processes in which DNA plays an important role.

Summary

Chemical reactions occur continuously in cells, the basic units of life. Cells carry out their biological functions through chemical reactions. Elements are materials that cannot break down into substances with new or different properties. They are the basic chemical form of matter. Atoms contain a characteristic number of positively charged protons, negatively charged electrons, and neutral neutrons. All elements are composed of atoms. Ions are atoms or molecules that have gained or lost electrons; ions have a positive or negative charge. Chemical bonds hold atoms together to form molecules. When those bonds involve the sharing of electrons, the bonds are called covalent bonds. When those bonds involve the attraction of two oppositely charged ions, the bonds are called ionic bonds. Weak bonds involving a partially positive hydrogen atom are called hydrogen bonds.

In chemical reactions, molecules interact and form different substances. Those interactions are accompanied by energy changes that drive the reactions living cells use for growth, reproduction, and all other functions. The membranes that surround cells enclose the chemical reactions of the cells and isolate them from the outside environment.

Organisms contain four major types of macromolecules. Carbohydrates store and transfer energy in cells and contribute to cell structure. Lipids store energy and form a major part of membranes. Proteins are structural components of cells, enzymes, and messengers and receivers of messages. The nucleic acids DNA and RNA store, transfer, and direct the expression of genetic information. Chemical compounds have biological activity because of their specific chemical structures. Chemical structure dictates biological function.

Key Concepts

Below is the beginning of a concept map. Use the concepts that follow to build a more complete map. Add as many other concepts from the chapter as you can. Include appropriate linking words.

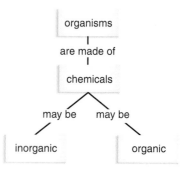

- atoms
- carbohydrates
- chemical bonds
- chemical reactions
- elements
- lipids
- macromolecules
- molecules
- nucleic acids
- proteins

Reviewing Ideas

1. What type of information does the periodic table of elements provide?
2. State in your own words the relationship among atoms, protons, neutrons, and electrons.
3. What are isotopes and why are they biologically important?
4. How does an ion differ from the atom from which it was formed?
5. Compare ionic, nonpolar covalent, and polar covalent bonds.
6. What is the importance of water in cells?
7. What is meant by pH, and how does pH affect cells and cellular processes?

 BSCSblue.com/vocabulary_puzzlemaker

8. Describe the structure of an amino acid. What type of bond forms the link between amino acids?
9. What are the building blocks of nucleic acids? Why are nucleic acids important in biology?
10. Why are lipids important to cell structure and function?

Using Concepts

1. Using simple molecular models, draw the electric attractions that allow sodium and chloride ions to dissolve in water. (Hint: Represent water as in Figure 1.9.)
2. Distinguish between amino acids and peptides and between simple peptides and polypeptides.
3. You are an analytical chemist working in a nutritional analysis laboratory. Someone brings you a tropical food made of only one type of macromolecule. How would you determine whether it is carbohydrate, fat, protein, or nucleic acid? (Reference: **Investigation 1B**)
4. You are given two samples of carbohydrates. One contains 6 carbon atoms per molecule (molecular formula $C_6H_{12}O_6$), and the other contains 12 carbon atoms ($C_{12}H_{22}O_{11}$). Which one is glucose and which one is sucrose? Explain your answer.
5. Animals usually store energy reserves as lipids; plants store them as polysaccharides. What is the advantage to animals of using lipids as storage molecules?
6. Margarine is produced by hydrogenating liquid vegetable oils so that they become solids like butter. What chemical change has taken place in the fatty acids?

Synthesis

Fat is more abundant in some cells of your body than in other cells. Explain how natural selection could account for that variation.

Extension

DNA is the source of genetic information for all living organisms, but the sequence of DNA nucleotides in organisms differs. In fact, the sequences of DNA can be characteristic for different species. If you were to compare the sequence of DNA nucleotides of similar and dissimilar organisms, what patterns would you observe? How is this evidence of evolutionary relatedness?

Web Resources

Visit BSCSblue.com to access
- Resources to help you understand important chemical concepts and the chemical basis of life
- Explanations and self-tests on biochemistry and an explanation of the chemical basis of your sense of taste and smell
- Web links related to this chapter

CONTENTS

LEARNING OUTCOMES

By the end of this chapter you will be able to:

A   Discuss why organisms need energy and how they obtain it.

B   Describe energy flow through an ecosystem.

C   Relate the first and second laws of bioenergetics to their implications for living systems.

D   Distinguish between synthesis and decomposition reactions in metabolism.

E   Summarize the importance of ATP in cellular energy transfer.

F   Describe how digestion breaks food into small molecules.

Energy, Life, and the Biosphere

■ *How does this scene demonstrate the relationship between life, energy, and the biosphere?*

■ *What is the source of energy that maintains these organisms?*

An alien spaceship orbits a planet making observations. Then it ejects a special robotic probe toward the planet's surface. Once retrorockets and parachutes slow the probe and allow it to land, a camera begins to scan the bleak landscape, and sensors measure the composition of the atmosphere and soil. The probe's mission: to determine whether life exists on the planet.

Does the mission seem simple to you? What do we mean when we say an object is living? No definition of life is completely satisfactory, but biologists agree that the ability to absorb and convert energy is a basic characteristic of life. Thus the energy required for the growth and organized maintenance of cells, organisms, and even communities is a central topic in biology. The study of the energy flow and energy transformations among living systems is called bioenergetics. This chapter examines the characteristics of living things and introduces the underlying topic of energy flow.

Organisms and Energy

2.1 Characteristics of Organisms

Try Investigation 2A ▽ **Are Corn Seeds Alive?**

If you were a member of an astronaut team sent to another planet, how could you tell whether objects on that planet were alive? You could gather clues and match them to your prior ideas of what constitutes life. To answer the question with confidence, you need to consider what characteristics living things have in common.

Movement and growth are two characteristics you might include in a definition of life, yet both living and nonliving things can move and grow. What makes living things different from nonliving things?

To help answer this question, consider not only *what* something does but also *how* it does it. Clouds move because they are carried by an external energy source, the wind. Clouds do not move on their own. Living things move by obtaining energy and transforming it into motion.

Table 2.1 lists characteristics generally found in living organisms (although not all organisms possess these characteristics to the same degree). Later chapters in this book consider these characteristics in detail. As you read these chapters during the year, you may wish to refer to the list. Compare the list with your own changing ideas about the characteristics of organisms. How are the two lists alike? How are they different?

TABLE 2.1
Characteristics of Organisms

- Take in and convert materials and energy from the environment; release wastes
- Have a high degree of chemical organization compared to nonliving objects (Figure 2.1)
- Have complex structural organization that is responsible for their appearance and activities
- Contain coded instructions (such as DNA) for maintaining their organization and activities
- Sense and react to changes in their environment
- Grow and develop during some part of their lives
- Reproduce others like themselves
- Communicate with similar organisms
- Move under their own power

a

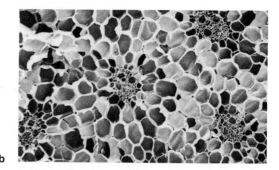
b

FIGURE 2.1

Organization of living and nonliving things. The arrangement of crystals in a granite rock **(a)**, shows less organization than the highly-ordered structure of a corn stem, *Zea mays* **(b)**, seen in cross section.

Biological Challenges

RESEARCH

Identification of Life

In 1976, two U.S. spacecraft, *Viking I* and *II*, arrived at Mars. Their primary purpose was to determine if life existed on the planet. Although no one imagined that Mars harbored higher forms of life, almost everyone agreed that microscopic forms might be present.

Each *Viking* lander carried instruments to detect soil microorganisms. Three experiments were designed to test for life in Martian soil. Each experiment exposed Martian soil to various substances and then monitored what happened. The assumptions were that if microorganisms were present in the Martian soil, they would take in food and give off waste gases or take in gases from the atmosphere and convert them to useful substances.

In one experiment, a nutrient solution was sprayed onto Martian soil (Figure 2.2). The carbon atoms in the nutrient solution were radioactively tagged so they could be followed. If Martian microorganisms "ate" any of the supplied nutrients and released the carbon atoms in a gas, the radioactive gas could be detected. This experiment seemed to give a positive result. Something in the soil chemically broke down the nutrient—almost like digestion.

Another experiment introduced gases from Earth into the Martian soil sample. The gases chemically combined with the soil—almost as if microorganisms synthesized organic material from the gases.

Some of the experiments suggested there might be life on Mars, but the results could not always be duplicated. In science, irreproducible results should be viewed with skepticism. The biggest drawback of these experiments was the lack of controls. How can you control for something you do not know about? Moreover, it is nearly impossible to repeat the experiments on Earth because we don't have soil from Mars, and duplicating conditions on Mars would be difficult. Still, the experiments had

too many unknown factors to say whether life was in Martian soils.

In early 2004, two U.S. mobile vehicles, "rovers," arrived at different parts of Mars to search again for life (Figure 2.3). This mission has been a major engineering success, and the rovers have found more evidence for water existing on Mars in the geologic past. Still, advanced cameras have not detected new evidence for life on Mars. Most scientists agree that a next step in the exploration of Mars will be to return soil samples to Earth for more careful analysis.

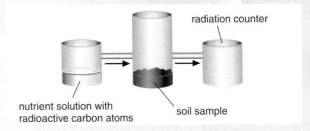

FIGURE 2.2

An experiment to detect life on Mars. The radiation counter is meant to detect gases produced by organisms as they digest the radioactive nutrients.

FIGURE 2.3

A U.S. mobile vehicle on Mars in 2004. Two mobile vehicles called "rovers" arrived on Mars from the U.S. in early 2004. The rovers were used to search for new evidence of life on the planet.

2.2 Energy and Nutrients

Living things need energy. Energy is the capacity to do work or to cause change. Organisms store energy in the organic molecules from which the organisms are made. Such energy is known as **chemical energy.** Organisms can use some of this energy as it is released during chemical reactions. The portion of this chemical energy that is available to do work is called **free energy.** Examples of free energy are the energy plants use for growing and producing food and the energy you use for exercise and thinking.

Living cells need a constant source of free energy for chemical and mechanical work and for transport. Chemical work includes constructing and breaking down large complex molecules, such as proteins and nucleic acids. Organizing these molecules into the larger structural components of cells, such as in muscle and skin, is also chemical work. Transport work involves the movement and concentration of the raw materials, or **nutrients,** needed to make complex molecules and to increase cellular organization during growth. Mechanical work includes movement, such as muscle contractions that enable you to kick a ball.

How do organisms obtain energy and nutrients? Some organisms, known as **heterotrophs,** obtain energy and nutrients from other organisms, either living or dead. Animals, fungi (mushrooms and molds), and most bacteria are heterotrophs. Other organisms obtain energy and

Try Investigation 2B ◆ Food Energy.

Focus On
Kilocalories

The unit of measure for energy traditionally has been the calorie, the amount of heat energy required to raise the temperature of 1 g of water from 15°C to 16°C. The unit of measure for food energy is the kilocalorie (kcal; also called the large Calorie), which is 1,000 calories (*kilo-* means 1,000). In the international system, the joule (J) is now the unit assigned to energy, but the kilocalorie still is used in biology. One kilocalorie is equivalent to 4.18 kilojoules (kJ).

The energy available from a specific food (Figure 2.4) can be measured by burning completely the food in a calorimeter. This chamber is designed to record energy changes while preventing energy exchange with the outside environment. For example, if 1 g of sugar is burned in a calorimeter, 4 kcal of heat energy are released. The mass of a teaspoon of sugar is about 5 g, so a teaspoon of sugar provides 20 kcal of energy. You use some of that energy for chemical and physical work and some to maintain your body temperature.

FIGURE 2.4

The "Nutrition Facts" label from a can of vegetable juice. The energy supplied by one serving is listed near the top in Calories, which are equivalent to kcals (kilocalories, or 1,000 calories).

nutrients from nonliving sources such as the Sun, minerals, and the air. These organisms are **autotrophs.** Autotrophs include plants, certain bacteria, and other organisms that capture energy from the Sun or from chemicals such as hydrogen sulfide (H_2S). Autotrophs use that energy to synthesize organic compounds from inorganic materials absorbed from their surroundings.

Many autotrophs capture energy from sunlight and use it to synthesize organic compounds from carbon dioxide and water, a process called **photosynthesis.** These organisms are called **photoautotrophs.** Some of the energy captured in photosynthesis is stored as chemical energy in organic compounds, which later serve as the source of free energy for cellular work. Other autotrophs, all of which are bacteria, can obtain free energy from inorganic chemicals in the environment. They use **chemosynthesis** to capture energy, which is stored as chemical energy and used for cellular work. This type of autotroph is a chemoautotroph. Autotrophs use the organic compounds they make as building blocks for maintenance, growth, and reproduction.

Heterotrophs consume autotrophs and other heterotrophs as food. Because only autotrophs can capture energy from inorganic sources, autotrophs directly or indirectly supply the energy and organic nutrients needed for the maintenance, growth, and reproduction of all heterotrophs (Figure 2.5). You are a heterotroph and must eat to obtain the nutrients and energy you need to grow and develop. The chemical energy that you obtain from your food can be traced back to its source—solar energy that plants and other autotrophs collected and converted to chemical energy. Even when you eat meat or other animal products, you are consuming energy sources that are derived from the plants those animals ate.

Both autotrophs and heterotrophs carry out chemical reactions that release the free energy of organic compounds. These reactions are known as **cell respiration** and are the subject of Chapter 5.

CHEMISTRY TIP

Organisms use the free energy released during chemical reactions. "Free energy" is free to be used, but it is not "free" in the sense of costing nothing. The molecules in which the energy is stored as chemical energy are broken down in the chemical reactions that make this energy available.

ETYMOLOGY

hetero- = other (Greek)
auto- = self (Greek)
-troph = feeding (Greek)
photo- = light (Greek)
chemo- = chemical (Latin, Greek)

Heterotrophs, such as animals, obtain food from other organisms. **Autotrophs** make their own food. Some autotrophs use light energy to do this in **photosynthesis.** Others use chemical energy to perform **chemosynthesis.**

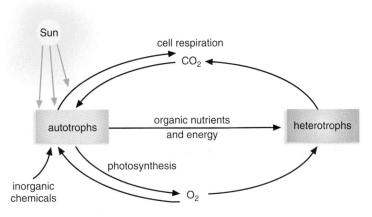

FIGURE 2.5

The relationship between autotrophs and heterotrophs. Energy passes from autotrophs to heterotrophs. Oxygen and carbon dioxide cycle repeatedly between them.

2.3 Energy and Ecosystems

The need for energy and nutrients links organisms in many complex ways. Autotrophs, which produce food other organisms use, are the **producers** in a community of living organisms, such as a forest or an ocean. Heterotrophs consume plants and other organisms for food; they are the **consumers.** For example, the trees in a forest or the seaweeds in an ocean capture energy from the Sun to make their own food from nutrients in the air, soil, and water. In turn, animals eat the plants, or they eat animals that have eaten plants. The plants are producers, and the animals are consumers. Bacteria, fungi, and other heterotrophs that break down and use dead plants and animals for food are **decomposers.**

Producers, consumers, and decomposers form a **food web** (Figure 2.6). Energy and nutrients flow from the environment through the producers to the consumers and finally to the decomposers. Decomposers complete the breakdown of organic nutrients and return inorganic nutrients to the soil or water, where they are available for reuse. These inorganic nutrients include mineral ions such as potassium (K^+) and iron (Fe^{+2}) and nitrogen-containing ions such as nitrate (NO_3^-) and ammonium (NH_4^+). Plants are especially important in returning these nutrients to the food web. They absorb nutrient ions dissolved in soil water through their roots. Heterotrophs obtain these nutrients by consuming plants.

The organisms in a food web depend on **abiotic,** or nonliving, factors, such as the soil, minerals, water, and weather. The organisms make up the **biotic,** or living, factors. The biotic and abiotic components of a particular place make up an **ecosystem** (Figure 2.6), such as a forest, a pond, or a prairie. Within each ecosystem are many **habitats,** places where particular organisms live. For example, in a pond, some organisms are bottom dwellers and others live along the shore; these are different habitats. All ecosystems combine to make up Earth's **biosphere.** The biosphere contains many ecosystems, such as coral reefs, deserts, marshes, and forests.

Check and Challenge

1. Automobiles use energy from gasoline to move. Is an automobile alive? Cite evidence to support your answer.

2. What is the relationship between energy and nutrients in an organism?

3. Distinguish between heterotrophs and autotrophs and give examples of each.

4. Describe the two processes by which autotrophs extract energy from the environment.

5. Describe the relationships among producers, consumers, and decomposers in an ecosystem.

Biology Online BSCSblue.com/check_challenge

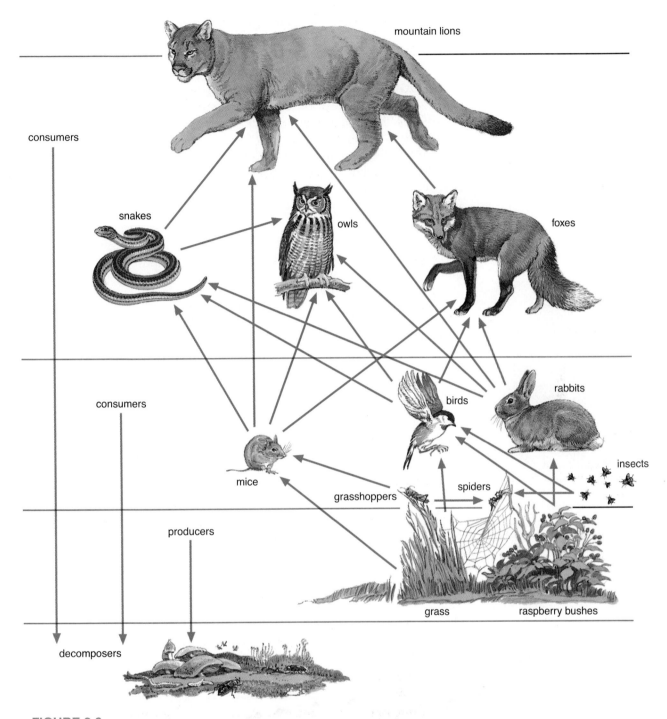

consumers

consumers

producers

decomposers

mountain lions

snakes

owls

foxes

birds

rabbits

insects

mice

grasshoppers

spiders

grass

raspberry bushes

FIGURE 2.6

A food web. Producers (autotrophs) support all the other organisms. Energy and nutrients (arrows) pass from producers to consumers and finally to decomposers. Decomposers include both microorganisms, such as many soil bacteria, fungi, and various other single-celled heterotrophs, and a number of animals, such as earthworms and many types of insects. Nearly all the organisms in this diagram are animals or plants. In aquatic environments, however, microscopic organisms are important as producers and primary consumers—organisms that consume producers directly.

Energy Flow

2.4 Energy Conversions

ETYMOLOGY

thermo- = heat (Greek)
dynamic = power or force (Greek)

Thermodynamics is the study of energy changes, especially the conversion of energy between heat and other forms.

Energy conversions are described by principles called the laws of thermodynamics. The **first law of thermodynamics** states that energy cannot be created or destroyed, but it can change form. On a broader scale, the first law is called the **law of conservation of energy.** This law states that the *total* energy of the universe is constant. Energy can exist in many forms, such as a log, an electric current, sunlight, a carbohydrate, or an ecosystem.

The chemical energy in a log is stored in its molecules. When a log burns, most of its energy changes to heat energy, which escapes into the surroundings. A small amount of the log's chemical energy remains in the ash and smoke. Eventually the heat energy spreads out into the surroundings, where it is no longer available to do work.

The first law of thermodynamics means that organisms cannot create their own energy, but must obtain it from an outside source. The source of energy for autotrophs is the Sun or inorganic chemicals; for heterotrophs, it is the chemical energy in food. For example, a wolf consumes a deer as a source of energy and nutrients (Figure 2.7). The wolf digests the food to simple compounds—such as glucose, fatty acids, and amino acids. The wolf's cells break down these compounds further. This releases free energy, which is used for cellular work, such as muscle contraction, growth, and tissue repair.

Like a burning log, the wolf releases some energy as heat energy (Figure 2.8). In the wolf, however, the conversion of chemical energy also releases free energy to do work in the living cells of the animal. That difference is a major distinction between living and nonliving things.

FIGURE 2.7

Biological energy conversion. A grey wolf *(canis lupus)* consumes a deer, a source of energy and nutrients.

 CHEMISTRY TIP

Reactions that absorb energy, such as photosynthesis, are called endothermic reactions. Reactions that release energy, such as the breakdown of carbohydrate or fat that releases energy for cellular activities, are called exothermic reactions.

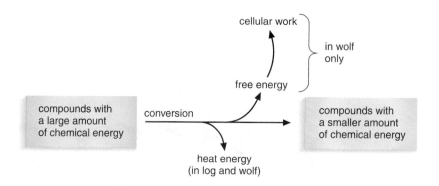

FIGURE 2.8

Energy conversion in a living and a nonliving system. An important difference between living and nonliving systems is the ability of living systems to conserve and use some of the energy released in chemical reactions.

2.5 Energy and Entropy

Note that, in the wolf, not all energy derived from food is useful. Like the log, a wolf releases energy to the surroundings as heat energy, which is no longer available to do work. Because energy tends to spread out into the surroundings, the free energy in a system is *slightly less* after each energy conversion than before. The **second law of thermodynamics** states that systems tend to change in a way that increases the disorder, or **entropy,** of the system plus its surroundings (Figure 2.9). The world becomes increasingly disordered as free energy is released. A burning log releases heat energy from its ordered molecules, which are reduced to ash, smoke, and water. The molecules in these materials are less ordered and contain less free energy than the macromolecules in the log. Free energy decreases and entropy increases as the log burns and the smoke disperses in the atmosphere.

What does the second law of thermodynamics mean for living systems? Organisms must be well organized to remain alive and to grow. For example, a pine seed contains stored energy and nutrients and a tiny immature plantlet called an embryo (Figure 2.10a). After the seed germinates, the embryo grows and develops into a tree. The mature pine tree is much more complex and organized than the embryo. Free energy was required for the embryo to grow into the organized tree. The source of this energy was sunlight, harnessed through photosynthesis.

A key to maintaining organization in all living systems is energy. In ecosystems, light or chemical energy flows from the environment (the

FIGURE 2.9

The tendency toward increasing disorder of a system and its surroundings. If no energy is added to maintain the system, it becomes disordered.

FIGURE 2.10

An example of biological organization. As an organism grows, order and organization increase. Compare the size and complexity of the embryo in the center of the seed **(a),** with the young tree **(b),** and the mature tree **(c).**

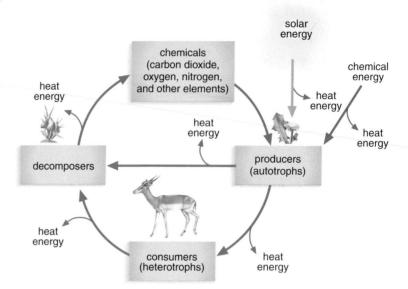

FIGURE 2.11

Energy flow in an ecosystem.
Organisms cannot recycle energy because their activities change it into a form (heat energy) that cannot be used to do work.

Sun or inorganic chemicals) to producers to consumers to decomposers. Producers, however, do not use all the energy they absorb in photosynthesis. (Photosynthesis and chemosynthesis, like all chemical reactions, are not 100% efficient.) Some of the light energy converts to heat energy, which escapes into the surroundings. Consumers convert part of the chemical energy from their food into free energy, which they use to do work. The remainder of the chemical energy becomes heat energy that escapes into the surroundings.

As energy flows through food webs, it eventually escapes into the surroundings in the form of heat energy (Figure 2.11). As a result, there is only a one-way flow of energy through food webs. The second law of thermodynamics states that all systems, including living ones, tend to become more disorganized. Living systems overcome that tendency toward entropy by constantly obtaining energy from their surroundings. Organisms stay organized and can function and grow only as the entropy of their surroundings increases. The total energy of the universe remains the same (first law of thermodynamics). It is, however, randomly dispersed as heat energy—an unusable form of energy for organisms—which increases the entropy of the universe (second law of thermodynamics).

Check and Challenge

1. How do the laws of thermodynamics apply to living systems?
2. How do organisms remain highly organized in spite of the universal tendency toward entropy?
3. What happens to a system as entropy increases?
4. What is chemical energy? How do organisms make energy available for work?
5. How are free energy and entropy related?

Metabolism and Energy Transfer

2.6 Enzymes and Energy

To release chemical energy to perform work, cells must have a way to break and form chemical bonds. Chemical reactions in most organisms take place within a fairly narrow range of temperatures. Those temperatures are not high enough to supply the activation energy (see Section 1.3) needed to start a reaction. How do organisms carry out the thousands of chemical reactions needed to survive?

All living cells contain specialized proteins called **enzymes** that lower the activation energy required to make a reaction proceed (Figure 2.12). In that way, enzymes greatly speed up chemical reactions that would otherwise occur too slowly to sustain life. The reactions do not consume the enzymes, which are thus reusable. Chemicals that lower activation energies are called **catalysts.** Enzymes are catalysts in living organisms.

Each type of enzyme catalyzes only one or a few specific reactions. The specific reaction catalyzed by an enzyme depends on a small area of its tertiary structure called the **active site** (Figure 2.13a). The active site has a shape that closely matches the shape of the starting molecule or molecules. The close fit of the starting molecule, called the **substrate,** into the active site of the enzyme brings the enzyme and substrate close together (Figure 2.13b). The

Try Investigation 2C Enzyme Activity.

CHEMISTRY TIP

Catalysts speed up chemical reactions by making it possible for them to occur in a different way that requires less activation energy.

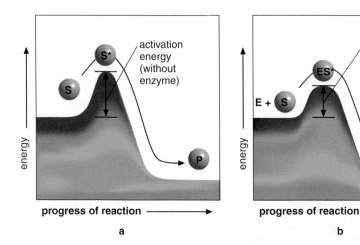

FIGURE 2.12

Activation energy without (a) and with (b) an enzyme. Consider the starting molecule, S, and the product molecule, P, which can be formed from S through a chemical reaction. To achieve an activated condition S*, S must acquire considerable energy. This activation energy can be represented as a hill that S must overcome before it will react to produce P. At normal temperatures and without an enzyme **(a),** most S molecules do not have sufficient energy to overcome the barrier. In an enzyme-catalyzed reaction **(b),** S combines temporarily with the enzyme E, forming a complex ES*, in which S requires less energy to form P (the barrier is lower). Because more molecules have enough energy to overcome the lower barrier, the reaction proceeds more rapidly than in the absence of the enzyme.

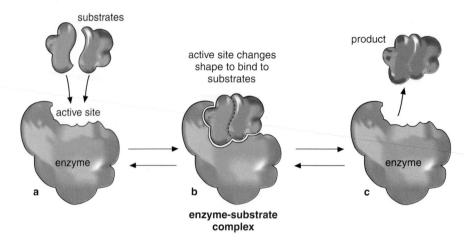

substrates

product

active site changes
shape to bind to
substrates

active site

enzyme

enzyme

a

b

c

enzyme-substrate
complex

FIGURE 2.13

The induced-fit model of enzyme action. (a), The substrates bind to the enzyme at the active site of the enzyme. **(b),** On binding, the enzyme changes shape to fit and hold the substrates. **(c),** The product of the reaction forms and the enzyme is recycled. The same enzyme can also catalyze the reverse reaction, a capacity of fundamental importance in biology.

resulting interaction lowers the activation energy, which allows the chemical reaction from substrate to product to proceed (Figure 2.13c). Because enzymes are not rigid, sometimes the binding of the substrate induces it to adjust its shape slightly, providing a better fit between enzyme and substrate. Once the chemical reaction occurs, the newly formed molecules break away, leaving the enzyme the same as it was before the reaction. Several thousand enzyme-catalyzed reactions may occur every minute in each of your cells.

Many reactions catalyzed by enzymes are reversible. With the aid of enzymes, two reacting molecules may combine to form a single molecule, or the single molecule may convert into the two smaller molecules. Thus an enzyme can speed up its specific reaction in *both* directions.

Enzyme reactions are faster at higher temperatures, but only within a narrow temperature range. Above certain temperatures, the enzymes break down. Enzyme activity also varies with the pH of the solution. Each enzyme is most active at a certain temperature and pH.

2.7 Chemical Reactions in Organisms

Chemical reactions occur continuously in all organisms. **Metabolism** consists of all the chemical activities and changes that take place in a cell or an organism. There are two types of metabolism: "building-up" reactions, or **synthesis,** and "breaking-down" reactions, or **decomposition.** Nearly all of these reactions involve enzymes.

Synthesis includes biosynthesis reactions that form larger, more complex biomolecules from small, less complex ones (Figure 2.14). Examples of biosynthesis reactions are the formation of starch from glucose and of DNA from nucleotides. Biosynthesis reactions build proteins from amino acids and, in turn, build tissues, such as muscle and blood, from the proteins. Photosynthesis builds sugars from carbon dioxide and water. Biosynthesis consumes free energy, because the products are more ordered and contain

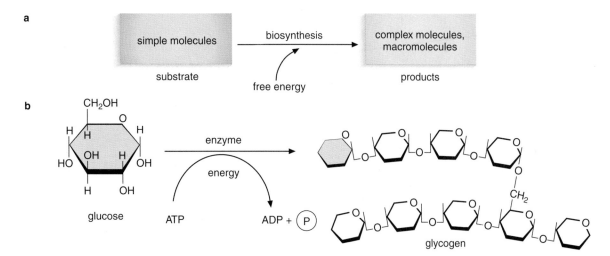

FIGURE 2.14

Biosynthesis of glycogen. Biosynthesis requires an input of free energy **(a).** The molecule ATP is used by all organisms as a storage form of cellular free energy. Decomposition of ATP releases some of its stored energy in a way that can perform cellular work, such as adding a glucose molecule to glycogen **(b).**

more chemical energy than the simpler, less ordered reactants. Biosynthesis enables organisms to grow and maintain their structure.

In decomposition reactions, larger molecules break down into smaller molecules. An example is the breakdown of glycogen to glucose in a muscle cell and of glucose to carbon dioxide and water during cell respiration. The energy stored in the glucose becomes available to the muscle cell or for other biosynthesis reactions (Figure 2.15).

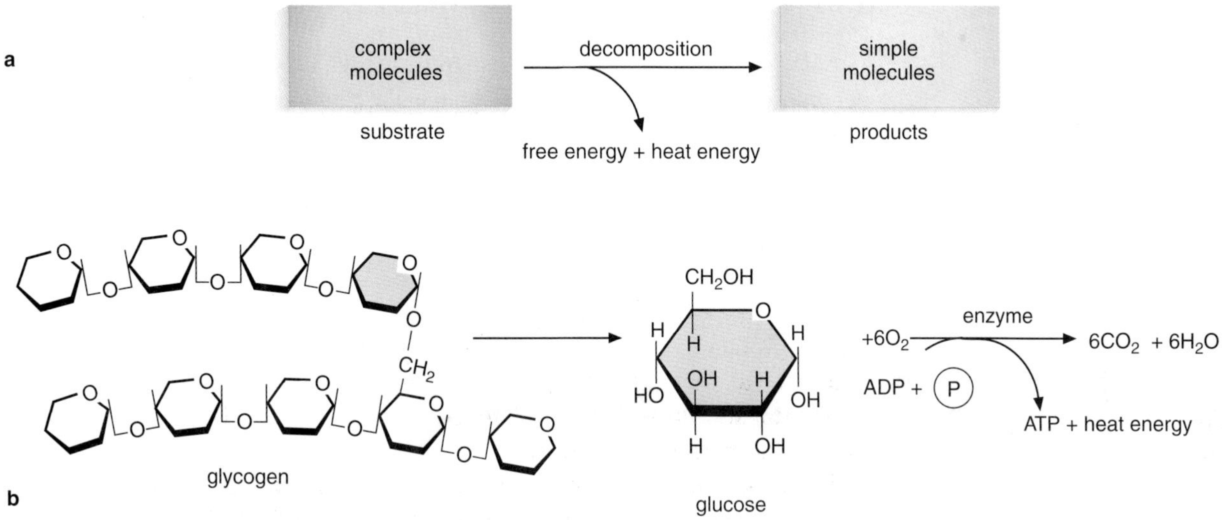

FIGURE 2.15

Decomposition of glycogen (animal starch). In decomposition reactions, free energy, and some heat energy, is released from the breakdown of complex molecules **(a).** An example of decomposition is the breakdown of glycogen in muscle cells **(b).** Glucose, from glycogen, also can break down in several steps to carbon dioxide and water. During this process, energy in the form of ATP is released and is used by the cell.

Decomposition reactions in cells release free energy that is stored in the structure of complex molecules, such as starch and glycogen. Almost invariably, the release of that free energy involves oxidation, the removal of electrons from a molecule. The available electrons are donated to another molecule, a complementary process called **reduction.** The oxidation of ethanol (and reduction of formaldehyde) is shown below.

Trapping energy from the environment involves similar processes. When a plant absorbs light energy from the Sun (photosynthesis), proteins in the leaf gain electrons from water. The water molecule splits and loses an electron. The protein receives an electron, so it is reduced. In the process, water is oxidized and becomes molecular oxygen, which is released into the atmosphere.

In biological systems, oxidation and reduction usually involve hydrogen transfer as well as electron transfer. When a molecule gains an electron (reduction), it often gains a hydrogen atom as well. Likewise, when a molecule such as water loses an electron, it often loses a hydrogen atom. The energy implications of hydrogen in reduced organic molecules will become clearer when you learn about photosynthesis in Chapter 4.

$$CH_3{-}CH_2{-}OH \underset{\text{reduction}}{\overset{\text{oxidation}}{\rightleftharpoons}} \overset{\overset{\displaystyle H}{|}}{CH_3{-}C}{=}O + 2\ e^- + 2\ H^+$$

ethanol   acetaldehyde

Decomposition reactions release free energy because they produce simple molecules from complex molecules, which increases entropy. Cells use some of the free energy and simple molecules released during decomposition for biosynthesis of other macromolecules. Thus synthesis and decomposition reactions in the cell are coupled to energy flow and the cycling of matter (Figure 2.16). This is an important characteristic of living cells.

2.8 Energy Transfer and ATP

How does decomposition release free energy? Through a process of decomposition called **oxidation,** which is the removal of electrons from a molecule, certain bonds are broken and rearranged. Some of the energy of the original molecule is released as heat energy and free energy. The free energy follows a series of electron transfers and ultimately ends up in a molecule called **ATP,** adenosine triphosphate. In this way, the free energy that is released

FIGURE 2.16

Coupling of synthesis and decomposition. Synthesis and decomposition are coupled through energy changes in the cell. Decomposition makes available some free energy that can be used for synthesis; most of the energy released is in the form of heat. For example, the synthesis reactions of photosynthesis convert light energy into chemical energy that can be used in the decomposition reactions of cellular respiration.

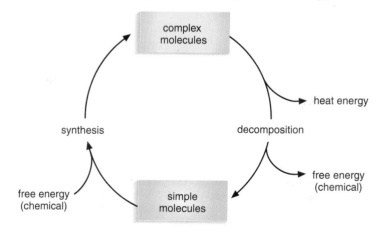

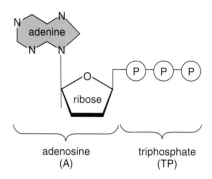

FIGURE 2.17

Structure of ATP. Where is the chemical-free energy stored in ATP?

through oxidation is available in chemical form (ATP) to drive the reactions of biosynthesis. ATP connects many energy-conversion reactions during metabolism and has been called the "energy currency" of living cells.

To see why ATP has been compared to money, imagine foreign tourists who arrive in New York City with only foreign currency. The tourists must pay a fee to change their foreign money into dollars, which they could use for their purchases. In a similar manner, a cell carries out chemical reactions that exchange the chemical energy of various organic compounds (foods) for the chemical energy of ATP. Then ATP pays most of the energy "debts" inside a cell. The "fee" is the energy lost as heat energy during the conversion.

ATP is a nucleotide consisting of adenine and ribose joined to a chain of three phosphate groups (Figure 2.17). The structure of ATP makes it an efficient and useful energy-transfer molecule in cells. Usually when an ATP molecule is involved in a chemical reaction, the bond between the second and third phosphate groups breaks and free energy is released.

In many reactions, the third phosphate group is transferred to an intermediate molecule, which carries some of the energy previously stored in ATP. The molecule that accepts the phosphate group from ATP gains free energy and is activated; it can then react usefully with other molecules in the cell. Figure 2.18 shows how the biosynthesis of glycogen from glucose uses ATP to form an activated intermediate compound, glucose-6-phosphate. Cells can use energy stored in ATP to supply activation energy or to drive energy-requiring reactions.

ATP is continually synthesized and broken down in cells. This synthesis and breakdown forms a cycle (Figure 2.19). When a molecule of ATP gives

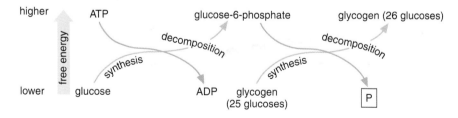

FIGURE 2.18

An example of biosynthesis using ATP. ATP is used to synthesize glucose-6-phosphate, an activated form of glucose, or blood sugar. Glucose-6-phosphate added to a chain of glycogen lengthens it by one glucose molecule. Glycogen is a useful way to store glucose. (⃞P is inorganic phosphate.)

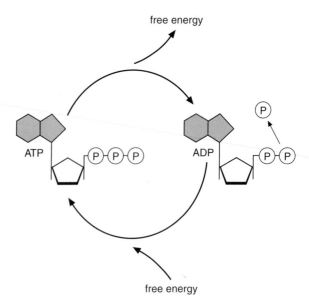

FIGURE 2.19

The ADP-ATP cycle. ATP molecules are continually rebuilt from ADP molecules, phosphates, and chemical energy.

CONNECTIONS

The use of ATP as an energy carrier in all organisms is evidence of a common evolutionary origin.

up one phosphate group, it becomes **ADP,** adenosine diphosphate. To form a molecule of ATP again, ADP must combine with one phosphate group. This requires free energy from the breakdown of organic compounds. ATP thus acts as an energy carrier, a go-between for the reactions in the cell that release energy and the reactions that consume energy. GTP (guanosine triphosphate) and UTP (uridine triphosphate) also are energy carriers in cells, but ATP is used much more often.

Cells need energy not only to fuel biosynthesis but also to remove wastes, take in nutrients, maintain internal ion concentrations, and move from place to place. ATP supplies much of the energy for this transport and mechanical work. ATP is the principal energy currency of living cells and was probably an energy carrier of primitive organisms. Today, ATP is used as an energy carrier in all known living cells—from single-celled heterotrophs to cells of humans and plants. This provides strong evidence for the unity of life on Earth.

Check and Challenge

1. What are the components and functions of metabolism?
2. How do enzymes help accelerate reactions?
3. What are the connections between synthesis and decomposition? How are they similar? How are they different?
4. Give an example of a reaction in which synthesis and decomposition are linked. Why are such reactions important?
5. Explain why ATP is referred to as the "energy currency" of the cell.
6. How are oxidation and reduction reactions related to energy conversion?
7. Describe several specific ways in which cells use the energy that is supplied by ATP.

Biology Online BSCSblue.com/check_challenge

Digestion

2.9 Digestion Inside and Outside Cells

Food provides organisms with a source of raw materials and cellular energy, which are essential for growth and maintenance. Autotrophs make their own food, fungi absorb nutrients, birds eat seeds and insects, and cows eat hay, grass, and grains. Each of these organisms has a digestive system that is adapted to the type of food the organism obtains.

As you will learn in Chapter 3, only relatively small molecules can pass through membranes and into cells, where energy release takes place. Most food particles that animals ingest, however, are much larger than molecular size. In order to use the raw materials and energy present in food, a heterotroph must break down large food particles into macromolecules, such as carbohydrates, proteins, and fats. It then must break down those macromolecules into sugars, amino acids, and other small molecules that can enter its cells. The processes that break down food are known collectively as **digestion.**

Digestion consists of two parts, one physical and the other chemical. The breakdown of large pieces of food into smaller ones is the physical part of digestion. In many animals, movements of the digestive cavity help accomplish this task. For example, the gizzard of a bird's digestive system is specialized to grind up food, as are human teeth (Figure 2.20).

Physical digestion, such as chewing and grinding, increases the surface area of food, making the chemical part of digestion easier. The large surface area of the food allows enzymes greater access to the food particles. Chemical digestion involves the breaking down of complex food molecules into simpler ones. Enzymes in the various organs of the digestive tract control these chemical reactions. Without chemical digestion, nutrients obtained from large food molecules could not be absorbed and used by the organism.

Most animals, including humans, rely on **extracellular digestion**—digestion that takes place outside the cells. Although it may seem odd to

CONNECTIONS

Digestive systems reflect evolution by means of natural selection. Each type of system is adapted to the diet, the environment, and the requirements of the organism.

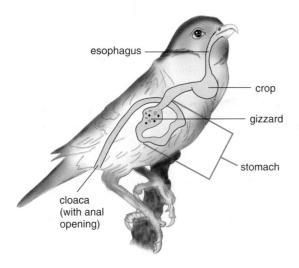

FIGURE 2.20

The digestive system of an animal. In some birds, food is temporarily stored in a sac called a crop. Farther along the digestive tract, a specialized part of the stomach—the gizzard—grinds up food to aid digestion. The walls of the gizzard are thick and muscular, an evolutionary adaptation to grinding. Some birds swallow sand and small pebbles that aid the grinding action.

esophagus

crop

gizzard

stomach

cloaca (with anal opening)

FIGURE 2.21

A Venus flytrap (Dionaea muscipula) in action. As soon as an insect lands on a leaf, the spiked blades start to close. Once trapped, the insect is digested by plant enzymes secreted into special leaf cavities. Such carnivorous plants obtain some of their nitrogen from insect prey and may have evolved in nitrogen-deficient soils.

refer to digestion as extracellular, consider that your digestive tract really is open to the outside environment, even though your body controls what enters and leaves. Most animals secrete digestive enzymes into a digestive cavity, where chemical digestion yields the simpler molecules that are then absorbed by the cells.

Most plant digestion takes place inside the cells with foods the plant has made itself. This type of digestion is called **intracellular digestion.** Digestive enzymes break down the food into small molecules that the cell can use. Digestion in single-celled organisms, such as *Paramecium*, occurs in the same way.

Some plants have the ability to capture insects and digest them in special cavities formed by the leaves. Figure 2.21 shows a Venus flytrap capturing an insect. After the insect is trapped, the leaf cells secrete enzymes that digest it.

Many organisms produce enzymes that digest food outside the organism itself and then absorb the nutrients into the cells. Most fungi, for example, digest materials from dead plants and animals. Bread mold is a fungus that secretes enzymes that diffuse out of the cells and digest bread (Figure 2.22). The mold then absorbs the products of digestion into its cells. Amino acids, sugars, minerals, water, and oxygen can diffuse into cells. Large molecules, however, cannot enter a cell by diffusion and are absorbed by other means.

Complex multicellular animals digest food in specialized cavities or digestive tubes with two openings. Food enters the mouth at one end of the tube, and material that cannot be digested passes out of the anus at the other end of the tube. The result is one-way movement of food and waste. The earthworm's digestive system, shown in Figure 2.23, is an example of such a digestive tube.

The digestive tube of most complex animals is divided into different regions with specialized functions. The specialization may depend on the diet of the animal. Goats and cows have four-chambered stomachs to handle

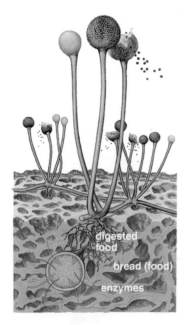

FIGURE 2.22

Bread mold. The expanded view at lower left shows that enzymes from the mold diffuse into the bread, digesting complex carbohydrates to simpler sugars. The mold then absorbs the sugars.

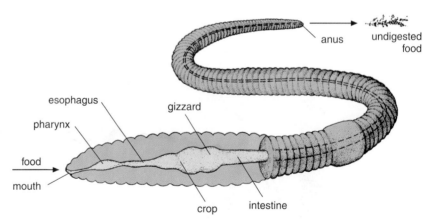

FIGURE 2.23

Digestive system of an earthworm. The earthworm's system is a complete digestive tube with two openings—the mouth at one end through which food is ingested, and the anus at the other end, through which wastes are eliminated. Digestion occurs extracellularly within the tube.

the hard-to-digest food that they eat. The stomachs of horses and rabbits have special side pockets in which microorganisms live and aid in the digestion of cellulose, a major component of grass. Animals that do not have such microorganisms, including humans, cannot digest cellulose. Carnivores consume mostly meat, which is more easily digested than grass. Consequently, a carnivore's digestive tract is relatively short.

2.10 An Overview of Human Digestion

Figure 2.24 shows the human digestive system. Refer to the diagram as you read about the events that take place during digestion. **Ingestion,** the process of taking food into the digestive tract, begins in the oral cavity. The chewing action of the teeth begins physical digestion, and the highly muscular tongue mixes food with saliva. **Saliva** is a watery secretion containing digestive enzymes that begin chemical digestion.

Try Investigation 2D ⬙ Starch Digestion.

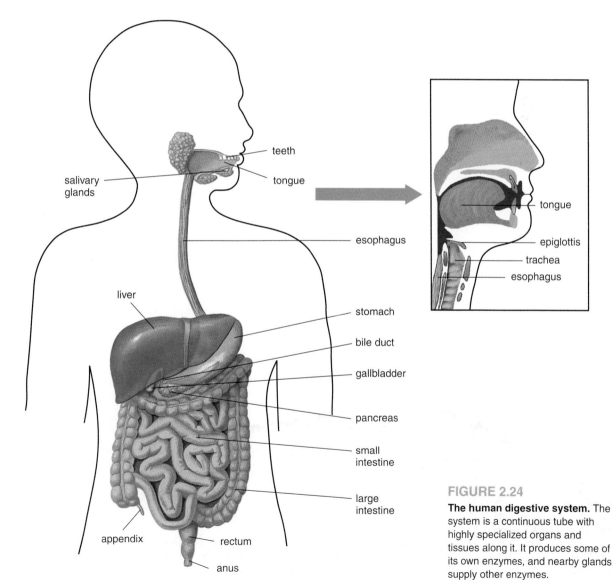

FIGURE 2.24

The human digestive system. The system is a continuous tube with highly specialized organs and tissues along it. It produces some of its own enzymes, and nearby glands supply other enzymes.

When you swallow food, it passes over the **epiglottis,** a trapdoor-like tissue that normally prevents food and liquids from entering the trachea (or airway). Food then enters the esophagus, a muscular tube connecting the oral cavity to the stomach. Wavelike contractions of the muscles of the esophagus move food to the stomach, a process called **peristalsis.** The same muscle action also moves food at later stages of digestion.

As the food moves along the approximately 9 m of digestive tract, it goes through various sections that function as a disassembly line. The next organ reached is the stomach. Contractions of muscles lining the stomach wall thoroughly break up and mix food with secretions from stomach glands. These secretions, called gastric juices, are composed of enzymes, mucus, and acid. As a result of their action, the contents of the stomach soon become souplike. From time to time, a ring of muscle that acts like a valve relaxes, releasing small amounts of partially digested food into the small intestine. After an average meal, the stomach usually empties in approximately 4 hours.

Food then enters the small intestine, a tube approximately 6 m long, where chemical digestion is completed and food molecules are absorbed. The pancreas and liver contribute digestive juices to the small intestine through ducts.

Food molecules are absorbed through the intestinal walls into the bloodstream. The blood carries the molecules to all the cells, where they are used in metabolism. Any undigested material eventually passes to the large intestine, where bacteria help produce several vitamins, gases, and other compounds. The vitamins and much of the water that was mixed with the food are absorbed through the walls of the large intestine. This absorption partly dries out the wastes, called **feces.** The feces are then eliminated through the anus.

2.11 Carbohydrates, Proteins, Fats, and Absorption

Carbohydrate digestion begins in the mouth with the action of an enzyme called **salivary amylase.** Amylase digests starch to shorter polysaccharides and maltose, or malt sugar, by breaking the chemical bonds in starch molecules and adding water molecules to the products of this breakdown. The following equation represents this reaction:

$$\text{starch + water} \xrightarrow{\text{salivary amylase}} \text{maltose}$$

Saliva has a pH between 6.0 and 7.4, and salivary amylase functions best in that pH range. The contents of the stomach, however, are very acidic (pH 1.0–3.5). Therefore no carbohydrate digestion takes place in the stomach.

Carbohydrate digestion is completed in the small intestine. (Figure 2.25 summarizes the functions of digestion.) The pancreas delivers pancreatic juices that convert the acidic food mixture to a basic pH again and contribute additional digestive enzymes. The maltose produced by salivary

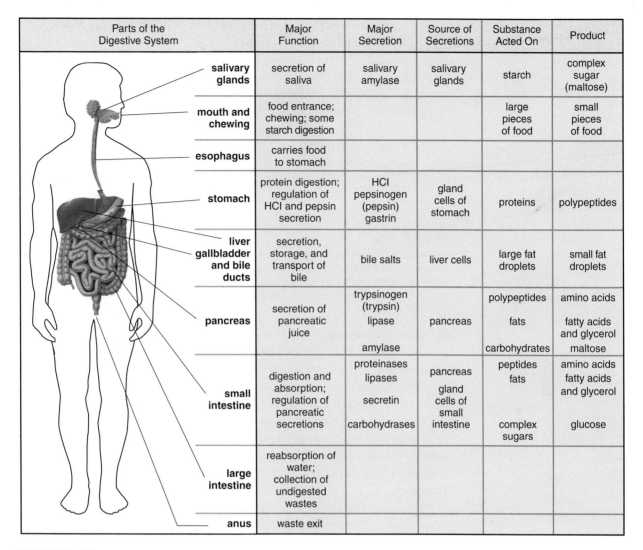

Parts of the Digestive System	Major Function	Major Secretion	Source of Secretions	Substance Acted On	Product
salivary glands	secretion of saliva	salivary amylase	salivary glands	starch	complex sugar (maltose)
mouth and chewing	food entrance; chewing; some starch digestion			large pieces of food	small pieces of food
esophagus	carries food to stomach				
stomach	protein digestion; regulation of HCl and pepsin secretion	HCl pepsinogen (pepsin) gastrin	gland cells of stomach	proteins	polypeptides
liver gallbladder and bile ducts	secretion, storage, and transport of bile	bile salts	liver cells	large fat droplets	small fat droplets
pancreas	secretion of pancreatic juice	trypsinogen (trypsin) lipase amylase	pancreas	polypeptides fats carbohydrates	amino acids fatty acids and glycerol maltose
small intestine	digestion and absorption; regulation of pancreatic secretions	proteinases lipases secretin carbohydrases	pancreas gland cells of small intestine	peptides fats complex sugars	amino acids fatty acids and glycerol glucose
large intestine	reabsorption of water; collection of undigested wastes				
anus	waste exit				

FIGURE 2.25

A summary of the parts and functions of the human digestive system. Note that trypsinogen and pepsinogen are the inactive forms of trypsin and pepsin, respectively.

amylase is further broken down to glucose by amylases added in the intestine. Most starch is digested in the small intestine, and the final result is usually glucose.

Protein digestion occurs in the stomach and in the small intestine. The enzymes that break down large protein molecules in the stomach require a strongly acidic environment. This condition is provided by stomach glands that secrete hydrochloric acid (HCl). The acid is so concentrated that it could destroy living tissue. The cells of the stomach lining are not harmed, however, because some of them secrete a thick, protective coat of mucus. Scientists now realize that many ulcers—wounds that form in the stomach wall—are caused by a bacterium (not acid) and can be treated with antibiotics.

As food enters the stomach, it stimulates certain cells to release a hormone called **gastrin,** which enters the bloodstream. When gastrin comes

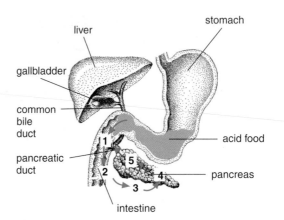

FIGURE 2.26

Regulation of digestion in the small intestine. Acidic food enters the small intestine **(1)** from the stomach and stimulates secretion of the intestinal hormone secretin **(2).** The secretin enters the bloodstream **(3)** and circulates to the pancreas **(4).** In the pancreas, secretin stimulates production of pancreatic juices **(5)** that flow into the intestine.

in contact with the glands that secrete hydrochloric acid, it acts as a signal that starts secretion. The nervous system is also a factor, so stress and tension can affect stomach-acid secretion.

The active protein-digesting enzyme in the stomach is **pepsin,** which is secreted by stomach gland cells in an inactive form called **pepsinogen.** Hydrochloric acid changes pepsinogen to active pepsin. Pepsin breaks large protein molecules into smaller polypeptides. Further digestion in the small intestine breaks down these polypeptides to amino acids. Humans can synthesize only about half of the amino acids necessary for making proteins. The others, called essential amino acids, must be provided by our diet.

When food enters the small intestine, pancreatic juice enters the intestine through the pancreatic duct and shifts the pH from acidic to basic (Figure 2.26). A basic pH is necessary for the intestinal enzymes to function. The intestinal enzyme **trypsin** breaks peptide bonds, producing amino acids from polypeptides.

Fats also are digested in the small intestine. Unlike carbohydrates and proteins, fats do not mix with water. Enzymes can digest only the fat molecules on the surface of the fat droplets. Fats are prepared for digestion by **bile,** a substance secreted by the liver and stored in the gallbladder. Bile salts physically break down fat droplets, increasing the surface area of the droplets available to the fat-digesting pancreatic enzymes. Bile does not contain digestive enzymes. The fat-digesting enzyme **lipase,** which is secreted in the pancreatic and intestinal juices, splits fats into fatty acids and glycerol.

The end products of digestion are amino acids, simple sugars, fatty acids, and glycerol (Figure 2.25). Cells can absorb these small molecules through their cell membranes and use them for free energy and as raw materials for building cellular structures.

These small molecules pass through the cells lining the small intestine. The surface area of the intestinal lining is increased tremendously by millions of small fingerlike projections called **villi** (singular: villus). Each

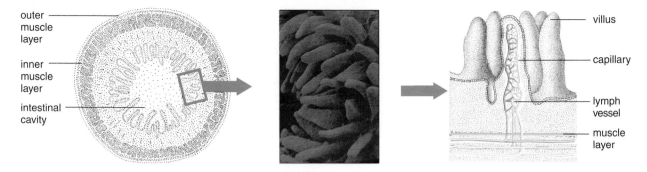

outer muscle layer

inner muscle layer

intestinal cavity

villus

capillary

lymph vessel

muscle layer

FIGURE 2.27

A cross section of the small intestine. Intestinal villi are shown enlarged in the micrograph (×136) and drawing. The products of digestion enter the blood and the lymph through the villi. Blood passes from arteries to capillaries to veins on its course through each villus.

villus contains **capillaries**—tiny, thin-walled blood vessels that serve as entry points to the bloodstream (Figure 2.27). Simple sugars, amino acids, fatty acids and some glycerol, minerals, and vitamins pass through the cells of the villi and enter the bloodstream through capillaries.

The blood carries the products of digestion to the cells. Inside the cells, the molecules are either broken down further to yield energy or used to synthesize substances the organism needs for growth and repair.

Check and Challenge

1. Trace the movement of food through the human digestive tract.
2. Explain the digestion of carbohydrates. Be sure to include the regions of the digestive tract where carbohydrate digestion occurs.
3. What is the role of the small intestine in the digestion of proteins and fats?
4. How does acid affect pepsinogen?
5. What is the role of gastrin in digestion?
6. Explain how digestive systems in different multicellular animals are adaptive.

Chapter
HIGHLIGHTS

Summary

All organisms require energy, which originates with the Sun and certain inorganic chemicals in the environment. Organisms use free energy to grow, move, reproduce, and maintain internal order.

Biotic and abiotic components together make up an ecosystem. Producers, consumers, and decomposers are the biotic components of an ecosystem. Energy flows in one direction from the Sun and inorganic chemicals to producers and then through a series of consumers and decomposers in the form of food. Living things cannot recycle energy because their activities convert it to heat energy, which they cannot use to do work.

The laws of thermodynamics allow us to predict the flow of energy in organisms and ecosystems. Organisms obey these laws. To stay alive, organisms must maintain and even increase a highly ordered state. This requires energy. Producers obtain energy from the environment. Consumers obtain energy from producers and other consumers. Decomposers obtain energy through the decay of dead producers and consumers.

Metabolism includes all the chemical reactions in an organism. Biosynthesis requires energy. Decomposition releases free energy. ATP connects these processes.

Energy changes drive metabolism. Enzymes speed up reactions in living cells that otherwise would not occur at adequate rates.

Heterotrophs break down food into molecules small enough to be absorbed and used by individual cells. Intracellular digestion takes place in the individual cells that will use the food. Most animals break down food by extracellular digestion, which takes place in a tube specialized for digestion. Specific enzymes break down fats, proteins, and polysaccharides in this system. The small molecules that are the products of digestion can be absorbed

and used by living cells as a source of energy or as nutrients for growth and repair.

Key Concepts

Use the concepts listed below to build a concept map of this chapter. Add as many other concepts from the chapter as you can. Include appropriate linking words for all concepts.

- abiotic factors
- producers
- food webs
- decomposers
- autotrophs
- heat energy
- enzymes
- entropy

- free energy
- chemical energy
- chemical reactions
- synthesis
- ATP
- metabolism
- absorption
- ecosystems

Reviewing Ideas

1. What is meant by "free energy" in organisms?
2. How do organisms use chemical energy to do work?
3. Discuss the functions and give examples of the living components of an ecosystem.
4. How are the biotic and abiotic components of an ecosystem related?
5. In your own words, state the first and second laws of thermodynamics.
6. What are the characteristics of enzymes? Why are enzymes essential to life?
7. Why is ATP the "currency" of metabolism?
8. How is ATP used as an energy carrier? Give an example.
9. What is the relationship among enzymes, energy, and reaction rates?
10. Why must most proteins, carbohydrates, and fats be digested?

11. Compare and contrast intracellular and extracellular digestion.
12. Explain how different organisms obtain food, and relate these methods to evolutionary adaptation.

Using Concepts

1. What would happen to ecosystems if there were no decomposers?
2. Figure 2.9 shows an example of entropy and the second law of thermodynamics. Think of another example and construct a model to illustrate it.
3. Organisms lose heat to their environment during metabolism. If this loss did not happen, would organisms still need an input of energy to survive? Explain.
4. You are part of a scientific team assigned to sample material brought back to Earth from Mars. What tests would you run on the samples to determine whether they contain life? Explain how you would run the tests and what information each test would provide.
5. Develop a concept map showing the structure and function of ecosystems.
6. How could you determine experimentally the number of kilocalories in a quarter-pound hamburger meal?
7. The first law of thermodynamics states that energy cannot be destroyed. What eventually happens to the energy from the Sun after it passes through an ecosystem?
8. When ATP breaks down, ADP forms. Does ADP contain energy? Explain. How could you answer this question in the laboratory?
9. Explain how synthesis and decomposition reactions are coupled in cells.
10. Why do cells use different enzymes to catalyze different reactions?
11. How is digestion affected by chewing food well as opposed to swallowing large bites?

Synthesis

1. How are chemical reactions, chemical bonds, and energy related?
2. How does ATP as a universal energy currency relate to the theory of evolution by natural selection?
3. Some of the enzymes in a particular organism are much more abundant than others. Explain how natural selection could select for that variation.
4. Relate what you know about human digestion to stomach and intestinal disorders, such as gas and heartburn. What might be the causes? How might antacids work?

Extensions

1. Describe how the first and second laws of thermodynamics might be used to explain the presence of some type of pollution (for example, smog, garbage, industrial waste) in your area.
2. Show in a cartoon the role of ATP in a cell.

Web Resources

Visit BSCSblue.com to access
- Additional information about energy flow in ecosystems, chemosynthesis (including deep-sea vent organisms), and thermodynamics
- Detailed information on the anatomy and physiology of digestion, including disorders that affect digestive function, such as ulcers and cystic fibrosis
- Web links related to this chapter

CONTENTS

LEARNING OUTCOMES

By the end of this chapter you will be able to:

A   Discuss the structure and function of membranes in living organisms.

B   Describe how materials are exchanged across membranes.

C   Explain how various organisms are adapted to maintain water balance while processing nitrogenous wastes.

D   Relate the structure of the human nephron to its function.

About the photo
This photo shows a
scuba diver breathing
under water with the aid
of an apparatus.

CHAPTER

3

Exchanging Materials with the Environment

■ *How do living organisms exchange materials with their surroundings?*

■ *What molecular processes are responsible for exchange?*

Consider this problem: You want to live in a safe, comfortable environment. So you build four thick walls and a sturdy roof around you. Now you are protected from wind, rain, and sun, and thieves cannot reach you. But you've got a new problem: You cannot bring in food, water, fresh air, or fuel. What if you want new clothes and visitors? How will you throw away your trash?

The solution to this problem is fairly simple. You build your room with doors and windows to let people and air move in and out. You have plumbing, gas, and electric lines. You may have a heater or air conditioner and telephone and cable lines. And you will probably put a lock on your door, so that you control who comes and goes. All these features control the world's access to your safe environment.

Living systems face similar problems. The surface of an organism is a barrier against destructive forces. That barrier, however, must allow the passage of food, water, waste, and communication signals if the organism is to survive.

Living Systems as Compartments

3.1 Exchanged Materials

ETYMOLOGY

cyto- = vessel or cell (Greek)
-plast or **-plasm** = molded shape (Greek)

Cytoplasm is the material that fills out a cell's shape.

This chapter will help you learn what materials an organism and its surroundings must exchange and how that exchange takes place. The simplest living organism is a single-celled bacterium. Its cell interior—its **cytoplasm**—is surrounded by a wall made of carbohydrates and proteins and a membrane made largely of phospholipids. The wall and membrane are barriers that separate this tiny compartment from the world around it (Figure 3.1). Even in large organisms that are made of many cells, such as humans, the cell is a protected compartment. A membrane encloses each cell like the walls of a room. Materials needed for life must enter each cell to be useful. What are those materials?

CHEMISTRY TIP

Ions have many roles in cells. Among other functions, they balance the charges of large organic molecules. For example, K^+ helps balance the negative charges of acid groups (—COO^-) of proteins.

First, and most important, organisms and their cells need water. Many chemical reactions essential for life use molecules that are soluble in water. Many cells also require oxygen for releasing the energy that powers cellular reactions. Cells also need the correct balance of ions, such as sodium (Na^+), magnesium (Mg^{+2}), calcium (Ca^{+2}), hydrogen (H^+), chloride (Cl^-), and potassium (K^+). The cells of autotrophs bring in carbon dioxide to build food molecules. Nutrients, including sugars and amino acids, must enter cells to supply energy and building material for cell components. Some

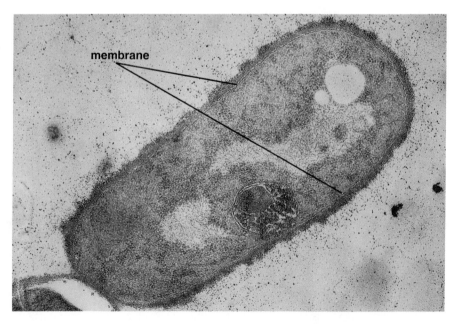

FIGURE 3.1

A *Bacillus megaterium* bacterium ($\times$30,500). A cell is a protected compartment separated from its surroundings by a barrier membrane. All cells are self-contained in this way.

hormones need access to the interior of cells to transmit messages. Besides all this traffic into a cell, wastes must exit. For example, ammonium ion (NH_4^+) is a potentially toxic waste. It constantly must be converted to a form that allows it to "go out with the trash" in the daily operations of a cell or a large organism such as a human.

3.2 Membrane as Barrier

Recall from Chapter 1 that membranes are composed of two thin layers of phospholipids and proteins. The lipids and proteins are fluid, not rigid, and move around each other like ice cubes in a punch bowl. How can this structure separate a cell from the external environment and at the same time regulate the flow of substances into and out of the cell?

Not all molecules are equally soluble in a membrane, as suggested by Figure 3.2. The nonpolar phospholipid tails of the lipid bilayer tend to repel charged particles such as ions but allow fat-soluble molecules to pass. Usually, the polarity, size, and electric charge of molecules determine whether they can pass through a membrane. For instance, water (H_2O) and ethanol (C_2H_5OH), small polar molecules, pass freely across the membrane. The small molecules of nonpolar gases such as nitrogen (N_2) and oxygen (O_2) also pass through.

Charged molecules such as the ions H^+ or Ca^{+2} can pass through only with the help of special proteins, called **transport proteins,** that are

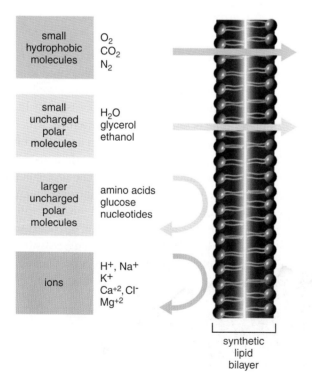

small
hydrophobic
molecules

O_2
CO_2
N_2

small
uncharged
polar
molecules

H_2O
glycerol
ethanol

larger
uncharged
polar
molecules

amino acids
glucose
nucleotides

ions

H^+, Na^+
K^+
Ca^{+2}, Cl^-
Mg^{+2}

synthetic
lipid
bilayer

FIGURE 3.2

A selectively permeable membrane. An artificial membrane composed purely of phospholipids would be permeable to some substances but would repel others.

outside

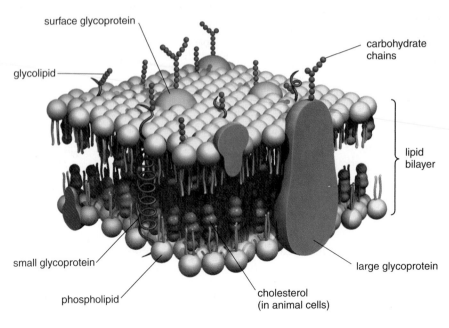

surface glycoprotein

glycolipid

carbohydrate
chains

lipid
bilayer

small glycoprotein

large glycoprotein

phospholipid

cholesterol
(in animal cells)

inside

FIGURE 3.3

The fluid-mosaic model of a membrane's structure. The bilayer of phospholipids provides fluidity. Proteins extending through the bilayer or embedded in the membrane's inner or outer surface make the surfaces asymmetric. Carbohydrate chains may be attached to both lipids and proteins.

ETYMOLOGY

per- = through (Latin)
meare- = to pass (Latin)

Only certain types of molecules can permeate, or pass through, a selectively **permeable** membrane.

CHEMISTRY TIP

Cholesterol molecules (See Chapter 1, Figure 1.20b) fit in the gaps between phospholipids and help regulate membrane fluidity and permeability.

embedded in the membrane. Their action is described in Section 3.4. Transport proteins also help fairly small polar molecules such as amino acids and sugars into the cell. Proteins and other very large molecules cannot pass through a membrane without special processes. By limiting entry in these ways, a membrane is **selectively permeable,** which means that membranes regulate the exchange of materials in a very specific way.

The structure of membranes is complex and allows them to perform many functions in the cell (Figure 3.3). For example, the proteins involved in ATP synthesis are situated in the membranes of specialized compartments within cells. Moreover, some proteins that are embedded in membranes have sugars attached to them. These proteins are called **glycoproteins.** Sugars also can be attached to the heads of membrane lipids **(glycolipids).** Glycoproteins and glycolipids act as antennae that receive chemical messages from other cells.

Check and Challenge

1. What materials need to be exchanged across a cell membrane?
2. In what way is an organism a protected compartment?
3. What is the significance of a membrane's being selectively permeable?

Biology nline BSCSblue.com/check_challenge

How Cells Exchange Materials

3.3 Diffusion and Osmosis

What happens when someone across the room opens a box containing hot pizza or freshly buttered popcorn? For one thing, that person suddenly is surrounded by hungry friends. How do these friends detect the food? The enticing aroma fills the room because a few molecules from the food migrate through the air and are detected by the nose of each hungry person. This movement of molecules is an example of **diffusion.** Diffusion refers to the movement of molecules from an area of higher concentration to an area of lower concentration.

Why does this movement happen? All molecules are in constant motion; they move and collide. The movement of each molecule is random, but there are more molecules in an area of higher concentration. In this initial situation, there are more chances that a molecule will move toward the less concentrated area than the other way around. Figure 3.4 illustrates this point. The result is that a substance moves from higher to lower concentration if it is not held back.

Try Investigation 3A ⬦ Cells and Movement of Materials.

Diffusion is a random process, and the entropy of the system increases as it occurs. In the final state, after a system has reached equilibrium or balance, there is less organization than when a concentration difference was present. Thus the increase in disorder is associated with an increase in entropy. (At equilibrium, molecules still move, but there is no longer a lower-concentration area into which to diffuse.)

When there is a difference in concentration of molecules across a distance, a **concentration gradient** exists. If you made a graph of the concentration versus the distance from a given spot, the graph would resemble Figure 3.5. Notice that the shape of this graph resembles a ramp. Going up the ramp requires energy; coming down is much easier. Because molecules diffuse from regions of higher concentration to regions of lower concentration, they are described as moving down their concentration gradient. Diffusion is a basic process underlying the movement of molecules into and out of cells.

CONNECTIONS

The tendency of diffusion to reduce concentration gradients is an example of the Second Law of Thermodynamics at work.

a   b   c

FIGURE 3.4

An example of diffusion. Molecules move from an area of higher concentration to an area of lower concentration until the concentration is the same throughout. In **(a)**, a crystal of potassium permanganate ($KMnO_4$) was dropped into a glass of water. The molecules diffuse through the water **(b)** until they are evenly distributed throughout **(c)**.

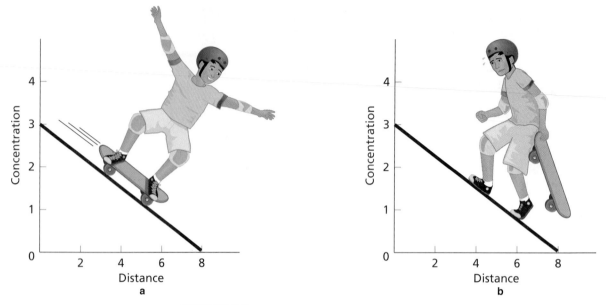

FIGURE 3.5

Energy and diffusion. The graphs represent a concentration gradient, a difference in concentration across a distance. **(a),** Entropy favors rapid diffusion down a concentration gradient, when the gradient is steep. **(b),** Moving against a concentration gradient takes energy. Diffusion in liquids effectively moves substances only for short distances.

Because membranes are selectively permeable and do not allow everything to pass through them, concentration gradients can build up across the cell membrane. For instance, the barrier formed by a membrane can act like a dam that holds back the water of a lake. In a cellular compartment, the membrane may hold back ions such as H^+. A great amount of potential energy is stored in this way, just as the water behind a dam has the potential to rush out if the dam is opened (Figure 3.6). The potential energy is based on the concentration gradient of substances. If the substance in question is charged (such as Na^+, Cl^-,

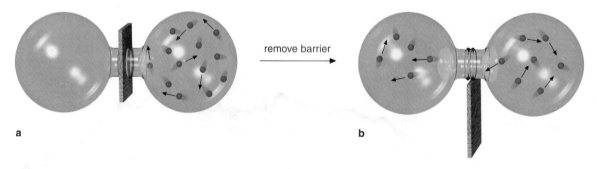

FIGURE 3.6

Diffusion of a gas. The motion of molecules in a glass container is random, but the net result is movement from an area of higher concentration to one of lower concentration. Initially **(a),** a barrier separates the two bulbs, with gas molecules (and potential energy) concentrated on the right side. When the barrier is removed **(b),** molecules begin to appear in the left-hand bulb. What happens to the concentration in the right-hand bulb? What happens to the gas pressure in each bulb?

or H+), an electric potential also forms across the membrane. For instance, H^+ gradients in membranes result not only in a difference in pH but also an electric potential. Nerve impulses rely on the buildup of electric potential. Concentration gradients provide potential energy to drive many cellular processes, including some types of membrane transport.

A special form of diffusion is the movement of water across a selectively permeable membrane. This movement of water down its concentration gradient is called **osmosis.** Osmosis is responsible for the movement of water across membranes. If an animal cell is placed in pure water, the concentration of water outside the cell is higher than inside. Water moves in, and the cell swells dangerously, possibly breaking open. If an animal cell is placed in a concentrated solution of a substance such as glucose, osmosis drives water out and the cell shrinks. Figure 3.7 illustrates how plant and animal cells respond to concentration differences. Cells of plants, fungi, and bacteria are surrounded by a fairly rigid cell wall (as well as a membrane), which helps protect them from too much swelling. The outward pressure of a cell against its cell wall is called **turgor.**

The rate of diffusion, including osmosis, depends on several factors. The size of the concentration gradient is important. A steeper gradient sends the molecules down faster. The surface area of the membrane also matters. A greater surface area relative to the enclosed volume results in a greater rate of diffusion. Both of those factors are critical to how organisms adapt to the challenges of exchanging the substances they need.

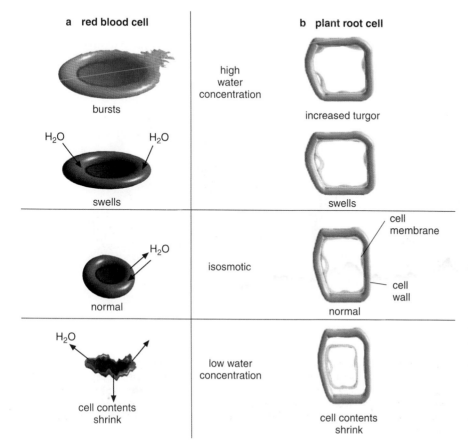

FIGURE 3.7

Effects of the environment on an animal cell (a) and a plant cell (b) in different water solutions. Initially the cells are in a solution with the same concentration of dissolved material as is found inside the cells. This is called an isosmotic solution. The animal cell can survive only fairly small variations from this concentration.

3.4 Passive and Active Transport

If osmosis and diffusion were the only processes of materials exchange, many substances needed by cells would soon become evenly distributed throughout the cells and their surroundings. Yet organisms must establish and maintain concentrations of materials inside their cells that may differ from concentrations resulting from diffusion. In addition, many substances cannot pass directly through lipid membranes by diffusion. Membranes are permeable to many of these substances but only with the help of transport proteins, which assist movement passively or actively. **Passive transport** involves diffusion without any input of energy. In contrast, **active transport** moves substances against their concentration gradients and thus requires energy.

Simple diffusion of neutral molecules such as oxygen or carbon dioxide into or out of a cell is a form of passive transport. Some molecules move down their concentration gradient with the help of transport proteins in the membrane. This transport is called **facilitated diffusion.** It is passive because the transported material moves down its concentration gradient (no energy is required). Facilitated diffusion makes transport more specific and speeds up the rate, but it does not work against the gradient. The transport proteins either form an open channel or attach to and carry specific molecules across the membrane (Figure 3.8a).

Active transport requires energy to move substances, in addition to the help of transport proteins (Figure 3.8b). One source of energy is the hydrolysis of ATP. Transport of Na^+, K^+, Ca^{+2}, and H^+ ions is directly linked to ATP hydrolysis. Energy also comes from coupling the movement of one substance against its gradient to the movement of another down its gradient. Some cells, for instance, need to build large concentrations of amino acids or glucose. Transport of these molecules is driven by the movement of Na^+ or H^+ down their concentration gradient.

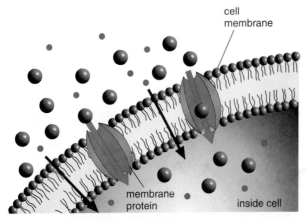

a   passive transport
(facilitated diffusion)

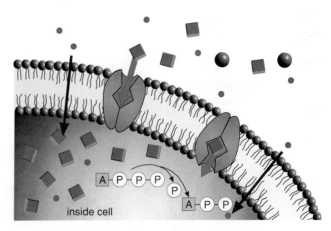

b   active transport

FIGURE 3.8

Passive and active transport. In simple diffusion, molecules pass directly through a membrane down their concentration gradient. In facilitated diffusion **(a),** a type of passive transport, membrane proteins transport molecules down their concentration gradient. Active transport requires the hydrolysis of ATP as well as membrane proteins **(b).** Active transport can move molecules against or with the concentration gradient.

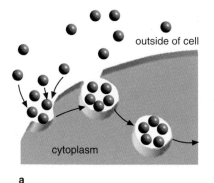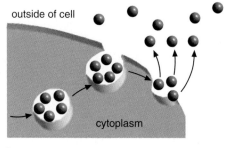

outside of cell   outside of cell

cytoplasm   cytoplasm

a   b

FIGURE 3.9

Endocytosis and exocytosis. Large molecules are transported into a cell by endocystosis **(a)**, and out of a cell by exocytosis **(b)**. Both processes require energy.

Maintaining specific gradients across cell membranes is essential to keep internal conditions in a range that permits life functions. For instance, many enzymes require a low concentration of Na^+, a high concentration of K^+, and a neutral pH, which depends on the concentration of H^+. Many necessary substances could not enter or leave cells without active transport. For example, the concentration of the nutrient ion known as nitrate (NO_3^-) is much greater inside the roots of a plant than in the surrounding soil. Plants need a constant supply of this nutrient, which is supplied by active transport.

Very large molecules such as proteins are moved into or out of a cell by a special active-transport mechanism that requires energy. The cell membrane actually folds around the substance to be transported, making a pocket to carry it in or out of the cell. **Endocytosis** (Figure 3.9a) is a useful way for unicellular organisms, such as amoebas, or very simple multicellular organisms to get food into their internal environment. **Exocytosis** (Figure 3.9b) helps cells remove waste materials. Exocytosis also moves specific molecules into the external environment, where they need to be to function. Some digestive enzymes, which could be harmful in the cytoplasm, and some hormones exit the cell in this way.

> **E T Y M O L O G Y**
>
> **exo-** = out (Greek)
> **endo-** = in (Greek)
> **cyto-** = cell (Greek)
>
> **Exocytosis** moves particles out of cells. **Endocytosis** moves them into cells.

Check and Challenge

1. What is a concentration gradient? How does it affect diffusion?
2. Explain the differences between active and passive transport. Why is each important to a cell?
3. In what way is a cell membrane selective?
4. Would you expect osmosis and diffusion to stop when substances are evenly distributed? Explain your answer.
5. What type of evidence demonstrates active transport?
6. How do membranes regulate the movement of molecules into and out of the cell?

Understanding Cystic Fibrosis

Cystic fibrosis (CF) is a common genetic disorder among Caucasian children, affecting about 1 in every 3,200 infants. Victims of CF suffer from chronic coughs, lung infections, and digestion difficulties. These problems usually become worse as patients become older, and only 25% of all CF patients survive into their thirties, although new treatments may help extend lifetimes.

CF is caused by mutations in the gene for a transport protein called the cystic fibrosis transmembrane conductance regulator protein, or CFTR protein. The protein normally has 1,480 amino acids and is found in the membranes of lung and intestinal cells. The normal form of the CFTR protein acts like a gate for chloride ions (Cl^-), controlling their exchange across membranes in the gas-exchange and digestive systems (Figure 3.10a).

In CF patients, the protein is different and does not function properly. For example, one particular mutation, which accounts for 70% of all CF cases,

causes the absence of one amino acid. This small change in structure has a large effect on function. Mutant versions of CFTR protein cannot regulate the exchange of chloride ions. Loss of chloride regulation means that the regulation of water balance is lost as well, because the concentration of chloride ions outside the cell affects the osmotic movement of water. As a result, thick mucus secretions build up outside the cells of the gas-exchange and digestive systems of people with CF (Figure 3.10b). The mucus then interferes with other processes. For example, mucus in the digestive ducts of the pancreas interferes with the release of certain digestive enzymes and consequently with digestion itself. Mucus in the lungs interferes with breathing and leaves CF patients vulnerable to lung infection. This serious disease illustrates the importance of regulating exchange across membranes.

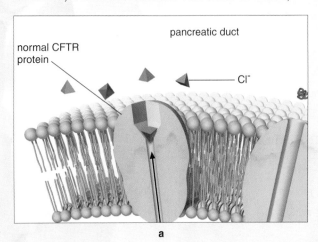

a

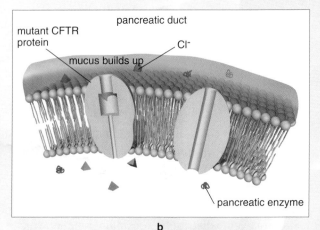

b

FIGURE 3.10

Choride transport in normal (a) and cystic fibrosis (b) cells. Normal CFTR protein regulates the flow of chloride ions across the membrane of pancreatic cells. Pancreatic enzymes secreted into the duct can aid digestion. Mutant CFTR protein interferes with chloride-ion movement across the membrane, which in turn interferes with the movement of water out of the cell. Pancreatic enzymes secreted into the duct get stuck in the thick mucus.

Exchange in Multicellular Organisms

3.5 Gas Exchange in Water

In most organisms, an important supply of energy for metabolism and other cell activities is cellular respiration (discussed in Chapters 2 and 5). Oxygen is essential for this process, and carbon dioxide is given off as a waste product. The correct balance of these two important molecules must be regulated carefully. The exact mechanisms of exchange depend in part on the external environment of the cell or organism. For example, a single-celled organism is directly in contact with its surroundings. In multicellular organisms, such as plants, animals, and fungi, some cells are buried deep within the organism. These interior cells also need nutrients, water, and oxygen or carbon dioxide and a way to remove wastes. These cells live in a more controlled environment than exterior cells or single-celled organisms, but they are more isolated from the environment as well.

Whether cells are in direct contact with the environment or in the interior of an organism, the basic processes of gas exchange are essentially universal. Gas exchange happens by diffusion across a membrane. Moreover, the gases involved, in this example oxygen and carbon dioxide, must be dissolved in water for diffusion to take place.

Water contains dissolved oxygen, but only in small amounts. To obtain that oxygen by diffusion, water dwellers such as fish, aquatic worms, and crustaceans need a very efficient gas-exchange mechanism. As with most exchange processes, efficiency requires a large surface area relative to volume. Breathing through gills is very efficient because gills have a large surface area made up of many fine, threadlike filaments (Figure 3.11). Each

> **Try Investigation 3B** ⬥ **Diffusion and Cell Size.**

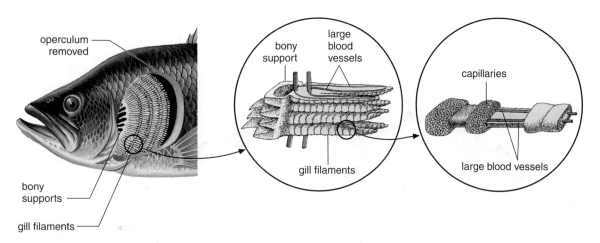

FIGURE 3.11

The structure of gills. Fish gills are thin filaments supported by bony structures and richly supplied with blood vessels. Each filament is made of disks that contain numerous capillaries. Water flows past these disks in directions opposite (countercurrent) to the flow of blood through the capillaries. A covering over the gills, called the operculum, protects the delicate filaments.

Focus On
Countercurrent Exchange

Gills are very efficient gas-exchange organs. Water comes in through the mouth of a fish, is forced over the gills, and passes out of the body through an opening in the body cavity that surrounds the delicate gill filaments. Gills are adapted to maximize the difference in oxygen concentration between blood in the capillaries and the water that passes over the gills. The filaments of gills are made up of thin disklike structures lined up parallel with the direction of water movement. Water flows over these disks from front to back. Within each disk, the blood circulates in the direction opposite to the movement of the water, from the back of the disk to the front. The concentration difference is maximized because the water flowing over the disks and the blood flowing within the disks run in opposite directions. This process is called countercurrent exchange (Figure 3.12).

As blood flows through the capillary, it takes up oxygen. It encounters water that has a higher concentration of oxygen because the water is just beginning its passage over the gills. This arrangement produces a diffusion gradient favoring the transfer of oxygen from the water into the blood along the entire length of the capillary. More than 80% of the dissolved oxygen in water diffuses into the blood.

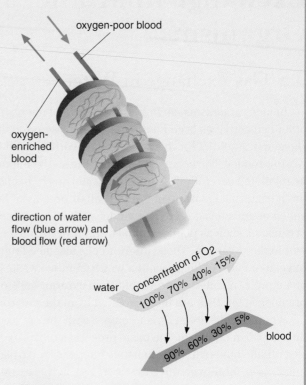

oxygen-poor blood

oxygen-enriched blood

direction of water flow (blue arrow) and blood flow (red arrow)

water

concentration of O2
100% 70% 40% 15%

90% 60% 30% 5%

blood

FIGURE 3.12

Blood and water flow in gills. With countercurrent flow, the blood continues to pick up oxygen from the water all along the length of the capillary. More than 80% of the dissolved oxygen in the water diffuses into the blood.

filament consists of a thin layer of cells surrounding a network of capillaries, which are short, narrow blood vessels. As water passes constantly over the gill surface, oxygen and carbon dioxide are exchanged between the blood circulating through these capillaries and the water surrounding the filaments. Because the gills have a huge surface area, diffusion easily can supply the small volume within the filaments. This large ratio of surface area to volume allows this exchange system to maintain a high rate of diffusion. Oxygen diffuses from the water into the blood down its concentration gradient and is carried to interior cells. Carbon dioxide, a waste gas, makes the reverse trip and diffuses from the gills into the water.

3.6 Adaptation to Life on Land

The concentration of oxygen is higher in air than in water (210 mL oxygen per liter of air as compared to 5 to 10 mL oxygen per liter of water, on the average). Although you might think this would give land organisms

CONNECTIONS

An organism's gas-exchange system reflects adaptation to the demands of its environment. Fish gills, for example, efficiently extract the limited amount of oxygen gas available in the water.

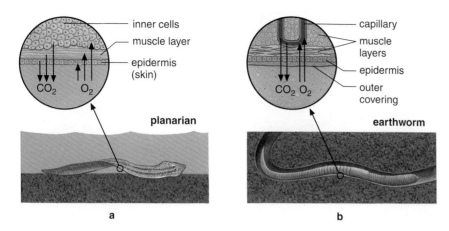

FIGURE 3.13

Simple air-breathing animals. Planaria (flatworms) **(a,)** and earthworms **(b),** have no special gas-exchange organs. Gases are exchanged directly through their skin.

a tremendous gas-exchange advantage, obtaining oxygen on land poses another set of challenges: Organisms living on land are constantly battling the tendency to dry out. With every breath, they lose precious water to the environment. In addition, in order for oxygen and carbon dioxide to enter or leave the cell, they must be dissolved in water, which they are not (for the most part) in the atmosphere. Thus land organisms must dissolve gases in water on the exchange membrane.

Minimizing the surface area of the exchange membranes would decrease water loss through evaporation. That solution, however, would work against the advantage of a large surface area for efficient exchange. To overcome these challenges, many species of land organisms have evolved exchange surfaces in an interior space. This protects the surface from excess evaporation caused by contact with the outside air and still allows a large area for exchange. Human lungs, discussed below, exploit the advantages of internalization, as do the internal surfaces of the leaves in plants.

Some land-dwelling organisms have no special gas-exchange organs, as seen in Figure 3.13. Others deliver oxygen to cells in interior spaces through complex transport systems (described in Chapter 7). Insects such as the grasshopper use a system of small, branched air ducts to carry oxygen throughout the body, shown in Figure 3.14. Extensive branching of the air ducts into smaller and smaller tubes provides a large surface area relative to volume that improves the rate of gas exchange. The smallest tubes are in direct contact with muscles and other body tissues. Cells can easily receive oxygen and give off carbon dioxide by diffusion through the cell membrane and the walls of the air tubes.

In many land animals, including humans, lungs are the organs of gas exchange. Lungs, which are located within your chest, minimize the effects of drying out by eliminating the one-way flow of oxygen that is so efficient in gills. With every contraction of the diaphragm, the muscle that controls breathing, the chest expands, and inhaled air moves into the lungs through

CONNECTIONS

Natural selection has favored many adaptations that allow a species to cope with the constraints imposed by the ratio of surface area to volume.

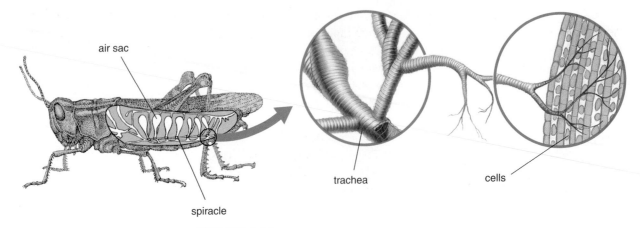

FIGURE 3.14

Gas exchange in a grasshopper. In insects, gas exchange occurs through branching air tubes called tracheae (singular: trachea). Air flows in and out of tracheae through openings called spiracles. The spiracles can close to retain water and keep foreign particles out.

a highly branched tubular passage. Exhaled air moves back out through the same passage (Figure 3.15). Because the diffusion surfaces of the lungs are exposed to a mixture of oxygen-rich and oxygen-poor air, the concentration difference is not great. Thus the gas-exchange efficiency of lungs is much less than that of gills, but so too is water loss.

Atmospheric air is usually dry, sometimes cold, and often dirty. The air you breathe passes through your nose, where it is filtered by hairs lining

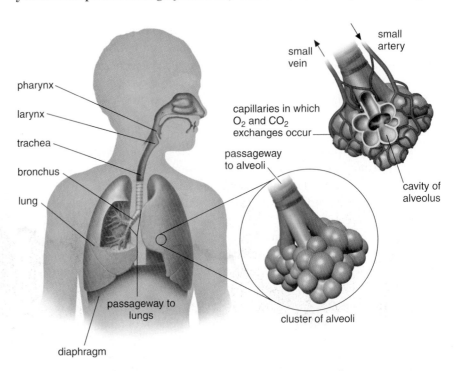

FIGURE 3.15

The human gas-exchange system. Two enlargements show clusters of alveoli and their blood supply.

the nasal cavities, moistened, and warmed. It then travels through branched passageways to reach microscopic cavities in the lungs called **alveoli.** Each lung has millions of alveoli, whose thin walls are richly supplied with capillaries. Figure 3.16 is a scanning electron micrograph of alveoli that shows the close association between alveolar walls and blood capillaries. Oxygen and carbon dioxide diffuse across the alveolar walls and the walls of the capillaries. A large volume of gas can be exchanged with the bloodstream in a very short time because the numerous alveoli of the lungs provide an enormous amount of surface area. If all the alveoli in an average person's lungs were spread out flat, they would cover an area of about 60 m^2 (about the size of two large parking spaces).

The internalization of exchange surfaces is just one example of an adaptation in terrestrial (land-dwelling) organisms that helps conserve water. Another water-conservation strategy involves barriers that limit the permeability of the outside of the organism itself. Such barriers are in place all the time. For example, air-breathing vertebrates and arthropods, plants, and fungi all have surface waxes and lipids that minimize water loss by evaporation. Glands in human skin secrete these oils and waxes as protective barriers. In plants, cells along the surface of a leaf secrete a waxy substance that forms a water-repellent covering called the **cuticle.** Not only does the cuticle reduce water loss from the leaf surface, it also blocks gas exchange between the air and these cells.

Specialized adaptations that can be regulated by the organism have evolved as well. In insects, water loss from the air ducts is minimized by spiracles, openings into the tracheae, which close whenever levels of carbon dioxide fall below a certain point. In plants, gases normally move into and out of the leaf tissue through openings known as **stomates** on the leaf surface. Each stomate is surrounded by a specialized pair of guard cells, which function as gates. When guard cells are swollen with water, they bend apart, opening the stomate (Figure 3.17). This opening allows carbon dioxide to diffuse in and water vapor and oxygen to exit. The loss of water by this

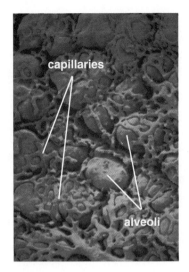

FIGURE 3.16

Scanning electron micrograph of alveoli, ×415. Capillaries in the alveolar walls provide a close relationship between blood and air.

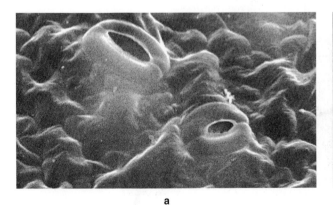

a

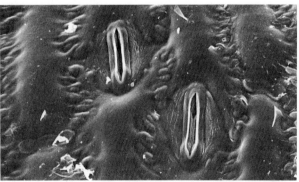

b

FIGURE 3.17

Stomates in a leaf. Guard cells act as gates around the stomates in the leaf surface. When open **(a)**, they allow water vapor to escape and carbon dioxide to enter the leaf. When water loss in the plant is higher than its replacement, the guard cells droop toward one another. This action closes the stomate **(b)**.

Focus On

Gas Exchange in Birds

There is a limit to the improvement in gas exchange that can result from increasing the total lung surface area for diffusion. With the evolution of birds, flying introduced gas-exchange demands that exceeded the capacities of saclike lungs. Many birds, such as hummingbirds, beat their wings rapidly for prolonged periods of time (Figure 3.18a). Such fast wing movement uses energy very quickly because it depends on rapid contraction of the wing muscles. Flying birds must oxidize large quantities of food molecules to replenish the ATP used in flight muscles. Thus they require a great deal of oxygen—more than a lung with a large surface area can deliver.

What system of gas exchange has evolved in birds? A bird's lung works like a two-cycle pump (Figure 3.18b). When the bird inhales, the air passes directly into a set of chambers called the posterior air sacs. When the bird exhales, the air passes into the lungs. On the following inhalation, the air passes from the lungs to a second set of air sacs, called the anterior air sacs. Finally, on the second exhalation, the air flows from the anterior air sacs out of the body.

What is the advantage of this complicated system? Air passes through the lungs in a single direction. Thus there is no mixing of oxygen-rich and oxygen-poor air as in the human lung; the air passing through the lung of a bird is always oxygen-rich. As in the gills of fish, the flow of blood past the bird lung runs in the direction opposite to the air flow in the lung (countercurrent flow). Thus bird lungs are very efficient at picking up oxygen from the air.

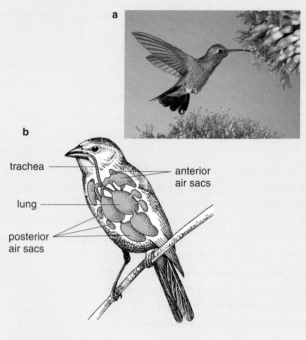

FIGURE 3.18

Rapid gas exchange in birds. (a), The energetic demands of flight and maintaining body temperature require hummingbirds to consume large quantities of oxygen and food. **(b),** Air sacs in birds ventilate the lungs, where gas exchange occurs. Two cycles of inhalation and exhalation are required for the air to pass all the way through the system.

pathway is called **transpiration.** If a plant loses more water than it can take in through its roots, the plant wilts. One protection against wilting depends on the regulated action of guard cells. As osmosis results in the loss of water from these cells, they shrink and draw toward one another, closing the stomate and minimizing further water loss. Thus photosynthesis requires an adequate water balance in plant cells in order to maintain a supply of carbon dioxide from the air.

3.7 Waste Removal

Organisms living in fresh water do not face the problem of drying out. Instead, they face the opposite challenge: They constantly must rid themselves of excess water. Figure 3.19 shows how unicellular organisms such as *Paramecium* living in freshwater use subcellular structures known

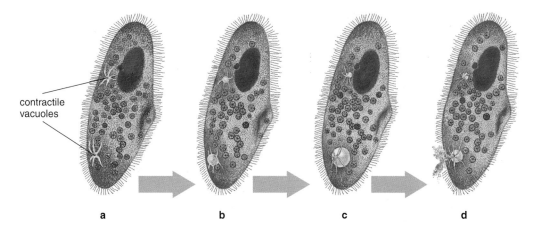

contractile
vacuoles

a   b   c   d

FIGURE 3.19

Waste removal in a one-celled organism. Contractile vacuoles in *Paramecium* rid the cell of excess water. The vacuoles **(a)** expand as water fills them through radiating canals **(b, c)**. The vacuoles then contract and eject the water from the organism **(d)**.

as contractile vacuoles to squeeze excess water from the cell. In addition to water, a variety of waste products must be removed from cells and organisms, including excess salts and carbon dioxide. The exchange of materials, including the removal of wastes, is essential to maintaining **homeostasis,** the balanced and controlled conditions in the internal environment of an organism. A variety of waste-removal mechanisms have evolved to help maintain homeostasis.

  In relatively simple organisms such as sponges (Figure 3.20) and *Hydra*, each cell simply excretes its wastes directly through the external surface. In larger and more complex animals, however, not enough of the body cells are in contact with the external environment for each cell to dispose of its own wastes. In other words, the surface area exposed to the environment is too low when compared to the volume of cells that require waste removal. Therefore, special organs have evolved for excretion and maintaining water balance in larger organisms. In fish, the excretion of carbon dioxide takes place in the gills, the same organs through which fish get oxygen. In saltwater fish, special cells in the gills also excrete salt, thus helping maintain the body's water-salt balance.

  Metabolism also produces by-products more toxic than waste gases, salts, and water. Perhaps the most troublesome by-product is the nitrogen-containing waste from the metabolism of proteins and nucleic acids. These nutrients lose nitrogen when they are converted to carbohydrates or fats. Amino groups removed from amino acids immediately become **ammonia** (NH_3) with the addition of a third hydrogen, a reaction that requires little energy. Some animals excrete ammonia directly, but it is very toxic to body tissues, so this form of nitrogenous waste removal can be used only if the organism lives in an environment with sufficient water to flush it from the body. The high solubility of ammonia makes it a safe excretory product in freshwater and saltwater protists and animals. *Paramecium* and other protists, as well as planarians and many other invertebrates,

ETYMOLOGY

homeo- = same (Greek)
stasis- = standing, not moving or changing (Greek)

Organisms normally maintain **homeostasis**, keeping their internal environment more or less constant.

FIGURE 3.20

A living sponge on the sea floor. Sponges are porous animals with thin tissues and a high ratio of surface area to volume. Their cells release wastes directly into the seawater they pump through their bodies.

excrete ammonia directly through their cell membranes or body coverings. Ammonia is also the chief nitrogenous waste of some vertebrates—the ocean fishes.

Some organisms convert their nitrogenous wastes to urea, which is much less toxic than ammonia. Urea can be excreted safely when diluted in a moderate amount of water, allowing body water to be conserved, an important advantage for land-dwelling animals. Humans and other mammals, some fishes, and amphibians excrete nitrogenous wastes chiefly as urea. Uric acid, an almost insoluble and nontoxic form of nitrogenous waste, is an adaptation of birds and many desert reptiles. The excretion of uric-acid crystals requires almost no loss of water.

The different forms of nitrogenous wastes show important evolutionary patterns. Animals that have evolved in water and land environments show differences in nitrogenous wastes related to the abundance or scarcity of water. Ammonia is more toxic than urea and is a common nitrogenous waste among aquatic organisms, which live where water quickly dilutes it. Organisms living in drier environments (where dilution is not possible) excrete primarily the less toxic compounds urea or uric acid.

3.8 Human Urinary System

The human urinary system is an example of how waste removal is critical to maintaining homeostasis. The excretory tubules of humans, the **nephrons,** are collected into compact organs, the **kidneys.** The two kidneys are the major organs in humans (as well as other mammals) responsible for processing the waste products of metabolism. Blood cycles through the kidneys, and nitrogenous wastes are removed. The removal of these wastes regulates the water balance by adjusting the concentrations of various salts in the blood. The kidneys, the blood vessels that serve them, and the plumbing that carries fluid formed in the kidneys out of the body compose the **urinary system,** illustrated in Figure 3.21.

Blood to be filtered enters the kidneys via the renal artery and leaves via the renal vein. The kidneys process the entire blood supply of the human body about once every five minutes. The waste fluid, **urine,** leaves the kidneys through a tube called the **ureter.** The ureter drains into a holding tank, the **urinary bladder.** The urinary bladder is periodically drained when the urine passes through a tube called the **urethra** during urination.

Each kidney contains approximately 1 million nephrons. A nephron (Figure 3.21c) is a long, coiled tube with one cuplike end that fits over a mass of capillaries. The other end of the nephron opens into a duct that collects urine. The tubular portions of the nephron are closely associated with a network of capillaries along the entire length. The cup of the nephron is called the **glomerular capsule,** or Bowman's capsule. The ball of capillaries within the cup is called a **glomerulus.** Collecting tubules from all the nephrons eventually empty into the ureter.

Nephrons have three functions: filtration, reabsorption, and secretion. These processes are summarized in Figure 3.22. Filtration occurs in the

Try Investigation 3C **The Kidney and Homeostasis.**

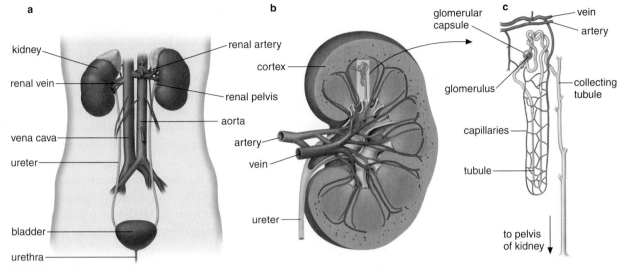

FIGURE 3.21

Waste removal in humans. (a), The human urinary system. **(b),** A section through the human kidney. **(c),** An enlarged view of one nephron with its surrounding capillaries.

glomerulus, where the fluid portion of the blood is forced into the glomerular capsule. Blood cells and most of the blood proteins are retained in the capillaries of the glomerulus. The filtrate—the materials that cross from the capillaries to the glomerular capsule—includes the blood plasma (the liquid portion of blood), nitrogenous wastes from cells, urea, salts, ions, glucose, and amino acids. Approximately 180 L of fluid enter the nephrons each day, yet only about 1.5 L of urine are eliminated from the bladder. The difference means that more than 99% of the fluid arriving in the glomerular capsule is returned to the blood.

Reabsorption and secretion take place in the tubule of the nephron. Cells of the tubule walls reabsorb substances needed by the body from the filtrate and return them to the blood. For example, salt and water in the tubule are reabsorbed into the capillaries. The water returns by osmosis, but active transport is necessary for the reabsorption of the sodium and potassium ions. Glucose, amino acids, and some urea also are reabsorbed by active transport.

Secretion occurs as the filtrate moves through the tubule. Cells of the tubule wall selectively remove from the surrounding capillaries substances that were left in the plasma after filtration or returned by reabsorption. The cells then secrete these substances into the filtrate. Excess potassium ions are excreted in this manner.

Reabsorption accounts for 85% of the salt, water, and other substances processed by the kidney. The remaining 15% is regulated by hormones or nervous-system controls. Excretion of sodium and potassium is regulated by **aldosterone,** a hormone secreted by the adrenal gland. When potassium levels in the blood are too high, aldosterone is released into the blood. That stimulates the secretion of potassium ions from the blood into the tubule near the collecting duct, which in turn lowers the blood potassium and

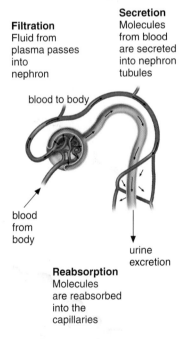

Filtration
Fluid from plasma passes into nephron

Secretion
Molecules from blood are secreted into nephron tubules

blood to body

blood from body

urine excretion

Reabsorption
Molecules are reabsorbed into the capillaries

FIGURE 3.22

Structure of a nephron, the structural subunit of the kidney. Filtration, reabsorption, and secretion in the nephrons result in the formation of urine.

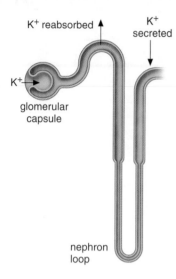

K⁺ reabsorbed

K⁺ secreted

K⁺

glomerular capsule

nephron loop

FIGURE 3.23

Potassium homeostasis in the kidney tubule. All of the potassium ions in the filtrate are reabsorbed into the blood by the time the filtrate passes the nephron loop. Under the influence of aldosterone, potassium ions are secreted back into the filtrate near the collecting duct, where they are excreted with the urine.

aldosterone levels (Figure 3.23). This is an example of **feedback regulation,** a process in which substances (such as aldosterone) inhibit their own formation. Feedback regulation helps maintain balance and stability.

Another example of feedback regulation involves water, which is important in regulating blood pressure. If a person becomes dehydrated, the water volume of the blood decreases and blood pressure drops. The hypothalamus in the brain detects this drop and stimulates the pituitary gland to release **antidiuretic hormone** (ADH) into the bloodstream. ADH causes the cell membranes of collecting duct cells to become more permeable to water. Water passes from the collecting duct to the bloodstream, and the proper volume of water in the blood is restored. The brain then detects the increased volume of the blood (and consequently higher blood pressure) and decreases the level of ADH, which in turn decreases the permeability of the collecting duct to water. Figure 3.24 illustrates this regulating mechanism.

The kidneys can also remove excess salt from the body, but only in small amounts. When humans eat or drink substances (such as seawater) that have high salt concentrations, their bodies actually lose water because, while trying to eliminate excess salt, the body excretes more water than was taken in. That is why shipwrecked people can easily perish from dehydration if they drink seawater.

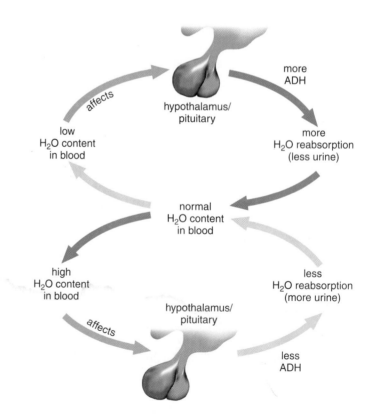

hypothalamus/ pituitary

affects

more ADH

low H₂O content in blood

more H₂O reabsorption (less urine)

normal H₂O content in blood

high H₂O content in blood

less H₂O reabsorption (more urine)

hypothalamus/ pituitary

affects

less ADH

FIGURE 3.24

Regulation of water loss in the kidney. Water content of the blood is controlled by ADH from the hypothalamus and pituitary gland.

Through filtration, reabsorption, secretion, and excretion, the kidneys remove nitrogenous wastes from the blood as urea, help regulate blood pressure, regulate water-salt balance, conserve blood glucose, and excrete excess salt, within limits.

Check and Challenge

1. Why are gills efficient organs for gas exchange in water-dwelling organisms?
2. What is the significance of the surface area of the gas-exchange membrane?
3. Why does oxygen diffuse into the capillaries around the alveoli?
4. What is the relationship between water conservation and urea as a waste product?
5. Describe how the structure of a nephron enables it to carry out its functions.
6. Distinguish among filtration, reabsorption, and secretion.
7. How does ADH act to regulate water balance?

Summary

A living system is a single or series of protected compartments. The internal conditions are usually different from conditions outside the organism. Internal conditions must be carefully balanced with regard to nutrients and wastes, a condition known as homeostasis. The cell membrane is selectively permeable, which helps it control an organism's exchange of substances with the environment.

The physical processes of diffusion and osmosis are responsible for movement of substances into and out of cells. Transport proteins in the membrane can help specific substances cross the membrane barrier. Transport is either passive or, if it requires energy, active. Exocytosis and endocytosis are responsible for exporting or importing large materials, respectively.

Gas exchange is an essential aspect of living processes. Exchange surfaces must be kept moist, and the ratio of surface area to volume affects the efficiency of exchange by diffusion. Land organisms must balance the need for large surface area of the exchange membranes against the danger of drying out.

Wastes must be expelled from all living systems. Nitrogenous wastes are particularly toxic and may be excreted as ammonia, urea, or uric acid. Each of these excretion compounds represents an adaptation to the environment in which a species lives.

Contractile vacuoles in unicellular organisms force wastes out of the cell. In humans, the kidneys are the major organs for removing waste products from the internal environment. The kidney removes nitrogenous wastes, regulates blood pressure, and controls water-salt balance. The nephron is the functional unit of the kidney. Hormones assist the urinary system in regulating ion balance, water levels in the blood, and blood pressure.

Key Concepts

Use the concepts below to build a concept map, linking to as many other concepts from the chapter as you can.

- diffusion
- gas exchange
- compartment
- waste removal
- environment
- adaptation
- endocytosis
- homeostasis
- transport protein
- water
- osmosis
- concentration gradient
- cell membrane
- evaporation
- active transport
- internal conditions
- exocytosis
- nitrogenous waste
- selective permeability
- evolution

Reviewing Ideas

1. Explain how the structure of a living organism resembles a protected compartment.
2. How does a cell membrane help regulate internal conditions and maintain homeostasis?
3. Explain the process of diffusion and why it is energetically favorable.
4. How do size, polarity, and charge affect the ways materials are exchanged across a cell membrane?
5. What is the key difference in active and passive transport of substances into and out of cells?
6. Describe how facilitated diffusion takes place.
7. What are two sources of energy for active transport? Explain your answer.
8. Draw a diagram to show how particles can enter a cell via endocytosis.
9. What is the significance of the surface area of an exchange membrane such as the one in the gills of fish?
10. The adaptations of land-dwelling organisms for getting oxygen show a trade-off of two needs. What are they?

Biology Online BSCSblue.com/vocabulary_puzzlemaker

11. Why is the excretion of nitrogenous wastes a critical issue for survival?
12. How does reabsorption in the nephron play a key role in maintaining homeostasis in a human body?

Using Concepts

1. Roots of plants are covered with fine root hairs that absorb water and nutrients from the soil. Why is this structure more efficient than a single large root would be?
2. When air enters the alveoli of human lungs, is the oxygen it contains truly in the internal environment of the human? Explain your answer.
3. Construct a curve on a graph that shows the relationship between the rate of diffusion and the ratio of surface area to volume.
4. What can you assume about water exchange in a plant that has wilted?
5. How does cystic fibrosis illustrate the importance of transport proteins in maintaining homeostasis?
6. A kidney dialysis machine cycles the blood through a filtering device using countercurrent exchange. Explain what this means and why the treatment must be done on a regular basis.

Synthesis

1. Human skin is coated with an oily substance called sebum. This substance slows water loss and inhibits growth of some bacteria and fungi. Given this information and what you know about surface area and exchange, explain why lungs are more vulnerable to infection than skin is.

2. In the context of evolution through natural selection, explain why multicellular organisms have specialized exchange systems and surfaces.

Extensions

1. Write a short essay using what you know about energy and chemistry to answer the following: Why is the bilayer the most stable arrangement in a watery environment?
2. Take a trip to a local wastewater treatment facility. Are any human wastes recycled? Are any other organisms involved in the treatment? Would you be willing to drink the water that comes from the treatment plant? Why or why not? Write a short paper on the importance of wastewater treatment.
3. Some limestone caves are being damaged by exposure to carbon dioxide exhaled by visitors. Write a short report on the damage and the chemistry involved.

Web Resources

Visit BSCSblue.com to access
- Information on cell membranes, transport processes, and kidney diseases
- Photographs, X rays, and CAT scans of human lungs and the excretory system
- Web links related to this chapter

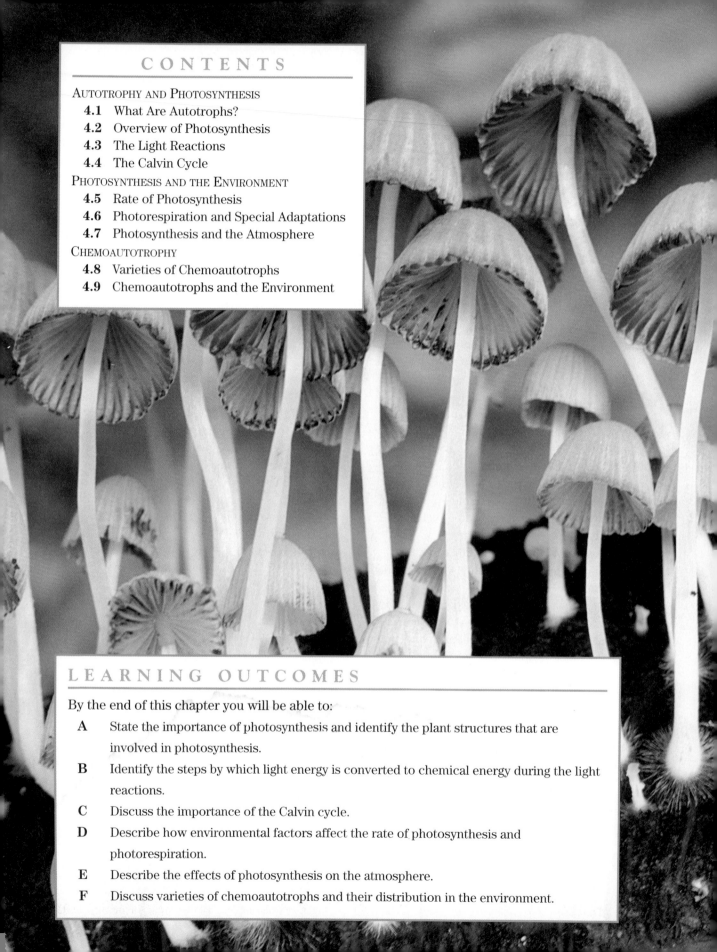

CONTENTS

LEARNING OUTCOMES

By the end of this chapter you will be able to:

A   State the importance of photosynthesis and identify the plant structures that are involved in photosynthesis.

B   Identify the steps by which light energy is converted to chemical energy during the light reactions.

C   Discuss the importance of the Calvin cycle.

D   Describe how environmental factors affect the rate of photosynthesis and photorespiration.

E   Describe the effects of photosynthesis on the atmosphere.

F   Discuss varieties of chemoautotrophs and their distribution in the environment.

◄ **About the photo**
This photo shows a
group of fungi.

CHAPTER

4

Autotrophy: Collecting Energy from the Nonliving Environment

■ *How do organisms obtain energy from their environment?*

■ *How could animals survive without plants?*

Photosynthesis can be compared to a "living bridge"—connecting the Sun with the organisms on Earth by providing the energy needed for life. Plants and other photosynthetic organisms harvest solar energy and use it to combine molecules of carbon dioxide into complex, energy-rich organic compounds. In a sense, plants are the "supermarket" that feeds much of the living world. But some environments are too dark or harsh for plants. In these places, bacteria that obtain energy from minerals such as iron or sulfur take their place. In this chapter, you will learn how organisms use the nonliving energy sources in their environment.

Autotrophy and Photosynthesis

4.1 What Are Autotrophs?

In Chapter 2, you learned the difference between autotrophs, which obtain energy from nonliving sources, and heterotrophs, which obtain energy from other organisms. All organisms need a source of energy and a source of carbon compounds for making sugars, amino acids, and other compounds necessary for life. Photosynthesis, performed by plants, some bacteria, and other small organisms called algae, uses the energy of sunlight to convert carbon dioxide to sugars. Enzymes convert the sugars to amino acids and other compounds. Autotrophs such as plants that depend on photosynthesis for both energy and carbon compounds are known as **photoautotrophs.** In most environments, photosynthesis supplies the energy and carbon compounds that photoautotrophs and the heterotrophs that depend on them need to survive and grow.

We depend on photosynthesis for the production of cotton and other fibers, wood, grains, vegetables, and fruits, as well as the animal feeds needed to produce meat, wool, and dairy products. We even rely on the products of ancient photosynthesis—petroleum and coal—to power cars and factories and to heat our homes. The oil and coal deposits we use formed millions of years ago from the partly decayed bodies of plants and the animals that fed on them.

But deep in the ground and under the ocean, there is not enough light for photosynthesis to occur. And some places, such as hot springs, are too hot, salty, or acidic for photoautotrophs to survive. Still, life exists in these extreme environments. Bacteria called **chemoautotrophs** obtain energy by oxidizing inorganic substances such as iron, sulfur, or other minerals, just as photoautotrophs obtain energy from sunlight. They use this energy to form sugars from carbon dioxide. These bacteria replace plants as the basis of food chains in many environments where photosynthesis is not possible (Figure 4.1).

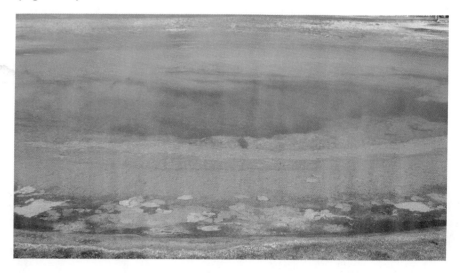

FIGURE 4.1

The upper geyser basin at Yellowstone National Park. Note the steam rising from the water. The water is too hot to support photoautotrophs. The colors are due to chemoautotrophic bacteria that can tolerate these extreme conditions.

TABLE 4.1
Nutritional Classification of Organisms

Source of Energy	Source of Carbon	
	CO_2 **(auto-)**	organic compounds made by other organisms **(hetero-)**
Light **(photo-)**	**photoautotrophs** (plants, algae, some bacteria)	**photoheterotrophs** (some bacteria)
Chemicals* **(chemo-)**	**chemoautotrophs** (some bacteria)	**chemoheterotrophs** (animals, fungi, some bacteria and other one-celled organisms)

*This category can be further divided according to whether organic or inorganic chemicals are used as a source of energy.

Table 4.1 summarizes the major types of autotrophs and heterotrophs. These categories are not absolute. For example, some algae and bacteria can obtain energy from either light or organic compounds, depending on which is more plentiful in their environment. These organisms are heterotrophs part of the time, and photoautotrophs at other times.

4.2 Overview of Photosynthesis

Photoautotrophs have adapted to take advantage of an energy source that they can never exhaust: sunlight. Light consists of a vibrating electric and magnetic field. The vibration of this field is like a wave (Figure 4.2). The length of the waves determines the light's color and energy; the shorter the wave, the greater its energy. Visible light waves have a certain range of energies. They have enough energy to cause small, reversible changes in the molecules that absorb them. When this occurs in our eyes, a chemical signal is generated that allows us to see. When it occurs in cells of a photoautotroph, the organism can capture the energy and use it. Photoautotrophic cells contain light-absorbing substances, or **pigments,** that absorb visible light. Although ultraviolet waves provide more energy

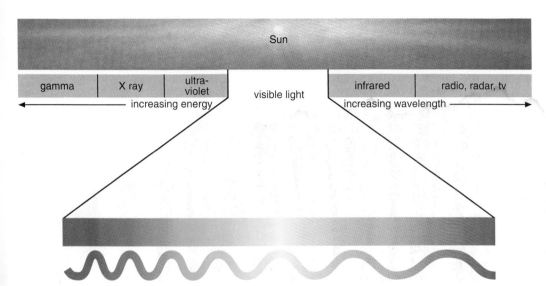

FIGURE 4.2

The electromagnetic spectrum. Energy that radiates from the Sun forms a continuous series of waves called a spectrum. The range of wavelengths that animals can detect with their eyes—visible light—is roughly the same range plants use in photosynthesis. Shorter wavelengths (blue light) have more energy than longer wavelengths (red light).

than visible light, few organisms have evolved the ability to absorb and make use of these high-energy waves without being damaged by them.

The light-absorbing pigments that function in photosynthesis are embedded in membranes within cells. These membranes, called thylakoids, form closed sacs. Some of the enzymes involved in photosynthesis are also embedded in the thylakoids. In the simple cells of bacteria, the thylakoids simply float inside the cell. In the cells of plants and algae, each thylakoid sac is part of an organized structure called a **chloroplast.** An outer chloroplast membrane separates the thylakoid from the cytoplasm and regulates the flow of materials into and out of the chloroplast. Chloroplast membranes have the same basic structure as other cellular membranes (see Section 1.8).

Figure 4.3 shows a chloroplast from a corn-plant leaf *(Zea mays).* The thylakoid membrane inside consists mostly of a series of flattened sacs, which increase the amount of membrane surface area that the chloroplast can hold. Many of the sacs are stacked like pancakes. These stacks are called grana (singular: granum).

The space surrounding the thylakoids is called the stroma. Enzymes in the stroma catalyze the formation of sugar from carbon dioxide and water, using the light energy captured in the thylakoids. Chloroplasts make a few of

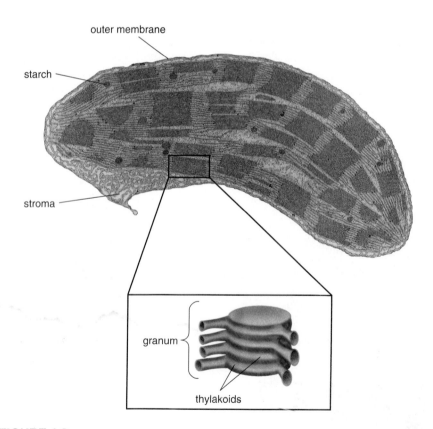

FIGURE 4.3

Electron micrograph of a chloroplast in a leaf of corn, *Zea mays,* ×24,000. The darker areas are stacks of thylakoids called grana; the drawing shows the structure of one enlarged granum. Photosynthetic pigments are embedded in the thylakoid membranes; DNA, RNA, and Calvin-cycle enzymes are in the stroma.

their proteins under the direction of their own DNA. The stroma contains the chloroplast's DNA, as well as the RNA and enzymes needed to make proteins encoded in the chloroplast DNA. Figure 1.35 summarizes the importance of DNA.

Most photosynthesis depends on the green pigment **chlorophyll,** found in the thylakoids. Plants contain two forms of chlorophyll, *a* and *b*, (Figure 4.4). Chlorophylls *a* and *b* absorb light in the violet/blue and orange/red ranges, but not in the green range (Figure 4.5). The green light that is not absorbed gives leaves their color. Other, accessory pigments absorb additional wavelengths of light; their absorbed light energy is transferred to a special type of chlorophyll *a* for use in photosynthesis. As the chlorophyll content of leaves declines in the fall, the accessory pigments become more visible (Figure 4.6). Some photosynthetic bacteria contain a form of a light-absorbing protein called rhodopsin instead of chlorophyll. Another form of rhodopsin occurs in the eyes of animals, where it is involved in vision.

The process of photosynthesis involves three energy conversions:

1. absorption of light energy,
2. conversion of light energy into chemical energy, and
3. storage of chemical energy in the form of sugars.

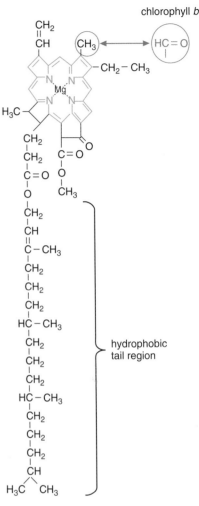

chlorophyll *a*

FIGURE 4.4

The structure of chlorophyll *a*. Chlorophyll *b* differs only in having a CHO—group in place of the circled CH_3—. Other forms of chlorophyll occur in algae and photosynthetic bacteria. The part of the molecule shown in green absorbs light; the hydrophobic tail helps to keep the molecule anchored in the lipid-rich thylakoid membrane.

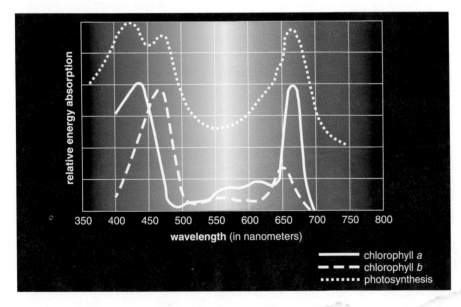

FIGURE 4.5

Absorption spectra for chlorophyll *a* and *b*. What wavelengths do these chlorophylls absorb most? least?

FIGURE 4.6

Autumn leaves. The colors are due to accessory pigments that become visible as chlorophyll is broken down.

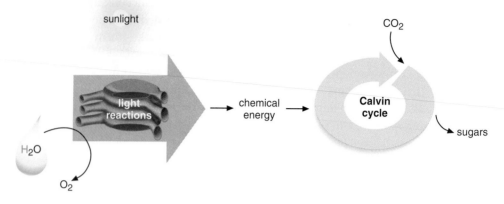

FIGURE 4.7

An overview of photosynthesis. Solar energy is converted to chemical energy in the thylakoid membranes. Enzymes of the Calvin cycle use this energy to reduce carbon dioxide, forming sugars.

These events occur in two groups of reactions, summarized in Figure 4.7. In the **light reactions,** pigment molecules in the thylakoids absorb light and convert it to chemical energy carried by short-lived, energy-rich molecules. The energy of these molecules is used to make 3-carbon sugars from carbon dioxide in a series of reactions known as the **Calvin cycle.** The chemical energy and carbon skeletons of the sugars are available to the plant for future growth.

The following equation summarizes the overall reactions of photosynthesis.

$$3\ CO_2 \ + \ 3\ H_2O \ \xrightarrow[\text{chlorophyll}]{\text{light energy}} \ C_3H_6O_3 \ + \ 3\ O_2$$

| carbon dioxide | water | | 3-carbon sugar | oxygen gas |

Note that, in this process, carbon dioxide is reduced and water is oxidized. This equation indicates only the major raw materials and final products of photosynthesis. The many steps and substances involved are explained in Sections 4.3 and 4.4. As you read the details that follow, keep in mind the overall process as it is described in this section. Remember that the products of the light reactions are used in the Calvin cycle. The two groups of reactions together accomplish the changes described by the previous equation.

Try Investigation 4A Photosynthesis.

4.3 The Light Reactions

The light reactions of photosynthesis convert visible light into the chemical energy that powers sugar production in the Calvin cycle. In these reactions, chlorophyll and other pigments in the thylakoid absorb light energy, water molecules are split into hydrogen and oxygen, and light energy is converted to chemical energy. Refer to Figure 4.9 as you read.

The light-absorbing pigments form two types of clusters, called photosystems (PS) I and II. The chlorophyll and other pigments in each

Biological Challenges

The Secret of Vegetation

People have long known that they and all other animals grow and develop by eating plants or by eating other animals that feed on plants. What is the "food" that plants eat, and how do they increase in size? In the seventeenth century, a physician named Jan Baptista van Helmont tried to answer these questions. He reached an amazing conclusion by doing something completely new. He did not just observe plants, but performed a quantitative scientific experiment on plant growth (Figure 4.8). In van Helmont's own words,

> I took an earthenware pot, placed in it 200 lbs of earth dried in an oven, soaked this with water, and planted in it a willow shoot weighing about 5 lbs. After five years had passed, the tree grown therefrom weighed 169 lbs and about 3 oz. But the earthenware pot was constantly wet only with rain or (when necessary) distilled water . . . and, to prevent dust flying around from mixing with the earth, the rim of the pot was kept covered with an iron plate coated with tin and pierced with many holes. . . . Finally, I again dried the earth of the pot, and found it to be the same 200 lbs minus 2 oz. Therefore, 164 lbs of wood, bark, and root *had arisen from the water alone.*

Van Helmont performed a beautifully simple experiment and tried to measure carefully, but he did not take into account the air. Neither van Helmont nor anyone else for another 100 years had any reason to suspect that the "food" that made plants grow was made in the leaves from carbon dioxide and water.

In 1772, Joseph Priestley discovered that plants affect air. In one experiment, he noted that when a burning candle is covered with a jar, its flame quickly goes out. However, when a sprig of mint is placed in the jar for a few days, and then the candle is covered, the candle burns for a short time.

By 1804, it had been shown that plants must be exposed to light for results such as Priestley's to occur. Plants had also been found to release oxygen and absorb carbon dioxide. Nicolas de Saussure showed that when a plant is exposed to sunlight, its weight increases by more than the weight of the carbon dioxide it absorbs. He concluded from this that plant growth results from the intake of both carbon dioxide and water.

Julius Robert von Mayer proposed in 1845 that plants absorb light energy and convert it into chemical energy, which is then stored in compounds. These compounds account for more than 90% of all plant substance.

5 YEARS

willow tree 5 lbs
soil 200 lbs

willow tree 169 lbs 3 oz
soil 200 lbs–2 oz

FIGURE 4.8

Van Helmont's experiment on plant growth in the late 1600s.
Over the next 200 years, scientists learned that plants manufacture energy from water, sunlight, and carbon dioxide.

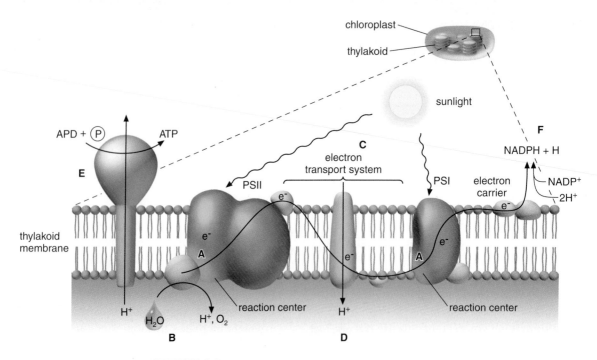

FIGURE 4.9

The light reactions of photosynthesis. In the thylakoid membrane, pigment molecules in each photosystem (PSI and PSII, green) absorb light and transfer its energy to the reaction center, which loses energy-rich electrons **(A)**. At the reaction center of PSII **(B)**, water (blue) is split, freeing electrons (e^-), protons (H^+), and oxygen (O_2). As electrons flow along an electron transport system **(C, yellow)**, protons are transported to the inside of the thylakoid **(D)**. ATP forms as the membrane enzyme ATP synthetase **(E, gray)** allows protons to diffuse back out of the thylakoid. The electrons and protons finally join with the hydrogen carrier $NADP^+$ **(F)**.

photosystem absorb light energy and transfer it from one molecule to the next. All this energy is funneled to a specific chlorophyll *a* molecule called the reaction center (Figure 4.10).

The reaction-center molecules accumulate so much energy that some of their electrons jump to other molecules, known as electron carriers. These molecules form an electron transport system between the two photosystems. The proteins and other molecules that serve as electron carriers are embedded in the thylakoid membrane and are organized in a way that allows easy movement of electrons from one carrier to the next. Some of them move through the membrane as they transport electrons to the next carrier in the system. Electrons from photosystem II move through the system to replace electrons lost from photosystem I. PSII receives replacements for these electrons from an enzyme near its reaction center that oxidizes water. This enzyme splits water molecules into protons, electrons, and oxygen.

$$2\,H_2O \longrightarrow 4\,H^+ + 4\,e^- + O_2$$

Recall from Chapter 1 that a hydrogen atom consists of a single proton in the nucleus and a single electron. Thus a hydrogen ion, H^+, is a single proton. When the enzyme oxidizes a water molecule, the oxygen is released as a gas, and the protons accumulate inside the thylakoid sac. The electrons reduce the PSII reaction center, replacing electrons that have already traveled to PSI. You might think of the transfer of electrons from water through PSII to PSI as a flow of reducing power.

Some photosynthetic bacteria obtain electrons from hydrogen sulfide gas (H_2S) instead of water. If you replace water in the above equation with H_2S, you can figure out what these bacteria produce instead of oxygen—solid sulfur. Bright yellow sulfur particles are visible in these cells under the microscope (Figure 4.11). The sulfur accumulates in quantities large enough that it can be profitably mined.

When electrons from water reach PSI, they receive an energy boost from the reaction center there. This gives them enough energy to reduce a molecule known as **NADP⁺** (nicotinamide adenine dinucleotide phosphate). The electrons, along with protons from water, combine with NADP⁺ to convert it to its reduced form, NADPH. This is the end of the electron flow in the light reactions. NADPH then provides the protons and electrons needed to reduce carbon dioxide in the Calvin cycle.

As electrons flow from carrier to carrier along the electron transport system, some of the solar energy they received from PSII powers the active transport of protons across the thylakoid membrane. A high concentration of positively charged protons accumulates inside the thylakoid. The difference in concentration and charge between the inside and outside of the thylakoid creates a difference in potential energy across the membrane, similar to the potential energy of a battery, with its positive and negative ends. And like a battery, this potential energy can do useful work. The concentrated protons diffuse out of the thylakoid through an enzyme complex in the membrane. As they do so, they transfer energy to the enzyme complex **ATP synthetase.** The enzyme then uses this energy to synthesize ATP from ADP and phosphate. Appendix 4A, "ATP Synthesis in Chloroplasts and Mitochondria," explains how the movement of electrons and protons results in ATP synthesis.

In summary, you can think of the energy from light as forcing electrons to flow from water to NADP⁺ in the chloroplast. The electrons retain this energy in NADPH. Some of the NADPH is used to synthesize ATP. Thus, photosynthesis converts light energy into the chemical energy of ATP and NADPH. This energy is used later to make sugars from carbon dioxide. ATP and NADPH, along with oxygen gas, are the products of the light reactions of photosynthesis.

CHEMISTRY TIP

During the oxidation of water, positively charged protons are separated from negatively charged electrons. Because these oppositely charged particles are attracted to each other, energy is required to separate them. This reaction converts solar energy into electric potential energy that helps to drive the flow of electrons.

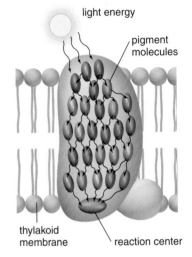

light energy

pigment molecules

thylakoid membrane

reaction center

FIGURE 4.10

Structure of a photosystem. Each photosystem consists of several hundred molecules of chlorophyll (green) and accessory pigments (orange and yellow) that absorb light energy and transfer it to a special chlorophyll *a* molecule, the reaction center. The reaction center is the only pigment molecule that can participate directly in electron flow in the light reactions.

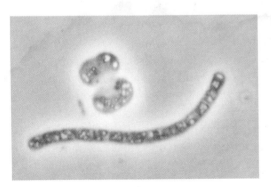

FIGURE 4.11

Photosynthesis without water splitting. The photosynthetic bacterium *Chromatium* oxidizes hydrogen sulfide gas instead of water, producing the yellow sulfur globules visible in its cells. Compare the water-splitting reaction ($2\ H_2O \longrightarrow O_2 + 4\ e^- + 4\ H^+$) to the oxidation of hydrogen sulfide ($2\ H_2S \longrightarrow 2\ S° + 4\ e^- + 4\ H^+$).

CONNECTIONS

The use of solar energy and oxidation-reduction reactions in the light reactions demonstrate how successful evolutionary adaptations are shaped by the laws of chemistry and physics.

Focus On

Protein Structure and Function

On the shores of San Francisco Bay, evaporation causes sea salt to accumulate in the wet, sandy soil. A pink scum on the sand reveals the presence of *Halobacterium*. This photosynthetic bacterium relies on a purple protein called bacteriorhodopsin, in place of chlorophyll and the light reactions of thylakoids, to carry out photosynthesis. Because bacteriorhodopsin is a fairly small protein molecule and its structure and action are similar to the rhodopsin that absorbs light in our eyes, biochemists have studied it intensively.

Biochemists have found that the bacteriorhodopsin molecule has seven sections that spiral across the bacterial cell membrane (Figure 4.12a). These seven coils cluster together to form a channel through the membrane (Figure 4.12b). Amino acids on the outside of this channel have hydrophobic (nonpolar) side chains, which associate with the membrane lipids. This helps to hold the protein in the membrane. Amino acids on the inside of the channel have hydrophilic (polar) side chains. Water and ions can pass through the polar interior of the channel between the inside and outside of the cell.

A molecule of retinal (a compound related to vitamin A) is weakly bound to one of the amino-acid side chains inside the channel, near the opening to the interior of the cell. When the retinal molecule absorbs light, it detaches from the protein and binds to a hydrogen ion. Retinal then binds to another amino acid at the outer end of the channel, releasing the hydrogen ion. As this cycle is repeated, light powers the transport of hydrogen ions out through the bacterial membrane. ATP synthetase then uses the difference in hydrogen-ion concentration between the inside and outside of the cell to power ATP production, just as it does in thylakoid membranes. The structure of bacteriorhodopsin is highly adapted to its function in photosynthesis.

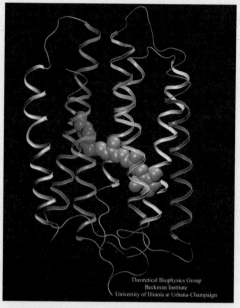

Theoretical Biophysics Group
Beckman Institute
University of Illinois at Urbana-Champaign

a

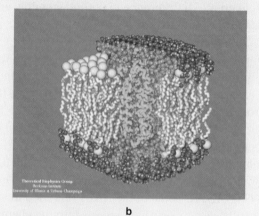

Theoretical Biophysics Group
Beckman Institute
University of Illinois at Urbana-Champaign

b

FIGURE 4.12

Bacteriorhodopsin, a photosynthetic pigment.
Bacteriorhodopsin and its attached retinal molecule (purple) **(a)**, form an ion channel, shown here in green **(b)**, through the cell membrane of *Halobacterium*.

Why does photosynthesis not stop with the synthesis of ATP and NADPH? There are two reasons. First, ATP and NADPH are not particularly stable compounds. A plant cannot conveniently store or transport energy in this form. Second, the light reactions do not produce any new carbon compounds that the organism can use to grow. Instead, the Calvin cycle uses the energy of ATP and NADPH to convert carbon dioxide into stable, easily transported sugars that provide energy and carbon skeletons for building new cells.

4.4 The Calvin Cycle

The Calvin cycle conserves the chemical energy produced in the light reactions in the form of sugars that the organism can use for growth. This completes the process of photosynthesis. Figure 4.13 summarizes the enzyme-catalyzed steps of the Calvin cycle as they occur in the stroma of a chloroplast. Follow these steps as you read the description.

At *a*, the beginning of the cycle, a molecule of carbon dioxide combines with the 5-carbon sugar-phosphate, ribulose bisphosphate (RuBP). This reaction is known as carbon dioxide fixation because it "fixes" carbon dioxide gas into an organic molecule. This produces an unstable 6-carbon molecule that immediately splits into two molecules of the 3-carbon acid, phosphoglyceric acid (PGA). At *b*, two enzymatic steps reduce each molecule of PGA to the 3-carbon sugar-phosphate, phosphoglyceraldehyde (PGAL). This requires one molecule each of ATP and the reducing agent NADPH from the light reactions.

As the Calvin cycle continues, a series of enzymatic reactions, *(c)*, combines and rearranges molecules of PGAL, eventually producing a 5-carbon sugar-phosphate. The final step, *d*, uses an ATP molecule from the light reactions to add a second phosphate group to the 5-carbon sugar-phosphate. This produces a molecule of the starting material, RuBP, thus completing the cycle. Three turns of the cycle, each turn incorporating one molecule of carbon dioxide, result in the formation of six molecules of PGAL. Of these, five molecules are required to regenerate RuBP. The sixth one is available for the organism to use for maintenance and growth.

CHEMISTRY TIP

A sugar-phosphate consists of a sugar molecule with one or more phosphate groups attached.

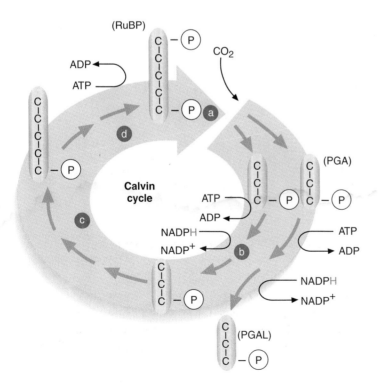

FIGURE 4.13

Reduction of carbon dioxide to sugars in the Calvin cycle. Enzymes in the stroma of the chloroplast catalyze each step of the cycle. Rubisco, the enzyme at **(a)**, catalyzes carbon dioxide fixation. The reactions require ATP and NADPH from the light reactions at **(b)** and **(d)**. ADP and $NADP^+$ return to the light reactions. A series of enzymes **(c)** catalyzes the recombination of five 3-carbon molecules of PGAL to form three 5-carbon molecules of RuBP.

Sugar-phosphates such as PGAL are removed from the Calvin cycle for use in other cellular functions, as shown in Figure 4.14. From these sugar-phosphates, plants can synthesize other compounds they need, such as carbohydrate polymers and amino acids. In plants, much of the sugar produced in photosynthesis is converted in the cytoplasm of leaf cells to the 12-carbon disaccharide sucrose. Leaf cells can consume sucrose or export it through the veins to supply the rest of the plant with energy and carbon skeletons. Chloroplast enzymes can also convert sugars to starch. Many plants accumulate starch in their chloroplasts during daylight. When photosynthesis shuts down at night, the chloroplasts break the starch down to supply the plant with energy and carbon skeletons.

An enzyme called rubisco catalyzes the reaction that incorporates (fixes) carbon dioxide into the Calvin cycle (Figure 4.13, step *a*). Because the product of this reaction is the 3-carbon acid PGA, plants that use only the Calvin cycle to fix carbon dioxide are called C_3 plants. In addition to providing energy for the synthesis of ATP and NADPH, light activates rubisco and several other enzymes of the cycle. For these reasons, and also because sufficient carbon dioxide is unavailable when the stomates are closed, the Calvin cycle cannot operate in the dark.

PGAL and other sugar-phosphates from the Calvin cycle are food for the plant. They supply energy and carbon skeletons to the entire plant, as shown in Figure 4.14. Some of the sugar-phosphates are made into lipids and others into amino acids and then proteins. (Chapter 5 discusses how these processes occur.) These changes can happen right in the chloroplast, elsewhere in the plant, or even in other organisms: Humans and other animals consume plants, using the products of photosynthesis as a source of carbon skeletons and energy.

Try Investigation 4B   Rate of Photosynthesis.

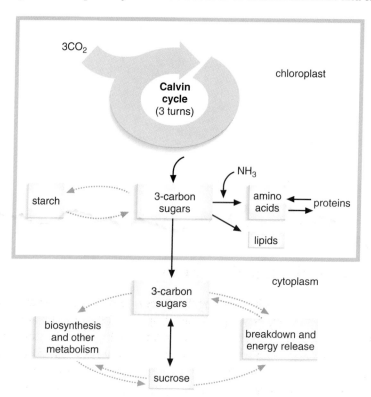

FIGURE 4.14

Fates of sugar produced in photosynthesis. Plants use the sugars produced in photosynthesis to supply energy and carbon skeletons for growth and other cell work. Much of this sugar is converted to sucrose or starch.

Biology Online BSCSblue.com/check_challenge

Focus On

Pathways and Compartments

Sometimes it is confusing to think about cycles, pathways, and molecules of metabolism in cells. Remember that the diagrams in this book are models based on the best available experimental data about cellular processes. Models help us understand how cells and their parts, such as chloroplasts, work, but there are no arrows, labels, or enclosing boxes in the actual cell. (There are, however, transport channels, preferred directions of reactions, and compartments within cells.)

Pathways and cycles are made of multiple copies of the enzymes and other molecules. The enzymes may be attached to a membrane, making the pathway or cycle more efficient. For example, the enzymes and electron-carrying proteins of the light reactions are embedded in sequence in the individual thylakoids. Calvin-cycle enzymes in the stroma may also be attached to the chloroplast membranes. The sugar molecules of the Calvin cycle are probably distributed evenly inside the chloroplast. The organization of enzymes and compartments in living cells is a result of millions of years of evolution that favor the most efficient arrangements.

PSI and PSII occur at many places along the thylakoids and make millions of molecules of ATP and NADPH. Because there are many molecules of each Calvin-cycle enzyme, many molecules of carbon dioxide can be fixed at the same time. Likewise, the activities of photosynthesis occur simultaneously in the thousands of chloroplasts in a single leaf. Do not let diagrams and models mislead you. They provide a way to represent and study biochemical pathways, but do not represent them exactly.

Check and Challenge

1. How is photosynthesis essential to life on Earth?
2. Why are photosynthetic organisms called producers in an ecosystem?
3. How is light used in photosynthesis?
4. What are the products of the light reactions, and how are they used?
5. Why does the Calvin cycle not operate at night?
6. What are the products of the Calvin cycle, and how are they used?

Photosynthesis and the Environment

4.5 Rate of Photosynthesis

Environmental conditions strongly affect photosynthesis. Therefore, they also affect the growth of photoautotrophs and the animals and other heterotrophs that depend on them. An understanding of environmental influences can help people protect and make more efficient use of Earth's resources. Environmental effects on organisms are usually described in terms of how they affect the **rate,** or activity per unit of time, of a

Materialism and Vitalism

Living things seem different from nonliving things. The ancient Greek philosopher Aristotle expressed this idea when he claimed that sperm gives a "vital force" to an egg, bringing it to life. Vitalism, the theory that a "life force" makes living matter different from nonliving matter, has come in and out of fashion over the centuries. During the Middle Ages and Renaissance, vitalism lost some of its appeal. Humans were thought to be unique in possessing a soul, but most people believed that our bodies, like everything else, were made of four elements (fire, earth, water, and air). Complex objects such as animals were just at higher levels on the same Great Chain of Being.

During the 1700s, chemists began to analyze living things. The materialists, who claimed that organisms and their products are composed of the same elements as nonliving things, began to oppose vitalism. In 1824, the German chemist Friedrich Wohler synthesized crystals of urea, a component of urine. He isolated urea from urine and then compared the crystals to it. They were identical! This seemed to contradict the idea of a vital spirit.

Because materialism claims to be able to explain something complicated (life) by completely reducing it to simpler ideas (chemical substances), it is called a reductionist theory. The chemical reductionist approach to biology gained ground during the 1800s. The major groups of biochemicals—carbohydrates, proteins, lipids, and nucleic acids—were identified. In 1899, Jacques Loeb discovered certain salts could force unfertilized sea urchin eggs to start developing. When he varied the salts, Loeb observed changes in the rate of growth. Loeb's experiments implied that chemistry, not a mysterious force, is the key to life.

By the 1950s scientific vitalism was dead. In an attempt to see how life could have originated, Stanley Miller passed electric current through gases such as methane and water vapor. Amazingly, the reaction produced amino acids. Around the same time, James Watson and Francis Crick uncovered the double helix structure of DNA. Their discovery revealed that the common link among all life is a particular chemical molecule.

Scientists are confident today in the power of physics and chemistry to explain the structure and function of living things. Some phenomena, such as consciousness, are too complex to explain easily in those terms. Scientists call consciousness an emergent property because it emerges from a complex brain composed of many simpler parts. As long as mysteries such as consciousness remain, vitalism will retain its appeal.

Experiments have demonstrated that everything is composed of chemical substances. Still, vitalistic thinking leads many people to believe that "natural" or "organic" products are safer and superior to "chemical" ones. Do these products contain chemicals?

biological process. For example, you can measure photosynthesis by how much carbon dioxide is consumed. Suppose you found that a plant consumed 2.5 g of carbon dioxide. Was it 2.5 g per second, per hour, or per day? These are very different rates of photosynthesis. The value expressed without the unit of time is meaningless. By expressing cell functions in terms of a rate, you answer the question, How fast did the process occur?

Light intensity, temperature, and the concentrations of carbon dioxide and oxygen all affect the rate of photosynthesis. You might expect an increase in light intensity to increase the rate of photosynthesis. Is that always the case? Study the graph in Figure 4.15. What happens to the rate of photosynthesis as light intensity increases? Note that before the light reaches the intensity of full sunlight, the rate of photosynthesis levels off. At this point, the light reactions are saturated with light energy and are proceeding as fast as possible. In still brighter light, chlorophyll accumulates energy faster than it can transfer that energy to the electron transport system. Some of this extra energy passes to oxygen molecules. The oxygen may react with water to form hydroxyl ions (OH^-) or hydrogen peroxide (H_2O_2). These substances can damage chloroplasts by reacting with pigments and proteins. A decline in photosynthesis, called **photoinhibition,** may occur.

How would light intensity affect the fate of a pine tree growing under a maple tree in a forest? Young pine trees need more light than young broad-leaved trees such as maple, beech, and oak. This gives young broad-leaved trees an advantage near the ground in a dense, shady pine forest. Eventually, they take over the forest as the pines die off. That, in turn, changes the types of animals found in the forest (see Chapter 25). This situation is an example of how a cellular adaptation can influence an entire ecosystem.

Temperature affects photosynthesis differently from light intensity. Study the graph in Figure 4.16 and explain the shape of the curve. Look carefully at

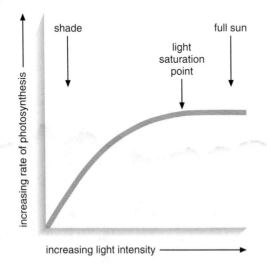

FIGURE 4.15

Effect of light intensity on photosynthesis. As light intensity increases, the rate of photosynthesis increases and then reaches a maximum rate. Data are generalized to show trends in C_3 plants.

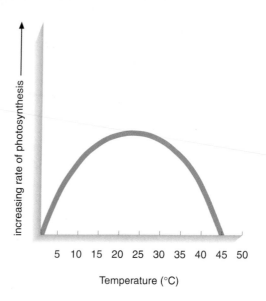

Temperature (°C)

FIGURE 4.16

Effect of temperature on photosynthesis. As temperature increases, the rate of photosynthesis also increases, and then declines. Data are generalized to show trends in C_3 plants that grow best between 20°C and 30°C.

each axis of the graph so you understand what the curve means. Develop a hypothesis to explain the changes in the rate of photosynthesis as the temperature increases.

An increase in carbon dioxide concentration increases the rate of photosynthesis to a maximum point, after which the rate levels off. A curve showing the effect of carbon dioxide on photosynthesis would be similar in shape to the one for light intensity in Figure 4.15, up to the light saturation point. Above the carbon dioxide saturation point, further increases in carbon dioxide concentration have no effect on photosynthesis.

The effects of light, temperature, and carbon dioxide all interact with each other. Any of these factors may be at an ideal level when another is far below optimum. In this case, the factors in shortest supply have the most effect on the rate of photosynthesis. This effect is called the principle of **limiting factors,** illustrated in Figure 4.17. Note that in maximum light, a higher temperature can increase the rate of photosynthesis.

FIGURE 4.17

Interaction of limiting factors. At high light intensity, the rate of photosynthesis is greater at 25°C than at 15°C. Thus, temperature can be a limiting factor when further increases in light intensity no longer stimulate photosynthesis. Data are generalized for typical C_3 plants.

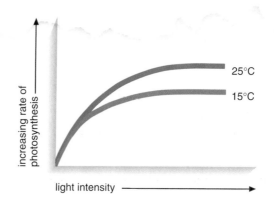

Brighter light will not do this. Thus plants grown in a greenhouse will not grow at the maximum rate, even in full sunlight, if the greenhouse is cold. In a forest, limiting factors include light, water, temperature, and nutrients. In the oceans, nutrients are the most important limiting factors for photoautotrophs in surface water. Below the surface, the availability of light limits photosynthesis.

4.6 Photorespiration and Special Adaptations

Figure 4.18 shows the effect of oxygen on photosynthesis by a C_3 plant. Normal atmospheric concentrations of oxygen (about 21%) can inhibit photosynthesis by up to 50%. How does this happen?

Recall from Section 4.4 that the enzyme rubisco incorporates carbon dioxide into sugars in the Calvin cycle. The molecular structures of oxygen and carbon dioxide help to explain how oxygen interferes with carbon fixation. Both molecules are held together by double bonds that keep the atoms about the same distance apart (Figure 4.19). This similarity allows rubisco to bind to either oxygen or carbon dioxide. When carbon dioxide binds to rubisco and combines with RuBP, two molecules of PGA form. But when oxygen replaces carbon dioxide in this reaction, the products include only one molecule of PGA, and one molecule of the 2-carbon acid glycolate (Figure 4.20). Glycolate is transported out of the chloroplast and partly broken down to carbon dioxide. The result is that the organism loses fixed carbon atoms, instead of gaining them. This pathway is called **photorespiration.** Unlike true cell respiration (see Chapter 5), the benefits of photorespiration are not yet clearly known. Photorespiration enables organisms to recover some of the carbon in glycolate. It may also help to reduce photoinhibition by providing a way for chlorophyll to release excess light energy.

CONNECTIONS

The effects of environmental conditions on photosynthesis and the adaptations of different plants to various environments are examples of how organisms interact with their environment as they evolve in it.

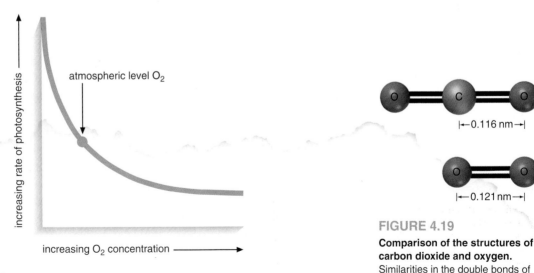

FIGURE 4.18

Effect of oxygen on photosynthesis. Increasing concentrations of oxygen inhibit the rate of photosynthesis in C_3 plants.

atmospheric level O_2

increasing rate of photosynthesis

increasing O_2 concentration

|←—0.116 nm—→|

|←—0.121 nm—→|

FIGURE 4.19

Comparison of the structures of carbon dioxide and oxygen. Similarities in the double bonds of these molecules allow rubsico to bind oxygen instead of carbon dioxide.

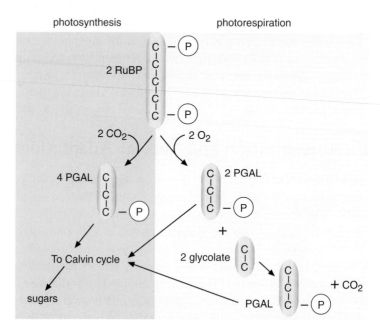

photosynthesis   photorespiration

FIGURE 4.20

Comparison of photosynthesis and photorespiration. Photorespiration occurs simultaneously with photosynthesis and results in the loss of previously fixed carbon dioxide. Both processes depend on the enzyme rubisco, which can react with either carbon dioxide or oxygen. High carbon dioxide levels favor photosynthesis over photorespiration. High oxygen levels promote photorespiration.

The loss of carbon from the Calvin cycle due to photorespiration slows the net rate of photosynthesis. Relatively high levels of carbon dioxide favor photosynthesis. Relatively high levels of oxygen favor photorespiration.

Weather also affects the balance between these two processes. Many plants reduce water loss during hot, dry weather by partly closing their stomates. When stomates are closed, however, carbon dioxide levels in the leaves may drop so low that photorespiration is favored over photosynthesis.

Two groups of plants have evolved adaptations that reduce photorespiration and aid survival in hot, dry environments. One group, including sugarcane, corn, and crabgrass, has evolved a system that first fixes carbon dioxide by incorporating it in a 4-carbon acid. For this reason, the process is called C_4 photosynthesis. C_4 plants have two systems of carbon dioxide fixation that occur in different parts of the leaves. Surrounding each vein in the leaves is a layer of tightly packed cells, the **bundle sheath** (Figure 4.21). Mesophyll cells surround the bundle sheath and extend into the air spaces in the leaf.

The mesophyll cells do not contain rubsico. Instead, they fix carbon dioxide by combining it with a 3-carbon acid. Unlike rubisco, the enzyme that catalyzes this reaction distinguishes well between carbon dioxide and oxygen. The resulting 4-carbon acid is rearranged and then transported to the bundle-sheath cells, as shown in Figure 4.22. There, carbon dioxide is released from the 4-carbon acid and refixed by rubisco, forming PGA by way of the Calvin cycle (Section 4.4).

CONNECTIONS

The binding of rubisco to both carbon dioxide and oxygen is an example of how chemical structures and reactions limit the range of possible adaptations.

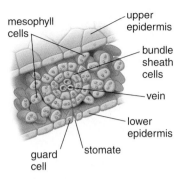

mesophyll cells
upper epidermis
bundle sheath cells
vein
lower epidermis
stomate
guard cell

FIGURE 4.21

Specialized leaf anatomy in a C₄ plant. Notice the tightly packed layers of cells—bundle-sheath cells and mesophyll cells—that surround the vein.

This arrangement delivers carbon dioxide efficiently to the bundle-sheath cells. By concentrating carbon dioxide in the bundle-sheath cells, the system favors photosynthesis and inhibits photorespiration. Although high temperatures raise the rate of photorespiration more than the rate of photosynthesis, the high concentration of carbon dioxide around rubisco in C_4 plants overcomes photorespiration. Thus C_4 plants can function efficiently at high temperatures while keeping stomates partly closed to reduce water loss. O_2 does not inhibit photosynthesis in C_4 plants.

By contrast, high temperatures can inhibit photosynthesis in C_3 plants by as much as 40 or 50%. In general, C_4 plants grow more rapidly than C_3 plants, especially in warm climates where C_4 plants evolve. Many C_4 plants can be

CONNECTIONS

C_4 photosynthesis is an adaptation by certain plants to a hot, dry climate. It is an example of evolution through natural selection.

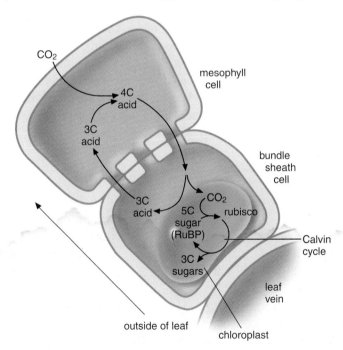

FIGURE 4.22

The C₄ photosynthetic pathway. Carbon dioxide first combines with a 3-carbon acid in the outside mesophyll cells. The resulting 4-carbon acid is then transported into the bundle-sheath cells, where carbon dioxide is released to the Calvin cycle and refixed by rubisco.

FIGURE 4.23

The jade plant (*Crassula*). This is an example of a CAM plant.

about twice as efficient as C_3 plants in converting light energy to sugars. Most of our important food crops, such as soybeans, wheat, and rice, are C_3 plants. Photorespiration may be a major factor limiting plant growth and world food production today.

Another specialization for photosynthesis was first discovered in desert plants such as cactus, jade (Figure 4.23), and snake plants. This system is called **CAM,** for crassulacean acid metabolism. CAM plants open their stomates at night and incorporate carbon dioxide into organic acids, just as C_4 plants do in daylight. During the hot, dry, desert days, the stomates close, conserving water. Enzymes then break down the organic acids, releasing carbon dioxide that enters the Calvin cycle just as it does in C_3 and C_4 plants. The CAM system is not very efficient. CAM plants can survive intense heat, but they usually grow very slowly.

Focus On
Biotechnology

One goal of biotechnology is to create more productive food crops. The production of C_3 crop plants in which oxygen does not inhibit rubisco has been the subject of two areas of genetic engineering research. The first was an attempt to change the gene for the enzyme rubisco, so that the enzyme would not bind oxygen. If the genetically engineered enzyme would still fix carbon dioxide and initiate the Calvin cycle, then we might greatly enhance photosynthesis and crop productivity. After unsuccessfully testing thousands of genetically altered plants for a form of rubisco that would bind carbon dioxide, but not oxygen, researchers concluded that the project was hopeless.

Some plant physiologists are trying to incorporate the genes for the C_4 system into C_3 plants, instead. If we could genetically engineer critical C_3 crops, such as soybeans, wheat, and rice, to photosynthesize as C_4 plants, food production might be increased, especially in hot, dry regions. This is not an easy task. There are many differences between C_3 and C_4 plants. Genetically altering photosynthesis will be extremely difficult, even with today's technology.

4.7 Photosynthesis and the Atmosphere

Photosynthesis supplies oxygen gas to Earth's atmosphere and food to Earth's organisms. Most organisms, including plants, use oxygen and release carbon dioxide. Photoautotrophs use the carbon dioxide again in photosynthesis, completing the cycle. These relationships are shown in Figure 4.24.

Photosynthesis produces enormous amounts of oxygen. Each year plants use as much as 140 billion metric tons of carbon dioxide and 110 billion metric tons of water in photosynthesis. They produce more than 90 billion metric tons each of organic matter and oxygen gas.

Because photosynthesis is the largest single biochemical process on Earth, any disruption of that process may have dramatic effects. The carbon dioxide content of the atmosphere has been increasing steadily at least since 1800. Each year, it reaches levels higher than any recorded in history. The increase is mostly due to large amounts of carbon dioxide released when people burn fossil fuels or clear land by burning rain-forest plants. Shrinking forests are also able to remove less and less carbon dioxide from the

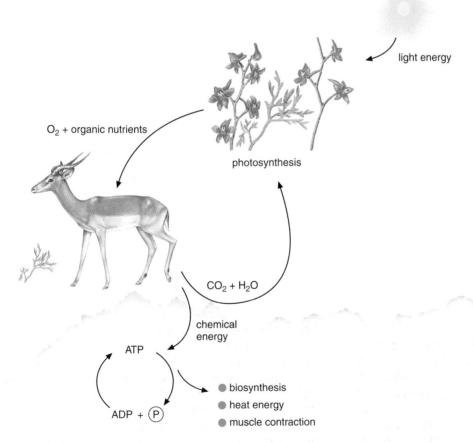

light energy

O_2 + organic nutrients

photosynthesis

$CO_2 + H_2O$

chemical energy

ATP

ADP + P

● biosynthesis
● heat energy
● muscle contraction

FIGURE 4.24

Summary of carbon, oxygen, and energy cycles in the biosphere. Note that energy flows in one direction—from the Sun to organisms—but substances such as carbon and oxygen cycle repeatedly through the biosphere.

atmosphere through photosynthesis. The rising carbon dioxide level may already have begun to heat up Earth's climates (see Chapter 25). Although the extra carbon dioxide could increase photosynthesis, preserving the balance of oxygen and carbon dioxide in the atmosphere may be vital to the future of all life on Earth.

Early in the history of life, when the Calvin cycle evolved, there was little oxygen in the air, so the binding of oxygen by rubisco was not a problem. After millions of years of photosynthesis, oxygen levels built up in the atmosphere and photorespiration became important. The C_4 and CAM pathways are examples of the adaptation of photosynthesis to an oxygen-rich atmosphere. Since the 1980s, scientists have noticed increased growth of C_3 plants in many parts of the world that were previously dominated by C_4 plants. Some biologists think this is happening because the increasing level of carbon dioxide in the atmosphere favors C_3 plants in their competition with C_4 plants.

Check and Challenge

1. Why is metabolic activity most accurately expressed as a rate?
2. How does increasing light intensity affect the rate of photosynthesis?
3. Describe the process of photorespiration. What is its effect on the rate of photosynthesis?
4. Describe the location and function of the two systems of carbon dioxide fixation in C_4 plants.
5. Which reaction in photosynthesis is inhibited by oxygen? Construct a graph that compares the effect of oxygen on C_3 and C_4 photosynthesis.
6. Explain the following statement: The C_4 pathway separates carbon fixation from the Calvin cycle in space, and the CAM pathway separates them in time.

Chemoautotrophy

4.8 Varieties of Chemoautotrophs

Chemoautotrophs are bacteria that obtain energy by performing chemical reactions, and fix their own carbon. The energy comes from oxidation of some substance in the environment, usually an inorganic mineral such as iron or sulfur. In general, this process does not provide as much energy as photosynthesis or heterotrophy. Many chemoautotrophs must oxidize large quantities of material to obtain enough energy to live. This may be one reason that chemoautotrophy occurs only in certain bacteria.

More complex organisms, such as plants, animals, and fungi, are all either photoautotrophs or heterotrophs. Chemoautotrophs do not compete well with other organisms. This is why they grow best in environments where other organisms cannot survive and light and organic compounds are in short supply.

Three questions are important in studying a chemoautotroph.

1. What is its source of energy?
2. What is its source of carbon?
3. What is its source of electrons for reducing carbon?

The answers to these questions are simple for most heterotrophs and photoautotrophs. Heterotrophs obtain electrons and energy by consuming carbon compounds produced by other organisms. Photoautotrophs obtain energy from sunlight, and carbon from carbon dioxide. Most of them obtain electrons by oxidizing water.

Chemoautotrophs fix carbon dioxide, usually with the Calvin cycle. Their sources of energy and electrons vary greatly, however (Table 4.2). Like photoautotrophs, chemoautotrophs have electron transport systems. They pass the electrons they extract from various substances through their electron transport systems to generate ATP and reduce NADPH (or a similar compound, NADH). The more reduced the electron source is, the more energy is released when it is oxidized. Chemoautotrophs that oxidize highly reduced substances such as hydrogen gas or elemental sulfur are able to produce more energy and grow faster than those that rely on partly oxidized electron donors such as ferrous iron and nitrite ions. Many chemoautotrophs can adapt to changing environments by switching electron donors or living heterotrophically when food is plentiful. Like photoautotrophy, chemoautotrophy is often a matter of degree.

Try Investigation 4C
Chemoautotrophs.

CONNECTIONS

Chemoautotrophs that are adapted to oxidize poor energy sources such as nitrite demonstrate that competition among organisms and variation in available resources both contribute to evolution.

TABLE 4.2
Some Common Energy-Yielding Oxidations Performed by Chemoautotrophs

Type of Bacteria	Electron Donor	Energy-Yielding Reaction
hydrogen bacteria	hydrogen (H_2)	$2\ H_2 + O_2 \longrightarrow 2\ H_2O$
sulfur bacteria	hydrogen sulfide (H_2S)	$H_2S + 2\ O_2 \longrightarrow SO_4^{-2} + 2\ H^+$
sulfur bacteria	elemental sulfur (S^0)	$2\ S^0 + 3\ O_2 + 2\ H_2O \longrightarrow 2\ SO_4^{-2} + 4\ H^+$
sulfur bacteria	thiosulfate ion ($S_2O_3^-$)	$S_2O_3^{-2} + H_2O + 2\ O_2 \longrightarrow 2\ SO_4^{-2} + 2\ H^+$
nitrifying bacteria	ammonia (NH_4^+)	$2\ NH_4^+ + 3\ O_2 \longrightarrow 2\ NO_2^- + 4\ H^+ + 2\ H_2O$
nitrifying bacteria	nitrite ion (NO_2^-)	$2\ NO_2^- + O_2 \longrightarrow 2\ NO_3^-$
iron bacteria	ferrous-iron ion (Fe^{+2})	$4\ Fe^{+2} + 4\ H^+ + O_2 \longrightarrow 4\ Fe^{+3} + 2\ H_2O$

Because many chemoautotrophs can use more than one electron source, you might wonder why any of them rely on the reactions that yield less energy. Some use different energy sources, depending on what is available. Others avoid competition by specializing in the use of resources that other organisms ignore.

4.9 Chemoautotrophs and the Environment

Because chemoautotrophs do not compete well with other organisms, they are not very important in the sun-lit, organic carbon–rich environments familiar to us. For example, the chemoautotrophic bacterium called *Deinococcus* survives on rocks near the South Pole, exposed to high levels of ultraviolet radiation and dry, frigid winds that would freeze-dry most organisms, without organic nutrients. However, underground and deep in the ocean where sunlight does not penetrate, chemoautotrophic bacteria are the carbon-fixing organisms on which others depend. Since the 1980s, studies of the deep ocean (below 1,000 m) and deep earth (below 500 m) have shown that chemoautotrophs are so common in these environments that they may make up the majority of life on Earth.

The oxidized end products of chemoautotrophs form important deposits of oxidized mineral ores, especially iron and sulfur. Chemoautotrophic bacteria also form nodules of manganese and other valuable minerals on the sea floor. These nodules may be mined in the future. Iron-oxidizing bacteria in bogs help to form "bog iron" (ferric hydroxide, $Fe(OH)_3$), an important source of iron for the early English settlers in New England. Metal-oxidizing bacteria are important in purifying copper metal from ores, and as contributors to water pollution near coal mines. They oxidize the copper in ore, converting it to soluble cupric ions (Cu^{+2}). The mine workers then collect the copper as metal. However, when the same bacteria attack coal that is rich in iron sulfide, or pyrite (FeS_2), they oxidize the sulfur in the pyrite, forming sulfuric acid (H_2SO_4). This strong acid dissolves aluminum and iron from the coal and rock. The acid and dissolved metals wash into local streams, killing many of the organisms living there (Figure 4.25).

Nitrogen-oxidizing bacteria are also important in the environment. Inorganic nitrogen ions are important plant nutrients. The supply of nitrogen in soil is often a limiting nutrient in plant growth. Chemoautotrophic bacteria contribute to plant growth by oxidizing ammonium ions (NH_4^+) to nitrite ions (NO_2^-), and nitrite to nitrate ions (NO_3^-). Most plants absorb and use nitrate more effectively than ammonium ions. The nitrogen cycle is discussed in more detail in Chapter 24.

In the deepest levels of the sea, chemoautotrophs are the primary producers that support heterotrophs such as animals. Deeper than 1,000 m, living things are increasingly rare. On most of the deep sea floor, cold water, high pressure, and scarce nutrients prevent the growth of most organisms. Around underwater volcanic vents, however, the hot water dissolves reduced forms of sulfur, manganese, and iron from rock. These minerals, as well as volcanic gases such as hydrogen, methane (CH_4), and carbon

FIGURE 4.25
Copper-rich, polluted water of the Queen River in Queenstown, Tasmania (Australia).
Chemoautotrophic bacteria oxidize copper ions in the acidic waste water that drains from copper mines upstream, coloring the water brown.

monoxide (CO), support a variety of chemoautotrophs. Among the animals that depend on these bacteria are worms that do not eat (Figure 4.26). Instead, their bodies contain large numbers of sulfur-oxidizing bacteria. Apparently the worms protect the bacteria inside their bodies, and the bacteria provide the worms with organic nutrients.

Hydrogen-oxidizing bacteria support another kind of biological community in the pores of rocks as deep as 2,800 m beneath the surface of the earth. Heterotrophic bacteria and fungi consume organic compounds that these chemoautotrophs produce. Many of these deep-earth dwellers are not adapted to survive under these conditions: Some are simply washed down by groundwater. They may grow so slowly in this nutrient-poor environment that they reproduce less than once per century. Eventually, they may die of starvation or be crushed as the pores they occupy fill up with mineral deposits.

Check and Challenge

1. What part of the light reactions of photosynthesis is similar to the oxidation of minerals by chemoautotrophs?

2. Why are chemoautotrophs rare among familiar organisms?

3. How do chemoautotrophs obtain organic nutrients?

4. Some deep-earth bacteria consume petroleum or natural gas. Are these organisms chemoautotrophs?

5. Some bacteria reduce metal ions or other inorganic substances. Is this behavior a clue that these organisms are chemoautotrophs? Explain.

FIGURE 4.26
Tube worms and other organisms living around an underwater volcanic vent on the Juan de Fuca ridge near the coast of Washington state. Minerals dissolved in the superheated water from the "black smoker" vent in the background form black solid particles when they reach the cooler ocean water. All these organisms depend on carbon fixation by mineral-oxidizing bacteria for carbon skeletons and energy.

Chapter
HIGHLIGHTS

Summary

Photosynthesis, the most common chemical process on Earth, transforms sunlight into the chemical energy that supports much of life on the planet. It also releases the oxygen that many organisms consume.

The cellular processes of photosynthesis include the production of ATP and NADPH by the combined actions of photosystems I and II of the light reactions. NADPH and ATP are used to fix carbon dioxide into sugars during the Calvin cycle. Both autotrophs and heterotrophs use the sugars for energy and the manufacture of cellular components. In eukaryotes, photosynthesis occurs in, and depends on, the structure of the chloroplast.

Environmental factors that directly influence the rate of photosynthesis in plants include light intensity, temperature, and the concentrations of carbon dioxide and oxygen. An increase of light intensity and carbon dioxide concentration tends to increase the rate of photosynthesis up to the point of saturation. After this, further increases do not stimulate the rate of photosynthesis. Increasing temperatures affect photosynthesis in the same way, except that high temperatures usually cause a decline in the rate of reactions. High concentrations of oxygen inhibit the rate of photosynthesis by fueling photorespiration. Environmental factors do not act individually but instead interact as limiting factors.

C_4 and CAM plants have evolved specializations that enable them to reduce photorespiration and water loss in hot, dry climates. C_4 plants reduce photorespiration by concentrating carbon dioxide at the Calvin cycle. In CAM plants, the stomates open at night and close during the day, greatly reducing transpiration. Both C_4 and CAM are effective adaptations to hot, dry climates.

Chemoautotrophs are bacteria that fix carbon, often by the Calvin cycle, and obtain energy by oxidizing substances in the environment, especially inorganic minerals. In general, this process provides less energy than photosynthesis or heterotrophy. Chemoautotrophs grow best in environments where other organisms cannot survive and light and organic compounds are in short supply. Many chemoautotrophs are able to use more than one electron source.

Chemoautotrophs support communities of organisms around underwater volcanic vents and deep in the earth. They are also important in the oxidation of ammonia in soil, and in the formation and mining of mineral ores.

Key Concepts

Use the concepts below to build a concept map, linking to as many other concepts from the chapter as you can.

- C_3
- C_4
- Calvin cycle
- chemical energy
- chemoautotrophs
- chloroplasts
- light energy
- light reactions
- oxidation
- photoautotrophs
- photorespiration
- photosynthesis
- reduction
- rubisco
- stroma
- thylakoids

Reviewing Ideas

1. During photosynthesis, how is light energy conserved in ATP and NADPH?
2. Does the Calvin cycle operate in the dark? Explain.
3. In what ways is photosynthesis important to humans?

4. What is the relationship between the light reactions and the carbon dioxide–fixing reactions of photosynthesis?
5. Describe how the structure of the chloroplast relates to its function in photosynthesis.
6. What happens to the sugars made during the Calvin cycle?
7. How does the special leaf anatomy of the C_4 plant support C_4 photosynthesis?
8. Compare photosynthesis in C_3 and C_4 plants.
9. Do chemoautotrophs gain energy by oxidizing or reducing substances in their environment? What step in photosynthesis is this like?
10. In which part of a pond would you look for chemoautotrophs?

Using Concepts

1. C_4 and CAM mechanisms of photosynthesis have evolved in some plants. What are the advantages of each type of photosynthesis?
2. Develop a concept map showing how the environment influences the rate of photosynthesis.
3. The curve in Figure 4.16 shows the effect of temperature increase on the rate of photosynthesis in a typical C_3 plant. Construct and explain a curve showing the response of a C_4 plant to the same increase in temperature.
4. Are the light reactions necessary in the mesophyll cells of C_4 plants? Explain the reasons for your answer.
5. Atmospheric levels of carbon dioxide have increased from 300 parts per million (ppm) to almost 355 ppm in recent years. What will be the effect of this increase on photosynthesis and growth in C_3 plants? In C_4 plants?
6. Why doesn't photorespiration occur in chemoautotrophs?

7. Could chemoautotrophs survive and grow on the rocky surface of the Moon, which has no atmosphere? Explain.

Synthesis

The level of carbon dioxide in the atmosphere continues to increase. How will this affect competition between chemoautotrophs and photoautotrophs? Explain the reasons for your answer.

Extensions

You have three plants—one C_3, one C_4, and one CAM plant. You also have a pH meter and a microscope. Describe how you could use only your pH meter and microscope (with the necessary supplies that go with each instrument) to identify each plant as C_3, C_4, or CAM.

Web Resources

Visit BSCSblue.com to access
- Information about chemoautotrophs and other organisms that live in extreme environments
- Research on the possibility of life on other planets
- Web links related to this chapter

CONTENTS

LEARNING OUTCOMES

By the end of this chapter you will be able to:

   A   Relate metabolism to cellular respiration.

   B   Distinguish among the three hydrogen-carrier molecules and their functions.

   C   Describe how the three stages of cellular respiration are related.

   D   Relate glycolysis and the Krebs cycle to compartmentalization in the cell.

   E   Explain the significance of the electron transport system to respiration.

   F   Discuss the importance of oxygen to respiration and photosynthesis.

   G   Explain the central role of the Krebs cycle in metabolism.

CHAPTER

5

Cell Respiration: Releasing Chemical Energy

■ *What problem do birds and other small animals face in winter?*

■ *What adaptations help them overcome this problem?*

All organisms require energy to carry out their life functions. Autotrophs collect this energy from the nonliving environment and store it in the form of organic compounds. Both autotrophs and heterotrophs use these compounds as energy sources. Evolution has produced a number of biochemical processes that enable organisms to obtain the energy stored in nutrients such as carbohydrates, fats, and proteins. The most efficient of those processes is cell respiration. In most organisms, the respiration reactions release usable energy while atmospheric oxygen oxidizes food molecules. This chapter explains how organisms convert the chemical energy stored in organic nutrients into free energy, which they use for cellular work.

An Overview of Respiration

5.1 Metabolism and Cell Respiration

Recall from Chapter 2 that metabolism, the chemical reactions in an organism, has two complementary parts—synthesis and decomposition. Synthesis reactions combine small, simple organic molecules to form more complex compounds, such as proteins and nucleic acids, for cell growth and maintenance. These reactions consume energy. Decomposition reactions release energy by breaking down organic food molecules to simpler forms. Organisms use some of this energy to make ATP, the major energy carrier in metabolism. Simple compounds produced by decomposition can also serve as carbon skeletons in biosynthesis. ATP molecules provide organisms with a ready supply of free energy in small, usable packets. Because ATP is made during decomposition and used during biosynthesis, it links these two processes, as shown in Figure 5.1.

Cell respiration is a decomposition pathway that provides the energy cells need to function. It is a series of reactions that release energy as they break down sugars and other substances to carbon dioxide and water. Each step in this decomposition is catalyzed by an enzyme. Respiration releases free energy by oxidizing sugars or other organic substrates. Some of this energy is conserved in ATP. ATP in turn provides the energy to power most life processes.

Cell respiration can occur in the presence or absence of oxygen. In **aerobic** respiration—occurring in the presence of oxygen—oxygen is the oxidizing agent that receives electrons from the decomposed substrates. In **anaerobic** respiration—occurring without oxygen—the substrate may be only partly decomposed, releasing less energy, or a nitrogen or sulfur compound may substitute for oxygen. This chapter will emphasize aerobic respiration.

The raw materials for aerobic respiration are carbohydrates, fats, and proteins. The 6-carbon sugar glucose ($C_6H_{12}O_6$) and glucose-phosphate ($C_6H_{11}O_6$—H_3PO_3) are important substrates for respiration. Animals produce them by digesting carbohydrates or by breaking down glycogen (animal

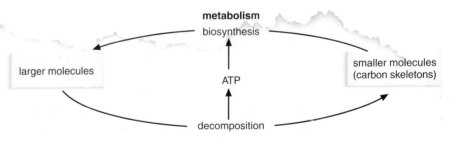

FIGURE 5.1

The importance of ATP in metabolism. Metabolism includes all decomposition and synthesis reactions in an organism. Cell respiration is part of decomposition. ATP mediates the transfer of energy in metabolism.

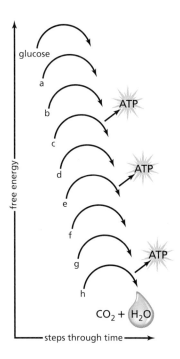

FIGURE 5.2

Energy transfer in cell respiration. A stepwise series of reactions transfers the energy of glucose to ATP. The letters *a* through *g* represent intermediate compounds in the decomposition of glucose to carbon dioxide and water.

starch) in the liver or muscles. Glucose is also the main form in which animals transport carbohydrate through the bloodstream. Sucrose fills this role in most plants. Plants produce glucose and glucose-phosphate by breaking down sucrose and starch.

During aerobic respiration, a great deal of energy released as glucose gradually oxidizes and breaks down to carbon dioxide. The overall reaction is summarized in the following equation:

$$C_6H_{12}O_6 \ + \ 6\,O_2 \ \xrightarrow{\text{enzymes}} \ 6\,CO_2 \ + \ 6\,H_2O \ + \ \text{energy}$$
$$\text{glucose} \quad\ \text{oxygen} \qquad\qquad \text{carbon dioxide} \quad \text{water}$$

Oxidizing one molecule of glucose releases much more energy than a single reaction needs, however, so glucose is not useful as a direct source of energy. You might compare glucose to the gas in your car's fuel tank. You might get to the store by dropping a match in the gas tank. The explosion might send some parts of you and the car to the store, but it would be rather inefficient (and foolish). A better way would be to start the engine and allow it to burn gasoline slowly. This approach would avoid wasting fuel and transport you to the store in one piece. The same is true for cells. To release all the energy in glucose at once would waste most of the energy and could heat the organism until it cooked itself. Instead, the "engine"—cell respiration—releases energy by oxidizing glucose in a series of small steps. Several of these steps release enough energy to drive the production of one molecule of ATP (Figure 5.2).

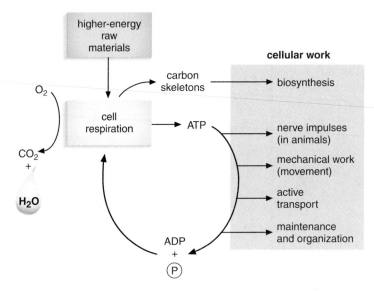

FIGURE 5.3

Products of cell respiration. Cell respiration breaks down sugars and other organic compounds. The products include carbon skeletons, used in biosynthesis, and ATP, which provides energy for various types of cellular work.

Cell respiration provides both ATP and the carbon skeletons needed for biosynthesis. Figure 5.3 summarizes these two major functions of cell respiration.

5.2 The Stages of Aerobic Respiration

The respiration of a simple carbohydrate such as glucose can be divided into three main stages—glycolysis, the Krebs cycle, and the electron transport system—as shown in Figure 5.4. Refer to the figure as you study each stage. Each stage involves a series of chemical reactions catalyzed by enzymes. The first stage is **glycolysis.** During this process, enzymes partially oxidize glucose and split it into two 3-carbon molecules. The partial oxidation releases enough energy to form a small amount of ATP.

An enzyme releases a molecule of carbon dioxide from each 3-carbon molecule that was produced in glycolysis. The resulting 2-carbon molecules are oxidized completely to carbon dioxide in the second stage, called the **Krebs cycle.** Additional ATP molecules form, conserving some of the energy released in this process.

Whenever one substance is oxidized, another must be reduced. As glucose is oxidized in glycolysis and the Krebs cycle, it loses electrons and protons. These pass to **NAD$^+$** (nicotinamide adenine dinucleotide). Hydrogen atoms released from glucose in stages 1 and 2 reduce NAD$^+$ to form NADH. NAD$^+$, an easily reduced molecule, is similar to NADP$^+$, which is reduced and oxidized in photosynthesis (see Section 4.3). NADH and

NADPH are identical except for the extra phosphate group in NADPH. Like NADPH, NADH is oxidized and reduced in a cycle (Figure 5.5).

NADH, in turn, is oxidized as it donates protons and electrons to the third stage of respiration, the **electron transport system.** This regenerates the supply of NAD^+. The protons and electrons release energy as the electron transport system transfers them to oxygen, forming water. This energy is available to form ATP. Most of the ATP is synthesized by the electron transport system.

Because oxygen must accept the electrons at the end of the electron transport system, stages 2 and 3 cannot proceed without oxygen. That is why this process is called aerobic respiration.

In one step in the Krebs cycle, two hydrogen atoms derived from glucose reduce a second hydrogen-carrier molecule, **FAD** (flavin adenine dinucleotide), instead of NAD^+ as follows:

$$FAD + 2\ H \longrightarrow FADH_2$$

Thus NADH, NADPH, and $FADH_2$ all carry hydrogen in cells. At the end of the electron transport system, hydrogen carried by NADH and $FADH_2$ reduces oxygen to form water. The energy released in this reaction is used to synthesize ATP.

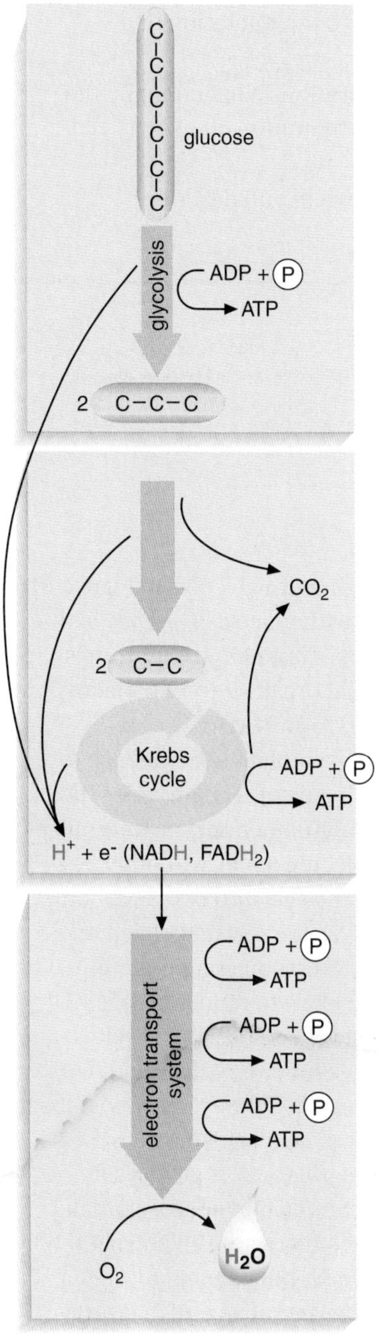

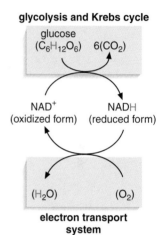

FIGURE 5.5

Role of NADH in cell respiration.
NADH acts as a carrier of reducing power. During cell respiration, NAD^+ is reduced as it receives electrons and protons from glucose; NADH is oxidized as it donates electrons and protons to the electron transport system. This replenishes the supply of NAD^+ to repeat the cycle.

FIGURE 5.4

The stages of cell respiration.
Aerobic respiration occurs in three stages—glycolysis, the Krebs cycle, and the electron transport system. As glucose and other substrates are oxidized to carbon dioxide and water, NAD^+ is reduced to NADH, and FAD is reduced to $FADH_2$. These reduced compounds carry hydrogen ions (H^+) and electrons (e^-) to the electron transport system.

> **Try Investigation 5A** ⬦ **How Does Oxygen Affect Cells?**

CONNECTIONS

FAD and NAD⁺ are derived from the B vitamins riboflavin and niacin. The consumption of foods rich in these vitamins is essential for energy metabolism.

Check and Challenge

1. How are carbon skeletons involved in biosynthesis and decomposition reactions?
2. What is the advantage of the stepwise nature of the reactions of cell respiration?
3. Why is glucose not a direct source of cellular energy?
4. Compare the process of cell respiration to the operation of a car's engine.
5. Breathing is also sometimes called respiration. What could be the connection between your need for cell respiration and your need to breathe?
6. Compare the three cellular hydrogen carriers and describe their functions.

The Reactions of Respiration

5.3 Glycolysis

Both aerobic and anaerobic respiration begin with glycolysis. Figure 5.6 shows the main steps of glycolysis, beginning with glucose. Glucose is the usual raw material for glycolysis in animal cells. Three important things happen during glycolysis—the glucose molecule breaks into two pieces, some ATP forms, and some NAD⁺ is reduced to form NADH.

Glycolysis begins when an enzyme converts a molecule of glucose to glucose-6-phosphate (Figure 5.6, step *a*). A molecule of ATP provides the phosphate and the energy to power the reaction. Another enzyme rearranges the glucose-6-phosphate, and a second ATP molecule donates another phosphate group (step *b*). The resulting molecule splits into two 3-carbon sugar-phosphates (step *c*). Other enzymes catalyze the rearrangement and partial oxidation of these molecules to form the 3-carbon compound **pyruvic acid** (step *d*). Oxidation of the sugar-phosphate leads to reduction of NAD⁺. Some of the energy released in this process is saved as ATP. The result is the formation of four molecules of ATP, two molecules of NADH, and two molecules of pyruvate. Recall that two molecules of ATP were needed to begin glycolysis. Thus, for each molecule of glucose, the net gain from glycolysis is two molecules of ATP. Table 5.1 summarizes the products of glycolysis.

In plant cells, glycolysis begins differently. Starch and sucrose break down to glucose or glucose-1-phosphate, which can enter glycolysis directly at step *a* in Figure 5.6. Three-carbon sugar-phosphates formed in photosynthesis can enter the process at step *c*. Several sugars and organic

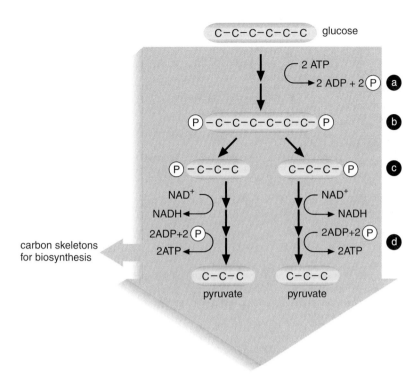

FIGURE 5.6

The partial oxidation of glucose in glycolysis. Glucose breaks down in small steps to two 3-carbon molecules of pyruvate. Many of the steps involve rearrangements of carbon skeletons. ATP and NADH form in these reactions. In plant cells, sugars from photosynthesis can enter glycolysis at step **(a)** or step **(c).**

acids formed during glycolysis can leave the pathway and serve as carbon skeletons for the biosynthesis of amino acids, lipids, and other compounds. Thus the biological roles of glycolysis in all organisms are the synthesis of ATP, NADH, and pyruvate and the formation of carbon skeletons.

At the end of glycolysis, the fate of pyruvate depends on whether oxygen is present (Figure 5.7). If insufficient oxygen is present, animal cells and some bacteria reverse the oxidation that produced pyruvate. They convert NADH and pyruvate into NAD$^+$ and **lactate,** another 3-carbon acid. NAD$^+$ cycles back to glycolysis, and glycolysis continues to provide a small amount

CHEMISTRY TIP

When a molecule of an acid (pyruvic acid, acetic acid, etc.) releases a positively charged hydrogen ion (H$^+$), it becomes a negatively charged ion with a similar name (such as pyruvate or acetate). Both types of names are commonly used to describe these substances. In living cells, many of the acid molecules exist as ions.

TABLE 5.1
Products of Glycolysis, per Molecule of Glucose

Substance	Molecules Formed
Pyruvate	2
NADH	2
ATP (net)	2

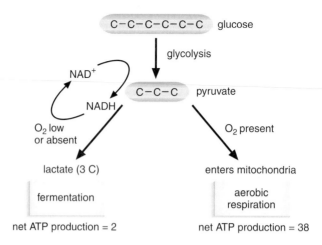

C–C–C–C–C–C glucose

glycolysis

NAD⁺

NADH

C–C–C pyruvate

O₂ low
or absent

O₂ present

lactate (3 C)

enters mitochondria

fermentation

aerobic
respiration

net ATP production = 2

net ATP production = 38

FIGURE 5.7

Oxygen and the fate of pyruvate. The oxygen concentration in a cell can affect the fate of pyruvate. If oxygen is present, then pyruvate enters aerobic respiration. If oxygen is absent, then pyruvate undergoes fermentation, forming lactic acid, ethyl alcohol, acetic acid, or other incompletely oxidized compounds.

of ATP until more oxygen becomes available. This anaerobic pathway is known as **lactic-acid fermentation.** Evolution has produced many other types of fermentation. For example, yeast and some bacteria ferment pyruvate to ethanol and acetic acid (vinegar), respectively. If sufficient oxygen is present, however, pyruvate enters the Krebs cycle. Note that glycolysis is the first stage in fermentation as well as in respiration.

Focus On
Fermentation

Wine makers have long known that the fermentation that produces wine is caused by living yeast. The special feature of fermentation is that it releases energy from sugars without using oxygen. In yeast cells, fermentation of glucose produces ethanol (ethyl alcohol) and carbon dioxide.

Certain bacteria also can carry out fermentation. In the 1850s, the French wine industry was troubled by spoiled wine. Louis Pasteur discovered that bacteria contaminating the wine were fermenting it. These bacteria spoil wine because they produce vinegar (acetic acid) instead of the ethyl alcohol produced by yeast. Fermenting bacteria and yeast are used to produce vinegar, alcohol, cheese (Figure 5.8), and yogurt.

FIGURE 5.8

Industrial fermentation. Bacteria ferment milk in the first stage of cheesemaking.

5.4 Mitochondria and Respiration

Cell respiration is a complex process. How are its many enzymes and chemical reactions organized and kept apart? The process of evolution has organized cell respiration and photosynthesis in similar ways. As in photosynthesis, the electron transport system of cell respiration is embedded in a membrane. In the simple cells of bacteria, the cell membrane serves this purpose. The enzymes of glycolysis, fermentation, and the Krebs cycle are suspended in the bacterial cytoplasm. Like all proteins, the enzymes of these pathways are synthesized according to instructions encoded in the cell's DNA (see Figure 1.33).

Other cells are more complex. Just as chloroplasts contain and organize the enzymes of photosynthesis in nonbacterial cells, specialized structures called mitochondria provide efficiency and organization to cell respiration. **Mitochondria** (singular: mitochondrion; Figure 5.9) are the organelles in which the Krebs cycle and the electron transport system occur. As in bacteria, glycolysis and fermentation occur in the cytoplasm. Mitochondria are called the powerhouses of the cell because they are the sites where most ATP is synthesized.

Some cells may have only ten to twenty mitochondria. Others that consume a great deal of energy (such as muscle cells) may have several thousand. Each mitochondrion is very small, usually only 2–3 μm long and about 1 μm thick. Large, stained mitochondria are visible under a compound microscope, but only the electron microscope reveals the detailed structure seen in Figure 5.9a.

Like a chloroplast, a mitochondrion has an outer and an inner membrane (Figure 5.9b). The two mitochondrial membranes consist of proteins and the usual double layer of lipid molecules. The inner membrane, however, contains so many enzymes that it is more protein than lipid. The inner membrane has many folds, or cristae, that extend into the inside of the mitochondrion. Organized in and on the cristae are all the enzymes of the electron transport system, the enzymes for ATP formation, and some of the enzymes of the Krebs cycle. Most of the enzymes of the Krebs cycle are within the mitochondrion's fluid-filled interior space, or matrix. The outer membrane regulates the movement of molecules into and out of the mitochondrion.

5.5 The Krebs Cycle

The Krebs cycle completes the decomposition and oxidation of glucose to carbon dioxide. The carbon dioxide is released as a gas. As the breakdown products of glucose are oxidized, NAD^+ and FAD are reduced, and a small amount of energy is saved as ATP. Within each mitochondrion, numerous copies of each enzyme catalyze the steps of the Krebs cycle. Follow Figure 5.10 as you read.

The process begins as pyruvate is transported into the mitochondria. Here, enzymes release a molecule of carbon dioxide from each 3-carbon pyruvate molecule, leaving a molecule of acetate, a 2-carbon organic acid (step *a*, Figure 5.10). This step also reduces one molecule of NAD^+ to NADH. A carrier

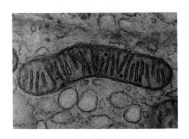

a

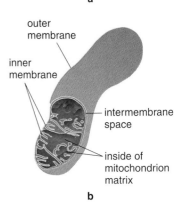

outer membrane

inner membrane

intermembrane space

inside of mitochondrion matrix

b

FIGURE 5.9

Mitochondria. (a), This transmission electron micrograph shows a mitochondrion in a human liver cell (×80,000; color added). **(b),** A three-dimensional cutaway drawing of the mitochondrion in **(a),** highlights its structures.

CONNECTIONS

The number of mitochondria in a cell is an example of the relationship between structure and function.

molecule, **coenzyme A** (CoA), binds to the acetate. The resulting complex is known as acetyl CoA. Coenzyme A delivers acetate to the Krebs cycle.

Acetate enters the Krebs cycle at step *b*, as an enzyme combines the acetate group of acetyl CoA with a 4-carbon acid (oxaloacetate) to form a 6-carbon acid (citrate). Coenzyme A is released and recycled to deliver more acetate. In steps *c* and *d*, other enzymes catalyze the rearrangement and oxidation of citrate. Two of the carbon atoms in citrate are oxidized to carbon dioxide. The hydrogen atoms that these carbon atoms lose reduce two molecules of NAD^+. These reactions produce a 4-carbon organic acid that is then rearranged and further oxidized (steps *e* and *f*). The result is a new molecule of oxaloacetate that begins another round of the cycle (step *a*).

The reactions at step *e* and step *f* that oxidize the 4-carbon skeleton reduce another molecule of NAD^+ and a molecule of FAD, producing another molecule of NADH and one of $FADH_2$. Some of the energy released in these reactions is saved, as a molecule of ATP forms from ADP and phosphate.

Steps *b–f* represent one turn of the Krebs cycle and result in the products shown in Table 5.2. Because each molecule of glucose produces

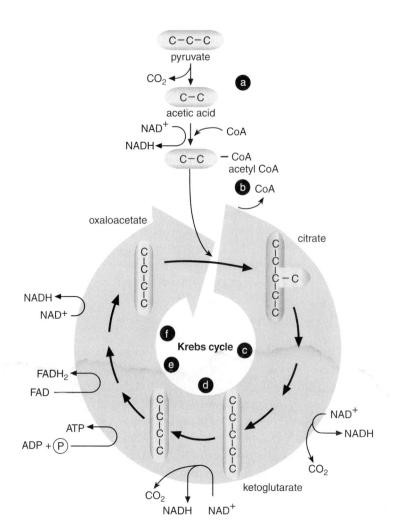

FIGURE 5.10

The Krebs cycle. The reactions of the Krebs cycle complete the breakdown of glucose to carbon dioxide. Each molecule of glucose provides two molecules of pyruvate to the Krebs cycle. The oxidation of pyruvate, formed from glucose during glycolysis, is linked to the reduction of NAD^+ and FAD and the formation of ATP.

TABLE 5.2
Products of the Krebs Cycle

Reaction	CO_2	NADH	$FADH_2$	ATP
Pyruvate to acetyl CoA	1	1	0	0
Krebs cycle	2	3	1	1
Total from 1 pyruvate	3	4	1	1
Total from 1 glucose = (1 glucose → 2 pyruvate)	6	8	2	2

two molecules of pyruvate in glycolysis, the products of the Krebs cycle are doubled in the last row of the table. The complete aerobic respiration of a molecule of glucose to six molecules of carbon dioxide can produce up to thirty-eight molecules of ATP. Stage three of respiration, the electron transport system, requires oxygen to produce most of that ATP.

5.6 The Electron Transport System

The oxidation of glucose in glycolysis and the Krebs cycle reduces NAD^+ to NADH and FAD to $FADH_2$. Reduced NADH and $FADH_2$ carry hydrogen atoms to the electron transport system. The electron transport system consists of a series of easily reduced and oxidized enzymes and other proteins known as **cytochromes.** These proteins are embedded in the inner membranes of mitochondria (Figure 5.11). The electron transport system

ETYMOLOGY

cyto- = cell (Greek)
-chrome = color (Greek)

Cytochromes are components of cells that often give them color.

area enlarged below

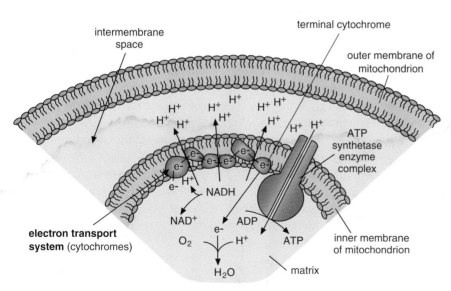

intermembrane space

terminal cytochrome

outer membrane of mitochondrion

H^+

ATP synthetase enzyme complex

NADH

NAD^+ ADP

electron transport system (cytochromes)

O_2 H^+ ATP

inner membrane of mitochondrion

H_2O matrix

FIGURE 5.11

The mitochondrial electron transport system. NADH and $FADH_2$ (not shown), produced in glycolysis and the Krebs cycle, donate electrons (e^-) to the electron transport system. These electrons are transferred through the system to oxygen. The oxygen is reduced to form water. Energy released by electrons as they pass through the system powers the active transport of protons out through the inner mitochondrial membrane. As highly concentrated protons outside the membrane diffuse back in, they pass through an enzyme complex that uses the energy of the concentration difference to synthesize ATP. This is how most ATP is generated.

CHEMISTRY TIP

Oxidizing agents such as oxygen have a strong affinity for electrons. The energy that electrons lose in passing from glucose to oxygen powers the synthesis of ATP.

separates hydrogen atoms into electrons and protons. The cytochromes transfer the electrons step by step through the system. The last, or terminal, cytochrome is an enzyme that combines the electrons and protons with oxygen, forming water. Only this final step of respiration uses oxygen.

At each transfer in the electron transport chain, the electrons release free energy. Some of the free energy enables enzymes in the inner mitochondrial membrane to actively transport protons from the matrix to the intermembrane space (see Figure 5.11). These protons become highly concentrated there. Such an unequal distribution of protons is unstable, and the protons tend to diffuse back into the matrix of the mitochondrion. As the protons cross the inner membrane, they pass through the ATP-synthetase enzyme complex, where ATP is synthesized from ADP and phosphate. (See Appendix 4A, "ATP Synthesis in Chloroplasts and Mitochondria," for more details of this process.) By providing energy to push protons out through the inner mitochondrial membrane, electrons from each molecule of NADH can drive the synthesis of up to three molecules of ATP. Each molecule of $FADH_2$ can drive the synthesis of up to two molecules of ATP.

Bacteria do not have mitochondria. Their cell membranes contain their electron transport systems. In some bacteria, electrons flow through the system to oxidizers other than oxygen, such as sulfate (SO_4^{-2}) or nitrate (NO_3^-). Instead of reducing oxygen to form water, these bacteria produce reduced sulfur or nitrogen compounds such as hydrogen sulfide (H_2S) and ammonia (NH_3). This process is called anaerobic respiration. Anaerobic respiration should not be confused with chemoautotrophy (Section 4.8), in which autotrophic bacteria obtain energy by oxidizing inorganic substances, including reduced forms of sulfur and nitrogen.

Some bacteria, known as **facultative aerobes,** can survive for long periods with or without oxygen. They are able to switch back and forth between fermentation and aerobic respiration, depending on the supply of oxygen. Other bacteria are actually poisoned by oxygen. These **obligate anaerobes** generate ATP entirely from fermentation or anaerobic respiration. Most organisms, such as animals and plants, are **obligate aerobes;** they cannot survive for long without oxygen.

To summarize, electrons and protons from NADH and $FADH_2$ reduce oxygen, forming water. The energy released in this process is used to synthesize ATP. As the electrons pass along the transport system to oxygen, they release free energy. Active transport proteins use this energy to pump protons out through the inner mitochondrial membrane. The difference between the proton concentration inside and outside of this membrane, in turn, supplies energy for the synthesis of ATP. Thus the role of the electron transport system is to synthesize ATP. The ATP can be transferred out of the mitochondrion and used by the cell. Figure 5.12 is a summary of the entire process of aerobic respiration leading to the synthesis of ATP. Table 5.3 (page 142) shows the number of ATP molecules synthesized from the complete respiration of one molecule of glucose.

CHEMISTRY TIP

Hydrogen sulfide (H_2S) is a reduced form of sulfur. It is a flammable toxic gas that smells like rotten eggs.

Try Investigation 5B ▽ Rates of Respiration.

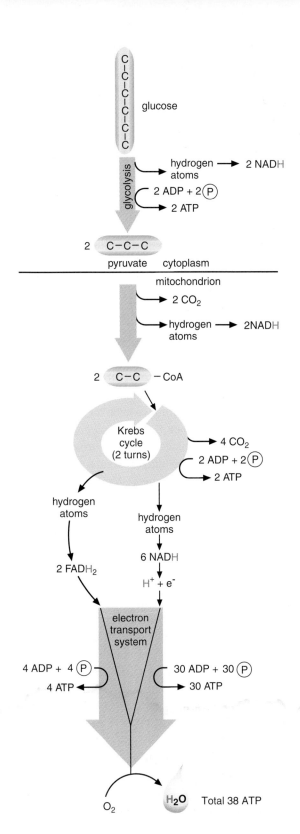

FIGURE 5.12

Summary of aerobic respiration.
As glucose is oxidized in glycolysis and the Krebs cycle, NAD+ and FAD are reduced to NADH and FADH$_2$. These carriers pass electrons to the electron transport system. ATP forms as these electrons lose energy in reducing oxygen. Each molecule of NADH generates three ATP molecules, and each molecule of FADH$_2$ generates two ATP molecules. The resulting oxidized NAD+ and FAD are recycled as more glucose is oxidized.

TABLE 5.3
ATP Synthesis in the Aerobic Respiration of a Molecule of Glucose

| Process | Number of Hydrogen Carriers | ATP Molecules Formed | | Potential Total ATP Molecules |
		Direct	From Hydrogen Carriers in ETS*	
Glycolysis	2 NADH	2**	6	8
Pyruvate to acetyl CoA	2 NADH	0	6	6
Krebs cycle	6 NADH	2	18⎤	24
	2 FADH$_2$		4⎦	
			Total per glucose	38

*ETS = electron transport system
**Four ATP molecules form, but two are consumed, for a net gain of two ATP molecules.

5.7 Oxygen, Respiration, and Photosynthesis

With all the importance placed on oxygen in this and the previous chapter, does it seem surprising that oxygen is involved only in the last reaction of respiration? Remember that oxygen is needed to oxidize glucose. Intermediates such as NADH only help cells to capture some of the energy released in this process as ATP. Without oxygen, cells must ferment glucose, forming only two ATP molecules per glucose molecule.

With oxygen present, organisms gain much more energy from their food. Humans and other animals have specialized organs and systems, such as lungs (Chapter 3) and circulatory systems (Chapter 7), that deliver oxygen to the electron transport system. This helps them release energy efficiently from food.

The atmospheric oxygen on which most life today depends is itself a product of life. Most photoautotrophs release oxygen to the air during photosynthesis. In general, the products of photosynthesis—oxygen and carbohydrates—are the raw materials for cell respiration. Cell respiration, in turn, provides the raw materials for photosynthesis (carbon dioxide and water). Thus the two processes complement each other (Figure 5.13). Both processes also provide carbon skeletons used in biosynthesis and use electron transport systems to form ATP, although their energy sources differ.

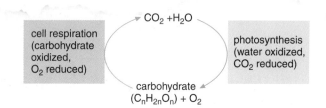

FIGURE 5.13

Relationship of cell respiration and photosynthesis. Respiration releases chemical energy by using the reduction of oxygen to water to drive the oxidation of sugar to carbon dioxide. Photosynthesis stores chemical energy by using the oxidation of water to oxygen to drive the reduction of carbon dioxide to sugar.

Focus On

Respiration during Exercise

What happens when you exercise? Enzymes in your liver cells break down glycogen, a form of starch, to glucose. The glucose travels through your bloodstream to your muscles, where it is respired to power your movements. But how can all this happen quickly enough for a racer to go from a standstill to an all-out effort in less than a second?

The rate at which a large polymer such as glycogen breaks down depends on the concentration and activity of the enzyme that decomposes it. When you prepare to exercise, the stress hormone epinephrine (also called adrenaline) binds to receptor proteins in the cell membranes of your muscle cells. This signal activates a pathway that stimulates an enzyme called a protein kinase. Protein kinases catalyze the addition of a phosphate group to other proteins. This chemical change activates the second protein, which is also a type of protein kinase. The second kinase activates the enzyme that breaks down glycogen. This system is called an enzyme cascade (Figure 5.14). It allows the rate of glycogen breakdown to rapidly increase over a millionfold.

When you exercise, you need to increase the supply of oxygen to your muscles (Figure 5.15). If you exercise at a moderate pace, your heart and lungs supply enough oxygen to your muscles to meet the demand of cell respiration. But most people can exercise so hard that muscle cells require more oxygen than the blood can deliver. Oxygen-starved muscle cells begin to ferment glucose to lactate. Accumulation of this acid causes muscle pain and cramps. Rapid breathing restores oxygen supplies. The bloodstream carries lactate out of the muscles and back to the liver, where it is slowly converted back to glucose. Soreness *after* exercise is usually due to slight tears in muscle fibers, not lactate accumulation.

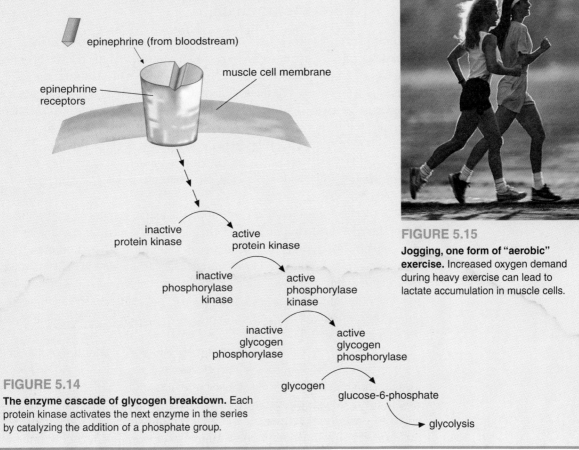

FIGURE 5.15

Jogging, one form of "aerobic" exercise. Increased oxygen demand during heavy exercise can lead to lactate accumulation in muscle cells.

FIGURE 5.14

The enzyme cascade of glycogen breakdown. Each protein kinase activates the next enzyme in the series by catalyzing the addition of a phosphate group.

Respiration and Cellular Activities

5.8 The Krebs Cycle in Fat and Protein Metabolism

Biologists originally thought the Krebs cycle was involved only in the breakdown of carbohydrates. Later research showed that the release of energy from fats and proteins also involves the Krebs cycle. When cells use the fatty acids of fats for energy, enzymes in the mitochondria break down the fatty acids to acetate. Coenzyme A transfers this acetate to the Krebs cycle, just as it does in the respiration of carbohydrates (see Figure 5.10). The cycle breaks down acetate from fatty acids and sugars in the same way. Because this process bypasses glycolysis, however, cells cannot ferment fatty acids as they do carbohydrates. Thus, without oxygen, most of the energy in fat cannot be transferred to ATP.

When cells use proteins in respiration, digestive enzymes first break down the proteins to amino acids. Other enzymes remove the amino groups and convert the ammonia this produces to safer nitrogen compounds. These compounds may be recycled or excreted. The carbon skeletons remaining from some amino acids can undergo reactions that form 4- or 5-carbon acids (oxaloacetate or ketoglutarate), which can enter the Krebs cycle. Figure 5.16 shows how carbohydrates, fats, and proteins can supply carbon skeletons to the Krebs cycle.

The Krebs cycle has a second important function in addition to its role in decomposition. The Krebs cycle and glycolysis also provide building blocks for biosynthesis. In autotrophs, these pathways, along with the Calvin cycle, lead to the synthesis of every organic compound the organism needs. In heterotrophs, these pathways lead to the synthesis of most, but not all, of the necessary organic compounds. Animals, for example, must consume the

Biology Online BSCSblue.com/check_challenge

organic compounds they cannot synthesize, such as vitamins, certain amino acids, and certain fatty acids.

It is important to note that most synthesis pathways are *not* the reverse of decomposition pathways. Separate enzymes and pathways for synthesis and decomposition help cells operate efficiently and control the activities of these pathways. For example, cells synthesize proteins by a very precise system that controls the position of each amino acid (see Chapter 9). Your digestion of the protein you eat, however, is anything but precise. Digestive glands supply enzymes to your stomach and intestines that break the bonds

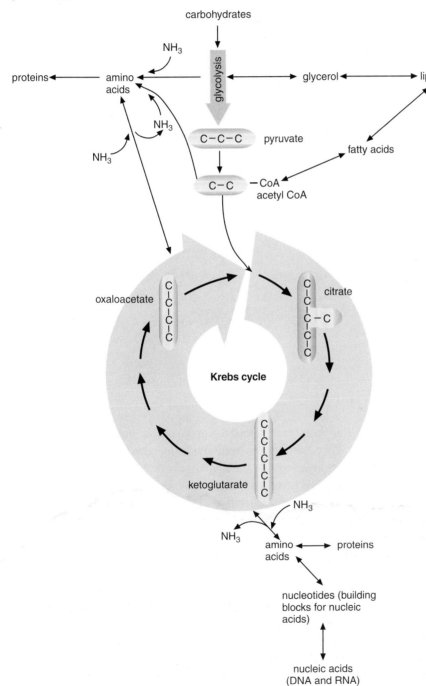

FIGURE 5.16

Other roles of the pathways of cell respiration. The reactions of cell respiration and particularly of the Krebs cycle contribute to both the decomposition and biosynthesis of carbohydrates, fats, and proteins. Certain amino acids can be synthesized from the carbon skeletons by adding amino groups ($-NH_2$) derived from ammonia (NH_3). Carbon skeletons can be formed from amino acids by removing the amino groups. Most organisms cannot convert fat to carbohydrate.

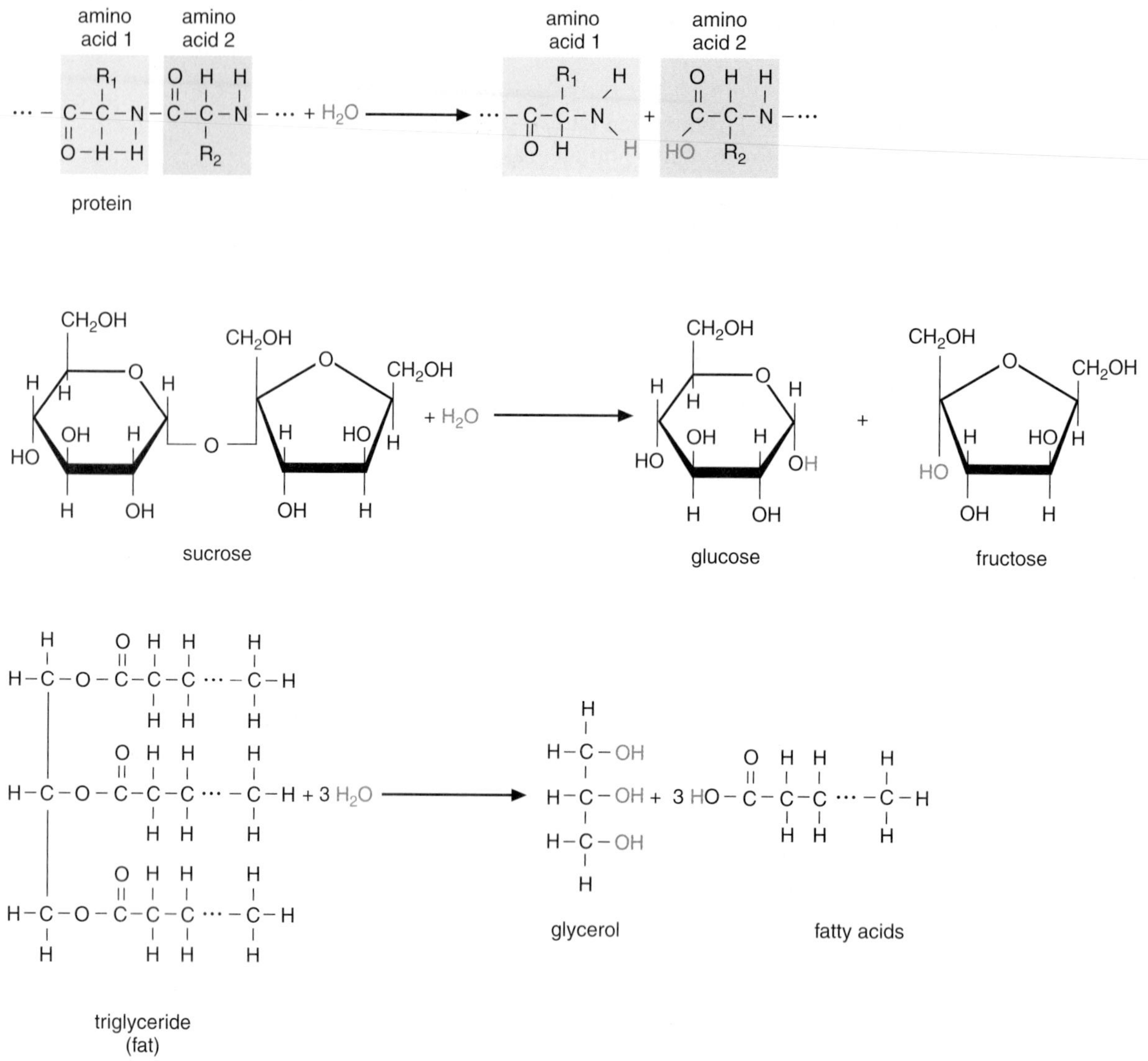

FIGURE 5.17

Examples of hydrolysis reactions.
In each case, bonds are broken in a water molecule and another molecule. Hydrolysis breaks down fats, polysaccharides, and proteins into smaller molecules that can undergo cell respiration.

ETYMOLOGY

hydro- = water (Greek)
lysis = loss or release (Greek, Latin)

Hydrolysis is a breakdown that involves water.

between neighboring amino acids until the protein is completely decomposed. This type of breakdown is known as **hydrolysis** because the components of water (H and OH) are inserted into the bond to break it (Figure 5.17). Most biological decompositions involve hydrolysis. Other digestive enzymes hydrolyze starch into simple sugars and fats into fatty acids and glycerol. Specialized cells that line your intestines actively transport the resulting amino acids, sugars, and other nutrients into your bloodstream.

5.9 Respiration and Heat Production

Cell respiration releases heat energy, which helps many organisms keep warm. Homeostatic mechanisms keep the internal temperatures of humans and many other animals nearly constant. Your own body temperature, for example, stays the same within one or two degrees. It may go up a few degrees when you are ill, but it drops back to normal when you are well.

Focus On

Energy Regulation

The way an organism regulates the use of its energy reserves is part of its overall adaptation. So it should not be surprising that different organisms manage their energy supplies in different ways.

For example, if you performed Investigation 5A, you learned that yeast cells, like many microorganisms, consume more sugar when oxygen is lacking. This is necessary because fermentation produces much less ATP than aerobic respiration does. But what makes glycolysis slow down when oxygen is available? PFK (phosphofructokinase) is one of the first enzymes in the pathway of glycolysis. The activity of PFK is stimulated by ADP and inhibited by ATP. When oxygen is available, cell respiration converts plenty of ADP to ATP. This inhibits PFK, slowing down glycolysis and, eventually, cell respiration. When oxygen is lacking, ATP production slows down and ADP builds up, stimulating PFK. This type of regulation, in which the product of a process inhibits its own synthesis, is called negative feedback. It occurs in many metabolic pathways.

Another type of feedback system operates in humans and many other animals. Two hormones with opposite effects—insulin and glucagon—cooperate to help regulate the level of glucose in the bloodstream. Insulin is produced by glandular cells in response to rising blood sugar levels. This hormone binds to proteins called insulin receptors on the surfaces of many types of cells. This signals the cells to absorb more glucose from the bloodstream and convert it to glycogen or fat. A deficiency of insulin or insulin receptors produces the disease called diabetes. Diabetic people have difficulty maintaining normal levels of blood sugar. When blood sugar levels fall too low, glucagon stimulates cells to break down glycogen to glucose and release it into the bloodstream. The balance between the actions of these two regulators helps to keep the level of glucose in the blood in a narrow range.

Some mammals (animals with hair and milk glands) have a specialized tissue called brown fat. Brown fat contains more mitochondria than any other body tissue and is adapted for rapid production of thermal energy. Respiration of stored fat in brown fat cells produces much heat energy and little ATP. This tissue is especially important for hibernating mammals and hairless or small mammals that tend to chill easily. The brown fat, located on the neck and between the shoulders, is especially active at the end of hibernation when the animal must quickly raise its body temperature to normal levels (Figure 5.18a). Why would brown fat be important for human infants?

Many plants have also evolved a form of respiration that produces a great deal of heat energy. The mitochondria of such plants contain an alternate branch of the electron transport system. In this pathway, some of the energy of electron flow results in the production of more heat energy and less ATP than in normal respiration. In one group of plants, the flowers become hot to the touch. At this temperature, organic compounds in the flower evaporate, producing an odor of rotting meat. The odor attracts certain flies and beetles.

FIGURE 5.18

Organisms that rely on cell respiration to keep warm. (a), The arctic ground squirrel, *Spermophilus parryi*, spends its summer on the Arctic tundra. In the winter, it hibernates in its nest. Brown fat enables the ground squirrel to quickly elevate its body temperature at the end of hibernation. **(b),** Skunk cabbage, *Symplocarpus foetidus*, releases heat energy that melts snow as it emerges from the ground.

a

b

starch (plants)
glycogen (animals)

cell
respiration

ATP

sucrose, PGAL (plants)
glucose (animals)

storage

fats

FIGURE 5.19

Energy regulation in animals.
Supply and demand determine how glucose and fatty acids are used. The rates of biosynthesis and decomposition vary in response to energy demand, but both types of reactions are always active to some extent.

CONNECTIONS

These mechanisms that regulate temperature are examples of adaptations that increase survival of the species.

When they land on the flower, pollen sticks to their bodies. As the insects fly from plant to plant, the pollen on their bodies fertilizes other flowers of the same species. (This process is pollination, discussed in Chapter 12.)

One plant that relies on the alternate pathway of respiration is aptly named skunk cabbage (Figure 5.18b). In addition to generating foul-smelling gases, the heat energy from alternate respiration melts snow cover over the skunk cabbage. The part of the stem bearing the flower grows through the hole in the snow, allowing pollination to occur while snow still covers the ground.

5.10 Control of Respiration

Organisms must control their rate of respiration in order to direct energy and carbon skeletons accurately to the pathways and cells where they are needed. Control is critical to organization, and cells must be organized to survive.

The mechanisms that control whether glucose is broken down in respiration or converted to starch or fat operate by supply and demand. For example, when animals such as humans use energy rapidly, their cells absorb glucose from the blood to make ATP (Figure 5.19). Liver cells then break down their stored glycogen to restore the blood sugar level. On the other hand, when the demand for energy is low, cells use excess glucose to synthesize glycogen, and then fat. Key steps in respiration are regulated to direct glucose and other substances to where they are needed.

CONNECTIONS

The control of respiration is an example of how cells regulate their activities to maintain homeostasis.

Check and Challenge

1. Why is the Krebs cycle so important in metabolism?
2. Why is oxygen necessary for the respiration of fat?
3. Of what adaptive value is the alternative pathway of respiration in plants?
4. How do plants and animals benefit from the heat energy released in respiration?
5. Discuss the control of glucose oxidation and ATP production in cells.
6. Describe how fats and proteins are brought into the Krebs cycle.

Biology Online BSCSblue.com/check_challenge

Energy Regulation and Obesity

In industrial countries such as the United States, obesity is one of the most common health risks, along with the use of alcohol and tobacco. When a hormone that helps to regulate appetite and body weight was discovered around 1996, many scientists and drug companies became interested in the possibility that it could be used as a treatment for severe, life-threatening obesity.

Fat cells (Figure 5.20a) produce the hormone called leptin and release it into the bloodstream. Leptin binds to a receptor protein on the surface of certain brain cells. This signals the brain that the stomach is full, so the person or animal stops eating. Other effects of leptin include increased cell respiration and reduced formation of storage lipids.

Obese rats and mice have been found to have reduced levels of leptin in their blood or defects in the sections of their DNA that code for leptin or its receptors (Figure 5.20b). Similar changes have been observed in some obese people. These observations raised hopes that leptin injections might help dangerously overweight people. Unfortunately, the story has become more complex, and the effects of leptin are still not completely understood. High leptin levels block the response of human liver cells to insulin. This may cause some people to become diabetic. In addition, high leptin levels have recently been found to reduce mineral density of bone. This in turn can weaken weight-bearing bones and increase the chance of experiencing a bone fracture. Mechanical stress on bones, whether due to exercise or body weight, tends to strengthen bones by stimulating mineral deposition. The role of leptin in regulating body weight and bone strength remains unclear.

Leptin now appears to be just one of many factors that affect appetite and body weight. These factors include other hormones, diet, exercise, and heredity. Defects in leptin and its receptor may help to explain why obesity seems to be inherited in some families. However, most obesity in the United States is due to lack of exercise and excessive consumption of fatty and sugary foods.

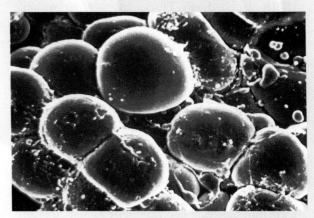

a

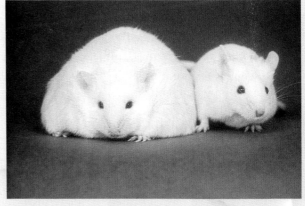

b

FIGURE 5.20

Regulation of metabolism by leptin. (a), This scanning electron micrograph shows human fat cells ($\times$110). Leptin produced by fat cells such as these helps regulate energy use and storage in animals. Leptin levels appear to be an important factor in the control of fat storage and body weight. Some obesity may be due to an inability to produce or respond to leptin. **(b),** Unlike the normal mouse *(right),* this obese mouse *(left)* is unable to respond to leptin because of a defect in its leptin-receptor proteins.

Chapter

HIGHLIGHTS

Summary

Metabolism consists of all the chemical reactions in an organism, including biosynthesis and degradation. Cell respiration is an important part of decomposition and involves stepwise reactions that oxidize carbohydrates, fats, or amino acids, with a release of energy that is conserved as ATP. Cell respiration also provides carbon skeletons for the biosynthesis of macromolecules the cell requires. Supply and demand in the cell determine whether carbon skeletons in this pathway are oxidized or used in biosynthesis.

Most types of organisms carry out aerobic (oxygen-dependent) respiration in three controlled, sequential stages: glycolysis, the Krebs cycle, and the electron transport system. Glycolysis, which occurs in the cell's cytoplasm, partly oxidizes sugars into pyruvate and produces a small quantity of ATP and NADH. Glycolysis is the first stage of both aerobic respiration and fermentation. In the absence of oxygen, fermentation occurs: NADH produced in glycolysis reduces pyruvate, converting it to lactate or alcohol. The resulting NAD^+ is cycled back to glycolysis. Fermentation results in the net synthesis of only two molecules of ATP per molecule of glucose.

In the presence of oxygen, aerobic respiration occurs. Pyruvate from glycolysis is transported into a mitochondrion (except in bacteria). NAD^+ oxidizes the pyruvate, forming acetate and NADH, and releasing carbon dioxide. Coenzyme A carries the acetyl group to the Krebs-cycle enzymes. That pathway completes the oxidation of the carbon skeleton to carbon dioxide. This produces ATP, NADH, and $FADH_2$. Hydrogen atoms carried by NADH and $FADH_2$ are used to synthesize ATP as electrons and protons move through the electron transport system to oxygen, forming water. The complete aerobic respiration of a molecule of glucose forms 6 molecules of carbon dioxide and a maximum of 38 molecules of ATP.

Cell respiration and energy production are closely regulated. Some plants and animals have special forms of cell respiration that release larger quantities of heat energy and produce less ATP.

Key Concepts

Use the concepts below to build a concept map, linking as many other concepts from the chapter as you can.

- glycolysis
- Krebs cycle
- electron transport system
- enzymes
- mitochondrion

- fermentation
- aerobic respiration
- energy
- carbon skeletons
- biosynthesis
- ATP

Reviewing Ideas

1. How are membranes involved in respiration?
2. Where in a cell does each part of cell respiration take place? Describe how the location of each part of the process is different in bacteria and in more complex cells.
3. Organisms can recycle some of the nutrients they require. What major requirement that cannot be recycled does cell respiration supply?
4. What are the biological functions of these processes?
   a. Cell respiration?
   b. Glycolysis?
   c. The Krebs cycle?
   d. The electron transport system?
5. Discuss the ways in which biosynthesis depends on cell respiration.
6. How does the Krebs cycle depend on glycolysis? How does the electron transport system depend on the Krebs cycle?

Using Concepts

1. The carbon skeletons of most amino acids come from the Krebs cycle. Would you expect aerobic or anaerobic organisms to have alternative pathways for amino-acid synthesis? Give the reasons for your answer.
2. Why would you not expect to find mitochondria in anaerobic organisms?
3. Use what you know about the effects of alcohol to propose a hypothesis that explains why animal cells evolved to produce lactate but not ethanol (ethyl alcohol) during fermentation.
4. Do plants contain mitochondria? Give the reasons for your answer.
5. How does respiration in plants depend on photosynthesis? How does respiration in animals depend on photosynthesis?
6. Some cells (other than bacteria) contain no mitochondria at maturity. If these cells receive ample glucose, could they use it for energy? Would they release carbon dioxide? Could these cells use fat as a source of energy? Explain your answers.
7. "Aerobic" exercise is activity that stimulates breathing and blood circulation, but does not exceed their ability to deliver oxygen to the muscles. Propose a hypothesis that explains why aerobic exercise is more beneficial than "anaerobic" exercise, which demands more oxygen than the heart and lungs can easily deliver.
8. The complete oxidation of glucose to carbon dioxide produces about 686 kcal of free energy. Oxidation of a molecule of glucose in a cell generates a maximum of 38 molecules of ATP. ATP provides about 7.3 kcal of energy. Calculate the efficiency of the ATP energy yield from the complete aerobic respiration of glucose. What happens to the rest of the energy?
9. Human cells cannot release energy from fat by first converting it to glucose and then respiring it. Propose a hypothesis that explains this situation.

Synthesis

1. What is the relationship between the Krebs cycle and the Calvin cycle?
2. The products of photosynthesis, 3-carbon sugar-phosphates, enter glycolysis at step c in Figure 5.6. How many molecules of ATP can be made from the complete aerobic respiration of one of those sugar-phosphate molecules?

Extensions

1. The cyanide ion (CN^-) is a poisonous chemical that blocks the flow of electrons to oxygen at the end of the electron transport system. Develop a hypothesis to explain why, as a result of this property, cyanide is lethal to most aerobic organisms.
2. Develop a game or a skit to explain the most important aspects of cell respiration.

Web Resources

Visit BSCSblue.com to access
- Explanations of mitochondrial structure and function
- Pathways of cell respiration
- Descriptions of mitochondrial diseases and their treatment
- Web links related to this chapter

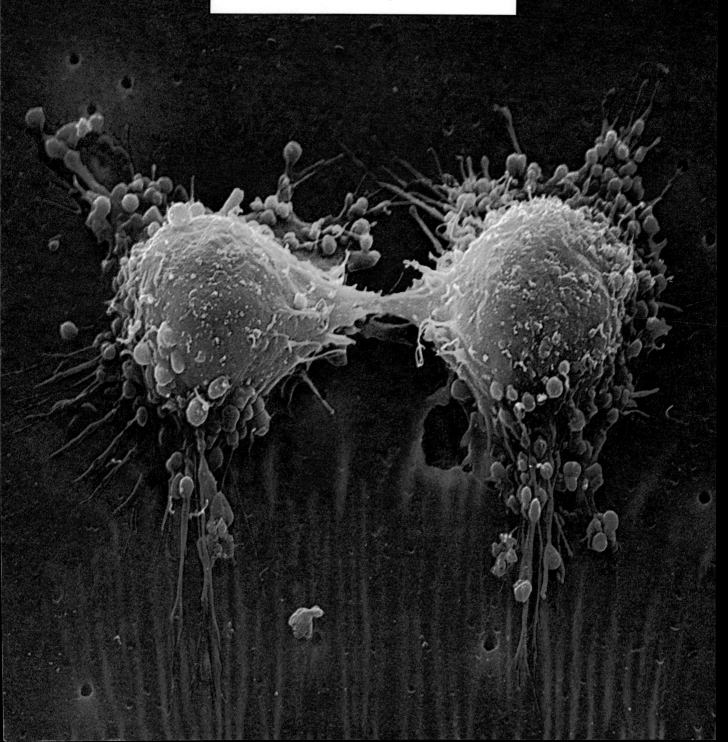

◀ **About the photo**
The photo to the left
shows a scanning
electron micrograph of
two prostate cancer cells
(×2000).

UNIT
2

The Cell: Homeostasis and Development

■ *What is happening to these cells?*

■ *How is heredity involved as these cells develop into a new organ?*

Cells are the basic units of life. All life begins as a single cell, and many organisms exist only as individual cells. Most organisms are composed of many cells, often trillions of them. All of an organism's cells carry the same hereditary information, encoded in their DNA.

Just as a car has many parts, a cell has many structures. Like the parts of a car, structures in a cell contribute to the cell's function. How are cells organized to perform the work they do? What coordinates the complex processes that occur in cells?

And if all of an organism's cells are controlled by the same genetic instructions, how can its cells develop into so many types? A cell's genes do not all act all the time. What activates skin-forming genes in some animal cells and muscle- or bone-forming genes in others? How do all these types of cells coordinate their activities to maintain homeostasis and ensure smooth, normal development?

These are all questions that you will explore in this unit.

CONTENTS

LEARNING OUTCOMES

By the end of this chapter you will be able to:

A   Explain the basic tenets of the cell theory.

B   Predict the possible effects of improved technology on the study of cells.

C   Distinguish between prokaryotic and eukaryotic cells.

D   Identify prokaryotic cell structures and explain functions of eukaryotic organelles.

E   Describe ways in which cells cooperate with each other.

F   Summarize the advantages of multicellular organization.

CHAPTER

6

Cell Structures and Their Functions

■ *How many different structures can you see in these cells?*

■ *What functions could they have?*

Cells are so complex that we are continually learning more about how they are made and how they work. The miniature world of cells may hold answers to most of our major health problems. Heart disease, cancer, some forms of mental illness and retardation, and viral diseases such as AIDS are all diseases of cells. Our attempts to increase food production depend on the ability of plant cells to convert the products of photosynthesis to edible, nutritious forms.

Two remaining important areas of exploration are the small worlds of cells and atoms and the huge world of the universe. Both stretch the limits of the imagination. Researchers who study either cells or stars encounter similar problems. They would like to be able to touch, sample, and observe directly the individual parts of these objects, but even with improved technology, much remains impossible. The parts that make up the cell are too small, and stars are too large. Both, in different ways, remain too inaccessible to study completely. In spite of the difficulties, biologists are making rapid progress toward unlocking the mysteries of the cell. In this chapter, you will learn what research has revealed about the cell so far.

The Basic Unit of Life

6.1 Cell Study and Technology

Like all living things, you are composed of tiny parts called cells, trillions of them. Other animals, plants, and fungi also are made of many cells. There are many billions of organisms such as bacteria, yeasts, and algae that are single-celled. Whatever their size, all are organisms and are composed of cells—the basic units of life. Figure 6.1 shows a few of the great variety of cell types that make up organisms.

The idea that cells are the basic units of life began to take shape in the early 1800s, as many biologists contributed data and ideas that led to the **cell theory,** which can be stated in two parts.

1. Cells, or products made by cells, are the units of structure and function in organisms.
2. All cells come from preexisting cells.

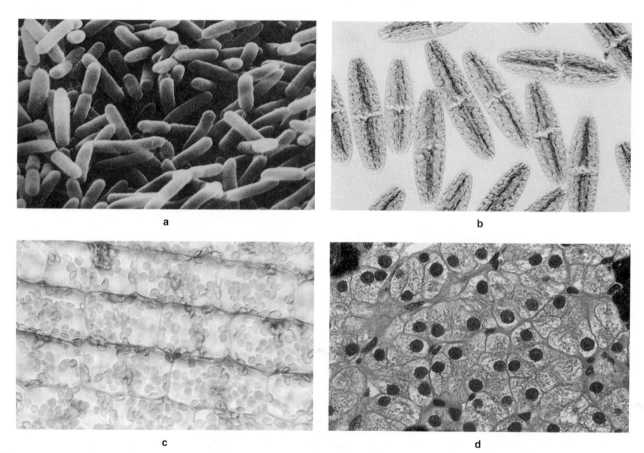

a

b

c

d

FIGURE 6.1

Cells. Examples of the variety of cells that make up all organisms (color added). Differences in cell structure and function account for many of the differences among living things. **(a),** Unicellular (single-celled) bacteria, ×20,000; **(b),** unicellular algae, ×150; **(c),** photosynthetic cells in a leaf, ×200; **(d),** cells from the liver of a salamander, ×400.

Once the cell theory was established, scientists began to study cell structure and function in detail. At first, biologists who studied cells saw them as just tiny blobs of jelly. These early scientists had no idea of the complex and detailed structures of cells. Progress depended on the technology of improved microscopes, better techniques to prepare cells for observation, and studies of cell function. Even modern light microscopes cannot uncover all the detailed wonders of the cell—some of its structures are too small to see without the electron microscope, which was developed in the 1930s (see Appendix 6A, "Preparing Cells for Study").

Electron microscopes (Figure 6.2) reveal very tiny cell parts and even some large molecules down to 0.5 nm—a magnification of more than a million. (How does that compare with the magnification of the microscope

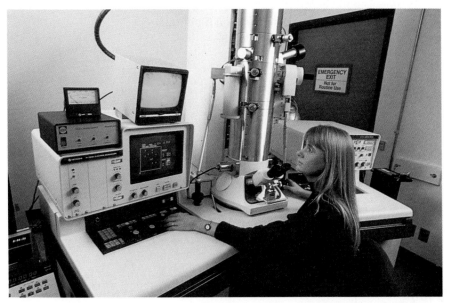

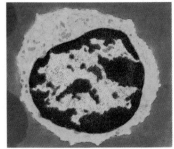

a

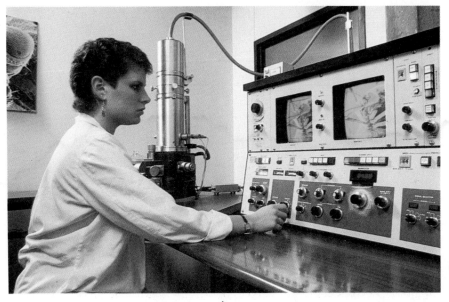

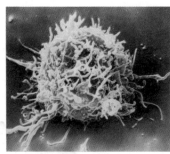

FIGURE 6.2

Transmission electron microscope (a) and scanning electron microscope (b), with typical images of similar white blood cells produced by each. Note the differences in the images produced of the same subject (color added).

b

you use in biology class?) The major drawback of the electron microscope is that the steps needed to prepare samples for examination kill any living cells before they can be observed. Therefore, the movements and chemical reactions of living cells cannot be studied in this way. Sample preparation can also alter cell structures, so the electron microscope does not always show them as they are in life. Scanning tunneling microscopes can be more powerful than electron microscopes and do not require such harsh treatment of samples (see Biological Challenges, Section 1.12). Like scanning electron microscopes, however, these devices can reveal only surface features.

Cells differ in size but average 10 to 20 μm in diameter. The smaller cells of bacteria may be only 1 μm long. Figure 6.3 compares the sizes of cells and cell parts to the units of measurement biologists use. Because cells are so small, models and diagrams, such as those in this chapter, are used to represent ideas about cell structure and function.

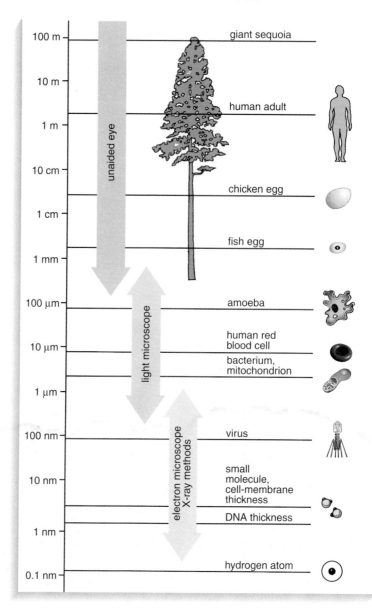

FIGURE 6.3

A comparison of sizes. Note that most cells are too small to be seen with the unaided eye. Each measurement (for example, 1 m, 10 cm) is one-tenth the size of the one above it.

Biological Challenges

DISCOVERIES

The Cell Theory

In the 17th century, Robert Hooke used his microscope (Figure 6.4a) to examine thin slices of cork. Cork is harvested from the dead bark of certain oak trees. Magnified many times under the microscope, these cork shreds seemed to be composed of many little boxlike units. The boxes looked empty; Hooke could see right through them. The rows of boxes reminded Hooke of the little rooms in which monks lived, so he named them cells.

After Hooke published his observations, others saw that the living parts of plants are also made up of cells. But these cells were filled with fluid and various smaller structures. Robert Brown saw a dense object in many cells in 1831 and named it the nucleus. M. J. Schleiden, a botanist, advanced the idea that plants are made of cells that contain nuclei and cell fluid.

In 1839, Theodor Schwann used a microscope to examine parts of animals. Animals also seemed to be made up of small units (Figure 6.4b). Often these units looked like fluid-filled sacs. Each sac contained a nucleus. Schwann suggested that animals are also made of cells.

Thirty years after Hooke discovered cells, Anton van Leeuwenhoek observed microscopic organisms, or microorganisms. Water from ponds, rain barrels, and rivers revealed a wealth of tiny creatures. Soon after van Leeuwenhoek published his papers, many microbiologists were looking for living organisms in soil, spoiled food, and other places.

Eventually, the structure of microorganisms was compared with the structures of plants and animals. Many microorganisms did not seem to be composed of cells. Instead, they appeared to be about the size of one cell from a larger organism. The idea quickly took hold that they were unicellular (single-celled). Nuclei and other parts were discovered in many of these microorganisms. A large group of scientists began to think of the cell as the basic unit of life—either a complete organism or part of a multicellular organism.

Cells were soon discovered in an increasing number of organisms. Numerous biologists observed cells dividing to produce more cells. A physician and biologist, Rudolf Virchow, saw in these events a principle—cells producing more cells through time. He stated the hypothesis simply: "All cells come from cells."

Today new technologies make possible extremely detailed studies of cell structure and function. New hypotheses about cell evolution are being investigated. The cell theory remains one of the unifying themes of biology.

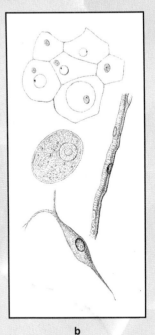

a   b

FIGURE 6.4

Early observations of cells. (a), One of Robert Hooke's microscopes. **(b),** The microscopic structure of small parts of animals as seen by Schwann: six cells from a fish *(top),* an oval cell from the nervous system of a frog, a long cell from the muscle of an unborn pig, and a spindle-shaped cell, also from an unborn pig.

6.2 Two Basic Types of Cells

Living cells can be separated into two groups, prokaryotes and eukaryotes, that differ in structure. This grouping is fundamental to the classification of organisms. Several types of strong evidence support the idea that the differences between prokaryotes and eukaryotes developed early in the history of life. These differences are essential to our understanding of the evolution of organisms. **Prokaryotes**—the bacteria—are the simplest living cells, and they are everywhere. They are common in soil, air, and water, and in or on every organism, including humans. Prokaryotic organisms are nearly always unicellular. Some prokaryotes inhabit extreme environments such as salt flats, hot springs such as those in Yellowstone National Park, and volcanic vents in the ocean floor. Others are so adaptable that you could find them almost anywhere on Earth. The smallest prokaryotes are only about 0.3 μm in diameter. Others are 1–5 μm across. Almost none are big enough to see without a microscope. Up to 700 million could fit side by side on the head of a thumbtack. Even the point of a pin can hold hundreds of bacteria (Figure 6.5.)

The cells of **eukaryotes** are larger (10–50 μm) and more complex than prokaryotes. These more complicated cells can form multicellular organisms: Plants, animals, and fungi are composed of eukaryotic cells.

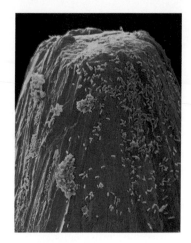

FIGURE 6.5

Prokaryotic cells. Prokaryotes are extremely small. Many bacteria are visible in this scanning electron micrograph (×290) of the point of a pin.

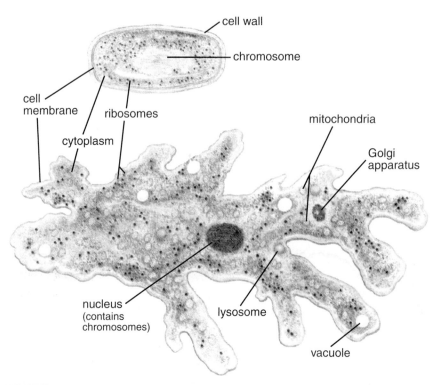

FIGURE 6.6

Two types of cells. A prokaryotic cell *(top)* and a eukaryotic cell *(bottom)*. Note the greater structural complexity of the eukaryotic cell (an amoeba) and its many membrane-enclosed parts, or organelles.

Unlike prokaryotes, eukaryotic cells have many parts, each with a specific function. This gives them the flexibility to develop into hundreds of specialized types that make up leaves, muscles, and other organs. Figure 6.6 compares a prokaryotic and a eukaryotic cell. You will learn about the structures shown in the diagram as you read this chapter.

Recall from Chapters 4 and 5 that prokaryotic cells lack mitochondria and chloroplasts. Eukaryotes contain many of these membrane-enclosed internal compartments. The most obvious difference between prokaryotes and eukaryotes is the **nucleus.** This membrane-enclosed structure contains the DNA of eukaryotic cells. Information encoded in this DNA directs the synthesis of the enzymes and other proteins that determine almost everything that happens in a living organism (see Figure 1.35 on page 47).

Check and Challenge

1. What is the cell theory?
2. In what ways does the study of cells depend on technology?
3. In centimeters, how long is your pencil? How does its size compare with that of a bacterium? of a eukaryotic cell?
4. Biologists have been able to keep isolated chloroplasts or mitochondria for a few hours or days. Can these structures be considered alive?
5. What is one advantage of using a light microscope rather than a more powerful electron microscope?

Cell Structure

6.3 Prokaryotic Cell Structure

The prokaryotes are bacteria. Most of them are unicellular but can associate in clusters, chains, and films. Nearly all prokaryotic cells have a rigid cell wall made of lipids, carbohydrates, and protein, but no cellulose. Figure 6.7 is a diagram of the structure of a prokaryote. Just inside the cell wall is a plasma membrane that encloses the cell. Prokaryotes usually have one chromosome made of a continuous, circular molecule of double-stranded DNA. The chromosome is attached to the plasma membrane in an area of the cell known as the nuclear region, or **nucleoid.** In addition, bacteria usually contain one or more smaller circular DNA molecules called **plasmids.** These extrachromosomal elements contain a few genes that help bacteria survive under specific conditions. The part of the plasma membrane attached to the chromosome may contain enzymes that aid in making a copy of the chromosome before the cell reproduces by dividing in two.

ETYMOLOGY

nucleus = kernel or core (Latin)
-oid = similar (Greek, Latin)
Early observers thought of the nucleus as the cell's "core." A bacterial cell's **nucleoid** is similar to a nucleus but lacks a selectively permeable nuclear membrane.

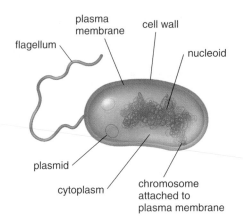

FIGURE 6.7

The structure of a prokaryotic cell. Electron micrographs revealed the details of the structure.

Most bacteria have one of three shapes—rod, sphere, or corkscrew (Figure 6.8). Some have **flagella** (singular: flagellum), long, whiplike extensions made of protein that rotate like propellers, enabling cells to swim through water or the body fluids of larger organisms. These swimming bacteria can sense substances in their environment. They swim toward food and away from harmful substances such as strong acids.

Prokaryotes are extremely diverse in their metabolism. Many of their metabolic processes, such as glycolysis, are similar to those of eukaryotes, but others are unique. For example, all ecosystems include many types of bacterial decomposers that help recycle nutrients such as carbon, nitrogen, and sulfur compounds. Some of the reactions these prokaryotes perform provide them with free energy or fixed carbon and nitrogen, as you learned in Chapters 4 and 5. These reactions also recycle nutrients that would otherwise remain unavailable in wastes and dead organisms. This function of prokaryotes is essential to other members of their ecosystems.

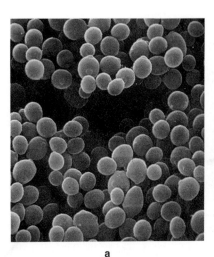

a

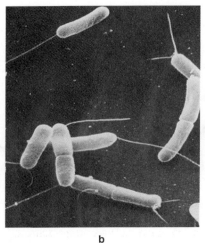

b

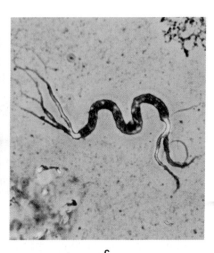

c

FIGURE 6.8

The three most common shapes of bacterial cells. (a), Cocci (spheres), ×45,000; **(b),** bacilli (rods), ×31,000; and **(c),** spirochetes (corkscrews), ×700. Note the flagella in **(b)**.

Many prokaryotes are autotrophs. They are important primary producers in lakes and oceans. Other bacteria are useful as the source of antibiotics and other compounds that are difficult to make artificially. Although some bacteria can cause human diseases such as skin infections, strep throat, rheumatic fever, and anthrax, most are beneficial. Bacteria that live in your intestines help digest your food and provide you with certain vitamins. Other bacteria are used in manufacturing fermented foods, such as cheese and yogurt, and life-saving drugs.

6.4 Eukaryotic Cell Structure

Eukaryotic cells are divided into small functional parts called **organelles.** This is an important difference from prokaryotes. Any part of a eukaryotic cell that has its own structure and function can be considered an organelle. Thousands of chemical reactions occur constantly in eukaryotic cells. Many of those reactions are not compatible, yet they can occur at the same time because membranes that surround many organelles divide the cell into compartments. The concentration of a substance may be very different in different compartments. Different reactions may produce a substance in one organelle and consume it in another. For example, a membrane separates the genes, which are made of DNA, from the rest of the cell. This separation protects the DNA from enzymes that might break it down. The selectively permeable membrane also helps the cell control which of its genes are expressed at any time. Thus compartmentation makes eukaryotic cells more efficient by separating specific processes and enabling a division of labor within the cell. Figure 6.9 shows the major organelles of most eukaryotic cells. Refer to this figure and locate the organelles as you read the descriptions that follow.

A plasma membrane (Figure 6.9a) encloses the contents of both eukaryotic cells and prokaryotic cells. The structure of cell membranes is described in detail in Section 3.2. The plasma membrane of plant and fungal cells and some unicellular eukaryotes is surrounded by a rigid structure, the **cell wall** (Figure 6.9f). The wall is composed of stiff fibers of cellulose and other complex carbohydrates (Figure 6.10). This enables it to support and protect the cell. Animal cells lack a rigid cell wall. The cell wall is the most consistent difference between plant and animal cells.

The most noticeable organelle in a eukaryotic cell usually is its nucleus (Figures 6.9b and 6.11). The nucleus is a cell's genetic control center because it contains the chromosomes. Looking for a nucleus is the most reliable way to decide whether a cell is a prokaryote or a eukaryote. A double layer of membranes forms the nuclear envelope, or nuclear membrane, that surrounds the chromosomes. Each eukaryotic chromosome consists of a single long DNA molecule wrapped around a series of protein "spools." One or more drops of concentrated RNA are usually visible in the nucleus. These bodies are called **nucleoli** (singular: nucleolus). The nucleoli are the sites where types of RNA that will become part of the cell's protein-synthesizing machinery are synthesized.

Within the plasma membrane, but outside the nucleus, is the cellular material, or cytoplasm. The cytoplasm was once thought to be mostly a

Try Investigation 6A ◇ Cell Structure.

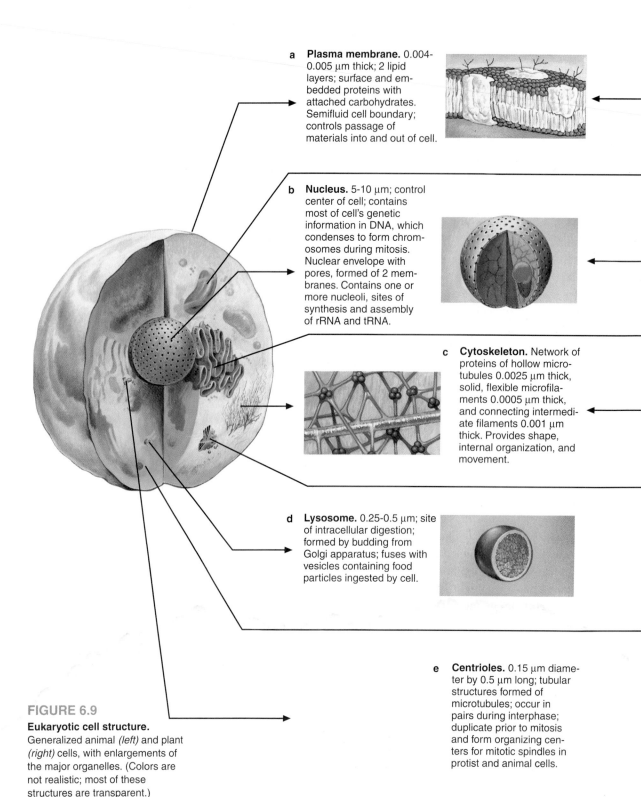

a **Plasma membrane.** 0.004-0.005 µm thick; 2 lipid layers; surface and embedded proteins with attached carbohydrates. Semifluid cell boundary; controls passage of materials into and out of cell.

b **Nucleus.** 5-10 µm; control center of cell; contains most of cell's genetic information in DNA, which condenses to form chromosomes during mitosis. Nuclear envelope with pores, formed of 2 membranes. Contains one or more nucleoli, sites of synthesis and assembly of rRNA and tRNA.

c **Cytoskeleton.** Network of proteins of hollow microtubules 0.0025 µm thick, solid, flexible microfilaments 0.0005 µm thick, and connecting intermediate filaments 0.001 µm thick. Provides shape, internal organization, and movement.

d **Lysosome.** 0.25-0.5 µm; site of intracellular digestion; formed by budding from Golgi apparatus; fuses with vesicles containing food particles ingested by cell.

e **Centrioles.** 0.15 µm diameter by 0.5 µm long; tubular structures formed of microtubules; occur in pairs during interphase; duplicate prior to mitosis and form organizing centers for mitotic spindles in protist and animal cells.

FIGURE 6.9

Eukaryotic cell structure.
Generalized animal *(left)* and plant *(right)* cells, with enlargements of the major organelles. (Colors are not realistic; most of these structures are transparent.)

f **Cell wall.** 0.1-10 µm thick; formed by living plant cells; made of cellulose fibers embedded in a matrix of protein and polysaccharides; provides rigidity to plant cells and allows for development of turgor pressure.

g **Mitochondrion.** 2-10 µm long by 0.5-1 µm thick; enclosed in double membrane; inner membrane much folded; most reactions of cellular respiration occur in mitochondrion; contains small amounts of DNA and RNA; may be several hundred per cell.

h **Endoplasmic reticulum (ER).** 0.005 µm diameter; tubular membrane system that compartmentalizes the cytosol; plays a central role in biosynthesis reactions. Rough ER is studded with ribosomes, the site of protein synthesis; smooth ER lacks ribosomes.

i **Golgi apparatus.** 1 µm diameter; system of flattened sacs that modifies, sorts, and packages macromolecules in vesicles for secretion or for delivery to other organelles.

j **Cytosol.** Semi-fluid material surrounding organelles; contains enzymes that catalyze cellular reactions.

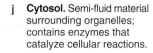

k **Chloroplast.** 5 µm long by 0.5-1 µm thick; enclosed by double membrane; this membrane system forms thylakoids in which light-absorbing pigments are embedded. All reactions of photosynthesis occur in chloroplasts.

l **Vacuole.** Variable size; large vesicle enclosed in single membrane; may occupy more than 50% of volume in plant cells; contains water and digestive enzymes; stores nutrients and waste products.

FIGURE 6.10

Cellulose fibers in the cell wall of an algal cell. Overlapping layers of these fibers and other materials make the cell wall rigid, ×18,000.

ETYMOLOGY

endo- = inside (Greek)
-plasm = a form (cell) (Greek, Latin)
reticulum = net (Latin)

The **endoplasmic reticulum** forms a network within the cell.

simple solution of salts and organic compounds in water. The organelles were said to float in this solution, also known as the **cytosol.** The protein-rich, semifluid material in the cell that surrounds and bathes the organelles is still sometimes called the cytosol (Figure 6.9j). The cytoplasm includes the cytosol and the organelles.

More recently, our model of the cytosol has changed from a simple solution to a more organized system. A network of several types of very fine protein fibers helps to shape the cell and organize the cytoplasm. Changes in this protein scaffolding also enable some cells to move or change shape. This system is known as the **cytoskeleton** (Figures 6.9c and 6.12). It includes hollow microtubules (25 nm thick), solid but flexible strands called microfilaments (5 nm thick), and connecting intermediate filaments (10 nm thick). The cytoskeleton may hold organelles in place or move them around. Much of the water in a eukaryotic cell may be loosely bound to the cytoskeleton or to various other proteins and solutes. This binding of water makes the cytosol more like an organized gel, in which many components have a specific place, and less like a simple solution.

Scattered throughout the cytoplasm of both eukaryotes and prokaryotes are many small bodies composed of RNA and protein, called **ribosomes.** Ribosomes catalyze the synthesis of a cell's proteins. In eukaryotes, some ribosomes are attached to a system of membranes called the **endoplasmic reticulum,** or ER (Figures 6.9h and 6.13). The ER membranes form tubes and channels throughout the cytoplasm. This system connects many of the organelles in the cell. In electron micrographs, the ER resembles wrapping paper, folded back on itself, running through the cell. Proteins that are synthesized at the ribosomes attached to the ER pass directly into the ER as they are formed. They are transported through the ER to their final destinations in the cell. The ER also carries other substances to places in the cell where they are needed.

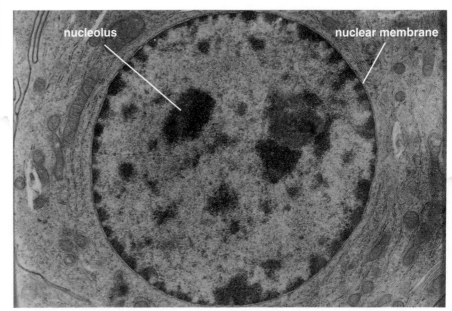

FIGURE 6.11

Cell nucleus. Transmission electron micrograph of the nucleus of a human liver cell, ×16,000. Note the double membrane that makes up the nuclear envelope. The large, dark, round body is the nucleolus.

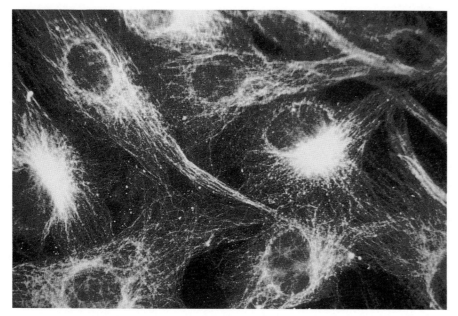

FIGURE 6.12

The cytoskeleton. Scanning electron micrograph image of the cytoskeletal protein network in hamster kidney cells, ×800. This internal protein network gives the cell shape and participates in cell movement.

Many substances that are exported from the cell pass through the ER to the **Golgi apparatus** (Figures 6.9i and 6.14). This organelle consists of a series of membranous sacs that look like a stack of pancakes. As material passes through these compartments, it is packaged in spherical, membrane-enclosed **vesicles** that appear to pinch off of the Golgi membranes. The vesicles can fuse with the plasma membrane, releasing their contents outside of the cell. Some vesicles deliver their contents to other organelles. Proteins synthesized on the ribosomes of the rough ER may be modified in the Golgi apparatus before they are released. For example, specific sugars are attached to some proteins before they are released from the cell. This system enables cells to release large polymers such as proteins and polysaccharides. These large molecules can then become part of an external

ETYMOLOGY

-icle = small (Latin)
A **vesicle** is a small "vessel," or container.

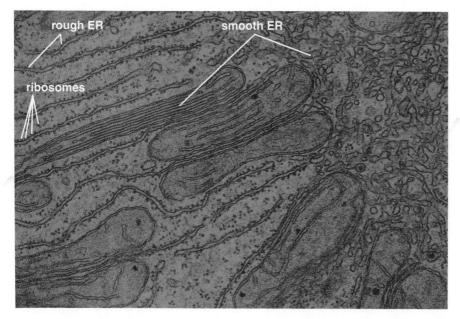

FIGURE 6.13

Endoplasmic reticulum. Transmission electron micrograph of endoplasmic reticulum in a rat liver cell, ×23,600. Note rough ER, with ribosomes, and smooth ER, without ribosomes. The ER carries newly made proteins to their destinations in or out of the cell.

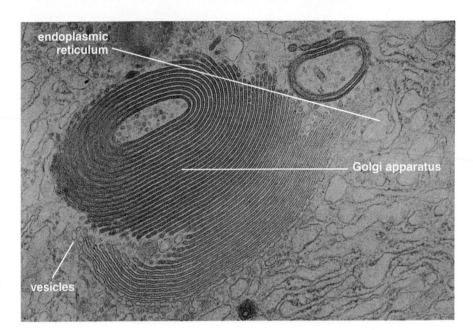

FIGURE 6.14

Golgi apparatus. Transmission electron micrograph of a Golgi apparatus from a hamster liver cell, ×94,000. Note the nearby ER and the vesicles formed at the Golgi apparatus.

structural framework. This process helps plant and fungal cells construct their cell walls. Animal cells rely on the same process as they build networks of protein fibers that strengthen organs such as your skin.

Together the ER, Golgi apparatus, and vesicles form a connected internal membrane system (Figure 6.15). Proteins made in this system become part of the membranes of the cell or end up inside various organelles. The lipid membranes of the vesicles that deliver these proteins also become part of the plasma membrane or the membranes of organelles.

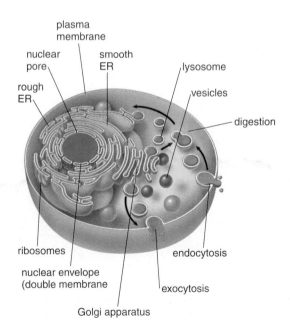

FIGURE 6.15

Internal membrane system of a eukaryotic cell. Proteins synthesized on the ribosomes of the rough ER pass through the ER, Golgi apparatus, and membrane vesicles on their way to their destinations.

The structure of the system enables it to act as a highway that directs proteins to target points inside the cell and to the plasma membrane for passage out of the cell.

Lysosomes (Figures 6.9d and 6.16) are special vesicles in animal cells and some other eukaryotes. They contain enzymes that break down the cell's old macromolecules for recycling. Lysosomes can also fuse with vesicles formed by endocytosis, digesting the food particles within. Some animal cells have lysosomes that fuse with the plasma membrane. Digestive enzymes are released outside the cell, where they break down bacteria and other foreign particles. For example, lysozyme, an enzyme found in lysosomes of tear-duct cells, protects the eyes from bacterial infection by breaking down bacterial cell walls. The **vacuoles** present in most plant cells (Figure 6.9l) are vesicles that enlarge as the cells mature. Vacuoles contain water, organic acids, digestive enzymes, salts, and pigments that give plant parts such as beet roots their characteristic color. Up to 90% of the volume of a mature plant cell may consist of its vacuole.

Chloroplasts (Figure 6.9k) and mitochondria (Figure 6.9g) are double-membrane organelles involved in energy reactions in the cell. Photosynthesis occurs in chloroplasts in plants and other photoautotrophic eukaryotes. Mitochondria are the major sites of ATP synthesis in most eukaryotic cells and have been called the power plants of the cell. These structures are described in detail in Chapters 4 and 5.

Centrioles (Figures 6.9e and 6.17) are tubular structures in the cells of animals and some fungi and algae. They participate in cell reproduction in

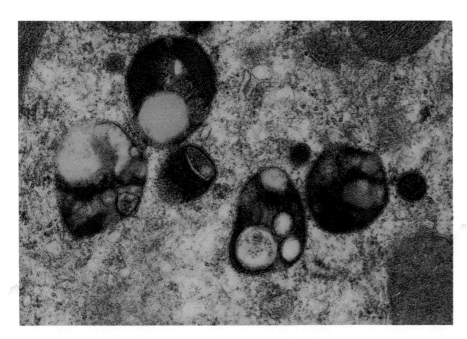

FIGURE 6.16

Lysosomes. Transmission electron micrograph of lysosomes in a mouse kidney cell, ×91,800. These vesicles contain enzymes that help break down wastes and foreign materials.

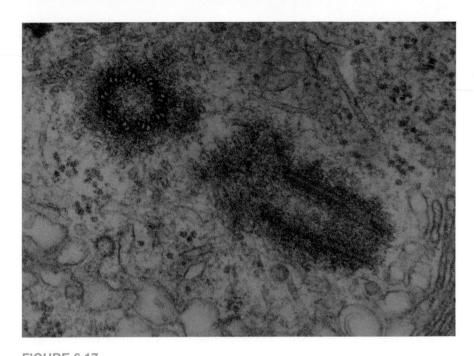

FIGURE 6.17

Centrioles. A pair of centrioles in a human cell, ×400,000. The cross section of the centriole at left reveals the nine triplets of microtubules that make up each centriole.

those organisms. Centrioles consist of a pair of cylindrical bundles of microtubules like those in Figure 6.9c. Mitosis, the primary form of cell reproduction in eukaryotes, is described in Chapter 8.

Some eukaryotic cells have flagella, but these structures are different from bacterial flagella (Figure 6.18). Bacterial flagella extend outside the plasma membrane. Eukaryotic flagella are covered by the plasma membrane. They consist of long bundles of microtubules. Enzymes associated with these microtubules provide energy for the motion of the flagellum by breaking down ATP.

Cilia (Figure 6.19) are short flagella. Cells that have cilia are often covered with hundreds of these organelles, arranged in rows. Eukaryotic flagella and cilia move cells along by whipping in an oarlike motion against the fluid surrounding a cell. Cilia can also help move material along a cell or tissue. For example, the breathing passages of humans and many other animals are lined with cilia-covered cells. The rhythmic beating of these cilia moves mucus and foreign particles such as dust out of the lungs. Substances in tobacco smoke paralyze these cells. As a result, cancer-causing smoke particles and disease-causing bacteria are more likely to enter smokers' lungs. Cilia on cells in your inner ear vibrate in response to sound waves. This is a critical step in hearing. Frequent exposure to loud sounds such as those at rock concerts or on aircraft runways can break off the cilia, causing permanent hearing damage.

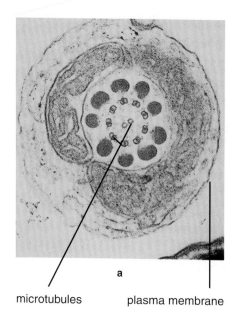

microtubules   plasma membrane

a

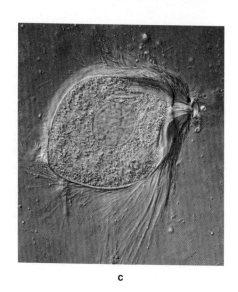

c

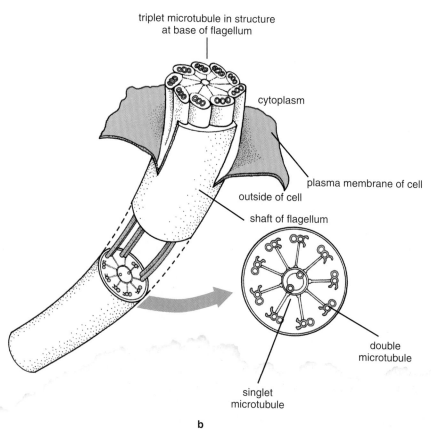

triplet microtubule in structure
at base of flagellum

cytoplasm

plasma membrane of cell

outside of cell

shaft of flagellum

double
microtubule

singlet
microtubule

b

FIGURE 6.18

Eukaryotic flagella. (a), Transmission electron micrograph, ×175,000 of a cross section of the
flagellum of a human sperm cell shows that it consists of nine pairs of microtubules arranged in
a cylinder around a tenth pair. **(b),** Parallel bundles of microtubules make up the internal
structure of the flagellum. **(c),** *Trichonympha* is a single-celled eukaryote that swims with
hundreds of flagella, ×135.

Table 6.1 compares various features of eukaryotic and prokaryotic cells. Use this table and Figure 6.9 to review this section.

FIGURE 6.19

Eukaryotic cilia. Scanning electron micrograph of rows of cilia on a hamster's inner ear cells, ×6,700.

TABLE 6.1
Some Differences between Prokaryotic and Eukaryotic Cells

Eukaryotic Features	Prokaryotic Features
• membrane-enclosed organelles, including nucleus	• no membrane-enclosed organelles
• multiple linear chromosomes, often in pairs	• single circular chromosome
• streaming in the cytoplasm (not always present)	• no streaming in the cytoplasm
• cell division by mitosis	• cell division without mitosis
• complex flagella	• simple flagella
• large ribosomes	• small ribosomes
• cytoskeleton	• no known cytoskeleton
• cellulose in cell walls	• no cellulose in cell walls
• DNA wrapped around proteins	• proteins bound to DNA

Focus On

The Cytoskeleton

The ability of eukaryotic cells to assume different shapes and to carry out coordinated and directed movements depends on the cytoskeleton (Figure 6.20), a complex network of protein filaments that extends throughout the cytoplasm. The cytoskeleton is responsible for the gelatinlike consistency of the cytosol. It makes possible the "crawling" movements of some cells, muscle contraction, and the changes in shape of developing vertebrate embryos. The cytoskeleton also helps organelles move from one place to another in the cytoplasm. The cytoskeleton may have been a crucial factor in the development of eukaryotic cells; prokaryotes do not seem to have cytoskeletons.

How do all these movements occur? Several types of proteins are involved. Motor proteins attach to various cell components and use the energy of ATP to move along the microtubules of the cytoskeleton. Some of these proteins can move in only one direction, while others carry their "cargo" the other way. There are also enzymes and other proteins that rearrange parts of the cytoskeleton and move cell components by adding or removing individual protein molecules at one end of a microtubule, making it longer or shorter. This internal transport system is responsible for the cytoplasmic streaming seen in some types of cells. In this process, part of the cytoplasm seems to flow through the cell in a specific direction, redistributing organelles and dissolved nutrients and wastes. Cell biologists are still exploring how these changes are controlled and coordinated. Many one-celled organisms get rid of wastes and excess water with a contractile vacuole. This structure absorbs water and wastes from the cytosol. Motor proteins then cause it to contract, squeezing its contents out through the plasma membrane.

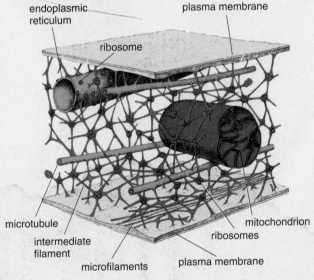

FIGURE 6.20

The cytoskeleton. This network of proteins in the cytoplasm of eukaryotic cells facilitates movement and helps the cell maintain its shape.

Check and Challenge

1. What are the important characteristics of prokaryotic cells?

2. Explain why metabolism in eukaryotic cells depends on the endoplasmic reticulum, but prokaryotes have survived quite well without such an internal transport system.

3. Which membrane gives eukaryotes an advantage over prokaryotes in protecting their genetic material from damaging chemical changes?

4. What three organelles enable eukaryotic cells to move?

5. Some medicines are delivered in capsules of artificial vesicles to be swallowed. Would this method be best for delivering medicines that act on target proteins in the plasma membrane, the mitochondria, or the cytosol? Explain your answer.

Multicellular Organization

6.5 Cooperation among Cells

When one-celled organisms divide, some new cells may remain together in a cluster (Figure 6.21). However, a cluster of cells is not necessarily a multicellular organism. Each cell has an individual life and may break away from the cluster at some point.

Try Investigation 6B From One Cell to Many.

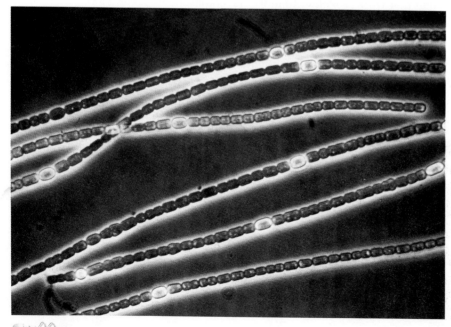

FIGURE 6.21

Chains of photosynthetic bacteria. Though prokaryotes such as these exist in clusters, chains, and films and as isolated cells, each bacterium is still an individual organism *Anabaena,* ×600, color added.

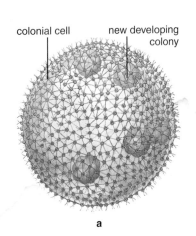
Some unicellular microorganisms live in groups called **colonies.** Members of a colony are usually related. They may interact in ways that give them advantages over isolated organisms, but each individual is still a separate organism. For example, some bacteria that attach to solid objects form colonies called biofilms. A biofilm may contain several unrelated types of bacteria that require similar environments. These organisms cooperate to change conditions in the biofilm, such as pH, to suit their own needs, but each cell remains a separate organism.

In some colonies, individual cells take on specialized roles. Such a colony may seem almost like a single organism. For example, *Volvox*, found in many ponds, is one of many types of colonial algae. During the spring and early summer, it forms a colony shaped like a hollow ball, as illustrated in Figure 6.22a. Inside the colony may be other developing colonies. Each colony has several hundred to tens of thousands of cells.

The cells of *Volvox* resemble the one-celled organism *Chlamydomonas* (Figure 6.22b). Each cell has a nucleus, two flagella, contractile vacuoles, a light-sensing structure, and a cup-shaped chloroplast. A gelatinlike layer surrounds each cell and separates it from its neighbors. Under the microscope, a *Volvox* colony looks like a gelatinous sphere with cells embedded in it (Figure 6.22c).

If you observe some types of *Volvox* under high magnification, you can see delicate strands of cytoplasm connecting the cells. You also can see that certain cells are larger than the rest. The colony appears to have a front end and a back end. As flagella move the *Volvox* through the water, the larger cells are usually at the back.

Volvox moves in a coordinated way. If the flagella of each individual cell of a *Volvox* colony were to move independently, the colony would move in a random, irregular way. Instead, the colony spins on its front-to-rear axis as it moves through the water. All the cells of a colony have to move their flagella in the same pattern to achieve this motion. Thus the cells are organized into a unit that works together to move the colony.

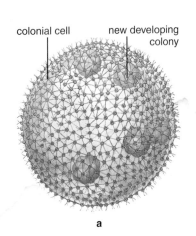

colonial cell   new developing colony

a

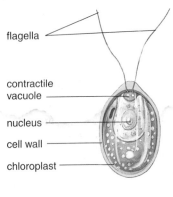

flagella

contractile vacuole

nucleus

cell wall

chloroplast

b

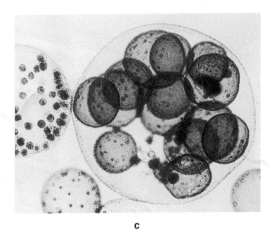

c

FIGURE 6.22

***Volvox*, colonial algae.** Individual cells of a *Volvox* colony, **(a),** look like the unicellular alga *Chlamydomonas,* **(b).** *Volvox* colonies, **(c),** are seen here through the light microscope, ×125.

Focus On

Biofilms

Biofilms have recently become the subject of great scientific interest. These bacterial colonies are important for both practical and basic scientific reasons. From a scientific point of view, they are interesting because they demonstrate how simple organisms such as bacteria can coordinate their activities for their own benefit, even when this involves a simple type of communication between different types of organisms. When different bacteria with similar environmental requirements attach to the same surface, they form a mixed colony. When the population of bacteria on the surface becomes crowded enough, the organisms begin to release a substance that acts as a chemical signal to other members of the colony. In a dense colony, the chemical signal accumulates to a level at which the bacteria begin to respond to it. They surround themselves with a mucuslike slime layer of protein and polysaccharide that makes the bacteria more difficult to dislodge (Figure 6.23). The slime layer also slows down diffusion between the cells and the surrounding fluid. Within this protected zone, the cells can regulate the surrounding pH and concentration of various foods and wastes.

Biofilms are also important from a practical point of view. They may help stabilize the colonies of helpful microorganisms that live on the internal and external surfaces of our bodies. The roots of many plants release substances into soil that encourage the formation of helpful biofilms on the roots and surrounding soil particles. This helps reduce the growth of other microorganisms that can cause disease. Some disease-causing bacteria also form biofilms that help

them stick to our bodies. (Dental plaque is a familiar example.) The film may also help protect the bacteria from antibiotics and our bodies' defense systems. Biofilms can contribute to corrosion in water pipes and on the hulls of ships. Advances in our understanding of biofilms may help promote health and reduce costly damage to ships, crops, and other property.

FIGURE 6.23

Biofilm. Bacteria in a dental plaque biofilm on an unbrushed tooth, ×2,000. Several types of bacteria are visible. At the center, a colony of cocci has completely covered a fibrous food particle.

Volvox is useful for investigating the functions of cells in simple multicellular organisms. If a *Volvox* cell is punctured, the cell dies, but the rest of the colony continues to live. Damage to or removal of cells can test whether those cells carried on any essential function for the whole organism. For example, only a few cells in the colony are capable of producing the offspring colonies. Different-sized cells in the colony may have different functions as well.

Is *Volvox* a colony or a multicellular organism? Many biologists consider it a borderline case. *Volvox* has some characteristics of a colony: Most of its cells are very similar. On the other hand, it does have some specialized reproductive cells, and the two ends of the colony are different. Perhaps *Volvox* should be considered just barely a multicellular organism.

6.6 Division of Labor

Most multicellular organisms, such as plants and animals, are much larger than *Volvox*. In Chapter 3, you learned that organisms must have enough surface area for the living cells within to exchange food, wastes, and other substances with their environment. *Volvox* is well adapted to this requirement: Its colonies are only one cell thick, so every cell is in contact with the surrounding water. Large plants and animals have evolved solutions to the surface-area problem that involve structures such as blood vessels, lungs, and leaves that add internal surface area. The development of these and other structures that enable large organisms to survive requires a number of specialized cell types.

Many different types of cells exist in multicellular organisms that are more complex than *Volvox*. All cells must carry on the basic activities of life, but each type of cell often takes on a special job as well. For example, a gland cell is specialized for making certain types of chemicals. A nerve cell is efficient in conducting electric signals, enabling the organism to respond to its internal and external environments. A muscle cell is specialized for movement. And the cells that form an organism's outer covering, or **epidermis,** may be specialized.

The drawing of a *Hydra* in Figure 6.24a shows cells with specializations. *Hydra* is a small, threadlike freshwater animal with a ring of tentacles at one end (Figure 6.24b). The animal's cells look slightly different and are different in specialization. In later chapters and in your laboratory work, you will see that cells of larger organisms are much more distinctive in appearance. For example, nerve and muscle cells are very long and thin, and human red blood cells are shaped like flattened spheres (Figure 6.25).

ETYMOLOGY

epi- = over (Latin, Greek)
derma = skin (Latin, Greek)

The **epidermis** of a plant or animal is its surface layer of cells and protective cell products.

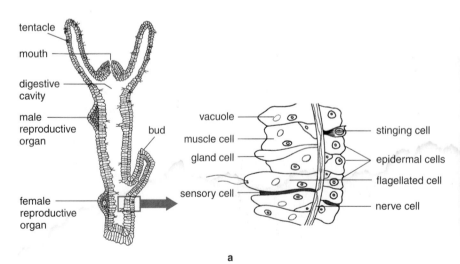

a   b

FIGURE 6.24

Cell types and tissues in *Hydra*, a small water animal. (a), A section through a *Hydra* and an enlarged view of some of the specialized cells in its body wall. The bud is the beginning of a new *Hydra*. **(b),** A *Hydra* viewed under low magnification, ×15. The animal is about 3 mm long.

In multicellular organisms, a group of cells with the same specialization usually work together. Each specialized mass or layer of cells is called a **tissue.** The epidermal cells in Figure 6.24a form part of a tissue. They form the covering, or epidermal, tissue of the *Hydra*, along with stinging cells and sensory cells.

Most animals and plants have many types of tissues. Different tissues may be organized into **organs** (eyes, hearts, and fruits, for example). Organs may be incorporated into **systems** of organs. For example, a circulatory system usually includes a heart, blood vessels, and blood. No matter how specialized the structures in an organism become, they are made from the same building blocks—cells and cell products (Figure 6.26).

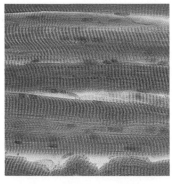

a

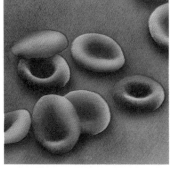

b

FIGURE 6.25

Specialized cells of a multicellular organism (human). **(a),** Muscle cells, ×124, color added. **(b),** Red blood cells, ×288, color added.

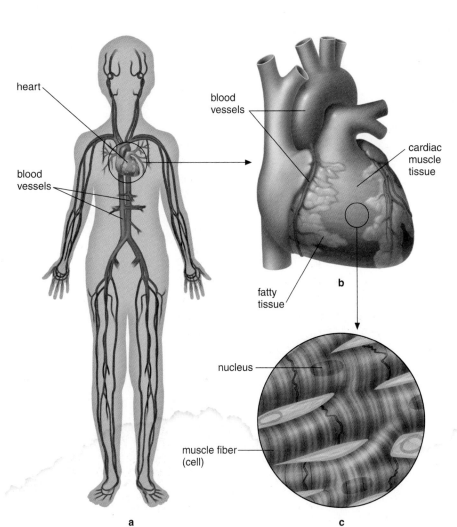

a

b

c

heart

blood vessels

blood vessels

blood vessels

cardiac muscle tissue

fatty tissue

nucleus

muscle fiber (cell)

FIGURE 6.26

Levels of structure in a multicellular organism. (a), The heart, blood, and blood vessels are the organs that make up an animal's circulatory system. **(b),** The heart is composed mostly of cardiac (heart) muscle tissue. **(c),** Specialized cells form this type of tissue.

Focus On

Junctions and the Extracellular Matrix

Cells in direct contact with each other often are connected at specialized regions of their plasma membranes, called junctions. Several types of junctions are present in animal cells. Some hold cells together. Others provide channels for communication between cells. In tight junctions (Figure 6.27a), specific proteins in the plasma membranes of adjacent cells make direct contact, so the plasma membranes are fused, and substances can't move through the spaces between the cells. Protein fibers anchored in the cytoplasm of neighboring cells form desmosomes (Figure 6.27b), which act like rivets that hold cells together but permit the passage of substances through the spaces between the cells. Gap junctions (Figure 6.27c) are channels formed by doughnut-shaped proteins embedded in the plasma membranes of adjacent cells. They allow small water-soluble molecules to pass directly from the

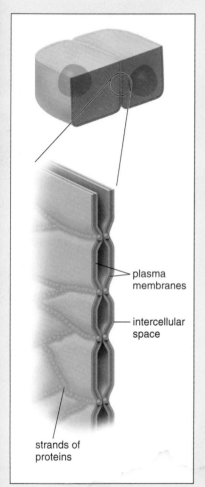

plasma
membranes

intercellular
space

strands of
proteins

a

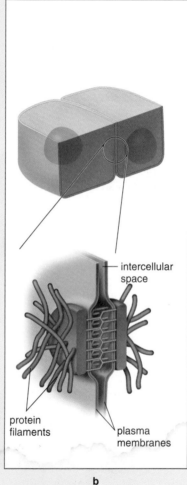

intercellular
space

protein
filaments

plasma
membranes

b

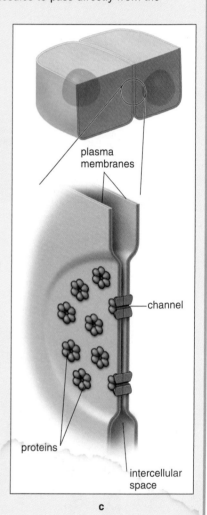

plasma
membranes

channel

proteins

intercellular
space

c

FIGURE 6.27

Examples of cell junctions in animal and plant cells. (a), Tight junction between animal cells. Here strands of protein stitch together the plasma membranes of neighboring cells. **(b),** Desmosome. This type of desmosome links the intermediate filaments of neighboring cells in animal tissues. **(c),** Gap junction. Tubular proteins embedded in plasma membranes connect the cytoplasm of neighboring animal cells.

cytoplasm of one cell to the next. Junctions are particularly important in epithelial tissue, which consists of cells tightly connected into sheets that cover internal surfaces such as breathing passages and digestive organs that come into contact with air, water, or food from outside the body.

In plants, cell walls are perforated by channels called pits (Figure 6.28a). Continuous strands of cytoplasm called plasmodesmata connect adjacent cells through the pits. Plasma membranes line the channels, and water-soluble materials can pass through them from cell to cell throughout the plant.

The cells of most multicellular organisms are at least partially surrounded by a solution of macromolecules called the extracellular matrix (Figure 6.28b). These proteins and carbohydrate molecules are secreted into the space outside cells (extracellular space) and form a network, or matrix, that holds cells together and allows them to migrate and interact with neighboring cells. The matrix may regulate the behavior of cells in contact with it. The extracellular matrix of plant and fungal cells is the cell wall. The matrix is currently the subject of considerable research.

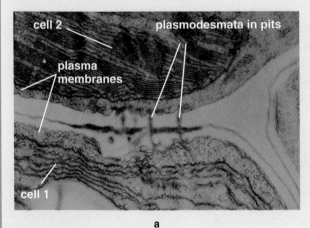
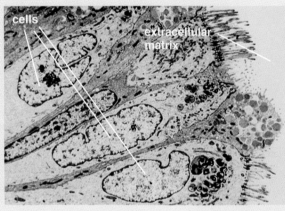

a

b

FIGURE 6.28

Examples of extracellular matrix. (a), Pits filled with plasmodesmata connect the cytoplasm of neighboring cells in a corn leaf, ×99,000. **(b),** These human cells are embedded in the extracellular matrix, ×15,000.

6.7 Systems

In most multicellular organisms, cells are small compared with the size of the entire organism. The inner cells cannot obtain nutrients directly from the outside environment or pass their wastes directly to the outside environment. Specialized systems are required to handle deliveries between the environment and the cells. In all but the smallest animals, a circulatory system is needed to transport materials. In fact, in most animals more than one system is involved in transport of materials through the organism's body.

Plants also have transport systems. They exchange gases with the atmosphere through the stomatal openings in their leaves. A transport system takes water and minerals from where they are absorbed in the roots to where they are needed in the leaves. Another transport system carries food produced in the leaves by photosynthesis to other parts of the plant. Both systems extend throughout the plant.

In general, specialized systems account for most of the complexity of multicellular organisms. Most of the systems, in turn, are necessary for these three reasons:

1. a division of labor occurs among cells,
2. many individual cells cannot work together without regulation and coordination, and
3. most cells are not in direct contact with the outside environment.

The different systems in an organism unite all its parts into a smoothly functioning whole.

In many organisms, additional specializations have developed within the organ systems. For example, some cells in animals' circulatory systems have nothing to do with transport. Instead, these cells have other specialized tasks such as protecting the body from infection. Thus continuing specialization has led to development of many different types of cells. Specialized systems, in turn, have contributed enormously to the diversity and complexity of multicellular organisms.

Figure 6.29 summarizes the increasingly complex organization of matter from atoms to the biosphere. Cells, the basic unit of life, are about midway

Levels of Structure in the Biosphere

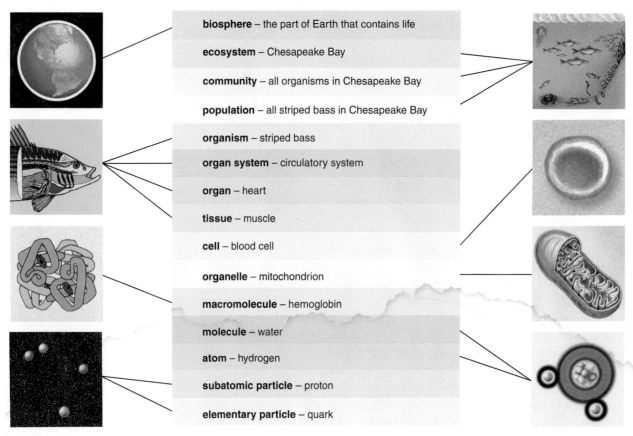

biosphere – the part of Earth that contains life

ecosystem – Chesapeake Bay

community – all organisms in Chesapeake Bay

population – all striped bass in Chesapeake Bay

organism – striped bass

organ system – circulatory system

organ – heart

tissue – muscle

cell – blood cell

organelle – mitochondrion

macromolecule – hemoglobin

molecule – water

atom – hydrogen

subatomic particle – proton

elementary particle – quark

FIGURE 6.29

Levels of structure in the biosphere. Matter in the biosphere is characterized by increasing levels of organization and complexity. Each higher level of order requires a greater input of free energy for its maintenance than the level below it. Where would you draw a line to distinguish between life and nonlife?

on the continuum of biological organization. They also are the lowest level of organization that truly can be considered living. Functions of cells, as well as those of the higher levels of organization, are considered in the remaining chapters of this book.

Check and Challenge

1. Give three reasons why specialized systems are necessary in large multicellular organisms.
2. Arrange these terms in increasing order of complexity: cells, systems, organs, tissues, organisms.
3. Can a single cell form a system or a tissue? Explain your answer.
4. What limits the size of cells?
5. How are cells organized in multicellular organisms? What are the advantages of that organization?

Chapter
HIGHLIGHTS

Summary

Prokaryotic and eukaryotic cells differ in their basic structures. Prokaryotic cells are smaller and less specialized than eukaryotic cells. The most distinguishing characteristic of eukaryotic cells is the presence of organelles: subcellular units of structure and function that increase efficiency. Organelles include the nucleus, mitochondria, chloroplasts, ribosomes, vacuoles, endoplasmic reticulum, and other compartments with specific functions in the eukaryotic cell. Many organelles are surrounded by membranes that divide eukaryotic cells into compartments with different functions.

Cells may exist alone as unicellular organisms. In fungi, plants, and animals, cells may be clustered and form multicellular organisms. Tissues, organs, and systems become more complex in larger multicellular organisms. In fact, most of the complexity of multicellular organisms results from their specialized systems.

Key Concepts

Use the following concepts to develop a concept map that describes a eukaryotic cell. Include appropriate linking words and add as many other concepts from the chapter as you can.

- structure
- function
- organelles
- nucleus
- prokaryotes
- unicellular organism
- multicellular organism
- division of labor

Reviewing Ideas

1. Compare the differences and similarities of prokaryotic and eukaryotic cells.
2. How do the walls of plants and bacterial cells differ? How are they similar?

3. How does the cytosol differ in prokaryotes and eukaryotes?
4. What cell part specializes in packaging materials for export? Give an example of how the plasma membrane also can package materials for export.
5. List three cell parts found in both prokaryotes and eukaryotes.
6. Define *unicellular, colonial,* and *multicellular.*
7. What difficulties are encountered in studying living cells of multicellular organisms?
8. What differences are likely to exist between unicellular organisms and cells of larger organisms?
9. Explain the difference between tissues and organs.

Using Concepts

1. Why is it important for cells to package some of their products in vesicles?
2. How does the environment of cells differ for a unicellular and a multicellular organism?
3. Compare the structure and action of flagella in prokaryotes and eukaryotes.
4. When eukaryotic cells reproduce by dividing in two, their nuclear membranes break down and then re-form around the nuclei of the two new daughter cells. Explain what is going on in this process and its importance to the cells.
5. Could there be an organism that has no ribosomes in its cells? Explain your answer.
6. Prokaryotes have no clearly defined cell nuclei. What other evidence could you look for in trying to determine whether they are cells?
7. Some biologists once wondered whether the endoplasmic reticulum was formed only after death by the process used to prepare cells for study. Why might they have asked such a question?

8. How is a biofilm different from a multicellular organism?

9. If cells are the units of structure and function in organisms, why would tissues and organs be needed?

10. Choose one of the differences between prokaryotic and eukaryotic cells from Table 6.1. Explain how the difference improves the efficiency of eukaryotic cells.

Synthesis

1. How might eukaryotic cells add material to their plasma membranes as they grow larger?

2. How are the characteristics of biological macromolecules important in the structure and function of cells?

3. How does the development of multicellularity demonstrate evolution through natural selection?

Extensions

Cell biologists are trying to determine whether material moves through the Golgi apparatus by passing from one membrane sac to the next, or whether the sacs move through the Golgi together with their contents, eventually breaking up into vesicles that travel to various destinations in the cell. Describe an experiment that could help answer this question.

Web Resources

Visit BSCSblue.com to access
- Web links to information and activities related to the structures and functions of various types of cells.

CONTENTS

LEARNING OUTCOMES

By the end of this chapter you will be able to:

   A   Summarize the adaptations made by plants to life on land.

   B   Compare the structure and function of xylem and phloem tissues.

   C   Describe the advantages offered by a closed circulatory system.

   D   Explain what blood pressure is and describe factors that affect it.

   E   Name the constituents of blood and describe the function of each.

   F   Explain how the circulatory system functions in homeostasis.

◄ **About the photo**
The photo to the left is
an arteriogram of a
human hand showing
the arterial structure
(enhanced).

CHAPTER

7

Transport Systems

■ *What system is responsible for the movement of blood throughout the body in humans?*

■ *How do transport systems contribute to the survival of multicellular organisms?*

Small, simple organisms living in water can obtain nutrients and eliminate wastes by diffusion and active transport through their surfaces. Larger, more complex multicellular organisms, such as most land plants and animals, cannot get by so easily. One key reason is the change in the ratio of surface area to volume with increasing size. Moreover, for organisms that live on land, the surface of the body is not normally in contact with liquid.

Consider your own body. Although we live on land, our cells are in contact with an internal liquid environment. In vertebrates, specialized systems have evolved that maintain this internal environment, provide cells with food and oxygen and remove carbon dioxide through the fluid around them, and eliminate the wastes resulting from metabolic processes.

Plants also have evolved systems for obtaining and eliminating materials and maintaining a stable internal environment. The transport system moves needed materials to different parts of the organism and carries away wastes for elimination. Just as in animals, the transport system plays a key role in maintaining the internal balance necessary for life.

Transport Systems in Plants

7.1 Adaptations for Life on Land

Fossils and other evidence indicate that the first land plants probably evolved from green algae about 430 million years ago.

Life out of water posed a new challenge: loss of moisture to the air. Thus, the first adaptations to land included a cuticle and protective structures for the gametes and embryos. Two groups of plants emerged during this period. One group had specialized tissue, called vascular tissue, consisting of cells joined into tubes that transported water and nutrients throughout the body of the plant. These were the first vascular plants—the ancestors of all plants except mosses and their relatives. Mosses and their relatives are descendants of the second group, nonvascular plants in which complex transport tissues did not evolve. These plants are restricted largely to damp environments and do not grow very large.

Life on land also presented other challenges. Soil contains water and minerals, but the light and carbon dioxide needed for photosynthesis must be obtained above ground. Vascular land plants adapted to this situation by differentiating into an underground root system that absorbs water and minerals and an aerial system of stems and leaves that makes food (Figure 7.1).

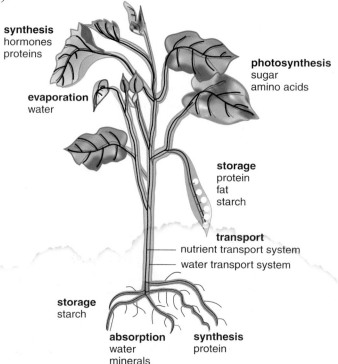

synthesis
hormones
proteins

photosynthesis
sugar
amino acids

evaporation
water

storage
protein
fat
starch

transport
nutrient transport system
water transport system

storage
starch

absorption
water
minerals

synthesis
protein

FIGURE 7.1

Transport in plants. Different parts of a plant have different activities, all of which require materials that must be transported where needed. Water is the material needed in greatest amounts. It also serves as the transport fluid, carrying minerals through one type of transport tissue and the products of photosynthesis through another.

This specialization required additional adaptations. The sections of roots that absorb water generally lack a cuticle. Note in the close-up view of the roots in Figure 7.2 that some cells have long, thin projections. These root hairs greatly increase the root's water-absorbing surface area.

Without the support provided by a water environment, how does a plant shoot stand upright in the air? An important adaptation of vascular plants is **lignin,** a hard material embedded in the cellulose matrix of the cell walls. These lignified walls support trees and other large vascular plants.

Water and minerals must be conducted upward from the roots to the leaves. The products of photosynthesis must be distributed from the leaves to the roots. These tasks are performed by a vascular system that extends throughout the plant. Hollow tube-shaped cells called **xylem** carry water and minerals up from the roots (Figure 7.2). Mature xylem cells are nonliving hollow tubes. Organic nutrients are distributed throughout

ETYMOLOGY

xyle = wood (Greek)
Wood is layers of **xylem** laid down during each year of a tree's growth.

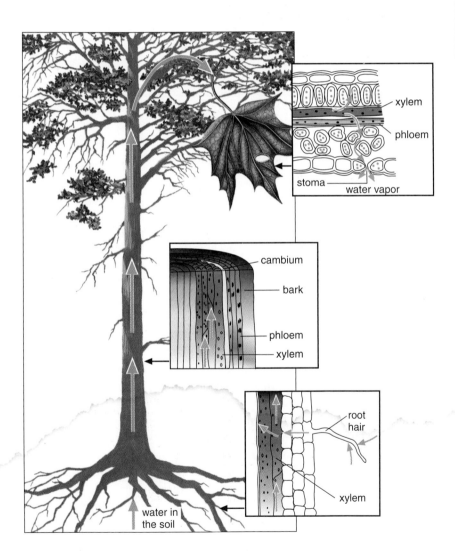

FIGURE 7.2

Movement of water through a plant. The arrows identify the path of water through this tree. Trace the path of the water from the root hairs through the xylem tissues to the leaves.

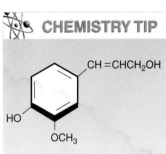

CHEMISTRY TIP

Lignin is a complex, irregular polymer made up of rigid subunit molecules containing phenolic rings (6-carbon rings with alternating single and double bonds). Coniferyl alcohol, shown here, is a subunit of lignin.

Focus On

Mycorrhizae

Many vascular plants interact with soil fungi to obtain what they need from the soil. This interaction causes slight modifications of the roots known as **mycorrhizae.** The fungi either form a sheath around the root or penetrate the root tissue, increasing the surface area of the root through outward extensions (Figure 7.3). The fungi are good at absorbing minerals, and they secrete an acid that converts the minerals to forms that can be used more readily by the plant. Minerals taken up by the fungi are transferred to the plant, and photosynthetic products from the plant nourish the fungi. In addition to enhancing mineral nutrition, mycorrhizae absorb water, and the fungi protect the plant against certain disease-causing organisms, or pathogens, in the soil.

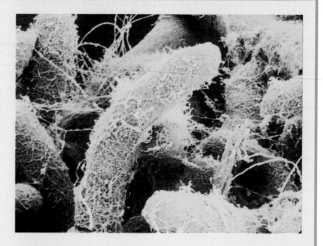

FIGURE 7.3

Root modifications. Scanning electron micrograph of mycorrhizae, ×50, between a *Boletus* mushroom and roots of an aspen tree, *Populus tremuloides*. The relationship benefits both the fungus and the plant.

---

the plant by active transport within the **phloem,** which consists of elongated cells arranged into tubes filled with streaming cytoplasm.

7.2 Water Transport

The xylem of flowering plants consists of two types of water-conducting cells, **tracheids** and **vessel elements** (Figure 7.4a), plus strong weight-bearing fibers. Tracheids have pointed ends and thick walls with pits that connect them to nearby cells. Water moves from cell to cell through these pits. Vessel elements are wider, shorter, thinner-walled, and less tapered. The end walls of vessel elements are perforated or missing. Water can flow freely through these openings. Columns of vessel elements form the xylem vessels through which water moves throughout the plant.

Tremendous amounts of water evaporate every day through the stomates as a plant exchanges gases with the air. For example, a typical red maple tree growing in a humid climate may lose as much as 2,000 L of water per day. This water must be replaced if the plant is to survive. Water entering through the roots of trees and other tall land plants often must be transported upward for great distances to reach the higher stems and leaves. Plants do not have pumping mechanisms like the hearts of animals, so how do water and the minerals it carries reach the upper parts of plants?

Scientists who have researched this problem have developed the cohesion-tension hypothesis as the likely mechanism for water transport through the xylem. The hypothesis is based on well-known phenomena—including the molecular properties of water and transpiration. Also, the root system exerts pressure that causes water and other materials to ooze out of

Try Investigation 7A   Water Movement in Plants.

a cut plant stem. Root pressure is a factor in transport, but careful measurements indicate that root pressure alone cannot account for the rise of water in taller plants and trees.

The properties of water cannot account completely for water transport either. Recall that hydrogen bonds form between water molecules. These bonds cause **cohesion**—the tendency of water to stick together. In addition, the positive and negative charges of water molecules (which are polar) form weak bonds to other charged molecules—the property called **adhesion.** Adhesion and cohesion cause water to rise up inside a glass tube placed in a container of water. This phenomenon is called capillary action. It occurs because the water molecules develop adhesion to charged groups on the walls of the tube, pulling them upward; additional water molecules are then drawn up by cohesion. The narrower the tube, the higher the liquid rises. This is because there is a larger charged surface area on the walls in contact with water. The walls of tracheids and vessels in plant xylem contain many charged groups, so they also take up water by capillary action. The rise of water by capillary action is not very rapid, however, and the height it can reach is limited by gravity and the diameter of the tube or vessel. Therefore, this process also cannot fully explain water transport in very tall plants, where water moves rapidly and to great heights.

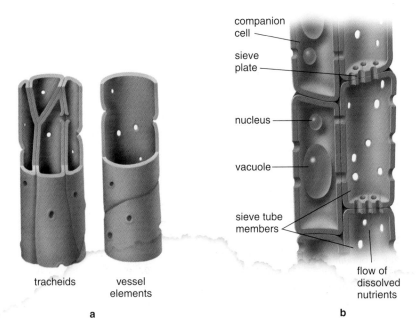

companion cell

sieve plate

nucleus

vacuole

sieve tube members

flow of dissolved nutrients

tracheids

vessel elements

a

b

FIGURE 7.4

Plant transport systems. (a), Xylem cells. Microscopic comparisons have shown that the xylem of nonflowering vascular plants that arose earlier in evolution, such as evergreens, is not as complex as that of flowering plants. The xylem of a pine consists mostly of tracheids with relatively thick walls and long tapering ends that fit into each other like fingers. **(b),** Phloem cells, consisting of sieve tube members, companion cells, and fibers, are found in broad-leaved, flowering trees considered hardwoods.

Due to cohesion, each water molecule that leaves the plant during transpiration tugs on the one behind it. This tugging is transmitted from water molecule to water molecule back into the leaf. Water molecules in the xylem of leaf veins replace those that left the mesophyll cells through the stomates. The result is that a long chain of water molecules is continually pulled through the xylem from root to leaf.

7.3 Nutrient Transport

In vascular plants, nutrients travel through living phloem cells joined end to end (see Figure 7.4b). Tiny pores in the walls at the ends of the phloem cells allow the contents of the cells to mix. The porous areas at the ends of these cells resemble tiny strainers, or sieves, so the phloem channels are often called **sieve tubes.**

Sugars and amino acids move through the phloem cells from the leaves to other parts of the plant. The rate at which the fluid moves is thousands of times faster than diffusion alone can account for. What is the possible mechanism of phloem transport?

The best explanation for the movement of sugars through the phloem is the pressure-flow hypothesis (Figure 7.5). According to this hypothesis, water and dissolved sugars move through the phloem from sources (areas of higher pressure) to sinks (areas of lower pressure). Sources include cotyledons and endosperm during germination (see Chapter 11), leaves

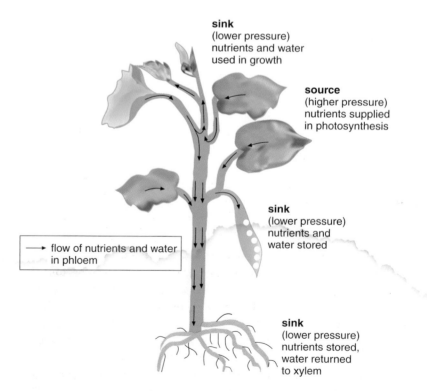

sink
(lower pressure)
nutrients and water
used in growth

source
(higher pressure)
nutrients supplied
in photosynthesis

sink
(lower pressure)
nutrients and
water stored

⟶ flow of nutrients and water
in phloem

sink
(lower pressure)
nutrients stored,
water returned
to xylem

FIGURE 7.5

Sources and sinks in phloem transport. The pressure-flow hypothesis, based on fluid pressure in phloem, accounts for the transport of nutrients and water.

during spring and summer, and some storage roots in early spring. Sinks are found in the many areas of a plant where water and sugars are used, including food-storage areas and growing leaf buds, root tips, flowers, fruits, and seeds.

The process begins at a source such as a leaf, where photosynthesis occurs. There sucrose is actively transported into the sieve tubes from the mesophyll cells. The companion cells, which lie on either side of the sieve tubes (see Figure 7.4b), produce a key protein involved in this transport. The resulting high concentration of sucrose draws water into the phloem cells. The higher pressure this produces forces the sucrose solution to move toward the lower pressure at the sinks. Thus, sucrose molecules move from phloem cell to phloem cell, from source to sink, through the sieve tubes.

At a sink, active transport removes sucrose from the phloem to be used or stored. As this occurs, water also leaves the phloem cells by osmosis; most of it returns to the xylem. The entire process depends on the uptake of sucrose and water by phloem cells in the leaves and active removal of sucrose and water from phloem cells in sink tissues.

Check and Challenge

1. Describe some adaptations plants have for living on land.
2. How does a plant benefit from the presence of mycorrhizae?
3. Describe the differences between xylem and phloem tissues.
4. Summarize the cohesion-tension hypothesis.
5. Explain how a plant's cells receive nutrients.

Transport Systems in Animals

7.4 Circulatory Systems

Like plants, all animals must exchange materials with the environment. Substances pass across the plasma membrane between each cell and a watery environment. A unicellular organism in a pond has such an environment. Diffusion and active transport carry substances across its plasma membrane. Figure 7.6 shows the transport systems of *Paramecium*, *Hydra*, an earthworm, and a grasshopper. One-celled organisms, such as *Paramecium*, take in food and form a vacuole that moves around inside the cell by the movement of the cytosol. A similar arrangement works for simple animals. For example, *Hydra* has a saclike body plan and tissues that are only two cell layers thick. Fluid containing food, oxygen, and carbon dioxide passes into and around its body as the *Hydra* moves. A specialized internal transport system is unnecessary. Exchange of materials in planarians and other flatworms occurs through a single opening of an internal cavity. The flatworm's thin body shape and the extent of the cavity keep its cells in contact with a watery environment.

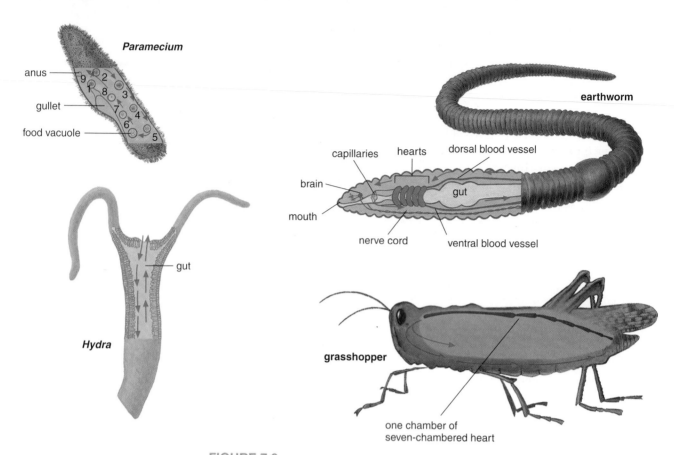

FIGURE 7.6

Transport systems in a unicellular organism and three animals. Compare the simple transport systems of *Paramecium* (a unicellular organism) and *Hydra* (a simple animal) with the more complex systems of the earthworm and grasshopper. Numbers indicate the pathway from endocytosis of food to exocytosis of indigestible waste in *Paramecium*. The grasshopper and earthworm both have contractile hearts. The earthworm has a closed system of blood vessels, and the grasshopper has an open circulatory system.

A single cavity is inadequate for internal transport in larger animals, especially if they live out of the water. Most such animals have digestive organs that break down food into nutrients their cells can use and excretory organs that collect wastes and eliminate them from the body. These are linked to the rest of the body by a circulatory system that transports nutrients, and often oxygen, to cells and carries wastes away. This transport is typically carried out by a pump (heart) and other organs and tissues, such as blood vessels and blood. Both the size of an organism and its level of activity play a role in how complex and efficient this system has to be.

Insects, crabs, and other **arthropods**—animals that carry their skeletons on the outside of the body and have jointed legs—have an **open circulatory system,** in which there is no separation between blood and other intercellular fluid. These fluids bathe the internal organs. Chemical exchange between the fluid and cells occurs as the blood oozes through sinuses, or spaces, surrounding the organs. Blood is also circulated by the heart's contraction and by body movements that squeeze the sinuses. For example, the grasshopper has seven simple heart chambers.

An earthworm, in contrast, has a **closed circulatory system,** which means that the blood is confined to vessels. Major vessels branch into smaller vessels that carry blood to or from the various organs. A number of tiny contractile hearts, illustrated in Figure 7.6, pump blood through this system to all the organs. The blood returns to the hearts through a second large vessel.

Blood travels through a closed circulatory system more rapidly than it flows through an open system. Open systems are common in animals that move sluggishly and in small animals such as insects that do not transport oxygen long distances in blood. Insects distribute oxygen through microscopic air ducts with branches that reach every part of the body. This system works well because of the small size of insects.

7.5 Circulation in Vertebrates

Humans and other vertebrates (animals with backbones) have a closed circulatory system, also called the cardiovascular system, which is far more complex than that of the earthworm. Figure 7.7 is a diagram of the human

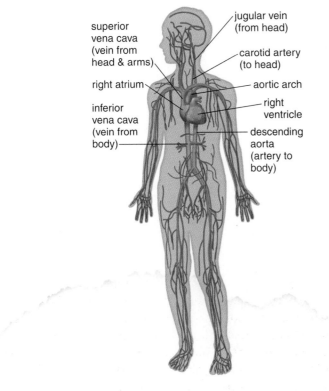

FIGURE 7.7

A simplified drawing of the human circulatory system. Blood with a high concentration of oxygen is shown in red. Blood with a low concentration of oxygen is shown in blue. Trace the path of the blood around the body.

circulatory system. The components of the cardiovascular system are the heart, blood vessels, and blood. The vertebrate heart consists of one or more **atria** (singular: atrium), chambers that receive blood returning to the heart, and one or more **ventricles,** chambers that pump blood out of the heart.

There are three types of blood vessels. **Arteries** carry blood away from the heart to organs throughout the body. The major artery leading from the heart and supplying blood to the body is the aorta. The major arteries that carry blood to the lungs from the heart are the pulmonary arteries. **Capillaries** are the network of microscopic vessels that infiltrate every tissue. The thin capillary walls allow chemicals to be exchanged between the blood and the intercellular fluid surrounding the cells and oxygen and carbon dioxide to be exchanged with air in the lungs. At their "downstream" end, capillaries rejoin to form **veins,** blood vessels that return blood to the heart. The relationship between arteries, veins, and capillaries is illustrated in Figure 7.8. Note that veins and arteries are distinguished by whether they carry blood to or from the heart, *not* by whether they carry oxygenated or deoxygenated blood.

During vertebrate evolution, increasingly elaborate closed circulatory systems have evolved to support higher levels of activity and cell respiration. Fish have a two-chambered heart with one atrium and one ventricle (Figure 7.9a). The ventricle pumps blood to the gills, where it picks up oxygen and gives off carbon dioxide. The blood next travels to the digestive system, where it picks up nutrients, then to the other tissues of the body, and finally back to the heart. Blood flow slows down in the capillaries because of their small size, but the flow of blood to the organs and back to the heart is increased by the fish's swimming movements.

Amphibians and most reptiles have a three-chambered heart with two atria and one ventricle (Figure 7.9b). The ventricle pumps blood into an

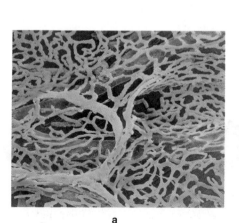
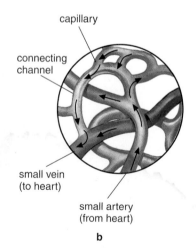

a   **b**

FIGURE 7.8

Blood flow in capillaries. (a), This scanning electron micrograph, ×100, of human capillaries lines the wall of the gall bladder. **(b),** Arteries and veins connect with capillaries by way of smaller vessels or connecting channels. Blood can pass from arteries to connecting channels to capillaries or directly through the connecting channels to veins.

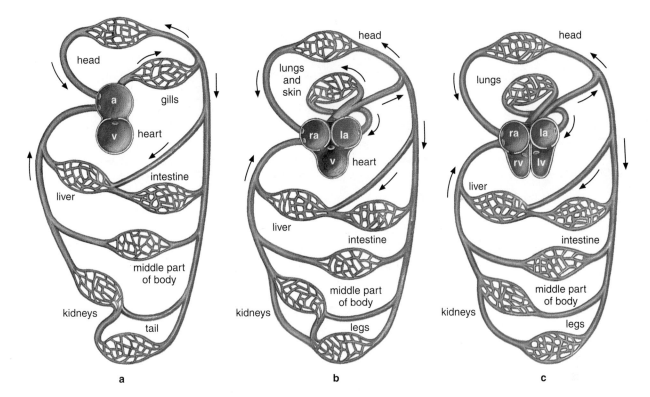

FIGURE 7.9

Vertebrate circulatory systems. Red represents oxygenated blood, and blue represents deoxygenated blood. **(a),** Fish have a two-chambered heart and a single system for blood flow. **(b),** Amphibians and most reptiles have three-chambered hearts with only one ventricle and some mixing of oxygenated and deoxygenated blood. **(c),** Mammals, birds, and crocodilians have four-chambered hearts and double circulation in which oxygenated blood is completely separated from deoxygenated blood.

artery with two branches. One branch leads to the lungs and skin, where the blood picks up oxygen while flowing through capillaries, then returns to the heart through veins. A second branch leads to capillaries serving all the organs except the lungs and returns to the heart through other veins. This pattern is called double circulation because the blood is pumped a second time after it has lost pressure from its trip through the capillaries, where it picks up oxygen. Double circulation ensures a vigorous flow of blood to the brain, muscles, and other organs. Oxygenated and deoxygenated blood continually mix in the single ventricle, however, so the level of oxygen reaching the organs of the body is lowered. Because of this, animals with three-chambered hearts tire easily.

The four-chambered heart found in mammals, birds, and crocodilians has two atria and two completely divided ventricles (Figure 7.9c). These animals have double circulation; the heart keeps oxygenated blood completely separate from deoxygenated blood. Because this system delivers high levels of oxygen, birds and mammals can maintain vigorous metabolisms. They are warm-blooded and can engage in high-energy activities such as flight or other rapid movement for long periods.

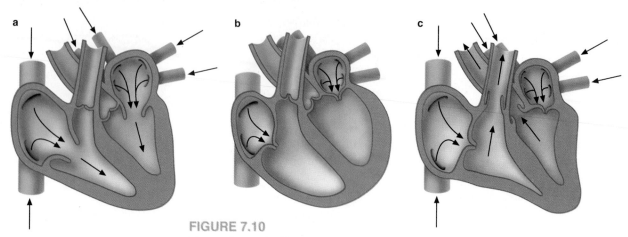

FIGURE 7.10

The cardiac cycle. Blood enters the atria, which contract **(a),** forcing blood into the ventricles. Then the atria relax and fill **(b),** and the ventricles contract **(c),** forcing blood into the pulmonary artery and the aorta. Then, the ventricles relax and the atria contract, repeating the cycle.

7.6 The Human Heart

Each heartbeat is a sequence of muscle contraction and relaxation called the **cardiac cycle** (Figure 7.10). In each cycle, the four chambers of the human heart go through phases of contraction, or systole, and relaxation, or diastole. The atria contract slightly before the ventricles.

When the atria are relaxed and filling, the ventricles are also relaxed. As pressure rises in the atria, the valves between the atria and ventricles (AV valves) are forced open, and the ventricles start to fill (Figure 7.11). Then the atria contract (atrial systole), forcing additional blood into the ventricles. Next, the ventricles contract (ventricular systole), causing the AV valves to

ETYMOLOGY

kardia = heart (Greek)

The **cardiac cycle** is the name given to the repeated contraction and relaxation of heart muscle that pumps the blood.

Try Investigation 7B ◇ **Exercise and Pulse Rate.**

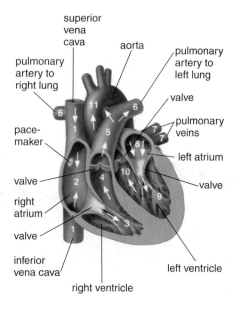

FIGURE 7.11

Blood flow. A drawing of a section through a human heart and the blood vessels leading to and from it. Trace the flow of blood into and out of the heart by following the numbers and arrows. Why do you think the wall of the left ventricle is thicker and more muscular than the wall of the right ventricle?

snap shut, and the pressure inside the ventricles rises sharply. With the increased pressure, the valves leading to the aorta and the pulmonary artery open, and blood flows out of the heart. After the blood has been ejected, the ventricles relax, and the cardiac cycle starts again.

Each heartbeat starts in the pacemaker, cells of specialized muscle tissue high in the inner wall of the right atrium. Without outside stimulation, this tissue creates electrical impulses through changes in ion channels and ion pumps in its cell membranes. These membrane changes alter the concentrations of sodium and potassium ions inside and outside the cells, creating an electrical potential across their membranes. The heartbeat impulse then spreads from the pacemaker throughout the heart muscle. Although the heartbeat starts in the heart itself, changes in the rate of the heartbeat are controlled by nerves outside the heart.

Figure 7.12 is a photograph of an AV valve in the human heart. These one-way valves prevent blood in the ventricles from backing up into the atria. Leaks in these valves cause the condition known as a heart murmur.

FIGURE 7.12

Human heart valve. This AV valve between the left atrium and left ventricle is called the mitral valve.

7.7 Molecular Basis of Muscle Contraction

Heart, or cardiac, muscle differs from skeletal muscle—the voluntary muscle that moves your arms and legs, for example—because it contracts involuntarily; that is, you do not control the beating of your heart as you do the movement of your arms. Heart muscle is branched but otherwise is similar in structure to skeletal muscle. The contractions of the heart and other muscles are produced by a fascinating molecular motor. This motor is largely composed of two proteins called **actin** and **myosin.** The patterns, or striations, formed by these molecules are visible with the electron microscope (Figure 7.13).

a

b

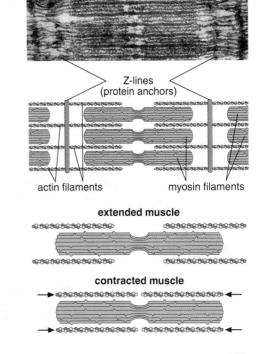

Z-lines
(protein anchors)

actin filaments   myosin filaments

extended muscle

contracted muscle

FIGURE 7.13

Mechanism of muscle contraction. (a), This transmission electron micrograph, ×40,000, shows a human striated muscle fiber. **(b),** Compare the micrograph of human muscle with the drawing showing how actin and myosin filaments enable muscle to contract and relax.

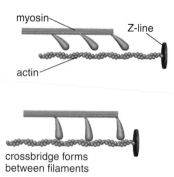

FIGURE 7.14

Muscle contraction at the molecular level. A simplified diagram of the sliding-filament model of muscle contraction is shown here. An enzyme uses the energy of ATP to form cross bridges between myosin and actin filaments, causing the muscle to contract.

The thin actin filaments are anchored to a structure called a **Z-line** at each end of the unit. The thicker myosin filaments contact the actin with rows of globular crossbridges, which bind to the actin. During contraction, these crossbridges "ratchet" along the actin filaments toward the Z-lines by continually releasing their attachment at one point, changing position, and reattaching at a point farther along (Figure 7.14). This motion causes the entire unit to contract. Each movement of the crossbridges requires ATP, so working muscles use a great deal of energy. Because heart muscle works constantly, a steady flow of nutrients and oxygen is essential. Thus, any interruption of blood flow to the arteries that serve the heart muscle itself can quickly cause a heart attack, possibly causing serious damage to the heart or death.

Check and Challenge

1. How does a vacuole in *Paramecium* serve some of the same functions as a circulatory system in animals?
2. Describe the differences between open and closed circulatory systems.
3. Why is the human heart described as a two-part pump?
4. Explain the differences in the structure of veins and arteries.
5. Why is a pulse taken at an artery and not at a vein?
6. Describe how the heart contracts and moves blood.
7. Explain the molecular basis for muscle contraction.

Biology Online   BSCSblue.com/check_challenge

Biological Challenges

Discovering the Secret of Circulation

From ancient times, the accepted theory about movement of the blood stated that it was pumped out to the body and then returned to the heart through both the arteries and veins, like the ebb and flow of the tides. This seemed the only logical explanation because capillaries could not be observed before the invention of the microscope.

William Harvey, trained as a physician and scientist, was taught the standard theory in medical school in 17th-century England. But his anatomy teacher showed him some tiny flaps that had been discovered in the heart and certain blood vessels of humans and some other vertebrates. Harvey's curiosity was aroused, leading him to investigate the circulatory system.

After careful observation and experimentation, Harvey presented a new hypothesis, that the blood circulates around the body like a "river with no end." One piece of evidence he used to support this idea was that the flaps he had seen as a student allowed a small instrument inserted into a blood vessel to move freely in one direction but not the other. Harvey reasoned that the flaps were valves that prevented blood from flowing backward (Figure 7.15). Therefore, blood could not "ebb" back into the heart by the same vessels through which it had flowed out. Harvey argued that blood must circulate from the heart through one set of vessels (arteries) and return through another set (veins). It was in the veins that the tiny valves had been observed.

A few years later, the Italian scientist Marcello Malpighi used a microscope to observe that the blood flowed from the smallest arteries into tiny capillaries and then into the veins. With the discovery of capillaries, the circulation of blood was established beyond doubt.

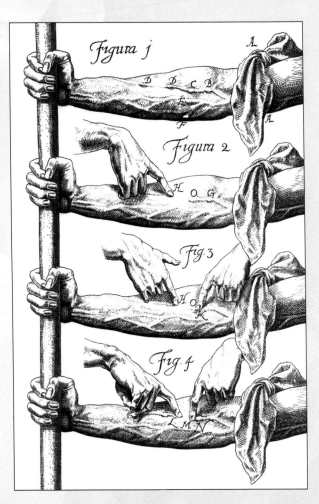

FIGURE 7.15

Harvey's explanation of the function of valves. "Let an arm be tied above the elbow as A (Fig.1). In the course of the veins certain large knots or elevations (B, C, D, E, F) . . . are all formed by valves. If you press the blood from the space above one of the valves, from H to O (Fig. 2), you will see no influx of blood from above; . . . . If you now apply a finger of the other hand upon the distended part of the vein above the valve O (Fig. 3), and press downwards, you will find that you cannot force the blood through or beyond the valve. If you press at one part in the course of a vein with the point of a finger (L, Fig. 4), and then with another finger streak the blood upwards beyond the next valve (N), you will perceive that this portion of the vein continues empty (L, N). That the blood in the veins therefore proceeds from inferior to superior parts of the heart appears most obviously."

Regulation and Transport

7.8 Blood Pressure

The circulatory system provides a good example of the interrelationship of structure and function. Blood vessels differ in the amounts of muscle and elastic tissue in their walls just as they differ in their functions in the system. For example, the largest arteries have walls made up largely of muscle and other elastic tissue (Figure 7.16). When the heart contracts, it forces blood into these arteries under great pressure, stretching their walls. This expansion allows more blood to enter and prevents the pressure from increasing greatly. During the relaxed phase of the heartbeat, the stretched elastic walls of the arteries contract, helping to push the blood along and maintain blood pressure.

The walls of smaller arteries are also made of muscle and elastic tissue. Signals from the nervous system can cause these arteries to contract and expand. Thus, blood pressure and the flow of blood into different parts of the body may be controlled.

Blood returning to the heart through the veins after passing through the capillaries is under much lower pressure than the blood flowing in the

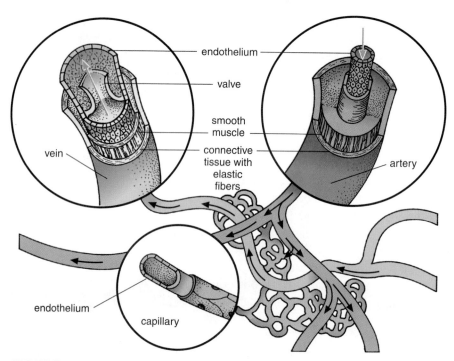

FIGURE 7.16

The structure of blood vessels. The walls of arteries and veins have three layers. The lining consists of endothelium, a simple layer of cells. The middle layer includes smooth muscle and elastic fibers. The outer layer is made of connective tissue with elastic fibers. Compare the thickness of the smooth muscle layer of the artery with that of the vein. Capillaries have only a single endothelial layer.

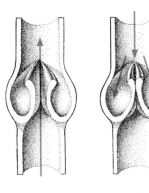

valve open   valve closed

FIGURE 7.17

Action of valves in veins. The valves regulate the flow of blood toward the heart. Note that back pressure on the valve tends to keep it closed until the pressure of blood on the other side opens it.

arteries. Veins have thinner walls with less muscle and elastic tissues than arteries. Valves in the veins prevent the blood from flowing backward (Figure 7.17). In some parts of the body, particularly those normally lower than the heart, contraction of the skeletal muscles around the veins helps to push the blood along (Figure 7.18). Gravity also helps the blood return from areas higher than the heart, especially the head.

A healthy blood pressure is maintained through complex interactions involving hormones and the nervous, excretory, and circulatory systems. Nerves connect pressure receptors in the aorta and in the artery leading to the head to cardiac-control centers in the brain. Other cardiac centers respond to sensory input, such as emotions, and to chemical input, especially the concentration of carbon dioxide. When an individual's blood pressure falls below normal range, or when exercise or danger call for

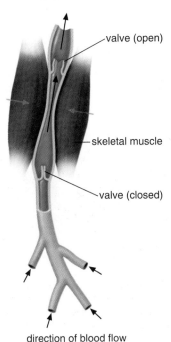

valve (open)

skeletal muscle

valve (closed)

direction of blood flow

FIGURE 7.18

Flow of blood through a vein. Movement of blood in veins is brought about by pressure from adjacent muscles. Compression forces blood in both directions, but valves prevent blood from flowing backward and away from the heart.

increased blood flow, the brain signals these receptors to increase the heart rate and constrict the blood vessels—changes that increase the blood pressure. The opposite adjustments occur when blood pressure is too high.

Various organs and tissues of the body respond differently to circulatory signals. As a result, the tissues that are most important during increased activity such as exercise or an emergency—the skeletal muscles and the heart itself—receive more of the increased flow of oxygen and nutrients (Figure 7.19).

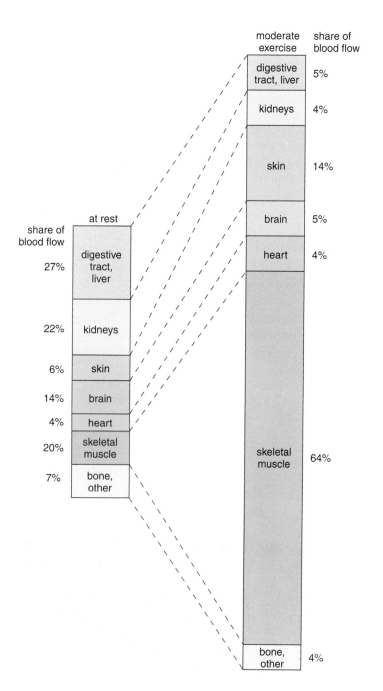

FIGURE 7.19

The distribution of blood supply during rest and exercise.
Contraction of smooth muscle in the walls of blood vessels reduces blood flow through specific organs.

Focus On

Genes and Cardiovascular Disease

Much cardiovascular disease is due to bad health habits such as smoking or eating a diet rich in fatty foods. Genes also can play an important role, however, because many of the factors that regulate the circulatory system are proteins encoded by specific genes. One example of a genetic cardiac disorder is **familial hypercholesterolemia.** This condition was discovered by investigating families that suffered many strokes and heart attacks. Research revealed that many of the people in these families had abnormally high levels of blood cholesterol (Figure 7.20), even if they ate a normal or low-fat diet.

The cause was eventually traced to a defect in a single gene. People with one defective copy of the gene and one normal copy had lower levels of an enzyme important in the regulation of cholesterol in the blood. The result was cholesterol levels well above normal and increased risk of cardiovascular disease. Those with two defective copies of the gene had cholesterol levels several times higher than normal. They often suffered severe heart attacks and even death in their twenties.

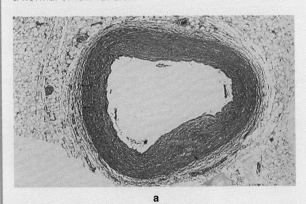

a

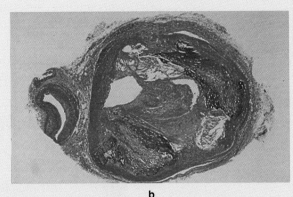

b

FIGURE 7.20

Cross sections of normal and diseased arteries. (a), This photomicrograph, ×10, shows a cross section of a healthy human coronary (heart) artery. **(b),** A cross section, ×3, of a human coronary artery is blocked by deposits of fatty plaque. Blockage of the coronary arteries can lead to a heart attack.

About 20% of the adult population in the United States has blood pressure constantly higher than the normal range, a condition called hypertension. Hypertension can cause serious health problems, even death. High blood pressure forces the heart to work harder, which can damage heart muscles; it can damage blood vessels in the brain so that they rupture, which can lead to a crippling or fatal stroke; and it can contribute to the development of atherosclerosis, further damaging the circulatory system.

The exact causes of 90% of hypertension cases are not known, and the condition often remains undiagnosed until health is seriously affected. However, hypertension can be controlled by medication prescribed by a physician. Regular physical examinations, proper diet, and exercise can help prevent or control the condition.

7.9 Composition of Blood

To keep the internal environment healthy, the body must closely regulate the composition of blood as well as its pressure. Vertebrate blood contains several types of cells suspended in a fluid. Specialized cells called red blood cells, or **erythrocytes,** transport oxygen. Erythrocytes contain an oxygen-carrying red protein called **hemoglobin.** This protein consists of four subunits, each of which carries an iron atom suspended in an organic molecule called a heme group (Figure 7.21). As erythrocytes begin to mature, the section of their DNA that encodes the structure of hemoglobin becomes active. Different groups of genes (DNA segments) are active in each of the types of cells that make up a plant or an animal (see Figure 1.35).

The iron of heme forms a temporary chemical bond with oxygen, which the erythrocyte transports to body cells. Hemoglobin binds oxygen strongly as the erythrocytes pass through the oxygen-rich environment of the lungs but releases the oxygen in the oxygen-poor environment of the capillaries. In other animals, substances of different chemical structure and color combine with oxygen in a similar manner.

Human erythrocytes live only about 120 days, so they must be produced constantly. Erythrocytes cannot reproduce in the bloodstream. In humans and other mammals, they lose their nuclei as they mature. Instead, new cells

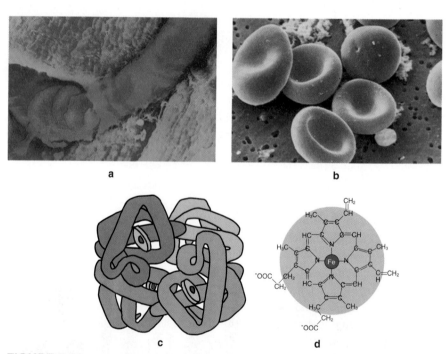

FIGURE 7.21

Circulation and function of erythrocytes. (a), This scanning electron micrograph shows erythrocytes moving single file through a human capillary, ×2000. **(b),** Each red erythrocyte contains many molecules of hemoglobin, ×6000. The small green objects are platelets, which participate in blood clotting. **(c),** Hemoglobin is a large molecule composed of four protein subunits. **(d),** Each subunit includes an iron-containing heme molecule, which can bind oxygen. Heme's oxygen-binding role in hemoglobin makes iron essential in the diet.

are manufactured in the marrow, the soft tissue in the long center of the bones of the body. This process is regulated by the kidneys, which monitor the level of oxygen in the bloodstream and release a hormone that stimulates the bone marrow to manufacture more erythrocytes whenever oxygen levels fall below normal range.

Specialized white blood cells, or **leukocytes,** circulate in the blood. They form the second line of defense against invading organisms such as bacteria and viruses. (Your skin and mucous membranes form the first line of defense.) Some types of leukocytes, called **macrophages,** surround bacteria and absorb them in much the same way as an amoeba takes in food. When there is an infection in the body, the number of leukocytes increases greatly. They help combat the infection by destroying the bacteria. Figure 7.22 shows some of the types of cells found in human blood.

The fluid portion of the blood, called **plasma,** consists of water, proteins, dissolved ions, amino acids, sugars, and other substances. Plasma carries most of the carbon dioxide generated as a waste product during cell respiration and delivers it to the lungs, where it is released during exhalation. The plasma absorbs digested food from the intestine, as described in Chapter 2, and carries it to all the cells of the body. Unusual variations in the levels of amino acids, sugars, or fats in plasma often indicates that the organs of the digestive system are not functioning properly. Plasma also carries hormones that are secreted by glands. Some of these hormones help maintain the proper composition of the blood. Insulin, for example, functions to keep blood sugar within the healthy range. Upsets in the hormonal system often can be detected by blood analysis.

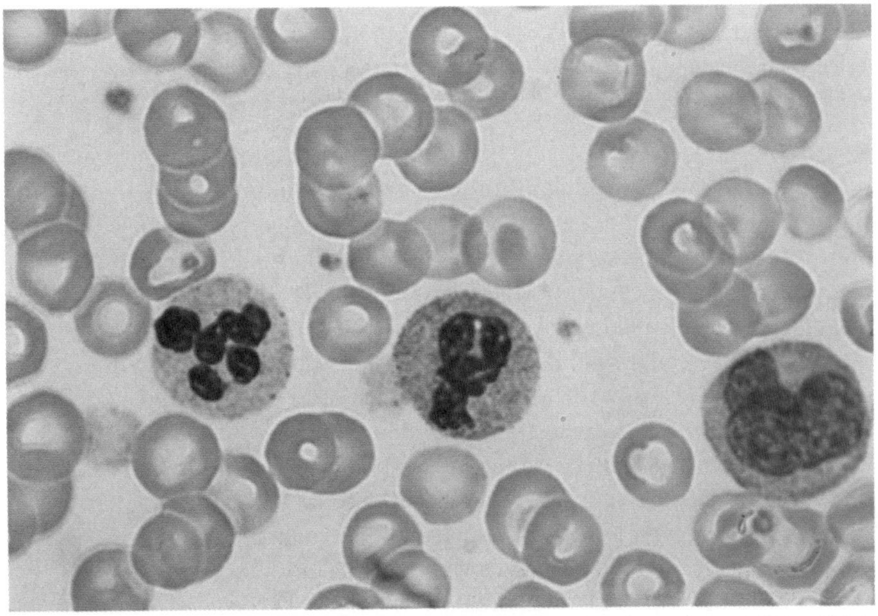

E T Y M O L O G Y

makros- = large (Greek)

phago = to eat (Greek)

A **macrophage** is a large white blood cell that "eats" foreign cells and substances in the blood by endocytosis.

FIGURE 7.22

Human blood smear. Cells visible here include erythrocytes, ×400, three types of leukocytes *(left, center,* and *right),* and platelets (the small particle, *upper left).*

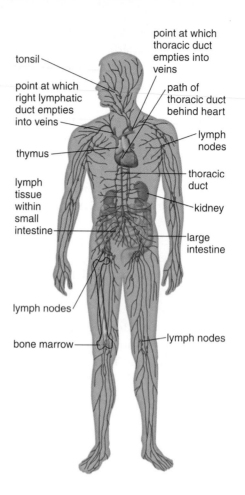

FIGURE 7.23

The human lymphatic system. The lymphatic circulation helps remove excess water from the tissues and filters out and destroys pathogens such as bacteria.

Dissolved ions in the plasma help maintain the osmotic balance between the blood and the intercellular fluid. Some of these ions also help maintain the normal pH of the blood. Normal functioning of nerves and muscles depends on the concentration of key ions in the intercellular fluid, which is determined by their concentration in the plasma. The kidneys maintain these plasma ions (also called electrolytes) at precise concentrations.

Some intercellular fluid is recycled into the circulatory system indirectly by the **lymphatic** system. The fluid in the lymphatic system, which contains certain specialized cells, water, large protein molecules, salts, and other substances, is called lymph. Lymph picks up intercellular fluid and passes through the walls of small lymph capillaries into larger lymph vessels. Finally, all of the lymph fluid empties from the two largest lymph vessels into the bloodstream at veins near the heart (Figure 7.23). Since the lymphatic system has no pump of its own, the flow of lymph (3 L/day) is extremely slow compared to the blood (6,200 L/day). Muscle contractions during body movement help push lymph through the lymphatic system.

A vital characteristic of blood is its ability to clot, or **coagulate.** Like a patch that seals a leaking garden hose, a clot can seal a wound and prevent further loss of blood. Coagulation begins when small cell fragments in the blood, called **platelets,** interact with a protein found in connective tissue that has been exposed at a wound site (Figure 7.24). This interaction makes the platelets become sticky and attract more platelets, forming a plug that partially seals the wound. The platelets also release enzymes that interact with plasma proteins known as clotting factors, beginning the chain of reactions diagrammed in Figure 7.24a. As a result of this enzyme cascade, an enzyme called active factor X (Roman numeral ten) forms. In the presence

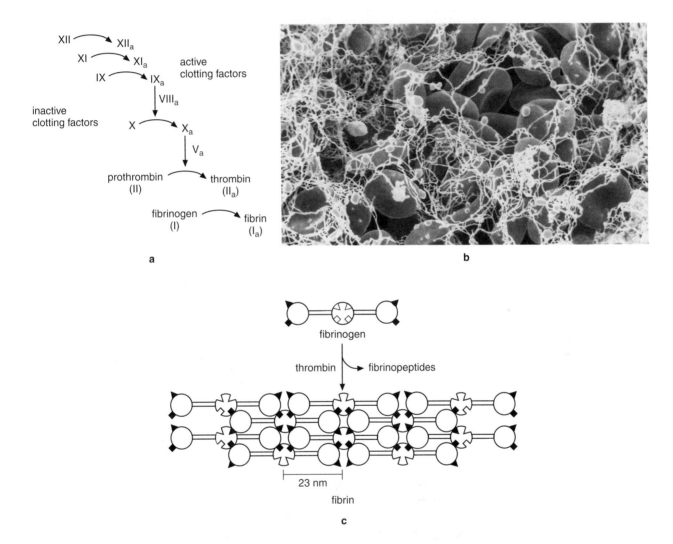

a

b

c

FIGURE 7.24

Formation of a blood clot. A blood clot is the result of a cascade of enzymatic reactions **(a),** that ends when the soluble protein fibrinogen is converted to insoluble fibrin strands. The fibrin strands provide a network in which platelets are trapped **(b),** forming a blood clot (human, ×2,200). The change from soluble fibrinogen to insoluble fibrin **(c),** involves the formation of a network of molecules.

In spite of the many advances in modern medicine, there still is no way to promote wound healing. The best approach is to keep the injury sterile and moist and let nature take its course in the healing process. This old method may soon change. Several companies, using recombinant DNA technology, are producing small proteins that appear to promote healing. These proteins, called growth factors, occur naturally in our bodies and stimulate cell migration and cell division, both of which are important in wound healing (Figure 7.25).

Growth factors applied experimentally to incisions in the skin of rats speeded the healing process. Five days after application, treated wounds were healing more than twice as fast as untreated wounds. The tissue-regeneration capabilities of growth factors may be valuable in treating areas that heal slowly, such as the transparent surface of the eye. Burn victims also may benefit from growth-factor research.

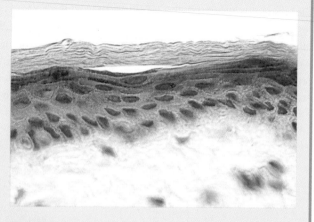

FIGURE 7.25

Section of normal human skin. This tissue section was stained with an antibody specific for EGF, epidermal growth factor, ×70. A chemical reaction produced a brown color where EGF was found. The dark blue oval structures are nuclei.

of calcium ions (Ca^{+2}), active factor X catalyzes the conversion of prothrombin, a plasma protein, to thrombin. Thrombin then acts as an enzyme to convert the soluble plasma protein fibrinogen to its insoluble form, **fibrin.** Fibrin forms a network of threads that trap additional platelets, erythrocytes, and other materials that form the clot (Figure 7.24b). Thrombin causes this change by clipping two peptides from fibrinogen. This change reduces the solubility of the fibrin molecule and exposes sites that can bind to other fibrin molecules. This causes them to link up and form the long strands found in the clot (Figure 7.24c).

Clotting is vital when wounds occur, and disorders of clotting can be very dangerous. If the cascade is triggered by internal damage to a blood vessel, a clot may form in it, blocking circulation and starving the tissues that the vessel serves of oxygen and nutrients. If the clot blocks one of the arteries supplying blood to the heart, a heart attack occurs. A clot that blocks an artery in the brain causes one type of stroke. Strokes can cause temporary or permanent paralysis of part of the body and may result in death. Problems with inadequate coagulation may be caused by hemophilia A, a serious genetic disorder. Affected individuals lack a normal copy of the gene that codes for factor VIII. Their clotting cascade is disrupted at the point where factor VIII should act, causing their blood to clot very slowly. Consequences can include serious injury or death from even minor cuts or bruises.

7.10 The Circulatory System and Homeostasis

Most organisms live in an external environment that is continually changing. Yet the internal environment of living things must be kept stable. An organism's ability to maintain homeostasis depends on the smooth interactions of its organ systems linked by the fluids of the transport system.

Different systems regulate, for example, the amount of hormones that are released, the level of sugar in your blood, and your metabolic rate. The transport system is the essential link in maintaining homeostasis because it carries hormonal signals and needed materials to all parts of the body. (Recall the example of insulin and glucagon from Focus On Energy Regulation, Section 5.9.)

Many control mechanisms have evolved that detect subtle changes in an organism's external environment and that make necessary adjustments to keep the internal environment constant. For example, when you become overheated, blood vessels just below the skin surface dilate, bringing more blood to your skin where excess heat energy can be radiated out of the body. Also, chemical messages carried in the blood stimulate the glands in your skin to begin producing and excreting sweat. As sweat evaporates, it removes heat energy from your body. Both of these responses help you cool down.

The gas-exchange system plays an important role in maintaining homeostasis. The rate of breathing is controlled by sensors that monitor the levels of gases in the transport system. Receptors in the aorta and in the artery that carries blood to the brain respond to the level of oxygen in the blood and to increases in the acidity of the blood caused by carbon dioxide. Another set of receptors in the brain also monitors increases in acidity, ensuring that the level of carbon dioxide in this important organ does not rise too high.

CONNECTIONS

The structures of plant and animal transport systems are diverse, yet they all display unity in their function—the maintenance of homeostasis.

CHEMISTRY TIP

When carbon dioxide dissolves in water, or plasma, it decreases pH. $CO_2 + H_2O \rightarrow HCO_3^- + H^+$

Check and Challenge

1. Explain how the structure of veins and arteries helps regulate circulation and blood pressure.
2. What is the difference between blood and erythrocytes?
3. What is the difference between blood and blood plasma?
4. Describe how a blood clot forms when the skin is broken.
5. How do blood and lymph differ?
6. What two organ systems work closely with the circulatory system in homeostatic control of the internal environment?

Chapter HIGHLIGHTS

Summary

In single-celled and simple multicellular aquatic organisms, diffusion provides a sufficient supply of materials necessary for life processes. Most multicellular organisms, however, cannot satisfy their needs through diffusion alone. Special structures have evolved in plants that transport raw materials to all the cells of the plant. Branching root systems obtain and transport water and dissolved minerals from the soil; xylem, with its tracheids and vessels, serves as the transport channels from roots to leaves. The cohesion-tension hypothesis, assisted by transpiration and root pressure, appears to account for the transport of water to the tops of tall trees. Phloem transports nutrients from the leaves to the rest of the plant. The pressure-flow hypothesis may account for how the food is transported through the phloem.

In animals, systems of pumps and vessels evolved, circulatory systems that transport food and oxygen to cells and remove waste products of cellular respiration. Some invertebrates have open circulatory systems in which blood passes through sinuses around the body tissues and then returns to a heart through small vessels. All vertebrates have closed circulatory systems with a heart, arteries, capillaries, and veins.

Arteries have muscular, flexible walls that withstand high blood pressure near the heart. Veins are less muscular and more flexible than arteries, and they are subjected to lower pressure. Exchanges of gases, wastes, and nutrients occur between the blood and the cells through capillaries, whose walls are only one cell thick.

Fish have two-chambered hearts; amphibians and most reptiles have three-chambered hearts; and mammals, birds, and crocodilians have four-chambered hearts. A four-chambered heart separates the systems for gas exchange and for circulation.

Blood consists of specialized cells, proteins, and plasma. Vertebrate erythrocytes contain the oxygen-carrying protein hemoglobin. Leukocytes provide the second line of defense against invading organisms. Some proteins function with platelets to form clots, which repair injuries and stop blood loss. Some plasma and other tissue fluids are picked up and returned to the circulatory system by the lymphatic system.

The circulatory system of animals plays a major role in maintaining homeostasis. Levels of blood sugar, hormones, and many other chemicals within the circulatory system must be maintained within certain limits or the body cannot function. The circulatory system functions along with the nervous, excretory, gas-exchange, and endocrine systems in maintaining homeostasis.

Key Concepts

Use the following concepts to develop a concept map of this chapter. Include additional terms as you need them.

- transport systems
- animals
- plants
- roots
- homeostasis
- stems
- leaves
- xylem
- phloem
- blood
- water and minerals
- nutrients
- oxygen
- waste products
- digestive system
- arteries
- veins
- capillaries
- lungs
- heart

Reviewing Ideas

1. How do root hairs help roots obtain water?
2. How do mycorrhizae help roots obtain water and minerals?

3. Explain the function of pits in the tracheids of xylem tissue.
4. Describe the role of the heart's pacemaker.
5. What is the difference in function between an atrium and a ventricle?
6. Describe the location and function of the different structures of the circulatory system that keep blood flowing in only one direction.
7. Describe the differences in function between erythrocytes and leukocytes.
8. How does the human circulatory system assist the endocrine system in completing its work?
9. How are the products of photosynthesis transported throughout a plant?

Using Concepts

1. Where does active transport function in the conducting tissues of plants?
2. How can transpiration help water reach the top of tall trees?
3. Explain the difference between water movement by capillary action and water movement explained by the cohesion-tension hypothesis.
4. Explain the pressure-flow hypothesis for food transport in phloem tissue.
5. What might be some advantages of a four-chambered heart over a three-chambered heart?
6. Explain why it would be difficult for gas and waste exchange to occur between cells and blood flowing in arteries.
7. Discuss the similarities and differences between the circulatory system and the lymphatic system.
8. How can some small multicellular organisms function without closed circulatory systems?

Synthesis

1. Review the process of osmosis described in Section 3.3. Explain how osmosis helps root hairs obtain water from soil.

2. Some modern land plants, such as the giant sequoias, reach tremendous size. Explain how a large size could be advantageous for these plants.
3. Dinosaurs were long thought to have been slow, sluggish animals. But skeletons discovered in the last 30 years indicate that some species were extremely vigorous and agile, suggesting that dinosaurs may have been warm-blooded animals with four-chambered hearts. Explain the logic behind this idea.
4. Certain world-class athletes who compete in endurance events such as a marathon have been suspected of "blood doping." To do this, the athlete has some blood withdrawn and stored several weeks before a competition. The erythrocytes from this blood are then concentrated and reintroduced into the athlete's blood just before the event. Explain how this practice might confer an unfair advantage.

Extensions

1. Research cardiovascular diseases and explain them to the class.
2. Research and report on the specialized adaptations in the blood and circulatory system of air-breathing mammals that perform deep dives, such as whales or elephant seals.

Web Resources

Visit BSCSblue.com to access
- Web links about plant and animal transport systems
- Web links about how your diet and behavior affect the health of your own heart and blood vessels

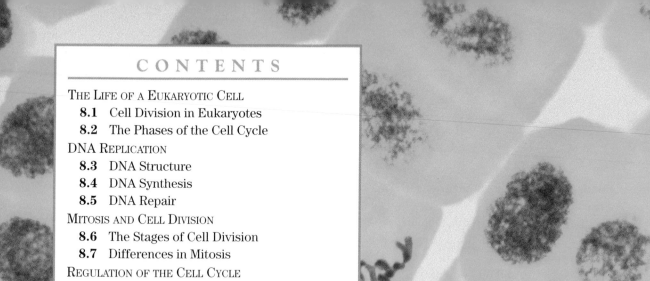

CONTENTS

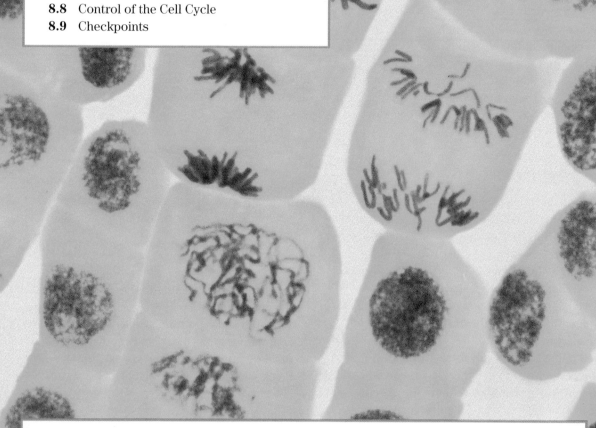

LEARNING OUTCOMES

By the end of this chapter you will be able to:

A   Compare the processes of cell division in prokaryotes and eukaryotes.

B   Describe the four phases of the cell cycle and how they are controlled.

C   Summarize the events of DNA replication and evaluate the importance of correcting DNA replication errors.

D   Describe the states of mitosis and compare and contrast mitosis in plant and animal cells.

E   Describe how the cell cycle is regulated.

CHAPTER

8

The Cell Cycle

■ *What is happening to these cells?*

■ *How is this event important in the growth and development of multicellular organisms?*

Like multicellular organisms, individual cells must grow and reproduce. Eukaryotic cells cycle through a series of ordered processes that result in duplication of the cell. First, the cell grows larger and duplicates its DNA. Then the cell divides into two cells, each of which receives a complete copy of the parent cell's genes. This mechanism of cell division evolved over millions of years as cells became more complex.

This chapter focuses on two critical events in the cell cycle—DNA replication and mitosis. Both of these processes involve intricate operations on the long, tightly wound DNA double helix. Errors in either DNA replication or mitosis can seriously damage or kill a cell or even a multicellular organism. In addition, you will examine the molecular events that regulate the cell cycle. This regulation ensures that both cell growth and division occur at appropriate rates and times. It also guarantees that each of the two daughter cells receives an exact copy of the genetic material from its parent cell.

The Life of a Eukaryotic Cell

8.1 Cell Division in Eukaryotes

In Section 6.3, you learned about the process of cell division in prokaryotes, which simply divide in two. Eukaryotic cells also reproduce by dividing in two. However, eukaryotic cell division is part of a more complex series of stages called the **cell cycle.** Unicellular eukaryotes, such as yeast and *Amoeba*, divide to produce two new identical organisms (Figure 8.1). Plants and animals, on the other hand, are multicellular organisms that usually develop from a single fertilized egg cell. The many types of cells that make up these organisms' tissues and organs develop through many cycles of division that began with that one fertilized egg cell.

By dividing into many cells, an organism's surface area can keep up with its growing volume. Plants have specialized regions at the tips of their roots and stems. Repeated cell divisions there produce the new cells that develop into the mature tissues of growing roots, stems, leaves, and other organs. During animal development, cell division produces many different types of cells that form the nerves, skin, and other organs. The timing of cell division is also critical to development. Cells in developing tissues pass through the phases of the cell cycle at various rates, coordinating their development with neighboring cells to produce organs and organ systems.

Eukaryotic cell division requires accurate replication and equal division of the genetic information encoded in the cell's DNA. Each new daughter cell must inherit an identical set of chromosomes. Reproduction is one of the most important cellular functions. In humans, an error in DNA replication or cell division can lead to birth defects, cancer, or other serious diseases.

Cell division also replaces cells that simply wear out or are damaged during the life of an organism. For example, cells in your skin are constantly dividing to replace those on the surface that have died or have been scraped off. Furthermore, cell division is necessary to form a new layer of skin at the site of an injury.

The cell cycle is remarkably similar in all eukaryotes. Evolution has produced little variation in this process among organisms. In fact, scientists have gained much of their knowledge of the cell cycle from studies of yeast, a simple fungus. These studies have found that yeast and human cells perform the cell cycle in a comparable fashion using very similar proteins.

8.2 The Phases of the Cell Cycle

Figure 8.2 shows the phases of the cell cycle. As a single cell completes the cycle, it becomes two new daughter cells. Early observations with light microscopes revealed that, when a eukaryotic cell divides, its nuclear membrane breaks down. The individual chromosomes separate and become visible as they are distributed to the daughter cells. This process of sorting and distributing the chromosomes is called **mitosis.** The period between divisions is called **interphase.** During interphase, the individual chromosomes are not

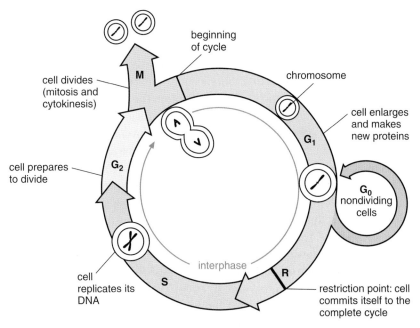

stages of the cell cycle

beginning of cycle

cell divides (mitosis and cytokinesis)

M

chromosome

cell enlarges and makes new proteins

G_1

cell prepares to divide

G_2

G_0 nondividing cells

cell replicates its DNA

interphase

S

R

restriction point: cell commits itself to the complete cycle

FIGURE 8.2

The cell cycle. G_1 (gap 1), S (DNA synthesis), and G_2 (gap 2) make up interphase. Cell division occurs in M (mitosis) phase. Cells in G_1 phase can stop dividing and enter G_0 (gap 0) or commit to perform an entire cell cycle by passing the restriction point (R). Note that DNA replication in S phase results in duplicated chromosomes.

visible in the nucleus. When cell biologists found that DNA replicates during part of interphase, they divided it into the three parts labeled in Figure 8.2. Eukaryotic cells are always in one of the five phases of the cell cycle. Cells pass in order through the phases known as **G_1** (Gap 1 or prereplication), **S** (DNA Synthesis), **G_2** (Gap 2 or premitosis), and **M** (Mitosis). During G_1, cells can also stop dividing and enter **G_0** (Gap 0 or nondividing cells).

During the gap stages (G_1 and G_2), cells grow and synthesize RNA, proteins, and other macromolecules. They either prepare for the next S or M phase or perform the cell's special function in a multicellular organism. Although the sequence of the cell cycle is the same for all cells, different types of cells spend widely different amounts of time in each phase.

The G_0 stage is a stopping point within G_1. Most cells in adult multicellular organisms are in G_0. These cells are metabolically active and specialized to perform the tasks necessary to sustain the life of the organism.

Cells constantly receive signals from their surroundings. Sometimes these signals stimulate a cell to divide. When a cell in G_0 or G_1 receives these signals, it passes through the **restriction point** (R). This "point of no return" commits the cell to a full round of the cell cycle. Once a cell passes this cell-cycle landmark, it cannot return to G_1 or G_0 without completing a full cell cycle.

Different types of cells vary in their ability to leave G_0 and commit to a cell-division cycle. For example, stem cells in your bone marrow are constantly dividing to produce replacements for worn-out red blood cells.

In contrast, the cells of your liver normally remain in G_0. The adult liver does not need to grow larger or to replace many of its cells. If part of the liver is removed through surgery, however, it is soon replaced by cells that leave G_0 and proceed through the restriction point. These cells divide until the liver reaches its previous size. At this point, the cells return to G_0, and growth ceases. Normal mature nerve cells rarely divide. Partly for this reason, damage to the brain or spinal cord usually cannot be repaired.

After passing the restriction point, a cell enters the S phase. During this phase, the DNA of each chromosome replicates to form a new identical set of chromosomes. This doubles the number of genes in the nucleus. Exact copying ensures that each new daughter cell will receive a complete copy of the parent cell's genetic information.

Cells pass from the S phase to G_2. During G_2, the cell prepares for mitosis by synthesizing specific types of RNA and proteins. These molecules will be required for mitosis during the upcoming M phase.

During interphase, the chromosomes spread out and fill up the nucleus. The M phase is the easiest phase to recognize because the chromosomes are condensed and visible through a light microscope. Mitosis is a series of events that ensures that each new daughter cell receives one copy of each chromosome. Mitosis is sometimes called **nuclear division** because the nucleus divides into two nuclei with identical sets of chromosomes. After mitosis, the whole cell divides. Division of the whole cell is called **cytokinesis.** After cytokinesis, each daughter cell enters G_1. They may then enter G_0 or commit to another round of the cell cycle.

ETYMOLOGY

cyto- = cell (Greek)
kinesis or **kinetic** = movement (Latin, Greek)

During **cytokinesis**, movement of the cytoplasm and plasma membrane divides a cell in two.

Check and Challenge

1. How does the importance of cell division differ in unicellular and multicellular eukaryotes?
2. What is the major event in M phase? S phase? G_0 phase?
3. What determines whether a cell stays in interphase or divides?
4. What are some factors that might influence the cell cycle?
5. Explain why DNA replication is important in cell division.

DNA Replication

8.3 DNA Structure

Mitosis provides each daughter cell with a complete set of chromosomes that are the same type and number as those of the parent cell. After scientists discovered that DNA is the cell's inherited genetic material, they began to study how cells accurately replicate their DNA molecules. Review Figure 1.35 for a summary of the roles of DNA in the life of a cell.

The process of DNA replication depends on the molecular shapes of DNA and its nucleotide bases. Base pairing depends on how many hydrogen bonds each nitrogen base can form with its counterpart. Adenine (A) pairs only with thymine (T) because these two bases can make two hydrogen bonds. Guanine (G) pairs only with cytosine (C) because three hydrogen bonds hold them together (see Figure 1.30). If you could uncoil and flatten a small part of the DNA helix, you would have a structure that looks like Figure 8.3. Notice that the sugar-phosphate backbones of the two strands have opposite orientations. Because the strands are parallel but run in opposite directions, the arrangement is called antiparallel (like a divided highway). The antiparallel structure of DNA is important in DNA replication, as you will see in the following section.

8.4 DNA Synthesis

During the S phase of the cell cycle, DNA replicates. The synthesis of new DNA is a multistep process. In eukaryotes, it involves more than 20 enzymes and other proteins. The process can be divided into three major parts (Figure 8.4):

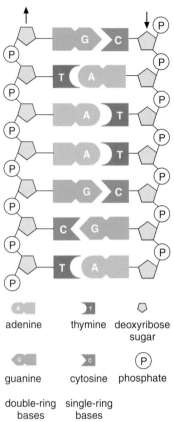

adenine   thymine   deoxyribose sugar

guanine   cytosine   phosphate

double-ring   single-ring
bases   bases

FIGURE 8.3

DNA base pairing. A short section of a DNA molecule, as it would appear if uncoiled and flattened, is depicted here. Sugar-phosphate bonds connect the nucleotides along each strand. Hydrogen bonds between the nitrogen bases connect the two strands. Compare the orientations of the sugar molecules in the two strands. Also compare how the sugar molecules are bonded to the bases and phosphate groups. Notice that the two strands have opposite orientations.

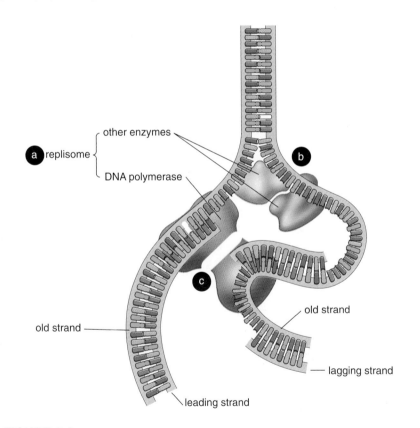

FIGURE 8.4

The three major parts of DNA replication. Enzymes and other proteins bind to the DNA at a replication origin, forming a replisome at **(a).** Enzymes use energy from ATP to unwind the DNA double helix at **(b).** DNA polymerase catalyzes formation of a new matching strand for each old DNA strand at **(c).** This occurs as one fast, continuous process on the leading strand and in discontinuous sections on the lagging strand.

1. binding of enzymes to existing DNA,
2. unwinding of the double helix, and
3. synthesis of a new matching strand for each existing strand.

First, enzymes and other proteins involved in DNA synthesis bind to specific regions of chromosomes called **replication origins.** Each chromosome has more than one replication origin. The proteins include an enzyme that unwinds the double helix, an RNA-synthesizing enzyme, and **DNA polymerase,** the enzyme that catalyzes the formation of the new DNA strands. The combination of DNA and proteins is called a **replisome.**

In prokaryotes, there is only one origin of replication. One is sufficient because bacteria contain only one small chromosome, which can replicate quickly. On the other hand, eukaryotes have several chromosomes (anywhere from three to dozens, depending on the organism). These chromosomes contain much more DNA than bacteria chromosomes do. Replication of a eukaryotic chromosome from a single origin would take an extremely long time.

DNA polymerase can add nucleotides only to the end of an existing nucleic-acid strand. Once the DNA double helix at the origin separates, an enzyme prepares each of the individual strands of DNA for synthesis of a matching strand. This enzyme synthesizes a short matching section of RNA that acts as a primer for DNA synthesis. The base sequence of the existing DNA determines the sequence of the matching strand, whether it is RNA or DNA. For example, wherever a thymine (T) occurs in the existing strand, an adenine (A) is added to the new strand.

Synthesis of the new matching strand occurs continuously on only one of the original strands. This is called the leading strand (Figure 8.5). The other original strand is called the lagging strand. On the leading strand, DNA

Try Investigation 8A ▽
DNA Replication.

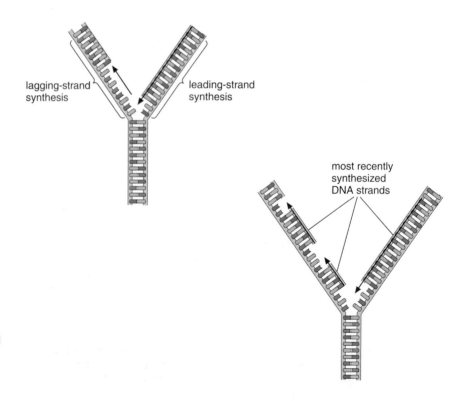

FIGURE 8.5

Synthesis of leading and lagging DNA strands. Synthesis of the leading strand is continuous and fast. Synthesis of the lagging strand occurs in short segments in the direction opposite to movement of the replisome.

lagging-strand synthesis

leading-strand synthesis

most recently synthesized DNA strands

E T Y M O L O G Y

replica = copy (Latin)
-some = body (Greek)

A **replisome** is a "body" that makes copies of DNA.

polymerase adds DNA nucleotides to the end of the RNA primer. As the replisome moves away from the replication origin along the DNA, it unwinds more of the double helix. At the same time, DNA polymerase extends the matching strand along the leading strand. Later, another enzyme replaces the small RNA primer with DNA.

Why is DNA replication discontinuous along the lagging strand? Recall that the double helix is antiparallel. DNA polymerase can extend a primer only in one direction. On the lagging strand, this is opposite to the direction in which the replisome moves. As a result, DNA synthesis on the lagging strand occurs in short, unconnected segments. Other enzymes follow the replisome, filling in the gaps on the lagging strand and joining the short segments.

Replisomes move away from the replication origin along the DNA in both directions (Figure 8.6). The result of the entire process is to replace the old DNA double helix with two identical ones. Each new double helix contains one strand of old DNA and one strand of new DNA. This type of replication is known as semiconservative (half conservative) replication because each of the two new double-stranded DNA molecules conserves one strand (half) of the original DNA, but adds one strand of new DNA. Once replication is complete, there are two new chromosomes in place of the original one.

The DNA of living cells is surrounded by different kinds of proteins that have many functions. Some of these proteins are involved in wrapping the DNA into the tightly condensed structure called the chromosome. The familiar shape of the chromosome shown in Figure 8.7 is visible only at mitosis. During interphase, DNA and proteins are loosely arranged in stringy masses throughout the nucleus.

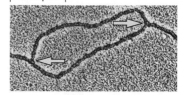

FIGURE 8.6

DNA synthesis. An electron micrograph of DNA replication in progress, ×100,000, is shown here. Replisomes are moving out from each replication origin in both directions, indicated by the arrows. This speeds up DNA replication.

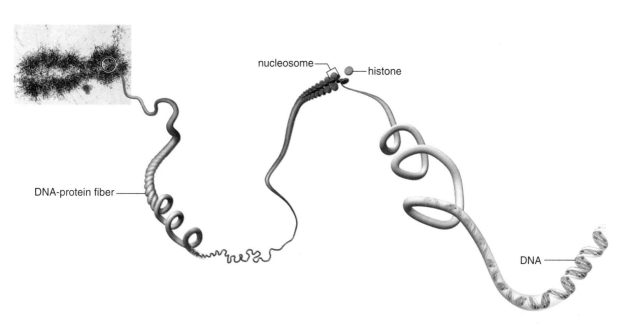

FIGURE 8.7

The structure of eukaryotic chromosomes. A DNA molecule wraps around histone proteins to form nucleosomes, the basic packing unit of eukaryotic chromosomes. The coiled, beaded chain of DNA with its nucleosomes forms still thicker coils that make up the chromosome.

Chromosome Structure

Each eukaryotic chromosome consists of a single DNA molecule that can contain as many as 300 million base pairs. If you could stretch out the DNA of such a chromosome, it would be almost 10 cm long. How can all that DNA fit in the nucleus without getting tangled? Moreover, because DNA has a negative electric charge, it repels itself. How can cells squeeze all that negatively charged DNA into a tiny nucleus?

The solutions to both of these problems lie in a DNA-protein complex called the **nucleosome.** Early observations of the DNA in chromosomes through electron microscopes revealed that, when the DNA was stretched out, it looked like a string with regularly spaced beads. These beads were later found to be nucleosomes. The core of each nucleosome is a disk made up of eight protein molecules—two each of four proteins called histones (Figure 8.8). Histone proteins are basic, having strong positive charges that balance the negative charges of DNA. As new DNA is synthesized during S phase, it quickly binds to histone proteins, forming nucleosomes. The DNA helix wraps twice around the edge of each protein disk. At each point where the histone proteins contact the DNA wrapped around the nucleosome, the amino acid arginine occurs in the histone protein molecules. Positively charged amino groups ($-NH_3^+$ or $=NH_2^+$) in the R group of this amino acid balance the negative charges of phosphate groups on the surface of DNA. DNA bound to nucleosomes is called **chromatin.** Eukaryotic DNA exists in this form most of the time.

Long strings of nucleosomes form thicker coils that pack together to make up the chromosomes that are visible through a scanning electron microscope (see Figure 8.7). This tightly packed structure "turns off" DNA by excluding the enzymes involved in gene expression. Enzymes can modify the histones bound to specific genes, loosening the chromatin structure and activating the DNA. In this way, different cell types in the same organism can activate specific groups of genes while keeping other genes silent.

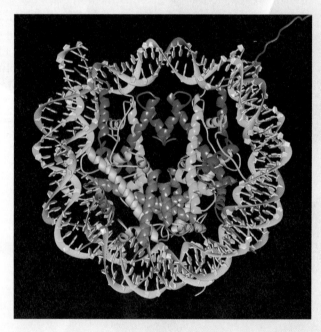

FIGURE 8.8

Structure of the nucleosome. Two molecules of each of the four core histones form a complex with two wraps of the DNA helix. Other histones bind to the DNA between nucleosomes. The four types of histone proteins are shown in different colors. Six of the eight histone molecules and one of the two loops of DNA are visible here. Each thick cylinder represents a protein secondary structure called an alpha helix (see Section 1.9, Figure 1.23). Each histone protein has three alpha-helix sections. The DNA molecule contacts each histone molecule at six points, once per turn of the DNA helix.

8.5 DNA Repair

The process of DNA replication must have as few errors as possible. The new DNA strands must be exact complements of the parental strands. Any change in the sequence of a cell's DNA is known as a **mutation.** Mutations can be nonharmful (silent), harmful, or even lethal to the cell. Mutations that persist to the next cell division are inherited by the daughter cells. This is important because many human diseases are caused by mutations. For example, mutations play a major role in the development of cancers (see Focus On Cancer, Section 8.9).

Cells have evolved intricate processes to detect and correct errors in replication as well as damage to DNA by environmental factors such as ultraviolet radiation. The first line of defense is the DNA polymerase that produces the DNA itself. About one in every 10,000 bases that this enzyme adds to DNA is incorrect. When DNA synthesis is complete, however, there is only about one new mutation per 10 million base pairs. The reason for this difference is that DNA polymerase proofreads its own work. After adding a new nucleotide to the growing chain, it checks to see if the resulting base pair is correct. If not, the enzyme pauses, removes the nucleotide, and replaces it with the correct one. The vast majority of replication errors are repaired in this way.

Although many cells in adult organisms are not actively dividing, they still need accurate instructions on how to function. Therefore, they need to detect and repair mutations introduced during replication or caused by environmental factors, such as **mutagenic** chemicals or radiation. Most mutations are known as mismatches because they consist of base pairs that cannot form hydrogen bonds (for example, adenine and cytosine). The process by which these mutations are repaired is known as **excision repair** (Figure 8.9). First, an enzyme recognizes the mismatch and binds to the

E T Y M O L O G Y

mutat- = to change (Latin)
gen- = creating or producing (Latin)

A genetic **mutation** is a change in DNA. **Mutagens** are chemicals or forms of radiation that cause mutations.

E T Y M O L O G Y

ex- = out (Latin)
cise- = cut (Latin)

Excision-repair enzymes cut out and replace mismatched nucleotides.

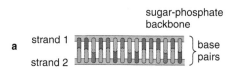

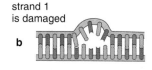

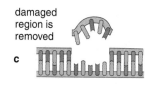

a strand 1 / strand 2 — sugar-phosphate backbone / base pairs

b strand 1 is damaged

c damaged region is removed

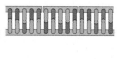

d DNA polymerase makes a new section of strand 1 that base-pairs correctly with undamaged strand 2

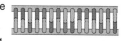

e another enzyme seals the backbone to restore strand 1

FIGURE 8.9

Excision repair. Normal base pairing is shown in **(a).** Mismatched bases in DNA cannot form base pairs **(b).** Enzymes remove part of one strand from the mismatched section **(c),** synthesize a replacement that matches the intact strand **(d),** and bond the new nucleotides to the rest of the strand **(e).**

DNA. It then breaks the sugar-phosphate bonds of the mismatched section and removes the damaged or mutant DNA. Next, DNA polymerase fills in the deleted DNA sequence. Finally, another enzyme forms sugar-phosphate bonds between the replacement piece and its neighboring nucleotides.

These mechanisms are only two of the many ways that cells repair mutations in their DNA. Without these mechanisms, cells would rapidly accumulate harmful or lethal mutations and cease to function properly.

Check and Challenge

1. What are the three different roles of DNA polymerase?
2. Explain how DNA synthesis can proceed in both directions from a replication origin, even though DNA polymerase can synthesize DNA only in one direction.
3. How is the number of replication errors in cells kept to a minimum?
4. How does the cell repair damaged DNA?
5. How are histones involved in gene expression?

Mitosis and Cell Division

8.6 The Stages of Cell Division

When DNA replication is complete, the cell passes from the S phase to the G₂ phase. The two copies of each chromosome made during the S phase are called **sister chromatids.** They are now ready to be separated and delivered to each new nucleus. As a cell enters the M phase, sister chromatids are still attached by proteins at a narrow point called the **centromere,** usually near the center of the chromosome (Figure 8.10).

FIGURE 8.10

Sister chromatids. A chromosome as it appears early in mitosis is shown here. **(a),** This scanning electron micrograph, ×87,000, shows a replicated chromosome in metaphase with its pair of sister chromatids joined at their centromere. **(b),** A drawing of **(a)** illustrates the chromosome's structures.

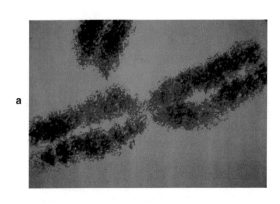

a

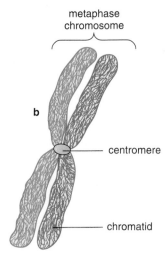

b

metaphase chromosome

centromere

chromatid

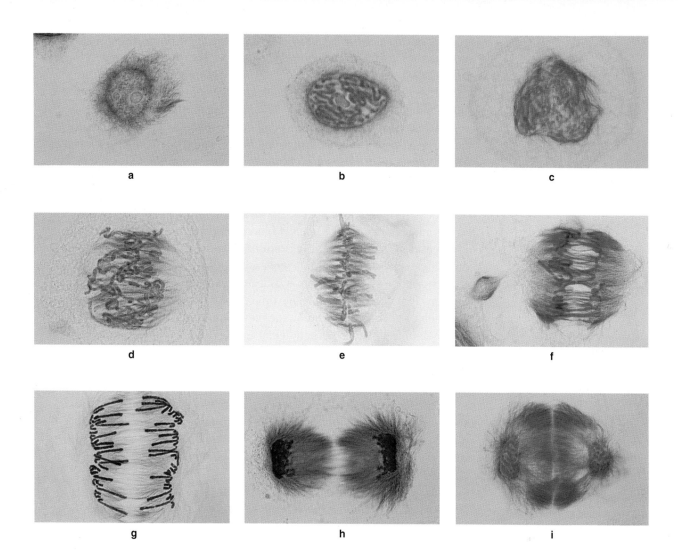

FIGURE 8.11

Interphase and stages of mitosis in the blood lily *Haemanthus*. Chromosomes are stained blue, and microtubules are stained red. Individual chromosomes are not visible during interphase **(a)**. During prophase **(b, c)**, the nuclear membrane breaks down, chromosomes become visible, and microtubules begin to assemble into a spindle. Approaching metaphase **(d)**, motor proteins in the kinetochores pull the chromosomes to the metaphase plate between the spindle poles **(e)**. Sister chromatids are pulled to opposite poles during anaphase **(f, g)**. Finally, a nuclear membrane forms around the new nucleus at each pole during telophase **(h, i)**, and cytokinesis separates the two daughter cells.

Separation of sister chromatids is called **chromosome segregation.** If segregation occurs correctly, each new nucleus receives one copy of each chromosome. A mistake at this stage will result in one nucleus with two copies of a particular chromosome and the other nucleus with none. Such daughter cells with abnormal numbers of chromosomes are called **aneuploid** cells.

The process of mitosis, once begun, is continuous (Figure 8.11) but is considered to have four distinct steps. Following a period of interphase (Figure 8.11a), the first step, called prophase, begins when the nuclear

ETYMOLOGY

se- = separate (Latin)
gregar- = group (Latin)
an- = not (Greek, Latin)
eu- = good (Greek)
-oid = form (Greek)

Segregation is separation from a group. **Aneuploid** cells do not have the normal number of chromosomes.

membrane breaks down into small vesicles. The chromosomes condense and become visible under a light microscope. Microtubules begin to form around the nucleus (Figure 8.11b) and join to form a **mitotic spindle** (Figure 8.11c and d). If the cell has centrioles, microtubules form between them and push them to opposite ends of the cell. (Centrioles are duplicated during interphase. Note that the plant cells shown in Figure 8.11 do not have centrioles; they have other structures to perform the function of centrioles.) The microtubules at the ends of the spindle are anchored to protein structures that surround the centrioles at each end of the cell. These sites, called the **spindle poles,** form around the centrioles, if they are present. Within each centromere is a protein complex called the **kinetochore.** Some of the microtubules in the spindle bind to the kinetochores of each chromatid so that a chain of microtubules connects each chromatid to a spindle pole. Sister chromatids move to opposite poles.

The second step of mitosis is called **metaphase.** By this time, motor proteins in the kinetochores have pulled the chromosomes into a ring between the two poles, forming the **metaphase plate** (Figure 8.11d and e). The metaphase plate is perpendicular to the spindle. By keeping the sister chromatids attached at their centromeres, in much the same way that you might pair up socks, the cell ensures that each daughter nucleus gets one copy of each chromosome.

In the third step, **anaphase,** enzymes break down the protein holding sister chromatids together. The sisters separate, and the motor proteins of their kinetochores pull them along the spindle microtubules to opposite spindle poles (Figure 8.11f and g). At this point, each segregated chromatid is called a chromosome.

Finally, the cells enter **telophase** (Figure 8.11h and i). The chromosomes begin to expand, and the nuclear envelope re-forms around them, producing two new nuclei. Soon after this, cytokinesis divides the cell in two, as the plasma membrane constricts between the nuclei and completes cell division. In cells that have cell walls, vesicles containing new cell-wall material fuse across the cell plate to complete the plasma membranes and cell walls of the two new cells. Both daughter cells now enter G_1. Each cell may enter G_0 or pass the restriction point and start another cycle.

8.7 Differences in Mitosis

Although the major events and molecular players of cell division are similar in all eukaryotic cells, there are some subtle differences. For example, cytokinesis begins during anaphase in most animal cells. The greatest difference between cell division in animal and plant cells reflects the fact that plant cells have cell walls. At cytokinesis in plants, vesicles containing cellulose begin to congregate between the two nuclei. These vesicles then begin to fuse, forming the plasma membranes of the two new daughter cells. The contents of the vesicles complete the cell wall between the two new cells. In some fungi, such as yeast, the nuclear envelope forms a bud instead of breaking down, and the spindle poles are embedded in the nuclear membrane.

Focus On

The Evolution of Mitosis

Prokaryotes demonstrate a simple type of cell division in which the DNA attaches to the plasma membrane. Cell division in some primitive unicellular eukaryotes is similar, but their chromosomes attach to the nuclear envelope instead of to the plasma membrane (Figure 8.12a). The spindle forms in the cytoplasm; however, as in yeast, the nuclear membrane never breaks down. Rather, it helps stabilize the spindle. In other unicellular eukaryotes, parallel bundles of microtubules penetrate the nucleus (Figure 8.12b). The ends of these bundles are not gathered at spindle poles.

In more complex eukaryotes, the nuclear envelope breaks down during prophase and re-forms only after chromosome segregation.

One hypothesis for the evolution of mitosis suggests that microtubules originally helped support the nuclear envelope. Over time, the microtubules became attached to the chromosomes themselves. As microtubules assumed a more active role in separating the chromosomes, the nuclear envelope became less important in mitosis.

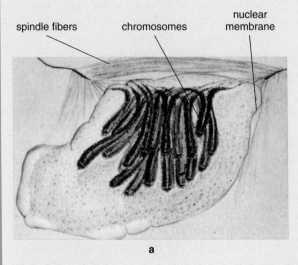

spindle fibers   chromosomes   nuclear membrane

a

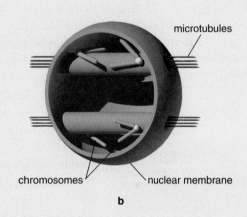

microtubules

chromosomes   nuclear membrane

b

FIGURE 8.12

Possible clues to the evolution of mitosis. (a), During mitosis in the unicellular eukaryote *Trichonympha grandis*, the spindle forms outside the nucleus, and the chromosomes attach to the intact nuclear membrane. **(b),** In other unicellular eukaryotes such as this dinoflagellate, the chromosomes also attach to the nuclear membrane. Microtubules do not form the typical tapered spindle shown in Figure 8.11f. Instead, they form parallel bundles that pierce the nuclear membrane.

Check and Challenge

1. What is the difference between mitosis and mitotic cell division?
2. What is the difference between chromosomes and sister chromatids in the M phase?
3. How do the chromosomes move during mitosis?
4. Briefly describe the major events of mitosis.

Regulation of the Cell Cycle

8.8 Control of the Cell Cycle

Once scientists had identified the major events of the cell cycle, many new questions arose. Why don't cells keep duplicating their chromosomes until the sister chromatids have segregated? What prevents mitosis from starting before the cell has completed S phase? If these steps were not performed in the correct order and at the right times, cells would begin to segregate DNA that was still replicating, or they would make huge numbers of copies of their chromosomes. The controls that regulate the order and timing of cell-cycle events are of major interest to scientists who study the eukaryotic cell cycle.

Early cell-fusion experiments showed that something in S-phase and M-phase cells could cause G_1 and G_2 nuclei to advance to the next phase (S or M). For example, when S-phase cells were fused with G_1 cells, the DNA in the G_1 nucleus began to replicate. When G_2 cells were fused to S-phase cells, however, no DNA synthesis occurred in the G_2 nuclei. These observations indicate that a factor or factors in S-phase cells can enter G_1 nuclei and initiate DNA replication. In the same way, M-phase cells can move G_2 nuclei into mitosis. Research on cells from organisms as diverse as yeast, clams, and frogs revealed an elegant mechanism that coordinates the timing of events in the cell cycle. It is interesting that, although yeasts are simple eukaryotes, they use the same basic mechanism for cell-cycle control that human cells use.

When cells leave G_0 and commit to a round of the cell cycle, different proteins within the cell begin to accumulate and then rapidly disappear as the cycle progresses. These proteins are called **cyclins** because they regulate progression through the cell cycle. The most important cyclins are the G_1 cyclins and the mitotic cyclins. G_1 cyclins begin to accumulate in late G_1 (Figure 8.13). They reach a peak during S phase. After DNA synthesis is complete, the mitotic cyclins accumulate until they peak at metaphase. At this point, the mitotic cyclins quickly disappear, and mitosis proceeds to completion. Cyclins act by binding to various kinases, which are enzymes that transfer a phosphate group from ATP to other enzymes. The phosphate group activates these enzymes. (Kinases are explained in Focus On Respiration during Exercise, Section 5.7.) The quantity of these kinases in the cell remains steady throughout the cycle. However, they are active only when bound to the appropriate cyclin. As the amount of a particular cyclin

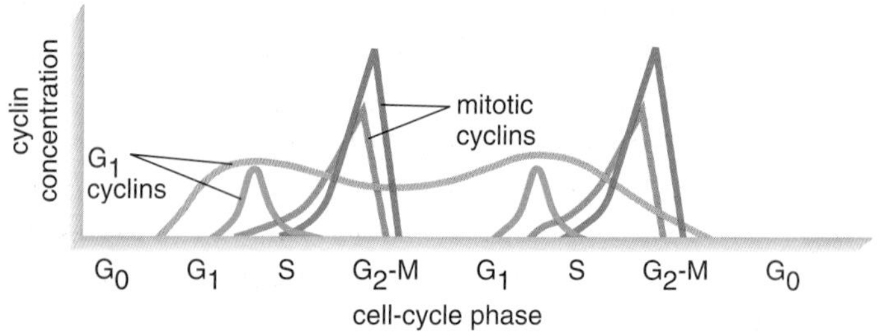

CONNECTIONS

The similarity of the cell cycle in a wide variety of diverse eukaryotes reflects their shared evolutionary origin.

CHEMISTRY TIP

Addition of a phosphate group to a protein causes a change in secondary structure that can either activate or inactivate the protein. During prophase, cyclin-dependent kinases add phosphate to proteins that form a network that supports the nuclear membrane. This chemical change causes the protein network to come apart, leading to breakup of the nuclear membrane.

FIGURE 8.13

Rise and fall of cyclin levels during the cell cycle. G_1 cyclins peak at S phase, and mitotic cyclins peak at metaphase in M phase.

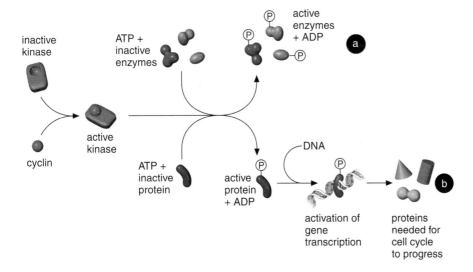

FIGURE 8.14

Regulation of progress through the cell cycle by cyclins. The abundance of different cyclins varies during the cell cycle. Each cyclin activates specific kinases. The kinases activate some enzymes directly at **(a)** and signal the cell to synthesize other proteins needed to progress to the next phase of the cycle at **(b)**.

rises, it activates more kinases. The kinases in turn activate various enzymes needed for progress through the cell cycle (Figure 8.14).

For example, during G_2, as mitotic cyclins begin to accumulate, they stimulate a few kinases. These kinases activate the pathway that leads to breakdown of the nuclear membrane and condensation of the chromosomes. As more cyclin accumulates, it activates other kinases that turn on the pathway to mitotic spindle formation, and so on. The gradual buildup of mitotic cyclins ensures that the events of mitosis will occur in the proper order and at the right times. The last pathway that the mitotic cyclins activate is one that breaks down specific proteins. Enzymes of this pathway break down the protein that holds sister chromatids together, allowing them to separate at anaphase. They also break down the mitotic cyclins themselves. The loss of cyclins inactivates the kinases, ensuring that the cell will pass from M phase to S phase. Mitotic cyclins do not reappear until the next G_2 phase.

8.9 Checkpoints

If something goes wrong during the cell cycle, what stops the cycle so that repairs can be made? Many things can go wrong during the cell cycle. For example, ultraviolet radiation may damage DNA. A pair of sister chromatids may fail to attach properly to the mitotic spindle. As cells multiply during the growth of multicellular organisms, uncorrected problems can cause large numbers of cells to die or malfunction. The reason this does not commonly occur is that eukaryotic cells have an elaborate system called checkpoint control that monitors the condition of the DNA, the chromosomes, and the mitotic spindle.

CHEMISTRY TIP

When a protein such as a cyclin or the centromere protein that holds sister chromatids together is to be broken down, an enzyme attaches a peptide called ubiquitin to it. This tag signals protein-digesting enzymes to attack the protein.

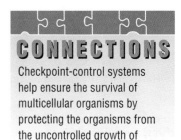

CONNECTIONS

Checkpoint-control systems help ensure the survival of multicellular organisms by protecting the organisms from the uncontrolled growth of damaged cells.

Focus On

Cancer

When checkpoint controls over cell division are damaged, a cell repeatedly divides, forming a mass of cells called a tumor. As the tumor grows, it may interfere with surrounding tissue or cells may break off and spread around the body in a process called **metastasis.** How do these cancers develop?

Growth signals from outside activate two pathways in a G_0 cell. One pathway promotes the start of the cell cycle; the other pathway inactivates checkpoint control. The cell then reenters the cell cycle. Genes that are involved in promoting cell division are called **protooncogenes.** Those that help inhibit cell division are called **tumor suppressors.** In cancer cells, both pathways are disrupted (Figure 8.15).

Signals transmitted from the cell surface to the nucleus promote growth and cell division. Because many proto-oncogenes are involved in these signal pathways, mutations in proto-oncogenes can convert them to oncogenes, or cancer genes, which stimulate cells to leave G_0 and divide whether or not there is a signal. For example, the normal Ras protein relays a growth signal from the plasma membrane to the nucleus. When there is no signal, *ras* is silent. However, certain mutations in the *ras* gene lead to proteins that are constantly active, even when there is no outside signal. Cells with these mutations are always preparing to leave G_0 and enter the cell cycle. Fortunately, these cells do not divide because no outside signal has turned off the checkpoint system.

The genes that encode the checkpoint proteins are called tumor suppressors because they suppress the development of cells into tumors. But if mutations inactivate these genes, the cell-cycle brake is removed with or without a signal from the outside. Many cancer cells have mutations in their tumor-suppressor genes.

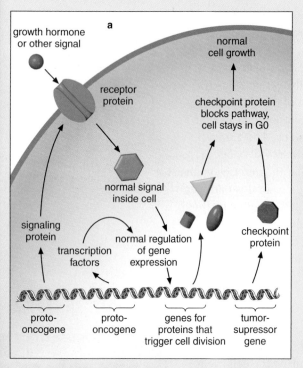

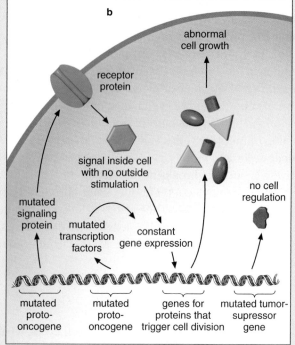

FIGURE 8.15

How a cell turns cancerous. Uncontrolled growth and cell division occur when both proto-oncogenes and tumor-suppressor genes are damaged by mutation. **(a),** Under conditions of normal cell growth, signals for cell division from both the cell's environment and proto-oncogenes enter the nucleus and activate expression of genes involved in cell division. Tumor-suppressor genes produce proteins of the checkpoint system that help keep cells in G_0. **(b),** Mutant proto-oncogenes become oncogenes that stimulate cell division whether or not they receive a signal from outside the cell. Mutant tumor-suppressor genes fail to block the progress of the cell cycle. Failure of these controls can result in development of a cancerous tumor.

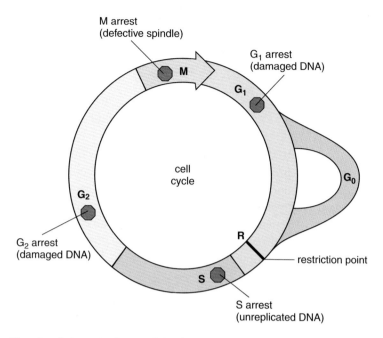

cell cycle

M arrest
(defective spindle)

M

G$_1$ arrest
(damaged DNA)

G$_1$

G$_0$

G$_2$

G$_2$ arrest
(damaged DNA)

R

restriction point

S

S arrest
(unreplicated DNA)

FIGURE 8.16

Cell-cycle checkpoints. Eukaryotic cells have checkpoint controls at several stages in the cell cycle. When these systems detect a problem, they arrest the cell cycle until the damage or mistake is repaired. This prevents production of daughter cells with genetic damage.

Checkpoint controls consist of proteins that detect mistakes and damage and quickly halt the cell cycle until repairs are made. When this occurs, the cell is said to be in **cell-cycle arrest.** For example, one human protein called p53 can detect DNA damage such as mismatched DNA base pairs (see Figure 8.9). When p53 detects a mismatch, it activates cell-cycle inhibitors that prevent the G$_1$ cyclin-kinase system from bringing the cell into S phase. Once the damage is repaired, p53 becomes inactive, and DNA replication can proceed. Yeast has a checkpoint system that monitors the condition of the mitotic spindle. If the spindle is damaged, checkpoint proteins inhibit the mitotic cyclin-kinase system and halt mitosis. Checkpoints throughout the cell cycle ensure that problems are corrected before the cycle progresses (Figure 8.16). Without checkpoint controls, mitosis could produce daughter cells with damaged or missing chromosomes.

Many other cell-cycle regulators are also involved in preventing cells from leaving the G$_0$ stage. When these regulators are inactivated, cells may divide at the wrong time. Therefore, mutations in the genes that encode these proteins can lead to uncontrolled growth. Research on these genes is critically important in the detection and prevention of the uncontrolled cell growth and reproduction known as **cancer** (Figure 8.17).

ETYMOLOGY

proto- = primitive or early form (Greek)
onco- = cancer (Greek)
meta- = change (Greek)
stasis = standing (Greek)

Proto-oncogenes can become oncogenes, or cancer-causing genes, if they are damaged by mutation. **Metastasis** involves a change in the state of cancer cells from a growing phase to one in which they spread throughout an organism and invade other tissues.

Check and Challenge

1. Explain why neither cyclins nor kinases alone can cause a cell to progress through the cell cycle.
2. How do controls on the cell cycle protect multicellular organisms from accumulating large numbers of damaged or defective cells?
3. What is the difference between a cancerous tumor and metastasis?
4. What are the functions of tumor-suppressor genes and proto-oncogenes in noncancerous cells?

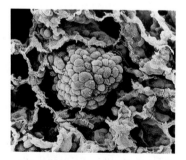

FIGURE 8.17

Lung cancer. This scanning electron micrograph shows a cancerous tumor, colored orange, in the alveoli (air sacs) of the human lung, ×200.

Chapter
HIGHLIGHTS

Summary

The eukaryotic cell cycle is a series of events leading to the formation of two offspring cells from a parent cell. There are five phases of the cell cycle: G_0, G_1, S, G_2, and M. Interphase consists of G_1, S, and G_2. From G_1, cells can also enter G_0, a nondividing phase. DNA synthesis occurs in S phase and begins at replication origins. Replisomes catalyze the synthesis of two new strands of DNA that are complements of the two parental strands. The leading strand is synthesized in a continuous process. The lagging strand replicates in short stretches of DNA that are then joined. Eukaryotic DNA replication is semiconservative; the resulting chromosomes each contain an old strand of DNA paired with a newly synthesized strand. Errors in replication are repaired by the DNA polymerase itself, while damaged segments may be repaired by the excision-repair system.

The newly replicated sister chromatids are segregated to the daughter nuclei during mitosis. In mitosis, arrays of microtubules form a mitotic spindle. Microtubules emanating from the spindle poles link to others that attach to the chromosomes at their centromeres. Motor proteins in the kinetochores then transport the chromosomes toward the metaphase plate. At anaphase, the sister chromatids separate and migrate to the spindle poles. The nuclear envelope re-forms around the new daughter nuclei in telophase. Cytokinesis usually follows mitosis.

Cells have elaborate systems that ensure the accurate replication and segregation of the chromosomes to the two new daughter nuclei. The timing and sequence of events are linked to the synthesis and disappearance of various cyclins throughout the cell cycle. Cyclins activate kinases. The kinases in turn activate enzymes needed for the cell cycle to progress. Eventually the cyclins signal their own destruction, allowing the cell to continue to the next phase. Errors that occur during this process are closely monitored by checkpoint control. Failure to complete each process of the cell cycle accurately leads to cell-cycle arrest until repairs can be made. Failure of the checkpoint-control system is one important step in the development of cancer.

Key Concepts

Use the following terms to construct a concept map of the cell cycle with appropriate linking words. Use additional terms from this chapter as needed.

- cell cycle
- mitosis
- metaphase plate
- DNA polymerase
- cyclin
- telophase
- G_1
- G_0
- kinetochore
- M
- checkpoints
- microtubule

Reviewing Ideas

1. What determines whether a cell will divide?
2. What is the biological importance of S phase of the cell cycle? of M phase?
3. Describe how errors are corrected in DNA.
4. What is the role of the spindle in mitosis?
5. There is an enzyme that forms sugar-phosphate bonds between neighboring sections of DNA. How is this enzyme essential to the replication of DNA?
6. How does DNA replication differ in prokaryotes and eukaryotes?
7. What two pathways are disrupted in cancer cells?
8. What are replication origins? How do they contribute to DNA replication?
9. When errors occur in DNA, what activity do checkpoint controls inhibit?
10. What is the role of the centromere in mitosis?

 BSCSblue.com/vocabulary_puzzlemaker

Using Concepts

1. Suppose a fellow student asked you to look through a microscope and observe the chromosomes in the nucleus of a leaf cell. Would you expect to see them? Give two reasons to support your answer.

2. Cells can be grown in a laboratory by a process called tissue culture. Imagine that you have two tissue cultures, one in G_1 phase and one in G_2 phase. Unfortunately, the cultures' labels have been misplaced. How would you determine which culture was in G_1 and which was in G_2?

3. Suppose you have yeast cells in a laboratory culture. To determine the rate of DNA synthesis in the cells, you plan to provide them with a radioactive nucleotide. The rate at which the cells incorporate the radioactive nucleotide into DNA will reveal the rate of DNA synthesis. Which radioactive nucleotide—A, T, G, C, or U—should you use? Explain your answer.

4. Based on the information in this chapter, can you predict what would happen if there were no DNA proofreading system in eukaryotic cells? Explain your answer.

5. What signal might be considered a growth-stimulating signal for a yeast cell?

6. How can you explain the fact that yeast cells need only a few cyclins, whereas animal cells have many?

7. How does the synthesis of new DNA differ from that of other molecules in the cell?

8. If a mutation arose in a G_1 cyclin, causing it to bind a kinase constantly, what would happen to the cell? Explain your answer.

Synthesis

1. Suppose you have animal cells growing in a laboratory tissue culture. Glucose containing radioactive carbon atoms is provided to the cells, and the radioactive carbon ends up in the DNA of the chromosomes. Trace the steps of how this happened.

2. What features of the structure of DNA help explain replication?

Extensions

1. Compare the cell cycle with a computer. What are their similarities and differences?

2. Do some research on the growth of cancer cells. Write a short report describing how their cell cycle differs from that of normal cells.

Web Resources

Visit BSCSblue.com to access
- Information about the cell cycle and cell division
- An explanation of how cyclins regulate the cell cycle
- An animated simulation of cell division in yeast
- A self-directed lesson on DNA and the cell cycle
- A summary of the stages of mitosis, with excellent detailed photographs of plant and animal cells in each stage
- Web links related to this chapter

CONTENTS

LEARNING OUTCOMES

By the end of this chapter you will be able to:

A Explain the connection between DNA and RNA in protein synthesis; describe the genetic code and its role in protein synthesis.

B Explain why proteins are important to biological systems.

C Identify the stages of transcription and explain what occurs during each stage.

D Summarize the events that occur in RNA processing.

E Identify the stages of translation and explain what occurs during each stage.

F Describe posttranslational modification and transport of proteins.

G Infer the consequences of RNA translation errors.

H Explain the relationship between viruses and host cells and describe the impact of viruses on living systems.

CHAPTER

9

Expressing Genetic Information

■ *How does an organism use the information stored in its genetic material?*

■ *Does a cell express all of its genetic information all the time?*

Living organisms store information in their genetic material. For an organism to survive, it must be able to read the encoded information and use it. This process is called gene expression. This chapter focuses on how organisms use the information stored in their genetic material.

Genes act by directing the synthesis of proteins. Some proteins are essential to cells' survival. Synthesis of these proteins is tightly controlled so that they are present when they are needed and not at other times. Cells synthesize other proteins in response to changing environmental conditions. Specific types of cells in multicellular organisms synthesize still other proteins. Your skin cells, for example, make some proteins that your muscle cells do not. Protein synthesis is also affected by viruses. When a virus infects a cell, the virus takes control of gene expression in the cell. In this chapter, you will learn how genes direct the synthesis of proteins, and how this process is regulated to ensure survival and normal development.

The Genetic Code: Using Information

9.1 Genetic Material

Genetic material consists of two types of biomolecules—DNA and RNA. Both are nucleic acids. Both are involved in gene expression, so both must store genetic information. This important function depends on two features of their molecular structure. First, nucleic acids consist of a long strand of repeating subunits that act as letters in a code. A second important feature is that the subunit bases of one strand pair with the bases of another strand. This base pairing is essential for sending genetic messages. One strand acts as a pattern or template for a new molecule. The new molecule is built according to the plan stored in the original strand. In Chapter 8, you saw how this complementary base pairing helps cells copy DNA during DNA replication. Base pairing also plays a role in expressing the genetic information stored in the master program of DNA.

How does DNA serve as a master program for organisms? Living cells store genetic information in DNA. For example, DNA specifies the primary structures of proteins. Recall from Chapter 1 that the primary structure of a protein—its sequence of amino acids—determines the protein's tertiary (three-dimensional) structure. By determining the primary structure of each protein, DNA indirectly dictates protein function. Proteins, in turn, carry out important cell activities. The genetic program of a cell or an organism thus goes through several molecular steps to direct cell activities.

DNA does not control protein synthesis directly, however. When a gene becomes active, an enzyme makes a temporary RNA copy of the information it contains. The process is analogous to making a disk copy of a software program. Figure 9.1 summarizes this flow of information.

RNA plays several roles in the expression of genetic information. **Messenger RNA** (mRNA) is the temporary copy of a gene that encodes a protein. The process of making an mRNA molecule is called **transcription.** The mRNA molecule provides the pattern that determines the order in which amino acids are added to the protein being made, a process known as **translation.** Protein synthesis takes place on ribosomes. The ribosomes themselves are made of proteins and another type of RNA, called **ribosomal RNA** (rRNA). Cells contain many ribosomes. About 80% of the RNA in a cell is rRNA. Each amino acid that will be used in making the protein is attached to a third type of RNA, **transfer RNA** (tRNA). In addition to these three types of RNA, eukaryotic cells have a variety of small nuclear RNA molecules that interact with specific proteins during RNA processing. Some RNA molecules act as catalysts, just as enzymes made of protein do. Many viruses store genetic information as RNA. This special situation is discussed in Section 9.8.

One of the most exciting chapters in the history of biological investigation was the cracking of the **genetic code** that describes how a

---

CHEMISTRY TIP

Recall that the subunits of a nucleic-acid strand are strongly connected through covalent bonds. In contrast, the pairing between bases of two strands depends on weaker hydrogen bonds. In the laboratory, many techniques for studying or cloning genetic material require separating the strands of paired DNA. Separation is carried out by raising the pH. When the pH is neutralized, the strands can pair once again.

ETYMOLOGY

trans- = across (Latin)
-scribe or **-script** = writing (Latin)
-late = side (Latin)

Transcription is a copying of information from DNA to RNA.
Translation is a conversion of information from one form (in this case, a nucleic acid sequence) to another (an amino acid sequence).

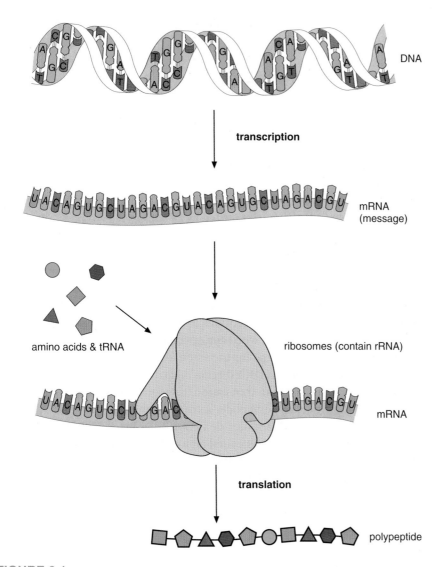

DNA

transcription

mRNA
(message)

amino acids & tRNA

ribosomes (contain rRNA)

mRNA

translation

polypeptide

FIGURE 9.1

The flow of information in the expression of genetic information. Information stored in DNA is copied to mRNA, which in turn directs the synthesis of a particular protein.

sequence of bases in DNA or RNA translates into the sequence of amino acids in a protein. A code is a system of symbols used to store information (Figure 9.2). Written language is a type of code developed by humans. For example, English contains 26 symbols, the letters of its alphabet. With these 26 letters, an unlimited number of words can be formed. The words, in turn, can store information. Yet, to people who do not know the code system, the stored information remains a mystery.

What are the symbols of the DNA code? The nucleotides serve as the four "letters" of the DNA "alphabet." The small size of this alphabet poses a problem. Proteins are large molecules built from smaller units, the amino acids. A genetic code requires at least 20 different code words—one for each amino acid.

FIGURE 9.2

Using a code. A language translator uses codes and symbols to translate information from one form into another.

What is the simplest way to combine DNA nucleotides to code 20 different words? If one type of nucleotide were to code a single amino acid, then only four amino acids could be coded with four nucleotides. Even two letters in a particular order would provide only 16 combinations, such as AA, AT, AC, AG, TA, TT, and so forth. When three nucleotides are grouped at a time, however, 64 triplet combinations, such as CTG, TAC, and ACA, are possible. In fact, using three nucleotides at a time provides extra code words. More than one genetic word can specify the same amino acid, just as in English we use *big* and *large* for the same idea.

Each nucleotide triplet in DNA directs a particular triplet to be formed in mRNA during transcription. In translation, a second base-pairing step is essential for reading the genetic code. A triplet in mRNA, called a **codon,** pairs with a triplet on a tRNA molecule carrying the correct amino acid (Figure 9.3). This tRNA triplet is called an **anticodon.** In this way, a codon in mRNA specifies a particular amino acid in the newly forming protein. Figure 9.4 shows codons for all the amino acids. Look for the "punctuation" codons that mean start and stop. Studies of organisms in all five kingdoms indicate that the genetic code is nearly universal.

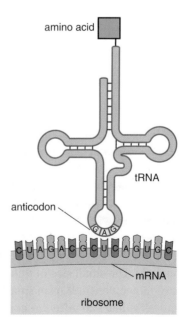

FIGURE 9.3

Translating the genetic code into protein. A molecule of transfer RNA (tRNA) with a specific amino acid attached reads each codon of a messenger RNA (mRNA) during protein synthesis (translation). Which amino acid is attached to this tRNA? (Hint: Use Figure 9.4 to find out.)

First Base	Second Base				Third Base
	U	**C**	**A**	**G**	
U	phenylanine	serine	tyrosine	cysteine	**U**
	phenylanine	serine	tyrosine	cysteine	**C**
	leucine	serine	stop	stop	**A**
	leucine	serine	stop	tryptophan	**G**
C	leucine	proline	histidine	arginine	**U**
	leucine	proline	histidine	arginine	**C**
	leucine	proline	glutamine	arginine	**A**
	leucine	proline	glutamine	arginine	**G**
A	isoleucine	threonine	asparagine	serine	**U**
	isoleucine	threonine	asparagine	serine	**C**
	isoleucine	threonine	lysine	arginine	**A**
	(start) methionine	threonine	lysine	arginine	**G**
G	valine	alanine	aspartate	glycine	**U**
	valine	alanine	aspartate	glycine	**C**
	valine	alanine	glutamate	glycine	**A**
	valine	alanine	glutamate	glycine	**G**

CONNECTIONS
The almost universal nature of the genetic code is strong evidence for the evolution of all living organisms from a common ancestral life-form.

FIGURE 9.4

The genetic code. The code is written in nucleotide triplets, or codons, in a strand of mRNA. Each triplet codon specifies an amino acid. For example, UGG codes for the amino acid tryptophan. Several amino acids have more than one codon. Some triplets are "punctuation" telling the system to start or stop translation.

9.2 Importance of Proteins

Why should protein synthesis be the way that genetic information is expressed? Proteins play a very important part in the function of any living system. Many proteins serve as the material that makes up cell structures or tissues. A protein known as keratin is the main structural material in skin, hair, and feathers (Figure 9.5). The protein collagen is a major component of the connective tissues of your body. Another protein, myosin, acts as a molecular motor to help muscles contract.

Other proteins are enzymes, as you learned in Section 2.6. Enzymes are essential catalysts that make the chemical reactions of living systems happen fast enough to be useful. Like enzymes, some other proteins also bind to specific molecules. For example, hemoglobin in your red blood cells binds to oxygen, providing a way for blood to deliver oxygen throughout the body.

FIGURE 9.5

An example of the importance of structural proteins. The feathers responsible for the unique appearance of this Raggiana bird of paradise, *Poradisaea raggiana,* are composed mostly of the protein keratin. Different patterns of keratin production help form feathers, horns, claws, and hair.

CONNECTIONS
Proteins demonstrate the relationship between structure and biological function.

Proteins also play a key role in communication within an organism. Insulin, for instance, is a protein **hormone** that helps maintain homeostasis. Hormones are chemical signals given off by cells in one part of an organism that regulate behavior of cells in another part of the organism. In vertebrates (animals with backbones), insulin circulates in the blood, regulating sugar uptake by cells. Cell signaling also involves protein receptor molecules attached to the cell membrane. They may pass along signals to a series of regulatory enzymes inside the cell, like relay runners passing a baton. Some regulatory proteins bind to specific DNA sequences or enzymes involved in transcription. These proteins help control gene expression.

A protein's structure determines its function, and information expressed from the code in DNA determines the structures of proteins. Collagen, for instance, exists as long fibers that bind cells together in tissues (Figure 9.6). Many enzymes have cavities or pockets that bind only specific substrate molecules. For example, the enzyme lysozyme, found in egg white and tears, helps destroy harmful bacteria by cutting a polysaccharide found in bacterial cell walls. Lysozyme has an active site that binds specifically to 6-sugar subunits of the target polysaccharide.

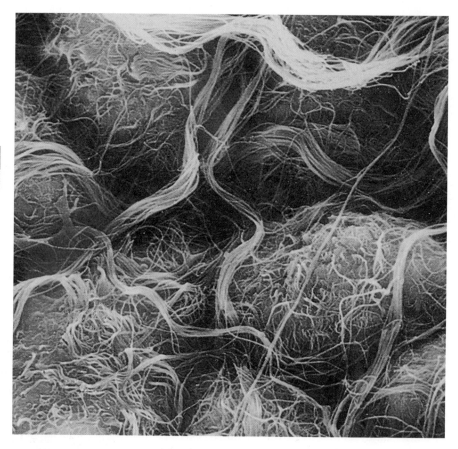

FIGURE 9.6

Collagen. A scanning electron micrograph of human pancreatic connective tissue, ×39,000, is shown here. These flexible protein fibers strengthen the tissues of many animals.

Biological Challenges

DISCOVERIES

Learning the Language of Genetics

What we know about genetic language has come a long way since the code was cracked. The first clue to the code was found by American biochemist Marshall Nirenberg in 1961 (Figure 9.7a). Working with a bacterial extract that contained ribosomes, Nirenberg found that adding mRNA from one cell type increased protein synthesis by ribosomes from another cell type. In other words, ribosomes could use instructions from a "foreign" mRNA. This observation suggested an experiment: Could ribosomes follow instructions from a custom-made RNA with a known sequence of nucleotides? If so, the code could be cracked.

Nirenberg made mRNA with only one base, uracil, repeated over and over again. The poly-U chain was added to 20 test tubes. Each tube contained ribosomes, enzymes, and other factors for protein synthesis. Each tube also contained a different radioactive amino acid. In 19 of the test tubes, nothing happened. In the 20th test tube, the radioactive amino acid phenylalanine was incorporated into polypeptide chains. Nirenberg had discovered the first codon. Gobind Khorana (Figure 9.7b), Severo Ochoa (Figure 9.7c), and other investigators soon discovered the mRNA codons for the rest of the amino acids.

In the years since then, scientists have analyzed the details of transcription and translation. There have been some surprises, of course. In the 1970s, genetic information was found to flow from RNA to DNA in special situations (see Section 9.8). Eukaryotic genes were found to be interrupted by noncoding regions, called introns (see Section 9.4). The ability to accurately read, cut, and rejoin DNA sequences has made gene cloning a reality.

Computer software is now used to analyze the enormous amounts of genetic information being discovered. Imagine looking for a specific sentence in a single book in a big library if you had no filing system and no page numbers. Scientists must have sophisticated ways to store, sort, and recognize genetic sequence data. Statistical methods that have worked with human language have been used to distinguish subtle differences in genetic data from different species. Different people speak the same language with different styles. In a similar way, biologists can recognize the DNA sequences of different groups of organisms by their characteristic "styles." In one case, this approach kept sequence data of yeast DNA from being erroneously recorded as human sequence data.

Software can also be used to identify similar sequences in different species. Comparisons to known genes help researchers determine the functions of newly discovered genes. They are also useful for estimating how closely various species are related.

a   b   c

FIGURE 9.7

Three biochemists who cracked the genetic code. Marshall Nirenberg **(a)** deciphered the first codon (UUU). Gobind Khorana **(b)** and Severo Ochoa **(c)** helped translate the rest of the genetic code.

Cells do not express all their genes at all times. Of the approximately 35,000 genes in a human cell, only a subset are active at any time. Each cell has a complete copy of the organism's DNA, but the activities of various types of cells are different. Think about the bird in Figure 9.5. Cells in the bird's feet are not expressing the particular combination of genes that would give rise to feathers.

Gene regulation is a key to how cells play different roles in a multicellular organism. Even in single-celled organisms such as bacteria, regulation of gene expression is important. The bacteria known as *Escherichia coli* live in your intestine, where they help digest your food. These bacteria have an enzyme that breaks down the disaccharide sugar lactose. However, they do not always make this enzyme. The gene that codes for this enzyme is switched on in response to a cellular signal that lactose is present (Figure 9.8).

Hormone signaling, availability of amino acids, and contact with other cells all help regulate gene expression. In multicellular organisms, specific groups of genes are active only in certain cell types and organs. Scientists are studying these tissue-specific genes to learn how genes control embryonic development.

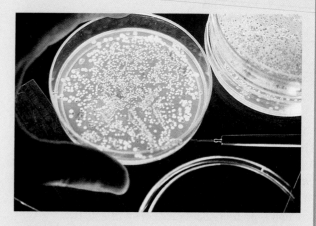

FIGURE 9.8

Regulation of gene expression in *Escherichia coli*. Each of the dots on this culture plate is a colony of many bacterial cells. In the culture medium, a chemical known as X-gal turns blue in the presence of beta-galactosidase, an enzyme that breaks down the sugar lactose. The gene that encodes this enzyme is active in the bacteria in the blue colonies.

Check and Challenge

1. What role does an mRNA molecule play in the flow of information for protein synthesis?
2. Distinguish between the roles of transcription and translation.
3. What is the genetic code?
4. Use Figure 9.4 to locate and list several code synonyms.
5. Examine Figure 9.4. Which is the least important of the three bases in a codon? Explain your answer.

Transcription

9.3 RNA Synthesis

Gene expression begins with RNA synthesis. The transcription enzyme **RNA polymerase** joins RNA nucleotides according to the base sequence in DNA. Prokaryotes have one type of RNA polymerase. Transcription is more

Try Investigation 9A Transcription.

complex in eukaryotes. Their nuclei have three RNA polymerases. Each is responsible for making different types of RNA. The basic process, however, is the same in almost all cells.

In eukaryotes, protein synthesis takes place outside the nucleus; however, mRNA, tRNA, and rRNA are built in the nucleus. The RNA molecules are then modified and move out to the cytoplasm through pores in the nuclear membrane. A specialized part of the nucleus, the nucleolus, is the site of rRNA synthesis. The nucleolus also is where rRNA and as many as 70 proteins are assembled into ribosomal subunits. During protein synthesis, two ribosomal subunits bind to each other and an mRNA to form an intact ribosome. Figure 9.9 summarizes the formation and function of the major types of RNA.

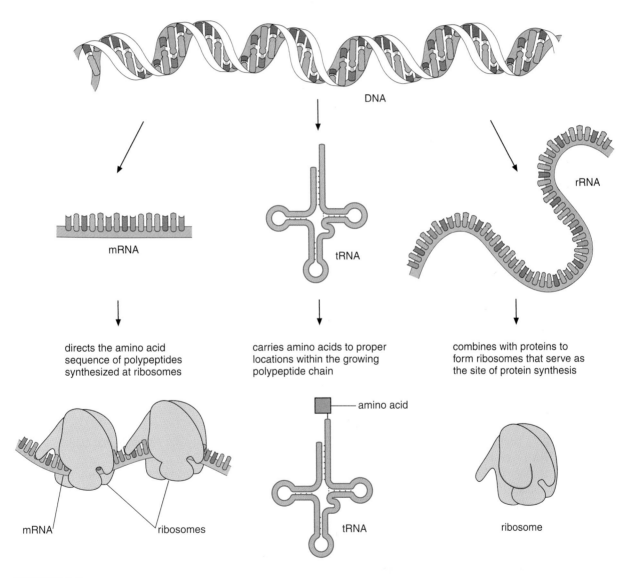

FIGURE 9.9

Transcription of the three types of RNA from DNA. Each type of RNA carries out a different function in protein synthesis. This figure uses a linear symbol for mRNA to emphasize that its sequence corresponds to the linear sequence of amino acids in a protein. In reality, the mRNA is folded and twisted rather than being straight or rigid.

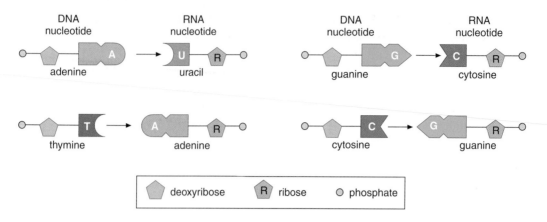

FIGURE 9.10

Each DNA nucleotide pairs with a particular RNA nucleotide. This pairing is the basis of the genetic code. Note that in RNA, uracil (U) replaces the thymine (T) of DNA.

Only one strand of the DNA, the coding or template strand, directs the synthesis of RNA. Recall that DNA replication involves the matching of complementary nucleotides—A with T and G with C (see Chapter 8). Similarly, in transcription, the sequence of nucleotides in new RNA is complementary to the sequence in the coding strand of DNA except that uracil (U) replaces thymine (T) (Figure 9.10).

DNA coding-strand nucleotides

A T T C G C A C C T A A

transcription

↓

U A A G C G U G G A U U

RNA nucleotides

Focus On

Molecular Motors

As RNA polymerase moves along a gene, it partially unzips the DNA double helix. This action puts an awkward twist in the DNA. How much force is required to do this? Sophisticated new experiments are testing the force involved in these minute molecular motors. For example, one end of a DNA molecule is attached to a glass slide, and the other end is tagged with a bead that can be moved by a laser beam. When a single molecule of RNA polymerase moves along the DNA, transcribing it, scientists determine the force exerted by measuring the movement of the bead. RNA polymerase was found to exert four times the force of another

molecular motor, the myosin molecule that contracts muscles (see Figure 7.13)

In addition to RNA polymerase, other enzymes are involved in unzipping the DNA double helix. Some of these enzymes unwind the helix and help separate the DNA strands. Others help stabilize regions of the DNA that have been opened up in this way. Figure 8.8 shows the complexity of chromosome structure. These enzymes are able to open up part of a tightly coiled chromosome during transcription and reassemble it afterwards without damage.

Transcription takes place in three stages. Initiation is the first stage (Figure 9.11a). It occurs when the enzyme RNA polymerase attaches to a specific region of the DNA. This attachment site is called the promoter region because it promotes transcription. It is located just before the segment of the DNA coding strand that will be transcribed. In eukaryotic cells, proteins known as initiation factors must be present for the RNA polymerase to attach to the promoter region.

CHEMISTRY TIP

The RNA-polymerase enzyme is moderately attracted to DNA in general. The enzyme has a much stronger attraction for the promoter site. This higher affinity causes the RNA polymerase to attach to this site.

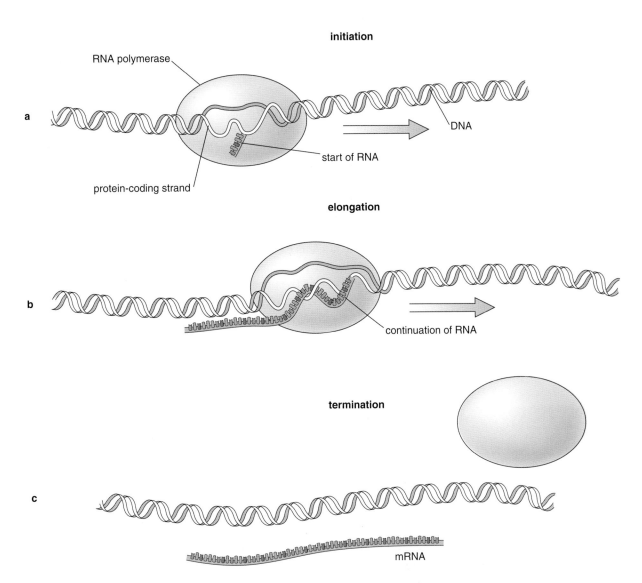

FIGURE 9.11

The three stages in transcription of RNA from a DNA template. RNA polymerase binds to the coding strand of the DNA at its promoter region **(a),** the RNA transcript elongates **(b),** and transcription ends when the RNA-polymerase molecule reaches the terminator region of the DNA **(c).**

The second stage is elongation of the RNA (Figure 9.11b). RNA polymerase partially unwinds the DNA, exposing the coding strand of the gene. The enzyme moves along the DNA away from the promoter site as it builds an RNA molecule. The sequence of DNA nucleotides determines the sequence of the RNA chain. A single complementary strand of RNA, called a primary transcript, is made.

The third stage of transcription is termination (Figure 9.11c). When RNA polymerase reaches the terminator region, or the end of the DNA to be transcribed, the enzyme and primary transcript are released from the DNA. This ends transcription.

9.4 RNA Processing

In prokaryotes, new mRNA is translated and broken down by enzymes within a few minutes (Figure 9.12). This rapid turnover of RNA helps regulate protein synthesis. In eukaryotes, mRNA can last from minutes to days. The life span of eukaryotic mRNA depends partly on how the primary transcript is processed.

A primary RNA transcript may contain as many as 200,000 nucleotides. (The average for human cells is 5,000.) Yet mRNA in the cytoplasm averages only 1,000 nucleotides. What happens to the rest? All three types of RNA are processed in the nucleus of eukaryotes before they leave the nucleus. Enzymes add additional nucleotides and chemically modify or remove others.

Figure 9.13 summarizes the processing of mRNA. First, enzymes attach a cap of chemically modified guanine nucleotides (methyl-guanine, or mG) to the starting end of the mRNA molecule (Figure 9.13a). Next, other enzymes replace part of the opposite end with a tail of 100–200 adenine nucleotides (Figure 9.13b). This addition is called a poly-A tail. The cap and the tail help protect mRNA from enzymes that break down nucleic acids. Generally, the

CHEMISTRY TIP

In methyl-guanine, one hydrogen atom is replaced by a methyl group ($-CH_3$).

CONNECTIONS

Transcription and mRNA breakdown in mitochondria and chloroplasts are similar to the pattern in prokaryotes. This evidence supports the theory that these organelles originated as bacteria.

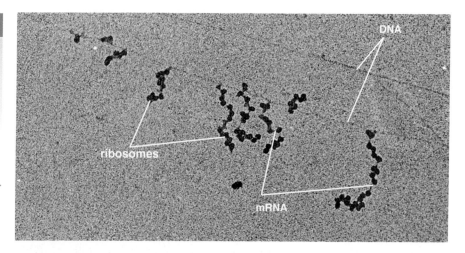

FIGURE 9.12

Rapid transcription and translation of bacterial genes. A transmission electron micrograph of an unidentified operon of the bacterium *Escherichia coli,* ×72,600. Ribosomes attach to mRNA, and protein synthesis begins even before transcription is complete. Note the series of lengthening mRNA molecules from top to bottom.

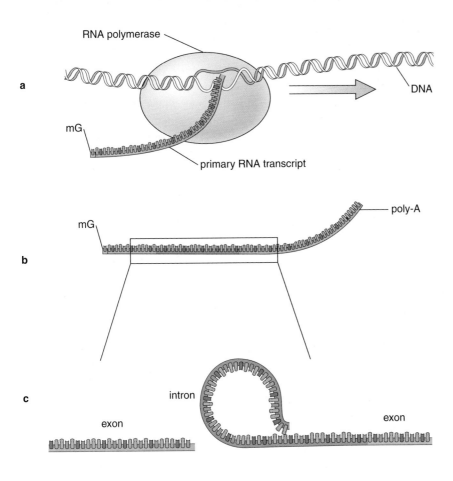

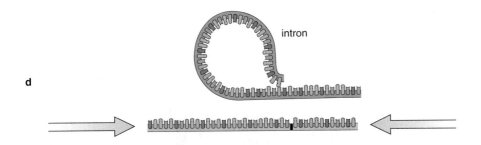

FIGURE 9.13

Processing of primary RNA transcript to mature mRNA. Changes to mRNA include the addition of an mG cap **(a)** and a poly-A tail **(b),** removal of introns **(c),** and splicing of exons **(d).**

longer the poly-A tail, the longer the life span of a particular mRNA. The tail may also help transport mRNA out of the nucleus. The cap apparently helps the mRNA attach to a ribosome and begin translation.

The final step in mRNA processing involves removal of some internal segments of the RNA that do not code for protein (Figure 9.13c). They are called **introns.** The parts of the transcript that remain (and code for protein) are called **exons.**

The process of removing introns and rejoining cut ends is called **splicing** (Figure 9.13d). Splicing requires precise recognition of the site to be cut. Splicing enzymes recognize the sequence GU at one end of an intron and AG at the other end. This can happen in several ways. Protein enzymes catalyze the splicing of tRNA in yeast and many other eukaryotes. In contrast,

ETYMOLOGY

inter- = between (Latin)
ex- or **exo-** = out or outside (Latin)

Introns are noncoding segments that intervene between coding segments (**exons**) of mRNA.

Most eukaryotic genes contain introns. During the chemical evolution that led to the first cells, RNA may have acted as both catalyst (as protein enzymes now do) and information-storage molecule (as DNA now does). Cutting and splicing of introns and exons by catalytic RNA may have been important in early evolution. Intron and exon genes may have been passed along to DNA as it became the major genetic molecule.

In contrast, most prokaryotes do not contain introns. They may have been lost through countless generations of evolution as prokaryotes became more streamlined. If so, introns would have been present in the first cells from which both prokaryotes and eukaryotes evolved.

Alternatively, introns may have evolved after the split between prokaryotes and eukaryotes. Evidence is less strong for this view. Either way, an advantage of introns is that they provide a way to shuffle exons. Exon shuffling rearranges the coding parts of genes to make new combinations. This produces new genes that could give organisms an evolutionary advantage.

catalytic RNAs splice RNAs produced by mitochondria, chloroplasts, and many unicellular eukaryotes. If introns are left in RNA, the consequences can be serious. For example, a change in one splice site of an intron in beta-globin, a component of the oxygen-carrying blood protein hemoglobin, results in defective hemoglobin. This causes a serious form of anemia called beta-thalassemia. In addition to splicing, an important step in the processing of tRNA is the chemical modification of several nucleotides and folding into a cloverleaf shape (Figure 9.14).

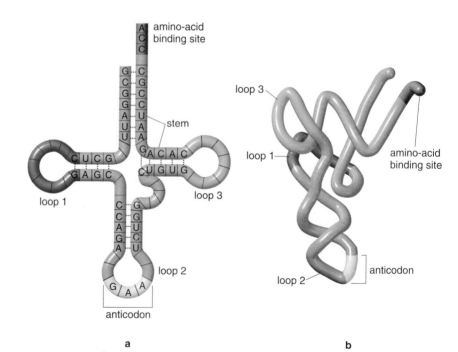

FIGURE 9.14

Structure of tRNA. Mature tRNA resembles a cloverleaf **(a)**, with the amino-acid binding site at the end of a stem and the anticodon at the loop on the opposite end. Base pairing between parallel parts of the tRNA molecule stabilizes the cloverleaf shape. The three-dimensional structure of the molecule is roughly L-shaped **(b)**.

Ribosomal RNA is not involved in coding. The primary rRNA transcript is spliced and modified to produce mature rRNA molecules. These molecules bind to proteins to form the large and small subunits of ribosomes. Often several ribosomes bind to the same mRNA molecule, forming a polyribosome (see Figure 9.12).

Check and Challenge

1. What is transcription? Where does it take place in prokaryotes? in eukaryotes?
2. What does RNA polymerase do?
3. Compare the roles of introns and exons in mRNA production.
4. Distinguish the functions of mRNA, tRNA, and rRNA.
5. Describe the differences between a primary RNA transcript and a mature RNA.
6. What types of molecules catalyze RNA splicing?

Protein Synthesis

9.5 Translation

Protein synthesis translates the codon sequence of mRNA into the amino-acid sequence of a protein. This happens on ribosomes, where tRNA acts as a molecular adapter. One end of a tRNA molecule carries a specific amino acid. The corresponding anticodon is at the opposite end of the tRNA molecule. In turn, the anticodon pairs with the mRNA codon that encodes this particular amino acid.

Attachment of the correct amino acid to its tRNA molecule is called tRNA charging. Twenty different enzymes carry out the tRNA charging reactions. Each enzyme bonds a different amino acid to its matching tRNA. A molecule of ATP provides the energy to form this bond. The accuracy of protein synthesis depends on the accuracy of tRNA charging.

Charged tRNA, mRNA, and the growing polypeptide chain come together at specific binding sites on a ribosome (Figure 9.15). At these sites, tRNA anticodons base-pair with mRNA codons. This positions the amino acids they carry so that they can bond to the growing polypeptide chain. Thus the sequence of codons dictates the amino-acid sequence.

One of the binding sites, the P site, holds the tRNA carrying the growing polypeptide chain. The A site holds the tRNA carrying the next amino acid to be added to the chain. Next to the P site is the exit site, or E site. An

Try Investigation 9B Translation.

CHEMISTRY TIP

The tRNA "charging" reaction attaches an amino acid to a tRNA molecule. It has nothing to do with electric charge.

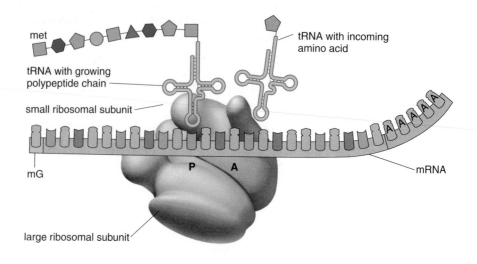

met

tRNA with incoming amino acid

tRNA with growing polypeptide chain

small ribosomal subunit

mG

P   A

mRNA

large ribosomal subunit

a

growing polypeptide chain

large ribosomal subunit

mRNA

small ribosomal subunit

b

FIGURE 9.15

The machinery of protein synthesis. (a), A charged tRNA sits in the A site of the ribosome, bound to the correct mRNA codon by base pairing. A second tRNA, carrying a growing polypeptide, is in the P site, bound to the previous mRNA codon. The E site is not shown. **(b),** A groove between the large and small subunits of the ribosome accommodates mRNA and the growing polypeptide chain.

uncharged tRNA leaves the E site after its amino acid is added to the growing chain. During translation, the ribosome moves along the mRNA strand one codon at a time.

Translation involves the same three stages—initiation, elongation, and termination—as transcription. All three stages require enzymes that are part

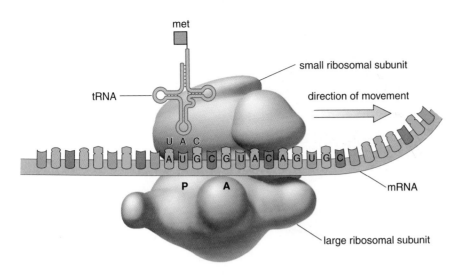

FIGURE 9.16

Initiation of translation. Two ribosomal subunits attach to an mRNA with an intact mG cap. This requires proteins called initiation factors and energy from GTP hydrolysis. A tRNA carrying the amino acid methionine (met) binds at the P site. The start codon, AUG, is positioned there. Most eukaryotic proteins do not begin with methionine. A polypeptide begins with this amino acid, and an enzyme later removes it.

of the ribosome. Initiation and elongation also require energy supplied by GTP (guanosine triphosphate), a molecule closely related to ATP. Figures 9.16 and 9.17 show the process of protein synthesis.

During initiation of translation, the ribosome attaches at a specific site on the mRNA (Figure 9.16). This site is the start codon, AUG. Which amino acid will begin the peptide chain?

During elongation, peptide bonds join each amino acid with the next in the sequence. A charged tRNA whose anticodon matches the next codon on the message enters the A site of the ribosome (Figure 9.17a). This positions the amino acid it carries to form a peptide bond with the amino acid attached to the tRNA at the P site (Figure 9.17b). When the bond forms, the polypeptide chain transfers to the tRNA at the A site (Figure 9.17c). The entire ribosome moves down the mRNA to position the next codon at the A site. The uncharged tRNA leaves the E site (Figure 9.17d). Now the tRNA that holds the growing polypeptide is at the P site. The A site is open and available for the next matching tRNA to bring in an amino acid. In this way, amino acids are added one at a time.

Translation terminates when a stop codon reaches the A site of the ribosome. Three codons can signal the end of translation—UAA, UAG, or UGA. No tRNA has an anticodon to complement these stop signals. Instead of a tRNA, a special protein known as a release factor binds to the stop codon in the A site. At this point, translation stops and the tRNA releases the polypeptide. The ribosome lets go of the mRNA, the tRNA, and the release factor; then the two ribosomal subunits separate. Figure 9.18 summarizes transcription and translation in eukaryotic cells.

 CHEMISTRY TIP

The P and A sites of ribosomes were named for the growing peptide chain and the next amino acid to be added to the chain, respectively. During formation of a peptide bond, the tRNA that carries the existing peptide chain sits in the P site of the ribosome, and the tRNA that will contribute the next amino acid occupies the A site. The new peptide bond forms between the acid group at the end of the peptide and the amino group ($-NH_2$) of the new amino acid.

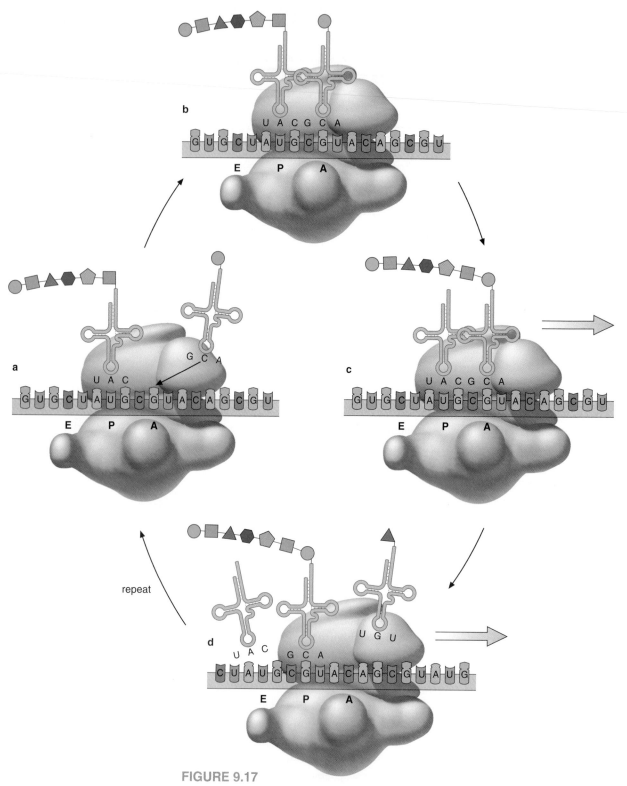

FIGURE 9.17

Elongation of a polypeptide chain on a ribosome. (a), A charged tRNA molecule approaches the A site. **(b),** It base-pairs with the mRNA, positioning its amino acid to form a peptide bond. **(c),** The bond forms, and the peptide chain transfers to the tRNA in the A site. **(d),** The ribosome moves one codon along the mRNA. The tRNA formerly in the P site leaves the E site, and the tRNA formerly in the A site occupies the P site. The cycle can now repeat.

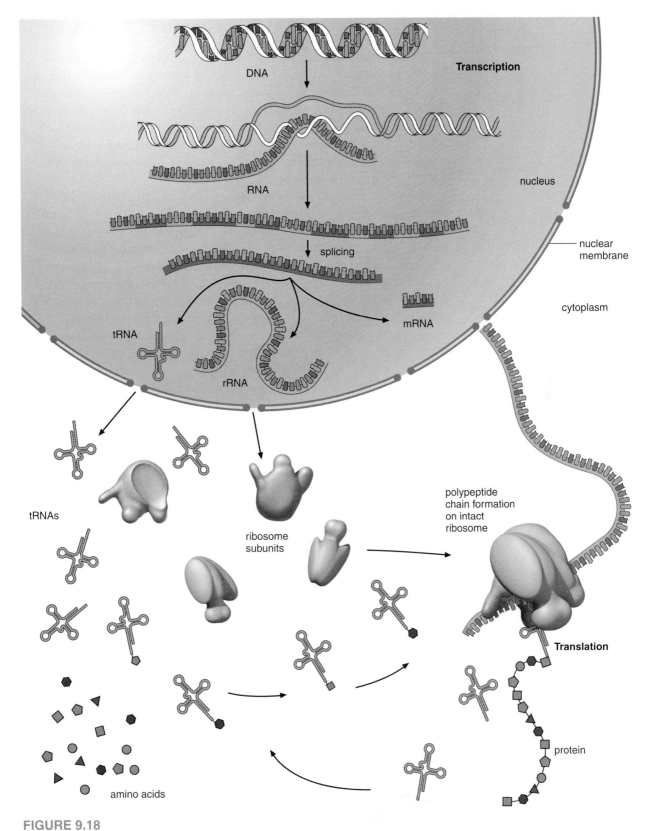

FIGURE 9.18

Summary of transcription and translation in eukaryotes. Transcription produces mRNA, tRNA, and rRNA. All three participate in translation.

9.6 Transport and Modification of Proteins

The new protein chain may not be functional. Many proteins must be chemically modified and folded into an active tertiary structure. Helper, or "chaperone," proteins often help stabilize the polypeptide as it is folded. Chemical modification frequently involves adding sugars to specific sites on the protein. In addition, enzymes may cut the polypeptide into smaller segments. For example, two enzymes, pepsin and trypsin, are responsible for the digestion of proteins you eat. These enzymes become active when part of their structure is removed.

After translation, the protein must be transported to where it will function. Sometimes the protein must move out of the cell, as in the case of hormones such as insulin. In such a case, a small membrane vesicle containing the protein fuses with the cell membrane. The protein is then released outside. Other proteins, however, become part of membranes.

Cells have many ways to send proteins to the correct location. Transport can start while the protein is still being translated. The process uses a signal that is part of the protein sequence, called the **signal sequence.** Proteins that will become part of membranes or leave the cell are made on ribosomes that are bound to the endoplasmic reticulum (see Section 6.4). The first few amino acids of these proteins are the signal sequence that directs the ribosomes making them to the ER. The signal sequence binds to a receptor protein in the ER membrane (Figure 9.19). As

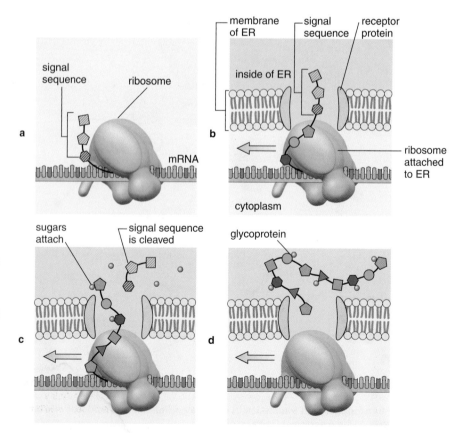

FIGURE 9.19

Synthesis of proteins for secretion or insertion in a membrane. The first few amino acids constitute a signal sequence **(a)**, which binds to a receptor molecule in the rough ER membrane **(b)** and threads the growing amino-acid chain through the membrane and into the ER as translation continues. Sugar molecules may be added to form glycoproteins **(c)** before the new protein is released to the inside of the ER **(d)**.

Several factors determine how much of a protein is in a cell, including how long the protein molecule survives. Proteins with errors often are recognized by their faulty structure. Enzyme complexes that break up faulty proteins, called proteosomes, are found in the nucleus and in the cytoplasm. Normal proteins also are destroyed at a regular rate. Each protein's life span may be different, depending on its function. For example, the hemoglobin molecules in red blood cells are constantly needed, and they survive for many days. Other proteins, such as those that regulate gene expression, may be needed only briefly. The life span of a given protein depends partly on the amino acids found at one end of the polypeptide chain. Data from studies of yeast show that proteins that end with the amino acid phenylalanine, for example, survive only a few minutes. Those with valine on the end survive for over 30 hours.

translation continues, the growing chain of amino acids threads through the membrane and into the space inside the ER. When translation is complete, the new protein is released from the ribosome into the inner ER. Often a sugar molecule is added to the protein, forming a glycoprotein. Proteins to be released from the cell pass from the ER to the vesicles of the Golgi apparatus.

9.7 Translation Errors

Errors sometimes occur during translation. Most are caught and corrected. The most common translation error results from misreading the nucleotide sequence. Initiation determines exactly where translation will begin. Starting from this point, the grouping of bases into codons is called the reading frame. If the start is shifted by one or two nucleotides in either direction, the frame changes. A different sequence of codons and amino acids will result. Figure 9.20 shows a frame shift.

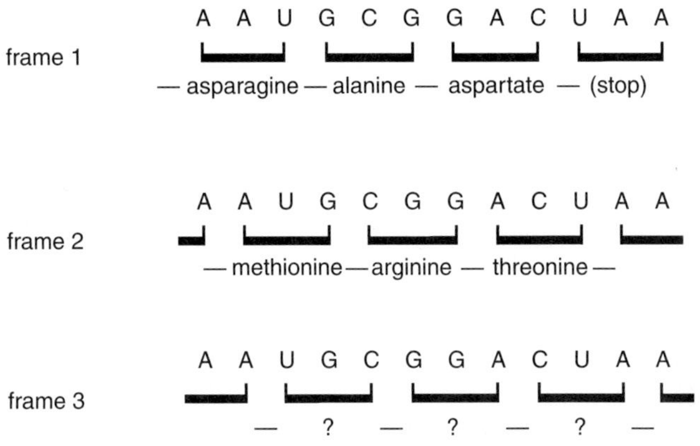

FIGURE 9.20

Three possible reading frames for a segment of mRNA. Each time the reading frame shifts, a different amino-acid sequence results. Use the genetic code (Figure 9.4) to identify the sequence of amino acids that would result in reading frame 3 when translation begins with the third nucleotide.

Some errors are due to splicing mistakes or changes in the DNA. For example, if a nucleotide in DNA is lost, a frame shift results. If a nucleotide changes so that a codon becomes a stop codon, translation can terminate partway through the message. The result is a partial polypeptide. Insufficient amounts of a particular amino acid also can disrupt translation.

In some cases, translational frame shifts or alternate initiation sites appear to be normal ways in which one mRNA can specify more than one polypeptide. These special situations are discussed in Chapter 14.

Check and Challenge

1. Describe the steps involved in adding amino acids to a polypeptide chain on a ribosome.
2. Why is accurate initiation of protein synthesis important?
3. What is a signal sequence, and what is its significance in protein synthesis?
4. What part do codons and anticodons play in translation?
5. What is a charged tRNA? What is its role in translation?

Viruses

9.8 Genetic Information and Viruses

Among the most important basic properties of life is the ability to replicate and to evolve. **Viruses** are tiny particles that have no cells, yet they replicate and evolve. Because they cannot do these things without help, viruses depend on the gene-expression machinery of the host cells they infect.

Viruses are much smaller and simpler than cells. They were discovered in 1892 by Russian botanist Dmitri Ivanovsky. He found that the infective agent of tobacco mosaic disease passed through a filter that would have retained bacteria. The agent was tobacco mosaic virus, TMV.

Viruses contain a small amount of genetic material. Some, such as the familiar viruses that cause colds, contain DNA. The T_2 **bacteriophage** that infects bacterial cells also is a DNA virus. Others, such as the influenza virus, store information in RNA. Most viruses consist of little more than a bit of DNA or RNA and a protective protein coat. Some contain a few enzyme molecules, such as those needed for transcription of their genes. Some viruses that infect animal cells have a membrane envelope, but they are not cells. They do not carry out metabolism or respond to stimuli, as cells do. Despite their simplicity and their dependence on host cells, viruses have an

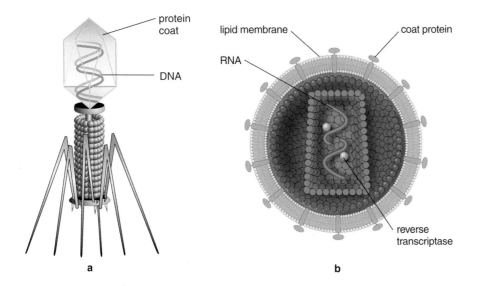

FIGURE 9.21

Two types of viruses. (a), Bacteriophage T$_2$, which infects bacterial cells, contains DNA surrounded by a protein coat. The elongated structure attaches to bacterial cells and injects DNA. **(b),** HIV (human immunodeficiency virus), which infects human cells, is surrounded by a protein and lipid membrane envelope. The genetic material is RNA. HIV also carries two molecules of the enzyme reverse transcriptase, ready to copy the RNA after entry into a host cell.

enormous impact on organisms from bacteria to humans. Figure 9.21 compares the structures of two types of viruses.

How do viruses replicate? The process varies among types of viruses, but the general principle of copying stored genetic information is the same as for cells. Viral replication falls into two patterns. In **lytic** infections (Figure 9.22a), the host cell's enzymes replicate the viral DNA. Viral genes are transcribed and translated on the host's ribosomes to make proteins for the outer capsule. New viral particles assemble. When there are many new viruses, the cell lyses (breaks open) and releases them to infect other cells.

In contrast, in **lysogenic** infections, the viral DNA (or a DNA copy of the viral RNA) inserts into the cellular DNA (Figure 9.22b). It is copied when the cell replicates. There is little or no production of new viruses. Instead, the genetic information for the virus is passed along to new cells. Sometimes an external stress such as starvation of the host cell activates a lytic cycle of replication. In animal cells, a few viral particles may be given off from time to time without lysing the cell. When the viral particle emerges, it can pick up part of the cell membrane. This membrane makes it more difficult for the host to recognize the viral particle as an invader. In the case of tumor viruses, the cell may lose control of normal growth and become cancerous.

ETYMOLOGY

lytic or lyso- = dissolving or destroying (Greek)
-genic = producing (Greek, Latin)

Lytic viral infections destroy cells. **Lysogenic** viruses can remain dormant in cells and may become lytic.

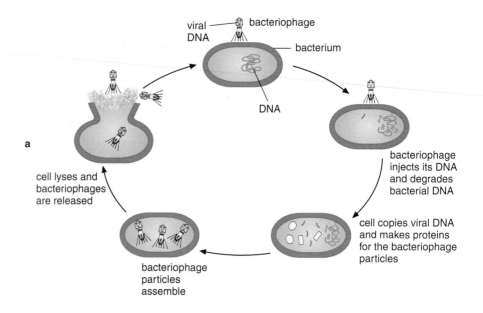

lytic infection

viral DNA — bacteriophage

bacterium

DNA

bacteriophage injects its DNA and degrades bacterial DNA

cell copies viral DNA and makes proteins for the bacteriophage particles

bacteriophage particles assemble

cell lyses and bacteriophages are released

a

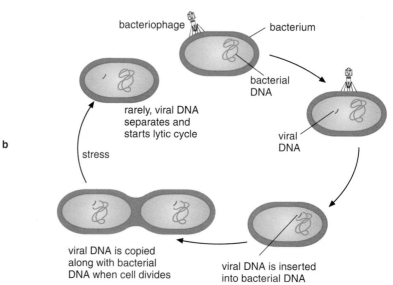

lysogenic infection

bacteriophage

bacterium

bacterial DNA

viral DNA

viral DNA is inserted into bacterial DNA

viral DNA is copied along with bacterial DNA when cell divides

stress

rarely, viral DNA separates and starts lytic cycle

b

FIGURE 9.22

Lytic and lysogenic viral reproduction. In a lytic infection **(a),** the viral genetic material takes over the cell machinery and directs it to make components that assemble into new viral particles until the cell breaks open. In a lysogenic infection **(b),** the viral DNA is incorporated into the cell DNA. There it goes along for the ride, being copied every time the cell divides. Lysogenic infections can become lytic.

9.9 Impact of Viruses

Viruses live at the expense of the host organism. They pose a serious threat to cellular life because they replicate quickly, evolve rapidly, and can remain hidden inside the host cells. Antibiotics are effective against bacterial infections because they interfere with aspects of bacterial metabolism that are different from their hosts' metabolism. They are useless against viruses because viruses do not have their own metabolism.

Focus On

Reverse Transcription

Some RNA viruses reverse the roles of DNA as template and RNA as copy. When these viruses infect a host cell, a viral enzyme called reverse transcriptase makes a DNA copy of the viral RNA. The viral DNA then joins the host cell's DNA and directs production of new virus particles. The process of reverse transcription was discovered in 1970 in viruses known as retroviruses. These include HIV (human immunodeficiency virus), the virus that causes AIDS (Figure 9.23).

Most vertebrate cells also contain retroviral genes. Some give rise to particles associated with reverse transcriptase activity. These particles generate new copies of DNA from RNA and insert them at random into the host chromosomes. This integration may disrupt or alter the function of a gene. In fact, many retroviruses induce tumors when their integration causes an ordinary gene to become a cancer-causing gene, or oncogene.

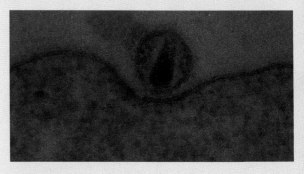

FIGURE 9.23

HIV (human immunodeficiency virus), entering a human leukocyte, or white blood cell (×900,000). Each virus particle is surrounded by a membrane vesicle taken from part of the host cell's plasma membrane.

Modern technologies have, in some cases, made the threat of viral diseases much greater. For example, increased air travel may discontinue the geographic isolation of the Ebola virus (Figure 9.24), which causes a highly contagious, usually lethal respiratory disease. Increased air travel also may have lead to the spread of severe acute respiratory syndrome (SARS). It first was reported in Asia in 2003, and spread to more than two dozen countries in just a few months.

Mechanical harvesting and international shipment of seeds, foods, and other agricultural products also can spread viruses that infect valuable crops and animals. Viruses also are exploited by advanced technologies. For example, scientists can disarm viruses by removing the genes that cause disease. Disarmed viruses are useful tools for delivering DNA in cloning experiments. Researchers are developing a vaccine for West Nile virus using components of the West Nile virus and other similar viruses.

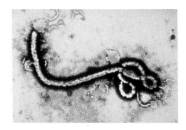

FIGURE 9.24

The Ebola virus (×26,400). This deadly virus occurs in isolated parts of East Africa, but air travel and human migration may cause it to spread to new regions of the world.

Check and Challenge

1. Bacteria and viruses are microscopic. In many cases, both can cause disease. Describe the similarities and differences between bacteria and viruses.

2. What two types of molecules are used by viruses to store genetic information?

3. What is a retrovirus?

4. Why are viral diseases often more difficult to treat than bacterial diseases?

5. What benefit do viruses provide for scientific research?

Summary

Genetic information serves as a master program to direct cell activities. Much of the genetic information encodes the primary structure for proteins. Proteins carry out numerous functions, including structural roles, cell signaling as hormones or cell-surface receptors, regulators of gene activity, and many catalytic functions. Genetic information is stored in DNA or, in the case of some viruses, as RNA. As the information is needed, it is expressed through transcription and translation. Regulation of gene expression is essential for different cells to carry out their particular activities.

In transcription, the coding strand of DNA is read as a template by RNA polymerases to build matching RNA molecules. Primary RNA transcripts are processed into tRNA, rRNA, or mRNA in the nucleus. Proteins combine with rRNA to form ribosomes. Amino acids are carried by their matching tRNAs to the ribosomes. Protein synthesis occurs as the sequence of codons in mRNA is translated into the sequence of amino acids in a protein. Newly transcribed proteins must fold into the appropriate three-dimensional structure in order to be functional. Often they are chemically modified, too. Proteins must travel to the appropriate location in order to do their job. Errors in transcription, RNA processing, or translation can result in poor function or absence of a particular protein.

A special exception to the usual flow of genetic information is found in RNA viruses. They use RNA as the long-term storage of information. One group of RNA viruses, the retroviruses, enter the host cell and make a DNA copy of their RNA genes. Viruses pose a serious threat to cellular life. They are exploited in biological research and for their potential as agents of gene therapy and vaccination.

Key Concepts

Use the concepts listed below and additional concepts from the chapter to build a concept map of the flow of information in gene expression. There is no single correct answer, but your map should show the most important logical connections among the major ideas in this chapter.

- amino acids
- transcription
- viruses
- codons
- protein transport
- ribosomes
- introns
- anticodons
- signal sequence
- translation
- information
- reverse transcription

Reviewing Ideas

1. Describe the function of tRNA, mRNA, and rRNA.
2. In what situation does RNA play a role similar to enzymes?
3. Are all genes expressed at all times? Explain your answer.
4. What are the major steps in processing primary RNA transcripts into mRNA, tRNA, and rRNA?
5. Why does each codon have three nucleotides instead of two or one?
6. Describe the major steps of protein synthesis on the ribosome.
7. What are the possible biological roles of the cap and poly-A tail of mRNA?
8. Describe the mechanism by which proteins that are destined to become part of a membrane or secreted are packaged and transported after translation.
9. Why is the primary structure of a protein so important in cell function?
10. Explain the difference between transcription and translation.
11. What are the steps in tRNA charging, and why is it needed for gene expression?

 **Biology** nline BSCSblue.com/vocabulary_puzzlemaker

Using Concepts

1. There are three forms of RNA polymerase in eukaryotes. The different enzymes make different forms of RNA. Explain the possible advantage of having more than one form of RNA polymerase.
2. Give the sequence of an mRNA molecule that is the complement of the following coded sequence of DNA: ATTACGCGGTCAGTA.
3. You are studying the rates of transcription and protein synthesis in plant cells growing in a tissue culture. You will do this by measuring the incorporation of radioactive isotopes into RNA and protein. What could you measure with radioactive phosphorus? Explain your answer.
4. A polyribosome consists of several ribosomes attached to an mRNA molecule. Would the proteins manufactured by a polyribosome be different or the same? Explain your answer.
5. How is the need for gene regulation in multicellular organisms different than in single-celled organisms?
6. What would happen to translation if a ribosome skipped one or more codons? Explain your answer.
7. What would happen to translation if a ribosome skipped one nucleotide? Explain your answer.

Synthesis

1. Explain how the primary structure of a protein can differ from the amino-acid sequence specified by the primary transcript.
2. What are the chief differences between DNA replication and transcription?
3. The genetic code is essentially universal. What is the significance of that observation in understanding the history of life on Earth? What are the implications for current research and medicine?

Extensions

1. Write a short essay explaining the statement "Genes control cellular respiration." Explain in detail how genes do this.
2. Show, as a cartoon or series of cartoons, the sequence of events that occur during translation on the ribosomes.

Web Resources

Visit BSCSblue.com to access
- Several chapters of an electronic biology textbook
- A computer simulation of protein synthesis
- Web links related to this chapter

CONTENTS

LEARNING OUTCOMES

By the end of this chapter you will be able to:

A Describe the stages of embryonic development in amphibians and humans.

B Describe how developmental patterns relate to evolutionary relationships.

C Describe human embryonic development.

D Describe methods used to understand mechanisms underlying differentiation.

E Explain the genetic-equivalence hypothesis and describe experiments performed to test it.

F Explain determination and the role the cytoplasm has in this process.

G Discuss examples of cell-cell interactions in differentiation.

CHAPTER

10

Animal Growth and Development

■ *What characteristics identify the different body forms among these eggs?*

■ *What biological processes must occur before the juvenile salmon resembles the adult form?*

A newborn baby forms through a series of fascinating biological events. For centuries, these events were mysteries. In the last 100 years, biologists have learned to do much more than simply describe developing embryos. Today biologists use the techniques of cellular and molecular biology to unravel the mysteries—to reveal the mechanisms that control an embryo's development. Very little of this work can be done directly on human embryos. However, work with many other animal species has given clear insight into the mysteries of human embryonic development.

This chapter describes the basic stages and events of animal development. You will learn how the cellular and molecular processes involved in development were discovered. The influence of an egg's cytoplasm on early development, the importance of interactions between cells, and the regulation of cell specialization are also explored.

Key Events of Development

10.1 Beginnings of the Embryo

Development of most new animals begins with **fertilization,** the union of two cells, one made by the male parent, the other by the female parent. These cells, the **sperm** and egg, are also called **gametes** (Figure 10.2). Animal sperm cells are usually very small and have a flagellum that they use to swim toward an egg. In contrast, egg cells are much larger. Their cytoplasm contains **yolk,** an energy-rich collection of lipids and proteins, and a variety of RNA molecules, many ribosomes, and mitochondria. Each gamete nucleus has one-half of the chromosome set found in each of the parent's cells (see Section 12.2).

To begin fertilization, the sperm cell touches the surface of the egg cell and fuses with it (Figure 10.1). The sperm nucleus enters the egg cytoplasm and meets the egg nucleus, and the two nuclei fuse. This joining of nuclei reestablishes the full chromosome set of the normal animal cell and the pairs of DNA gene sequences expected in every cell. A fertilized egg, or **zygote,** is the earliest stage of the **embryo.**

Before fertilization, the eggs of many animals are metabolically inactive: They use little energy and make almost no new protein or RNA. Fertilization turns on the egg's metabolism. This **activation** usually occurs within seconds of egg-sperm fusion. Cell respiration increases, and soon new proteins are made, using messenger RNA molecules already present in the cytoplasm. When the zygote begins to divide, it is as metabolically active as most adult cells.

Activation has two other major effects. One is a rapid change in the plasma membrane, which blocks fertilization by a second sperm. (Animal embryos nearly always develop abnormally when their cells contain more than two sets of chromosomes.) The other result of activation is rearrangement of the zygote cytoplasm by movements in the cytoskeleton (Figure 10.3). This helps produce differences among cells when they divide. The events of activation start the process of changing the zygote into a complex multicellular organism.

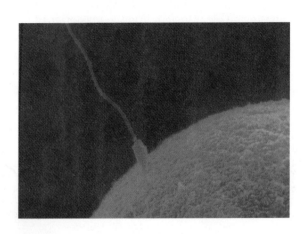

FIGURE 10.1

Egg-sperm fusion. A scanning electron micrograph of contact between a purple sea urchin (*Arbacia puretulata*) sperm and egg, ×20,000, is shown here. The nucleus of the sperm and the nucleus of the egg fuse to initiate formation of the zygote.

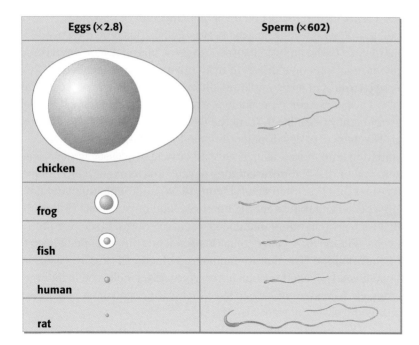

Eggs (×2.8)	Sperm (×602)
chicken	
frog	
fish	
human	
rat	

FIGURE 10.2

Eggs and sperm of several organisms. Note that the sperm are enlarged ×602, much more than the eggs. In the chicken, frog, and fish, the ova are surrounded by other materials (shown in outline).

10.2 Growth, Differentiation, and Form

How does a zygote become a complex multicellular organism composed of many types of cells, tissues, and organ systems? Animal development includes growth, cell specialization, and formation of tissues and organs

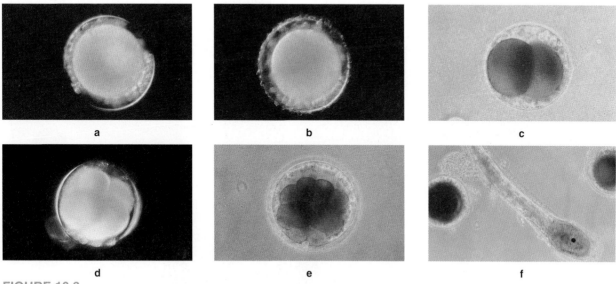

FIGURE 10.3

Rearrangement of cytoplasm during cleavage. After the unfertilized egg of the sea squirt, *Styela clava*, **(a)** is fertilized, polarity is established **(b)** when pigments cover a portion of the cytoplasm, resulting in a yellowish region and a gray, non-pigmented area. These regions are evident at the 2-cell stage **(c)**, where the yellowish lower hemisphere (which will become the muscle cells of the tail) is distinct from the dark upper half of the embryo. At this stage, the embryo is divided into right and left halves. Development continues through the 4-cell stage **(d)** to the 32-cell blastula stage **(e)**. The last photograph **(f)** shows the tadpolelike larva.

(Figure 10.4). Through many cell divisions, the embryo becomes multicellular. The new cells enlarge between divisions. The increased number of cells in the embryo causes it to grow larger. As the embryonic cells divide, some become different from others, a process called **differentiation.** A cell is completely differentiated when it possesses all the features of a specific cell type, such as a muscle cell or a skin cell. As cells differentiate, they organize to form the tissues and organs of a complete animal. This period of development is called **morphogenesis.** Growth, differentiation, and morphogenesis are all critical processes in development.

Each type of cell that differentiates during development has a unique structure and function. Skin cells (Figure 10.5a) are tough, thin, and flat. They are specialized to protect the body. Skeletal muscle cells (Figure 10.5b) are filled with protein fibers that enable them to contract (see Section 7.7). Nerve cells (Figure 10.5c) have long, thin branches that are specialized to transmit information. The mature human red blood cell (Figure 10.5d) lacks a nucleus. It is specialized to transport oxygen. Each cell type in an organ— your stomach, for example—has a specific location, and each performs a specialized and essential role.

ETYMOLOGY

morph- = form or shape (Greek)
genesis = forming or creating (Greek)

Morphogenesis is the forming of the major organs and tissues of an embryo.

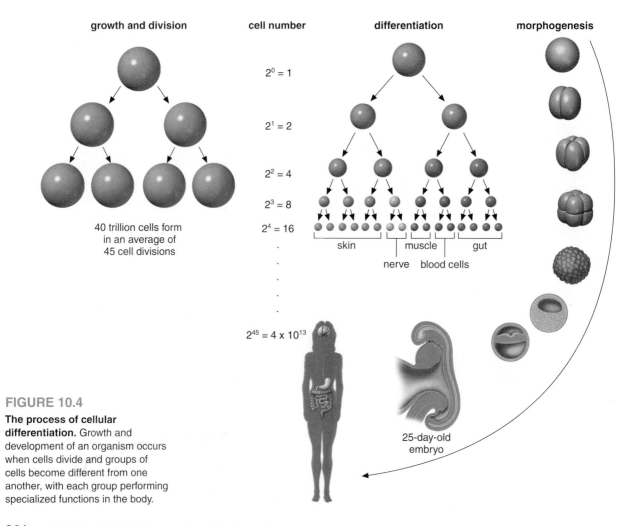

growth and division

40 trillion cells form in an average of 45 cell divisions

cell number

$2^0 = 1$

$2^1 = 2$

$2^2 = 4$

$2^3 = 8$

$2^4 = 16$

$2^{45} = 4 \times 10^{13}$

differentiation

skin

nerve   blood cells

muscle   gut

morphogenesis

25-day-old embryo

FIGURE 10.4

The process of cellular differentiation. Growth and development of an organism occurs when cells divide and groups of cells become different from one another, with each group performing specialized functions in the body.

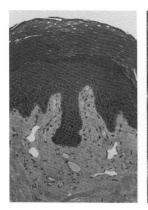

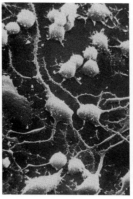
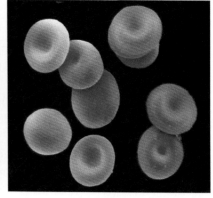

a   b   c   d

FIGURE 10.5

Four types of differentiated human cells. (a), Epidermal cells, ×99, cover the outside of the body and line internal cavities. **(b),** Each cell in skeletal muscle, ×320, has many nuclei. **(c),** Cerebral cortex nerve cells, ×3900, receive, integrate, and transmit information about the cell's environment. **(d),** Mature red blood cells, ×9000, have no nucleus, function for about 120 days, and are continually being replaced.

CHEMISTRY TIP

Keratin is rich in the sulfur-containing amino acid cysteine, which has the R group —CH_2SH. Disulfide bonds (—S—S—) between the cysteine of neighboring keratin molecules help form a tough cytoskeleton that is responsible for the strength of skin, hair, feathers, and claws. The more disulfide bonds that form between the cysteine molecules in hair, the curlier the hair becomes.

CONNECTIONS

Control of differentiation by gene expression is an example of the genetic and chemical regulation of biological structure and function.

All of an animal's cells carry the same genetic information, but how do they become different from one another? Furthermore, how is development controlled and organized? Proteins are the keys to differentiation in animal cells. Even cell movement in morphogenesis involves proteins on each cell's surface. Specific groups of genes are expressed in each type of cell (see Section 9.2). This leads to the production of specific proteins. For example, each type of cell in Figure 10.5 contains unique molecules (Table 10.1). The cells of the outer skin layers produce an extracellular matrix of keratin, a tough protective protein. The banded fibers in muscle cells contain the proteins actin and myosin, which enable the cells to contract and relax. Nerve cells contain small molecules called neurotransmitters, which carry nerve impulses from one cell to another (see Section 21.7). The major protein in red blood cells is hemoglobin, which transports oxygen to and from tissues (see Section 7.9). Differences between cells in gene expression lead to differences in cell form and function.

TABLE 10.1
Examples of Differentiated Cells in the Human Body

Cell Type	Cell Function	Characteristic Structures	Characteristic Molecules	Molecule Function
epidermal skin cell	protection of body surface and retention of fluids	flat, thin shape; several layers; tough, flexible	keratin	protection of body against abrasion and desiccation
skeletal muscle	contraction	long, cylindrical shape; more than one nucleus; banded cell (fiber); many mitochondria	actin, myosin	contraction
nerve	transmission and reception of stimuli	branched extensions from cell body	neurotransmitters	transmission of nerve impulses
red blood cell	transport of oxygen to and from tissues	biconcave; disk shape; lack of nucleus	hemoglobin	transport of oxygen

10.3 From One Cell to Many: Making the Organism

Try Investigation 10A
Development in Polychaete Worms.

ETYMOLOGY

blast- = bud (Greek)
The **blastula** is a budlike ball of cells at the end of cleavage; an early nerve cell is a neuroblast.

After fertilization, the zygote divides into two cells. During this period of development, called **cleavage,** the cells usually divide simultaneously, doubling in number with each cycle. The embryo divides into smaller and smaller cells (Figure 10.6).

By the end of cleavage, the embryo consists of a mass of many cells called a **blastula.** The cells are usually all of the same general size and appearance. The shape of the blastula depends on the structure of the original egg and how its yolk is arranged. The blastula of a sea star is a hollow sphere; a duck's blastula is a thin, hollow disk on the surface of the huge yolk mass. Many simpler animals, such as snails, have solid blastulas.

Differentiation and morphogenesis first become obvious as some cells move from the surface to the interior of the blastula. The embryo now becomes a three-layered **gastrula.** The three cell layers, called the **primary germ layers,** will form all the body's tissues. The outer layer, the **ectoderm,** will form the skin, nervous system, and related structures. The inner layer, called the **endoderm,** is usually a tube and will become the lining of the

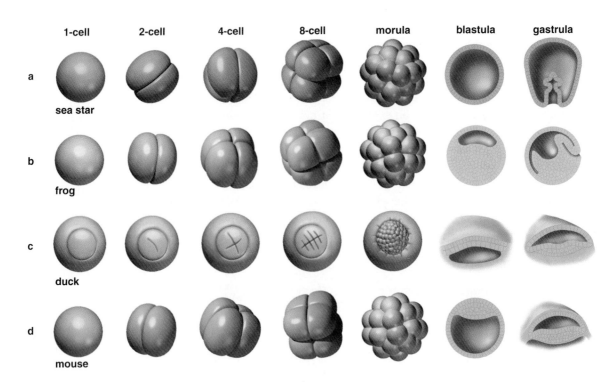

FIGURE 10.6

An overview of early development. Early cell divisions in these organisms are directed by the distribution and amount of yolk in the fertilized egg (not to scale). The invertebrate sea star **(a)** has a sparse, evenly distributed yolk, similar to that of the vertebrate frog **(b)** and mouse **(d).** In contrast, the duck **(c)** has a dense yolk limited to one area of the egg. From the 16- to 64-cell stage, an embryo is referred to as a morula.

digestive system. In between lies the **mesoderm,** which will produce the skeleton, muscles, heart, blood, and many other internal organs. Different animals have different ways to form the three germ layers of the gastrula. In sea stars, part of the blastula surface moves inward to form the endoderm tube, much like what happens when you press your thumb into a balloon. In reptiles, birds, and mammals, the layers separate, and cells migrate into the space between them.

Development of a blastula into a gastrula, or gastrulation, involves major changes. Morphogenesis includes coordinated movements of individual cells and tissues, changing cell shapes, folding or splitting of cell layers, formation of tissue masses by local cell division, and even shaping of organs by genetically timed death of some cells. Different proteins hold the surfaces of neighboring cells together. Cell movement during morphogenesis involves the controlled breaking and remaking of these chemical bonds.

In vertebrates the general shape, or **body plan,** of the organism appears during gastrulation. The first mesoderm becomes the **notochord,** a stiff rod that develops into part of the backbone. The notochord runs down the middle of the embryo just beneath the dorsal ectoderm. This development establishes the anterior-posterior axis, a line running from head to tail (Figure 10.7). At this time, the dorsal-ventral direction (from the back to the belly) and right and left become obvious. Later, a large head, segmented backbone, and limbs complete the vertebrate body plan.

ETYMOLOGY

gaster = stomach or belly (Greek)
ecto- = outer (Greek)
endo- = inner (Greek)
meso- = middle (Greek)
derma = skin (Greek)

The **gastrula** has three cell layers, the **ectoderm, mesoderm,** and **endoderm.** The endoderm becomes the lining of the stomach and other digestive organs. Mesoderm forms bones, skeletal muscles, and other internal organs. Ectoderm becomes the skin and nervous system.

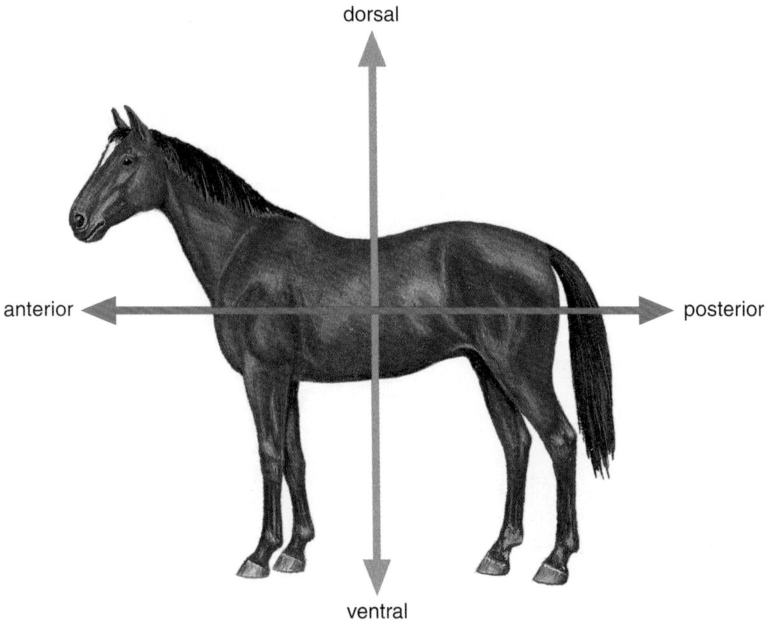

FIGURE 10.7

Anterior-posterior and dorsal-ventral axes in the animal body plan. During the early stages of development, the structure of this body plan is directed by a genetic program such as the one depicted in Figure 15.2b.

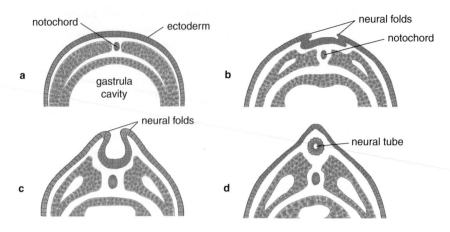

FIGURE 10.8

Development of the neural tube. A vertebrate gastrula possessing a notochord **(a)** begins the process of neurulation (formation of the neural tube) when the ectoderm begins to fold **(b).** The edges of the U-shaped fold arch toward the midline of the embryo **(c),** where the ectoderm fuses to form the neural tube **(d).** Note the reorganization of the primary germ layers in the resulting embryo, called a neurula.

Above the notochord, the dorsal ectoderm folds up to become the **neural tube** (Figure 10.8). This tube will form the brain, spinal cord, and nerves. The tissue interactions that produce the neural tube establish the foundation on which the later stages of development are based (Figure 10.9).

Some animals, such as birds and mammals, develop directly into young that are like the adult. Others—such as frogs, sea stars, and insects—first form a **larva** (plural: larvae), a feeding individual that looks nothing like the adult. The larva later goes through **metamorphosis,** a series of changes that transforms the larva into an adult. A maggot is the larva of a fly; a caterpillar is a larval butterfly. A swimming tadpole is a frog's larva; during metamorphosis it grows legs, loses its tail, and develops lungs.

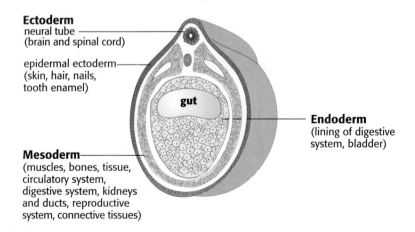

FIGURE 10.9

Cross section of a vertebrate embryo. Note the approximate locations of the three primary germ layers and the lists of some mature tissues and organs into which the layers develop.

Developmental Diversity

10.4 Developmental Patterns and Evolutionary Relationships

The basic developmental pattern varies greatly among animals. In some animals, such as snails and worms, the first two cells are unequal in size. The small cell and its descendants become the ectoderm, and the large cell leads to the mesoderm and endoderm. If the two cells are separated experimentally, each develops only into its limited types of tissues. Fly eggs divide differently, but each part of their cytoplasm contains different RNA and protein molecules. During cleavage, each cell receives molecules that control its fate. In all these embryos, the opening of the gastrula becomes the mouth.

On the other hand, when a sea star or vertebrate zygote cleaves, the first two cells are identical. Cell-separation experiments show that both cells can produce all three kinds of tissue and develop into a complete embryo. Accidental separation of these two cells produces identical twins. In both of these animals, the first opening of the gut forms the anus, while the mouth forms later at the other end of the gut.

Developmental patterns are a clue to relationships among living groups of animals. Even when adult animals are very different, embryonic similarities reflect relatedness. Differences can suggest a more distant relationship or adaptation to different environments. The examples above suggest that sea stars and vertebrates are more closely related to each other than they are to snails, worms, and flies. Each group shares features that the other does not possess. Figure 10.10 illustrates some developmental differences and

CONNECTIONS

Similarities and differences in developmental patterns provide evidence of evolutionary relationships among major groups of animals.

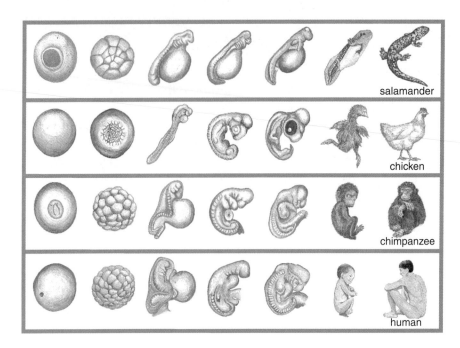

FIGURE 10.10

Developmental stages of several vertebrate animals. Note the similarities that suggest the close relationship of animals whose adult forms are quite different.

similarities among the vertebrates. Cleavage seems very different in birds and mammals. The large bird egg forms a small disk of cells on top of the huge yolk, whereas the tiny, yolk-free mammal egg divides completely and forms a hollow blastula. However, both form disk-shaped gastrulas and continue through morphogenesis in very similar ways.

Charles Darwin was among the first biologists to compare developmental patterns to help determine relationships among animal species. Today molecular biology is beginning to reveal how this works. Related species possess many of the same genes. Even though birds and mammals start development quite differently, morphogenesis in both follows a similar genetic program inherited from a common ancestor.

For example, similar genes in many animals are responsible for **segmentation,** division of the body into a number of similar sections. Grooves on the surface make the segments of worms and fly larvae visible. The segments of vertebrates are clearest in the skeleton. The body-pattern genes were first discovered in fruit flies that carried errors in these genes (Figure 10.11). They developed legs on their heads in place of antennae, had extra wings, or produced larvae that were missing several body segments. Errors in these **homeotic genes** can transform one organ into another.

To study these genes, biologists compared DNA from the abnormal flies with DNA from normal flies. They found 11 homeotic genes close together on one chromosome. Each gene acts in a different body segment. Every one of these genes contains one or more copies of a 180-base-pair sequence that codes for a protein segment (called a homeodomain) that is 60 amino acids long. This DNA sequence is called a **homeobox.** Each of the genes encodes a protein that includes the 60-amino-acid homeodomain. This part of the

ETYMOLOGY

homeo- = similar (Greek)

A **homeotic** transformation is one in which one body structure replaces another. **Homeotic genes** direct the development of animal body parts. The **homeobox,** a sequence of 180 DNA base pairs, is extremely similar in all genes in which it is found.

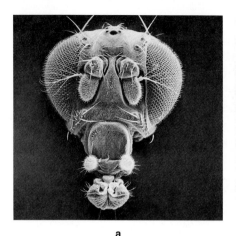

a

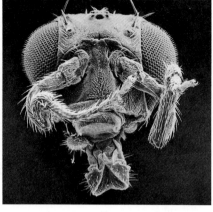

b

FIGURE 10.11

A homeotic mutation in the fruit fly *Drosophila melanogastor*. Scanning electron micrographs of the heads of a normal fly with antennae **(a)** and an *Antennapedia* mutant fly **(b)** are shown here. Note that the homeotic mutation is expressed by legs forming in the segments that should form antennae.

protein binds to DNA, regulating the transcription of important genes. Homeotic genes are located close together, in the same order as the body segments whose development they control. Figure 10.12a illustrates this relationship in the fruit fly.

Nearly identical gene sequences were found in mice. These mouse genes were named ***Hox* genes** for the *h*omeob*ox*es they contain. Biologists were amazed to find that the mouse genes are arranged in the same order as the corresponding ones in flies (Figure 10.12b).

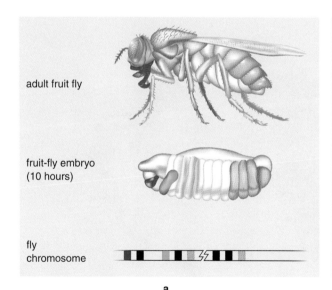

adult fruit fly

fruit-fly embryo (10 hours)

fly chromosome

a

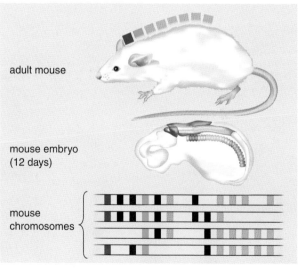

adult mouse

mouse embryo (12 days)

mouse chromosomes

b

FIGURE 10.12

Expression of body-pattern genes. Homeotic genes **(a)** in the fruit fly and *Hox* genes **(b)** in the mouse are color-coded to indicate which gene is active in which body part and to identify which genes have similar DNA sequences in these two species. Note the similar arrangement of genes on the chromosomes and its relationship to the embryo segments expressing those genes. The break in the fly chromosome is meant to show that the cluster of genes active in anterior structures is located away from the cluster of genes that are active in posterior structures. The mouse has four *Hox* gene clusters on different chromosomes.

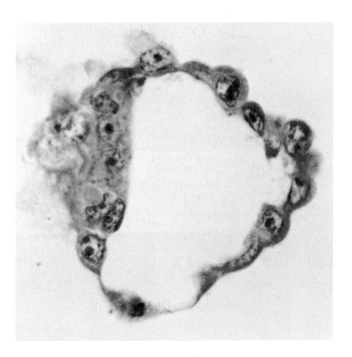

CONNECTIONS

The similar roles of homeotic genes in distantly related animals such as mice and flies are evidence of the shared ancestry of all life.

Homeotic genes and similar regulators of morphogenesis in other animals show that most animals share the same basic genetic program for the body plan. These genes have changed very little during evolution. Changes in homeotic genes often lead to embryonic death or severe abnormalities. Occasionally, the change in a body part might persist. A mutation in a homeotic gene could produce a sudden evolutionary change. The molecular analysis of embryonic development has provided much evidence of the evolutionary relationships among animals.

10.5 Human Development

The development of humans and most other mammals (animals with hair and milk glands) is unique (see Section 12.7). Their embryos develop within the mother. The mother's body provides a warm, protected environment. Her blood circulation provides nutrition and oxygen to the embryo and takes away wastes and carbon dioxide.

The human egg is very small, 0.1 mm in diameter, and contains almost no yolk. The fertilized zygote cleaves as it moves down the duct from the ovary into the uterus. About 5 days after fertilization, the embryo, called a **blastocyst,** resembles the hollow blastula of other animals. The blastocyst sinks into the wall of the mother's uterus to develop and grow.

Part of a thick mass of cells inside the blastocyst forms the disk that becomes the embryo (Figure 10.13). In this disk, gastrulation occurs just

ETYMOLOGY

cyst = bladder or sac (Greek)
A **blastocyst** is a blastula that resembles a small sac.

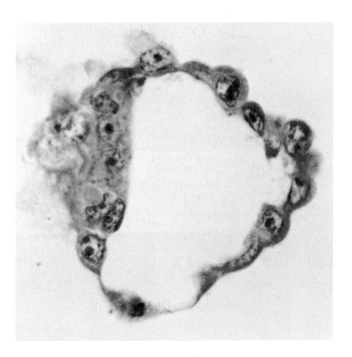

FIGURE 10.13

Section of a human blastocyst. The embryo develops from the inner cell mass, the thicker part of the blastocyst seen at the left.

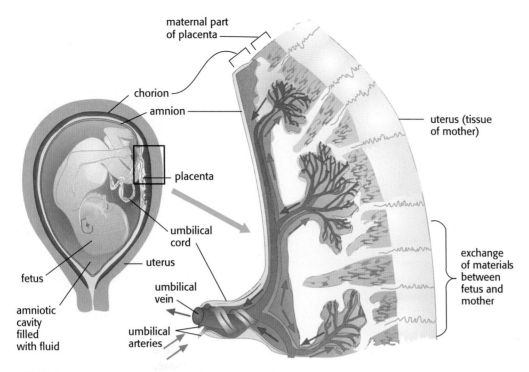

FIGURE 10.14

Embryonic membranes. A developing human fetus in the uterus, the embryonic membranes, and the placental connection. Part of the placenta is enlarged to show both fetal and maternal circulations. The mother's blood does not mix with that of the fetus; exchange of materials occurs across the thin membranes that separate the fetal capillaries from small pools of maternal blood.

as it does in birds and reptiles. The rest of the blastocyst develops into membranes that surround, nourish, and protect the embryo (Figure 10.14). The **amnion** immediately surrounds the embryo. The **chorion** encloses all the other membranes and forms from the blastocyst's thin outer wall.

As gastrulation begins, the chorion extends fingerlike projections, or **villi** (singular: villus), into the lining of the uterus. The chorionic villi and uterine lining form the **placenta,** which exchanges nutrients, wastes, oxygen, and carbon dioxide between mother and embryo. Blood vessels in the umbilical cord connect the embryo to the placenta. Arteries carry wastes and carbon dioxide from the embryo to the placenta; veins return nutrients and oxygen. The mother's blood flows through cavities in the placenta. The chorionic villi extend into these cavities. Nutrients and wastes pass through the villi and embryonic blood vessels, but the two blood supplies remain completely separate.

A human takes about 40 weeks to develop in the uterus (Figure 10.15). After the beginning of the eighth week, the embryo is called a **fetus.** After 3 months, or the first trimester, most of the organs have begun to form, and the skeleton can be seen in ultrasound images. Most of the last 3 months, or third trimester, is a period of rapid growth and maturation of organ systems.

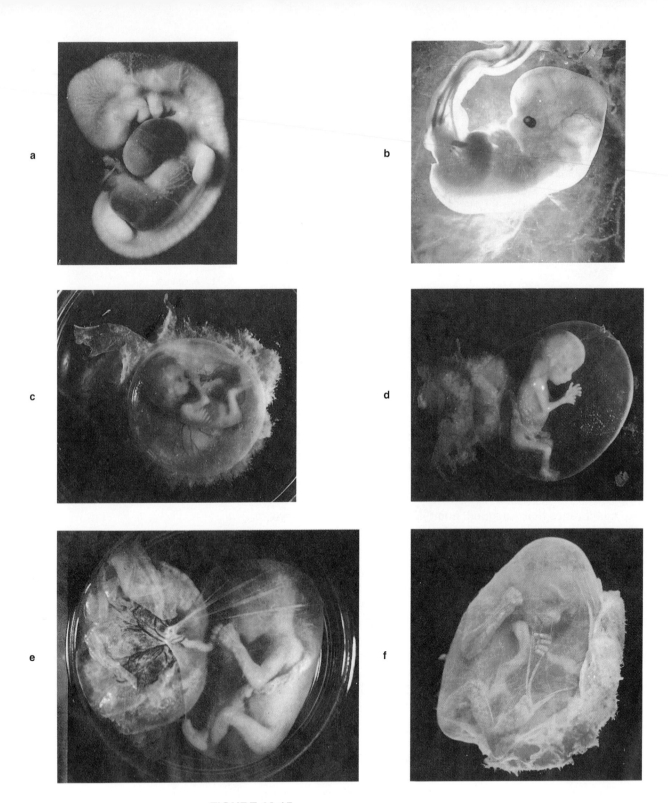

FIGURE 10.15

Stages in the early development of a human embryo and fetus (not to scale). At 4 weeks after fertilization **(a),** the human embryo, like all vertebrates, has a tail. At 6 weeks **(b),** retinal pigments mark the eye, and the head is growing rapidly. By 8 weeks **(c),** the embryo looks human and has fingers and toes, which are even more distinct at 12 weeks **(d).** By 14 weeks **(e)** and 16 weeks **(f),** the fetus is moving, making facial expressions, and thumb-sucking.

10.6 Birth Defects

Unfortunately, noticeable and sometimes serious birth defects are not uncommon. Some are caused by defective genes and others by environmental factors acting on normal or abnormal developmental genes. The causes of many birth defects are still unknown, yet all stem from disrupted development.

Polydactyly, the condition of having extra fingers or toes, is caused by an altered gene. Biologists do not yet know how this gene alters human limb development, but an idea has come from experiments on chicken embryos. As the first bud of a chick's limb begins to grow, a particular gene is active only in the posterior part of the bud. It produces a protein that seems to control the pattern of digits in the limb by regulating the transcription of other genes. Experimenters exposed cells in the anterior part of a limb bud to this protein. As the embryo developed, the treated limb developed extra digits in a mirror-image pattern. Research work with the chicken limb may help answer our questions about human limb development.

Neural-tube defects occur when part of the neural tube does not close completely (Figure 10.16). In spina bifida, the posterior end of the neural tube fails to close, and even the body wall remains open. Surgeons can close the opening, but other severe problems persist through life. When the anterior part of the tube fails to close, a large part of the brain fails to develop. In this condition, called anencephaly, the exposed brain degenerates, and the top of the skull fails to form. These individuals usually do not survive after birth.

Both genes and environmental factors affect neural-tube development. Experiments with mice suggest that up to half of neural-tube defects could be prevented if pregnant women consumed sufficient folic acid and vitamin B_{12}. For this reason, physicians recommend that women of childbearing age take a daily folic-acid supplement. Researchers have also identified at least three genes involved in neural-tube development. For example, for the tube to close, its cells must be able to stick to each other. In mice with inherited

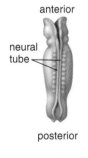

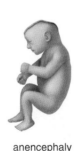

anterior

neural tube

posterior

a

normal

b

anencephaly

c

spina bifida

d

e

FIGURE 10.16

Human embryo neural-tube closure and congenital neural-tube defects. Neural tube closing at 23 days of development **(a)** leads to normal development **(b).** Failure of the anterior end of the neural tube to close results in anencephaly **(c).** Failure of the posterior end of the neural tube to close leads to spina bifida **(d).** Because the muscles of the legs are normally controlled by nerves that extend from the posterior end of the spinal cord, many people with spina bifida are unable to walk **(e).**

THEORY

Engineering Life

Giraffes' long necks allow them to reach leaves high off the ground. This specialization gives them access to food that is unavailable to zebras and other grazers. Could this adaptation evolve in another species? If an antelope developed a very long neck but did not change in other ways, it would fall over. As giraffes evolved, natural selection ensured that their muscles, circulatory systems, and skeletal joints were balanced and coordinated to support a longer, heavier neck. Uncoordinated changes in body form interfere with successful adaptation. Hence, natural selection tends to eliminate poorly adapted individuals.

Can scientists predict the changes that are most likely to evolve in the future? Structural adaptationism is a theory that assumes that living organisms are well adapted. Because organs must work well together, any changes in one requires changes in others. Like engineers studying a building design, structural adaptationists build theoretical models using ideas from mechanics to study the structural integrity of organisms. As adaptationists, they assume that organisms evolve in ways that make them more efficient. Consider car designs as an analogy. In designing new cars, auto engineers try to consider desirable factors such as interior space, fuel consumption, engine power, and handling. In general, the improvement of one factor requires sacrifices in another. As the photos here suggest, structural adaptationists imagine that natural selection works like engineering. Like the best-designed cars, only the most efficient organisms succeed in the competitive "market" of survival.

Scientist can use structural engineering models as theories that may predict future evolutionary changes. The models weigh efficiency of body form in a given environment with the likelihood that the form will arise and survive among offspring. Since future changes are not observable, scientists look to the past to test the models. For example, one model predicts the evolution of the cheetah from its ancestors. Based on specific assumptions about the environment, competitors, and prey at the time, the theoretical model assigns a high rate of success to the efficient cheetah body type.

Critics of these models point out that many organisms have useless traits, features that persist without contributing to survival. Examples of such traits include the human appendix and the production of light by some marine microorganisms. Others claim that engineering models oversimplify the picture since we do not know the details of past or future environments that make a change beneficial or harmful. Even if these objections are valid, models based on concepts of engineering help us understand the reasons that some body plans are so common.

A cheetah and a racecar. Both are structured for speed on a certain type of surface. Imagine a different surface, such as thick mud. How would these structures need to change to move quickly on mud? Would the cheetah and the car change in the same way?

neural-tube defects, some of the cell-surface adhesive molecules are different, so the closing cells fail to stick to each other. In summary, healthy genes and environments both are needed for normal development.

Check and Challenge

1. How do developmental patterns help biologists determine evolutionary relationships among animals?

2. What events in early development suggest that frogs are more closely related to sea stars than either group is related to snails or worms?

3. What kinds of genes that regulate development suggest that flies and mice are related? How do these genes' structure and organization on the chromosome suggest their importance in all types of animals?

4. In your own words, describe the human blastocyst. Which of its parts contributes cells to the embryo? to the two major embryonic membranes? to the placenta? Which of the mother's structures also contributes to the placenta?

Mechanisms of Cell Differentiation

10.7 Exploring the Mechanisms of Differentiation

Just describing the development of an embryo cannot tell us what cellular and molecular processes control this orderly series of events. To find an explanation, a scientist must propose a testable hypothesis and then design an experiment to test it. To propose a hypothesis, the scientist must first have a question that she or he is trying to answer.

Developmental biologists ask questions like these: How does a zygote produce many types of cells? Do differentiating cells lose some of their DNA as they specialize to make only certain proteins? If not, what turns some genes on and others off? Does a differentiating cell follow a rigid genetic program or do cells communicate to regulate each other's development? Our understanding of development comes from experiments that test hypotheses based on such questions.

Early experiments with embryos involved surgery. Scientists removed certain cells or moved tissues to new locations. These experiments revealed much about the influence of neighboring tissues on development. A later method involved replacing the nucleus of an unfertilized egg with the nucleus of a differentiated cell. The development of this new cell showed how much development that nucleus could control.

Molecular methods now help determine which genes are active in a particular cell. Scientists can make large quantities of a particular gene's

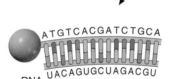

FIGURE 10.17

DNA-RNA hybridization. Single-stranded DNA is prepared by heating or chemical treatment, and a chemical tag is added. In cells with complementary RNA, tagged DNA hybridizes. The chemical tag reveals the location of the complementary RNA.

ETYMOLOGY

hybrid = of mixed origin (Latin)

DNA-RNA hybridization produces a mixed double helix with one DNA strand and one RNA strand.

DNA and use chemicals to separate the DNA's two strands. Then they attach dye molecules to the DNA to make it visible. These tagged DNA molecules are used as probes to detect RNA with a matching nucleotide sequence (Figure 10.17). This method is called **DNA-RNA hybridization.** Scientists treat embryos with DNA probes and examine them with microscopes. Cells that have transcribed the gene contain mRNA that matches the probe's sequence. The dye bound to the probe marks these cells.

10.8 The Genetic Equivalence of Differentiating Cells

When a cell differentiates, what happens to the genes it will not use? The selective-gene-loss hypothesis (Figure 10.18a) proposes that differentiating cells lose some genes. The genetic-equivalence hypothesis (Figure 10.18b), however, states that all cells contain the same genes, but some genes become inactive during differentiation. Following are the results of some experiments designed to test these hypotheses. Does this evidence indicate that differentiated cells are genetically identical or different?

In 1952, Robert Briggs and Thomas King injected the nuclei of differentiated cells from leopard frogs *(Rana pipiens)* into unfertilized frog eggs. They replaced the egg's nucleus with the nucleus of a differentiated

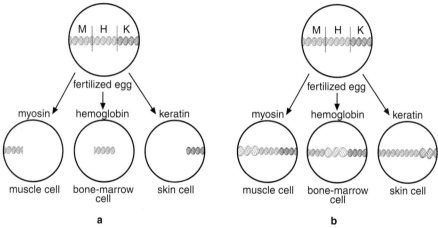

FIGURE 10.18

Two possible explanations for differentiation. (a), Selective-gene-loss hypothesis: Differences between the proteins of various cells are due to selective loss of genes as cells divide during development. Each cell type retains only a specific set of genes. **(b),** Genetic-equivalence hypothesis: All cells have the same genes, but control mechanisms turn on expression of only specific genes in each cell type.

cell. These eggs never met a sperm; the injection activated the egg. A nucleus from a blastula, an early stage of embryo development (Figure 10.19a), supported development of the egg all the way to becoming a tadpole. When the researchers used a nucleus from a skin cell (Figure 10.19b), development stopped soon after gastrulation. The more differentiated cell still had all of the genes needed for early development, yet its nucleus could not support production of all kinds of cells.

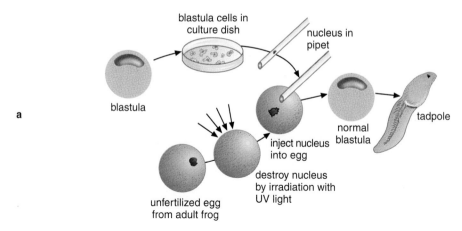

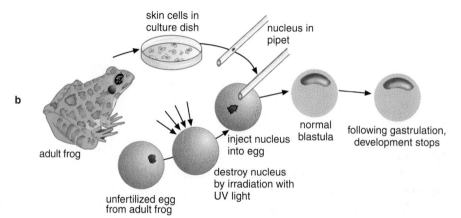

FIGURE 10.19

Diagram of an experiment showing that the nucleus of a differentiated cell contains the same set of genes as the zygote. Cells from an early developmental stage retain the ability to direct formation of a complete tadpole **(a),** while the differentiated skin cells **(b)** do not.

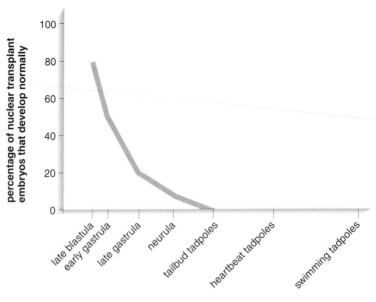

FIGURE 10.20

Graph of the percentage of frog embryos created by nuclear transplant that develop normally in relation to the developmental stage of the donor nucleus. As cells become more fully differentiated, their nuclei are less able to direct development to the swimming-tadpole stage.

Soon after this, John Gurdon's experiments with South African clawed frogs *(Xenopus laevis)* showed that an egg with the nucleus of a differentiated tadpole cell could develop into a reproducing adult frog. Nuclei from adult skin cells did not support development past the tadpole stage, and no genes were lost as cells differentiated. The results support the hypothesis that all cells in an individual are genetically equivalent, but differentiation does restrict the expression of some genes (Figure 10.20).

Focus On

A Sheep from a Differentiated Cell Nucleus

Since the mid 1990s, experiments have produced a few normal sheep, cattle, monkeys, donkeys, mice, and pigs from eggs given adult cell nuclei. Like the frog experiments, this is an example of cloning, the production of a group of genetically identical offspring. To clone a sheep, Ian Wilmut and his colleagues replaced the nucleus of a sheep's egg with the nucleus of a cell from a female sheep's udder (see Biological Challenges: Cloning in Section 12.2). This cell was specialized to make milk proteins. Wilmut's group grew udder cells in their laboratory. To remove the genetic blockages created during differentiation, they used the nucleus of a cell in the G0 phase of the cell cycle (see Section 8.2). This nucleus and all of its descendent nuclei supported the development of an adult sheep born in 1996 (Figure 10.21). This result supports the genetic-equivalence hypothesis.

Farmers and scientists see the agricultural value of cloning animals by nuclear transfer because one prized animal could be replicated many times without the random changes that can occur with two parents. A major obstacle is that the method is very inefficient. Currently, nuclear transfer produces a live offspring in less than one of every 100 attempts.

FIGURE 10.21

Lamb 6LL3, known as Dolly, the first reported cloned mammal. Dolly, cloned from a mammary cell of a Finn-Dorset ewe, is genetically and physically distinct from her surrogate mother, a Scottish Blackface ewe.

10.9 Determination and Differentiation

What controls the differences in gene expression between different types of cells? Many embryonic cells are committed to their fates long before they differentiate. **Determination** is the process by which a cell commits to a particular course of development. In some animal embryos, determination occurs independently in each cell. Other species have a less rigid determination system. Their cells communicate, affecting each other's differentiation.

Experiments with two-cell embryos of a snail and of a frog demonstrate two extremes of determination (Figure 10.22). The two cells were separated and allowed to develop separately. Each frog cell produced a complete tadpole. On the other hand, the snail cells did not produce normal larvae. Both cells continued to cleave, but the smaller cell produced only ectoderm, and the larger cell made a mass of mesoderm and endoderm. The snail embryo's cells were already irreversibly determined. In contrast, the frog embryo's two cells were not yet determined.

Eventually, the frog embryo's cells do become determined. If one or two cells are removed from a 32-cell frog embryo, the remaining cells will adjust and development will continue normally. But if a certain cell is removed from a 32-cell snail embryo, no mesodermal organs will be produced. The embryo cannot adjust to make up for the missing cell.

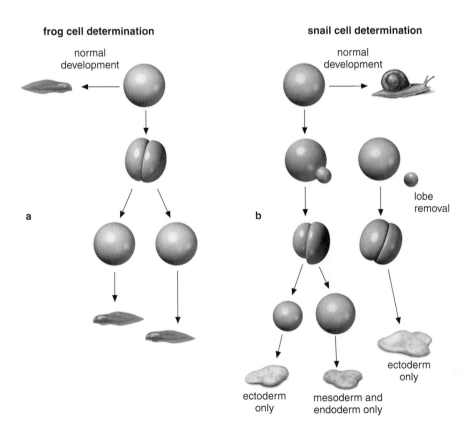

frog cell determination

snail cell determination

FIGURE 10.22

Determination experiments. (a), Each cell of a two-cell frog embryo can develop into a complete tadpole. **(b),** The snail small cell develops into an abnormal ectodermal mass; the large cell forms abnormal mesoderm and endoderm. Removing the polar lobe produces an embryo with two equal cells that make no mesoderm or endoderm. The mesoderm-endoderm determinants are in the lobe.

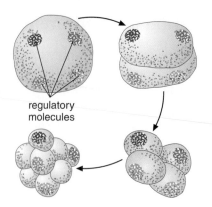

FIGURE 10.23

Regulatory molecules are not equally distributed in the cytoplasm of an egg. As the egg divides, different molecules may be incorporated in each offspring cell, activating different genes.

What determines the fates of snail embryonic cells so early in development? Some proteins and other molecules are distributed unevenly in the egg cell (Figure 10.23). Cleavage tends to leave each cell with different cytoplasmic components that later influence gene expression.

10.10 Cytoplasmic Determination

What makes a snail's first two cells different? Figure 10.22b shows that a lobe of cytoplasm forms near one pole of the zygote before it divides. The cytoplasm in the lobe looks different from the rest and becomes part of the larger cell. Could factors in the lobe cytoplasm determine each cell's fate? To find out, scientists used microneedles to remove the lobe from a zygote. The first cleavage produced two cells of equal size that developed into an abnormal larva with no heart or intestine. This evidence supported the idea that the large cell becomes mesoderm and endoderm because of something it receives in the lobe. What regulator is in the lobe?

The control factors in the snail lobe are still unknown, but in the embryos of tunicates (sea squirts), they include RNA and proteins. Sea squirts are spineless relatives of the vertebrates. Their embryos develop into microscopic larvae. Some of their eggs have a layer of brightly colored yellow or orange pigment right below the cell surface. Just after fertilization, the pigment streams into a crescent-shaped area that will become the posterior end of the embryo (see Figure 10.3b). At the same time, light gray pigment granules stream to the anterior end. After several divisions, only a few cells are either yellow or gray (Figure 10.24). The gray cells form the notochord. The two yellow cells produce the muscles in the larva's tail. Removing both yellow cells stops tail-muscle development completely. Removing one of them produces a larva with half the normal number of muscle cells. Thus, differentiation in tunicates seems to begin with movements of the zygote's cytoplasm that carry pigment granules and regulatory molecules into different regions of the cell. Cleavage distributes these molecules to different cells.

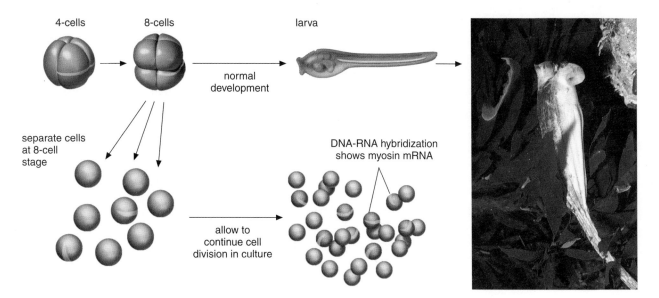

FIGURE 10.24

Cytoplasmic determination and larval tail-muscle differentiation in the tunicate _Styela_. Yellow crescent pigment persists in the tail-muscle cells. When cells are separated and allowed to divide, those containing the crescent cytoplasm express myosin (muscle protein) genes. The others do not.

Yellow cells grown in isolation produce molecules found in muscle—the enzyme cholinesterase and the mRNA for the muscle protein myosin. However, the yellow pigment does not regulate differentiation. DNA-RNA hybridization has shown RNA molecules attached to the cytoskeleton, moving with the pigment. Some of these RNAs encode regulatory proteins that turn on the genes that lead to muscle differentiation.

RNA and protein molecules are also distributed unevenly in fruit-fly eggs. DNA-RNA hybridization revealed one mRNA at the embryo's anterior end and a different one at the posterior end (Figure 10.25). These RNAs encode proteins that regulate differentiation by controlling either the transcription or translation of genes that are active during development of various types

CHEMISTRY TIP

Acetylcholine ($C_7H_{17}NO_3$) is the molecule that transmits a nerve impulse to a muscle, starting contraction. Cholinesterase is the enzyme that destroys acetylcholine to prevent continuous muscle contraction.

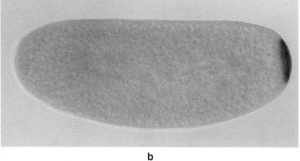

a   b

FIGURE 10.25

Anterior and posterior determinants in the fruit-fly _Drosophila_ embryo. Different mRNAs detected by DNA-RNA hybridization are shown here. **(a),** The mRNA that encodes a protein called Bicoid occurs in the anterior tip of the embryo. **(b),** The mRNA that encodes a protein called Nanos is most concentrated at the posterior tip. While Bicoid protein activates specific gene transcription in the anterior part of the embryo, the Nanos protein inhibits translation of the same gene transcript in the posterior cells.

of cells. The mRNAs from the egg are translated during early development. Cells form anterior, middle, or posterior tissues, depending on the amount of each protein that they possess. One kind of abnormal fly produces headless larvae. They have two posterior ends and do not survive. These flies' eggs have neither the anterior RNA nor the anterior protein. They develop normally when the anterior end of the zygote is injected with anterior cytoplasm taken from a normal embryo.

All these experimental results support the hypothesis that RNA and proteins in egg cytoplasm help control differentiation by regulating gene expression. This pattern seems to occur among all types of animals.

10.11 Cell-Cell Interactions

Differentiation is more flexible in vertebrate embryos than in sea squirts and flies. Hans Spemann and Hilde Mangold demonstrated this in the 1920s, using salamander embryos. They hypothesized that a signal from the notochord shifts the neighboring dorsal ectoderm cells from skin to neural-tube differentiation. They transferred pieces of ectoderm between two gastrulas. When cells that normally form skin were moved to the dorsal region, above the notochord, they produced a neural tube instead. When dorsal cells that should later form the neural tube were placed in the ventral area, they produced skin (Figure 10.26a). This experiment showed that the cells respond to other cells nearby as they become determined. If the dorsal tissue is taken from a very late gastrula, it develops as a neural tube in the skin-forming area (Figure 10.26b). By now, the signals to become neural tissue have been received, and this tissue does not turn back to become skin. This process is called **embryonic induction.**

FIGURE 10.26

Spemann and Mangold's first transplant experiments. (a), When the donor is a blastula or very early gastrula, both dorsal and ventral ectoderm are not yet determined. Ventral ectoderm becomes a neural tube in the new dorsal location; dorsal ectoderm becomes belly skin in its new ventral location. **(b),** A late-gastrula donor provides ventral ectoderm that can still be induced to become neural, but its dorsal ectoderm is now determined to make a neural tube regardless of its location.

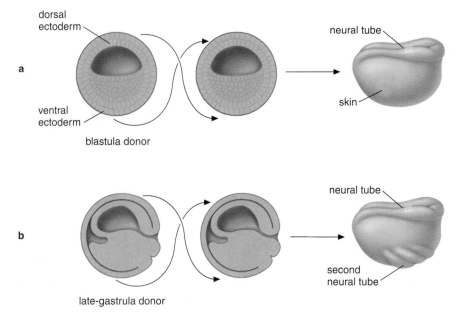

dorsal ectoderm

neural tube

a

ventral ectoderm

skin

blastula donor

neural tube

b

second neural tube

late-gastrula donor

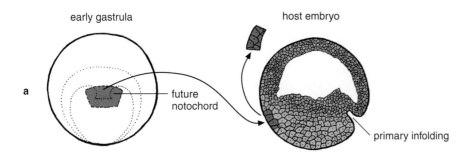

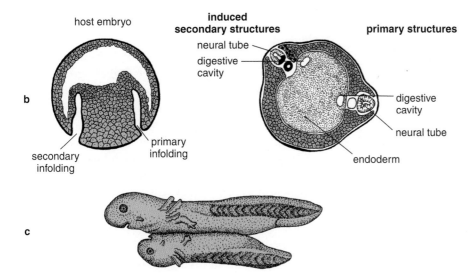

FIGURE 10.27

Spemann and Mangold's embryonic induction experiment. **(a),** Future notochord tissue of an early gastrula was removed and transplanted into a host embryo as shown. **(b),** The transplanted tissue caused a second point of infolding and a second notochord to appear in the host embryo, inducing a second neural tube. **(c),** This transplant produced two embryos joined at the abdomen.

Spemann and Mangold's next experiment showed that the notochord is the source of the inducing signal. They took from a blastula or early gastrula a piece of tissue that would later become the notochord (Figure 10.27a). When transplanted to another embryo, this tissue formed an extra notochord. In turn, the extra notochord induced the neighboring ectoderm to form a second neural tube (Figure 10.27b). The second notochord started a series of events that led to a nearly complete second larva attached to the main individual (Figure 10.27c).

Communication between notochord and ectoderm is only one of many kinds of cell-cell interactions in development. Spemann and Mangold's results so excited biologists that the search for the molecular mechanisms of neural induction has continued ever since. When a piece of notochord is cultured with ventral ectoderm, the ectoderm cells are soon induced to become neural cells. To determine whether induction requires cells to be touching, scientists cultured these tissues on opposite sides of a plastic filter. The pores of the filter were so small that no cell contact would occur, but molecules could pass through. Induction occurred through the filter. This experiment showed that contact is not required. The notochord communicates with the ectoderm by releasing a substance that can pass through the filter.

Molecules and Muscle Differentiation

When muscles do not differentiate normally, the fetus or newborn is seriously abnormal. Muscles are composed of contracting fibers; each fiber consists of many individual cells. Muscle cells, or myofibrils, have several nuclei and are filled with arrays of the proteins actin and myosin that enable them to contract (see Section 7.7). Figure 10.28a illustrates how a myofibril forms from undifferentiated mesoderm cells that become myoblasts. Several myoblasts fuse with each other, forming a large cell with several nuclei. This cell, the myotube, synthesizes actin and myosin and matures into a myofibril.

Research has found that two genes, *myoD* (Figure 10.28b) and *myf-5*, are active in locations where muscles develop. When extra copies of these genes are added to cartilage- or nerve-forming cells, the cells turn into myoblasts. The MyoD and Myf-5 proteins activate myoblast differentiation. The gene *myogenin* is important in the fusion of myoblasts to form myotubes. The gene *MRF4* controls the maturation of myotubes into myofibrils. The protein products of these genes regulate transcription from similar sites on DNA.

Scientists have genetically engineered embryos (see Section 15.3) in which each of these genes has been destroyed. Did each embryo make normal muscle? No. When *myoD* and *myf-5* were both knocked out, no myoblasts developed. If only the *myogenin* gene was destroyed, normal myoblasts appeared, but they did not fuse to make myotubes. Cells in embryos that lacked *MRF4* matured only to the myotube stage. Thus, each gene's role in muscle differentiation became clear.

There is much more to learn about the intricate steps of muscle development. These discoveries pave the way for future hypotheses and experiments. For instance, what regulates the genes that make these regulators? An answer to this question will help us better understand muscle development.

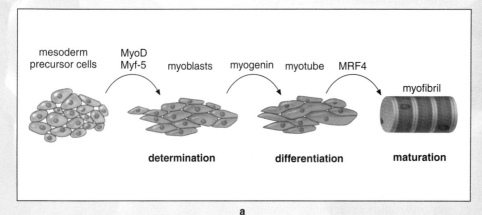

a

b

FIGURE 10.28

Muscle-cell commitment and differentiation. (a), Cell commitment, myoblast differentiation, and fusion to form myofibrils is the process shown here. Important regulatory proteins are indicated next to the steps they influence. **(b),** A mouse embryo stained for expression of *myoD* gene is shown at right. Note that the *myoD* gene is active in muscle segments and embryonic limbs.

Experiments of this sort started biologists searching for the inducing substance. More than 20 years later, molecular techniques finally revealed that induction involves a number of genes and proteins. Several of the DNA sequences and their proteins were found in embryos of the South African clawed frog. Two inducing proteins that the notochord produces are called chordin and noggin. DNA-RNA hybridization shows that, as early as gastrulation, the *chordin* and *noggin* genes are active in cells that will later form the notochord. Scientists injected chordin protein into the ventral side of frog embryos before gastrulation. The chordin induced the ventral cells to form a second anterior-posterior axis, somewhat like the one Spemann and Mangold's embryo formed (see Figure 10.27). This showed that chordin is important in notochord formation and in establishing the anterior-posterior axis of the frog embryo. Abnormal frog embryos with no dorsal tissues formed dorsal structures when treated with noggin RNA. Noggin protein also causes ventral ectoderm in laboratory tissue cultures to become nervous tissue instead of skin. All these results confirm the importance of noggin and chordin in neural induction.

The cell interaction story is not simple. The chordin and noggin proteins do not regulate gene transcription directly. Each interferes with the action of another protein that controls the production of a family of other proteins. These latter proteins regulate the transcription of many specific genes involved in the development of nerve cells. Scientists have found more than 10 developmental control proteins in notochord cells alone. Some of them help determine which part of the neural tube forms the brain and various types of nerves.

These cell-cell interactions and their associated genes and proteins are only a small part of the developmental picture. Molecular methods are helping scientists to better understand the events that bring an embryo from fertilization to birth.

Check and Challenge

1. Compare determination with differentiation. Which process occurs first? Which process produces visible changes in the cell?

2. Compare the selective-gene-loss hypothesis with the genetic-equivalence hypothesis. Explain what each hypothesis says. Which is supported by more evidence? Give examples of this evidence.

3. Explain molecular hybridization in your own words. If a single strand of DNA hybridizes with an RNA molecule, what does this suggest about their nucleotide sequences?

4. How can differences in cytoplasm explain determination? Give an example and explain how it relates to differentiation that occurs later.

5. In your own words, summarize the meaning of the phrase *cell-cell interaction*.

6. Draw and label a picture that shows the steps and the results of Spemann and Mangold's second experiment. Draw a diagram of an experiment that tests the hypothesis that a soluble molecule, and not cell-surface contact, is the key factor in neural induction.

Chapter
HIGHLIGHTS

Summary

The development of most animals begins when gametes come together at fertilization, creating the zygote. The cell divisions of cleavage produce a multicellular blastula, and the first signs of cell differentiation are seen as the blastula transforms into a gastrula. Gastrulation establishes the three primary germ layers (ectoderm, mesoderm, and endoderm). Further differentiation and morphogenesis produce the tissues and organs of the larva and adult. Developmental biologists focus their experiments on hypotheses about the control of cell determination, differentiation, and morphogenesis.

Developmental differences and similarities provide one kind of evidence of evolutionary relationships among types of animals. In flies and mice, genes that control morphogenesis have similar DNA structure and organization, showing the shared evolutionary ancestry of animals.

Experimental analysis of cell determination and differentiation in embryos depends on microsurgical manipulations, cell and tissue culture, and molecular-biology techniques. All of an animal's cells contain a complete set of genes. Determination and differentiation involve selection of the genes that are expressed in each cell. Cytoplasmic determination can distribute gene-regulation factors to different cells during cleavage, determining the cells' fates. Interactions between cells may influence cell determination throughout development. Cells and tissues exchange molecular signals, stimulating further specialization of structure and function.

Key Concepts

Use the following concepts to develop a concept map. Add other terms from the chapter as you need them.

- differentiation
- proteins
- genetic equivalence
- cell-cell interactions
- determination
- rearrangement of cytoplasm

Reviewing Ideas

1. How are growth, differentiation, and morphogenesis related to each other? Can any of the three processes occur independently of the others? Explain your answer.
2. Explain how mammals differ from other vertebrates in their developmental processes. In what ways are the developments of mammals and of birds similar?
3. What are two ways in which observations of early embryos relate to the theory of evolution?
4. What kinds of experiments demonstrate that nuclei become more restricted in their developmental potential as they differentiate? Which experiment (or experiments) demonstrates that unused genes in differentiated nuclei can still be expressed, given the proper environment?
5. Describe how regulatory molecules in the cytoplasm might become segregated into different cells during development. How might these regulators affect the subsequent developmental fate of each cell?

Using Concepts

1. List the ways that a zygote differs from a gastrula of the same species. Use information from the chapter and consider characteristics such as the number and size of cells, the contents of the nucleus and cytoplasm, and the normal capacity for gene expression in each nucleus. Under normal conditions, which cell has more developmental potential—the zygote or a cell from a gastrula? Explain your answer.

 BSCSblue.com/vocabulary_puzzlemaker

2. Explain the two alternative hypotheses about the genetic basis for differentiation. Which hypothesis is more strongly supported by the evidence presented in the chapter? Explain your answer.

3. Housekeeping proteins are present in almost all of an organism's cells because they are vital for cell survival. Specialized proteins are found only in certain cells and are required for the cell's special functions. Examine Table 10.2. Name three possible housekeeping proteins and three possible specialized proteins. What might need to occur in a skin cell's nucleus before it could produce a daughter cell that would differentiate into a pancreatic cell?

TABLE 10.2
Two Cell Types and Some of Their Proteins

Cell Type	Proteins Synthesized
skin	DNA polymerase, phosphoglucokinase, keratin, actin
pancreas	actin, phosphoglucokinase, amylase, DNA polymerase, trypsin

4. Name the two mechanisms by which embryonic cells become committed to particular fates. What is the importance of cleavage to each of these mechanisms? What is the importance of gastrulation?

5. This chapter discusses a number of transcription-factor proteins that are important in determination and morphogenesis in mice, fruit flies, and frogs. Name some of those proteins and explain in which organ or tissue systems they control development. What sorts of experimental techniques would you use to determine whether a particular embryonic tissue was actively making the RNA for one of these proteins?

Synthesis

1. What is the relationship between the processes of reproduction and development? In what way is development part of the process of reproduction? Explain your answer.

2. Is there any embryological evidence supporting the theory of evolution? Explain your answer.

3. What is the role of development in the structure and function of an animal?

Extensions

1. Write a short essay about the mechanism by which cytoplasmic molecules (those in the egg or those made by embryo cells) contribute to differentiation of the animal. How might such regulatory molecules function?

2. Research and sketch the development of a specific animal. Some possible choices are a bird, opossum, kangaroo, platypus, butterfly, spider, shark, and bony fish. What relationship, if any, exists between the animal's environment, its way of life, and its development?

Web Resources

Visit BSCSblue.com to access
- Resources related to animal development
- Web links to activities and information about embryonic development in humans and other animals
- Web links to a collection of images that allows you to explore the development of a human embryo in detail from conception to birth

CONTENTS

LEARNING OUTCOMES

By the end of this chapter you will be able to:

A Describe the structures involved in seed germination.

B Explain primary and secondary growth in plants.

C Discuss the factors that affect plant germination and growth.

D Discuss the actions of various plant hormones.

E Describe how plants respond to light, gravity, and day length.

◀ **About the photo**
This photo shows a
seedling emerging from
the soil.

CHAPTER

11

Plant Growth and Development

■ *How would you design an experiment to learn if this seedling is responding to a source of light?*

■ *What other forces affect the growth of this plant?*

In the story of Jack and the beanstalk, Jack's mother threw some bean seeds out of the house in disgust after Jack had traded the family cow for the seeds. The seeds sprouted and grew high into the sky. Although this is only a fairy tale, it illustrates a real aspect of plant growth and development—some plants grow extremely fast. Bamboo shoots, for example, can grow more than 1 m per day. Giant sequoia and redwood trees can grow extremely tall—almost 100 m. Certain plant tissues develop and grow throughout the life of the plant.

What is the difference between growth and development? Growth is an increase in size. Increases in both the number and size of cells contribute to growth. Development is the process by which the cells of a new organism become specialized to perform different functions. Various cells develop into a plant's many tissues, specialized for photosynthesis, nutrient transport, and other necessary tasks. This chapter examines how plants grow and develop and some factors that affect these processes.

Plant Development

11.1 The Embryo and the Seed

Try Investigation 11A
Seeds and Seedlings.

Sexual reproduction begins with fertilization in both plants and animals. The embryo that develops into a new plant forms from the zygote. In most species, the embryo remains dormant inside a seed until the environment is suitable for growth. Asexual reproduction also occurs in plants. The details of reproduction are described in Chapter 12.

The major stages of development of the embryo of a flowering plant are shown in Figure 11.1. About 95% of all plant species are flowering plants. The process is somewhat different in nonflowering seed plants such as conifers (needle-leaved trees). In both kinds of plants, mitotic cell divisions of the zygote form a spherical mass of cells that develops into the embryo (Figure 11.1a).

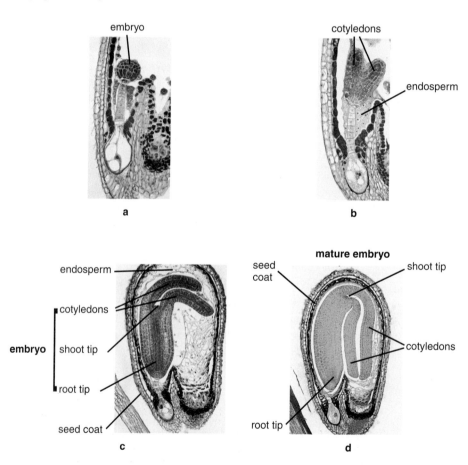

FIGURE 11.1

Development of a plant embryo. The embryo develops **(a)** from the zygote and includes one or more cotyledons, a shoot tip, and a root tip, ×125. Endosperm, a food-storage tissue, surrounds and nourishes the developing embryo **(b, c).** It may be completely consumed before the seed is mature **(d).** The core of densely stained cells in the center of the embryo in **(c)** and **(d)** is beginning to differentiate into future vascular tissue (xylem and phloem).

As the embryo develops, it is surrounded by a tissue called **endosperm.** The endosperm helps transfer nutrients from the mother plant to the developing embryo. In some species, the embryo completely digests and consumes the endosperm during seed development. In others, the endosperm persists in the seed, serving as a source of nutrients for the young plant until photosynthesis can begin.

Differentiation begins as small bumps form on the developing embryo (Figure 11.1b). These bumps become the **cotyledons,** or seed leaves, of the embryo. Cotyledons are not true leaves, but in many plants, once above ground, they carry on photosynthesis until the first leaves of the newly sprouted plant develop. The cotyledons absorb nutrients from the endosperm that nourish the embryo. The cotyledons of many species are also specialized to store some of these nutrients as protein, lipids, or starches that are broken down later to support the growth of the embryo into a seedling.

During seed formation, genes that encode storage proteins and the enzymes of starch and fat formation are highly active in the endosperm or cotyledons. The nutrient reserves will nourish the embryo when it begins to grow into a new plant.

Next the embryo elongates quickly. Cells in the embryo divide rapidly. The new cells begin to differentiate into specialized structures. Cells between the cotyledons become the embryonic shoot, which will later produce the stem and leaves (Figure 11.1c). At the opposite end of the embryo, the embryonic root develops. A zone of undifferentiated cells remains at the tip, or apex, of both shoot and root, even in mature plants. These cells form the **apical meristems.** Meristem cells divide and produce new cells that differentiate into all the specialized tissues of a mature plant.

Maternal flower tissues form a tough **seed coat** enclosing the endosperm and the embryo (Figure 11.1d), which stops growing and remains dormant until the seed sprouts. The maturing seed becomes so dry that enzymes cannot function effectively. Transcription and translation stop, and cell respiration slows down to an extremely low rate. Dry, dormant seeds can remain alive for many years.

Plant and animal development are similar in some ways and different in others. Development of an animal embryo depends on movements of cells. The cell walls of neighboring plant cells are connected, however, so plant cells must differentiate where they are formed. Thus the position of an embryonic cell determines its future development. For example, a cell produced in the stem apical meristem will become part of the plant's protective surface layer if it is on the surface of a stem or leaf. Cells deeper in the stem become part of the phloem or xylem (see Sections 7.1–7.3).

The cytoplasm of plant and animal egg cells is not homogeneous. Organelles and molecules that were unevenly distributed in the cytoplasm of a plant or animal egg cell are divided unequally among the embryonic cells as the zygote divides. The resulting differences in the cytoplasm of the embryonic cells can signal the genes, helping determine how each cell will develop. Embryonic development in many animals, however, does not include a period of dormancy, as in seeds.

E T Y M O L O G Y

endo- = inside (Greek)
sperm = seed (Greek)
The interior of some seeds consists mostly of **endosperm.**

11.2 Seed Germination

When the environment is suitable, **germination,** or sprouting of the seed, occurs. If water and oxygen are plentiful and the temperature is favorable, the embryo resumes metabolism, growth, and development. It produces enzymes that begin to digest the stored food in the endosperm and cotyledons. The released nutrients then move to the growing regions of the embryo. Cell respiration speeds up in these metabolically active tissues. The absorbed water restores turgor to the embryo's cells. This outward pressure enables them to expand, contributing to growth. The seed coat fractures, and the embryonic root emerges. The shoot grows up through the soil surface. Figures 11.2 and 11.3 show that the details of germination differ in dicots (plants that have two cotyledons in their embryos) and monocots (plants with only one cotyledon).

Germination is a critical stage in a plant's life cycle. Intricate mechanisms have evolved that favor germination only when survival of the seedlings is most likely. For example, if the seeds of plants that grow in cold climates (such as pines or apple trees) germinated in the fall, the seedlings would not be mature enough to survive the winter. These seeds are genetically programmed to remain dormant until they experience several weeks of cold followed by warmer temperatures. Many desert plants germinate only after heavy rainfall ensures sufficient soil moisture. Some forest plants germinate only after intense heating by fire. This ensures that the seedlings will receive mineral nutrients from the ash of plants that have burned in forest fires and that their supply of sunlight will not be blocked by dense, unburned foliage. Some seeds, such as avocado, germinate only in total darkness, which serves as a cue that the seeds are well covered by soil. Many small seeds, such as

FIGURE 11.2

A germinating wheat seedling. The endosperm and cotyledon are still in the seed coat. Typical monocots, wheat seedlings push their green, photosynthetic embryonic leaves out of the soil.

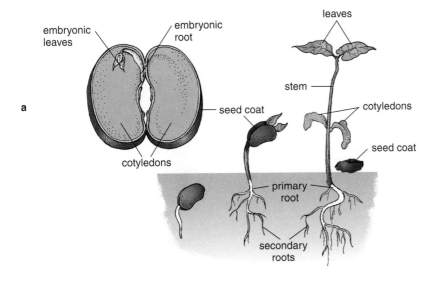

a

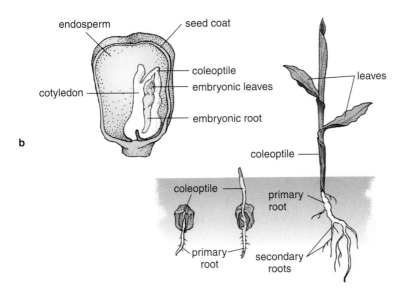

b

FIGURE 11.3

Seed germination. Germination of a dicot **(a)**, bean, and a monocot **(b)**, corn, is shown here. Unlike dicots, the monocot cotyledon stays buried with the endosperm in the seed coat. The first leaves of the dicot seedling emerge as the stem lengthens. In monocots, a protective sheath, the coleoptile, covers the first leaves to emerge. These leaves grow up from the stem apex, which remains protected near the soil surface until the reproductive phase of the plant's life begins.

those of lettuce, do not have enough stored nutrients to enable deeply buried seedlings to grow to the soil surface. These seeds can germinate only when faint light signals that the seeds are covered by a shallow layer of soil.

11.3 Primary and Secondary Growth

During germination, the root and then the stem begin their **primary growth**—growth from the meristems present in the embryo. Cell divisions in the apical meristems provide a steady supply of new cells. These cells expand mostly along the direction of the root and stem, making these organs longer (Figure 11.4). Meristems at each **node,** or point at which a leaf emerges, also contribute cells to stem growth. Starch and lipids stored in the cotyledons or endosperm support the growing embryo until the first leaves expand and begin to carry on photosynthesis. The root penetrates the soil, anchoring the plant, and begins to absorb the water and minerals needed for photosynthesis and

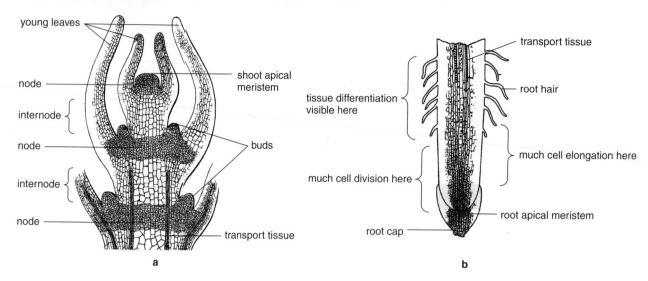

FIGURE 11.4

Primary growth. In shoots **(a)**, apical and nodal meristems provide new cells. Expansion of these cells elongates the internodes (stem segments between nodes). In roots **(b)**, elongation of cells produced in the root apical meristem lengthens the root, pushing the root tip through the soil. (Meristems are in yellow.) Notice the growth similarities and differences in roots and shoots.

growth. A tough tissue mass called the **root cap** covers and protects the apical meristem as the root grows through the soil.

Growth and development go hand in hand. While a new stem or root segment is completing its growth, its cells begin to differentiate into three major tissue types. Surface cells make up the **epidermis** (Figure 11.5a) that covers the plant. Most epidermal tissue is covered by a hydrophobic **cuticle** of wax and other lipids that keeps moisture in and pathogenic microorganisms out. Phloem and xylem consist of **vascular tissue** (Figure 11.5b). The other tissues that fill up the plant body, giving it shape and internal support, are called **ground tissue** (Figure 11.5c). Some ground tissues become specialized. The mesophyll cells of leaves are specialized for photosynthesis. Other ground tissues contribute to nutrient storage, mechanical support, or other functions.

CHEMISTRY TIP

The waxes in plant cuticles melt at higher temperatures than most other plant lipids. This ensures that they will remain solid in hot weather and not melt and drip, creating gaps in the cuticle.

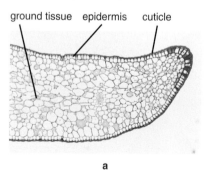

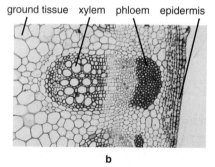

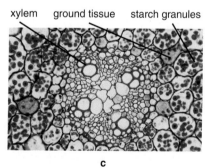

FIGURE 11.5

The three major categories of plant tissues. (a), Epidermal tissue in the upper surface of a lily *(Clivia)* leaf is shown, ×100. Note the cuticle (stained red) on the surface. **(b),** Vascular tissue as it appears in a cross section of a bundle of xylem and phloem in the stem of sunflower, *Helianthus annuus,* is shown, ×400. (c), Ground tissue in a developing root of a buttercup, *Ranunculus,* is shown, ×90. Note the many plastids containing starch granules, which are stained purple.

Two important factors in plant development are the growth of the cell wall and the rate and orientation of cell division. As plant cells mature, they add material to their walls. The thicker, stronger wall resists cell expansion, so growth slows down. The final size of a plant organ is the result of a race between cell growth and cell-wall hardening. Mitosis produces small, thin-walled cells that are capable of expanding. During primary growth, most cell divisions in stems and roots are horizontal. This produces vertical columns of cylindrical cells. Long cellulose molecules in the cell walls of these cells wind around the cells, keeping them from expanding sideways, so the cells grow by elongating vertically, lengthening the stem and root. Different patterns of cell division and expansion produce organs with other shapes. What would be the shape of an organ that developed from a mass of cells that divided with random orientations and had cell walls containing randomly oriented cellulose molecules?

Leaves are a good example of the growth and development of a plant organ. A leaf begins to form as randomly oriented cell divisions produce a bump on the side of a shoot apex. At first, cells in the center of each bump divide, producing small, fingerlike growths. Then meristem cells on the sides of the bud begin to divide at right angles to the leaf surface. The new cells expand, and the leaf becomes flatter (Figure 11.6). Leaves become broader, but not thicker, because their cells divide only perpendicular to the surface. The number of layers of cells in the leaf, and thus its thickness, are limited.

Growth and development reflect both genetic and environmental influences. Genetic factors strongly influence the shape of the leaf bud, the distribution and orientation of cell divisions, and the amount and distribution of cell enlargement. The transcription of specific genes in each meristem cell is affected by chemical signals and physical forces created by neighboring cells. All of these factors help determine whether each cell will become part of a leaf, a petal, or another organ. For example, if ground tissue and veins grow at the same rate, a leaf with a simple outline results (Figure 11.7a). On the other hand, if the growth is more vigorous near the veins than in the other regions, a lobed shape forms (Figure 11.7b). Leaf shape is also affected by the environment. For example, the submerged leaves of some aquatic plants are very different from the leaves above the water surface (Figure 11.7c).

FIGURE 11.6

Early leaf development. New leaves are shown as they begin to form on a shoot tip of a sugar maple, *Acer saccharum,* ×135. Repeated divisions perpendicular to the surface of each young leaf will be followed by horizontal cell expansion.

CONNECTIONS

The thin, flat form of leaves demonstrates adaptation through natural selection. This shape maximizes the surface area available for light absorption and photosynthetic gas exchange with the atmosphere.

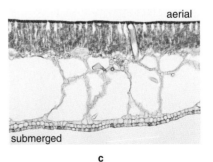

a   b   c

FIGURE 11.7

Genetic and environmental influences on leaf form. (a), Uniform growth of ground tissue produces an elm *(Ulmus rubra)* leaf with a simple, rounded shape. **(b),** Rapid growth of ground tissue near veins produces a lobed maple *(Acer saccharum)* leaf. **(c),** Water lily *(Nymphaea odorata)* leaves growing in air are relatively compact. Submerged leaves are thick and spongy with additional internal air spaces, ×100.

Complete development of the leaf, however, requires exposure to light so that chlorophyll and other photosynthetic pigments can be synthesized. The hormonelike plant-growth regulators (see Sections 11.4–11.7) also appear to play an important role in leaf development.

In some plants, older parts of stems and roots that have completed primary growth continue to increase in diameter. This **secondary growth** comes from the **vascular cambium,** another type of meristem. The cambium is a cylindrical layer near the outer surface of roots and stems. It produces cells that differentiate into two types of transport tissue, shown in Figure 11.8. The inner surface of the vascular cambium provides cells that differentiate into xylem. Xylem cells have thick walls. They are the main component of wood. Wood produced in successive growing seasons supports trees and results in annual rings that indicate a tree's age. Cells produced on the outer surface of the vascular cambium develop into phloem. Phloem transports dissolved sugars and amino acids from the leaves throughout the plant.

Trees and other woody plants develop a meristem called a cork cambium that produces their bark. Bark protects the internal plant tissues from dehydration and attack by animals and microorganisms. The xylem cells at the center of a large tree no longer carry water, but their thick, tough cell walls help support the tree. Even if the inner part of its wood is destroyed, a tree often can continue its normal growth because the functioning tissue is near the bark. The Chimney Tree of California (Figure 11.9) is a giant redwood whose interior was burned out, leaving only the outer shell. You can walk inside the base of the trunk and look straight up through the center of the tree and see the sky. From the outside, the tree looks normal and continues to grow like other healthy redwood trees.

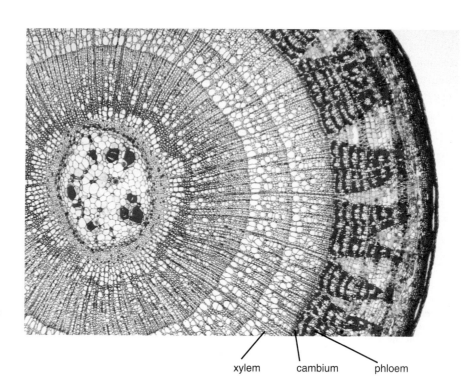

xylem   cambium   phloem

FIGURE 11.8

Secondary growth in a three-year-old basswood sapling. The newly formed cambium has begun to produce layers of cells that differentiate into xylem toward the inside and phloem toward the outside, ×40.

FIGURE 11.9

The Chimney Tree of California. The nonfunctional xylem in the center of this tree has burned out. The tree survives because the cambium, phloem, and water-carrying xylem in the outer part of the trunk are intact.

Figure 11.10 shows the locations of the major meristems. Note the meristems in buds on the sides of stems. Buds grow into branches, leaves, and flowers. Just as the bud at the tip of the main stem contains an apical meristem, a bud that produces a branch contains the branch's apical meristem. Root branches, or secondary roots, arise from the **pericycle,** a cylinder of meristem tissue that surrounds the xylem and phloem in the root. Figure 11.11 summarizes the process of plant differentiation.

As most plants mature, they begin to produce reproductive structures such as flowers and, eventually, seeds and fruits (see Section 12.5). The timing of flowering is strongly affected by environmental factors such as the length of the night. The regulation of flowering is discussed in Section 11.9.

ETYMOLOGY

peri- = surrounding (Latin, Greek)
cycle = circle (Greek)

The **pericycle** is a cylindrical tissue layer that surrounds the inner part of the root.

FIGURE 11.10

Location of major meristems in a typical plant. Meristems are plant tissues that remain embryonic, producing new cells that are responsible for growth. Apical meristems bring about primary growth; cambium is responsible for secondary growth. Meristems in buds produce cells that differentiate into all the tissues of leaves, branches, and flowers. Red marks the positions of meristems.

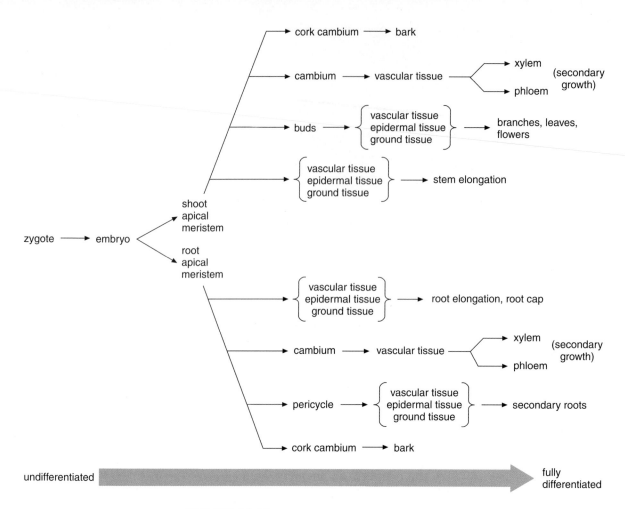

FIGURE 11.11

Stages in plant-tissue differentiation. Each organ contains examples of all three types of tissues. Note the similarities in development between the tissues of the root and the shoot.

Check and Challenge

1. Describe the development of the plant embryo inside the seed of a flowering plant.
2. What is the role of the endosperm for the developing plant embryo?
3. How do cotyledons function in the growth and development of a young seedling plant?
4. Distinguish between primary growth and secondary growth in plants.
5. What specialized tissues function in primary and secondary growth in plants?
6. What conditions contribute to germination? Describe what happens when a seed germinates.

Control of Growth and Development

11.4 Factors Affecting Plant Growth

What determines when and how plant organs grow? Genes provide the primary control, but how do different bits of genetic information direct a cell's growth and differentiation? Factors that act as cues for expression of different genes at different times include temperature, night length, nutrition, chemical signals from other parts of the plant, and activities of neighboring cells. Figure 11.12 summarizes the interactions of the many internal and external factors that may affect plant growth and development.

Many of the effects of these factors are signaled by substances called **plant growth regulators** (PGRs). These compounds function somewhat as hormones do in animals. Botanists have identified five major classes of interacting PGRs that influence growth and development. Their production is under genetic control, but environmental cues play critical roles in

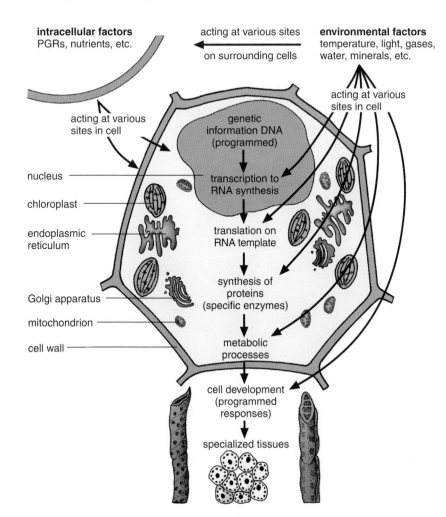

FIGURE 11.12

Interaction of factors affecting plant cell development. Two important cell processes in plant development are the synthesis of cell wall components and their export from the cytoplasm via the Golgi apparatus. Another is the regulation of genes that affect the development of plastids into chloroplasts or more specialized types of plastids.

regulating the production of a PGR and determining which cells are sensitive to it. PGRs are produced by tissues that have other functions, and they often are produced in one place and transported to different locations within the organism. These compounds may cause different effects in different parts of the plant, at different times, or in different concentrations. Each PGR may have a primary effect on each cell type, from which a variety of responses can follow. These compounds occur in extremely small concentrations. They affect many metabolic pathways and other cell activities, usually through effects on gene expression.

11.5 Auxins

Auxins were the first PGRs to be identified. They are produced in apical meristems and move through the plant by active transport. Auxins can stimulate receptive cells in the growing regions of the plant to elongate, but the effects depend on a number of factors, especially the concentration of the auxins. At extremely low concentrations, for example, auxins promote elongation of roots, but at higher concentrations, they inhibit elongation. Auxin produced in a stem apex inhibits the growth of other buds on the same branch. Removal of the apical bud often results in increased branching, as other buds begin to grow out.

You could apply auxins to plant cuttings to stimulate the formation of roots when you plant a cutting (Figure 11.13). Auxins also promote the development of fruits from flowers. Usually, the developing seeds provide the necessary auxin. Spraying some flowers with auxin can cause them to develop into fruits without fertilization, so no seeds develop inside. Seedless tomatoes and cucumbers can be produced this way.

A synthetic auxin called 2,4-D is a herbicide used to kill dicot weeds. At the concentrations generally used, 2,4-D provides a deadly overdose of auxin

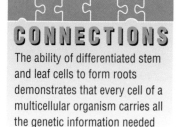

CONNECTIONS

The ability of differentiated stem and leaf cells to form roots demonstrates that every cell of a multicellular organism carries all the genetic information needed to develop into any type of cell present in the organism.

CHEMISTRY TIP

2,4-D is an artificial PGR that has some auxinlike effects. Molecules of artificial auxins resemble natural ones in structure and the distance between their polar and nonpolar regions. These features, required for auxin activity, help biochemists determine the geometry and chemical characteristics of the auxin-receptor protein.

FIGURE 11.13

Practical use of auxin. The application of auxin induced this cutting to form roots.

to dicots but does not affect the less-sensitive monocots, such as members of the grass family. Therefore, it can be used to control weeds such as dandelions in lawns and in grain fields.

11.6 Other Growth Stimulants: Gibberellins and Cytokinins

Gibberellins were first discovered in the 1920s by the Japanese scientist Eiichi Kurosawa when a fungal infection caused rice plants to grow abnormally tall and weak. The fungus *Gibberella* produces a chemical that caused the abnormal growth. The newly discovered compound was called gibberellin. More than 100 gibberellins have been discovered in fungi and plants since then, though usually only a few are present in any species.

Gibberellins synthesized in the apical parts of stems and roots stimulate stem elongation. This effect can be enhanced by auxins. Germinating embryos produce gibberellins that stimulate the transcription of genes that encode digestive enzymes in endosperm. These enzymes break down the stored nutrients in endosperm. The embryo then uses the nutrients as it grows.

Gibberellins are also involved in reproduction. Plants treated with gibberellins may produce flowers that develop into seedless fruits. Gibberellins also cause fruits to grow (Figure 11.14) and may counter the effects of herbicides. The height gene that Gregor Mendel studied in pea plants encodes an enzyme required for gibberellin synthesis. Treatment with gibberellins can cause some dwarf plants to grow to the height of normal varieties (Figure 11.15). Some dwarf varieties that do not respond to gibberellin lack a functional receptor protein for one of these PGRs.

The **cytokinins** are a third group of naturally occurring PGRs that promote cell division and organ development, such as the growth of lateral

CHEMISTRY TIP

Gibberellins have a complex structure. The best known of these compounds is gibberellic acid (GA). The many slight variations in the structure of GA are given numbers rather than names: Gibberellic acid is also known as GA_3. Other gibberellins include GA_1, GA_2, and so on.

FIGURE 11.14

Effect of gibberellin treatment on grapes. The grapes on the left served as the control in this experiment; they were not treated with gibberellin. The grapes on the right were sprayed with a gibberellin solution early in their growth to increase their mature size.

FIGURE 11.15

Effect of gibberellin (GA₃) on dwarf corn plants. Each dwarf plant shown here was treated with the indicated dosage of GA₃ and allowed to continue growing for 7 days. Note the increase in height of the plants with increased dosage. The plants treated with 10 or 100 μg GA₃ have grown to the height of normal corn plants.

CONNECTIONS

PGRs are part of the adaptation of plants to life as multicellular organisms.

buds or stems. Cytokinins usually work in combination with auxins and other hormones to regulate the total growth pattern of the plant. They are found mostly in plant parts with actively dividing cells, such as root tips, germinating seeds, and fruits. Cytokinins oppose aging of plant tissues. Evidence indicates that cytokinins are produced mainly in the roots and then transported throughout the rest of the plant. They also are produced in developing fruits.

Cytokinins are necessary for stem and root growth, as well as chloroplast development. Cytokinins stimulate the growth of lateral branches and inhibit the formation of lateral roots. In contrast, auxins inhibit the formation of lateral branches and promote the growth of lateral roots.

11.7 Growth Inhibitors: Abscisic Acid and Ethylene

Two naturally occurring PGRs inhibit cellular activities and help plants survive environmental stresses such as drought. **Abscisic acid** ($C_{15}H_{20}O_4$) is synthesized in response to dry conditions. Abscisic acid stimulates the closing of stomata. This action protects the plant against water loss, though it also reduces photosynthesis. Buds and seeds become dormant when abscisic acid accumulates in them. During embryo formation, abscisic acid in seeds stimulates synthesis of storage protein and prevents embryos from germinating until they are mature and the environment is favorable for seedling growth.

Ethylene is a PGR that is a simple gas (C_2H_4). It promotes aging of tissues, such as the ripening of fruits. Ethylene opposes many effects of auxins and cytokinins. For example, you learned earlier that auxins produced in an apical bud stimulate the growth of the stem or branch below. In other buds lower on that branch, however, the auxin stimulates the

Focus On

Development of New Plants

PGRs make it possible to grow whole plants from individual cells. To do this, biologists grow plant cells in sterile containers with PGRs. At equal concentrations of auxins and cytokinins, many cells continue to divide, forming an undifferentiated tissue called callus. Different concentrations of auxins and cytokinins induce callus to form roots or buds and, in some cases, to grow into a complete plant (Figure 11.16).

Plants grown from callus are genetically identical to the parent plant. This quickly produces many offspring from a plant with desired traits. Most house ferns, for example, are produced in this way. It is also possible to change the cells' DNA and then use them to form new plants with useful traits. For example, biologists inserted into cells of cotton plants a gene that encodes a protein that is toxic to many crop-eating insects. These cells were then used to produce insect-resistant cotton plants. Selecting a useful individual from a million cells can be much faster and more efficient than screening a million plants.

Other techniques for developing new plants use protoplasts—cells that have lost their cell walls. Researchers use enzymes to digest away the cell walls (Figure 11.17). Protoplasts of different species may fuse, forming a hybrid cell. In such a hybrid cell, only some of the chromosomes of each original cell survive. This method may be useful for combining genes from different species. Protoplasts or thin tissues can also be bombarded with microscopic gold particles to which DNA is attached. The particles pass though the cells, leaving bits of DNA behind. Crop plants produced using these techniques carry genes from bacteria and other organisms that provide resistance to insects and herbicides that are used to kill weeds.

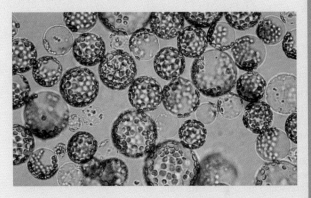

FIGURE 11.17

Plant protoplasts. The cell walls have been removed from these cells in preparation for fusion or genetic manipulation.

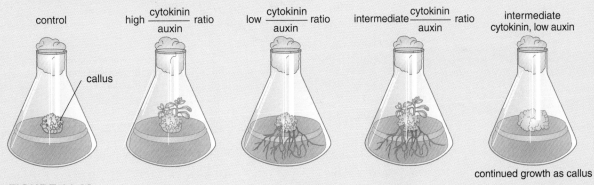

FIGURE 11.16

Interaction of auxins and cytokinins in plant tissue culture. The culture media were supplemented with various levels of these two PGRs.

production of ethylene. The ethylene, in turn, suppresses the development of those lower buds. Similarly, as a mature leaf begins to age, its cytokinin content declines, and it produces ethylene. The ethylene stimulates breakdown of proteins and other components of the old leaf. The amino acids and other nutrients that are released travel through the phloem to young, growing tissues. Ethylene also makes leaves, flowers, and fruits drop from an aging plant (Figure 11.18).

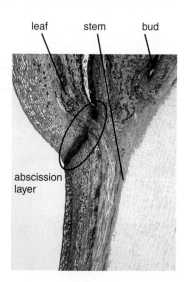

leaf stem bud

abscission
layer

FIGURE 11.18

The beginning of leaf shedding.
The circled vertical band in the
micrograph is the abscission layer,
which forms at the attachment of a
leaf or fruit to the stem before the leaf
or fruit falls. The abscission layer
seals off the vascular tissue
connecting the organ to the stem. An
increase in the ratio of ethylene to
auxin in this tissue triggers the
process.

Hundreds of years ago, Chinese farmers unknowingly produced ethylene
when they burned incense to ripen fruit stored in ripening rooms. Taking
advantage of this property of ethylene, many farmers pick delicate fruits,
such as tomatoes, when they are green and less susceptible to damage. The
fruits are shipped in an atmosphere of carbon dioxide, which blocks the
action of ethylene. When the tomatoes arrive at their destination, they are
treated with ethylene to speed their ripening so they can be sold.

Figure 11.19 presents some of the actions of the five major groups
of PGRs.

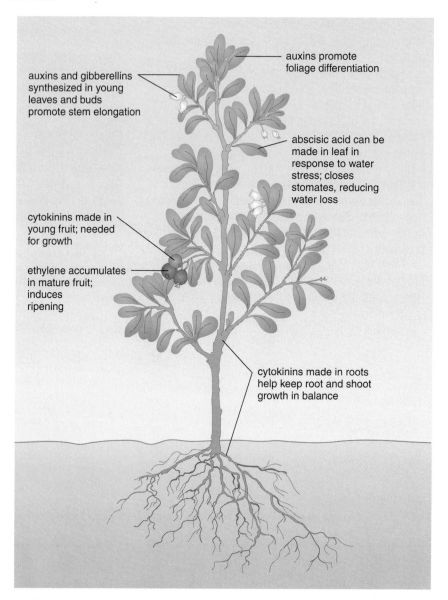

auxins promote
foliage differentiation

auxins and gibberellins
synthesized in young
leaves and buds
promote stem elongation

abscisic acid can be
made in leaf in
response to water
stress; closes
stomates, reducing
water loss

cytokinins made in
young fruit; needed
for growth

ethylene accumulates
in mature fruit;
induces
ripening

cytokinins made in roots
help keep root and shoot
growth in balance

FIGURE 11.19

Some effects and interactions of PGRs in plant organs. PGRs typically affect tissues in
many organs. However, the same PGR can have different or opposite effects in different organs.

Control of Organ Formation

One of the remaining great mysteries of plant biology is exactly how changes in the metabolism of PGRs and the regulation of various genes combine to produce organs such as flowers. These intricate structures are accurately recreated many times in the same plant, yet differ in specific ways among varieties and species. The control system that directs plant growth and development is slowly being worked out by a number of research groups.

Elliot Meyerowitz of the California Institute of Technology and Enrico Coen of the John Innes Institute in England have studied many snapdragon (*Antirrhinum majus*) and wild mustard (*Arabidopsis thaliana*) plants with genetic changes (mutations) that lead to abnormal flowers. They found that these plants carry altered homeotic (organ-forming) genes (Chapter 10). Some of these genes appear to encode protein kinases—enzymes that regulate the activity of specific proteins by attaching phosphate groups to them. These kinases help regulate gene expression by controlling the activity of RNA polymerase. Different forms of this enzyme, in turn, may transcribe groups of genes involved in the development of specific organs or tissue types, though the details of this system are not yet clear.

At a certain point in the life of a shoot apical meristem, the bumps that form on its edges stop developing into leaves and instead become the parts of a flower. In the plants that Coen and Meyerowitz studied, the apical meristem produces four whorls (rings) of these bumps when it forms flowers. The outermost whorl develops into the sepals, or small leaves on the base of the flower. The second whorl forms the petals, and the third and fourth whorls become the male and female parts of the flower, respectively.

Many of the mutations that Coen and Meyerowitz investigated seem to affect only two neighboring whorls. For example, at least five mutations were found to affect only whorls 2 and 3, changing petals to sepals and male parts to female parts. Coen and Meyerowitz showed that the type of organ formed in each whorl is determined by three groups of regulatory genes. Group A genes are transcribed only in whorls 1 and 2, group B in whorls 2 and 3, and group C in whorls 3 and 4. Thus a different combination of groups A, B, and C is active in each whorl. These combinations determine the fate of each whorl of developing organs (Figure 11.20).

Other experiments have provided hints about other details of this system. Some of the genes are expressed only for a short time in a developing flower part. Auxin, gibberellin, and cytokinin levels in each flower part also rise and fall at various stages of development. Differences in PGR concentrations between the tip and base of a petal or other flower part may help shape growth and development of these organs as well, by activating or repressing the activities of various genes and enzymes.

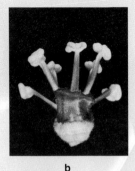

a　　　　b

FIGURE 11.20

Flower development mutants. Flowers of snapdragon (*Antirrhinum majus*) in **(a)** have green, leaflike sepals on the outside (whorl 1) surrounding large petals (whorl 2). Mutant *Ovulata* snapdragons in **(b)** have a defect in one of the group A genes. Their sepals become female flower parts, and their petals become male parts.

1. Would it be more accurate to say that each PGR controls specific cellular processes or that it controls development of specific organs?

2. Compare the effects of auxins and cytokinins on plant growth. What is the primary effect of each PGR?

3. How can one type of fruit ripening in a room affect how a different type of fruit ripens in that same room?

4. Gibberellins and auxins both influence elongation of cells. How do their effects differ?

Plant Responses

11.8 Plant Movements and Growth Responses

Try Investigation 11B ◇ Tropisms.

Plant survival depends, in part, on movements and changes in growth in response to a stimulus. Most plant movements are responses to changes in the environment. For example, when any part of the leaf of the sensitive plant, *Mimosa pudica*, is touched, the leaflets droop together suddenly (Figure 11.21). This movement results from changes at the cellular level—the plasma membranes allow potassium and other ions to leak out of cells. This leads to rapid outward osmosis of water from certain cells at the bases of the leaves and leaflets. These cells lose turgor pressure and collapse, making the leaflets droop. This movement may make the leaflets less attractive to grazing animals or protect them from damage by rain or hail.

a

b

FIGURE 11.21

Response of the sensitive plant to contact. Leaflets of a mimosa plant, *Mimosa pudica*, **(a)** droop when they are touched **(b).**

Other plant movements are really changes in the type or direction of growth. Growth toward or away from a stimulus is called a **tropism.** Tropisms result from differences in growth between parts of an organ. You probably have observed that most plants grow toward light. Charles Darwin studied the growth of plants toward light, called **phototropism.** Cells on the lighted side of the shoot stop growing, while those on the shaded side continue to elongate (Figure 11.22). This differential growth bends the plant toward the light.

What causes phototropism? Many experiments designed to answer this question have been carried out on oat seedlings. In uniform lighting and in the dark, oat seedlings grow straight up. With one-sided lighting, they bend. Evidence from the experiments indicated that in the light, the tip of a growing shoot produces an auxin that causes bending. Darwin's experiments showed that if the tip is covered, the shoot will not bend. Other researchers found that a shoot would bend even in the dark if they applied auxin to only one side of the shoot.

Eventually it was found that auxin synthesized in the tip of the shoot moves down the stem equally on all sides, but it is also transported horizontally from the lighted side to the shaded side, where it stimulates elongation of the cells. Increased auxin on the shaded side stimulates synthesis of enzymes that loosen the cell-wall structure, enabling cells to expand.

Gravitropism is growth toward or away from Earth's gravitational pull. Stems are negatively gravitropic; that is, they grow away from gravity. If stems are placed in a horizontal position, they bend upward, although this response can be variable. Auxins may play a critical role in gravitropism, as do calcium ions. Roots are positively gravitropic, and the evidence indicates that sensitivity to gravity occurs in the root cap. There, plastids filled with dense starch grains fall to the bottom of certain cells (Figure 11.23). Contact between the plastids and plasma membranes signals the direction of gravity. Again auxins appear to be involved, along with abscisic acid and perhaps other PGRs, in stimulating downward growth.

FIGURE 11.22

Seedlings growing toward light. Auxin transport away from light reduces growth on the lit side and promotes growth on the shaded side of each stem.

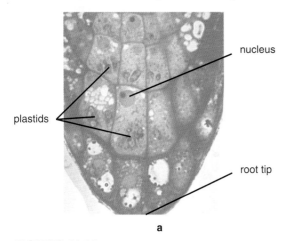

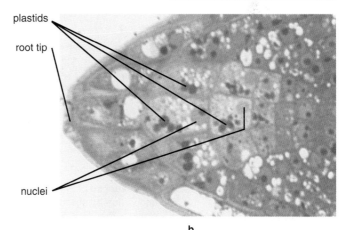

a

b

FIGURE 11.23

Gravitropism. Detection of gravity in root cap cells of thale cress *(Arabidopsis thaliana)* is shown here. When the root is vertical **(a),** starch-containing plastids (dark circles) reside at the bottom of the cells. When the root is turned on its side **(b),** gravity makes the plastids settle quickly to the new bottom of the cells.

11.9 Photoperiodism

The changing seasons affect many organisms. Some plants respond to the lengthening nights of summer and autumn by producing reproductive organs (flowers, seeds, and fruits). A response to the relative length of light and darkness in a 24-hour period is called **photoperiodism.**

Many spring-flowering plants, such as daffodils, bloom only when day length exceeds a certain number of hours. They are called long-day plants. Fall-flowering plants, such as asters, on the other hand reproduce only when day length is shorter than a certain number of hours and so are called short-day plants. Day-neutral plants flower whenever they become mature, regardless of the day length, although temperature may play a role in their flowering. Florists use knowledge of photoperiodism to "force" short-day plants such as chrysanthemums to bloom year-round.

In the 1940s, biologists learned that the length of the night, rather than the length of the day, controls photoperiodism. They found that a brief period of darkness during the day has no effect on flowering. Even a few minutes of light during the night, however, prevents flowering, showing that photoperiodic responses depend on the length of uninterrupted darkness. Long-day plants flower when the night is *shorter* than a critical length; short-day plants flower when the night is *longer* than a critical length.

How do plants measure night length? Plants contain a pigment known as **phytochrome** that has two slightly different chemical structures. One form (P_r) absorbs red light; the other form (P_{fr}) absorbs far-red light (a wavelength in the farthest red part of the visible spectrum—see Figure 4.2). Phytochrome is synthesized as P_r and remains in that form as long as the plant is in the dark. In sunlight, which is richer in red light than in far-red light, the P_r absorbs red light and is converted to P_{fr} (Figure 11.24).

After sunset, P_{fr} gradually converts back to P_r. At sunrise, the level of P_{fr} rapidly increases again as P_r absorbs red light. In this way, the conversion of

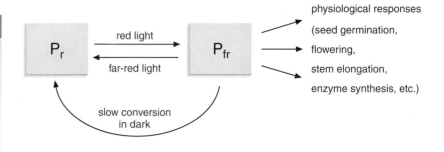

FIGURE 11.24

The response of phytochrome to light. Phytochrome occurs in two forms: P_r (red absorbing) and P_{fr} (far-red absorbing). Absorption of red light converts P_r to P_{fr}; absorption of far-red light converts P_{fr} to P_r.

phytochrome from one form to the other marks the beginning and end of the dark segment of the photoperiod. This conversion also acts as a switch that controls many events, such as flowering, germination, and bolting (the rapid stem growth that often precedes flowering).

Studies of genes involved in flowering in wild mustard (*Brassica kaber*) suggest that, in at least some plants, night length and temperature do not activate genes that are needed for flowering. Instead, they may block the action of genes that inhibit flowering.

Check and Challenge

1. How do phototropism and geotropism affect plant growth? Predict the effects on growth if a plant responded abnormally to light and to gravity.

2. Describe how phytochrome acts in plant photoperiodism.

3. What is the survival value of photoperiodism to a plant?

Chapter
HIGHLIGHTS

Summary

Plants grow when the number and size of their cells increase. Development is the process by which the cells of a new organism form the differentiated tissues and organs of a complete individual. A seed contains the embryo with its cotyledons and a reserve energy supply. When conditions are appropriate, the seed germinates. The initial growth of the root tip and then the stem tip uses the energy reserves inside the seed. Once the leaves emerge from the buds on the stem, the plant uses photosynthesis to harvest the energy it needs for continued growth, development, and reproduction. Growth occurs at specific areas in the plant called meristems. Primary growth—increase in length—occurs in roots and stems at apical meristems; secondary growth—increase in diameter—occurs at the cambium and, in some plants, at the cork cambium. Additional leaves and branches arise from the meristem tissue in the buds.

Five major groups of plant growth regulators (PGRs) have been identified—auxins, gibberellins, cytokinins, abscisic acid, and ethylene. PGRs act directly on various enzymes and metabolic pathways and indirectly by influencing the activity of various genes that are involved in growth and development. Protein kinases are an important part of the signaling pathways through which PGRs produce their effects. PGRs interact in response to environmental cues, such as light and gravity, to affect plant growth and development. For example, tropisms are variations in plant growth that are directed toward or away from environmental stimuli such as light intensity. Plants also exhibit photoperiodism, a response to relative length of the dark period that involves the pigment phytochrome. Much remains to be learned about the mechanisms by which PGRs interact with each other and with the environment to regulate growth and development.

Key Concepts

Below is the beginning of a concept map for plant development. Use the concepts listed and other concepts from the chapter to build a more complete map.

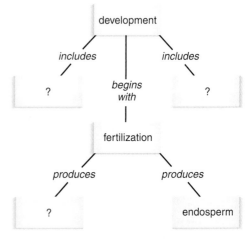

- differentiation
- embryo
- germination
- growth
- apical meristem
- night length
- leaf formation
- root growth
- seed
- stem growth
- auxin
- bud

Reviewing Ideas

1. Some seeds survive and germinate; others do not. How is this an example of a contribution of the environment to natural selection?
2. Could a seed produce a plant if its endosperm were removed before it germinated? Why or why not?
3. What is the difference between a cotyledon and a true leaf?
4. Why is it advantageous for the root tip to grow from the embryo before the stem tip starts growing?

5. What is the difference between an apical meristem and a cambium?
6. How are PGRs used to cause a new plant to differentiate and grow from callus?
7. Contrast the action of auxins and cytokinins on the growth of lateral stems and roots.
8. How are xylem and phloem tissues involved in the growth and development of a plant?

Using Concepts

1. Most cell division in plants occurs in or near meristems, and not throughout the entire organism. What are the advantages of this for plants?
2. Some seeds must undergo freezing temperatures followed by warmer temperatures before they can germinate. How does this help plants survive?
3. What are some of the environmental factors that influence plant growth and development? What are their effects?
4. How does putting green fruit in a paper bag promote ripening?
5. What might be some beneficial uses of an artificial PGR that functions only in certain species of plants and not in others?
6. How could a short-day plant and a long-day plant growing in the same area flower on the same day of the year?
7. How might day length, phytochrome, and PGRs interact in the germination of seeds planted just below the soil surface?
8. Deciduous trees, such as oaks and maples, lose their leaves in the fall. Describe how PGRs and environmental stimuli might bring about this response.
9. Ten years ago, a farmer built a fence 1.5 m high and attached one end of it to a tree that was 7 m high. Now the tree has grown to a height of 14 m. How far above the ground is the attached end of the fence? Explain your answer.

Synthesis

1. Explain how the formation of a plant embryo is similar to and different from the formation of an animal embryo.
2. Genes from plants can be inserted into bacteria, making them produce plant proteins. Explain how this capability supports the theory of evolution by natural selection.
3. Plant development appears to be much more closely linked to environmental cues than animal development is. Explain how this adaptation would enhance survival of plants.

Extensions

1. Assume auxins have the same effect on humans as they do on plants. Write a short story about how you would use auxins and the effects they would have.
2. Diagram the life cycle of a plant. Show the PGRs likely to be most active at each stage in the cycle.
3. Abscisic acid occurs in some animals. Use what you have learned about the functions of this compound in plants to predict whether abscisic-acid levels in the body of a hibernating animal would rise at the beginning or end of its winter hibernation. Give the reasons for your prediction.

Web Resources

Visit BSCSblue.com to access
- Notes and information on plant biology
- Activities and other resources for learning about plant growth and development
- Web links to information and activities for plant-tissue culture at home

Heredity: Continuity of Life

■ *What evidence do you see that these organisms are related?*

■ *How can you explain the resemblances among close relatives?*

W hy do children resemble their parents? Parents pass a copy of their DNA to their offspring. This reproductive inheritance is the blueprint for the development of the new organism. It acts in each of an organism's cells throughout its lifetime. In sexually reproducing species, both parents contribute genes to the offspring. What happens when the parents' genes are different? Genetics is the branch of biology that seeks to identify the rules of inheritance.

Genetics enables scientists to predict some characteristics of an organism if they know its ancestry. These traits are determined by fairly simple interactions between parental genes. Other traits are determined by more complex interactions among many genes and environmental factors. Variations of the rules that describe simpler forms of heredity can explain more complex genetic patterns. Advances in analyzing and manipulating DNA have greatly accelerated the pace of genetic discovery. As a result, geneticists can now manipulate specific traits in agricultural plants and animals. Genetic research has led to tests for forms of human genes that are likely to cause disease. In the next few years, treatments for inherited diseases that involve changing a patient's DNA will probably become a reality.

CONTENTS

LEARNING OUTCOMES

By the end of this chapter you will be able to:

A   Describe reproduction in plants and animals.

B   Explain the importance of meiosis in maintaining chromosome numbers and identify the stages of meiosis.

C   Infer the advantages of a dominant diploid stage in the life cycle of plants and animals.

D   Relate the process of fertilization in flowering plants to their successful domination of land environments.

E   Compare external and internal fertilization.

F   Discuss the influence of hormones on the human male and female reproductive systems.

G   Discuss causes of infertility and methods of contraception.

CHAPTER

12

Reproduction

■ *What benefit does the flower gain from the activity of this bee?*

■ *How might a cluster of flowers give this species a reproductive advantage?*

Reproduction is a basic function of living things. Unlike other life functions—eating, sleeping, breathing—reproduction has value only for the survival of a species as a whole. Through reproduction, a species passes its unique genetic material to future generations.

For a species to survive, it must reproduce successfully in every generation. Natural selection has produced a variety of adaptations that greatly increase the chance that offspring will be produced and survive. Some organisms reproduce individually; their offspring are genetically identical to the parent. Others reproduce sexually, producing offspring that receive genetic information from two parents. In this way, sexual reproduction produces new combinations of genes.

Reproduction is tightly regulated. Study of reproduction helps explain how species survive. It also provides useful information for protecting the health of human mothers and children and making decisions about reproduction, which involve values and ethics.

Cell Division and Reproduction

12.1 Asexual Reproduction

Asexual reproduction requires a single parent. In this process, one or more of an organism's cells form a genetically identical offspring. A group of genetically identical cells or organisms produced through asexual reproduction is called a **clone.** Many organisms reproduce asexually. Figures 12.1 and 12.2 show some examples of asexual reproduction. As you read in Section 6.3, prokaryotes reproduce by simply dividing in two, a process called binary fission. One-celled eukaryotes such as *Paramecium* (Figure 12.1a) also reproduce by dividing in two. Other eukaryotic organisms that reproduce asexually include many fungi and simple animals such as *Hydra* (Figure 12.1b). *Hydra* reproduces by budding: The buds grow to adult size and detach as new individuals. The flatworm *Planaria* reproduces by fragmentation (breaking into pieces), as shown in Figure 12.1c. Each piece then grows into a new worm.

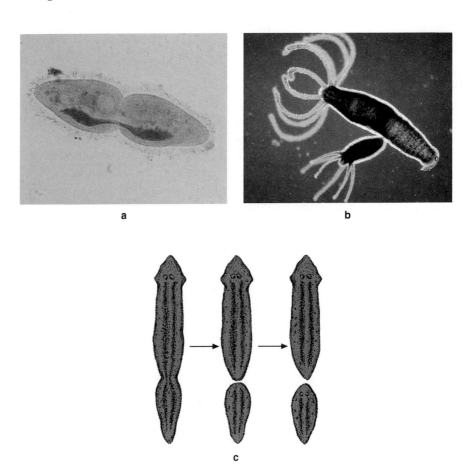

a   b

c

FIGURE 12.1

Forms of asexual reproduction. (a), Mitotic cell division in *Paramecium caudatum* (color added, ×400); **(b),** budding in *Hydra viridis* (color added, ×8); **(c),** fragmentation in *Planaria.*

a   b

FIGURE 12.2

Asexual reproduction in plants. (a), Fragments of the cactus known as jumping cholla
(Opuntia bigelovii) break off easily and are carried by wind or animals to new locations. The
fragments form new plants where they land. **(b),** Leaves of *Bryophyllum* produce little plantlets,
each of which may fall to the ground near the parent and grow as a new plant.

Some plants also reproduce through fragmentation. Small fragments of
some cacti, such as prickly pear and jumping cholla (Figure 12.2a), spread
over great distances in the wind or in the fur of an animal. Stored water and
nutrients sustain the fragment until it can take root at a new site.

Asexual reproduction in plants is also called vegetative reproduction. This
type of reproduction is efficient in filling an area with plants, but it is less
successful in quickly spreading plants to new locations. For example,
strawberries and some grasses grow new plants along horizontal stems called
runners. At various points along the runner, roots grow down, and a new plant
develops. Potatoes are enlarged underground stems. New potatoes can grow
from each bud, or eye. Aspen trees reproduce when new trees sprout from the
roots of older trees. Cuttings from a parent plant can be rooted and then grown
into mature plants. Figure 12.2b shows a less familiar form of vegetative
reproduction. The parent plant produces new plants along its leaf edges.

12.2 Chromosome Numbers

Each species has a characteristic number of chromosomes. Prokaryotes
generally have only one major chromosome, consisting of a single circle of
DNA. One or more smaller plasmids (see Section 6.3) carry only a few genes,
such as those responsible for resistance to antibiotics.

The number of chromosomes varies among eukaryotes. Humans and the
tropical fish known as black mollies have 46 chromosomes, turkeys have 82,
and giant redwoods have 22. Even with the same number of chromosomes,
however, different species carry different genetic information.

Cells of most organisms that reproduce sexually have pairs of similar
chromosomes. Each parent provides one member of each pair. Cells that
carry a double set of chromosomes are called **diploid.** Cells with just one set
of chromosomes are called **haploid.** The haploid number of chromosomes

> **ETYMOLOGY**
>
> **diplo-** = double (Greek)
> **haplo-** = single (Greek)
> **-oid** = form (Greek)
> **Diploid** cells have two sets of
> chromosomes. **Haploid** cells have a
> single set.

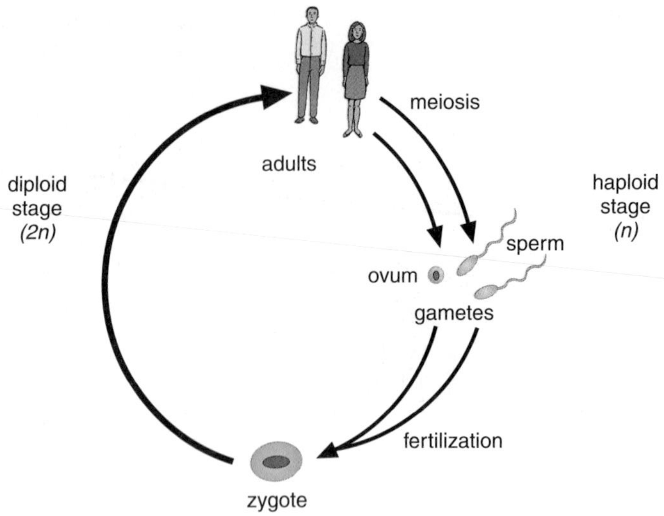

FIGURE 12.3

Changes in chromosome number in a typical sexual life cycle. In meiosis, diploid somatic cells produce haploid gametes. Gametes unite during fertilization, restoring the diploid chromosome number. The new individual has a combination of chromosomes from the two gametes.

Diagram labels: adults — meiosis — diploid stage (2n) — haploid stage (n) — sperm — ovum — gametes — fertilization — zygote

for a species is represented by the symbol *n* (Figure 12.3). The diploid number is represented by *2n*. For example, human cells contain 46 pairs of chromosomes. What are the diploid and haploid numbers for humans?

In diploid organisms, the two chromosomes of a pair are called **homologous.** Homologous chromosomes, or homologues, are similar in structure. (An exception is the pair called the sex chromosomes, which determine the sex of an individual. The sex chromosomes may be very different in size and shape and may carry different genes.) Homologues carry the same genes, though their DNA sequences may be slightly different. These differences produce the variety you see among members of the same species.

In asexual reproduction, the cells of parent and offspring carry identical sets of chromosomes. In sexual reproduction, however, *two* parents contribute chromosomes to offspring. For this reason, the reproductive cells that fuse during sexual reproduction must each carry *half* the normal number of chromosomes. If these cells contained the same number of chromosomes as the parents' **somatic** cells (body cells), the number of chromosomes would double in each generation. Consequently, the sexual reproductive cells of diploid organisms are haploid.

In sexual reproduction, each parent produces haploid gametes. Male gametes are sperm, and female gametes are ova (singular: **ovum**), or eggs. During fertilization, male and female gametes join and their nuclei fuse. A new individual develops from the diploid fertilized egg, or zygote.

A special cell-division process, **meiosis,** produces the haploid gametes. Generally, a gamete must either fuse with another gamete or die. In fungi and simple plants, meiosis produces different types of haploid cells called **spores.** Most spores can develop into haploid organisms without fertilization. In sexual reproduction, meiosis and fertilization are complementary processes: Meiosis produces haploid gametes, while fertilization restores the diploid chromosome number. Interestingly, a few species of plants and animals have lost the ability to reproduce sexually. For example, whole populations of desert whiptail lizards (Figure 12.4) are entirely female clones. Their eggs develop without being fertilized.

FIGURE 12.4

Natural cloning. Populations of desert whiptail lizards, *Cnemidophorus neomexicanus,* consist entirely of females. The factors that led to natural selection for asexual reproduction in these populations are not yet clear.

Cloning

Why is asexual reproduction most common among simple organisms? Sexual reproduction increases variety by producing new genetic combinations. This diversity provides raw material for natural selection. This effect may be an important factor in evolution.

In the 1970s, scientists developed ways to insert specific DNA sequences in bacteria and other cells. These modified cells, grown for many generations, were clones that carried the foreign genes. This process became known as cloning a gene. Cloning the human insulin gene in bacteria made it possible to mass-produce human insulin. Genetically engineered crop plants produce much of our food.

Artificial cloning is not a new idea. People have cloned crops by planting cuttings since ancient times. As early as the 1930s, biologists cloned carrot plants from single cells. The first cloned animal was a frog, produced in the 1950s by transplanting a nucleus from a cell of an early-stage frog embryo into a frog egg whose nucleus was removed. However, the ability to clone mammals is relatively new, and still crude.

In the 1980s, scientists began to separate the cells of young cattle embryos to clone the offspring of high-producing animals. Each cell developed into a complete embryo and was implanted in a separate cow for development and birth. In 1997, Ian Wilmut's research group in Scotland cloned the now famous sheep Dolly, (Figure 12.5). Researchers first removed the nucleus from a sheep's ovum and replaced it with the nucleus of a somatic cell from a different sheep. At this point, the ovum contained its original cytoplasm, including mitochondria and their DNA, but it had the nucleus of another sheep. The scientists then used an electric shock to stimulate the ovum to divide and develop. The young embryo was inserted in another sheep's uterus. This sheep gave birth to Dolly, who resembled the sheep that donated the somatic-cell nucleus. Thus, the genetic information transplanted into the clone determined the outcome.

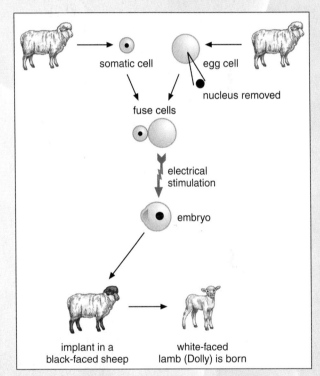

FIGURE 12.5

Artificial cloning. Scientists transferred a somatic cell's nucleus from one sheep to the ovum of a different sheep to clone Dolly. What evidence shows that the donor of the somatic cell nucleus is Dolly's genetic parent?

12.3 Meiosis and the Production of Gametes

Meiosis (Figure 12.6) differs from mitosis in three important ways. First, cells divide twice during meiosis, but the chromosomes are not duplicated after the first division. This halves the chromosome number from $2n$ to n. Second, meiosis distributes a random mixture of maternal and paternal

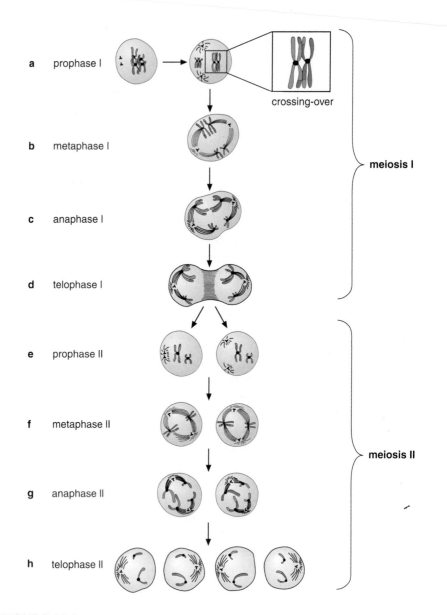

FIGURE 12.6

Meiosis. The stages of meiosis for a cell with $2n = 4$. As in mitosis, each chromosome enters meiosis as a pair of sister chromatids joined at the centromere. During prophase I **(a)**, homologues pair and cross over. Homologues attach to spindle fibers from opposite poles during metaphase I **(b)** and migrate to opposite poles during anaphase I **(c)**. Following telophase I **(d)**, each daughter cell contains a single mixed set of paternal and maternal chromosomes. These chromosomes are not replicated before meiosis II begins **(e)**. During metaphase II **(f)**, sister chromatids attach to spindle fibers from opposite poles, as in mitosis. Sister chromatids separate during anaphase II **(g)**. The result **(h)** is four haploid cells.

chromosomes to each gamete. This results in new genetic combinations. Third, homologous chromosomes pair up side by side during the first meiotic division. In this position, they often exchange corresponding pieces of DNA, a process called **crossing-over.** Recall that sexually reproducing organisms inherit a complete set of chromosomes from each parent. Crossing-over changes each chromosome into a mixture of maternal and paternal genes. This adds to the genetic variety of the gametes.

Meiosis involves two nuclear divisions—meiosis I and meiosis II—that produce four haploid cells. How does meiosis reduce chromosome number? Each division includes the same four stages—from prophase to telophase— as mitosis (see Section 8.6). As in mitosis, the process begins with duplicated chromosomes consisting of joined pairs of sister chromatids. However, meiosis I differs from mitosis in two ways. First, homologous chromosomes pair and cross over during prophase I (Figure 12.6a). Second, sister chromatids do not separate during anaphase I. Instead, homologous chromosomes migrate to opposite poles (Figure 12.6c). The result is a pair of cells, each containing a single set of chromosomes (Figure 12.6d). Note that each chromosome still consists of a pair of sister chromatids. When each cell divides in meiosis II, these chromatids separate (Figure 12.6g). Each of the four daughter cells receives a haploid set of chromosomes. Meiosis is a complex process because it does three important things: reduces chromosomes to the haploid number, provides genetic variation, and ensures the correct distribution of chromosomes into the resulting cells.

Meiosis does not always divide the cytoplasm equally between daughter cells. Figure 12.7a shows that in most male animals, including humans, meiosis produces four equal-sized sperm. In females, however, most of the cytoplasm remains in one cell (Figure 12.7b). This cell becomes the ovum. In

CHEMISTRY TIP

During prophase of meiosis I, several proteins form a zipperlike complex between homologues, ensuring that corresponding genes are side by side. Some of these proteins are enzymes that catalyze crossing-over by cutting and rejoining segments of DNA.

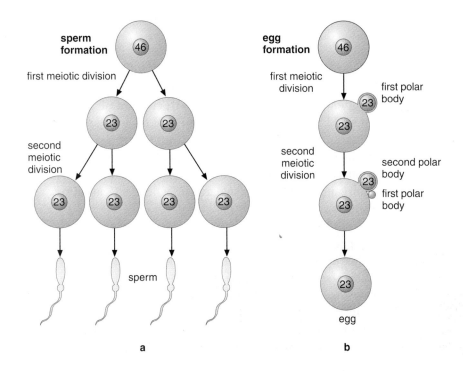

FIGURE 12.7

The results of meiosis in humans. **(a),** In males, meiotic division usually results in four equal-sized sperm, each with 23 chromosomes. **(b),** In females, two polar bodies form, and only one egg cell results.

animal species, the two small cells, called **polar bodies,** usually break down and disintegrate. In flowering plants, one polar body survives and remains diploid. Both this cell and the ovum are fertilized. The fertilized polar body develops into the endosperm that nourishes the embryo (see Section 11.1).

Check and Challenge

1. Why does asexual reproduction result in a clone?

2. Section 12.2 gave several examples of the numbers of chromosomes in the cells of various species. Explain the significance of the fact that all of them have even numbers of chromosomes.

3. What do the symbols $2n$ and n mean? How do these symbols correspond with the terms *haploid* and *diploid?*

4. Why is the reduction in chromosome number important in sexual reproduction?

5. What is crossing-over?

6. What are three important outcomes of meiosis?

Sexual Reproduction

12.4 Sexual Reproduction in Microorganisms

Prokaryotes reproduce asexually through cell fission. They do, however, exchange genetic information. A tube of cytoplasm can temporarily connect some bacterial cells. Some DNA passes through this tube. This process is called **conjugation.** Like sexual reproduction, it promotes genetic variation. However, conjugation does not produce offspring.

Conjugation also occurs in many unicellular eukaryotes (Figure 12.8a). Microorganisms that reproduce both asexually and sexually include many unicellular or colonial green algae. The life cycles of these organisms include both haploid and diploid stages, a pattern called **alternation of generations** (Figure 12.8b). These organisms are haploid during most of their lives. Sometimes they produce gametes, which fuse into a diploid zygote. Eventually meiosis produces more haploid cells. Alternation of generations is common in parasitic microorganisms. Many, such as *Plasmodium,* the organism that causes malaria, infect one species during the diploid stage and another during the haploid stage.

Many fungi and other microorganisms switch from asexual to sexual reproduction in response to changes in their environment. Stresses such as poor nutrition induce sexual reproduction in many such organisms. The mushrooms you may see sprouting from a wet lawn are just the fruiting structures that bear the fungus's reproductive cells. Most of the fungus consists of filaments of haploid cells in the soil. Cells of opposite mating types join to form the mushroom. The nuclei of these fused cells remain separate, however. On the underside of the mushroom are thin sheets of

CONNECTIONS

Sexual reproduction produces genetic variation. Natural selection acts on genetic variation. The result is the evolution of adaptations and of new species.

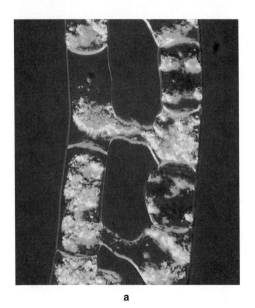

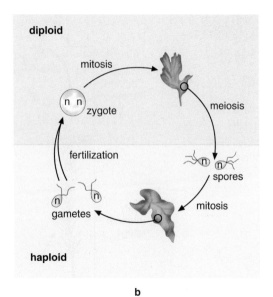

a   b

FIGURE 12.8

Conjugation and reproduction in green algae. (a), Two strands of the conjugating colonial alga *Spirogyra elegans* exchange DNA through temporary cytoplasmic bridges, ×190. **(b),** A diagram of alternation of generations shows that the haploid and diploid stages of the same organism often look completely different.

fungal tissue called gills. Here the nuclei fuse, meiosis occurs, and haploid spores are produced. The spores then disperse and grow into new filaments. Some fungi have particularly complicated reproductive cycles.

12.5 Sexual Reproduction in Plants

Plants generally reproduce sexually. However, many plants can also reproduce asexually, and some have lost the ability to reproduce sexually. Plant life cycles involve alternation of generations. Simple plants, such as mosses, spend most of their lives in the haploid stage (Figure 12.9). Like their aquatic ancestors, the green algae, these plants require a moist environment to reproduce. Their sperm swim through wet soil to reach the plants' female reproductive structures. If you examine the underside of a fern frond, you may see brown spots such as those in Figure 12.10. These spots are reproductive structures in which meiosis produces haploid spores. The spores fall to the ground and grow into tiny haploid plants that produce male and female gametes. The sperm require water to reach the ova. After fertilization, the zygote grows into a new diploid fern.

More complex plants are adapted to live in a wider variety of environments (see Chapter 11). These plants are large diploid structures. Their haploid stage is just a small tissue in their reproductive organs. The ova are protected inside these structures. Unlike fern sperm, the sperm of seed plants do not need to swim through wet soil. Wind or symbiotic animals, such as bees, bats, and butterflies, carry the sperm that has been packaged inside tough protective **pollen** grains to the female organs.

The most successful plants are the flowering plants. Haploid cells in the flowers produce the gametes. Each flower may produce sperm, ova, or both.

> **Try Investigation 12C**
> **Reproduction in Mosses and Flowering Plants.**

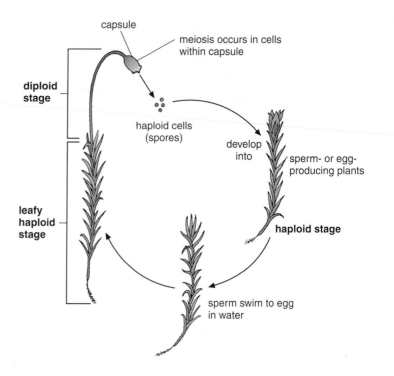

FIGURE 12.9

Life cycle of a moss. All plant life cycles involve alternation of generations. Mosses and other simple plants pass most of their life cycles in the haploid stage. The diploid plant quickly grows and produces cells that undergo meiosis to form haploid spores. The diploid plant then dies.

FIGURE 12.10

Spore-producing structures. On the underside of a frond of a licorice fern, *Polypodium vulgare,* the mature spores will be released and fall to the ground. There they will develop into tiny haploid plants.

At the center of an ovum-producing flower, one or more modified leaves called **carpels** fuse edge to edge, forming a hollow structure (Figure 12.11a). The base of this structure is the **ovary,** which contains one or more small structures called **ovules.** The ova develop in the ovules. Within each ovule, a specialized cell undergoes meiosis (Figure 12.11b). Four haploid cells result, three of which disintegrate. The fourth cell divides by mitosis to produce seven cells. One of these cells becomes the ovum. Another is a large cell containing two nuclei known as polar nuclei.

Cells in the **anther** undergo meiosis, producing four haploid cells (Figure 12.11c). Each haploid cell divides mitotically to produce a pollen grain containing two cells—a tube cell and a second cell that divides to produce two haploid sperm nuclei.

Sexual reproduction begins as the anthers shed pollen. In nonflowering seed plants such as pines and other needle-leaf trees, wind carries the pollen to the carpel of the same or other flowers. Insects, bats, and other animals that eat flower parts carry the pollen of flowering plants. The transfer of pollen from anther to carpel is called **pollination.** Cross-pollination (pollination between two different plants of the same species) increases genetic variation by combining chromosomes from two parents. Many intricate mechanisms for pollination have evolved. Some insects and flowering plants have become so completely dependent on each other that neither can reproduce without the other (see Appendix 12A, "Pollination by Insects Aids Fertilization").

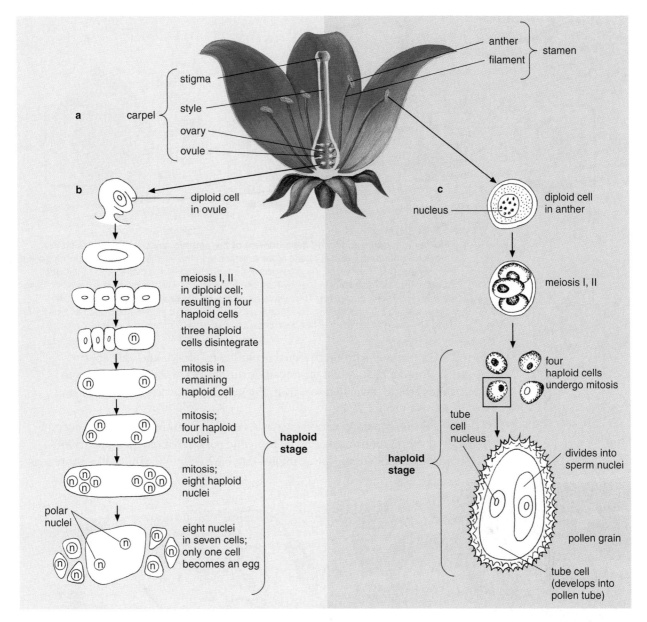

FIGURE 12.11

Parts of a flower. The carpel and stamen **(a)** make up the female and male parts, respectively, of a typical flower. The haploid stage in the life cycles of flowering plants is a small part of the flower. Gametes in flowering plants form from specialized cells. In the development of both ova **(b)** and sperm **(c),** haploid cells formed by meiosis divide by mitosis before gamete production is complete.

When pollen lands on the tip of the carpel or stigma, the pollen grain germinates and forms a pollen tube that grows toward the ovule, carrying the sperm nuclei (Figure 12.12a). Fertilization occurs when one sperm nucleus fuses with the egg (Figure 12.12b). The resulting diploid zygote divides mitotically and eventually develops into an embryo (Figure 12.12c). The second sperm nucleus fuses with the two polar nuclei, forming a triploid (3n) cell that develops into the endosperm. Fertilization ends the short haploid stage in the life cycle of a flowering plant. The ovule then becomes a

CHEMISTRY TIP

The female reproductive structures of many plants release chemical signals that attract sperm or animals that carry pollen. The disaccharide sucrose ($C_{12}H_{22}O_{11}$) attracts moss sperm. Flowering plants produce aromas that attract bees and other pollinating animals.

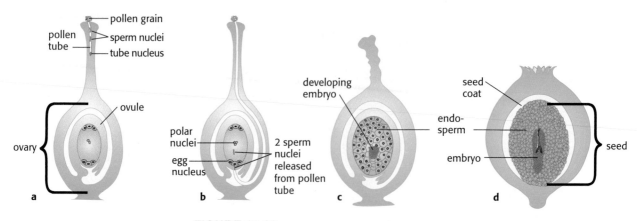

FIGURE 12.12

Pollination, fertilization, and development of the embryo, endosperm, and seed in a flowering plant. (a), Nutrients and other substances in the stigma stimulate the pollen grain to germinate. **(b),** The pollen tube grows down through the style and ovary wall to the ovule. Double fertilization occurs in the ovule as one sperm fuses with the ovum and the other fuses with the two polar nuclei. **(c),** The zygote develops into the embryo, and the triploid cell develops into the endosperm. The stigma and style shrivel up and may fall off after fertilization. **(d),** The ovule wall becomes the seed coat, and the ovary wall becomes the fruit.

seed, forming a protective coat around the embryo and endosperm (Figure 12.12d). Auxin produced by the seeds stimulates the ovary to enlarge and develop into a fruit. Figure 12.13 shows some of the many forms of fruits.

Seeds spread in various ways. A coconut drifting on an ocean current may travel thousands of miles from one island to another. Many fruits—those of dandelions, for example—are carried by the wind. Others, such as

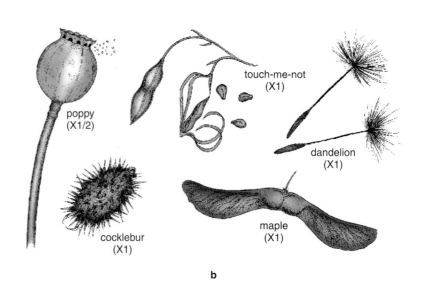

FIGURE 12.13

Examples of the diversity of fruits. (a), Each edible fruit is a mature ovary containing the seeds of the plant. Think about how edible fruits are adaptive for plants. **(b),** Many fruits have structural adaptations for seed dispersal. How do these fruits help disperse the seeds they contain?

a   b   c

FIGURE 12.14

Adaptations of flowering plants to diverse environments. (a), Mountain dogwood, *Cornus nuttallii,* in a pine-sequoia forest; **(b),** hedgehog cactus, *Echinocereus,* in a desert; **(c),** water lily, *Nymphaea odorata.*

cockleburs, are adapted to stick to animal fur. Many seeds remain inside fruits, which may be eaten by animals and transported great distances in their digestive systems. Eventually the animals deposit the undigested seeds along with other organic wastes, which help nourish the young plants in their new location.

Flowering plants are found in widely different environments (Figure 12.14). Several adaptations contribute much to their success: (1) the dominance of the diploid stage in the life cycle, which allows development of complex structures; (2) the evolution of pollen, which allows transfer of sperm from plant to plant without the need for water; (3) the evolution of the seed, which protects the dormant embryo and provides food and protection for the young plant; and (4) a variety of adaptations that promote pollen and seed dispersal.

> ## CONNECTIONS
>
> The large, starchy endosperm of wheat, rice, corn, and other grains makes them useful as foods. The use of nutrient reserves in seeds and fruits by both germinating seeds and plant-eating animals reflects the shared metabolic pathways all organisms have in common.

12.6 Sexual Reproduction in Animals

A large majority of animals reproduce sexually. Most animals that reproduce sexually have organs called **gonads** that produce gametes. These organs include ovaries, which produce ova, and **testes** (singular: testis), which produce sperm.

In many simple animals, such as *Hydras,* and some sponges, each individual produces both eggs and sperm. Most of these animals, however, are not usually self-fertilizing. Instead, one individual's sperm fertilizes another's eggs. A few vertebrates produce both eggs and sperm, but none are known to produce both simultaneously.

The simplest animals are aquatic. Most of them release large numbers of gametes into the water, a process known as spawning, or **external fertilization.** Many of these gametes fail to meet and fertilize. Most of the gametes, embryos, and immature offspring are consumed by other organisms.

In contrast, land animals and a few aquatic animals, such as whales, depend on internal fertilization. During **internal fertilization,** a male releases sperm into a female reproductive organ. Fertilization occurs within the body of the female. This protects the gametes from the hazards of the outside environment and requires fewer gametes than external fertilization.

Amphibians—vertebrates such as frogs that reproduce in water and spend part of their lives on land—are intermediate between internal and external fertilization (Figure 12.15). The male deposits sperm over the eggs as the female releases them into the water. This greatly increases the chance of fertilization. Like spawning animals, amphibians lay large numbers of eggs, and most of the young do not survive to reproduce.

Internal fertilization makes reproduction more efficient. Because the egg is sheltered within the female's body, fewer eggs are required. Spawning animals, such as fish, may lay more than 6 million eggs, although few are fertilized. This large number increases the chance of fertilization in an otherwise inefficient setting. In addition, because most aquatic organisms do not care for their eggs or their young, predators eat most of them. Internal fertilization is part of an evolutionary trend among larger, more complex organisms toward fewer gametes and offspring and greater parental care for each offspring. Mammals—animals that have hair and produce milk—include many of the most extreme examples of this trend. Humans and elephants usually produce only one offspring at a time and only a few in their lifetimes. The offspring develop within the mother's body and receive food and protection from parents and other adults for several years after birth.

Primitive mammals provide evidence of how mammalian reproduction evolved. Only two species of mammals—the spiny anteater, or echidna, and

FIGURE 12.15

Wood frogs, *Rana sylvatica*, mating. The male deposits sperm over the eggs as the female releases them.

the duckbilled platypus—lay shelled eggs. A few species, called marsupials, give birth to immature embryos that complete development in an open pouch on the outside of the mother's body (Figure 12.16). Older offspring return to the pouch to nurse and when they are threatened.

In most animals, sperm are much smaller than eggs (Figure 12.17). The large, immobile egg is adapted as a storehouse for the nutrients and organelles that support the embryo's development. In contrast, the tiny sperm cells contain little but a nucleus, a flagellum, and a few mitochondria that provide energy for movement. Because an animal sperm contains little stored nourishment, it cannot live long after being released. If a sperm is to reach an egg, both must be shed at about the same time and place. Many patterns of behavior have evolved that help ensure this. Reptiles, birds, and other animals perform elaborate movements as they prepare to mate (Figure 12.18). The movements usually do not attract animals of a different species, so these rituals also help restrict breeding to individuals of the same species.

Although internal fertilization uses eggs more efficiently, many sperm are still needed. There are three reasons for this. First, fertilization must occur in a brief time. Sperm live only a short time because they contain little stored food. For example, the life span of a human sperm cell is about 3 days. The human egg usually remains fertile for only about 1 day. For fertilization to take place, egg and sperm must be released at about the same time.

A second reason that large numbers of sperm are needed is that the microscopic sperm must swim to the egg. In large vertebrates, they must

FIGURE 12.16

Marsupial reproduction. This young grey kangaroo, *Macropus,* begins its development within its mother's body. After birth, a young marsupial crawls into its mother's pouch, attached to a nipple, and completes development.

a

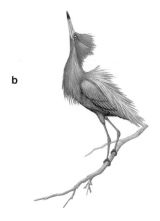

b

FIGURE 12.18

Mating behavior. Many animals have evolved elaborate behavior patterns that ensure that mating will occur only between members of the same species. **(a),** The male fiddler crab, *Uca,* attracts the female by waving its one large claw. **(b),** The male reddish egret, *Egretta rufescens,* faces the female in this position. Mating occurs only if she responds correctly.

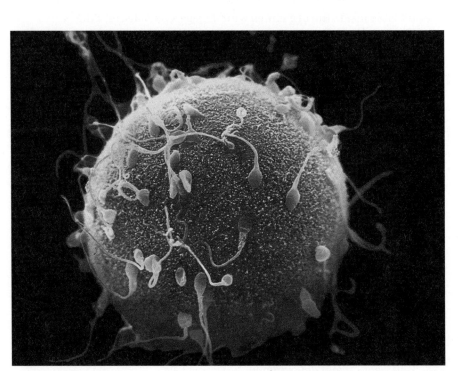

FIGURE 12.17

Internal fertilization. Scanning electron micrograph of human sperm penetrating an egg (color added, ×1,210). Only one sperm will fuse with this mature egg, initiating the growth and development of a new person.

Liquid is necessary for the movement of sperm during sexual reproduction. Life evolved first in the oceans, so this requirement became a problem only as life moved onto land. Plants and animals adapted to reproduction on land in similar ways. Simpler land plants and animals grow only in damp environments. Ferns can grow in slightly drier places, but they can reproduce only when a thin film of water is present, as after a rain. Similarly, adult amphibians can seek out food on land but must return to water to reproduce.

Pollen, produced by seed plants, protects sperm from dehydration. This protection enables seed plants to reproduce in drier environments. Internal fertilization serves a similar function in land vertebrates and insects. Among these organisms, the reproductive organs provide the liquid necessary for movement of sperm. The environment inside a plant's carpels or an animal's reproductive organs somewhat resembles the conditions ancient organisms found in the oceans. Flowering plants and land animals have other parallels, including nutrition of the offspring (by endosperm in seeds and by nutrients supplied by the mother's body or stored in the egg) and protection of the embryo (by the seed coat in flowering plants and by egg shells or the mother's body in land animals).

travel several centimeters. Relatively few sperm ever reach the egg. Most of the 200 million sperm that human males release die before reaching the egg.

The third reason is that the enzyme activity of many sperm is required to penetrate the membrane and layers of cells around the egg. Once one sperm enters the egg, chemical and physical changes immediately take place that make it impossible for other sperm to penetrate the egg. The nucleus of the sperm that is inside the egg unites with the egg's nucleus, and the development of a new individual begins.

Many insects avoid these problems by storing sperm in a pouch inside the female. The female releases some of the sperm as she lays eggs. Female bees, for example, may store sperm from a single mating for an entire life span.

Check and Challenge

1. Some plants, animals, and fungi can reproduce both sexually and asexually. How can you explain the fact that many of these organisms tend to switch from asexual to sexual reproduction under stresses such as starvation?
2. Distinguish between pollination and fertilization in plants.
3. Describe the formation and function of the plant embryo, endosperm, seed, and fruit.
4. In what ways are flowering plants better equipped than other plants to reproduce in a wide range of environments?
5. What are the functions of ovaries and testes in animals?
6. How is the number of eggs an animal produces related to its form of fertilization and development?
7. Explain the importance of animal premating rituals.

Reproduction in Humans

12.7 Egg Production and the Menstrual Cycle

The details of reproduction vary greatly among animals. Differences in reproduction among animals reflect their adaptations to various environments. This section describes human reproduction as one example.

In human females, the ovaries are located inside the body cavity (Figure 12.19). The ovaries usually release eggs, which are barely big enough to see (about 0.1 mm), one at a time. The egg travels through one of two tubular structures, called **oviducts,** to the **uterus.** This muscular, pear-shaped organ is where the embryo develops. During birth, strong muscular contractions push the baby out of the uterus through the **vagina.** This muscular passageway also functions as the route through which the male deposits sperm during sexual intercourse and through which menstrual fluid passes out of the body.

The egg-releasing cycle, called the **menstrual cycle,** usually lasts about 28 days, but this varies among individuals and from cycle to cycle. The inner lining (endometrium) of the uterus builds up in preparation for receiving a fertilized egg. If the egg is not fertilized, the lining and its rich blood supply disintegrate and flow out through the vagina. The first day of menstrual flow marks the first day of the menstrual cycle.

ETYMOLOGY

mensis = month (Latin)

duct = leader or channel (Latin)

The **menstrual** cycle is about 1 month long. The **oviduct** is the passageway through which the ovum passes.

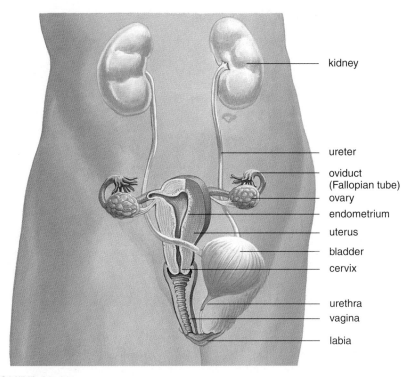

kidney

ureter

oviduct
(Fallopian tube)

ovary

endometrium

uterus

bladder

cervix

urethra

vagina

labia

FIGURE 12.19

The human female reproductive system. The oviducts are tubes that conduct the immature and fertilized eggs from the ovaries to the uterus where the embryo will develop. The excretory system is shown in gray for reference.

The nervous system, several glands and other organs, and a variety of hormones interact to regulate the menstrual cycle. Figure 12.20 illustrates the events of this cycle. The hypothalamus, a glandular part of the brain, acts in the menstrual cycle the way a thermostat acts in a house. A thermostat does not keep a house at a constant temperature. The temperature fluctuates slightly as the thermostat turns the furnace on and off. Just as a thermostat monitors and adjusts temperature, the hypothalamus monitors and helps adjust the level of hormones in the circulatory system.

At the start of the menstrual flow, **estrogen** and **progesterone,** hormones released into the bloodstream by the ovaries, are at low levels. These low levels cause the hypothalamus to secrete gonadotropin-releasing hormone (GnRH) into blood vessels that supply the pituitary gland. This structure lies just underneath the hypothalamus. GnRH stimulates the pituitary to release two hormones, follicle-stimulating hormone (FSH) and luteinizing hormone (LH), that act on the ovary. FSH causes an egg to start maturing inside a fluid-filled sac, or follicle, in the ovary.

FSH and LH also stimulate the maturing follicle to release more estrogen into the bloodstream. Estrogen signals the lining of the uterus to thicken and grow more blood vessels. Around day 14 of the cycle, a sudden increase of LH from the pituitary gland causes the follicle to burst and release the egg.

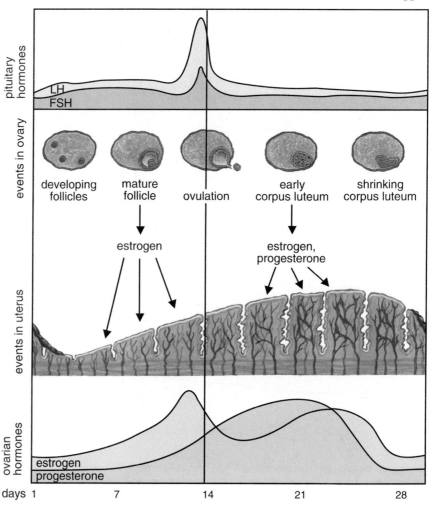

FIGURE 12.20

The menstrual cycle. Interactions among the nervous system and several organs, glands, and hormones regulate the cycle.

This process is called **ovulation.** The ruptured follicle becomes the **corpus luteum.** The corpus luteum continues to release estrogen and progesterone. These hormones stimulate additional buildup of the uterine lining.

The hypothalamus detects the increased levels of estrogen and progesterone and slows down the release of FSH and LH by the pituitary gland. If the egg is not fertilized while traveling down the oviduct toward the uterus, the corpus luteum starts to degenerate, and the levels of progesterone and estrogen decline. The decrease in estrogen and progesterone signals the lining of the uterus to break down. Menstruation begins as the blood-filled uterine tissues disintegrate and flow through the vagina. The onset of the menstrual flow marks the beginning of another cycle.

If the egg is fertilized in the oviduct, it undergoes mitosis on its way to the uterus and becomes an embryo. When the embryo arrives in the uterus, it implants in the lining and begins to form a placenta (see Section 10.5). The placenta begins to release a hormone called human chorionic gonadotropin (HCG). (Many pregnancy tests detect the presence of HCG in the mother's urine.) HCG signals the corpus luteum to continue releasing high levels of progesterone and estrogen. These hormones support the uterine lining in which the embryo develops and indirectly prevent the start of another menstrual cycle (Figure 12.21a). The corpus luteum functions for the first 3 months of **gestation,** or development of the embryo within the uterus. After 3 months of gestation, the placenta stops releasing HCG and becomes the main source of estrogen and progesterone (Figure 12.21b).

Another hormone, the small peptide known as oxytocin, is released at the time of birth. It causes the muscles of the uterus to contract, expelling the baby. After birth, oxytocin stimulates milk release and contraction of the uterus to its prepregnancy size. Shortly after birth, the placenta separates from the uterine wall and is expelled as the afterbirth. Without the placenta,

ETYMOLOGY

corpus = body (Latin)
luteum = yellow (Latin)

After an ovarian follicle releases an ovum, it becomes a **corpus luteum,** or "yellow body."

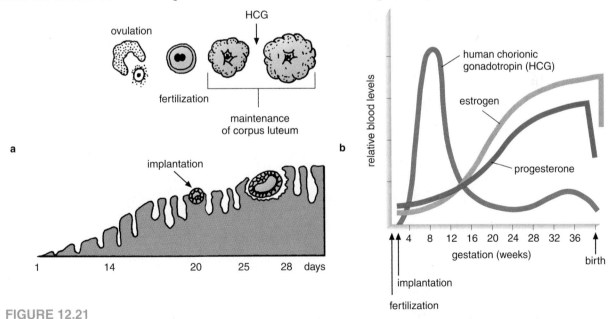

FIGURE 12.21
Hormonal changes during pregnancy. (a), At first, HCG secreted by the placenta causes the corpus luteum to produce the estrogen and progesterone necessary to maintain the pregnancy. **(b),** Later the placenta produces these hormones.

estrogen and progesterone levels drop. The low levels of these hormones signal the hypothalamus once again to release GnRH, which causes the menstrual cycle to resume.

Menstruation is a characteristic of female primates (humans, apes, and their relatives). Most other mammals do not menstruate, but they do have estrus, or heat, cycles when they ovulate and mate. Animals such as deer and elk enter estrus once a year, and most dogs go into estrus twice a year. In other mammals, such as mice and rabbits, mating provokes ovulation. Menstrual cycles, estrus cycles, and ovulation at intercourse are all adaptations that help ensure reproductive success.

12.8 Sperm Production

In human males, the testes are located in the **scrotum,** an out-pocketing of the body wall. In this location, the testes remain cooler than the rest of the body. Human sperm need this slightly cooler environment to develop normally. Sperm are produced by meiotic cell division in the highly coiled tubes (seminiferous tubules) of the testes (Figure 12.22). Sperm are stored in the epididymis, a coiled part of the sperm duct (vas deferens) adjacent to the testes.

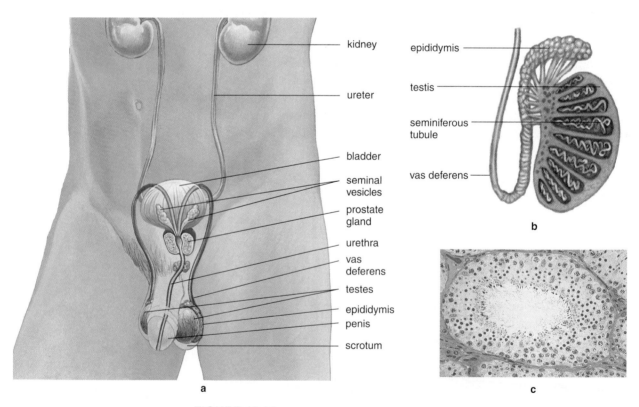

kidney

ureter

bladder

seminal vesicles

prostate gland

urethra

vas deferens

testes

epididymis

penis

scrotum

epididymis

testis

seminiferous tubule

vas deferens

b

a

c

FIGURE 12.22

The human male reproductive system. (a), Sperm produced in the testes travel the vas deferens, mix with seminal fluid, and are ejaculated through the urethra. The excretory system is shown in gray for reference. **(b),** Each testis is composed of many tightly packed coils of seminiferous tubules in which sperm develop. **(c),** In this cross section through a seminiferous tubule, ×100, notice the darkly stained heads (nuclei) of the sperm along the edge of the central canal. Their flagella extend into the canal.

Additional structures, the prostate gland and seminal vesicles, produce seminal fluid, which transports sperm. The seminal fluid also contains much fructose, a sugar that provides energy for the sperm. During **ejaculation,** sperm and seminal fluid, together called semen, travel through the sperm duct and out of the body through the urethra. If sperm are not ejaculated, they eventually are reabsorbed by the tissues in the male reproductive tract.

Males and females produce the same regulatory hormones. GnRH from the hypothalamus stimulates the pituitary to release FSH and LH. LH stimulates cells between the tubules to secrete **androgens,** a group of male hormones. The major androgen secreted by the testes is **testosterone.** FSH stimulates other cells in the testes to produce sperm. Although both sexes share the same regulatory hormones, there is no evidence that males have a cycle of reproductive hormones similar to that of females.

12.9 Secondary Sex Characteristics

There are prominent differences in the appearance of male and female animals of the same species (Figure 12.23). Many of these differences are due to hormonal differences between the sexes. Both males and females produce estrogen and androgens. Females produce more estrogen, and males produce more androgens. These hormones control the development of **secondary sex characteristics,** such as facial hair in males and growth of breasts in females. Estrogen controls the development of secondary sex characteristics in females, causing changes in the breasts, bone structure, and fat deposits under the skin. These changes begin at **puberty,** the beginning of sexual maturation. In females, puberty begins at about the same time as the first menstrual period. In North America, this usually occurs between the ages of 10 and 12. However, the age at which puberty begins is affected by many factors, such as heredity, percentage of body fat, and nutrition.

Androgens control the development of secondary sex characteristics in males. Beginning at puberty, the androgens cause changes, for example, in the voice and in body-hair distribution. Puberty in North American males usually begins around the age of 12, but this varies widely among individuals. Most males enter puberty later than most females.

FIGURE 12.23

Secondary sex characteristics in a mating pair of lions, *Panthera leo.* Think of other examples in which the female and male of a species do not closely resemble each other.

12.10 Infertility and Contraception

Some couples find it difficult or impossible to conceive and bear children. The fertility problem may lie with either partner or depend on the combination of the two individuals. For example, irregular menstrual cycles can prevent ovulation. Some conditions, such as an overactive thyroid gland, can interfere with embryo implantation. Blocked oviducts can prevent fertilization. A man may produce too few sperm or sperm that cannot swim well enough to reach the egg. Sometimes the combination of the environment of the vagina and reduced mobility of sperm makes conception unlikely. Hormone treatments can sometimes establish regular ovulation. Other treatments, including taking specific vitamins, may improve sperm mobility.

Some fertility problems require medical intervention. For example, **in vitro** fertilization (Figure 12.24) can help women with blocked oviducts to conceive. If the oviducts are blocked—by malformation or by scarring from infection—sperm cannot reach the ovum to fertilize it. To bypass the blockage, physicians remove several ova directly from the woman's ovaries and mix them in a culture dish with the husband's sperm. After fertilization, the embryos are implanted in the mother's uterus.

Couples may choose to limit the size of their families. Some methods of contraception provide a physical barrier between the egg and sperm (Figure 12.25). The condom is a sheath that is slipped over the penis prior to intercourse. When used correctly, latex condoms also reduce the spread of AIDS and other sexually transmitted diseases. A diaphragm is a flexible cap that is inserted through the vagina to cover the cervix. Because the cervix varies in size, an effective diaphragm must be fitted by a health care professional. Both the condom and the diaphragm are quite effective when used properly.

Other methods of contraception prevent ovulation. Norplant is a device that is implanted surgically in the upper arm. It releases progesterone, preventing ovulation for up to 5 years. Depo-Provera, a progesterone-based

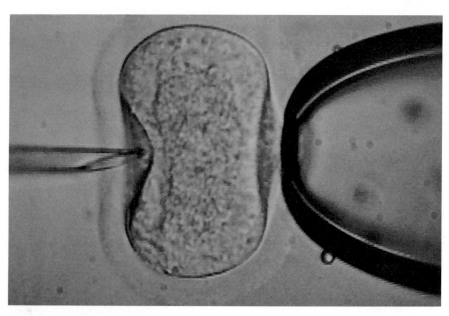

FIGURE 12.24

In vitro fertilization. The needle at left is puncturing the coating of the egg (×250). This allows sperm to enter more easily.

FIGURE 12.25
A variety of contraceptive devices.
Starting clockwise from top left are a diaphragm, a contraceptive pill dispenser, a sterile package containing Norplant, and condoms.

injectable drug, suppresses ovulation for 3 months. Similarly, taking pills with carefully regulated doses of estrogen and progesterone inhibits the release of the pituitary hormones that stimulate ovulation. Some low-dose contraceptive pills alter the timing of egg release and may prevent implantation. Because these hormone-based methods can have serious side effects or health risks for the women who take them, they must be prescribed by a physician.

Other contraceptives include foams or jellies that immobilize sperm. Foams or jellies are inserted into the vagina before intercourse. They are most effective when used together with a diaphragm or condom.

Surgical sterilization is yet another form of contraception, but it is usually permanent. The sperm ducts can be cut and tied in a process called a vasectomy. The oviducts can also be cut and tied in a procedure called a tubal ligation. Both procedures are difficult to reverse.

Effective contraception requires planning. Once sperm have entered the vagina, there is little chance of preventing fertilization. The only 100% reliable method of birth control is abstinence from intercourse.

Check and Challenge

1. Develop a concept map that shows the relationships among the following female hormones: GnRH, estrogen, progesterone, FSH, LH, and oxytocin.
2. What is the relationship between a follicle and a corpus luteum?
3. What are the effects of FSH and LH in the male reproductive system?
4. Describe the relationship between male and female hormones and the onset of puberty.
5. Describe and compare naturally occurring and artificial barriers to contraception.

Summary

Many organisms reproduce asexually, producing offspring identical to the parent. Prokaryotes reproduce by binary fission. Asexual reproduction in eukaryotes usually involves mitotic cell division. Examples of asexual reproduction include budding and fragmentation among animals and the growth of runners among plants. In some animals, an unfertilized egg can develop into an adult organism.

Sexual reproduction requires two gametes and the complementary processes of meiosis and fertilization. Two important events occur in meiosis. First, meiosis reduces the number of chromosomes from diploid to haploid during the production of gametes. If meiosis did not occur, the number of chromosomes would double in each generation. Second, crossing-over, which occurs during the early stages of meiosis, exchanges genes between homologous chromosomes. The genetic recombination that results from meiosis and crossing-over is important in maintaining genetic variation in species.

Sexual reproduction in flowering plants requires pollination followed by double fertilization, which results in a triploid endosperm and a diploid embryo. Evolution has produced many elaborate schemes through which pollination occurs.

Animals with external fertilization are mainly aquatic and produce large numbers of gametes and young. Most of them do not survive. Animals with internal fertilization usually live on land and produce small numbers of eggs but large numbers of sperm. Elaborate behavior patterns have evolved that result in simultaneous release of eggs and sperm.

Most mammals have an estrus cycle in which females ovulate and become receptive for mating at specific times during the year. The menstrual cycle is found only among the primates and is produced by an elaborate interaction among the brain and various glands, other organs, and hormones. The cycle coordinates the release of ova with preparation of the uterus to receive a fertilized egg. If the egg is not fertilized, the uterine lining degenerates, resulting in menstrual flow.

Sperm production in male animals is also under control of hormones. In addition, sex hormones cause males and females to exhibit secondary sex characteristics. Some couples have difficulty conceiving. A lack of ovulation, sperm blocked from reaching eggs, or low sperm activity are all factors that can result in infertility. Contraceptive methods use some of these same factors to prevent conception.

Key Concepts

Use the concepts below to build a concept map, linking as many other concepts from the chapter as you can.

- meiosis
- gametes
- fertilization
- haploid
- asexual reproduction
- sexual reproduction
- pollination
- diploid

Reviewing Ideas

1. How is the structure of the ova and sperm related to their function in fertilization?
2. Explain three important functions of meiosis.
3. Describe the differences between metaphase I and metaphase II in meiosis.
4. Why does moss need a moist environment to reproduce?
5. Why is fertilization in flowering plants called double fertilization?
6. Explain how insects help pollination in flowering plants.
7. What adaptations in aquatic animals help ensure that external fertilization will occur?

 Biology Online BSCSblue.com/vocabulary_puzzlemaker

8. What adaptations in land animals help ensure that internal fertilization will occur?
9. Explain the relationship between the follicle and ovum in a human female.
10. Explain how estrogen and progesterone influence the uterus in the menstrual cycle.
11. What could be concluded from a blood test indicating the presence of human chorionic gonadotropin (HCG) in a human female?
12. What is the action of HCG, and what organ does it affect in human females?
13. Explain how LH and FSH function in human reproduction.
14. Describe the physical changes that occur at puberty in human females and males.
15. What is oxytocin, and what role does it play in human reproduction?

9. Compare the processes of mitosis and meiosis.
10. How are the organs of the human male and female reproductive systems adapted for the continuation of the species?

Synthesis

1 Why is maintaining reproductive processes important in protecting endangered species?
2. In what way is the first cell division in meiosis unlike cell division associated with mitosis?

Extensions

1. Write a short paper about an imaginary human population that reproduces asexually. Describe how this might make a difference in the structure of society.
2. Make a drawing that depicts the three most important results of the process of meiosis.
3. Write a short essay to explain the reasons that species show a wide variety of reproductive methods.

Using Concepts

1. Why is it important to maintain a constant chromosome number within a species?
2. What would prevent the pollen from one plant species from pollinating flowers of another species?
3. Compare the evolutionary advantages of asexual and sexual reproduction.
4. Explain why chromosomes cannot be referred to strictly as maternal or paternal after meiosis has occurred.
5. Explain how double fertilization in flowering plants differs from fertilization in humans.
6. What is the importance of reproduction to individual organisms? to populations? to a species?
7. What advantages for land animals does internal fertilization have over external fertilization?
8. Explain the difference between estrus (heat) and menstrual cycles.

Web Resources

Visit BSCSblue.com to access
- Resources about reproduction
- Information about photomicrographs of pollen tubes and cells at various stages of meiosis
- More detailed explanations of meiosis and the practical importance of bacterial conjugation
- Web links related to this chapter

Genes and Chromosomes

13.1 Heredity and Environment

ETYMOLOGY

frater = brother (Latin)

Genetically, **fraternal** twins are as similar and as different as other brothers and sisters.

⚛ CHEMISTRY TIP

Most chemical reactions, including those catalyzed by enzymes, proceed faster at higher temperatures. However, as enzymes absorb more energy at higher temperatures, their secondary and tertiary structures can change. This can reduce their effectiveness as catalysts, as in the Siamese cat.

Organisms are products of their heredity and of their surroundings. Your height and build are probably similar to those of your parents, but they are also influenced by environmental factors such as nutrition and exercise. A common example of the interaction of heredity and environment is the coloring of Siamese cats (Figure 13.1). These cats inherit genes for enzymes that produce a dark pigment in their fur. These enzymes function best at temperatures below the cat's normal body temperature. Siamese cats have dark markings on their ears, nose, paws, and tail, all areas with a low temperature. If the cat's belly is shaved and an ice pack is applied, the replacement hair will be dark. Likewise, a tail that is kept warm by a cloth wrapping would soon be covered with light-colored fur. These changes are temporary, however, unless the ice pack or wrapping is kept in place. This example shows that the environment has a strong impact on gene expression.

Studies of twins can help separate the effects of inheritance and the environment. Identical twins develop when a zygote forms two complete embryos. These embryos have exactly the same genetic information. Fraternal twins arise from separate eggs and sperm cells. Fraternal twins are no more genetically similar than other siblings. If identical twins exhibit the same trait more often than fraternal twins, then the trait is probably heavily influenced by genetic factors. If the trait differs in identical twins, then the environment must have a strong influence on the trait. Such studies show that both genes and environment are important.

The idea of "blending" inheritance was once a popular explanation of heredity. According to this theory, an individual's genetic makeup was formed when the parents' genes mixed at fertilization. The result was a sort of averaging of the parents' genes. Once the parents' contributions were blended, they could not be passed on separately to future generations.

Genetics no longer includes the theory of blending inheritance, but it was a reasonable idea. People's height and coloring are often intermediate between those of their parents, and many dogs look like combinations of their parents. Using pollen from a snapdragon with white flowers to fertilize one with red flowers produces a plant with pink flowers. Blending inheritance was a logical explanation for many visible differences.

Still, there were always events that blending inheritance could not explain. Some individuals resemble their grandparents or other relatives more than their parents. Healthy parents sometimes produce sickly offspring, and some sickly parents have healthy children. These apparent exceptions were the clues that led to a better theory—one that could explain more observations. Section 13.2 explains that theory.

13.2 Mendel and the Idea of Alleles

In the 1860s, Gregor Mendel (Figure 13.2) used garden peas to study heredity. Genetics has its foundations in Mendel's work. His choice of the garden pea as an experimental organism showed thoughtful planning. Peas are very easy to grow, and Mendel could study many generations during his eight-years of experiments. Peas are also self-fertilizing; Figure 13.3 shows how fertilization usually occurs in pea plants. When Mendel wanted to cross-fertilize his plants, he removed the stamens and dusted the stigmas with pollen from a different plant.

Mendel decided not to study traits, such as seed size, that show a continuous range of variation. Instead, he concentrated on traits that did not fit the blending theory. Mendel found seven different characteristics of pea plants, shown in Figure 13.4, that he could study in an either-or form. He classified peas as either smooth and round or wrinkled; plants were either tall or short; pods were either green or yellow; and so on. Mendel worked with his plants for several years to be sure he had true-breeding varieties for each of the traits he selected. True-breeding plants produce offspring identical to themselves generation after generation. Mendel then

FIGURE 13.2

Gregor Mendel (1822–1884). Mendel was a monk, gardener, high school science teacher, mathematician, and scientist. By experimenting with peas in his monastery garden, Mendel developed the fundamental principles of heredity that became the foundation of modern genetics.

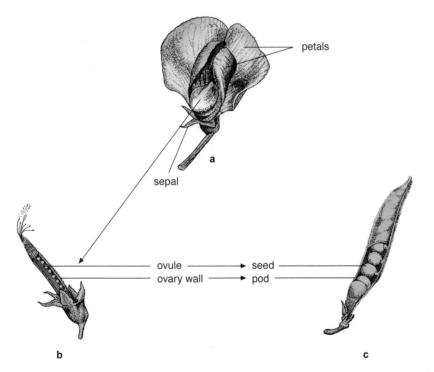

FIGURE 13.3

Self-fertilization in a pea flower. The petals of the pea flower **(a)** completely enclose the reproductive organs. As a result, the pollen from the anthers falls on the stigma of the same flower **(b).** Pollen tubes grow down through the female reproductive organ to the ovules in the ovary. The ovules develop into seeds, and the ovary wall develops into the pea pod **(c).**

seed shape	embryo color	flower and seed coat color	pod shape	pod color	flower position	stem length
round	yellow	colored	inflated	green	axial	tall
wrinkled	green	white	constricted	yellow	terminal	short

FIGURE 13.4

The seven characteristics of garden peas. Mendel's findings contradicted the blending theory of heredity.

CONNECTIONS

Most genes have different alleles, resulting in the genetic variation that is the basis of evolution through natural selection.

crossbred his plants and classified all the offspring. As he analyzed the data from many breeding experiments, Mendel looked for patterns of inheritance.

Consider a mating between a plant with round seeds and a plant with wrinkled seeds (Figure 13.5). Would the offspring plants produce partly wrinkled seeds? Or would some seeds be round and others wrinkled? Mendel found that the next generation of plants produced only round seeds. When those plants were allowed to self-fertilize, however, some plants in the next generation produced round seeds and other plants produced wrinkled seeds. What conclusions can you draw from these results?

The blending theory could not explain these results. Mendel demonstrated with pea plants that both parents pass on to their offspring genetic *factors* that remain separate generation after generation. Even when the wrinkled seed factor remained hidden in the second generation, it was still present and able to act in the third generation. It was not blended into the round seed factor.

Today the concept of genes has replaced Mendel's vague idea of factors. But the concept of genes has changed drastically during the past century as methods for studying genes and their interactions have improved. A gene is now defined as a segment of DNA whose sequence of nucleotides codes for a specific functional product. These products include ribosomal and transfer RNA and proteins.

Most genes exist in more than one form, or **allele.** Each allele of a particular gene has a different base sequence. Today we would call Mendel's

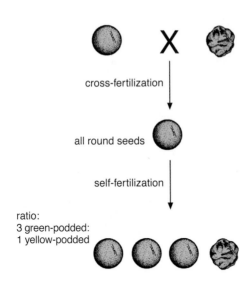

cross-fertilization

all round seeds

self-fertilization

ratio:
3 green-podded:
1 yellow-podded

FIGURE 13.5

Results of a mating between pea plants breeding true for round seeds and wrinkled seeds. Mendel relied heavily on results such as these in developing his theory of inheritance.

FIGURE 13.6

A woman with the widow's peak trait. Note how her hairline tapers to a point above her nose.

factors alleles. Thus, the allele for round seeds encodes an enzyme that produces round seeds. The allele for wrinkled seeds is a different form (actually an inactive form) of the same gene. The wrinkled seed allele has an insertion of an extra bit of DNA sequence. This extra DNA prevents the gene from encoding an active enzyme. Wrinkled seeds are the result.

All organisms have genes that exist as several different alleles. Human examples of alleles are found in the characteristics known as widow's peak and male pattern baldness. The allele for widow's peak is one form of a gene affecting the shape of the hairline. Individuals who have this allele often exhibit the trait (Figure 13.6). Conversely, individuals who have a different allele of the gene have smooth hairlines. In male pattern baldness, there is a gradual loss of hair on the top of the head (Figure 13.7). This trait also results from one allele of a single gene. Most traits scientists once thought were determined by single genes—such as hair color, skin color, nose shape, and handedness—actually result from the complex interaction of several genes with each other and the environment.

13.3 Genes and Chromosomes

The arrangement of genes in chromosomes differs in eukaryotes and prokaryotes. Eukaryotic chromosomes are long molecules of DNA wrapped around proteins (see Biological Challenges: Chromosome Structure in Section 8.4). Only part of this DNA codes for proteins. Noncoding DNA is not translated. Some noncoding DNA consists of short sequences of bases

FIGURE 13.7

Male pattern baldness. The expression of the trait is influenced by sex hormones. It is much more common in men, hence its name.

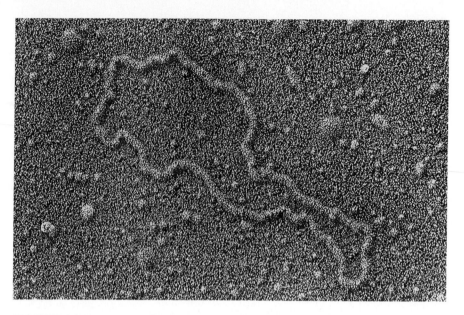

FIGURE 13.8

Bacterial plasmid. A scanning electron micrograph (×79,000) of a bacterial plasmid. These small circular DNA molecules are common in some types of bacteria. Their ability to transfer genes between cells makes plasmids useful in genetic research.

CHEMISTRY TIP

The flat molecules of some stains bind to DNA by fitting between the paired bases of the double helix. Others reveal banding patterns by binding to regions of repetitive noncoding DNA.

repeated thousands of times. About 1.5% of human DNA is expressed as protein. The importance of most of the rest is still unclear.

The situation is different in prokaryotes. Bacteria have a single circular chromosome with little associated protein. Their genes generally do not have introns, and an estimated 90% of their DNA is translated. Many bacteria also have plasmids—small circles of DNA that contain additional genes (Figure 13.8). Plasmids may move from one bacterium to another. Genetic engineers use plasmids to introduce modified genetic material into bacterial cells. Plasmids also can carry genes that provide resistance to antibiotics.

You have seen stained chromosomes in eukaryotic cells, but those chromosomes probably all looked alike. Homologous chromosomes carry the same genes, although their genes may be present as different alleles. How can biologists distinguish homologous chromosomes from other chromosomes in the same cell? The size of a chromosome and the position of its centromere provide some clues, but several chromosomes of the same organism may appear similar. Stains help by binding to specific regions of chromosomes. This creates banding patterns that are specific to each of an organism's chromosomes (Figure 13.9a). Chromosome painting (Figure 13.9b) is another technique that can be used to identify quickly all of an organism's chromosomes.

Chromosomes are easiest to study in the condensed form they have during cell division. For this reason, white blood cells are used most frequently to study human chromosomes. These cells can easily be made to divide and grow in culture. Chemicals are added to stop cell division during metaphase. The cells are placed on microscope slides and treated with water. This causes them to swell, and their chromosomes spread apart. Stains applied to the resulting chromosome spread produce the banding

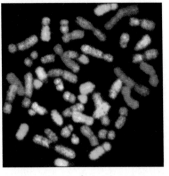

7   X

a   b

FIGURE 13.9

Techniques for identifying chromosomes. (a), Human chromosomes 7 and X are similar in size and shape. The banding patterns evident in the photographs show more clearly in the drawings. These two chromosomes, which have almost the same length and centromere location, can be distinguished by their banding patterns. **(b),** Fluorescent dyes of different colors are chemically bonded to short pieces of DNA (probes) that bind to genes on different chromosomes. This process stains each homologous pair of chromosomes a different color.

patterns seen in Figure 13.10. This technique helped determine that human cells (except gametes) have 46 chromosomes.

Stained chromosomes can be photographed under the microscope. Individual chromosomes can be cut out of an enlarged photograph and arranged by size and shape to form a display called a **karyotype.** Figure 13.11 shows a karyotype of a human cell. Karyotypes of fetal cells can be used to check for suspected chromosomal abnormalities in developing fetuses.

FIGURE 13.10

Human chromosome spread. The banding patterns result from differences in stain absorption by different regions of the chromosomes. Banded chromosomes enable biologists to detect missing or extra chromosome parts more easily than uniformly stained chromosomes do. The banding patterns also have made the mapping of genes on chromosomes more accurate.

ETYMOLOGY

karyo- = nut or nucleus (Greek)

A **karyotype** reveals the types of chromosomes present in a cell's nucleus.

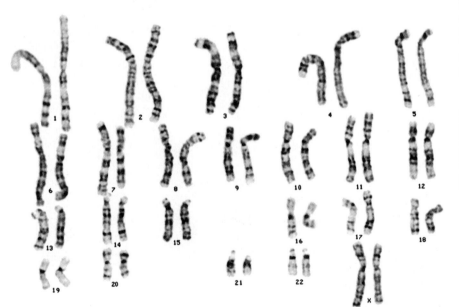

FIGURE 13.11

A karyotype of a human cell. The chromosomes are arranged as homologous pairs. The larger groups (for example, pairs 1–3) include chromosome pairs with similar size and centromere position. A karyotype is prepared by cutting out individual chromosomes from a photograph and matching them, pair by pair. The pair of X chromosomes determines this individual's sex.

Biological Challenges

The Human Genome Project

The Human Genome Project (HGP), involving thousands of scientists and technicians, had a single goal. Its objective was to determine the sequence of the estimated 3 billion base pairs of DNA in the human genome (Figure 13.12). An organism's entire DNA makes up its genome. The human **genome** consists of 22 numbered pairs of chromosomes, the two sex chromosomes (X and Y), and the mitochondrial chromosome. Starting in 1988, the HGP was an international effort that was projected to take 15 years and cost $3 billion. The HGP was completed in 2003 and cost more than $1 billion. It provided a map showing where in our chromosomes each of the estimated 35,000 human genes are. It also provided the complete DNA sequence of each chromosome.

The HGP sequenced sections of DNA from many people. Together these samples span our entire genome. Sequence data alone, however, cannot explain how genes work. This would require comparisons of human genes with those of other well-studied organisms. So far, the complete genomes of more than 100 species of bacteria, and dozens of eukaryotes have been sequenced.

A new specialty in biology, called bioinformatics, has been created to analyze the vast amount of sequence data generated by the HGP. It uses computers to compare human DNA sequences with those of other organisms. If a human gene is similar to a fruit-fly gene, for example, then the two sequences may have similar functions. This approach is being used to help develop diagnostic tests and treatments for genetic diseases. The HGP identified genes associated with single-gene disorders, such as Huntington's disease and cystic fibrosis. It also discovered genes that influence more complicated disorders, such as cancer and heart disease, which involve many genes and environmental factors.

As with any powerful technology, many ethical concerns were associated with the HGP. Some people feared that they would be unable to obtain jobs or health insurance if they were found to carry alleles that may contribute to disease, even though they were healthy. Part of the project's budget was used to study its ethical, legal, and social implications. The intent was to anticipate any social problems from the HGP and to develop strategies for dealing with them before they could cause harm.

The vast amount of data produced by the HGP will keep scientists working for many years and impact society for generations to come.

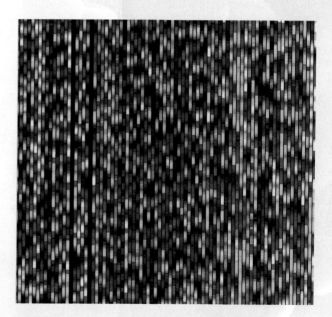

FIGURE 13.12

Electrophoresis results from a human DNA-sequencing experiment. Each band corresponds to one base (adenine, guanine, thymine, or cytosine) in the DNA sample. The color of each band indicates which base is present. Automated machines produce these images and convert them into DNA base sequences in a computer's memory.

Mendelian Patterns of Inheritance

13.4 Probability and Genetics

Diploid organisms usually carry different alleles of many genes. The distribution of those alleles among the gametes in meiosis and the combination of alleles that come together in fertilization are a matter of chance. **Probability** is a branch of mathematics that predicts the chances that a certain event will occur. Geneticists use probability to predict the results of matings.

You are probably familiar with the idea of probability. What are the chances that a tossed coin will turn up heads? What are the chances that you will draw an ace from a deck of cards (Figure 13.13)? Probability is usually expressed as a fraction. The chance of the coin landing heads up is one out of two, or $\frac{1}{2}$. Your chance of drawing an ace is 4 out of 52, or $\frac{1}{13}$. A batting average in baseball also is a statement of probability. A player who has a .250 batting average has about a 25% probability ($\frac{1}{4}$, or 1 in 4 chance) of making a hit.

This kind of calculation works when each event (a coin toss or a turn at bat) has no effect on other events. The events are said to be independent of one another. For example, if you get three heads in a row when flipping a coin, the probability that the next toss will land heads up is still $\frac{1}{2}$. The coin does not remember the first three tosses—chance events have no memory. You can, however, predict the probability of getting all heads when tossing four coins simultaneously. The chance that two or more independent events will occur together is the multiplication product of their chances of occurring separately. Thus to determine the chance that all four tosses will land heads up, you multiply $\frac{1}{2} \times \frac{1}{2} \times \frac{1}{2} \times \frac{1}{2}$. The result is a $\frac{1}{16}$ chance.

Geneticists use probability to predict the alleles of the offspring of various crosses, or matings. These predictions can be compared with the results of breeding experiments. Mathematical tests can help show whether the difference between the observed and predicted results is significant—in

Try Investigation 13A ⬦ Probability.

FIGURE 13.13

Probability in card drawing. The chance of drawing an ace is $\frac{1}{13}$. What is the chance of drawing a spade? What is the chance of drawing the ace of spades?

Biology Online BSCSblue.com/check_challenge

other words, whether the difference is due to chance or to some other factor (see Appendix 13A, The Chi-Square Test). Genetic counselors use probability to help parents assess the risk of passing on a genetic disorder to their children.

Remember that genetic ratios are estimates of probability, not absolute numbers. However, the larger the sample size, the less deviation you would expect from predicted ratios. In the next section, you will learn to use probability to explain the behavior of genes.

13.5 Inheritance of Alleles

The principles of probability enable you to explain the results of Mendel's experiment (see Figure 13.5). In this experiment, Mendel crosed plants that differed in only one trait: seed shape. Such a cross is called a **monohybrid cross.** Mendel dusted pollen from true-breeding, round-seeded plants onto the stigmas of true-breeding, wrinkled-seeded plants. The plants involved in this first cross are called the **parental (P) generation.**

The plants that grew from the seeds of the parental crosses produced all round seeds. This generation of plants is called the **first filial, or F_1, generation**. Mendel then allowed the F_1 generation plants to self-fertilize. He obtained 5,474 plants with round seeds and 1,850 plants with wrinkled seeds in this **second filial, or F_2, generation,** approximately a 3:1 ratio. Mendel's actual data are in Table 13B.1 in Investigation 13B.

For each of the seven characteristics Mendel studied, only one trait of each pair was visible in the F_1 generation. He termed this trait the **dominant** trait. The alternative trait, which was not visible in the F_1 generation, he called the **recessive** trait. In Figure 13.4, the dominant traits are in the top row and the recessive traits in the bottom row. Round seeds are dominant, and wrinkled seeds are recessive. Note that the recessive trait reappeared unchanged in the F_2 generation. In the F_2 generation produced by self-fertilization of F_1 plants, the two traits appeared in a ratio of approximately 3:1, dominant to recessive.

Mendel was not the first to observe a 3:1 ratio of traits, but his use of mathematics in biology was unusual for his time. He used mathematics to conclude that each true-breeding plant has two identical copies of the factor for a particular trait. When gametes formed during meiosis, only one copy of the factor went into each pollen or egg cell. At fertilization, the F_1 generation received a round seed factor from one parent and a wrinkled seed factor from the other parent. Only one of these factors, either round or wrinkled, went into each gamete formed by the F_1 plants. Mendel called this separation of factors the **principle of segregation.** Today Mendel's factors are called alleles. The alleles of each gene segregate (separate) during meiosis when homologous chromosomes are divided among the gametes.

The allele for a dominant trait is commonly represented by a capital letter (such as *R* for round seeds), and the allele for the recessive trait is represented by the lowercase letter (such as *r* for wrinkled seeds). Remember that pea plants, like most multicellular organisms, are diploid, so they have two alleles for each gene. Therefore, the **genotype,** or genetic makeup, of the true-breeding, round-seeded plant in the P generation would

Try Investigation 13B ◇ Seedling Phenotypes.

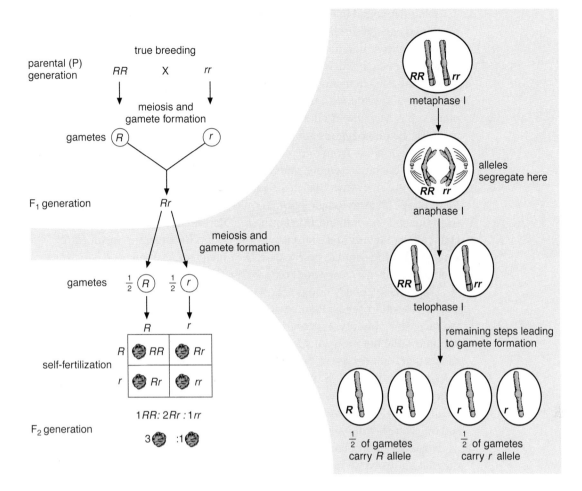

FIGURE 13.14

A monohybrid cross that demonstrates Mendel's principle of segregation. Each plant has two factors (alleles) for each trait, which segregate during the formation of gametes. Fertilization restores paired alleles. The grid, or Punnett square, shows how gametes of the F_1 generation combine to form the genotypes of the F_2 generation.

have been *RR*. The wrinkled-seeded parental plant had the genotype *rr* (Figure 13.14). During meiosis, each plant produced haploid gametes with only one allele for each gene. At fertilization, each F_1 plant received a chromosome that carried the *R* allele from its round-seeded parent and a homologous chromosome with the *r* allele from its wrinkled-seeded parent. Therefore, the genotype of the F_1 plants was *Rr*. When the F_1 plants produced their gametes, the two alleles *R* and *r* segregated during meiosis. Half of the gametes received the *R* allele. The other half received the *r* allele.

If both alleles are the same (*RR* or *rr*), as in the parental plants, the genotype is **homozygous.** If the alleles are different (*Rr*), as in the F_1 individuals, the genotype is **heterozygous.** The genotype of each individual is responsible for its **phenotype**—its appearance or observable characteristics. Because the round-seeded phenotype is dominant, both the homozygous *RR* and the heterozygous *Rr* genotypes produce plants with round seeds.

ETYMOLOGY

phene- or **phane-** = appear (Greek)
hetero- = different (Greek)
homo- = same (Greek)
zygo- = joined (Greek, Latin)

An organism's **phenotype** describes how it appears. If different alleles of the same gene (such as *R* and *r*) occur together in the cells of an organism, the organism is said to be **heterozygous.** If the alleles are identical (such as *R* and *R*), the organism is **homozygous.**

What happened in the F_2 generation? Half the gametes that the F_1 plants produced carried the R allele and half carried the r allele. Recall that the chance that two independent events will both occur is the product of the chances that each will occur separately. Thus the chance that an F_2 plant will receive the r allele from both parents is $\frac{1}{2} \times \frac{1}{2} = \frac{1}{4}$, or 25%. This would produce an F_2 plant with the homozygous genotype rr. The chance that an F_2 plant will receive the R allele from both parents is also 25%. This would produce the homozygous genotype RR. What is the chance that the plant will receive the R allele from one parent and the r allele from the other, producing the heterozygous genotype Rr?

There are two ways that this could happen: The male parent could supply the R allele, and the female parent could supply the r allele. The reverse—r from the male parent and R from the female—is also possible. There is a 25% chance that each of these situations will occur. Because the result (a Rr genotype) is the same in both cases, the chance of an F_2 plant having the genotype Rr is $\frac{1}{4} + \frac{1}{4} = \frac{1}{2}$, or 25% + 25% = 50%.

The homozygous genotype RR will produce the dominant phenotype: a plant with round seeds. Plants with the homozygous genotype rr will have the recessive phenotype, wrinkled seeds. Because the round seed trait is dominant, the heterozygous Rr plants will have round seeds. Figure 13.14 shows that, on average, 75% of the F_2 plants will have round seeds, and 25% will have wrinkled seeds. A 3:1 ratio of phenotypes in the F_2 generation often indicates that the trait is determined by a single gene with two alleles. Ideally, three-fourths of the F_2 generation will have the dominant phenotype, and one-fourth will have the recessive phenotype.

The diagram in Figure 13.14 is a Punnett square. This tool is useful for calculating probable ratios of genotypes and phenotypes. All the possible genotypes of the gametes one parent can produce are at the top of the square, and the genotypes of the other parent's gametes are at the side. The square is then filled in like a multiplication table with the genotypes that would result from the union of those gametes. The Punnett square in Figure 13.14 shows how the 3:1 ratio of genotypes results in the F_2 generation of a monohybrid cross.

To follow the inheritance of two characteristics at once, Mendel made **dihybrid** crosses—crosses between individuals that differ in two traits. Earlier he had found that the traits of round seeds and yellow embryos are dominant and that the traits of wrinkled seeds and green embryos are recessive. Imagine a cross of plants that are true-breeding for round seeds with yellow embryos ($RRYY$) with plants that are true-breeding for wrinkled seeds and green embryos ($rryy$). Can you predict the genotypes and phenotypes of the F_1 generation?

Mendel did not know about chromosomes and meiosis, but the behavior of chromosomes during meiosis explains his results. The two genes involved in this cross are on different pairs of homologous chromosomes. The genotype of all the F_1 plants was $RrYy$. When meiosis occurred in these plants, half the gametes received the R allele and half received the r allele. At the same time, half the gametes received the Y allele and half received the y allele. Because the seed-shape and embryo-color genes are located on different chromosomes, their inheritance is not connected. Whether a

Try Investigation 13C ▽ A Dihybrid Cross.

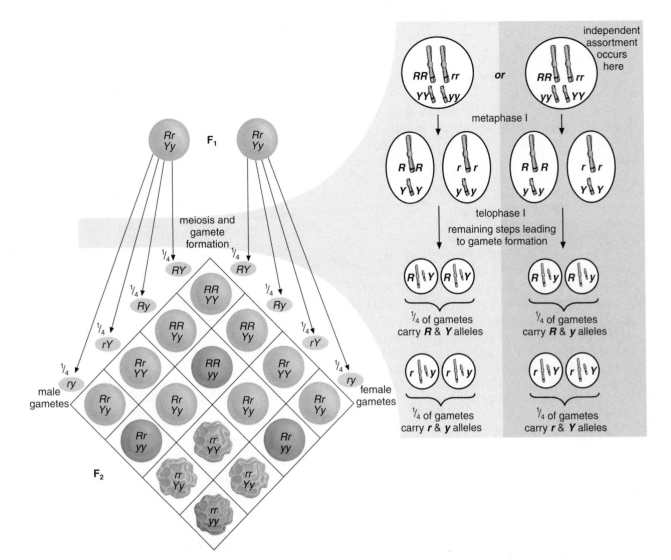

FIGURE 13.15

A dihybrid cross that illustrates the principle of independent assortment. What phenotypes are present in the F₂ generation and in what ratios? These results demonstrate that the traits for seed shape and embryo color assort separately. (Note that there are 12 round to every 4 wrinkled seeds and 12 yellow to every 4 green seeds. Each trait independently shows the 3:1 ratio typical of a monohybrid cross.) *R* = round seed; *r* = wrinkled seed; *Y* = yellow embryo; *y* = green embryo.

particular gamete receives a chromosome that carries the *R* allele or the *r* allele has no effect on whether it will also receive a chromosome that carries a *Y* or *y* allele.

If you count the distribution of phenotypes in Mendel's F₂ generation (Figure 13.15), you will see that their ratio is 9:3:3:1. This ratio is characteristic of the F₂ generation in a dihybrid cross. The genes for seed shape and embryo color separated from each other and were inherited independently. The results of dihybrid crosses formed the basis for Mendel's **principle of independent assortment:** Alleles for one characteristic assort, or divide up among the gametes during meiosis, independently of alleles for other characteristics.

Focus On

The Chromosome Theory of Inheritance

Theodor Boveri and W. S. Sutton each made the connection between chromosomes and Mendel's factors in 1902. Boveri showed experimentally that although egg and sperm cells are different, they make equal genetic contributions to the new organism (Figure 13.16). Mendel had already found evidence for this conclusion. He found that the ratio of dominant to recessive phenotypes was always the same, regardless of which parent carried the factor for the dominant trait.

Boveri and Sutton reasoned that if the genetic contributions of sperm and eggs are the same, the genes ought to be located in the same place in the two types of gametes. Where could this be? Sperm cells are composed mostly of a nucleus and just a small amount of cytoplasm. The nucleus of the egg is very similar to the nucleus of the sperm. However, the cytoplasm of the egg is very different from that of the sperm. Because of the similarities of nuclei and differences in cytoplasm, Boveri and Sutton both concluded that the nucleus contains the genes.

Careful observations of meiosis showed that the chromosomes in the nucleus behave like genes:

Homologous chromosomes segregate among the gametes like Mendel's factors. Nonhomologous chromosomes assort independently, just as Mendel claimed the alleles of different genes do. Table 13.1 compares the activity of chromosomes with the behavior of Mendel's factors. The work of Sutton and others supported these ideas, which became known as the chromosome theory of inheritance.

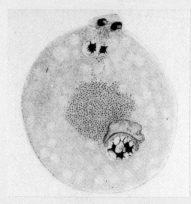

FIGURE 13.16

A drawing by Theodor Boveri of his observations of fertilization of the egg of a parasitic nematode (roundworm). The chromosomes of the egg cell and a polar body are visible at the top. The sperm nucleus is at lower right. Boveri's sharply detailed drawing appears as realistic as a photograph.

TABLE 13.1
Comparison of Genes and Chromosome Behavior

Hypothesis of Gene Behavior	Observations of Chromosome Behavior
Gametes have half the number of genes that body cells have.	Gametes have half the number of chromosomes that body cells have.
Gene pairs separate during gamete formation.	Chromosome pairs separate during gamete formation.
In fertilization, gametes unite, restoring the original number of genes.	In fertilization, gametes unite, restoring the original number of chromosomes.
Individual genes remain unchanged from one generation to the next.	Individual chromosomes retain their structure from one generation to the next.
The number of possible gene combinations can be calculated.	The number of possible chromosome combinations can be calculated.

13.6 Sex Determination

Chromosomes come in matching pairs except for the sex chromosomes, which may be different. This pair of chromosomes determines the sex of the individual. In humans, other mammals, and the fruit fly *Drosophila melanogaster*, the sex chromosomes are labeled *X* and *Y*. Females have two

X chromosomes; males have one X chromosome and one Y chromosome. Therefore, all the eggs produced during meiosis have an X chromosome. Half of the sperm produced by a male have an X chromosome, and the other half have a Y chromosome. Thus sperm determine the sex of the offspring. If an egg is fertilized by a sperm with an X chromosome, the zygote develops into a female. If the sperm has a Y chromosome, the offspring is a male. Look at Figure 13.11. Is it the karyotype of a male or a female? Figure 13.17 shows how the sex chromosomes determine the sex of human offspring.

Some insects, such as grasshoppers, crickets, and roaches, have a different system: Females have two X chromosomes, males have one, and there is no Y chromosome. Thus males have one less chromosome than

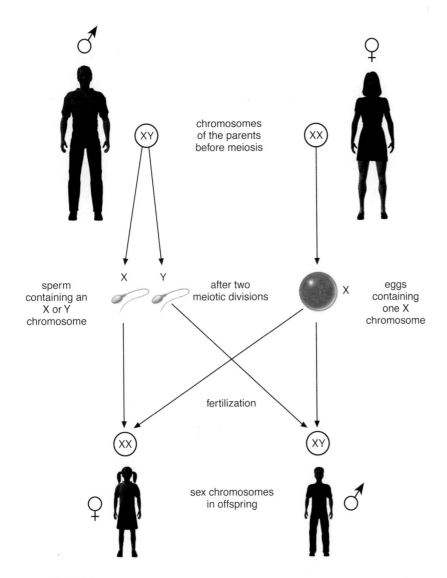

FIGURE 13.17

Sex determination. Sex in humans is determined by the sex chromosomes of the egg and the sperm at the moment of fertilization. XX produces a female; XY produces a male. The symbol ♂ represents male; ♀ represents female.

females. Birds, some fish, and some insects have what is called a Z-W system of sex determination. Here the male has two matching sex chromosomes (ZZ), whereas the female has one Z chromosome and one W chromosome. Some plants, such as spinach and date palms, have separate female and male plants and have sex chromosomes that follow the X-Y system of sex determination. Most plants and some animals have no sex chromosomes.

Check and Challenge

1. Describe Mendel's two principles of inheritance. How do these two principles relate to the behavior of chromosomes during meiosis?
2. What are the chances that a family with three children has only sons? After two sons, what is the probability that the next child will be a boy?
3. Distinguish between genotype and phenotype and between homozygous and heterozygous.
4. Using a Punnett square, work out the probable ratios for the phenotypes of offspring from a cross between a homozygous (TT) tall pea plant and a heterozygous (Tt) tall pea plant. (The tall phenotype is dominant.)
5. What types of gametes are produced by the following plants: $GGRr$ (green pods and round seeds) and $TtYy$ (tall plants with yellow embryos)?

Other Patterns of Inheritance

13.7 Multiple Alleles and Alleles without Dominance

Mendel worked with traits that were clearly either dominant or recessive. Some genes do not follow this pattern, however. When red-flowered snapdragons are crossed with white-flowered snapdragons, all the F_1 plants have pink flowers. This phenotype is intermediate between those of the parents. This type of inheritance is known as **incomplete dominance.** When the F_1 plants self-fertilize, the F_2 plants show a 1 red: 2 pink: 1 white ratio (Figure 13.18).

Another pattern of inheritance is seen in human blood types. Blood type depends on the presence or absence of type A or type B carbohydrates on the surface of red blood cells. The alleles I^A or I^B code for different forms of an enzyme that add different sugars to the carbohydrate bound to the plasma membrane. When a person has the $I^A I^B$ genotype, both types of

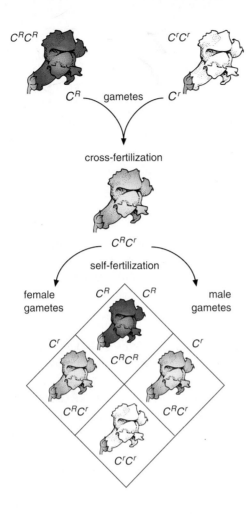

FIGURE 13.18

Incomplete dominance of flower color in snapdragons. Note that the symbols for alleles used here are different from the system used earlier for simple Mendelian genes. C refers to the flower-color gene. Superscripts indicate the alleles for red (C^R) and white (C^r) flowers. The phenotype of the F_1 heterozygote ($C^R C^r$) is intermediate between those of the true-breeding parents. The ratios for both the genotypes and phenotypes of the F_2 generation are 1:2:1.

carbohydrates are produced and the individual's blood type is AB. When both phenotypes appear in heterozygous individuals, as in this case, the alleles are said to be **codominant.**

Blood types also involve **multiple alleles.** In addition to the I^A and I^B alleles, the allele i codes for no active enzyme. An ii genotype produces type O blood. The symbols I^A, I^B, and i are used to show that the A and B traits are codominant and the O trait is recessive. The genotypes, phenotypes, and surface carbohydrates for the four different blood types are shown in Table 13.2. Although there are three alleles for blood type, no individual has more than two of the alleles.

Blood types are important in transfusion. If someone with type A blood receives a transfusion of type B blood, the donated red cells clump together, clogging the blood vessels. The clumping is caused by **antibodies,** which are defensive proteins found in the blood. Antibodies bind to foreign substances and are an important defense against infection (see Section 23.8). People do not produce antibodies for carbohydrates on their own red blood cells, but they do produce antibodies for foreign carbohydrates. Thus a person with type B blood produces anti-A antibodies. If this person receives a transfusion

CHEMISTRY TIP

Both type A and type B carbohydrates contain a molecule of the simple sugar galactose, but in type A cells, there is an acetyl group ($-CH_2-COOH$) and an amino group ($-NH_2$) attached to the galactose molecule. In type O blood, no galactose is added to the cell-surface carbohydrate.

linked genes

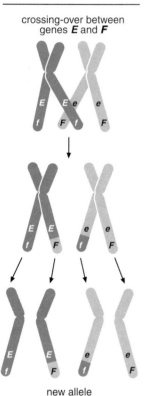

crossing-over between
genes **E** and **F**

new allele
combinations

FIGURE 13.19

Crossing-over. Crossing-over frequently occurs between homologous chromosomes during the early stages of meiosis. This results in genetic recombination (new combinations of genes). Here the symbols *E* and *e* stand for the alleles of one gene, and *F* and *f* are the alleles of another gene on the same chromosome. Because these two genes are close together, crossing-over probably would not occur between them very often.

TABLE 13.2
Genetics of ABO Blood Types

Phenotype (blood type)	Genotype	Carbohydrate on Surface of Red Blood Cells	Antibody Response
type A	$I^A I^A$ or $I^A i$	A	anti-B
type B	$I^B I^B$ or $I^B i$	B	anti-A
type AB	$I^A I^B$	A and B	neither anti-A nor anti-B
type O	ii	neither A nor B	anti-A and anti-B

of type A blood, the anti-A antibodies that are produced attack the A surface carbohydrates. Table 13.2 lists the antibodies and red-blood-cell carbohydrates found in individuals of different blood types.

13.8 Linked Genes

Each chromosome carries many genes. Genes on the same chromosome are said to be linked. Linked genes are often inherited together. Most of the traits Mendel studied were not due to linked genes. If they were, he might not have developed the principle of independent assortment.

Alleles of linked genes do not always stay together. Recall from Section 12.3 that homologous chromosomes often exchange pieces when they pair in meiosis (Figure 13.19). That exchange produces new allele combinations that greatly increase variation. The farther apart two genes are on a chromosome, the more likely a break will occur between them. If two genes are far enough apart, breaks between them may be so frequent that they assort independently.

The frequency with which linked traits become separated reflects how far apart on the chromosome the genes for those traits are. This information can be used to map the positions of genes on chromosomes (Figure 13.20). Important genes that are hard to identify can also be mapped by observing traits produced by nearby marker genes. Researchers have identified many disease-related genes in this way, including the allele responsible for Huntington's disease, a serious inherited disorder of the nervous system.

13.9 X-Linked Traits

An understanding of linked genes came from studies of fruit flies (Figure 13.21). At Columbia University in the 1910s, Thomas Hunt Morgan and his students Alfred Sturtevant and Calvin Bridges found an unusual white-eyed male fly. When they mated this fly with a normal red-eyed female, all the offspring had red eyes. The F_2 generation showed a 3:1 ratio of red-eyed flies to white-eyed flies (Figure 13.22a). These results confirmed that the white-eye trait is recessive. However, only males had white eyes. Why should this trait be linked to a fly's sex?

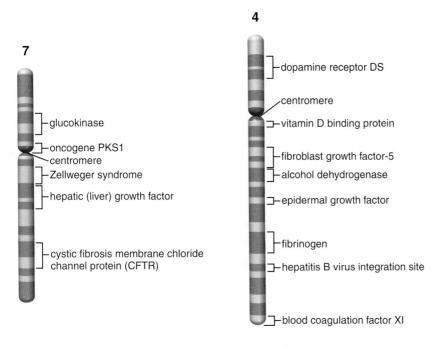

7

- glucokinase
- oncogene PKS1
- centromere
- Zellweger syndrome
- hepatic (liver) growth factor
- cystic fibrosis membrane chloride channel protein (CFTR)

4

- dopamine receptor DS
- centromere
- vitamin D binding protein
- fibroblast growth factor-5
- alcohol dehydrogenase
- epidermal growth factor
- fibrinogen
- hepatitis B virus integration site
- blood coagulation factor XI

FIGURE 13.20

Gene maps. A few of the genes that have been mapped to human chromosomes number 7 *(left)* and 4 *(right)*. Linked genes are inherited together because they are on the same chromosome.

Morgan's group eventually found white-eyed females. When they crossed these flies with red-eyed males, the F₁ offspring included only red-eyed females and white-eyed males (Figure 13.22b). Morgan's explanation for these results was that the gene for eye color is carried on the X chromosome. There is no eye-color gene on the Y chromosome. Males have only one X chromosome. Therefore, whichever eye-color allele they receive is expressed. A male fly that carries the recessive white-eye allele will have white eyes. Females have two X chromosomes, so the red-eye trait is dominant in female flies. White eye color was the first known example of an **X-linked trait,** a trait whose gene is carried only on the X chromosome. Since then, many other X-linked traits have been found in *Drosophila,* and more than 300 have been identified in humans. Well-known examples of human X-linked, recessive alleles included those responsible for red-green color blindness and hemophilia, a disease in which blood does not clot normally.

The British geneticist Mary Lyon added to Morgan's work. She suggested that, early in the development of a normal female, one X chromosome becomes inactivated in each body cell. This process occurs at random: In some cells, the X-linked genes inherited from the mother are expressed. In other cells, the X-linked genes inherited from the father are active. Thus the cells of a female express a mixture of X-linked traits (Figure 13.23).

FIGURE 13.21

Adult fruit flies, *Drosophila melanogaster* (×10). The top fly is male; the bottom fly is female. Note the difference in their coloring.

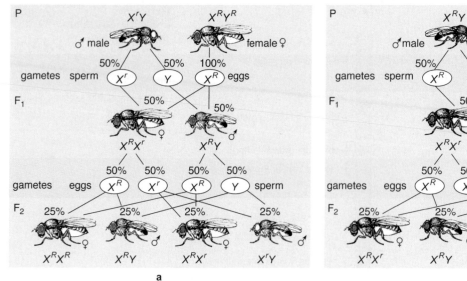

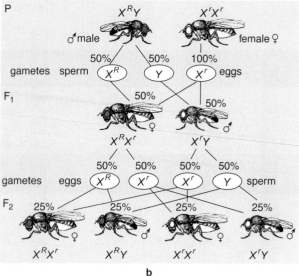

a

b

FIGURE 13.22

X-linked inheritance. (a), When Morgan crossed a white-eyed male with a normal red-eyed female, all the F_1 flies had red eyes. The F_2 flies showed the expected 3:1 ratio of red to white eyes, but only the male flies showed the recessive, white-eye trait. **(b),** In the reciprocal cross, a white-eyed female was crossed with a red-eyed male. All male offspring had white eyes because they inherited their X chromosome from their mother. The ratio of F_2 flies also is shown. Compare the F_1 generations of the two crosses.

Lyon proposed that X inactivation could explain a darkly staining mass called a Barr body (Figure 13.24) that normally appears in the nucleus of female cells but not in male cells. She suggested that the Barr body is the tightly condensed, inactive X chromosome. Errors in meiosis (see Section 13.10) cause some females to receive an extra X chromosome. Cells of these XXX females have two Barr bodies. Lyon interpreted this observation to mean that only one X chromosome in a cell remains active. Embryos that receive extra chromosomes usually do not survive. But females with an extra X chromosome do survive. Lyon's X-inactivation hypothesis may help to explain why an extra X chromosome is not as disruptive as an extra copy of another chromosome.

FIGURE 13.23

A calico cat. The coloring of this female cat is a visible indication of X inactivation. The cat carries two different alleles for coat color on its two X chromosomes. Random X inactivation during embryonic development results in this patchwork-colored coat in some female cats.

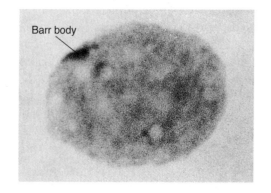

Barr body

FIGURE 13.24

Barr body in the nucleus of a female cell. The darkly stained Barr body is the inactivated X chromosome.

13.10 Nondisjunction

One of the X-linked traits found in *Drosophila* is the recessive eye-color trait known as vermilion (bright red). Morgan's student Calvin Bridges crossed many vermilion-eyed females with normal red-eyed males (Figure 13.25a). About once in every 2,000 flies, a vermilion-eyed female or a red-eyed (but infertile) male appeared in the F₁ generation. How could this happen?

A female in the F₁ generation with vermilion eyes must have two X chromosomes carrying the recessive v allele. But her red-eyed father's only X chromosome carried the V allele for the dominant red-eye trait. How could a vermilion-eyed female be produced? By the same token, how could a red-eyed male appear in the F₁ generation?

Bridges suggested that these flies must have developed from gametes produced by abnormal meiosis. If for some reason the sex chromosomes failed to separate in meiosis, then the offspring developing from those gametes would have unusual numbers of sex chromosomes. Bridges found that the cells of these exceptional flies did have extra or missing sex chromosomes. He called the failure of homologous chromosomes to separate in meiosis **nondisjunction.** Figure 13.25b shows the results of nondisjunction of the X chromosomes in the female parent. Nondisjunction also can occur between the X and Y chromosomes in males.

Nondisjunction also occurs in the sex chromosomes and other chromosomes of humans. The effects may be severe. Certain syndromes, or

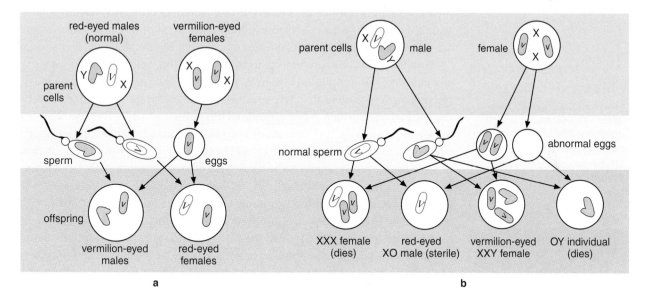

a   b

FIGURE 13.25

The effect of nondisjunction on inheritance of an X-linked trait. (a), The gene for eye color is carried on the X chromosome. V represents normal red eye color; v represents vermilion. Although this gene affects eye color, it is not the same one represented by the symbols R and r in Figure 13.22. Males inherit their eye-color gene on the X chromosome they receive from their mothers. **(b),** Nondisjunction of the X chromosome can produce red-eyed, sterile males and vermilion-eyed females. Bridges explained these rare individuals by suggesting that sometimes the X chromosomes fail to separate during meiosis in egg cell development. When Bridges examined the cells of these exceptional flies, he found the abnormal number of chromosomes that he had predicted. Nondisjunction also can occur in males.

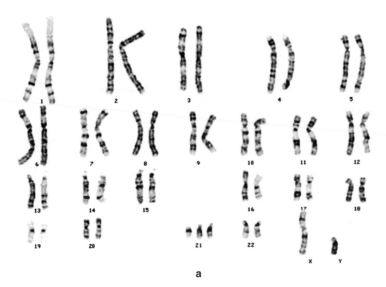

a

b

FIGURE 13.26

Down syndrome. (a), Chromosome number 21 identifies this male Down syndrome karyotype. **(b),** This girl shows typical facial features of Down syndrome.

typical combinations of symptoms, can result from abnormal numbers of sex chromosomes. Individuals with only one X chromosome and no Y chromosome (an XO genotype) are usually short, sexually underdeveloped, infertile females. The condition is known as Turner syndrome. Females with an XXX genotype often have limited fertility and may be slightly disabled mentally. An XXY genotype results in Klinefelter syndrome. These males often are tall, sexually underdeveloped, and may have a slight mental disability.

Nondisjunction can produce an individual with three copies of a chromosome. This condition is known as trisomy. Most trisomies of autosomes (all chromosomes other than the sex chromosomes) disrupt development, and the embryo usually dies. A few autosomal trisomies are viable. Trisomy of human chromosome 21 (Down syndrome) causes limited mental abilities, short stature, characteristic facial features, and heart defects, although severity varies (Figure 13.26). Some people with Down syndrome are severely retarded; others are only mildly so. Approximately 95% of them reach adulthood. Of those, 80% live into their early 50s. Down syndrome is the most common serious birth defect in the United States. It occurs approximately once in every 700 live births. Mothers of age 35 or older are more likely to give birth to infants with Down syndrome. Some pregnant women now choose to have the karyotype of the fetus checked for trisomy 21 and other chromosomal abnormalities.

13.11 Multigene Traits

Most traits are affected by more than one gene. This is typical of traits such as height and weight that show a continuous range of variation. For example, there are many variations of eye color in humans. Several genes are involved in the production and distribution of pigment in eyes.

Like eye color, most human traits are **multifactorial**—they are affected by several genes and environmental factors. Multifactorial traits, such as

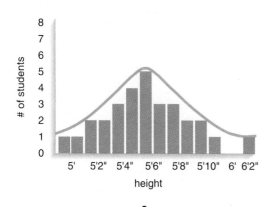

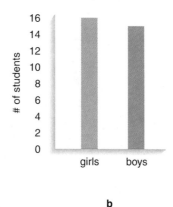

FIGURE 13.27

Continuous and discontinuous variations. (a), This graph of the heights of students in a high school biology class shows a continuous distribution. Height is a quantitative, or continuous, characteristic. **(b),** This graph shows the distribution of the sexes in the same class. Sex is a qualitative, or discontinuous, characteristic.

height, are also known as quantitative traits. The environment includes both external factors, such as light and temperature, and internal ones, such as the organism's metabolic activities. Most multifactorial traits, such as height, intelligence, color, and the control of metabolic processes, vary in finely graded steps. For example, production of the skin pigment melanin is thought to be controlled by four major genes, none of which demonstrates simple dominance. Independent assortment of the alleles of these genes results in continuous variation in human skin color. Exposure to sunlight also affects skin color.

A graph of a multifactorial trait usually resembles a bell-shaped curve (Figure 13.27a). A discontinuous distribution is typical of characteristics such as sex that exist in only two forms and are affected by few genes (Figure 13.27b). Compare the graphs of the two types of distributions. What type of distribution do most human characteristics demonstrate? Because most familiar traits are multifactorial, it is easy to see why the theory of blending inheritance was once popular.

Check and Challenge

1. What are multiple alleles? Can a person have more than two alleles for a single gene? Explain your answer.

2. How is a trait that is determined by multiple genes different from one that is determined by multiple alleles?

3. Should a person with type A blood receive a transfusion of type O blood? Why or why not?

4. How does crossing-over affect linkage?

5. What offspring would result from a cross of a homozygous red-eyed female fruit fly and a vermilion-eyed male fruit fly? How might vermilion-eyed male offspring be produced?

6. A tall, sexually underdeveloped human male has cells with one Barr body each. What sex chromosomes would you expect to see in a karyotype from this individual?

7. How does the inheritance of X-linked traits differ from that of other traits?

Chapter
HIGHLIGHTS

Summary

The concept of a gene has changed during the past century as research techniques have allowed more detailed analysis. A gene is a region of DNA that codes for a product. Most genes exist as several alleles that code for different products. Studies of twins demonstrate the effects of environment on gene expression.

Chromosomes are complex structures, each containing many genes. In addition to coding DNA, chromosomes contain other segments of DNA that are never transcribed. Using techniques such as staining to produce banding patterns, scientists can identify homologous chromosomes. Homologous chromosomes carry the same genes, but these genes may be different alleles.

Mendel's work forms the basis of modern genetics. His principles of segregation and independent assortment can be explained by the behavior of homologous chromosomes during meiosis. Probability can be used to predict the outcome of breeding experiments.

Working with *Drosophila*, Morgan and his students identified many X-linked traits; helped to establish the chromosomal basis of inheritance; and explained nondisjunction, crossing-over, and recombination.

Key Concepts

Use the following concepts to develop a concept map for this chapter. Include additional concepts as you need them.

- alleles
- codominant
- genes
- multifactorial
- homologous chromosomes
- multiple alleles
- nondisjunction
- recessive
- trait
- X-linked

Reviewing Ideas

1. How is a karyotype made and used?
2. How does sex determination differ in humans, birds, and grasshoppers?
3. Identify some of Mendel's experimental procedures. How did they contribute to his success as a geneticist?
4. Explain Mendel's two principles of inheritance.
5. What sort of evidence indicates that genes for two traits are linked?
6. What are X-linked traits? What experimental evidence first established their existence?
7. What is the Lyon hypothesis? What observations support this hypothesis?
8. Distinguish between the contributions of genes and the environment to phenotypes.
9. What is meant by nondisjunction? What sort of conditions can it produce?

Using Concepts

1. Could you establish a true-breeding variety of pink snapdragons? Why or why not?
2. What are the possible blood types of children in the following families?
   a. Type A mother, type A father
   b. Type A mother, type O father
   c. Type B mother, type AB father
   d. Type AB mother, type AB father
3. In paternity lawsuits, blood typing often is used to provide genetic evidence that the alleged father could not be related to the child. For the following mother-child combinations, indicate which blood types could not have been the father's.

Mother	Child	Impossible Blood Types for Father
O	B	
B	A	
AB	B	

4. A brown mouse is crossed with a heterozygous black mouse. If the mother has a litter of four, what are the chances that all of them will be brown? (Black is dominant; brown is recessive.)

5. A cross is made between true-breeding tall red-flowered plants and true-breeding short white-flowered plants. Tall and red are dominant traits. What would be the expected ratios of phenotypes in the F_1 and F_2 generations? What is the expected phenotypic ratio of the F_2 generation if red and white are codominant?

6. Radish roots may be long, round, or oval. In a series of experiments, crosses between radishes with long roots and radishes with oval roots produced 159 plants with long roots and 156 plants with oval roots. Crosses between round and oval produced 199 round and 203 oval. Crosses between long and round produced 576 oval. Crosses between oval and oval produced 121 long, 243 oval, and 119 round. How is root shape inherited? (It may help to make a table of the four crosses and their results.)

7. In tomatoes, red fruit color is dominant to yellow. Round-shaped fruit is dominant to pear-shaped fruit. Tall vine is dominant to dwarf vine.
   a. Suppose you cross a true-breeding tall plant bearing red round fruit with a true-breeding dwarf plant bearing yellow pear-shaped fruit. Predict the appearance of the F_1 generation.
   b. Assuming that the genes controlling the three traits are on three different pairs of chromosomes, what are the possible genotypes in the F_2 generation?
   c. What are the expected phenotypic ratios?

8. In which sex would you expect X-linked traits to be found most frequently in each of the following species: dogs, grasshoppers, robins?

9. State the goals of the Human Genome Project.

10. Often the results of breeding experiments do not follow expected Mendelian ratios. Instead of 3:1 or 9:3:3:1 ratios in F_2 generations, there may be results such as 1:2:1, 9:7, 9:3:4, or 7:1:1:7. How can you explain such ratios?

Synthesis

1. Alleles of the same gene segregate; alleles of different genes assort. Explain this statement as it relates to meiosis.

2. Suppose you examined the cells of a plant and found 12 chromosomes: a long straight pair, a short straight pair, a medium-length straight pair, a long bent pair, a short bent pair, and a medium-length bent pair. You then breed plants of this species for several generations.
   a. At the end of this time, would you expect to find some plants with all the straight chromosomes and none of the bent ones? all bent ones and no straight ones? Explain your answer.
   b. What proportion of the gametes should have three straight chromosomes and three bent ones? four straight and two bent? six bent?

Extensions

1. How will a complete gene map for humans be useful? How might it be a problem?

2. Imagine an allele for a recessive X-linked trait. Diagram or describe the situations in which it would be evident through five generations, beginning with a heterozygous female.

Web Resources

Visit BSCSblue.com to access
- Web links to explanations and activities related to human genetics, karyotyping, and the role of genes in health and disease
- Resources about genetics, including Gregor Mendel's original paper, a database of human genes, help in analyzing genetics problems, and a simulated fruit-fly genetics lab

CONTENTS

LEARNING OUTCOMES

By the end of this chapter you will be able to:

A   Explain how genes function.

B   Explain cytoplasmic inheritance.

C   Explain genomic imprinting.

D   Apply the principle of epistasis to genetics.

E   Describe genetic anticipation and its effects.

F   Explain the behavior of transposable elements.

CHAPTER

14

Other Forms of Inheritance

- *What could cause the varied colors of these wild horses?*

- *Does stable nuclear inheritance explain all patterns of inheritance?*

How different will explanations of inheritance be in 20 years? It would be very surprising indeed if we found out nothing more than we now know about genetics. Scientists have repeatedly confirmed Mendel's basic principles. However, they also have found that inheritance and gene expression are often more complex than Mendel thought. Some genes are inherited from only one parent, or act differently depending on which parent contributed each allele. Some genes affect the expression of other genes. There are also genes that change from one generation to the next or move from one chromosome to another.

In this chapter, you will be introduced to the present state of knowledge about how the passing of genes from one generation to the next can contribute to the health and appearance of an organism. The examples you study may raise questions in your mind about other genetic conditions that haven't been explained yet. Perhaps you will help extend this story by discovering additional pieces of the puzzle of inheritance.

New Explanations of Inheritance

14.1 Understanding Gene Function

Research in genetics, as in any field of science, builds on the work of previous scientists. New evidence is used to retest and adjust existing scientific explanations. For example, the discovery of introns required scientists to expand their description of the structure of genes (Figure 14.1). When Mendel published his explanation of inheritance patterns in 1865, he provided experimental evidence to back up his claims. Unfortunately, his paper went unnoticed for many years. When scientists in the 20th century consulted published research, they rediscovered Mendel's findings. This knowledge was almost 50 years old, but it quickly advanced our understanding of how inheritance works.

You can expect most of what you know now about genetics to be consistent with future explanations. The reason for this consistency is that scientific explanations are built on a firm foundation of evidence and

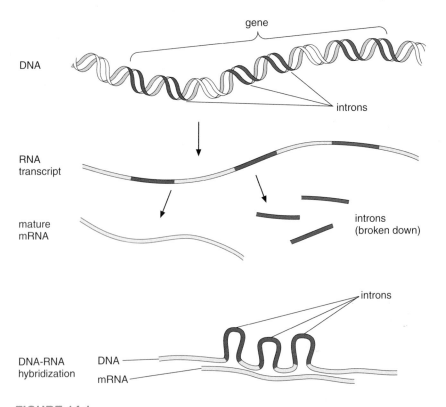

FIGURE 14.1

A discovery that helped change ideas about genes. Removal of introns from RNA makes many mRNA molecules shorter than the genes from which they are copied. When mRNA binds to its gene, introns in the DNA form loops because the RNA contains no matching segment. This discovery showed that proteins are not just direct translations of DNA sequences.

Other hormones regulate genes by a similar but more complex sequence. For example, glucagon regulates production of glucose by liver cells (see Focus On Energy Regulation, Section 5.9). The hormone itself does not enter the nucleus. Instead, it sets in motion a cascade of biochemical reactions in the cytoplasm and in the nucleus. A protein kinase, an enzyme that activates other enzymes by adding a phosphate group, serves as the transcription factor in this case. Many eukaryotic genes are regulated by the action of other genes or by various transcription factors. Genes are highly regulated during development of a mature organism from an embryo.

CHEMISTRY TIP

ATP often is the source for the phosphate group that is transferred to other molecules such as proteins. ADP is left after the phosphate has been donated.

14.2 Cytoplasmic Inheritance

Four familiar Mendelian patterns of inheritance are summarized in Table 14.1. Which, if any, of these patterns accurately describes the inheritance that would produce family histories like those shown in the human pedigrees in Figure 14.4? As you study the figure, think about the particular characteristics that do and do not fit these data.

The inheritance patterns described in Table 14.1 are based on nuclear inheritance. Genetic information carried on the chromosomes in the nuclei

TABLE 14.1
Four Patterns of Inheritance

Simple Dominant	Simple Recessive
Males and females are equally likely to have the trait.	Males and females are equally likely to have the trait.
Either parent can transmit the trait to offspring of either sex.	Either parent can transmit the trait to offspring of either sex.
The trait does not skip generations (usually).	The trait often skips generations.
The trait is present whenever the corresponding gene is present (usually).	Often, both parents of offspring who have the trait are heterozygous; they carry at least one copy of the allele.
	Only homozygous individuals have the trait.
	The trait may appear in siblings without appearing in their parents.
	If a parent has the trait, those offspring who do not have it are heterozygous carriers of the trait.

X-Linked Dominant	X-Linked Recessive
All daughters of a male who has the trait will also have the trait.	All daughters of a male who has the trait are heterozygous carriers.
There is no male-to-male transmission.	There is no male-to-male transmission.
A female who has the trait may or may not pass the gene for that trait to her son or daughter.	The trait is far more common in males than in females.
	The son of a female carrier has a 50% chance of having the trait.
	Mothers of males who have the trait are either heterozygous carriers or homozygous and express the trait.
	The daughters of a female carrier have a 50% chance of being carriers.

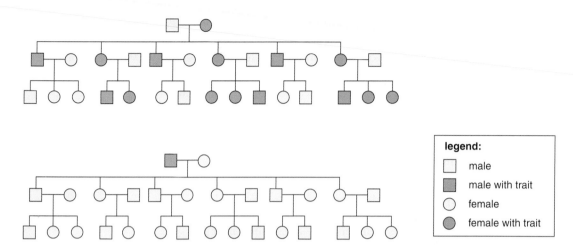

FIGURE 14.4

Two puzzling pedigrees. Can you identify the inheritance pattern that explains each of these pedigrees? Both pedigrees illustrate inheritance of the same trait.

of eukaryotic cells exhibits these inheritance patterns. Genetic information is not, however, found *only* in the nucleus. Mitochondria and chloroplasts have their own DNA. Because these organelles are found in the cytoplasm, their DNA is the source of cytoplasmic inheritance.

Each eukaryotic cell has hundreds to thousands of mitochondria. Each mitochondrion has several copies of **mitochondrial DNA (mtDNA).** Mitochondrial DNA is a double-stranded, circular molecule. Scientists finished reading the sequence of all 16,569 base pairs of human mtDNA in 1981. It codes for 13 polypeptides, 2 rRNAs, and 22 tRNAs. This information is not sufficient for all the activities of the mitochondrion in its role as a site of cell respiration. The rest of the molecules needed by mitochondria are encoded by nuclear genes. In this way, the action of nuclear and mitochondrial genes is connected.

Despite this connection, mtDNA is copied independently. It does not depend on replication of nuclear DNA during mitosis or meiosis. New mitochondria are made through fission, similar to the process of reproduction in prokaryotes. In addition, mitochondria do not sort evenly between daughter cells during meiosis or mitosis. For this reason, even different cells in a single organism can contain mitochondria that differ slightly in the information carried on their mtDNA.

What is the source of mitochondria in a multicellular organism? To answer this question, examine the diagrams showing fertilization in Figure 14.5. Notice that an ovum (egg) is a very large cell with many mitochondria that help meet its great energy needs for growth. In a sperm cell, the mitochondria are in the tail. They provide energy to help the sperm move. At fertilization, the tail and mitochondria of the sperm disintegrate. The mitochondria in the fertilized egg and developing embryo all come from the ovum. In most sexual animal species, mitochondrial inheritance is maternal. This observation has made comparisons of mtDNA sequences a useful way to trace human ancestry. A family tree built from these data shows the maternal lines for a particular population or individual. Now can you provide an explanation for the pedigrees shown in Figure 14.4?

CONNECTIONS

Mitochondria are found in nearly all eukaryotic cells, including those that have chloroplasts, so mitochondria must have appeared first. Both organelles may be descended from free-living bacteria that became part of early primitive eukaryotic cells. This evolutionary pattern demonstrates the unity of eukaryotic life.

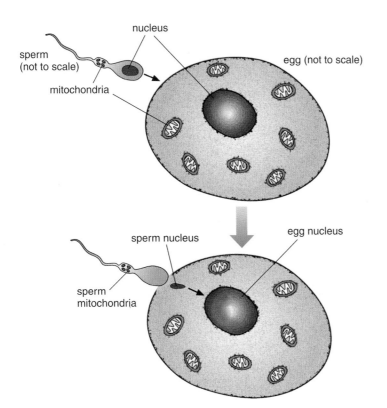

FIGURE 14.5

The fate of mitochondria in human fertilization. During fertilization, a sperm penetrates an ovum. The tail of the sperm, which contains mitochondria, disintegrates. The mitochondria in the developing zygote all come from the ovum. Mitochondrial inheritance is entirely maternal in humans.

Focus On
Maternal Inheritance

Mitochondria are not the only source of cytoplasmic inheritance. Like mitochondria, chloroplasts carry their own DNA (Figure 14.6). Most plants inherit chloroplasts only from their female parent, but there are many exceptions to this pattern. Traits encoded by chloroplast or mitochondrial genes are usually said to be inherited maternally. Chloroplast DNA is double-stranded and encodes about 100 different polypeptides. However, about 90% of chloroplast proteins are encoded by nuclear genes. This cooperation between organelle and nucleus underscores their mutually dependent relationship.

Chloroplasts are somewhat larger than mitochondria. Both types of organelles in many ways resemble prokaryotes. Chloroplasts and mitochondria may have evolved from free-living prokaryotes that parasitized other cells (see Section 17.7).

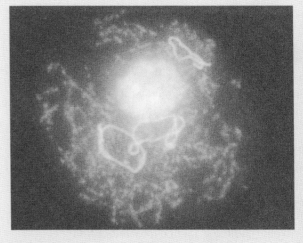

FIGURE 14.6

Cytoplasmic inheritance. This cell, a chrysophyte alga, *Ochromonas danica,* has been treated with a fluorescent stain that binds to DNA. The nucleus is the large light disk. Around the nucleus are four plastids, which are visible as bright loops because the DNA is arranged in a ring around the inside of the plastid. This arrangement is typical of brown algae and chrysophytes. The tiny scattered fluorescent dots represent mitochondrial DNA.

Biological Challenges

Medicine and Mitochondria

Humans rely on energy from oxidative cell respiration to survive—that is why people die if deprived of oxygen for even a few minutes. With this in mind, it is not surprising that mutations that affect mitochondria can be quite serious. Mutations can affect mitochondrial function in two ways. First, a mutation in mtDNA obviously can affect mitochondrial function directly. Second, a mutation in a nuclear gene that encodes some product essential for mitochondrial structure or function will also affect the mitochondria.

In the 1960s and 1970s, scientists focused on finding the biochemical changes that caused the symptoms of mitochondrial diseases. The symptoms of a certain disease included droopy eyelids, weak muscles, inability to tolerate exercise, abnormal eye movement, and even brain disease. In 1963, W. K. Engel developed a stain that showed abnormal mitochondria in the skeletal muscles of people who had this disease. The odd appearance of affected tissue was called ragged red fibers (Figure 14.7). It is still used as a sign that skeletal-muscle mitochondria are abnormal.

In 1988, a research group at the Institute of Neurology in London found the first known mtDNA mutation linked to human disease. The deletion in mtDNA was present in 9 of 25 patients with a mitochondrial disease of skeletal muscle called mitochondrial myopathy. Later, D. Wallace at Emory University found evidence to connect a single nucleotide mutation in mtDNA with an eye disease called Leber's Hereditary Optic Neuropathy. Since then, many other mtDNA defects have been linked to disease. Some nuclear mutations that affect mitochondria also are known.

A difficulty in analyzing disease associated with mitochondrial inheritance is that each cell has hundreds of mitochondria. Each mitochondrion has two to ten copies of its mtDNA. The most complicating factor is that, when cells divide, the mitochondria are not evenly sorted into the new cells. Thus, cells or populations of cells can have a mixture of normal and abnormal mtDNA. People with only a few mutant mitochondria may not show symptoms. Consequently, the diseases can appear to skip a generation.

One explanation for the large number of mtDNA defects that have been identified is that mitochondria have a less effective DNA repair system than do nuclear genes. As a result, the mutation rate in mitochondria is 10 times that of nuclear genes. For cells that last a long time in the body, such as nerve or muscle cells, aging may result in an accumulation of mtDNA errors so great that the individual becomes ill or dies.

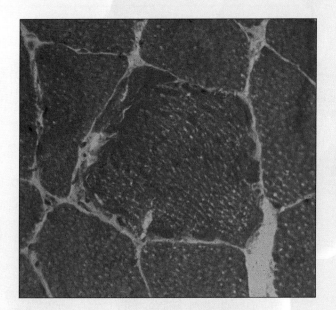

FIGURE 14.7

Normal and abnormal muscle fibers. Normal muscle fibers, a bundle of strands called myofibrils, surround a ragged red fiber. This tissue was treated with the dyes hematoxylin and eosin, which stain muscle cytoplasm pink and show the disorganized myofibrillar pattern that suggests a "ragged" appearance.

14.3 Genomic Imprinting

When Mendel's ideas about inheritance became widely accepted, scientists assumed that it makes no difference which parent donates which allele to a heterozygote. However, the expression of some traits does not follow this pattern. For example, Huntington's disease is caused by a dominant allele. However, people who inherit that allele from their fathers tend to develop symptoms at an earlier age than do people who inherit it from their mothers. On the other hand, the muscle disease myotonic dystrophy is usually more severe in people who inherit it from their mothers. What makes the difference in these cases?

Part of the answer came from studies published in 1991 of dwarf mice. Normal growth in mice depends on a growth hormone called insulinlike growth factor II (Igf_2). Dwarf mice are homozygous for a recessive defective Igf_2 allele. Scientists were surprised to find that even heterozygous mice are dwarfs if they received their normal Igf_2 allele from their mothers. The reason that the parental origin of each allele matters is a normal process known as **genomic imprinting** (Figure 14.8). Imprinting of certain genes can affect their behavior.

Here is how genomic imprinting appears to work. In most multicellular organisms, somatic cells contain two copies of each chromosome. One is

CONNECTIONS
Genomic imprinting extends the ways in which sexual reproduction produces genetic variation.

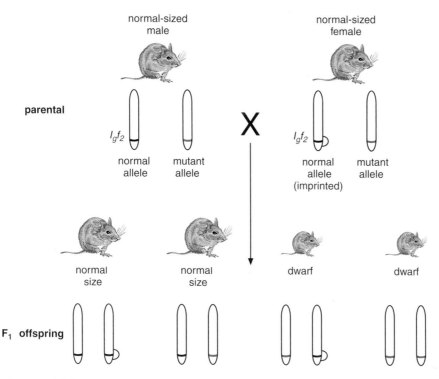

FIGURE 14.8

Genomic imprinting in mice. Heterozygous mice with only one copy of the dominant normal allele of the Igf_2 gene grow to normal size (parental generation). However, in female mice, imprinting inactivates the Igf_2 gene during meiosis. Only the offspring that inherit the normal allele from their father can grow to normal size. Heterozygous offspring that inherit the imprinted normal allele from their mother are dwarfs.

inherited from the mother (maternal) and the other from the father (paternal). For each chromosome, particular genes occur at specific locations (loci). Thus, an individual has a maternal copy and a paternal copy of each gene (except for genes on the sex chromosomes). Until about 1990, geneticists assumed that it does not matter whether you inherit a particular allele from your mother or your father. It will behave the same way. However, the evidence from Huntington's disease and dwarf mice shows that this assumption is not correct for every gene.

The behavior of some genes changes when the chromosome regions that contain these genes are marked by imprinting. The nature of the change depends on whether the imprint is from the female or male parent. Mice need to inherit a normal *Igf2* allele from their fathers or else dwarfing results. Normally an active copy of the *Igf2* gene is inherited from the father, while imprinting inactivates the maternal copy. Keep in mind that imprinting does not cause the mutation in dwarf mice. The mutation is simply a clue that helped scientists discover imprinting of this gene.

Genomic imprinting affects at least 100 genes in humans. Imprinting inactivates many genes, but fewer genes are activated by imprinting. Genes probably become imprinted during meiosis, as shown in Figure 14.9. During meiosis, the inherited imprint is removed and a new imprint is imposed. The imprint corresponds to the sex of the parent. The imprint probably remains with the gene throughout the life of the individual. The only time it is removed or changed is during the production of gametes in meiosis.

During meiosis, the imprint changes to match the sex of the individual. For example, look at the pedigree in Figure 14.9a. Individual 2 is a man. He has inherited an imprinted allele from his mother. This allele then bears a *female* imprint in all the somatic cells of his body. However, when meiosis produces sperm of individual 2, the original imprint is removed or altered. Now individual 2's sperm will carry genes with a *male* imprint (Figure 14.9b). Therefore, the same allele will produce the different phenotypes seen in individuals 2 and 3.

Focus On

X-Inactivation

X-inactivation is another process that appears to involve methylation. In the cells of female mammals, one of the two X chromosomes is largely inactive. Which X chromosome gets inactivated—the maternal copy or the paternal copy—is random. In addition, it may vary from cell to cell. The X chromosomes of cats carry pigment genes. The patchwork coloring of a calico cat (see Figure 13.23) is an example of the mosaic effect of X-inactivation varying in different cells. X-inactivation is different from genomic imprinting because X-inactivation involves the coordinated regulation of an entire chromosome. Genomic imprinting affects individual genes or parts of a chromosome.

The mechanism of X-inactivation is under much scrutiny. It appears to involve the product of a gene that is expressed only from the inactive X chromosome. Methylation is a likely candidate for the biochemical process involved. X-inactivation is removed as gametes are formed. For this reason, each gamete receives an active X chromosome to pass along to offspring.

a

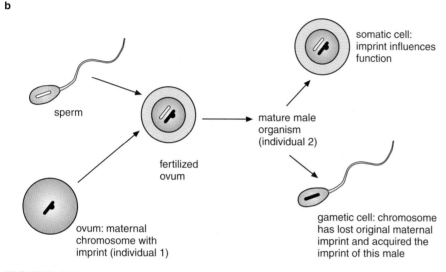

Individual 1

Individual 2
inherits a maternal
imprint

Individual 3
inherits a paternal
imprint

legend
☐ male
◼ male with trait
○ female
● female with trait

b

sperm

fertilized
ovum

ovum: maternal
chromosome with
imprint (individual 1)

mature male
organism
(individual 2)

somatic cell:
imprint influences
function

gametic cell: chromosome
has lost original maternal
imprint and acquired the
imprint of this male

FIGURE 14.9

Genomic imprinting during meiosis. Imprinting marks certain genes. The imprint
corresponds to the sex of the parent. **(a),** This process leads to an inheritance pattern in which
the behavior of an imprinted allele depends on the sex of the parent who contributed it.
Individuals 2 and 3 inherited the same allele but have different phenotypes. **(b),** The imprint
survives in somatic cells throughout the life of the individual who inherits it. In the gametes that
an individual produces, however, the imprint changes to reflect the individual's sex.

Genomic imprinting is a topic of much investigation. Finding out the
exact molecular mechanisms involved is one goal of the research. Initial
evidence suggests that methylation of certain DNA bases may be part of the
process. Methylation is a chemical tagging process that regulates genes.
During development, methylation patterns change in different cells, helping
various tissues differentiate. The temporary pattern of methylation during
development, however, is *not* the same process as imprinting. Genomic
imprinting remains the same in somatic cells throughout the life of the
individual. Imprinting may function to control the dosage of a particular
gene, making one copy inactive. In some cases, imprinting may also be
essential to signal that the developing zygote received a set of chromosomes
from each parent. For some species, contribution of genetic information
from two parents may be an absolute requirement for survival and normal
development of an embryo.

CHEMISTRY TIP

Methylation is the addition of a
methyl group (—CH$_3$) to a
molecule. The nitrogenous bases
of DNA, for example, can be
methylated.

14.4 Epistasis

ETYMOLOGY

epi- = upon (Greek)

stasis = stopping or standing (Greek)

In **epistasis**, the effects of one gene "stand on," or depend on, the effects of other genes.

Genes do not always act independently. In some cases, one gene cannot exert its phenotypic effect unless a second gene also is expressed. This situation is called **epistasis.** There are many ways in which epistasis can occur. For example, consider two genes, gene 1 and gene 2, that encode different enzymes, enzyme 1 and enzyme 2. These enzymes are needed for subsequent steps in a biochemical pathway (Figure 14.10a). The starting material is molecule A. It is converted by enzyme 1 to an intermediate molecule B. Next, enzyme 2 converts molecule B to the final product, C. Obviously, if there is a mutation in gene 2 that inactivates enzyme 2, there will be no product C. But what if there is a mutation in gene 1? No intermediate molecule B will be available for the action of enzyme 2 (Figure 14.10b). Even if gene 2 is normal, the lack of enzyme 1 prevents normal function of enzyme 2, encoded by gene 2. Thus, the phenotypic expression of gene 2 is masked by the defective expression of gene 1. In this example, gene 1 is said to be epistatic to gene 2.

A typical example of epistasis is found in the coloring of Labrador retriever dogs. A dominant allele, *B*, leads to a black coat (Figure 14.11). Homozygous recessive dogs *(bb)* are brown. A second gene, *E/e*, affects the deposition of the dark pigment in the hair. The dominant allele *E* enables the expression of *B/b*. In homozygous recessive dogs *(ee)*, pigment deposition is

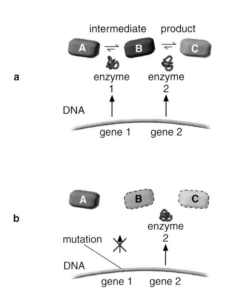

FIGURE 14.10

Epistasis at the molecular level. (a), Gene 1 codes for enzyme 1, and gene 2 codes for enzyme 2. Hence, gene 1 is epistatic to gene 2. **(b),** A mutation in gene 1 could interfere with expression of gene 2. Although gene 2 is active, lack of enzyme 1 prevents enzyme 2 from producing molecule C.

phenotype

BbEE   Bbee   bbEE   genotype

FIGURE 14.11

Epistasis and coat color in Labrador retrievers. A dominant allele, *B*, produces a black coat. The homozygous recessive genotype *bb* produces a brown coat. A second gene influences expression of this coat-color gene. The dominant allele *E* permits the allele *B* or *b* to be expressed, while the homozygous recessive genotype *ee* always produces a yellow coat. The genotype shown below each figure is only one of several genotypes that could produce each coat color. What is the color of a dog with the genotype *BbEe?*

defective and the hair accumulates little pigment. These dogs are light yellow no matter which alleles of the *B/b* gene they carry. *E/e* is epistatic to *B/b*. The *e* allele masks the expression of the *B/b* gene that encodes black or brown coloring. Epistasis does not contradict Mendel's observations. It is one of several modifying conditions that can shift the expected ratios of phenotypes in a genetic cross. However, the same genetic principles still apply.

Check and Challenge

1. Why do scientific explanations change?
2. Compare the prokaryotic operon with eukaryotic gene regulation.
3. Explain the connection between genomic imprinting and parental inheritance.
4. What is the minimum number of genes involved in epistasis?
5. Is inheritance in eukaryotes strictly nuclear? Explain your answer.

Special Mechanisms of Inheritance

14.5 Genetic Anticipation

Genetic mechanisms such as cytoplasmic inheritance, imprinting, and epistasis are fairly widespread. They are normal mechanisms of inheritance. Generally, the patterns are stable from one generation to the next. However, that stability is missing in some special genetic mechanisms discussed in Sections 14.5 and 14.6.

For years doctors and geneticists observed that, with certain inherited disorders, symptoms showed up at an earlier age in some patients than was true for their parents or grandparents. With each generation, the onset of symptoms generally was earlier, and the severity of symptoms generally increased. This shift toward earlier symptoms of an inherited disorder is called **genetic anticipation.** One of the most famous inherited disorders showing this phenotypic pattern is Huntington's disease. Examine the pedigree shown in Figure 14.12. Can you identify any pattern that can help explain the earlier age of onset of symptoms?

Although most DNA sequences are transmitted from parent to child as exact copies, there are exceptions. In certain genes, the number of copies of a repeated sequence of three nucleotide bases can increase or decrease within a normal range when the gene passes from parent to offspring. If a threshold number of copies of the repeat is reached, however, the gene becomes very unstable. The number of repeats then increases steadily with each generation. This situation is called **trinucleotide repeat expansion** and is responsible for genetic anticipation.

Many disorders show genetic anticipation. Among these are Huntington's disease (a neurological disorder), myotonic dystrophy (a muscular disorder), and Friedreich's ataxia (a neurological disorder). The first disease in which

Focus On
Genetic Counseling

Genetic anticipation in disorders such as myotonic dystrophy or Huntington's disease raises some difficult questions for genetic counselors. A laboratory examination of chromosomes and DNA sequence can reveal not only whether an individual has inherited a mutant allele but also how many copies of the critical trinucleotide repeat are present. This is true even for people who do not have symptoms. Should counselors recommend testing when there is no early treatment or prevention? The molecular data will predict if and when a person is likely to develop symptoms. For some people, the test is a welcome relief, either to know they did *not* inherit the mutation or to know they *did* and be able to plan accordingly. For others, the stress of finding out is too great—they prefer to wait and see if symptoms appear later in life.

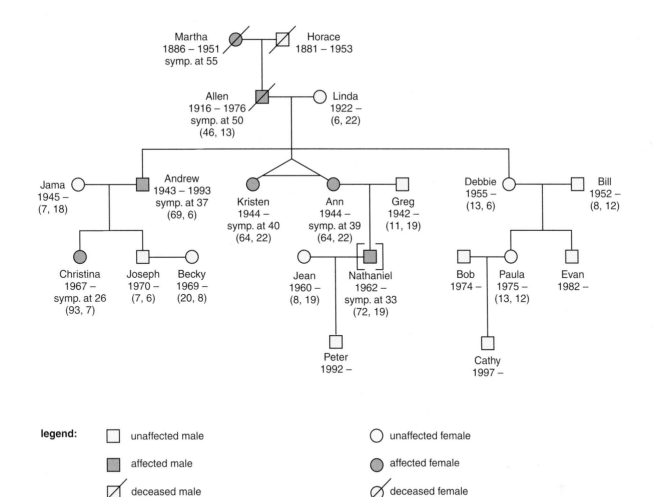

legend:

□ unaffected male   ○ unaffected female

■ affected male   ● affected female

⧄ deceased male   ⊘ deceased female

▣ biological son, placed for adoption

△ monozygotic ("identical") twins

symp. = onset of Huntington's disease

FIGURE 14.12

Huntington's disease pedigree. The gene associated with this disease includes a repeated sequence of three nucleotides. The numbers in parentheses indicate the number of repeats of this sequence in each chromosome. Is there any connection between the age at which symptoms appear and the number of trinucleotide repeats? Can you determine from these data whether the Huntington's disease allele is dominant or recessive?

this phenomenon was observed is an X-linked form of mental retardation known as fragile-X syndrome. The mutant genes associated with these genetic-anticipation disorders have unstable regions of trinucleotide repeats (Figure 14.13).

Figure 14.13 shows that the gene associated with Huntington's disease, located on chromosome 4, contains CAG as a long trinucleotide repeat. When the number of copies of CAG in the gene is less than 39, the person usually does not show symptoms (Figure 14.14). Above

. . .AGCTAGCAGACTGATCGATGTACGTACGTTAGCTAGTGCATGAGCGATGCTAGCTTAGCTAGT
CTATGCATTAGCAT**CAGCAGCAGCAGCAGCAGCAGCAGCAGCAGCAGCAGCAGCAGCAGCAG**
CAGCAGCAGCAGCAGCAGCAGCAGCAGCAGCAGCAGCAGCAGCAGCAGCAGCAGCAGCAGCAG
CAGCAGCAGCAGCAGCAGCAGCAGCAGCAGCAGCAGCAGCAGCAGCAGCAGCAGCAGCAGCA
GCAGCAGCAGCAGCAGTGGCATCGATGCATGATCTAGCATAGGACTCTAGAGACCCCATGCA
TTACGATTACGATTATCGACCCCATAGGGATCGTACGATGCATCGATGCAGCATG. . .

FIGURE 14.13

Part of the DNA sequence of a mutant allele for Huntington's disease. The CAG
trinucleotide repeat region can become unstable and expand as this gene passes to offspring.

40–50 copies, the gene becomes very unstable. Some patients have
more than 120 copies per gene.

The trinucleotide repeat associated with each disorder is different.
For instance, trinucleotide repeat expansion in myotonic dystrophy
involves CTG. Men with symptoms of fragile-X syndrome have 200 or
more copies of the CGG trinucleotide near the coding region of the
associated gene. The unstable trinucleotide in Friedreich's ataxia is GAA.
It appears that the instability and expansion of repeats is the problem,
rather than the particular trinucleotide sequence. The exact mechanism
that causes expansion is currently under study.

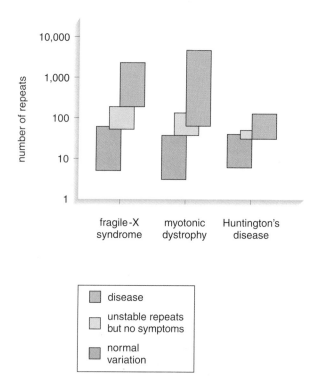

FIGURE 14.14

Unstable trinucleotide repeats and genetic disorders. The number of trinucleotide repeats
is related to phenotype for certain genetic disorders. The overlap of symbols suggests that
other factors are also involved. Note that the scale on the vertical axis (number of repeats) is
exponential, not linear.

14.6 Transposable Elements

For many years, geneticists assumed that DNA sequences remained at stable locations in the genome. Barbara McClintock was the first person to suggest that genetic sequences could move. She observed spotty or streaked coloring of kernels of corn, *Zea mays* (Figure 14.15). To explain this pattern, McClintock suggested that pieces of genetic material—which became known as **transposable elements**—moved from one chromosome to another. When a transposable element is inserted in a gene, it interferes with the expression of that gene. McClintock hypothesized that, during the development of corn kernels, transposable elements move in and out of genes involved in pigment production. This results in the uneven coloring of the kernels. It took more than 30 years of research for McClintock and others to find enough evidence to confirm this hypothesis.

One line of evidence that supported McClintock's hypothesis came from the study of bacterial genes. In bacteria, regions of DNA with repeated sequences at each end can appear at new sites in the bacterial genome. The repeats are inverted—that is, they run in opposite directions at either end of the transposable elements (Figure 14.16a). The repeats are involved in the mechanism to break and rejoin DNA as the piece inserts at a new site. In both bacteria and eukaryotes, some transposable elements do not move. Instead, a copy moves to a new location.

Transposable elements of 1,000 to 5,000 base pairs have been identified in bacteria. They carry another gene or genes between the inverted repeats,

Try Investigation 14A ⬥
Jumping Genes.

FIGURE 14.15

The effects of transposable elements. The color spotting on these kernels of corn *(Zea mays)* results from the movement of transposable elements that can interrupt expression of genes involved in the production of pigment.

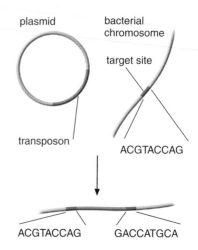

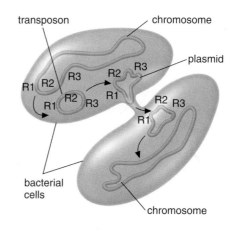

b

FIGURE 14.16

Bacterial transposons. (a), A transposon carrying one or more genes can insert itself in the chromosome at a target site. The gene encoding the enzyme transposase usually is located on the transposon. This enzyme is involved in inserting the transposon at its target site, creating inverted repeats. **(b),** Transposons can carry genes back and forth between plasmids and the bacterial chromosome. In this example, genes *(R1, R2, R3)* that provide resistance to an antibiotic move from the chromosome to a plasmid. The plasmid then carries the resistance genes into other bacterial cells.

making it possible to insert a variety of genes at new sites. This combination of genes is called a **transposon.** Some bacterial transposons carry antibiotic-resistance genes between a bacterial cell's chromosome and plasmids. Once on a plasmid, these genes can be transferred to other bacteria (Figure 14.16b).

Molecular data supporting the existence of these "jumping genes" in bacteria also lend support to McClintock's theory about transposable elements in corn. When a transposable element appears in a coding region or control region for a gene, a mutation results. This can be a problem, but it also is a source of genetic variation. In corn many copies of transposable elements are unstable between generations. Some even change position in somatic cells during the life of an individual plant. This instability produces the streaked color of corn kernels, such as those in Figure 14.15. Transposable elements also occur in humans. For instance, one form of the blood disease hemophilia results from the movement of a transposable element to the middle of a gene involved in blood clotting.

In many cases in eukaryotic cells, the transposable element neither moves nor makes a DNA copy that goes to the new site. Instead, an RNA intermediate is made. The enzyme reverse transcriptase (see Focus On Reverse Transcription, Section 9.9) makes a DNA copy from the RNA intermediate. The new DNA inserts itself in the genome. Retroviruses such as HIV may have evolved from such **retrotransposons.**

Check and Challenge

1. What is genetic anticipation?
2. Describe the connection between trinucleotide repeat expansion and genetic anticipation.
3. Was the action of transposable elements predicted by Mendel's experiments? Explain.
4. What effect can movement of transposons have on gene expression?

Chapter HIGHLIGHTS

Summary

Mendel's work and that of other early geneticists laid the foundation to explain inheritance, but additional research has greatly expanded our understanding. Once a gene is inherited, its expression depends on many factors, including the action of gene regulation and, in some cases, the effects of genomic imprinting or the influence of other genes. Some DNA sequences are unstable, and their movement can alter the expression of a variety of genes. Despite all these complications, most genes are inherited and expressed accurately.

Gene regulation can result from the action of other genes or external influences, including hormones. The phenotype also is influenced by environmental factors. In bacteria, some sets of genes share a common set of control elements (operator, promoter, and repressor sequences). This combination is an operon. In a typical operon, the action of RNA polymerase is blocked by the binding of a repressor protein to the operator. In response to a stimulus, this barrier is removed and transcription begins. Eukaryotic genes are regulated in a variety of ways. Often regulation involves the binding of transcription factors to a DNA sequence known as a response element. With the transcription factor bound, the response element can influence initiation of transcription at a nearby gene.

Not all inheritance in eukaryotes is nuclear. Mitochondria and chloroplasts have their own DNA that replicates independently. Cytoplasmic inheritance is a normal phenomenon, as is genomic imprinting. Many eukaryotic genes acquire an imprint during meiosis. This imprint results in a difference in gene function in the offspring that reflects the parental origin. Furthermore, the action of some genes is directly influenced by one or more other genes. In epistasis, the action of one gene can mask the expression of another gene.

Certain inherited disorders show increasing severity of symptoms or earlier onset of symptoms in each new generation. This change is called genetic anticipation. The underlying molecular mechanism is the expansion of unstable trinucleotide repeats in the affected gene. Other unstable DNA sequences are the transposable elements of prokaryotes and eukaryotes. These elements can insert in new locations in the genome, sometimes changing the function of a gene at the site of insertion.

Key Concepts

Use the concepts below to build a concept map, linking to as many other concepts from the chapter as you can.

- epistasis
- mitochondria
- genomic imprinting
- gene expression
- regulation
- transposon
- operon
- transcription factors
- cytoplasmic inheritance
- genetic anticipation
- trinucleotide repeat expansion

Reviewing Ideas

1. Explain the importance of the statement "Most genes are inherited and used correctly."
2. What is an operon, and how does it regulate gene expression?
3. Describe the role of response elements and transcription factors in eukaryotic cells.
4. What is cytoplasmic inheritance?
5. Describe genomic imprinting. Is this a normal function or a mutation? Explain your answer.
6. If a gene is said to be epistatic to another gene, what do you know about the action of the two genes?
7. What are the characteristics of genetic anticipation?

 Biology Online BSCSblue.com/vocabulary_puzzlemaker

8. Why would a patient want to know the number of copies of a trinucleotide repeat in the gene associated with Huntington's disease, myotonic dystrophy, or other disorders that show genetic anticipation?
9. What is X-inactivation?
10. Explain the behavior of transposons and the effect they have on inheritance.

Using Concepts

1. Why is the genetic information carried on mtDNA said to be inherited maternally?
2. Can children inherit mutations in their mitochondrial DNA from their father? Explain your answer.
3. What is the expected ratio of black, brown, and yellow phenotypes in a cross between two Labrador retrievers with the genotype *BbEe?*
4. Huntington's disease is inherited as a dominant trait that can show genetic anticipation. Suppose a parent knows that she or he has inherited a copy of the mutant allele associated with Huntington's disease. What information would the parents need to make an informed prediction about whether their child will develop the symptoms of the disease?
5. Describe two different genetic mechanisms in which the parent of origin for a particular gene makes a difference to the results of a cross done with mice.

Synthesis

1. In what ways does cytoplasmic inheritance support the idea that chloroplasts and mitochondria are evolved from prokaryotic cells?
2. Why does a peptide hormone act through a cascade of intermediate steps to stimulate gene

transcription while a steroid hormone actually enters the nucleus? (Hint: Consider the chemical nature of steroid hormones, which are complex lipids, as compared with proteins.)
3. Some bacteria carry antibiotic-resistance genes in transposons. Why is this an important issue for medicine?

Extensions

1. Draw a diagram to show two examples of inheritance that are dependent on the mother: inheritance of a mitochondrial gene and inheritance of a nuclear imprinted gene.
2. Research and report on one of the following inherited diseases: Friedreich's ataxia, fragile-X chromosome, or Philadelphia chromosome. Describe the disorder and explain how it is inherited.
3. Discuss ethical implications of genetic medicine.

Web Resources

Visit BSCSblue.com to access
- Resources for understanding and experimenting with genetics
- Information on the inheritance and biology of Huntington's disease, instructions for experiments in yeast genetics, and an animated explanation of how transposable elements produce variable kernel color in corn (*Zea mays*)
- Web links related to this chapter

CONTENTS

LEARNING OUTCOMES

By the end of this chapter you will be able to:

A Describe the Human Genome Project.

B Explain how functional genomics helps scientists use sequence data.

C Describe new technologies that have contributed to the genome project.

D Describe mechanisms of mutation and relate them to disease.

E Distinguish between gene therapy and traditional treatments for genetic disease.

F Discuss ethical, legal, and social implications of genetic technology.

CHAPTER

15

Advances in Molecular Genetics

■ *Why is human DNA placed in bacterial colonies?*

■ *What practical uses can you think of for this technology?*

Imagine having to go through a stack of 20 different encyclopedias to find the one word in all of those books that is misspelled. You might think this work would take you and all your classmates more than the whole year to complete. This is the kind of large-scope research that human geneticists are facing as they try to identify the causes of inherited diseases.

The Human Genome Project was completed in 2003. It established that human DNA consists of over 3 billion base pairs and contains information for approximately 35,000 genes. This data is being analyzed to help researchers find the causes of genetic diseases and to aid in their treatment and prevention.

In this chapter, you will read about some of this research and its applications, the kinds of organisms and human diseases that are being studied, and some of the medical discoveries and ethical issues that arise as a result.

Studying Genomes

15.1 The Genome Projects

An organism's genome consists of all of its genetic information. Research on various organisms' genomes is being conducted around the world. By the late 1980s, projects with the name Genome had begun in the United States under the sponsorship of the National Institutes of Health and the Department of Energy. As other countries became involved, an international organization called the Human Genome Organization was formed to coordinate all the research. The pharmaceutical industry and nonprofit research organizations also play an important role in developing new genome information. Many companies are using this genome information to improve plant and animal products, such as grains and livestock. They are also trying to apply this knowledge to the development of better treatments for genetic diseases, such as muscular dystrophy and cancer.

The Human Genome Project (HGP) has been called biology's "moon shot." The HGP was an international effort to study the genomes of humans and other organisms. Scientific organizations in the United States and the United Kingdom produced most of the data.

The HGP's central goal was to determine the sequence of the approximately 3 billion base pairs of DNA that make up the 24 different human chromosomes (numbers 1–22, X, and Y; see Figure 15.1). Other important goals include the study of how the genes function and the ethical, legal, and social issues related to genetic research.

The HGP began with efforts to map genes to specific chromosomes. Geneticists have already mapped the positions of many human genes (Figure 15.1), a first step in determining which region of DNA to begin sequencing. Mutations in genes that cause disease are also being discovered.

Genome scientists are studying many different organisms besides humans. In fact, the first genomes sequenced were those of viruses, not humans. Viral genomes are much smaller than the human genome. The first ones that were sequenced had only a few thousand base pairs. Next, bacteria that cause human infections were sequenced. These bacterial sequences help reveal how certain bacteria cause disease. This information is being used to help design new and more effective antibiotics. For example, genetic mutations in *Staphylococci*, *Streptococci*, and enteric (living in the intestine) bacteria have made these germs resistant to many forms of penicillin. A new synthetic compound, Zyvox (manufactured by Pharmacia & Upjohn), works so early in the pathway of bacterial protein synthesis that the bacteria will be slow to alter their own genes to resist this drug.

One of the first bacterial genomes sequenced was that of *Escherichia coli*. This common bacterium lives symbiotically in human intestines, where it helps digest our food. Scientists have used *E. coli* as an experimental

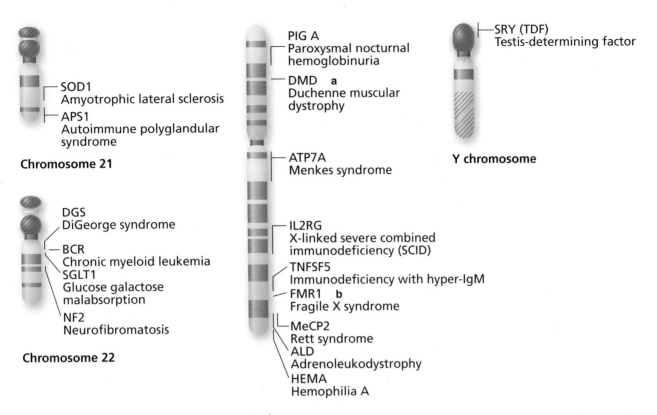

FIGURE 15.1

Locations of genes on some human chromosomes. The dark and light bands shown on each chromosome represent banding patterns seen with specific stains. These bands do not identify actual genes but serve as landmarks to help researchers label large regions of chromosomes that may contain genes. These diagrams are highly simplified representations of chromosomes that include many, many more genes than are being shown here.

organism for about 100 years. Much of our knowledge of cell biology and the functions of DNA are based on studies of *E. coli*. Publication of *E. coli's* genome sequence enabled scientists to relate their knowledge of the biology of this organism to its complete DNA sequence.

The first eukaryotic genome to be sequenced was that of brewer's yeast, *Saccharomyces cerevisiae*. This yeast is an important model of more complex eukaryotic cells, even though *S. cerevisiae* is a single-celled organism. Like *E. coli*, this yeast is often used in biological research because it grows quickly and because deleting, modifying, or adding genes is relatively simple. Completion of the yeast genome helped advance understanding of eukaryotic cells.

The first multicellular organism sequenced was a tiny worm, *Caenorhabditis elegans* (Figure 15.2a). This animal has only 959 cells, and the fate of each cell during development from zygote to adult worm has been mapped out (Figure 15.2b). The ability to relate a genetic-sequence map to the development of the tissues and organs of this small animal was a powerful step toward understanding how genetic programs direct the

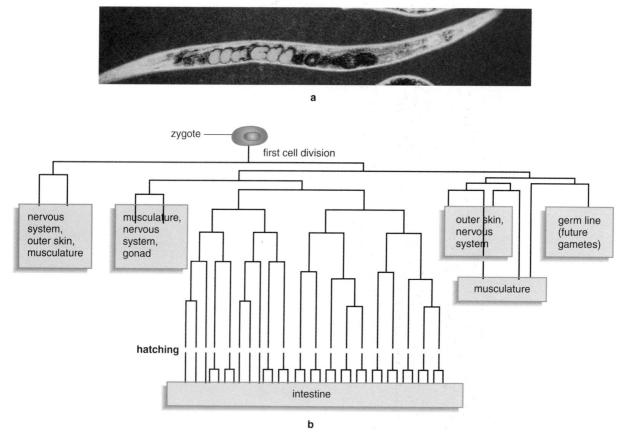

FIGURE 15.2

Development of a simple multicellular organism. The embryonic origin of each cell of *Caenorhabditis elegans* **(a)**, a nematode worm, has been identified, ×200. A fate map **(b)** of its cells shows how the worm normally develops. The effects of mutations in various genes on the development of specific cells reveal the genetic basis of development in this simple animal.

development of more complex organisms such as humans.

Other genomes now under study include those of organisms with substantial commercial value, such as corn, rice, cotton, pigs, and cattle. Intensive gene-mapping efforts are under way to identify the position of genes in these organisms that may be important for disease resistance or improving production.

These genetic studies have also led to controversy. Breeders have begun to use the new genetic knowledge to modify the genomes of crops and livestock, especially those of corn, soybeans, and cotton. These changes have increased crop yields and created more effective disease resistance in these plants. However, they have also generated concern about whether people might be harmed by eating products made from genetically modified organisms. Most scientists believe that these risks are very small. Nonetheless, these concerns have been widely discussed and, in some cases, have resulted in the banning of certain foods.

15.2 Functional Genomics

Functional genomics is the study of DNA sequence information to help explain cell functions. The genome projects are providing enormous amounts of information for functional genomics to interpret. An important tool in functional genomics is the computer analysis of DNA sequences (Figure 15.3) to predict the structures and functions of the proteins they encode.

For example, diseases such as muscular dystrophy are caused by mutations in genes that encode various muscle proteins. Functional genomics specialists can use the DNA sequence of a mutant muscle-protein

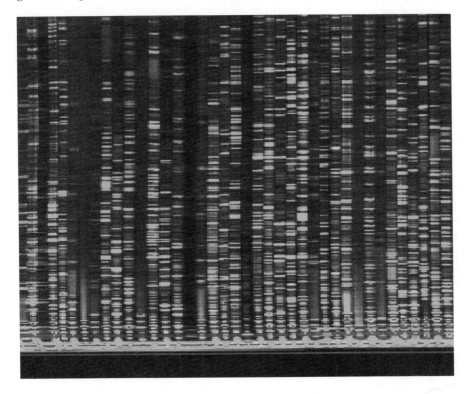

FIGURE 15.3

Computer image of DNA sequence data. Each band contains DNA that is one base longer than the band below it. The color of each band indicates the terminal nucleotide in that band's sequence.

Focus On

DNA Sequencing

Automation has sped up the process of DNA sequencing. It has reduced the time required from days to hours. DNA sequencing is based on the elongation of DNA chains by DNA polymerase.

To begin, the DNA to be sequenced is divided into four batches—one for each base, A, C, G, and T. After separating the two DNA strands, the researcher adds a short strand of DNA that is complementary (or matching) to one end of the sample DNA (Figure 15.4). This short strand is called a primer and is required to initiate DNA synthesis. The primer is labeled with a fluorescent dye; a different colored dye is used in each batch. In the presence of a supply of nucleotides, DNA polymerase extends the primer using the sample DNA as a template so that the DNA becomes double-stranded. In addition to the enzyme and nucleotides, a small amount of a modified nucleotide is added to each batch. Whenever a modified base is incorporated into the growing chain, it stops elongation of the primer strand whenever it is incorporated. The modified nucleotide is the last one added to the DNA. A different nucleotide (A, C, G, or T) is modified in each batch. The result is a mixture of partly double-stranded DNA molecules that differ in the length of the double-stranded region. In each batch, the last base in the double-stranded region of every DNA molecule is the same.

After DNA synthesis is complete, the researcher combines the four batches and separates the double-stranded DNA. Next, the DNA molecules are sorted by electrophoresis. This technique separates molecules by size—the shorter DNA molecules move faster and travel farther through a porous gel. The DNA forms a series of bands in the gel as shown in Figure 15.3. Each band contains DNA that is one base shorter than the one above it. Because a different dye and modified base was used in each batch, the color of each band identifies the last base in its DNA. Thus, the series of colored bands reveals the base sequence of the original DNA. A computer equipped with an optical scanner can read the sequence quickly and store the information for future study.

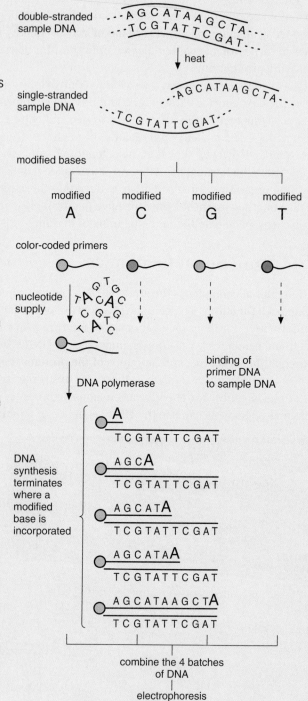

FIGURE 15.4

DNA sequencing procedures. Four batches of sample DNA are heated to separate the two DNA strands, and a differently labeled primer is added to each batch. Addition of a supply of nucleotides and DNA polymerase leads to synthesis of the complementary strand.

a . . . TGTCGATGCCAATCTCTCAGA . . .

b . . . cysteine – arginine – cysteine – glutamine – serine – leucine – arginine . . .

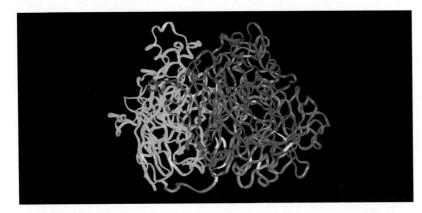

c

FIGURE 15.5

An example of functional genomics at work. Once the DNA sequence of a gene has been determined **(a)**, it is stored in a computer database. A computer program identifies start and stop codons and introns and converts the DNA sequence into a predicted amino acid sequence **(b)** for the encoded protein. In a different example **(c)**, a complete amino acid sequence is analyzed by a sophisticated program that predicts the three-dimensional structure of the protein. Comparison of the structures of normal and mutant proteins can reveal the nature of a genetic defect.

allele to predict how the mutation alters the shape and function of the protein (Figure 15.5). This approach enables scientists to move very quickly from a DNA sequence to the specific cause of a disease. As computer analysis of DNA sequences improves, researchers hope to use it to study how thousands of genes interact to produce either a healthy or diseased cell.

An example of the use of functional genomics is the study of mutations in the *cytochrome P450* gene. This gene encodes the enzyme Cytochrome P450, which breaks down medicines and toxins that enter our bodies. Variation in this gene causes individuals to differ in their ability to break down foreign substances. Scientists can use computer models of these variations (Figure 15.6) to determine whether some people may have difficulty in metabolizing a new medicine. For some, the medicine may prove to be toxic. In this way, computer models can help reduce the need or dangerous drug testing on human beings.

Most genes do not act by themselves. Instead, they interact in conjunction with many other genes. As the expression of one gene increases, expression of another gene may decrease or stop. Many diseases, such as cancer, involve changes in the expression of many genes. Scientists who study these diseases have developed a method to look at the expression of thousands of genes at once. Several thousand short DNA sequences are bound to a glass chip in an arrangement called a microarray (as shown in the Prologue opener photo).

CHEMISTRY TIP

Cytochrome P450 (also known simply as P450) is an enzyme that helps break down a wide variety of substances such as the cancer-causing compounds found in cigarette smoke. P450 metabolizes these substances by adding oxygen (from O_2) and hydrogen and electrons (from NADPH) to a substrate molecule, represented here as A.

$$A + 3\ NADPH + O_2 \longrightarrow AOH + 3\ NADP^+ + H_2O$$

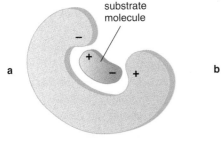

normal enzyme

mutant enzyme

FIGURE 15.6

Model of variations in enzyme and substrate interactions. Interactions of normal **(a)** and mutant **(b)** enzymes with a substrate molecule. If the altered amino acid sequence of the mutant enzyme interferes with binding of the substrate, the enzyme's activity will decrease, and the substrate will not be metabolized as quickly or efficiently.

Each of these sequences matches part of a different gene. Binding of RNA (or a more stable DNA copy of the RNA) in a cell extract to its matching spot in the DNA microarray indicates that the corresponding gene was active in the cell being tested. Computer analysis of microarray binding can identify groups of genes that are turned on and off together. Comparison of these patterns of gene expression in normal cells and cancer cells helps identify genes that may be involved in the development of cancer. By using DNA microarrays and computers in this way, scientists can quickly analyze data from thousands of genes to focus on those few that are most likely to be important in a particular disease.

15.3 Technologies

Humans have a long history of employing the beneficial characteristics of living organisms to perform practical tasks, as in using bacteria to produce cheese and antibiotics, or yeast to make bread and wine. Charles Darwin pointed out that humans have been practicing the selective breeding of plants and animals for centuries, manipulating traits that were part of an organism's normal genetic potential. Current methods developed to study the molecular biology of bacteria and viruses, however, enable the manipulation of individual genes that can be isolated and enable the alteration of their sequences. These manipulated genes can then be returned to the original organism or incorporated into the genetic makeup of a different organism.

Using genetic engineering, as these techniques are called, biologists can prepare recombinant DNA–DNA whose segments are from different sources. Genetic technology allows biologists to isolate specific genes, recombine genes from different organisms, and transfer this recombinant DNA into host cells. In the host cells, the recombinant DNA can be replicated, or cloned, to provide multiple copies of the gene for study. Alternatively, the cloned DNA may be used to express huge quantities of proteins, such as human growth hormone. Several practical applications of recombinant DNA were discussed in Sections P.2 and P.3 of the Prologue.

Creating a recombinant organism requires several conditions. The first requirement is identifying a specific gene and excising it from a chromosome. The second is finding a DNA carrier, or vector, that can be introduced into living host cells and can replicate there. Third, a method of joining the gene to the vector DNA is required. Finally, a method is needed to detect cells that have replicated the recombinant DNA, cells that may be synthesizing the desired protein.

As an example, the following procedure might be used to produce human growth hormone (GH) by genetic engineering (Figure 15.7). A DNA fragment containing the GH gene is liberated from the chromosome using **restriction enzymes.** This family of enzymes cuts DNA by recognizing and cleaving at specific nucleotide sequences. If the gene of interest and the vector are cut with the same restriction enzyme, their free ends can be joined by an enzyme called DNA ligase, an enzyme also involved in DNA repair, which functions to "glue" the strands together at their cut ends, forming recombinant DNA.

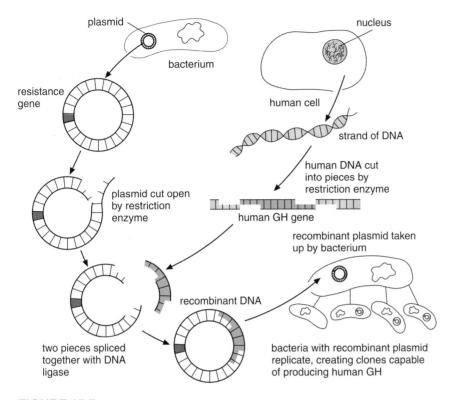

FIGURE 15.7

Genetic engineering. Recombinant DNA can be prepared by cutting plasmids from one source and DNA from another source with the same restriction enzyme. The complementary cut ends base pair with each other, and the backbones are joined by DNA ligase. The recombinant plasmids are taken up by host bacteria and are then replicated many times.

The most commonly used DNA vectors are plasmids, small circles of DNA found outside the chromosome in prokaryotes (see Section 6.3). In a culture medium, bacteria can be induced to take up plasmid DNA (usually *Escherichia coli*). After entering the host cell, the recombinant DNA replicates along with the host cell's DNA. This process is known as cloning—the reproduction and growth of genetically identical cells or organisms. The product that the recombinant DNA codes for, such as human GH, is synthesized by the genetically engineered host cells. Cloning can also furnish multiple copies of a gene for further research.

One method of identifying bacterial cells that have taken up the recombinant DNA is to use plasmids that carry what is known as a resistance gene. This gene confers resistance to various antibiotics. Bacteria containing plasmids with this resistance gene can grow on a culture medium that contains antibiotics, whereas bacteria that lack those plasmids will die.

Advances in the genome project have also benefited greatly from improved technologies. A fast, efficient machine can sequence almost 100 different fragments of up to 1,000 base pairs of DNA at a time in a few hours. Large sequencing laboratories may have dozens or even hundreds of such machines.

Preparing samples for DNA sequencing requires a combination of technologies to isolate pure DNA and to mix it with the appropriate chemicals to carry out the sequencing reaction. These reactions benefit from the use of robot technology, which can move tiny amounts of liquid with great speed and precision (Figure 15.8). Many of these reactions also use the **polymerase chain reaction** (PCR), a method of producing many copies of a tiny sample of DNA.

Recall that each chromosome consists of one very long DNA molecule. Before the DNA can be sequenced, enzymes must be used to cut it into more manageable pieces. This is accomplished by treating the DNA with restriction enzymes that cut DNA wherever a certain short base sequence

FIGURE 15.8

A laboratory robot. DNA is being loaded into plastic plates via this robotic arm for lab research to map the human genome. This device can operate more quickly and accurately than a human technician can. It can also repeat the same tasks for many hours without getting tired or bored.

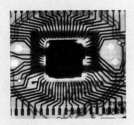

Biological Challenges

The Polymerase Chain Reaction

The development of PCR in the early 1980s by Kary Mullis revolutionized the way scientists analyze DNA and earned him the 1993 Nobel Prize for Chemistry. Since the development of PCR, as little as a single molecule of a DNA fragment can be copied many times to provide sufficient amounts for sequence or mutation analysis. This technique is called polymerase chain reaction because it uses a DNA polymerase to produce exact copies of a DNA sequence. The term *chain reaction* refers to the fact that the DNA it produces becomes the template for additional DNA synthesis in the next cycle of reactions.

An automatic temperature-control device repeatedly warms and cools the reaction mixture. This device repeats the procedure shown in Figure 15.9 about 30–40 times. First, heating separates the double-stranded DNA molecules. Next, DNA polymerase acts at a lower temperature to synthesize matching strands for the single-stranded DNA. The cycle then repeats. The result of all this DNA synthesis is several million copies of the targeted DNA sequence.

Each cycle requires only 3–4 minutes, so the entire procedure can usually be completed in a few hours. The success of PCR requires use of DNA polymerase from *Thermus aquaticus*, a bacterium that lives in hot springs. Unlike the enzymes of most organisms, its DNA polymerase is stable at the high temperatures needed to separate double-stranded DNA. The secondary structure of most proteins is destroyed by even moderate heating, a result quite evident from the frying of an egg. The albumin protein that makes up most of the egg white quickly congeals into a rubbery mass well before the lipid-rich yolk is cooked. The use of DNA polymerase from *Thermus aquaticus* in PCR is a good example of the practical application of biological knowledge.

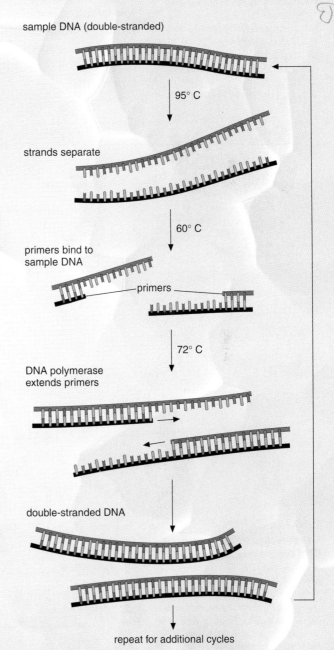

FIGURE 15.9

Polymerase chain reaction (PCR). Each cycle roughly doubles the number of DNA molecules.

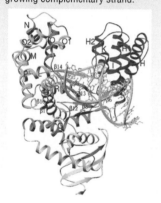

happens to occur. Once scientists have determined the sequence of each piece, they must assemble the sequences in the correct order and determine whether the sequences include any errors. To help determine the order of the pieces, scientists, using two different restriction enzymes, can perform an experiment to map the DNA. Because each enzyme cuts the DNA at a different sequence, the experiment will produce overlapping sets of pieces. This experimental data helps identify the order in which the pieces were originally arranged (Figure 15.10). Computer technology plays an important role in both assembling the sequence and determining the possibility of errors. To minimize errors, both strands of each DNA fragment are sequenced several times. Robots, very small samples, and computer technology all help make DNA sequencing a relatively inexpensive procedure.

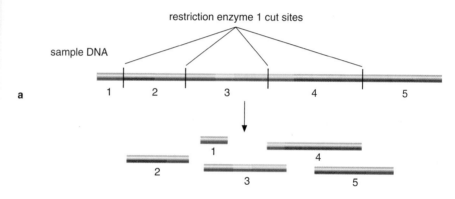

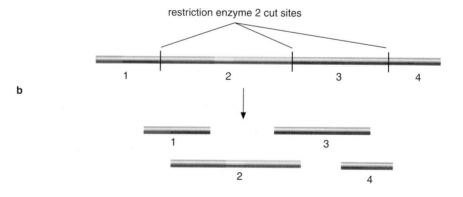

FIGURE 15.10

Use of restriction enzymes in DNA sequencing. (a), Restriction enzymes cut DNA into pieces that are short enough to be sequenced individually. The sequences of the pieces must then be arranged in the correct order. **(b),** Treating a second sample of the same DNA with a different restriction enzyme results in a different series of cuts. **(c),** Overlaps between the two sets of pieces reveal their correct order.

Focus On

Single Nucleotide Polymorphisms

Genetic markers that allow us to detect allelic variation among individuals have a long history in genetics, beginning with the identification of the ABO blood groups in the early 1900s. Most recently, **single nucleotide polymorphisms** (SNPs) have generated enthusiasm because they are common and easy to detect.

SNPs are genetic variations in which alleles differ by only one or a few scattered nucleotides (Figure 15.11). Most SNPs have little or no effect on phenotype, but they can be detected when DNA is sequenced. SNPs occur in about one of every 1,000 base pairs of DNA. Therefore, there are millions of SNPs in each person's genome.

Your unique set of SNPs distinguishes you from everyone else on the planet unless you have an identical twin. This variation can be used to study inheritance patterns and mutations that cause disease. It can also help determine paternity.

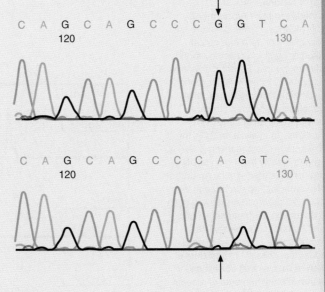

FIGURE 15.11

A single nucleotide polymorphism (SNP) revealed by DNA sequencing. The two sets of graphs show the corresponding DNA sequences of two individuals. A SNP can be seen at position 127, where one person has adenine (green) and the other has guanine (black).

The ability to copy tiny amounts of DNA with PCR to produce samples large enough for analysis has provided an important tool for examining very small amounts of DNA that may be present in blood or tissue samples left at a crime scene. It is also a useful technique for studying the minute amounts of DNA that remain in some fossils. Comparisons of the DNA sequences of ancient and modern organisms help reveal evolutionary relationships.

Another powerful tool for DNA analysis is based on genetic polymorphism and the use of restriction enzymes. Because of genetic variation, any particular restriction enzyme will cut the DNA of individuals of the same species in different places. Therefore, the resulting DNA fragments will have different lengths. When the fragments of cut-up DNA are separated by electrophoresis, these differences can be made visible. Comparison of the lengths of these fragments can be used to identify individuals or mutations that create or eliminate sites where the enzyme binds to and cuts the DNA. This technique is called **RFLP** (restriction fragment length polymorphism) **analysis.** A problem in RFLP analysis comes from the fact that most restriction enzymes recognize DNA sequences that are only six bases long. As a result, the DNA in a human cell contains over 1 million sites where a restriction enzyme will cut. The

FIGURE 15.12

Short Tandem Repeat (STR) analysis in law enforcement. This technique relies on four-base sequences of DNA that repeat different numbers of times among different people. Forensic scientists examine 13 different STRs found on different chromosomes. DNA samples coming from the same individual match at all 13 STR locations. To analyze an STR, the polymerase chain reaction (PCR) is used to amplify the region of DNA that contains the repeats. Amplified DNA samples are run through thin straw-like gels (called capillary gels) to determine the fragment lengths (and number of repeats). The fragments are compared to an allelic ladder that consists of a mixture of DNA molecules that correspond to all the different numbers of repeats for that STR (gel 1). Most individuals inherit different numbers of repeats from their parents and show two DNA peaks (heterozygous, gel 3). Other individuals inherit the same number of repeats from both parents (homozygous, gel 2). The presence of more than two peaks indicates that a sample comes from more than one person (mixed sample, gel 4).

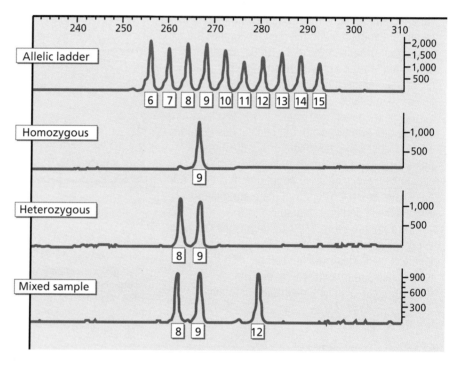

mixture of variously sized fragments that this produces forms an unreadable continuous smear of DNA in an electrophoresis gel.

When RFLP analysis is used to identify individuals (for example, in a legal trial), a short piece of synthetic DNA, called a **probe,** is added to the

Focus On

Forensic Testing

The testing of biological samples, such as blood left at a crime scene, has a long history in law and genetics. Since the 1990s, the PCR technique has helped forensic scientists—those who work in legal settings—use very tiny amounts of materials left at crime scenes for identity testing. This technology has been aided by the discovery of large numbers of individual genetic differences called short tandem repeats (STRs; Figure 15.12). It is now possible to determine with a high degree of reliability whether an individual matches a particular sample.

Besides use in law enforcement, DNA typing has also been used to determine identity in other situations. Some of these uses have generated considerable scientific and political discussion. For example, it has long been thought that one of our first presidents, Thomas Jefferson, may have fathered a child with one of his slaves, Sally Hemmings. In 1998, DNA testing of known descendants of Hemmings's son Eston and Jefferson showed that their Y chromosomes share certain rare DNA sequences. This evidence strongly suggests that Jefferson was Eston's father. However, it leaves open the possibility that Eston's father was another Jefferson relative who would have shared the same DNA sequence. Testing like this can help resolve arguments about events that happened hundreds of years ago.

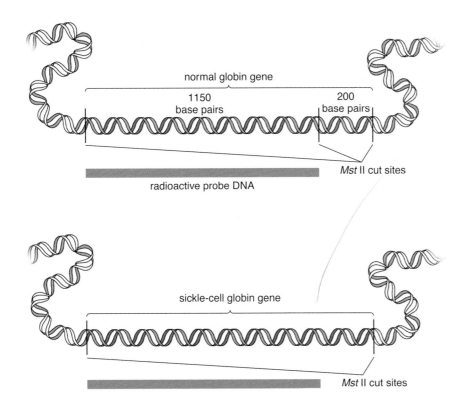

FIGURE 15.13

RFLP analysis in medical diagnosis. The restriction enzyme *Mst* II cuts the globin gene at three sites. The sickle-cell mutation eliminates one of these sites. The gene fragment that binds to the radioactive probe is longer than normal in the case of sickle-cell DNA.

In the top diagram: "normal globin gene", "1150 base pairs", "200 base pairs", "*Mst* II cut sites", "radioactive probe DNA". In the bottom diagram: "sickle-cell globin gene", "*Mst* II cut sites".

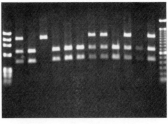

FIGURE 15.14

PCR-RFLP. PCR is used to amplify the region of the N-acetyltransferase 2 gene that contains a single restriction site for *Kpn* I. The amplified DNA is then cut with the enzyme, and the DNA fragments are size-separated on a gel and stained. Lane 1 (far left) is the control, a molecular size standard. Lane 2 shows a heterozygote, a person carrying one copy each of the two alleles (three bands, one long uncut fragment and two small pieces of the cut fragment). Lane 3 shows a homozygotic person with two copies of the DNA that is cut by the enzyme (two smaller bands). Lane 4 shows a person with two copies of the alternate allele, DNA that is not cut by the enzyme, and shows only one larger band. The last lane (far right) is another molecular size marker.

DNA after electrophoresis. The probe is selected so that it will bind to a sequence that occurs only a few times in the DNA being tested. A radioactive atom that will expose X-ray film is attached to the probe. The X-ray film is then used to reveal the location of the DNA bands that contain the probe-binding sequence, which may or may not have been cut by the restriction enzyme. Because of variation among individuals in the size of these DNA fragments, the positions of these bands can be used to identify the source of the DNA.

A variation of this technique can be used to determine whether a person carries a disease-causing allele, increasing his or her chances of developing certain diseases. For example, the sickle-cell mutation occurs in the gene that encodes a type of globin, the protein hemoglobin. A probe that binds to the globin gene can reveal the presence of the sickle-cell allele in the DNA of a healthy heterozygote (Figure 15.13). Polymorphisms in another gene, N-acetyltransferase 2, are associated with an increase in urinary bladder and colorectal cancers due to the body's decreased ability to inactivate certain carcinogens. Alternate alleles can be detected by first amplifying the chromosomal region containing the gene, then trying to cut these DNA fragments with a useful restriction enzyme (Figure 15.14). This method, called PCR-RFLP, avoids the use of radioactively labeled probes.

Check and Challenge

1. Why might scientists want to obtain the DNA sequence of organisms other than humans?
2. About how many base pairs of DNA are on the average human chromosome?
3. How do scientists determine DNA sequences?
4. How does functional genomics lead to predictions about biology?
5. What advantages does PCR provide for scientists studying DNA?
6. How common are single nucleotide polymorphisms?

Applications and Issues in Molecular Genetics

15.4 Mutations and DNA Repair

Try Investigation 15A ⬥ Determining Mutation Frequency in Bacteria.

ETYMOLOGY

mis- = wrong or bad (Old English, Old French)

non- = not (Latin)

sense- = meaning (French)

Missense mutations cause the wrong amino acid to be inserted in a protein. A **nonsense** mutation occurs when a stop codon replaces an amino acid codon. The truncated protein is generally not functional.

Mutations are changes in the DNA sequence. Mutations have many causes, including exposure to certain chemicals and radiation. Most mutations, however, probably result from failure of the DNA copying and repair mechanisms to work with 100% accuracy. A mutation can affect protein structure directly, or it can occur in the regulatory regions of the genome that control when and where each protein is synthesized. The most common type of mutation is the **point mutation,** in which one base pair changes into a different one (Figure 15.15). Most point mutations result in SNPs that have little or no effect on phenotypes. In some cases, the mutation changes an amino acid that is important to protein structure or function. These cases are called **missense mutations. Nonsense mutations,** on the other hand, change the codon for an amino acid into a stop codon, resulting in an abnormally short protein.

Type of mutation		Effect on sentence	Effect on DNA	Effect on protein structure
original sequence		The red cat ate the fat rat.	C TAT C A G T G T A C	–leucine–serine–valine–tyrosine–
point mutations	missense	The red hat ate the fat rat.	C TAT **T** A G T G T A C	–leucine–leucine–valine–tyrosine–
	nonsense	The red cat ate the fat.	C TAT C A G T G T A **A**	–leucine–serine–valine–[stop]
frameshift mutations	addition	The reb dca tat eth efa	C TA **A** T C A G T G T A C	–leucine–isoleucine–cysteine–valine–
	deletion	The rec ata tet hef	C TA C A G T G T A C . . .	–leucine–glutamine–cysteine–. . .

FIGURE 15.15

Mutations in DNA and in an English sentence. The effects of point mutations vary widely. Some missense mutations have little or no effect on phenotype; others have significant effects. Most nonsense and frameshift mutations eliminate normal gene function.

 BSCSblue.com/check_challenge

In addition to point mutations, there are **frameshift mutations,** in which one or two base pairs are inserted or deleted from the DNA, thereby disrupting the pattern of three-base codons (reading frame). Frameshift mutations change every subsequent codon. In most cases, this also creates a new stop codon. The result is a shortened, nonfunctional protein in which all amino acids after a certain point are changed.

Different mutations in a gene can produce the same phenotype. For example, many different mutations have been found in the human gene that encodes the enzyme phenylalanine hydroxylase. Normally this enzyme converts the amino acid phenylalanine to tyrosine, another amino acid. Any mutation that reduces the activity of this enzyme results in a disorder called phenylketonuria (PKU). If PKU is not detected in early infancy, toxic by–products build up from the breakdown of excess phenylalanine. These toxins damage developing nervous systems and result in mental retardation. Since the 1960s, a simple blood test for this common condition has been used on all newborn infants. For children born with these mutations, a diet low in phenylalanine largely protects them from developing PKU.

In contrast to PKU, some genetic disorders, such as sickle-cell anemia, are caused by one specific point mutation. Another common example of such a mutation is a form of dwarfism called achondroplasia (Figure 15.16). More than 98% of the cases of this condition are caused by the same point mutation in the gene that encodes the receptor for a growth hormone. When only one mutation causes all or most cases of a disorder, detection of the condition is easier because a single genetic test can be used. When many

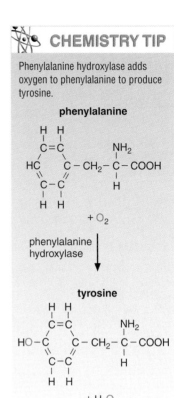

CHEMISTRY TIP

Phenylalanine hydroxylase adds oxygen to phenylalanine to produce tyrosine.

phenylalanine

$+ O_2$

phenylalanine
hydroxylase ↓

tyrosine

$+ H_2O$

When the enzyme phenylalanine hydroxylase is lacking or defective, other enzymes break down phenylalanine by replacing its amino group (—NH_2) with oxygen to form toxic waste products called phenylketones.

FIGURE 15.16

Achondroplasia, a dominant form of dwarfism. Over 98% of cases of this growth disorder are caused by the same mutation in the gene for a growth factor receptor protein. Many cases occur because new mutations arise during gamete formation in families with no history of dwarfism.

 CHEMISTRY TIP

Achondroplasia is usually due to a mutation that changes a specific guanine base to adenine (G to A point mutation). The result is a protein with the highly polar, basic amino acid lysine in place of the neutral amino acid glycine.

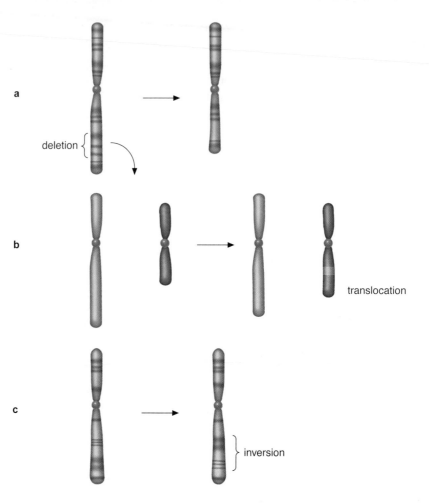

FIGURE 15.17

Chromosome rearrangements.
These mutations include deletions
(a), translocations **(b)**, and
inversions **(c)**.

different mutations in a gene can lead to the same disease—as is the case for
PKU, cystic fibrosis, and breast cancer—genetic testing may be more
difficult. Testing is especially difficult if the gene is very large and sites of the
mutations vary widely.

Like the DNA in individual genes, large regions of chromosomes can be
mutated by the deletion or movement of genetic material (Figure 15.17).
Chromosomal mutations are especially common in cancer cells. A clue to
how chromosomal rearrangements might lead to the development of cancer
came from the study of a form of leukemia (cancerous growth of white
blood cells). About 70% of people with this form of leukemia were found to
have cancer cells with a mutation called the Philadelphia chromosome
(Figure 15.18). In this mutation, parts of chromosomes 9 and 22 break off
and are **translocated** (exchange positions). Because the breaks in both
chromosomes occur within genes, the result is a fused gene that consists
of the promoter and beginning of the protein-encoding part of a gene on
chromosome 22 and the end of a protein-encoding region from chromosome
9. The promoter ensures that the mutant protein is continuously expressed.
This protein has been found to activate the *ras* pathway that regulates cell
division (see Focus On Cancer, Section 8.9). This activity probably accounts
for the cancer that results.

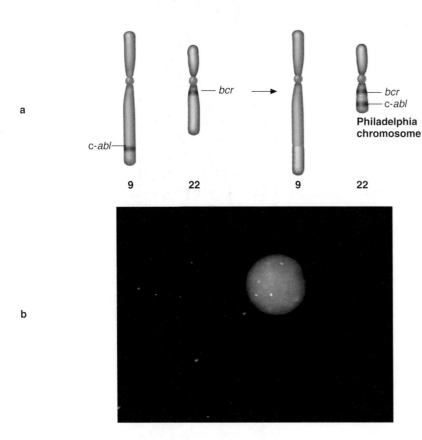

a

b

9   22   9   22

bcr → bcr
c-abl

c-abl

Philadelphia
chromosome

FIGURE 15.18

The Philadelphia chromosome.
(a), A translocation transfers a
small piece of chromosome 9 to
chromosome 22, which is now
referred to as the Philadelphia
chromosome. **(b),** A fluorescent
probe reveals the presence of this
out-of-place DNA. Look closely to
notice that the c-*abl* gene is pink,
the *bcr* gene is green, and the
fusion gene (c-*abl*/*bcr*) shows up as
yellow. The translocation joins parts
of two genes, resulting in a fused
protein.

In multicellular organisms, mutations can occur in somatic cells. These
mutations are not inherited. Some somatic-cell mutations are not important;
at worst, they result in the death of one mutant cell. Others cause
uncontrolled cell growth, eventually leading to cancer. These more
dangerous mutations usually occur in tumor-suppressor genes or in genes
involved in the repair of mismatched bases or other types of damaged DNA
(see Section 8.5).

One kind of somatic-cell mutation is a normal part of development in
many animal species. This process, called **gene amplification,** creates
extra copies of specific genes. In multicellular organisms, some specialized
cells produce large quantities of a few specific proteins. For example,
glandular cells in the human pancreas produce large quantities of insulin.
During differentiation, some types of cells produce extra copies of
chromosomal regions that contain the genes for these proteins. This is
especially common in insect larvae. The extra copies may be inserted in the
chromosomes or float freely in the nucleus, like bacterial plasmids. Gene
amplification may have contributed to the evolution of genes such as
ribosomal RNA genes that normally exist as multiple copies. Oncogenes (see
Focus On Cancer, Section 8.9) are often amplified in cancer cells. This
mutation contributes to the development of cancer. Gene amplification also
has been shown to contribute to the development of resistance to
medications in some cancer cells.

E T Y M O L O G Y

ampli- = large (Latin)
Gene **amplification** enlarges the
effect of a gene by increasing the
number of copies of the gene.

CONNECTIONS

The likely contribution of gene
amplification to the evolution of
multicopy genes such as rRNA
demonstrates the close
relationship between evolution
and development.

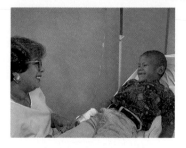

FIGURE 15.19

A cancer patient receiving chemotherapy to treat the disease. Chemotherapy, a treatment for some cancers, involves one or more drugs administered according to precise medical protocols.

ETYMOLOGY

germ = bud or seed (Latin)

Gametes are also known as germ cells. The term **germ line** refers to gametes and their ancestral diploid cells.

15.5 Genetic Disorders and Gene Therapy

The most important result of the HGP will be new ways of treating or preventing inherited disease. Genetic disorders and their treatments are as varied as any other medical condition. Treatments can include surgery, medication, and other approaches. For example, tuberous sclerosis is an autosomal (not sex-linked) dominant condition in which small, noncancerous tumors (or tubers) can form in the skin and brain. In the brain, they can result in seizures. Treatments include surgery to remove the tumors in addition to medication to control the seizures. Most types of cancer, whether genetic or not, are treated with medicine (chemotherapy, Figure 15.19), surgery, or radiation.

All of these treatments, however, deal with the symptoms of the disease and not its genetic cause. **Gene therapy** attempts to treat the genetic defect itself rather than its results. It is one of the most exciting areas in medical research today. Gene therapy can be divided into two general categories. In one, called **germ-line therapy,** the DNA of an affected individual's gametes would be changed so that the abnormal allele would not pass on to any future offspring. Because this form of therapy changes the genes of future generations, it has been controversial. The other form of gene therapy, which treats somatic cells, is less controversial and more likely to be applied in the near future. This form of gene therapy could correct the disease-causing allele in the cells that express the gene.

For example, some forms of diabetes are caused by a deficiency of insulin (see Focus On Energy Regulation, Section 5.9), and some people are abnormally short because they lack sufficient growth hormone. Both insulin and human growth hormone are now produced in large quantities by genetically engineered bacteria and used by millions of people every year. Even so, a better treatment would enable cells in the body to increase and decrease insulin production in response to need, as it does in healthy people. Scientists are trying to make this possible by replacing the defective insulin gene with an effective gene in the pancreatic cells of diabetic individuals. Similarly, if a normal gene for growth hormone could be placed in the pituitary cells that normally make this protein, abnormally short children might be able to grow to average adult height.

Gene therapy requires that genes be brought into cells and then integrated, or moved, into the cells' DNA. There are several methods of doing this. One way involves placing the DNA in a lipid vesicle that will allow it to move through the plasma membrane. Another way is to insert the gene in the DNA of a virus that is adapted to bring its own DNA into the cell (Figure 15.20). Each of these approaches has advantages and drawbacks.

In some cases, when foreign DNA moves into a cell, it becomes part of the cell's chromosomes but not necessarily in the same place as the cell's defective gene. Occasionally, the new DNA may be inserted in the middle of another gene and inactivate that gene, creating a new problem. Similarly, using viruses to carry DNA into cells can lead to unintended viral diseases.

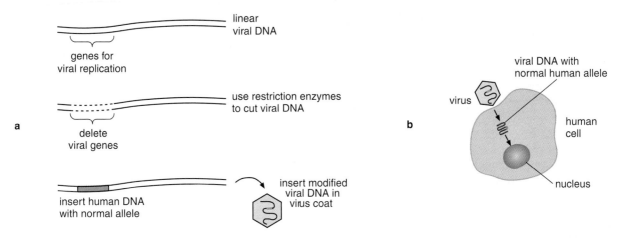

FIGURE 15.20

One approach to gene therapy. (a), Scientists remove the disease-causing genes from a virus and insert a copy of a normal human gene in the virus's DNA. **(b),** If the treatment is successful, the virus carries the human gene into the patient's cells, where it is permanently integrated into the chromosomes.

Finally, gene therapy is easier in some tissues and organs than in others. Bone marrow cells, which are easy to remove and replace, are good targets for gene replacement. Brain and heart cells, on the other hand, are more difficult to treat. Even slight damage to these tissues can be dangerous. Some successes have already been achieved in treating cystic fibrosis patients by inserting the normal **CFTR** (Cystic Fibrosis Transporter) gene into cells of the nose. The cells were genetically transformed and began to transport chloride ions normally (see Biological Challenges: Understanding Cystic Fibrosis in Section 3.4). Patients with an inherited defect in their immune systems have also shown some improvement following replacement of the gene that encodes the missing enzyme. Gene therapy will increase in importance as we learn more about how to integrate new genes into host-cell DNA, how cells regulate gene expression, and why some organs are more or less receptive than others to foreign DNA.

15.6 Ethical, Legal, and Social Issues

The availability of the human DNA sequence is helping scientists identify the genes that play an important role in human health and disease. Along with DNA sequence information, the HGP is revealing much human variation, such as SNPs. The technology is now at hand to cheaply and efficiently identify normal variations as well as those alleles that may increase a person's risk of developing disease. These advances have created a situation that challenges our current legal, ethical, and social views of genetics. Some people argue that it is unethical to identify healthy people who carry alleles that may predispose them to disease in the future. These individuals may be targeted by employers, schools, or insurance companies for discrimination.

CONNECTIONS

Medical treatment of genetic disorders can relieve an unhealthy condition. This demonstrates the interaction of genetic and environmental factors in determining phenotypes.

THEORY

What is a Gene?

Neither Gregor Mendel nor Charles Darwin used the term *gene*, but both wrote about individual, heritable factors that determine the characteristics of organisms. Where are these factors? Most people thought that when genes from each parent combined in a zygote, they blended. But if the traits of offspring are just a blend of those of the parents, then differences between members of a species should disappear over time. This idea did not agree with observations of plants and animals, however, nor with Mendel's ratios of inherited traits. Mendel's work was important in developing the idea of a gene as a factor responsible for a specific trait.

Walter Sutton's and Theodor Boveri's careful observations of dividing cells provided the first strong evidence that the hereditary factors are located on chromosomes. They called these factors genes. Thomas Hunt Morgan's observations of fruit flies confirmed Sutton's and Boveri's conclusions. Morgan also demonstrated that genes do not blend in offspring but remain distinct structures in chromosomes.

Biologists still think of a gene as an individual unit of heredity, but what counts as a gene is not always clear. Since James Watson's and Francis Crick's studies of DNA, biologists have also thought of genes as segments of DNA or RNA that code for particular polypeptides. But in some cases, as shown in the model of hemoglobin, several DNA segments contribute to the synthesis of a single protein. Even when a single DNA segment encodes a whole protein, the protein may not be synthesized unless other genes (such as RNA polymerase and tRNA genes) are active. Expression of many genes also depends on environmental factors, such as temperature, pH, and the presence of certain hormones. Since a gene is defined in terms of its function in coding proteins and its heritability, should these other factors be considered part of the gene? What about segments of DNA that regulate gene expression but do not encode protein or those for which we have not yet discovered a function?

Researchers continue to link segments of chromosomes with the presence of traits, but the chain of links gets more complicated every day. New discoveries in genetics make the one gene–one trait model seem less and less useful. Perhaps the idea of a gene will evolve like the terms *heat* and *electricity* from a physical object to a somewhat loose name for a general topic.

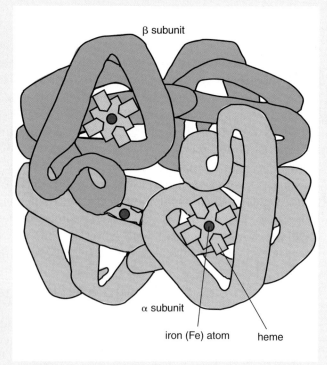

A model of hemoglobin. Hemoglobin is the protein that carries oxygen through the bloodstreams of vertebrate animals. The complete protein consists of four peptide subunits, two each of two different types (α and β). The DNA sequences that encode these peptides include introns that are not translated. Each peptide is bound to an iron-carrying nonprotein molecule called a heme ring. Many enzymes are needed for the biosynthesis of heme and the absorption and metabolism of iron. What is the hemoglobin gene? Does it exist?

In addition, the identification of the genetic basis of the normal variation in characteristics, such as height and skin color, could create new opportunities for discrimination. The history of discrimination against individuals based on their skin color or ethnic background has heightened our awareness of these concerns.

On the other hand, identification of alleles that predispose people to disease allows individuals who are at risk to be identified at an early stage in their lives when prevention and early diagnosis are possible. For example, individuals who might be at genetic risk for developing lung cancer could be encouraged to avoid alcohol, tobacco, or airborne pollutants in the workplace. Those at risk for developing colon cancer could modify their diets to reduce their risk. Individuals who have an elevated risk for high blood pressure or breast or prostate cancer could receive diagnostic tests on a regular basis, such as blood-pressure screening or breast or prostate exams. Early detection of these disorders can lead to safer and more effective treatments that can greatly prolong the lives of the at-risk individuals. Information about genetic risk factors is also useful for people who are considering having children.

As technology advances, tests for dozens or even hundreds of genes will become relatively cheap. As a result, it will be possible to create a very complex genetic profile for each individual. As more and more tests become available, patients, physicians, and other health-care workers will have more decisions to make. An understanding of the role that genes play in health and disease will be essential in helping you decide which tests to have and when. The role of the environment in modifying genetic risk factors will also be an important variable to consider in determining how likely it is that you or a member of your family might be affected by a particular condition. These challenges will provide you with the opportunity to make decisions based on your knowledge of biology in general and your own environmental and genetic profile in particular.

Check and Challenge

1. Describe the different types of mutations.
2. How can several different mutations cause the same genetic disease?
3. What are the main difficulties in gene therapy?
4. What are the important potential benefits and risks of genetic screening?

Summary

Advances in technology have made the mapping and sequencing of genes faster and cheaper. The genomes of a number of viruses, bacteria, and multicellular organisms have been sequenced. The Human Genome Project (HGP) has moved from the determination of DNA sequences to the analysis and interpretation of that information. Computers help scientists predict the structures of normal and mutant proteins from their gene sequences. Computers also help identify functional relationships among genes by analyzing large bodies of data generated from DNA microarrays. Other powerful tools for the analysis of gene sequences include PCR, SNPs, laboratory robots, and RFLP analysis. These techniques have made DNA analysis an important source of evidence in medicine and law.

Alterations to DNA include nonsense and missense point mutations and additions and deletions that cause frameshift mutations. Many different types of mutations can inactivate a gene, resulting in the same disorder. This variety makes the design of DNA-based tests for inherited diseases difficult. Not only individual genes and nucleotides but also large regions of chromosomes are subject to mutation. Chromosome-level mutations include deletions, translocations, and inversions. Some mutations in multicellular organisms are not inherited because they occur in somatic cells. Somatic-cell mutations can lead to cancer and other disorders. Gene amplification, however, is a type of somatic-cell mutation that is a normal part of embryonic development. In some cases, such as the ribosomal RNA genes, additional copies of amplified genes have become a permanent part of the genome.

Gene therapy for germ-line cells is unlikely to be important in the short term but may become more common in the future. Techniques for somatic-cell gene therapy, now an area of intense activity in medical research, are currently being developed. Some limited success has already been achieved in somatic-cell therapy. The many ethical, social, and legal issues raised by genetic technologies are still being debated. These issues include the privacy of people who have been found to carry alleles that may cause disease in the future and the risk of discrimination on the basis of genetic background. However, widespread genetic profiling may have a positive effect if it leads to advances in individualized plans for disease prevention and treatment.

Key Concepts

Use the following concepts and others from this chapter to construct a concept map.

- single nucleotide polymorphisms
- RFLP
- DNA sequencing
- DNA microarray
- DNA probes
- frameshift
- deletion
- gene therapy
- polymerase chain reaction
- Human Genome Project
- restriction enzymes
- genome
- missense
- translocation
- gene amplification
- somatic cells

Reviewing Ideas

1. What is the main goal of the Human Genome Project?
2. What was the reason for sequencing the genome of a simple multicellular organism (worm)?
3. What is the reason for sequencing the genomes of important crops and food animals?
4. How are electrophoresis and computers used in DNA sequencing?
5. How can functional genomics speed up the design and testing of new medications?

 Biology Online BSCSblue.com/vocabulary_puzzlemaker

6. What problem does PCR solve?
7. Is it likely that you carry some SNPs? Explain your answer.
8. Is RFLP analysis more useful for determining a DNA sequence or for identifying an individual? Explain your answer.
9. Distinguish between a missense mutation and a nonsense mutation.
10. Would deletion of three nucleotides cause a frameshift mutation? Would a deletion of five nucleotides cause a frameshift? Explain your answers.
11. Why would it be difficult to design a genetic test for a disease that was caused by mutations in a large gene?
12. Describe the results of gene amplification.
13. Distinguish between germ-line therapy and somatic-cell gene therapy.
14. What are the major likely risks and benefits of gene therapy?

Using Concepts

1. How could one frameshift mutation compensate for another?
2. Gene amplification in cancer cells has been shown to lead to resistance to cancer-killing medications when the dose of medication is increased gradually. Starting with a high dose is a more effective way to kill cancer cells. Describe a hypothesis that explains these results.
3. How could gene therapy cause new genetic damage?
4. How could restriction enzymes be useful in gene therapy?

Synthesis

1. How could a mutation in a tRNA gene help overcome the effect of a point mutation in a gene that encodes a protein?
2. Would people with chromosomal translocations or with point mutations be better candidates for gene therapy? Explain your answer.

Extensions

1. How could a mutation in a tRNA gene help overcome the effect of a nonsense mutation in a gene that encodes a protein?
2. Research the current status of genetic screening and counseling for cystic fibrosis, sickle-cell anemia, Tay-Sachs disease, or another common genetic disorder. Report to the class on the technology available today for your topic and the current ethical, legal, and social issues involved.

Web Resources

Visit BSCSblue.com to access
- Resources about current research in molecular genetics and its applications
- Web links to information on the inheritance and biology of many normal human traits and diseases, a database of SNPs and their phenotypes, news updates on the HGP and the sequencing of other genomes, and current information about oncogenes and the interactions between environmental influences and genetic variation

CONTENTS

LEARNING OUTCOMES

By the end of this chapter you will be able to:

A Relate the study of genetics to that of population genetics and discuss the factors that can affect gene-pool equilibrium.

B Explain the Hardy-Weinberg model.

C Discuss evolution through natural selection.

D Explain genetic drift and contrast its effects on large and small populations.

E Discuss the role of quantitative traits in microevolution.

CHAPTER

16

Population Genetics

■ *What differences can you see among these sea lions?*

■ *What are the sources of this variation?*

No two individual organisms are exactly alike. Even members of the same species show differences in their anatomy, their behavior, and their genes. For example, some breeds of dogs are tall and sleek and bred for speed. Others are short with long snouts and claws suited to digging. Some breeds are active, whereas others are docile. Some are long-haired and suited to cold climates; others are short-haired and adapted to tropical temperatures.

As Charles Darwin pointed out almost 150 years ago, variation among individuals is the raw material for evolution. Variation is what allows populations to adapt to new environmental conditions or to be selectively bred for desirable traits. As you read about variation in this chapter, ask yourself these questions: Where do the differences among individuals come from? What are the forces that affect variation? What would the biological world be like without variation?

Genetic Variation in Populations

16.1 Populations and Gene Pools

Evolution is defined as change in populations through time. How much time? Biologists usually think of two time scales of evolution. **Microevolution** is change within species, which can occur over dozens or hundreds of generations. **Macroevolution** usually involves much longer periods of time and includes the origin of new species. This chapter focuses on microevolution.

The definition of evolution emphasizes change, but what type of change? The biologists of the early 20th century would probably have defined this as "a change in the appearance of organisms" or "a change in species." However, modern investigators have a more precise answer. They think of microevolution as change in the genetic composition of populations. The field of biology that studies microevolution is called **population genetics.** Population geneticists use Mendel's laws of inheritance, biochemical analysis of genes and proteins, mathematical models, and other techniques to understand microevolution.

A central concept in population genetics is the **gene pool** (Figure 16.1). A gene pool consists of all the genes of a local population of organisms. For example, if a population of plants consists entirely of homozygous purple-flowered plants, then its gene pool has many copies of alleles that make purple flowers and no alleles that make white flowers. In this case, the frequency of the allele for purple flowers is 100%, and the frequency of the allele that makes white flowers is 0%. If the population consists entirely of

FIGURE 16.1

Imaginary representation of a gene pool. The gene pool is a collection of all of the genes and their different alleles that are in a population. In simple population-genetics models, the gene pool is imagined to be an infinitely large body of water containing all the alleles. The gene pool pictured here has two alleles that affect flower color, represented by the symbols *p* (for white) and *P* (for purple).

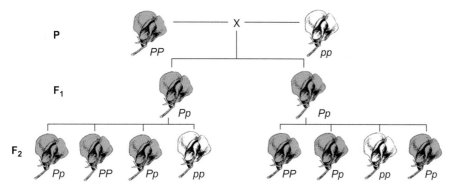

FIGURE 16.2

Allele frequencies in three generations of a population. Homozygous purple-flowered plants are crossed with homozygous white-flowered plants to produce heterozygotes in the F₁ generation. Note that the frequency of each allele is the same (50%) in all three generations.

homozygous white-flowered plants, then the frequencies of alleles in the gene pool are reversed—100% alleles for white flowers and 0% alleles for purple flowers. It is also possible that both types of alleles are present in one gene pool. For instance, one particular gene pool might have 70% alleles for purple and 30% alleles for white. A different population of the same species might have a gene pool with 50% alleles for purple and 50% alleles for white. In each case, the frequencies of the two types of alleles add up to 100% (Figure 16.2).

Notice that a gene pool refers to one population, not to an entire species. Organisms that are together in one geographical region make up a local population. The organisms in the local population interbreed. Because they share the same environment, these organisms experience the same pressure from natural selection. For these reasons, the local population is the unit of evolution. Population genetics focuses on the gene pools of local populations.

16.2 Genetic Variation

All humans are members of the same species, *Homo sapiens*, but no two individuals are exactly alike. There are many similarities and differences among people. For example, eye color and blood type differ among individual *Homo sapiens*. Differences in these traits are due to genetic differences. The human gene pool carries alternative alleles that affect blood type and many other traits. Other species also have variation in their gene pools. For example, apple trees are all members of one species, but the fruit produced by different trees can be red or yellow, hard or soft, sweet or tart, large or small. These differences are caused partly by genetic variation. When two or more alleles of a gene for a trait are present in a gene pool, the population is said to be **polymorphic** for that trait.

Population geneticists have studied the gene pools of many species of plants and animals. They have examined variation in obvious traits such as

ETYMOLOGY

poly- = many (Greek)

-morph = form (Greek)

A **polymorphism** is one of several forms of an allele.

shape and color. In many cases, they have also found genetic variation in the amino-acid sequences of proteins and the nucleotide sequences of DNA. For example, the fruit fly *Drosophila melanogaster* has an enzyme called alcohol dehydrogenase. There are two slightly different forms of alcohol dehydrogenase that can be distinguished by electrophoresis. The two forms differ by only one amino acid. The amino-acid difference is caused by one nucleotide difference in the DNA. In natural populations all over the world, *D. melanogaster* gene pools are polymorphic for the alcohol dehydrogenase gene. The populations have both types of alleles and produce both varieties of alcohol dehydrogenase.

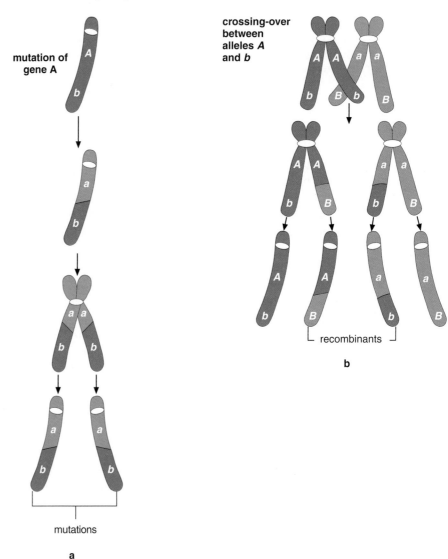

FIGURE 16.3

Mutation and crossing-over, two sources of genetic variation. Mutation **(a)** gives rise to new alleles, such as when *A* mutates to *a*. Before mutation, only *A* alleles were present. Crossing-over **(b)** generates new combinations of alleles at different genes. Before meiosis, only the combinations *Ab* and *aB* occur. Crossing-over between the two genes results in the recombinant types *AB* and *ab*.

Not all genetic variation has an effect on the organism. Because of the redundancy of the genetic code, some mutations that change a nucleotide cause no change in amino-acid sequence (see Figure 9.4). For example, the codons CAA and CAG in mRNA both specify the amino acid glutamine. This means that two individuals could have slightly different genes but produce identical proteins. Many mutations in noncoding, nonregulatory regions of DNA have no effect on the organism. Also, some mutations cause an amino-acid change that has no effect on protein function. For example, a change from the codon AAA to AAG would substitute the amino acid arginine for lysine. These amino acids have similar R groups that can roughly substitute for one another in a protein. Most gene pools probably carry many such examples of hidden polymorphisms.

Observation of protein sequences and DNA sequences in the gene pools of many species has led scientists to the conclusion that almost all large populations of sexually reproducing organisms have much variation in their gene pools. How much variation? Typically, about 0.5% of the DNA bases are variable. The chromosomes of *D. melanogaster* contain 165 million base pairs. Therefore, almost a million nucleotide sites differ among fruit flies. The human genome contains less variation than that of the fruit fly. The genomes from two different individuals are 99.9% the same. This means that the 3 billion base pair human genome is expected to contain about 3 million variable sites. The ultimate source of all that variation is mutation (Figure 16.3a). The second source of variation is the recombination of alleles during meiosis (Figure 16.3b). Crossing-over and independent assortment in meiosis produce new combinations of alleles.

16.3 The Hardy-Weinberg Model

Scientists sometimes use simple models to understand nature. Population geneticists use the **Hardy-Weinberg model,** an idealized mathematical model of gene pools. This model was proposed in 1908 by the mathematician Godfrey H. Hardy and the physician Wilhelm Weinberg. They recognized a simple mathematical relationship in the allele frequencies in gene pools. Their insight remains a fundamental part of population genetics.

The Hardy-Weinberg model makes several simplifying assumptions (Table 16.1). Imagine a large population of diploid, sexually reproducing plants. The plants are assumed to have nonoverlapping generations, which means that the parents die before the offspring reproduce. This kind of life

CONNECTIONS

The Hardy-Weinberg model is an example of how scientific theories often work. Imperfect models lead to experiments that help scientists improve their models.

TABLE 16.1
Assumptions of the Hardy-Weinberg Model

• The organism is diploid.	• Reproduction is sexual.
• Generations are nonoverlapping.	• Gametes unite at random.
• Population size is very large.	• Migration is negligible.
• Mutation is negligible.	• Natural selection does not operate.

cycle occurs in annual plants (plants that live 1 year or less) and in insects that have only one generation per year. The model also assumes that mutation and migration are negligible. No natural selection operates. Finally, reproduction involves random union of gametes, or random mating.

Now suppose that there are two alleles in the gene pool of this imaginary population that affect flower color. Some gametes carry an allele that produces purple flowers. Other gametes carry an allele for white flowers. Each generation of plants is created by randomly sampling the gamete pool. What does random sampling mean? Imagine reaching into an infinitely large pool of gametes. Pull out two gametes at random, and unite them to form a diploid zygote. Repeat this process over and over, uniting pairs of gametes, until all the zygotes for the new generation have been formed. Each time the random sample yields two alleles for purple flowers, a homozygous genotype is formed. Each time the random sample yields two alleles for white flowers, a different homozygous genotype is formed. Each time the sample gives one allele for purple flowers and one allele for white flowers, a heterozygote is formed.

The Hardy-Weinberg model predicts a simple relationship between the allele frequencies in the gamete pool and the genotype frequencies in the next generation. If the allele for purple flowers has a frequency p, the allele for white flowers has a frequency q, and $p + q = 1$, then the frequency of plants that are homozygous for the allele that causes purple flowers in the next generation will be p^2. The frequency of plants that are homozygous for the allele that causes white flowers will be q^2. The frequency of heterozygous plants will be $2pq$. The Hardy-Weinberg model enables you to use the allele frequencies to calculate all of the genotype frequencies. This is the first important result of the model.

The second main result of the Hardy-Weinberg model is that the allele frequencies are stable over time. Under the assumptions of this model, the frequency of the allele for purple flowers remains constant generation after generation. The genetic variation does not disappear; if a gene pool is polymorphic, then it will stay polymorphic. This prediction would have excited Charles Darwin, who did not know about Mendel's laws of inheritance. Darwin thought that blending inheritance would cause populations to lose genetic variation. This was a serious problem for his theory of evolution by natural selection because the theory requires genetic variation. The Hardy-Weinberg model shows that under some simple circumstances, with no evolutionary forces operating, genetic variation tends to remain in populations.

Keep in mind that the Hardy-Weinberg model is highly idealized. No population exactly meets all of the assumptions of the model. Even so, the model is a useful tool for thinking about gene pools. Populations are often observed to have allele frequencies that are stable from generation to generation. Genotype frequencies are often very close to the proportions predicted by the model. The predictions of the Hardy-Weinberg model show a good fit to the real world, even though the assumptions are not completely realistic.

CONNECTIONS

Mendel's theory of inheritance provided a way to explain how inherited variation could lead to evolution.

Changes in Gene Pools

16.4 Microevolution in Large Populations

Natural selection is one of the most important factors that change gene pools. One of the best-known examples of natural selection involves the English peppered moth (*Biston betularia*). The moth flies at night and rests on tree trunks during the day. Examination of museum collections has shown that in 1850, most *B. betularia* in central England had light-colored wings. There were only a few moths with dark wings. At that time, the gene pool had about 95% alleles for light color and 5% alleles for dark color. By 1900, the situation was reversed—almost all the moths were dark, and there were very few light moths. The gene pool of this species changed dramatically in 50 generations. Population geneticists hypothesized that natural selection and environmental change were responsible for this dramatic shift. Before 1850, the tree trunks were covered by light-colored lichens. Light-colored moths resting on light-colored tree trunks were well camouflaged (Figure 16.4a). This protected them from being eaten by birds. After 1850, industrialization created much pollution in central England. The pollution killed the lichens and exposed the dark tree trunks. Light-colored moths resting on dark tree trunks then became easy prey for birds (Figure 16.4b). In the new environment, the dark moths were better camouflaged. Natural selection changed the frequencies in the gene pool and helped the population adapt to its changed environment.

Another well-known example of selection is the sickle-cell polymorphism in human hemoglobin. Hemoglobin is the protein in red blood cells that carries oxygen throughout the body (see Section 7.9). Most people have normal hemoglobin, but there is an alternative form that differs by one amino acid. The amino-acid difference is caused by a change in one nucleotide in

> Try Investigation 16A ▽ Sickle-Cell Disease.

FIGURE 16.4

Natural selection in the peppered moth, *Biston betularia.* (a), The light form of the moth is camouflaged on light tree trunks but visible on dark tree trunks. **(b),** The dark form of the moth, called *carbonaria,* is camouflaged on dark-colored tree trunks but visible on light, lichen-covered tree trunks.

 CHEMISTRY TIP

Hemoglobin molecules contain iron, which gives blood its red color. The hemoglobin protein contains four polypeptides. In sickle-cell carriers (heterozygotes), hemoglobin molecules are made from either normal or mutant polypeptides.

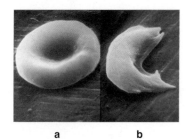

a   b

FIGURE 16.5

Red blood cells. Normal and sickle-cell human red blood cells, ×26000. **(a),** Normal cells retain their disk shape under low oxygen concentration. **(b),** The red blood cells of individuals who carry two copies of the sickle-cell allele collapse under low oxygen concentration and take on a sickle shape.

the DNA. Homozygotes for this alternative allele, called the sickle-cell allele, have serious health problems. Their hemoglobin does not do a good job of carrying oxygen to the tissues. Under certain conditions, their red blood cells become deformed. The deformed cells are sickle shaped instead of being the usual disk shaped (Figure 16.5). The sickle-cell homozygotes tend to be anemic, which means that they do not have enough red blood cells. In spite of its bad effects, the frequency of the sickle-cell allele is as high as 20% in certain regions of Africa. Why would such a harmful allele be so frequent?

The population geneticist J. B. S. Haldane was the first to suggest a reason for the high frequency of the sickle-cell allele in Africa. He noticed that the sickle-cell allele was most frequent in areas where people were often infected by parasites that cause malaria.

Haldane hypothesized that the sickle-cell allele is advantageous in certain environments because it protects carriers against malaria. Research has confirmed this hypothesis. Homozygotes for the normal hemoglobin allele have no anemia, but they can catch malaria. Sickle-cell homozygotes have anemia, but they are resistant to malaria. Heterozygotes who carry one normal allele and one sickle-cell allele are resistant to malaria and only slightly anemic. Natural selection favors the heterozygote where there is malaria and keeps both alleles in the population.

Natural selection is the main factor that changes the gene pools of large populations, but several other factors can also cause change (Table 16.2). For example, animals sometimes migrate great distances. If they are able to find mates in their new locations, then they can change allele frequencies in the local gene pools. They can even introduce new alleles to their new

TABLE 16.2
The Main Mechanisms that Affect Gene Pools

Mechanism	Effect on Gene Pool
Natural selection	Increased reproductive success leads to increased allele frequency; poor reproductive success leads to decreased allele frequency.
Genetic drift	Random changes in small populations
Gene flow	Change in the gene pool resulting from migration of individuals between populations
Mutation	Direct conversion of one allele to another

populations. The gametes of many plants and aquatic animals are also mobile and can influence distant gene pools. The term **gene flow** is used to describe the effects of migration between gene pools. When continued over many generations, gene flow between geographically separated populations tends to make the gene pools similar.

Mutation is another factor that can change allele frequencies. Mutations arise spontaneously at a very low rate. They usually result from slight errors during the replication of DNA. Most mutations have bad effects or no effects on their carriers, but some rare mutations are beneficial. The beneficial ones are favored by natural selection and gradually increase in frequency in the gene pool. For example, mutant alleles that provide resistance to common insecticides spread quickly through a population of houseflies.

CONNECTIONS
Gene flow and other broad patterns of biological change depend on specific local factors such as geographical barriers to migration.

CHEMISTRY TIP

DNA polymerases, the enzymes that replicate DNA, also have proofreading functions. Polymerases that do little proofreading cause a relatively high mutation rate.

Focus On
Founders of Population Genetics

The founders of population genetics were J. B. S. Haldane (1892–1964), Ronald Fisher (1890–1962), and Sewall Wright (1889–1988). All three had exceptional mathematical skills. They are remembered for their contributions to population genetics theory, which developed after the rediscovery of Mendel's work in 1903. Haldane and Fisher worked in Britain, while Wright (Figure 16.6) worked in America. Haldane developed much of the basic theory of natural selection and wrote books on biochemistry and political philosophy. Fisher concentrated on evolution in large population and emphasized the similarities between theoretical physics and population genetics. He also did important work in theoretical statistics. Wright was most interested in evolution in small populations subject to both genetic drift and natural selection. Wright also developed the basic theory of inbreeding.

FIGURE 16.6
Sewall Wright. The only American founder of population genetics, Wright is best known for his influence on the development of quantitative genetics and evolutionary biology.

16.5 Microevolution in Small Populations

All of the factors that affect the gene pools of large populations also affect small populations. In addition, small populations are subject to **genetic drift.** Genetic drift is random change in allele frequencies. Genetic drift can have a substantial effect in small populations but has little effect in large populations.

To understand why genetic drift affects small populations more than large populations, think of a coin-tossing experiment. If you toss a coin 1,000 times, it is fairly certain that there will be about 500 heads and 500 tails. It would be extremely unlikely to get only 300 heads and 700 tails. If you tossed the coin only 10 times, you might get 5 heads and 5 tails, but it is also fairly likely that there would be 4 heads and 6 tails, or even 3 heads and 7 tails. The 1,000-toss experiment will be close to the expected value, but the 10-toss experiment can be way off just by chance.

The same kind of chance effect can work in gene pools. Imagine a randomly mating population in which no selection operates. For large populations, the gene pool is sampled many times to produce the next generation of organisms. With hundreds or thousands of samples, it is likely that the allele frequencies in the progeny generation will be very close to the allele frequencies in the gamete pool. On the other hand, in small populations, the gamete pool is sampled only a few times. The allele frequencies in the progeny generation might be quite different from those in the gamete pool just by chance. In small populations, the allele frequencies jump around, changing randomly from generation to generation. In large populations, however, there is little or no change in allele frequencies from generation to generation.

Sometimes a small number of migrants starts a new population. For instance, Darwin's finches dispersed to the Galápagos Islands from the South American mainland (see Section P.5). They also dispersed among the islands. Sometimes a few migrants landed on islands where there were no other finches, and they started new populations. The gene pools of those new populations were sometimes quite different from the mainland populations just by chance. Genetic drift that influences new populations is called the **founder effect.** A new population can have allele frequencies that are very different from its source population if the number of organisms that establish the new gene pool is small (Figure 16.7).

In addition to causing random changes in allele frequencies, genetic drift in small populations causes a gradual loss of genetic variation in the gene pool. Genetic drift reduces the frequency of heterozygotes over time. The gradual increase in homozygosity is called **inbreeding.** Populations that are limited to only a few organisms every generation eventually become very inbred. For example, the California condor (Figure 16.8) is an endangered species that has been reduced to small numbers. Because there are very few of these animals, all matings occur between close relatives. The population has become highly inbred. In the most extreme case, if a population were limited to just two organisms each generation, then its whole genome would become completely homozygous in about 20 generations.

a

b

Large populations can also become inbred if there is a **population bottleneck.** A bottleneck means that the number of organisms is drastically reduced for a few generations. The American bison (*Bison bison,* commonly called a buffalo) went through a severe population bottleneck in the 19th century because of hunting. Now the bison is protected, and its population has increased, but the population bottleneck has left the buffalo populations partially inbred.

FIGURE 16.8

The California condor, ***Gymnogyps californianus.*** This endangered species has become inbred because of small population numbers. An adult condor typically has a wingspan measuring 2.5–3 m. Condors were once abundant in the western United States and northern Mexico, but by the early 1980s, there were only about 20 animals left in the wild. In 1987, all known California condors were captured and brought to zoos to prevent the extinction of the species. They have been successfully bred in captivity. There are now more than 100 California condors, some of which have been released into the wild in an attempt to reestablish natural populations.

There is evidence that the wild cheetah populations of south and east Africa (Figure 16.9) suffer from inbreeding depression. Males have sperm counts that are only one-tenth those of other cat species. About 70% of their sperm have abnormal shapes. Electrophoresis of cheetah proteins reveals very little genetic variation. In other mammalian species, including humans, the genes that control the immune system are highly polymorphic. Cheetahs are so similar in their immune genes that a typical animal can accept a skin graft from an unrelated cheetah. The recipient's immune system does not reject the donor's skin as foreign.

In captivity, cheetahs reproduce poorly and suffer high infant mortality. Because of their genetic uniformity, most cheetahs are susceptible to the same viruses. It is likely that cheetahs became highly inbred about 10,000 years ago as a result of a severe population bottleneck.

FIGURE 16.9

A cheetah, *Acinonyx jubatas.* Natural populations of cheetahs in Africa have gone through population bottlenecks. The reduction in population size led to inbreeding depression and a loss of genetic variation.

Inbreeding can have bad effects on populations. Most gene pools have some recessive alleles that are harmless as single copies in heterozygotes but are harmful or even lethal when homozygous. The typical human is estimated to have seven alleles that would be lethal if they were homozygous. Inbreeding increases the frequency of those harmful and lethal alleles. Because of the increased homozygosity of harmful recessive alleles, inbred populations suffer from **inbreeding depression,** which means that their fertility and survival are reduced compared with populations that are not inbred. Some isolated human populations are relatively inbred and have high frequencies of inherited diseases. Animals in zoos, endangered species, and certain registered breeds of dogs are subject to inbreeding depression because their populations are small. To minimize inbreeding, breeders often try to arrange matings between the most distantly related animals in managed populations.

There is one useful aspect of inbreeding. By mating brothers and sisters generation after generation, geneticists have produced highly homozygous lines of mice, fruit flies, and other common research organisms. Having genetically identical organisms makes it possible to repeat experiments with great accuracy. It also makes it possible for different laboratories to use the same experimental material.

16.6 Quantitative Traits

Many of the phenotypes of organisms that are important in microevolution are multifactorial traits that are influenced by more than one gene (see

Section 13.11). Multifactorial traits are also called **quantitative traits** because of the way they are measured. For example, body size is measured in grams or kilograms. Life span is measured in days, months, or years. In contrast, single-gene traits are usually of the either-or type. The color of English peppered moth wings is a single-gene trait. The wings are either light or dark. Their size, however, is a quantitative trait. Size is measured on a continuous scale and is affected by multiple genes. The genes that affect quantitative traits are called **quantitative trait loci** (singular: locus), or **QTLs.**

Like other kinds of genes, QTLs are variable in most populations. Evidence of this variation comes from **artificial selection** experiments. In these experiments, a breeder changes a plant or animal population by selective breeding. For instance, if a breeder wants larger chickens, only the largest hens and roosters are allowed to mate each generation. The medium-sized and small chickens are not allowed to breed. A few generations of controlled matings will increase the frequencies of alleles that cause larger body size and reduce the frequencies of the alternative alleles. Breeders have done hundreds of artificial selection experiments. The usual result is that artificial selection can change virtually any quantitative trait (Figure 16.10). The genetic variation in populations makes it possible for breeders to change their gene pools. This variation also allows populations to adapt to environmental changes.

Quantitative traits also vary in human populations. The most obvious evidence for this is that biological relatives look more like each other than like other non-related people. Relatives share alleles that influence height,

E T Y M O L O G Y

locus = place (Latin)

In genetics, a **locus** is a specific place in or part of a chromosome, such as a gene or the part of a gene where a mutation has occurred.

pouter

jacobin

rock dove

fantail

FIGURE 16.10

Artificial selection in pigeons. Pigeon breeders have produced a wide variety of shapes, colors, and behaviors through artificial selection. The ability of the population to respond to artificial selection indicates the presence of genetic variability in the gene pool. Variation among breeds was of great interest to Charles Darwin, who raised many varieties of pigeons on his estate in England.

Focus On

The Jackson Laboratory

The Jackson Laboratory is an independent, nonprofit research institution dedicated to improving human health, primarily through genetic research on mice. Founded in the 1920s, the laboratory is located on Mt. Desert Island, near Acadia National Park on the coast of Maine (Figure 16.11). This remote location was chosen to avoid contamination of valuable mouse stocks with city mice.

The Jackson Laboratory produces about 3 million inbred mice per year. Some of the inbred lines have been maintained by brother-sister mating for more than 100 generations. The inbred mice are sold to research labs all over the world. The proceeds from these sales and donations are used to support basic research at the Jackson Laboratory on the genetics of osteoporosis, aging, immunology, cancer, and other areas of study.

FIGURE 16.11

The Jackson Laboratory on Mt. Desert Island. This lab on the coast of Maine is a worldwide supplier of inbred mice for research and is the site of much research on the genetics of disease.

head shape, pigmentation, and other quantitative traits. Close relatives are most likely to have the same alleles and look the most similar. Distant relatives are likely to share fewer alleles and look less similar. Shared alleles and shared environments both contribute to similarities among relatives. Family members experience the same diet, health care, and housing conditions. They usually grow up with the same parents and attend the same schools. They acquire similar attitudes about exercise, social interaction, reading, and sports. Consequently, the similarity of family members is partly genetic and partly due to their shared environment.

The closest relatives are **identical twins** (Figure 16.12). Their entire genomes are identical. Pairs of identical twins arise from a single fertilized egg that splits into two embryos early in development. Fraternal twins share only half of their alleles, on average, and are no more closely related than ordinary brothers and sisters. Fraternal twins arise from two fertilized eggs that develop at the same time. Geneticists can compare the two types of twins to answer questions about the relative importance of genetic and environmental factors for various human quantitative traits. If the genes are more important for determining a phenotype, then identical twins will be more similar, on average, than fraternal twins. If the environment is more important, then identical twins will be about as similar as fraternal twins, on average. Twin studies have shown that for many anatomical traits, the genetic factors are more important. For many behavioral traits, environmental factors are more important.

Some scientists have proposed that variation in human intelligence depends mostly on genetic factors. These claims are controversial. Some of the studies that made these claims examined too few people to give accurate

FIGURE 16.12

Identical twins. The similarities and differences between twins give clues about the effects of genes and environment on human characteristics.

results. Other studies used only adopted identical twins who grew up in different homes. In principle, it is possible to estimate the importance of genetic factors by studying identical twins raised in different environments, but in practice, there are many problems with this kind of study. Even if they were adopted as children, the twins shared prenatal and early childhood environments. For instance, if the mother smoked or drank alcohol while she was pregnant, both fetuses would be affected. Also, the adopted homes of twins raised apart are sometimes not very different. Some of the raised-apart identical twins used in studies of intelligence grew up in the same neighborhoods and went to the same schools. As a result, the shared environmental factors tend to make the twins more alike, regardless of genetic effects.

Focus On

The Danish Twin Study

The government of Denmark maintains detailed records on all Danish citizens, including dates of birth, death, marriage, and the occurrence of twins. Researchers have used this information to study the degree of genetic determination of life span and other traits. Examining the records of several thousand deceased twin pairs showed that identical twins were slightly more similar in their life spans than were fraternal twins. Cases of infant mortality and childhood mortality were excluded from the analysis. The results suggest that about 20% of the variation in life spans is explained by genetic factors. The remaining 80% is explained by environmental factors.

QTL Mapping

One method of QTL mapping involves crossing two inbred lines that differ in a quantitative trait. Imagine that you have two inbred lines of *Drosophila* that differ in life span (Figure 16.13a). Inbred line 1 has an average adult life span in the laboratory of 75 days, while inbred line 2 has an average life span of 40 days. Your goal is to determine the chromosomal locations of the genes that cause the life-span differences.

Your first step is to find genes that are easy to study and have different alleles in the two lines. These genes are called markers. For instance, a DNA-sequence difference that is easily detected by restriction enzymes could serve as a marker. In a typical *Drosophila* QTL mapping experiment, 80–100 markers are used. The markers are scattered throughout the genome on all the chromosomes, and their chromosomal positions are known. Your second step is to cross the two lines to produce an F_1 generation (Figure 16.13b). The F_1 flies are crossed to produce F_2 flies and so on. Because of crossing-over, after a few generations each fly will have a mixture of line 1 and line 2 genetic material in each chromosome.

Your next step is to make two kinds of observations on the recombinant population. You determine the genotype of each fly at each of the markers and measure the fly's life span. The marker genotypes tell which parental line is the source of each chromosomal segment. Genes that are physically close on the chromosome are seldom separated by crossing-over. Therefore, if a life-span QTL and a particular marker are close to one another, then they will tend to be transmitted together. If there is a QTL that extends life span near, say, marker 4, then the marker allele and the QTL will tend to be transmitted together from inbred line 1 to all of the descendants that inherit that chromosome segment from line 1. As a result,

flies that have the marker 4 allele from inbred line 1 tend to live longer than recombinant flies that carry the marker 4 allele from line 2 (Figure 16.13c). This indicates that there is a QTL affecting life span near marker 4. Experiments of this sort have identified a few chromosomal segments carrying QTLs that extend life span. With similar mapping on a finer scale, it is possible to find the specific genes within a chromosome segment that alter the phenotype.

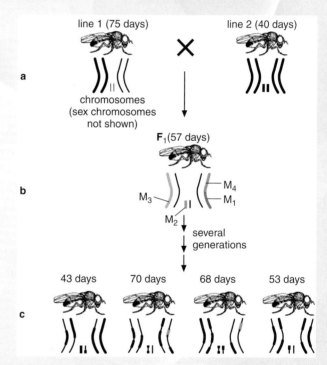

FIGURE 16.13

QTL mapping experiment. An experiment to find the chromosomal locations of genes (QTLs) that influence life span. **(a),** Repeated crosses between the descendants of a cross between long-lived fruit flies (line 1) and short-lived flies (line 2) produce flies with a mixture of genes from both lines. **(b),** Easily identified differences between the two lines in alleles of marker genes (M1–M4) indicate the origin of each chromosomal segment. **(c),** In this example, long-lived descendants have inherited the allele for marker gene M4 from long-lived line 1. The close relationship between this allele and life span indicates that M4 is located near a QTL that affects life span.

The most recent development in the study of quantitative traits is **QTL mapping.** This is a set of procedures for finding the chromosomal locations of genes that cause variation in quantitative traits. The procedures are similar to those used to map single genes but use many genetic markers scattered throughout the genome. In the near future, it is likely that QTL mapping studies will identify specific genes responsible for variation in many quantitative traits, including blood pressure, obesity, and tumor resistance.

Check and Challenge

1. Why is genetic drift most important in small populations?
2. What are the main forces that change gene pools? Explain how they work.
3. How is genetic drift like a coin-tossing experiment?
4. Are studies of twins more important in understanding single-gene traits or quantitative traits? Explain your answer.

Chapter
HIGHLIGHTS

Summary

Population genetics is the science of microevolution, which is defined as "change in the gene pools of local populations." Gene pools of sexually reproducing species with large populations tend to be highly genetically variable. Under the simplest circumstances, described by the Hardy-Weinberg model, gene frequencies are unchanging.

The main factors that cause change in gene pools are natural selection, genetic drift, mutation, and gene flow. Classic examples of natural selection include the color variants of the English peppered moth and the sickle-cell polymorphism in humans. Genetic drift is especially important in small populations. Drift causes random changes in allele frequencies and leads to increased homozygosity and inbreeding depression.

Quantitative traits are influenced by many genes, called QTLs. Quantitative traits can be altered by artificial selection. Human populations are genetically variable for many quantitative traits. The relative roles of genes and environments in determining human traits can be estimated by twin studies. Variation in anatomical traits is explained mostly by genetic factors, whereas variation in behavioral traits is explained mostly by environmental factors. The specific genes that influence quantitative traits can be identified using QTL mapping.

Key Concepts

Construct a concept map of this chapter using the following terms. Add linking words and other terms from the chapter as needed.

- artificial selection
- gene flow
- genetic recombination
- mutation
- QTL
- genetic drift
- twin study
- natural selection
- quantitative trait
- genetic variation
- gene pool
- inbreeding

Reviewing Ideas

1. What is a gene pool?
2. Describe two techniques that can be used to detect genetic variation.
3. What are the assumptions of the Hardy-Weinberg model?
4. What is natural selection?
5. What are some of the similarities and differences between artificial and natural selection?
6. What are the factors that affect survival of the English peppered moth?
7. Why is the sickle-cell allele present at high frequencies in certain regions of Africa?
8. What is the original source of all genetic variation?
9. What is genetic drift?
10. Describe the founder effect, and explain how it works.
11. What is an inbred line?
12. What are the two main factors that cause individuals of the same species to appear different?
13. What is the difference between identical twins and fraternal twins?

Using Concepts

1. What do you think would happen to the frequency of the sickle-cell allele in sub-Saharan Africa if malaria were eliminated?
2. Which gene pool do you think would have a higher degree of heterozygosity: a self-pollinated or an open-pollinated plant? Explain your answer.

 Biology Online BSCSblue.com/vocabulary_puzzlemaker

3. Many species of animals avoid mating with close relatives. Explain this behavior as an adaptation that helps the species survive.
4. If you were managing a captive population of an endangered species, such as the California condor, what would you do to increase its numbers?
5. Why are founder effects most visible on islands?
6. Explain why populations evolve although organisms within the population do not.

Synthesis

1. Explain how modern ideas about microevolution combine the ideas of Charles Darwin and Gregor Mendel.
2. How has the development of modern techniques of molecular biology influenced the study of microevolution?
3. Use what you have learned in this chapter to explain why microorganisms can adapt to environmental changes more quickly than vertebrates.
4. How does population size and degree of gene flow influence a gene pool?

Extensions

1. Write a letter to Charles Darwin, and tell him about how modern ideas in genetics relate to his theory of evolution by natural selection.
2. Suppose a pair of parakeets arrives on an uninhabited island in the Pacific Ocean. Write a short story or poem about the traits of the pair and their descendants on the island.
3. Make up an imaginary environmental change, and suggest three ways that populations might adapt to the change.

Web Resources

Visit BSCSblue.com to access
- Information about the Hardy-Weinberg model, molecular evolution, museum exhibits, scientific societies, and scientific journals on evolution and population genetics
- Web links related to this chapter

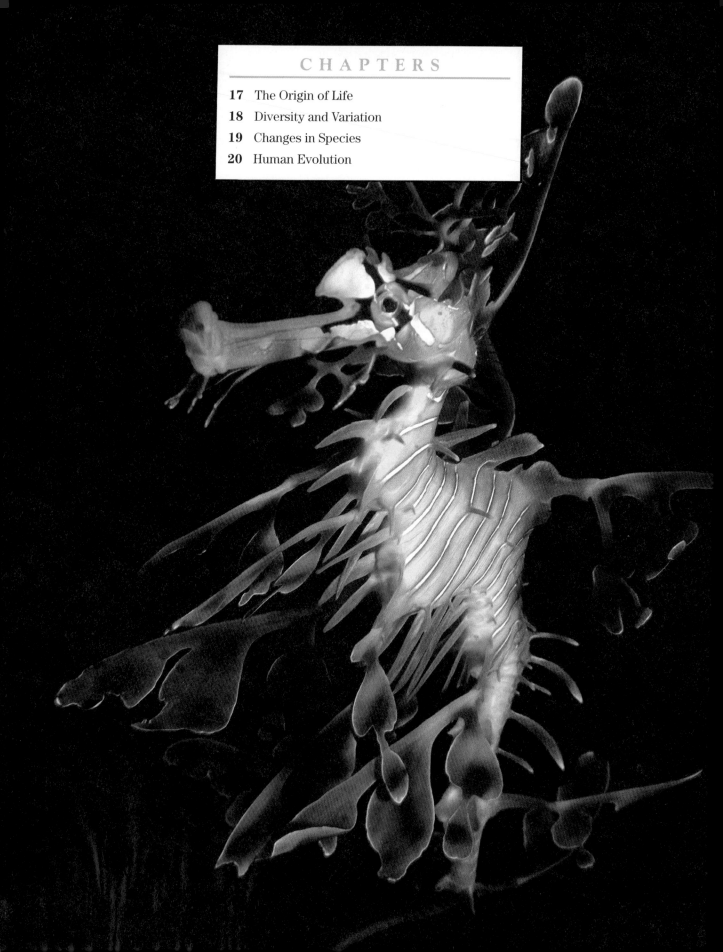

About the photo
This photo shows a leafy sea dragon (*Phycodurus eques*).

UNIT
4

Evolution

■ *How could you decide which organisms are related to this one?*

■ *How can you account for the differences between this organism and its relatives?*

Early in the history of Earth, the first living things developed. Today, their descendants include millions of types of bacteria, plants, animals, and other organisms. Evidence from fossils shows that at least as many organisms have become extinct. Some fossils closely resemble living organisms. Others seem bizarre in comparison to familiar organisms and probably have no living descendants.

How did all this diversity arise? What maintains it? What determines which organisms survive and how they differ from their ancestors? These are some of the questions you will study and explore in this unit.

The many types of organisms differ not only from each other; there is also great variation among individuals in populations of the same type of organism. Differences among dogs, humans, or oak trees are easy to see. Variation in populations of bacteria, fungi, and other organisms also exists. However, these distinctions are not always obvious without chemical analysis of DNA or other cell components. When and how do differences among individuals in a population become so great that they are no longer all the same species or type of organism? This unit will help you answer this important question in the study of evolution.

Biology Online BSCSblue.com/webquest

437

CONTENTS

LEARNING OUTCOMES

By the end of this chapter you will be able to:

A   Describe the origin of the universe and probable conditions on early Earth.

B   Evaluate hypotheses about the origin of life and identify the probable characteristics of early life-forms.

C   Distinguish between chemical and biological evolution.

D   Describe the fossil record for prokaryotes and for eukaryotes.

◀ **About the photo**
This photo shows
stromatolite structures
formed by cyanobacteria,
in Shark Bay, Australia.

CHAPTER

17

The Origin of Life

■ *What is required for life to begin?*

■ *How did life arise on Earth?*

Of all the questions scientists investigate, the origin of life on Earth probably evokes the most public curiosity and controversy. Where did life come from? When did it begin? How did the first living things look? These questions are interesting but difficult to answer. There are no witnesses or direct evidence to investigate. But it is possible to study the questions and find plausible answers. Scientists cannot always investigate the distant past by experiments and direct observations. It is not possible, for example, to repeat the formation of the solar system, the extinction of the dinosaurs, or the evolution of humans.

Science does, however, successfully investigate events that occurred long before there were humans to observe and record them. Astronomers, for example, use tools such as the Hubble Space Telescope to develop reliable explanations for the origin of the universe and the formation of stars, galaxies, and planets. By examining deposits in Earth's crust and by creating computer simulations of a collision between a large meteor and Earth, scientists have produced strong evidence that a meteor may have killed off the dinosaurs. Anthropologists and molecular biologists have used fossil remains and DNA analysis to reconstruct many aspects of human evolution. It is possible, then, to develop reliable inferences about ancient events such as the origin of life. Used carefully, inference can help scientists draw sound conclusions about things that they cannot observe or measure directly.

FIGURE 17.1

The Hubble Space Telescope.
This robotic telescope, launched in
1990, orbits 600 km above Earth. It
has allowed astronomers to look
deeper into space than is possible
with telescopes on Earth's surface.
Observations with the Hubble and
other space telescopes have
confirmed the hypothesis that the
universe is expanding.

The Origin of Earth

17.1 The Big Bang

Life on Earth cannot be older than Earth itself. What does science tell us about the origin and age of Earth, and about the universe in which it resides? Measurements of light coming from deep space indicate that the stars and galaxies were much closer to each other in the past than they are now. They are still moving rapidly away from each other.

How do astronomers know that the universe is expanding? In the 1920s, Edwin Hubble, the scientist for whom the Hubble Space Telescope (Figure 17.1) was named, studied light from distant galaxies to test the idea that the size of the universe is stable. Light consists of waves (see Figure 4.2, page 103). The length of any wave changes if the object that is producing it moves toward or away from the object receiving the wave. For example, think about the sound waves produced by the siren on an ambulance. If the ambulance moves away from you, the sound waves spread out (the wavelength increases), and the siren has a lower sound. If the ambulance moves toward you, the sound waves are compressed (the wavelength decreases), and the sound is higher. Light waves behave the same way. When a source of light moves away from the observer, the wavelength of the light increases and the visible light is a bit redder. This is called a redshift. (When the source of light moves toward the observer, the wavelength of the wave decreases, and the visible light is a bit bluer—a blueshift.)

Hubble measured the wavelengths of light from distant galaxies (Figure 17.2) and discovered that they were consistently shifted toward

FIGURE 17.2

Spiral galaxy. Photographed by the Hubble Telescope, the galaxy NGC 4414 is 60 million light-years away. Measurements of light from such distant galaxies provide evidence that the universe is expanding.

the red end of the spectrum. He concluded that those objects were moving away from Earth. Hubble also found that distant objects were moving away more quickly than closer objects. His theory of the expanding universe has been confirmed by repeated observations.

Because scientists can calculate the rate of the universe's expansion, they can work backward to determine the time at which the universe was much smaller. Evidence indicates that, about 15 billion years ago, the whole universe was concentrated in one superdense mass that exploded. That explosion, called the big bang, hurled matter and energy into space. Gravity pulled some of the matter together to form galaxies and stars. Gravity also pulled matter into orbit around stars. The clumps of matter around our own star, the Sun, became planets. Some planets were large enough to have enough gravity to attract smaller masses that became moons or other satellites.

The best current estimate is that Earth formed about 4.6 billion years ago. Meteorites and the oldest rocks from the Moon confirm that estimate. Analysis of rocks brought back from the Moon indicates that it formed at about the same time as Earth. The Moon may have formed when a meteor collided with Earth, sending a large chunk of Earth into space. Meteors are thought to be bits of material left over from the formation of our solar system. As illustrated in Figure 17.3, geologists have worked out a history of Earth from evidence in its rocks.

era	million years ago	first evidence of
Cenozoic	7-5	humanlike apes
	65	primates
Mesozoic	140	flowering plants
	220	mammals
	235	dinosaurs
Paleozoic	300	reptiles
	360	amphibians
	400	land animals
	430	land plants
	520	vertebrates
Precambrian	2,100	eukaryotes
	2,500	free oxygen released into atmosphere by prokaryotes
	3,500	prokaryotes
	4,600	formation of Earth

FIGURE 17.3

Geologic timescale. This timeline relates several major biological events to the history of Earth.

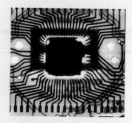

Biological Challenges

TECHNOLOGY

Using Radioactivity to Date Earth Materials

Accurate dating of rocks and fossils provides important information about the history of Earth and its organisms. Scientists use the decay of radioactive isotopes as a "clock" to determine the ages of rocks and fossils.

Relatively young fossils or other organic material are dated by measuring the decay of radioactive carbon-14 (Figure 17.4). The ratio of carbon-14 to carbon-12 in dead material indicates how long it has been since the organism died. A limitation of the technique is that the half-life of carbon-14 is only 5,730 years. If a specimen is 57,300 years old, 10 half-lives have elapsed. As a result, so little of the original carbon-14 remains that it is difficult to measure accurately, thus the method is only applicable to materials that are younger than about 50,000 years old.

Most rocks contain little carbon and are too old to date with carbon-14. These rocks and the fossils they contain must be dated with isotopes that decay more slowly. Potassium-40 has a half-life of 1.3 billion years. Potassium-argon (as well as argon-argon) dating methods have been very useful in determining the age of volcanic rocks. The ages of fossils in sedimentary rocks sandwiched between layers of volcanic

deposits can be determined once dates for the surrounding volcanic rocks have been established. Potassium-argon techniques are one of several methods that NASA scientists used to determine the age of rocks from the Moon.

Another key radioactive "clock" depends on the decay of uranium into lead. When magma (molten rock) solidifies, traces of uranium are trapped in crystals of the mineral zircon (zirconium silicate, $ZrSiO_4$). As time passes, uranium-238 decays into lead-206 in the zircon. Measurement of lead isotopes in rocks can provide ages with an error of less than 1%. Dating methods with uranium are now used in geologic systems from only thousands of years old to materials as old as the solar system (4.6 billion years). Overall, the uranium-lead dating system has been the most important method for determining the ages of rocks, calibrating the history of life on Earth, and measuring when the solar system formed.

Sophisticated equipment and accurate methods are required for dating rocks and minerals. The accuracy of ages is confirmed by using a method properly and by reproducing the results. Using two or more dating methods on the same rock or mineral is another way to test and confirm the initial results.

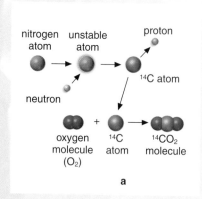

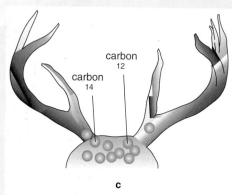

FIGURE 17.4

The basis of carbon-14 dating. (a) A small fraction of the carbon dioxide in the atmosphere contains carbon-14 formed by cosmic rays colliding with nitrogen atoms. **(b)** Plants and other autotrophs absorb some of the radioactive carbon-14 and incorporate it into their tissues. Animals and other heterotrophs absorb the carbon-14 by consuming autotrophs. **(c)** When an organism dies, its carbon-14 (red atoms) continues to decay at a known rate, decreasing in amount relative to carbon-12 (blue atoms). The amount of carbon-14 remaining can be measured to determine the approximate age of the organism's remains.

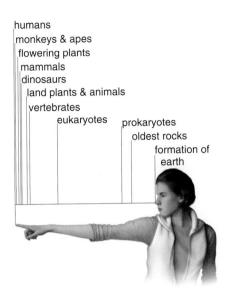

FIGURE 17.5

Earth's history. The history of Earth—and life—compressed into the distance between the tip of a person's nose and the tip of the fingernail on the index finger. The history of our species, *Homo sapiens,* would disappear with one pass of a nail file.

Figure 17.5 compares the 4.6 billion years of Earth's history to a timescale based on the distance between the tip of a person's nose and the end of the nail on the index finger. Note where the first signs of life appear and where humans first appear.

17.2 Early Earth

Plenty of evidence indicates that Earth's interior is hot. Miners, for example, encounter higher temperatures deep underground. Volcanoes and hot springs are also evidence of Earth's hot interior (Figure 17.6). The decay of radioactive elements such as uranium (U), thorium (Th), and some isotopes of potassium-40 is the primary source of heat energy within Earth. The entire planet was probably hot when it first formed. As the outer crust cooled, gases escaping from the planet's hot interior formed a primitive atmosphere.

The early atmosphere must have been involved in the origin of life. For this reason, the composition of that atmosphere has been the topic of much experimentation and debate. It likely originated from volcanic gases: mostly nitrogen (N_2), carbon dioxide, water vapor, and hydrogen (H_2), with small amounts of carbon monoxide (CO). Early hypotheses suggested that the young atmosphere included ammonia (NH_3) and methane (CH_4), but later sunlight was found to rapidly break down ammonia into nitrogen and hydrogen, so there was probably very little of it.

All geologic evidence indicates that oxygen gas (O_2) probably was not present in the early atmosphere. Thus the first organisms could not have been aerobic. Early on, oxygen was bound in compounds such as water and carbon dioxide. Oxygen began to accumulate in the atmosphere only after the first photosynthetic organisms started to produce it, about 2.1–2.4 billion years ago.

CHEMISTRY TIP

Atoms of radioactive isotopes release energy to their environment as they spontaneously break down into the smaller atoms of other elements.

FIGURE 17.6

Eruption of a volcano in Hawaii. The molten lava that pours from such volcanoes is evidence of Earth's hot interior. Volcanoes produce a mixture of gases that is probably similar to the mixture in Earth's early atmosphere.

At first, the oxygen released by photosynthesis would have combined with other elements, such as iron, instead of remaining free in the atmosphere. Oxygen-containing compounds such as iron oxides are present only in small quantities in the oldest rocks and did not become abundant until much later in Earth's history (see Figure 17.3). The atmosphere probably reached modern levels of oxygen about 360 million years ago as plants became abundant on land.

Check and Challenge

1. How can scientists investigate processes or events that occurred in the very distant past?

2. What was the big bang? What evidence supports the theory of the expanding universe?

3. What evidence supports the conclusion that Earth has a hot core?

4. What gases were probably in Earth's early atmosphere? What is the evidence?

5. Explain why free oxygen gas probably was not in the early atmosphere of Earth.

Evolution of Life on Earth

17.3 The Beginnings of Life

The surface of Earth 4.6 billion years ago would have been hostile to modern life. First, organic compounds do not form easily in an atmosphere rich in nitrogen and carbon dioxide. Second, ultraviolet radiation from the

Biology Online BSCSblue.com/check_challenge

Sun bathed the surface of Earth. Ultraviolet radiation is particularly harmful to DNA. Today, a layer of ozone (O_3) in the atmosphere (Figure 17.7) protects life from ultraviolet radiation. The ozone layer formed only after about 1 billion years of photosynthesis added oxygen gas to the atmosphere. Ultraviolet radiation converted some of this oxygen to ozone. But when life began, the atmosphere had little or no oxygen, (O_2).

How could life have originated in this chaotic environment of volcanoes, intense radiation, extreme temperature variations, and scarce oxygen? Popular scientific explanations include the following:

1. *Life originated on some planet of another star and traveled to Earth through space.* This idea does not explain life's origin. If life came to Earth from someplace else, we still must explain how it originated in that other location. It is very difficult to investigate that possibility.

2. *Life originated by unknown means on Earth.* The earliest life may have been very different from today's organisms. Because no such remains have been found of those early organisms, their detection is extremely difficult. Like explanation 1, this explanation is very difficult to investigate.

3. *Life evolved from nonliving substances through interaction with the environment.* Organisms are chemical systems that maintain themselves and undergo biological evolution. These characteristics of organisms originated from complex chemical changes through time, or chemical evolution. Because the chemical principles at work then still apply, we can study those changes now, in nature and in the laboratory.

CHEMISTRY TIP

Oxygen molecules (O_2) in the stratosphere (the upper atmosphere) absorb solar ultraviolet radiation. This energy breaks the bonds in the oxygen molecule: $O_2 \rightarrow 2\ O$. Free oxygen atoms then react with other oxygen molecules to form ozone (O_3): $O + O_2 \rightarrow O_3$. When ozone molecules absorb ultraviolet light, the process is reversed. By intercepting this component of solar radiation, the ozone layer in the stratosphere protects organisms from its hazardous effects, including damage to DNA that can result in genetic mutations and cancer.

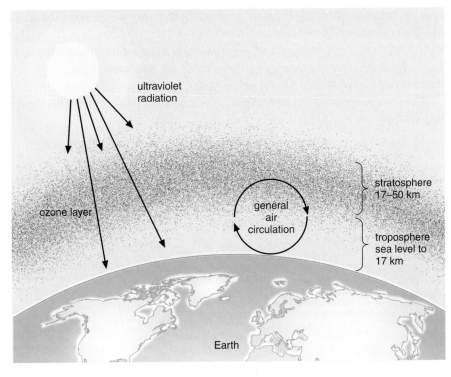

FIGURE 17.7

Earth's atmosphere. Ozone (O_3) in the stratosphere absorbs most of the ultraviolet radiation from the Sun. Early Earth did not have this protection.

CONNECTIONS

The dynamic, spirited exchange of ideas in research on the origin of life is an example of the investigative nature of science.

Each of these explanations for the origin of life has both advocates and critics. The third explanation is the most suited to scientific investigation because it can be stated in the form of a hypothesis. Section 17.4 describes some of the experiments designed to test that hypothesis.

Many people believe that a supernatural force or deity created life. That explanation is not within the scope of science. That is, we cannot use the methods of science to confirm or reject it. Therefore, such beliefs are not part of scientific debates about the origin of life.

17.4 Chemical Evolution

Most scientific research on the origin of life has focused on the third explanation from the previous section: Life evolved from nonliving substances by interacting with the natural environment. In the 1920s, the Soviet scientist Alexander Oparin and the British scientist J. B. S. Haldane separately described this hypothesis in detail. Oparin and Haldane started with the idea that the early atmosphere consisted of methane, ammonia, hydrogen, and water vapor. According to their hypothesis, energy sources such as radioactivity, lightning, cosmic radiation from space, and heat energy from volcanoes caused gases in the atmosphere to react, forming organic compounds. Those compounds then accumulated in the oceans, which came to resemble a hot soup. Figure 17.8 shows how such chemical evolution might have occurred.

Oparin and Haldane hypothesized that life evolved by further chemical reactions and transformations in the complex organic soup. They differed over the details of how life actually evolved. Both, however, thought the first life-forms were heterotrophs that fed on the organic compounds in the soup. With this food supply and few competitors, early autotrophs would have had little advantage over heterotrophs. These ideas are the basis of the **heterotroph hypothesis** for the origin of life.

The Oparin-Haldane version of the heterotroph hypothesis requires three major steps for the origin of life:

1. There had to be a supply of organic molecules, produced by nonbiological processes.
2. Some processes had to assemble those small molecules into polymers such as nucleic acids and proteins.

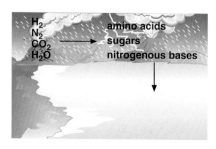 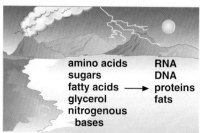 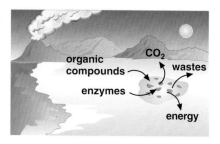

FIGURE 17.8

The heterotroph hypothesis. This hypothesis assumes that life processes developed gradually. The first organisms were heterotrophs that lived on the organic compounds in the oceans. Autotrophs evolved later, and photosynthesis produced today's oxygen-rich atmosphere.

3. Other processes had to organize the polymers into a system that could replicate itself, using the organic molecules produced in step 1.

Although there are only three main steps, each is a major change in moving from nonliving to living. How might these steps have occurred?

The first step requires organic compounds to form in an inorganic environment. Is that possible? As early as the 1820s, chemists had synthesized the organic substance urea (CH_4N_2O) from the inorganic compound ammonium cyanate (NH_4CNO). But that was in the laboratory. Could this happen in a natural setting?

Harold Urey and Stanley Miller worked on this problem in the 1950s at the University of Chicago. In laboratory experiments, they recreated conditions that might have existed on Earth 4.6 billion years ago. Miller built and sterilized an airtight apparatus (Figure 17.9). In the apparatus, methane, hydrogen, and ammonia gases circulated past an electric spark. A container of boiling water in the apparatus added hot water vapor. As the water vapor circulated, it cooled and formed "rain." Thus Miller created some of the conditions (hot gases, rain, and lightning) that might have been present in the early atmosphere.

While circulating the gases for a week, Miller examined the liquid in the apparatus. There were two visible changes. First, there was a small mass of black tar in one part of the apparatus. Second, the water turned red.

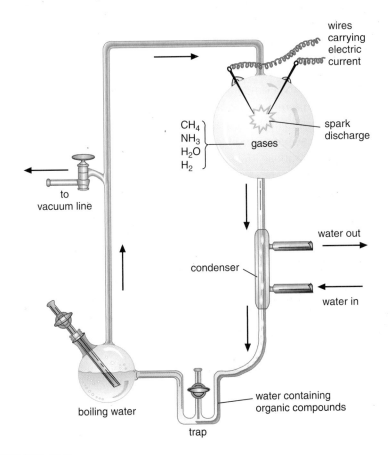

wires carrying electric current

CH₄
NH₃
H₂O
H₂

gases

spark discharge

to vacuum line

water out

condenser

water in

boiling water

water containing organic compounds

trap

FIGURE 17.9

A drawing of Miller's apparatus. Miller used this equipment to recreate conditions thought to exist in Earth's primitive atmosphere. What was he trying to find out?

FIGURE 17.10

The search for organic compounds from space. Cyril Ponnamperuma **(a)** and his associates were the first to detect amino acids in a meteorite when they analyzed a sample of the Murchison meteorite **(b).**

a

b

Chemical tests showed that many compounds, most unidentifiable, had accumulated in the water. Some of the gas molecules had combined to form new and more complex molecules.

When Miller and Urey analyzed the liquid, they found many organic compounds, including some amino acids. That was an exciting discovery, because amino acids are essential to every known life-form.

More recent evidence indicates that only small amounts of methane and ammonia were present in the early atmosphere. Miller and others repeated his experiment using only carbon dioxide, water, nitrogen, and hydrogen. The new system produced simple amino acids, such as glycine, as long as oxygen was absent and the ratio of hydrogen to carbon dioxide was the same. After those later experiments, Miller noted that at least a little methane is needed to produce more complex amino acids. More recent experiments have produced at least 13 of the 20 common amino acids found in proteins. Those experiments also have produced all the bases found in DNA and RNA. Even small amounts of ribose, the sugar in RNA, have been produced.

Perhaps some organic compounds arrived on Earth from space. Evidence from meteorites indicates that organic compounds do form in space. In 1970, at the Ames Research Center in California, Cyril Ponnamperuma and others found amino acids and other organic compounds in a meteorite that fell to Earth near Murchison, Australia, in September 1969 (Figure 17.10). Of the seven amino acids found, two do not occur in organisms on Earth. Since 1970, organic compounds have been found in many other meteorites. In 1986, several spacecraft flew through the tail of comet Halley (Figure 17.11) when it passed near Earth. Their instruments detected several organic compounds, including formaldehyde (CH_2O). This compound is needed for the formation of ribose, the sugar in RNA.

Some organic molecules also may have formed at volcanic vents deep in the ocean (Figure 17.12). These vents, which are actually cracks in Earth's crust, release gases from Earth's interior at very high temperatures. Some biochemists think that simple organic molecules might have formed in these vents first, rather than on the surface of the planet.

From Miller's work, the study of deep-sea vents, and the discovery of organic compounds in meteors and comets, we can infer that organic compounds could have formed on early Earth or come to Earth from space.

FIGURE 17.11

Comet Halley, seen on March 8, 1986. Instruments flown through the comet's tail detected organic compounds. The blue tail at left indicates the presence of carbon monoxide (CO).

In either case, there is a plausible explanation for step 1 as we test the Oparin-Haldane hypothesis about the origin of life.

With a reasonable explanation for step 1 in the Oparin-Haldane hypothesis, we now must explain step 2—the formation of more complex molecules such as nucleic acids and proteins from the organic materials in the oceans. For this to happen, the smaller molecules must become concentrated enough that they are likely to collide and react with each other. These reactions can form larger polymers. The small molecules could have become concentrated in a solution, or on a surface to which they could stick.

In 1985, A. G. Cairns-Smith, of the University of Glasgow, Scotland, proposed an interesting hypothesis to explain the second possibility. He suggested that clay particles might have helped to form the first organic polymers. Clay consists of tiny, layered crystals with a specific, repeating structure (Figure 17.13) that can attract and concentrate certain molecules. Ionized amino acids and other small, charged organic molecules could have bound to the repeated pattern of silicon, iron, and oxygen atoms on the surfaces of such crystals. The crystals could then have catalyzed the bonding together of these small molecules concentrated on their surfaces to form proteins and other polymers.

Many questions remain, but the work by Cairns-Smith and others since allows us to infer that polymers could have formed from simple organic compounds in the prebiological soup. This means we have a plausible explanation for step 2 in the Oparin-Haldane hypothesis.

Self-replication, a central feature of living systems, is the third and final step for the origin of life according to the Oparin-Haldane hypothesis. This part of the hypothesis presents an interesting "chicken-and-egg" problem. DNA is the universal information molecule for all life on Earth. Replication of DNA, however, requires proteins. And the structure of proteins is encoded

FIGURE 17.12

A deep-sea vent at a mid-ocean ridge in the Atlantic Ocean. The hot gases from Earth's interior that escape through these vents may have produced organic materials central to the origin of life.

CONNECTIONS

Evolution is central to the heterotroph hypothesis. As heterotrophs exhausted the supply of nutrients in the ocean, natural selection would have favored autotrophic life-forms able to use other sources of energy, such as inorganic compounds and light.

FIGURE 17.13

Structure of clay. Repeating crystal structures in clay particles (scanning electron micrograph). Crystalline clays may have served as a blueprint for assembling the first biochemical polymers.

by DNA. It seems impossible, therefore, to produce proteins (the "chicken") without DNA (the "egg"), and vice versa. Experiments with RNA have produced a possible answer to this problem. This work has led to the hypothesis that life began in an "RNA world."

How is RNA related to self-replication—step 3 in the Oparin-Haldane hypothesis? As we have seen, the components of RNA can be produced nonbiologically. They also can join spontaneously. Some RNA molecules can catalyze their own partial replication. In the laboratory, scientists have made RNA assemble short strings of bases that are a halfway point to self-replication. If RNA can be shown to reproduce itself completely, we will have an answer to the chicken-and-egg problem of evolving an information molecule and a catalytic molecule at the same time. This would provide a very plausible explanation for step 3. According to the RNA-world hypothesis, RNA served as both information molecule and catalyst at first. Later, DNA became the main information molecule and protein enzymes the primary biological catalysts. So far, scientists have found no RNA that can replicate itself completely. Much more work is necessary to determine whether RNA could have been the first self-replicating polymer.

RNA molecules can also undergo simulated Darwinian evolution in the laboratory (Figure 17.14). As with all evolution, the process begins with variation in a population. The experiment starts with a collection of many

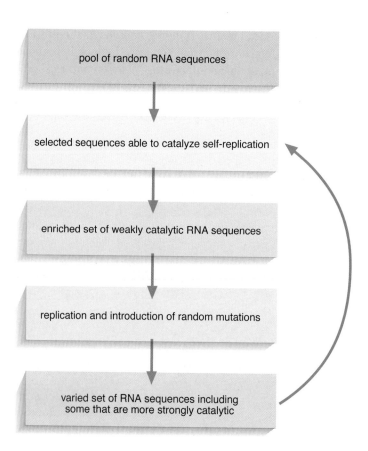

FIGURE 17.14

Directed evolution of RNA molecules. Beginning with a mixture of RNA molecules with various sequences, directed evolution allows a scientist to develop a ribozyme (RNA enzyme) that can catalyze its own replication or another desired reaction. Think about ways in which the evolution of RNA could have led to the first living systems.

Focus On

Catalytic RNA

For decades, scientists thought that proteins were the only catalysts in cells. In the early 1980s, however, biologists learned that RNA is the catalyst in a chemical reaction in some single-celled eukaryotes. Additional experiments found that RNA can catalyze the same reaction in other cells. In this reaction, RNA cuts and rejoins itself in the absence of any of the protein enzymes that normally carry out those reactions. RNA molecules that act as enzymes have been given the name **ribozymes.** Some ribozymes catalyze changes in other molecules, while others act on themselves. Because a ribozyme that acts on itself is changed in the reaction, these RNAs are not true catalysts.

The ability of an RNA molecule to serve as a catalyst depends on its three-dimensional structure (Figure 17.15). That structure results from the sequence of bases in the RNA. In the example shown in Figure 17.15, the ribozyme has two sections that can bind to complementary sections of a substrate RNA. In the substrate RNA, however, the two base-pairing sections are slightly closer together than the matching sections of the ribozyme. When the two RNA molecules bind together, the substrate molecule is stretched until it breaks apart.

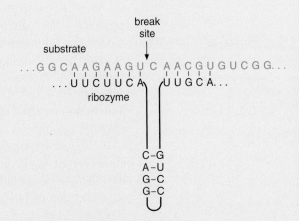

FIGURE 17.15

The structure of a ribozyme. Ribozymes are RNA molecules that act as enzymes. The ability of an RNA molecule to act as a catalyst depends on the molecule's three-dimensional structure. The linear sequence of the bases determines the molecule's shape. A double-stranded region at least three base pairs long is needed to stabilize the loops that give the ribozyme its catalytic structure. Note that the substrate RNA in blue has one more nucleotide (C) than the ribozyme between the two base-paired regions. This strains the structure of the substrate molecule so that it breaks at this point.

different RNA molecules in a solution of RNA nucleotides. The investigator mimics natural selection by allowing the RNA molecules to replicate themselves, if they can. Some of the RNAs may join the nucleotides in the solution into duplicates of their own structures. After some time has passed, those RNA molecules with shapes well suited to this task will be more numerous. They have been "selected" from the original population of various shapes. Scientists have used such simulated evolution to produce RNA molecules that catalyze specific, desired reactions. This process is similar to the use of artificial selection to produce plants and animals with desired traits.

The discovery that RNA can act as an enzyme reminds us that biology is an experimental science. It is a process of discovery, not a body of facts. The ideas and hypotheses provided in this and other textbooks are examples of knowledge that can change as research produces new information. The discovery that RNA is a catalyst in cells reminds us that science is a constantly evolving attempt to understand how nature works.

> Try Investigation 17A ◇ Molecular Evolution in the Test Tube.

CONNECTIONS

Ribozymes demonstrate the importance of structure to biological function.

17.5 Biological Evolution

Try Investigation 17B ▽ Coacervates.

The boundary between chemical evolution and biological evolution is not easy to define. Some specialists on the origin of life think of the formation of self-replicating polymers as marking the end of chemical evolution and the beginning of biological evolution.

Chapter 1 discusses the characteristics of life, but what is life? There is no single agreed-upon definition, but NASA's Exobiology Program, which studies and searches for life on other planets, has adopted the following working definition: "'Life' is a self-sustained chemical system that is capable of undergoing Darwinian, or biological, evolution." Thus the origin of life was the beginning of biological evolution.

Biological evolution consists of three processes: self-reproduction, mutation that can be inherited, and natural selection. In other words, for something to be considered alive, it must be able to reproduce itself, acquire mutations that can be passed on to its offspring, and be subject to natural selection. All life on Earth meets those requirements.

Life on Earth also consists of cells. Cell theory holds that all life is made of cells and that all cells come from preexisting cells. How did cells arise? To answer that question, we have to explain the evolution of membranes, which separate the contents of the cell from the external environment. Scientists have proposed that cell membranes provided an advantage for early life-forms as they competed for survival. For example, the plasma membrane protects DNA and RNA from agents that might harm them. In addition, as you learned in Chapter 3, membranes help establish the concentration differences required for chemical reactions such as ATP synthesis to occur in cells.

Just explaining the advantage of a structure does not, however, tell us how it might have evolved. The origin of cells and membranes is still not clearly understood. It is not even clear at what stage in the origin of life cells arose. Perhaps some reactions required for chemical evolution could only occur within cells. Or perhaps cells arose after nucleic acids or other polymers began to replicate themselves.

Many people have wondered about the origin and structure of the first cells. In the early 1980s, Carl Woese, at the University of Illinois, proposed a model for the formation of the first cells. Woese suggested that life began on Earth before the planet was fully formed. High levels of carbon dioxide in the atmosphere caused a strong, warming "greenhouse effect." Conditions on Earth were similar to those that exist now on Venus: That planet's atmosphere is mostly carbon dioxide, which retains heat energy like a greenhouse. Meteorites that hit primitive Earth provided large quantities of dust that were spread by the wind. Woese suggested that water vapor condensed around the dust particles, forming droplets. Each droplet, warmed by the atmosphere, then acted as a primitive cell in which chemical reactions occurred that allowed life to begin. Other scientists have proposed their own models of how the first cells formed. We are still a long way from having a complete picture of the process.

Besides cells, other possibilities for the first life-forms include proteins, DNA, and RNA. Amino acids in the oceans may have joined spontaneously to form proteins (Figure 17.16). Perhaps genetic information was encoded in the amino-acid sequence of a protein. Many proteins are enzymes that catalyze the synthesis of other substances. If there is a way for proteins to replicate themselves without the help of RNA, however, it has not been discovered.

Could DNA or RNA have been the first life-form? As we have seen, short strands of RNA can self-replicate. RNA also stores information, directs the synthesis of proteins, and has limited ability as a catalyst. In certain viruses, such as the virus that causes AIDS, RNA also directs the synthesis of DNA. The experiments of Miller and others have shown how the components of RNA nucleotides may have formed from atmospheric gases and simple organic compounds present in meteorites and comets.

A few scientists maintain that the first life-forms could have functioned only if they contained both proteins and nucleic acids. Perhaps the first life-form was similar to a virus. Viruses consist of DNA or RNA surrounded by a protein coat (Figure 17.17). They form crystals (Figure 17.18) and can synthesize proteins, use energy, and reproduce only by taking control of living cells. Viruses are so dependent on their host cells that they probably originated after their hosts. Until more data become available, their role in evolution remains unclear.

All the ideas about the origin of life and the nature of the first life-forms have supporters and opponents. More experiments and further exploration of other planets and moons in our solar system may add new pieces to this puzzle.

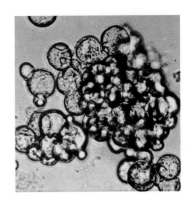

FIGURE 17.16

Proteinlike microspheres made in a laboratory. Each sphere is 5 microns in diameter. These structures, composed of amino acids, resemble living cells in some ways. Microspheres are selectively permeable and appear to divide by budding when they reach a certain size, but they cannot grow, reproduce, or evolve.

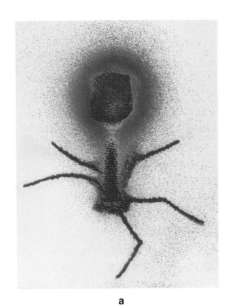

a

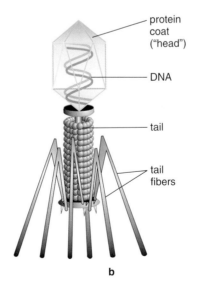

protein coat ("head")

DNA

tail

tail fibers

b

FIGURE 17.17

Structure of a virus. Viruses are made of DNA or RNA and a protein coat. **(a),** Electron micrograph of a virus that infects bacteria, ×94,500. **(b),** Diagram of the same type of virus.

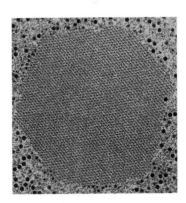

FIGURE 17.18

Crystallized virus in the nucleus of an animal cell. This crystal (×35,000) consists of many complete, infectious adenoviruses that can cause disease in humans.

The Record of the Rocks

17.6 Microfossils and Prokaryotes

Some scientists have tried to investigate early life-forms by searching for fossils. What may be the oldest known microfossils, or fossil microorganisms, were discovered in 1993 in northwestern Australia by a team led by J. William Schopf of the University of California, Los Angeles. These fossils include 11 types of microorganisms.

The fossils are of tiny, single-celled filaments (Figure 17.19). Their shapes resemble certain modern bacteria capable of photosynthesis. Because of this similarity, scientists hypothesize that the organisms might already have developed the pathways of photosynthesis.

The microscopic fossils were embedded in mineral grains. These grains were encased in a type of rock that formed almost 3.5 billion years ago. The fossils suggest that life was thriving and diversified at that time. Finding them showed that life appeared on Earth much earlier than scientists had thought. This discovery could focus more attention on the hypothesis that life originated elsewhere in the universe and somehow reached Earth.

The Australian fossils were found with domelike structures called **stromatolites.** The domes, about 30 cm high and 150 cm across, are composed of many wafer-thin layers of rock (Figure 17.20a). Microfossils are also found in stromatolites at Gunflint Lake in western Canada and around the shores of ancient Lake Bonneville in Utah. Today, limestone-secreting bacteria form similar structures at Hamelin Pool, an extremely salty arm of Shark Bay, Australia (Figure 17.20b).

Older microfossils will be very difficult to find. Some of the oldest known rocks on Earth, in Greenland, are about 3.8 billion years old. Older rocks have melted, eroded, or otherwise changed until they are no longer

ETYMOLOGY

stroma- = bed or layer (Greek, Latin)
litho- = rock (Greek)
A **stromatolite** is a kind of layered rock produced by bacteria that deposit minerals they collect from seawater.

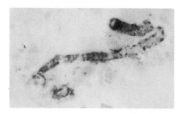

FIGURE 17.19

Fossil filament. A 3.5-billion-year-old prokaryote fossil in a stromatolite from western Australia. This fossil contains the oldest known cells.

FIGURE 17.20

Fossilized and living stromatolites. (a), Approximately 530 million-year-old fossil stromatolites from Wyoming. Note the characteristic layered structure. The image is approximately 5 cm across. **(b),** Modern living stromatolites in Shark Bay, Australia, built by cyanobacteria. Each dome is about 30–100 cm in diameter.

recognizable. Thus any fossils of the first life-forms have been destroyed. The Australian, Canadian, and African fossils are thought to be prokaryotes, but their structures were probably less organized than modern prokaryotes. Their genetic information may have been scattered in pieces throughout the cell, rather than organized in a single chromosome.

Section 17.4 explained why the first organisms probably were heterotrophs. Because the early atmosphere lacked oxygen gas, those organisms were also anaerobic. Carl Woese suggests that the first organisms may have been **methanogens**—anaerobic bacteria that obtain energy by using carbon dioxide to oxidize hydrogen (Figure 17.21). This reaction produces methane gas, which the bacteria release into the environment. Methanogens and related bacteria live today in conditions thought to be like those of early Earth. Some live near **hydrothermal** volcanic vents deep in the ocean where temperatures are much higher than the boiling point of water. Other bacteria thought to be relatives of methanogens have been found in acidic environments, such as rock wastes from mining, and in extremely salty waters, such as the Dead Sea between Israel and Jordan and the Great Salt Lake in Utah.

ETYMOLOGY

-gen = to produce (Greek)
hydro- = water (Greek)
thermal = hot (Greek)
Methanogens are methane producers. **Hydrothermal** vents are volcanic openings in the sea floor.

CHEMISTRY TIP

Methanogens obtain chemical energy by catalyzing the oxidation of hydrogen: $2 H_2 + CO_2 \rightarrow CH_4 + O_2$. Methane produced by these bacteria is the main component of natural gas.

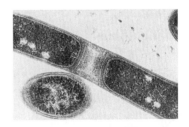

FIGURE 17.21

Modern methanogenic bacteria. Methanogens similar to these may have been the first organisms. They are anaerobic and obtain energy by combining carbon dioxide and hydrogen to produce methane. The cells are approximately 0.6 microns in diameter.

17.7 Eukaryotes

Prokaryotes probably were the only organisms on Earth for more than 1 billion years (Figure 17.22). The first microfossils that may have been eukaryotes are about 2.1 billion years old. Fossil evidence indicates that eukaryotic cells became common by 750 million years ago. Eukaryotic cells contain membrane-enclosed organelles such as a nucleus, mitochondria, and chloroplasts.

Lynn Margulis, of the University of Massachusetts, Amherst, hypothesizes that mitochondria and plastids (such as chloroplasts) originated as free-living prokaryotes. Margulis proposes that eukaryotes originated from a **symbiosis** between large anaerobic prokaryotes and smaller aerobic or photosynthetic prokaryotes. The large cells absorbed the smaller ones (or alternatively, small parasitic cells bored into larger ones). Instead of being digested, however, the smaller cells survived, protected inside the host cells (Figure 17.23). Photosynthesis and aerobic respiration in the small cells produced sugars and ATP that may have benefited the host cells. This relationship may have allowed all the partners to survive in higher numbers than those cells that lacked such relationships. Eventually, the internal partners, or **endosymbionts,** lost the

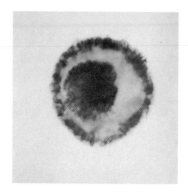

FIGURE 17.22

Fossil prokaryote. A 1.4-billion-year-old structure believed to be the fossil of a single-celled eukaryotic organism. This fossil contains one of the oldest known eukaryotes.

ETYMOLOGY

sym- = together (Greek)
bio- = life (Greek)
endo- = inside (Greek)
-on = thing (Greek)
Symbiosis is a relationship in which different types (species) of organisms live together. An **endosymbiont** is an organism that lives inside its symbiotic partner.

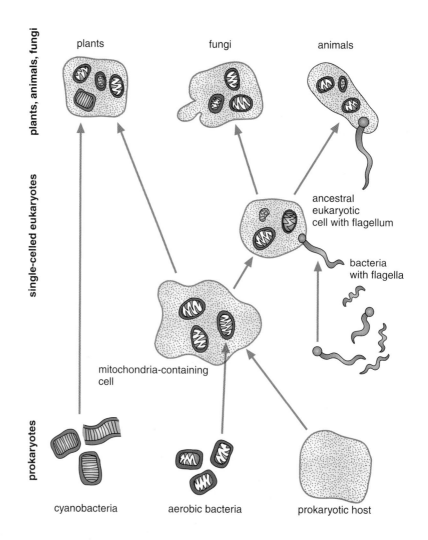

FIGURE 17.23

The endosymbiont hypothesis. The mitochondria and plastids (such as chloroplasts) of eukaryotic cells may have originated from free-living prokaryotes.

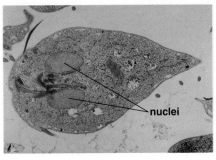
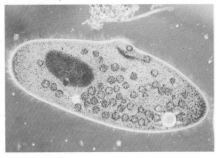

a   b

FIGURE 17.24

Unusual eukaryotes. (a) *Giardia lamblia,* ×16,000, has two haploid nuclei and no mitochondria.
(b) *Paramecium,* ×160, has one large nucleus (stained brown here) and several small nuclei
(stained green here). The first eukaryotes may have resembled these single-celled organisms.

ability to live independently. Aerobic purple bacteria likely were the ancestors
of mitochondria. Photosynthetic bacteria—cyanobacteria—were the likely
ancestors of plastids.

A large body of evidence supports Margulis's **endosymbiont hypothesis**
that mitochondria and plastids were once free-living prokaryotes. For
example, both organelles have their own DNA and ribosomes. Their genes
and ribosomes are similar to those of bacteria. Both organelles also have
double membranes. The outer membranes of these organelles may have
evolved from the vacuoles that surrounded them when their host cells first
took them in.

Some single-celled organisms don't have all the organelles usually found
in eukaryotes (Figure 17.24). Some of them may be survivors of a
transitional phase when the first true eukaryotes were developing. It is also
quite possible that the ancestors of these organisms were more typical
eukaryotes that lost certain structures as they evolved.

Check and Challenge

1. What is the evidence that the oldest fossils were autotrophs?
2. What characteristics make methanogens and related bacteria good
   candidates for being the first organisms?
3. How might mitochondria and plastids have originated?
4. What evidence supports the idea that mitochondria and plastids
   originated from free-living prokaryotes?

Summary

It is not possible to repeat historical events, such as the origin of life, but inference helps scientists study such events and draw plausible conclusions about how they occurred. The big bang probably occurred about 15 billion years ago. Planets, including Earth, formed around the Sun about 4.6 billion years ago. Gases escaping from within Earth formed a primitive atmosphere. Scientists hypothesize that chemical evolution, driven by a variety of energy sources, led to the origin of life on Earth. Catalytic RNA may have served as both information molecule (before DNA) and functional molecule (before proteins).

According to one definition, living systems must be able to reproduce, must be subject to mutations that can be passed on to offspring, and must be subject to natural selection. The first living thing may have been a "naked" information molecule such as RNA, DNA, or protein. It may also have been a cell-like structure.

The first life-forms were probably heterotrophs that lived on organic compounds in Earth's oceans. The oldest fossils appear similar to modern bacteria. Photosynthesis probably evolved very early, but significant levels of oxygen did not accumulate in the atmosphere until about 1 billion years ago. The oldest known fossils of eukaryotes are about 2.1 billion years old. Eukaryotes became abundant by 750 million years ago. Mitochondria and plastids may have arisen as free-living prokaryotes that occupied host cells and increased their energy yield.

There are more questions than data relating to the origin of life on Earth. Creative thinking and additional research may one day yield more definite answers.

Key Concepts

Copy the concept map shown here, and fill in the missing linking words. Use the concepts listed and other concepts from the chapter to expand the map.

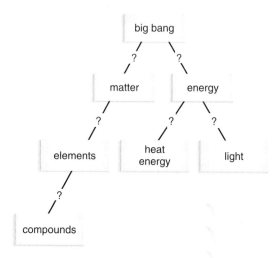

- chemical evolution
- biological evolution
- catalytic RNA
- heterotroph
- information molecule
- inference
- photosynthesis
- eukaryote
- endosymbiont
- functional molecule

Reviewing Ideas

1. Explain the evidence for the theory of an expanding universe.
2. How is the origin of life related to the origin and physical evolution of planet Earth?
3. How do the ages of meteorites help determine the age of Earth?
4. What is the evidence that free oxygen gas was not present in the primitive atmosphere?
5. Explain the limitations of science in investigating some of the ideas about the origin of life.
6. Explain the use of inference to study historical events and processes in science.
7. Why was there no ozone layer on Earth when life first originated?
8. Describe the steps in the Oparin-Haldane hypothesis for the origin of life.
9. What properties of a mineral crystal make it a candidate for an early information molecule?

Biology Online   BSCSblue.com/vocabulary_puzzlemaker

10. Why would a "naked" organic molecule be a less likely candidate for the first life-form than an organism with a membrane?
11. Why has no one found fossils older than 3.485 billion years?
12. Describe the results obtained by Miller in his experiments with and without methane and ammonia. What is the significance of the difference?

Using Concepts

1. Describe how Miller's experiment either supported or refuted the Oparin-Haldane hypothesis.
2. Explain why the identification of formaldehyde in the tail of comet Halley is important to the hypothesis of chemical evolution for the origin of life.
3. Why is RNA a more likely candidate than DNA for the first life-form?
4. Compare the suggested conditions of early Earth with the present conditions on Venus and Mars.
5. Why is the first life-form more likely to have been a prokaryote than a eukaryote?
6. Which is better understood—how organic compounds necessary for life accumulated on Earth or how the first cells formed?

Synthesis

1. Describe the differences between Darwinian evolution, as described in the Prologue, and chemical evolution.
2. A characteristic of living things is their need for energy. How was that need met by the earliest organisms, and in what way did that need drive the evolution of early life-forms?

3. Use the discussion in the Prologue about pseudoscience to explain why claims of "creation science" about the origin of life do not meet the test of scientific explanations.
4. What evidence indicates that the origin of life was tied directly to the physical evolution of planet Earth?

Extensions

1. Describe possible interactions between early life-forms and the materials in the seas around them.
2. Make a drawing that shows how you think primitive Earth appeared as it was cooling from its molten state.
3. If intelligent life exists on a planet in another galaxy, describe the problems you would have in trying to communicate with the extraterrestrials by radio signals.
4. Explain what inferences an extraterrestrial could draw about the presence or absence of life on Earth by observing the planet from another solar system.
5. Explain how the position of Earth in the solar system is related to the presence of life on the planet.

Web Resources

Visit BSCSblue.com to access
- Resources related to the origin of life
- NASA research about the origin of life
- NASA research about the possibility of life elsewhere in the universe
- Web links related to this chapter

CONTENTS

LEARNING OUTCOMES

By the end of this chapter you will be able to:

A   Discuss the difficulty of defining a species and explain the biological meaning of species.

B   Explain homology and give examples of homologous structures.

C   Describe the classification hierarchies used to categorize organisms and how they relate to one another.

D   Describe three ways to classify species.

E   Describe how the general characteristics of the five kingdoms differ.

F   Predict the effect of new knowledge on classification systems.

◀ **About the photo**
The photo to the left
shows a coral reef.

CHAPTER

18

Diversity and Variation

■ *How many different organisms can you identify here?*

■ *What important biological phenomenon is illustrated?*

We humans share this planet with an incredible number of different types of living things. Scientists have identified more than 3 million types of organisms, and there are probably several million more that have not yet been discovered or classified. This incredible variety—a basic characteristic of life on Earth—is referred to as **biodiversity.** Some types of living things resemble each other, but most do not. For example, dogs do not look like cats, and cats do not look like humans. Even organisms of the same type, such as cats, show a great deal of variation. Persian cats, for example, have different coloring and longer hair than Siamese cats.

To deal with the tremendous biodiversity, we need a logical way to classify organisms. Many different systems of classification are possible. Flowering plants, for example, could be classified according to the color of their flowers: yellow-flowered plants in one group, blue-flowered plants in a second, red-flowered plants in a third, and so on. This type of system might be useful, but it has serious limitations because it relies on one superficial characteristic. For example, it does not recognize the differences between woody trees and small herbs. The classification system used most widely today considers how organisms are related to one another. It groups them on the basis of their evolutionary history, to the extent that it is known. This chapter explains how classification systems bring order to biodiversity and how they make the great diversity of life-forms more understandable.

Bringing Order to Diversity

18.1 The Species Concept

ETYMOLOGY

taxo- = to arrange or put in order
(Latin, Greek)
-nomy = law or form of organization
(Greek)

Taxonomy is the science of
organizing classifications according
to a system of rules.

For thousands of years, humans have tried to classify organisms in ways that reflect relationships and help distinguish one type of organism from another. The science that grew out of these efforts is **taxonomy.**

The basic grouping used in biological classification is the **species.** A species is a group of organisms that is capable of breeding, or mating, with one another in nature to produce fertile offspring. Members of one species usually cannot produce fertile offspring with members of another species. For example, all humans are members of a single species. If a Texan and an Australian meet, they could produce children who would grow into fertile human adults.

Individual members of a species may look very different from each other. For example, not all dogs look alike, but all of them make up one species. Some dogs are recognizable breeds, such as collies and beagles. Others are intermediate mongrels, produced when members of different breeds mate. The mongrels can also mate and produce young that are even more varied.

Such differences among members of a species are known as variations (Figure 18.1). Inherited variation is the raw material of evolution. Natural selection acts on variation, resulting in changes in species or the evolution of new species (see Section 19.4).

The variations in a population include polymorphism, or variation in form; geographic variation; and individual variation. Polymorphism occurs when two or more forms, or morphs, exist in the same population. The light and dark forms of the peppered moths in England are an example (see

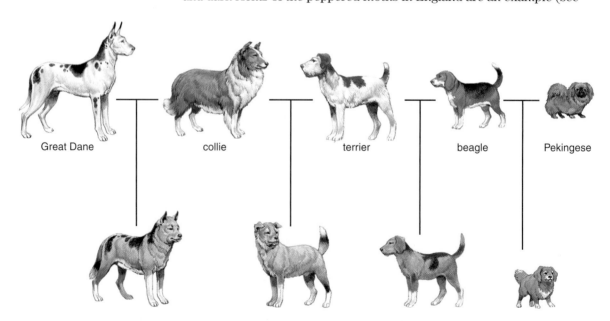

Great Dane   collie   terrier   beagle   Pekingese

FIGURE 18.1

Five breeds of dogs. Their mongrel offspring are intermediate between the breeds. Note the variation in size, coloring, and body form.

FIGURE 18.2

Geographic variation in the human species. In the past, climatic factors such as temperature, humidity, sunlight, and altitude contributed to human variation via natural selection. Today human populations intermingle more easily, and many distinctive geographic variations are disappearing. Nevertheless, a few variations remain, such as height and body form.

Figure 16.4). The differences between males and females of the same species are another example of polymorphism.

Geographic variation occurs when a species occupies a large geographic range that includes distinct local environments. Populations in those local environments often have unique physical characteristics. Geographic variation accounts for many of the differences in the human species (Figure 18.2).

Individual variation occurs in all populations of organisms that reproduce sexually. This type of variation is less obvious than geographic variation. We are very much aware of individual variation in our own species. We are less sensitive to individuality in populations of other animals and plants (Figure 18.3).

FIGURE 18.3

Variation within a population. Individual variation in a cattle population.

a

b

FIGURE 18.4

Two species of canines. Though related, the domestic dog, *Canis familiaris* **(a)**, and the coyote, *Canis latraus* **(b)**, remain separate because of differences in behavior and rare interbreeding.

Members of related species may interbreed occasionally. If the two groups fail to produce a significant number of hybrids, however, they remain separate species. Domestic dogs and coyotes (Figure 18.4) are one example. Behavioral differences in the care and feeding of the young limit the number of dog-coyote hybrids (coydogs). Because the hybrid offspring are so rare and scattered, two coydogs are unlikely to meet each other and mate.

A species then is a population of individuals that breed and produce fertile offspring *under natural conditions* (as opposed to unnatural conditions, such as captivity). Species remain separate from one another in three basic ways: (1) potential mates do not meet; (2) potential mates meet but do not breed; (3) potential mates meet and breed but do not produce fertile or viable offspring.

An example of the first case involves the grizzly bear and the polar bear (Figure 18.5). Although these two species have reproduced together in captivity, no hybrids have been discovered in the wild. The reason is simple: Grizzly bears live in forests and eat fish, berries, and small animals. Polar bears live on snowfields and ice floes and depend on seals for food. These two species do not meet in the wild.

Species that share the same habitat but remain separate are examples of the second case, in which potential mates meet but do not breed. A Colorado blue spruce cannot breed with a Colorado blue columbine growing in its shade (Figure 18.6). Similarly, a giraffe and an ostrich cannot mate even though they may come in contact with each other. These species are too different to breed together.

The dog and coyote exemplify the third case—potential mates meet and breed but with poor results. In some cases, the offspring of two species may be sterile (incapable of reproducing). The mule, a cross between a horse and a donkey, is almost always sterile.

The species concept has certain limitations. First, it does not apply well to organisms that do not reproduce sexually. Second, even species that reproduce sexually may be only partly separated. Some populations are intermediate between subspecies, or races, which can interbreed, and species, which cannot. Third, the species concept does not easily accommodate the slow and usually unmeasurable changes that occur in species through time.

a

b

FIGURE 18.5

Two species of bears. The grizzly bear, *Ursus arctos* **(a)**, and the polar bear, *Ursus maritinius* **(b)**, are related but distinct species. What keeps these species separate?

a   b

FIGURE 18.6

Distinct species. Species that come in contact but do not interbreed. The Colorado blue spruce, *Picea pungens* **(a),** and the Colorado blue columbine, *Aquilegia caerulea* **(b),** do not crossbreed even though they share the same habitat.

18.2 Classification and Homologies

Classification is important for both practical and scientific reasons. Taxonomists identify wild relatives of crop plants that provide breeders with useful alleles. They also identify weeds and help develop wild species into new crops. Taxonomy also benefits public health. Accurate identification of parasites and the organisms that carry them (such as certain mosquito species) is critical for disease control.

Classification plays a role in ecology too. The ecosystems that sustain life on Earth depend on millions of species, many of which are not yet identified. To preserve these ecosystems, we must identify their most important species. Some species are especially sensitive to chemical pollution. Changes in their populations warn us of changes in environmental quality.

Classification also provides a key for understanding the unity of life. Taxonomists group species according to how closely they think the species are related by ancestry. This approach produces a huge family tree that unites every living thing on Earth (see Figure 18.19).

Focus On

Species

Horses and donkeys can produce hybrid offspring called mules. A mule is a cross between a male donkey and a female horse. The offspring of a female donkey and a male horse is called a hinny. Normally mules and hinnies are sterile because they are a hybrid of two species. However, strange things happen. In 1984, a mule named Krause gave birth to a colt (Figure 18.7). Geneticists from the San Diego Zoo determined the number of chromosomes of both the mule and the new colt. The tests showed that Krause and her offspring were mules. (Chester, the father, was a donkey.) Horses have 64 chromosomes, donkeys have 62, and mules have 63. Krause did not stop there. In 1987, she gave

birth to another colt. Such occurrences illustrate how difficult it can be to define a species.

FIGURE 18.7

Fertile mule. The mule Krause with her two colts, Blue Moon and White Lightning.

rose

red alga

sea urchin

earthworm

summer tanager

FIGURE 18.8

One way of classifying organisms. One similarity among these organisms is obvious—their color. This is a poor basis for classification.

CONNECTIONS

DNA is the universal information molecule in all kingdoms. It is an example of diversity of type and unity of pattern.

E T Y M O L O G Y

homo- = same (Greek)
ana- = according to (Greek)
logos = proportion or logic (Greek)

Homologous structures develop following the same "logic," or ancestral genetic plan. **Analogous** structures serve similar functions but do not reflect shared ancestry.

What characteristics are most helpful in classifying organisms? The organisms in Figure 18.8 are all red, but they have little else in common. Taxonomists use many characteristics to classify organisms, including structure, biochemistry, behavior, and genes. Stable characteristics are the most useful. For example, structures such as skeletons (in animals) and flowers (in plants) vary less within a species than do size and color. For this reason, structure provides much information that is useful in classifying organisms.

Classification focuses on structures that indicate *a related evolutionary ancestry*, not just similarity. These structural resemblances are called **homologies.** For example, as Figure 18.9 shows, the flipper of a whale, the wing of a bird, and the arm of a human are similar in their bone structure. Fish, amphibians, reptiles, birds, and mammals all share this limb pattern. All these limbs are homologous. On the other hand, insect wings have a very different structure and contain no bones at all. The wings of birds and insects have the same function, but they do not reflect a shared ancestry between birds and insects. Structures that are similar in appearance and function but are not the result of shared ancestry are **analogies.**

Two factors make anatomical (structural) characteristics particularly important in classification. First, they are easy to observe in organisms and fossils. Second, fossils are the only evidence we are ever likely to have of extinct species. An extinct species may be the ancestor of a living one. Knowledge of past organisms helps us determine the relationships of living ones.

Chemical homologies, such as similarities in the structures of cellular polymers, are also evidence of close evolutionary relationships. The sequence of amino acids in a protein molecule or of nucleotides in DNA is now relatively easy to determine. Comparisons of these sequences in

Focus On

Chemical Homology

Differences among species in the nucleotide sequences of homologous genes provide clues to evolutionary relationships. The more closely two species are related in their evolutionary history, the fewer nucleotide differences show up in their genes. A comparison of the DNA sequences of different species can yield evidence of how closely the species are related (Tables 18.1 and 18.2). Usually, chemical evidence from studies of protein or DNA sequences supports the same family tree of species that has been built up on the basis of evidence from anatomy and behavior. This agreement strengthens our confidence in both kinds of evidence. In some cases, the picture is more complex. For example, bacteria exchange bits of DNA quite easily. This makes it more difficult to determine their relationships from their DNA sequences.

An interesting finding in protein studies is that some positions in a particular type of protein molecule are occupied by the same amino acid in every species analyzed. These invariant amino acids may be part of an enzyme's active site or may help maintain the protein's tertiary structure. Changing one of those amino acids would alter the protein's function. An individual with an altered protein would be at a disadvantage compared with other members of the population. Therefore, natural selection acts against variations in these critical amino acids.

TABLE 18.1
Nucleotide Sequence of Part of a Gene in All Apes and Monkeys

Source	DNA Sequence
Human	* * * * C A C A A T A
Chimpanzee	* * * * C A C A A T A
Gorilla	T A A T C A C A A T A
Orangutan	T A A T C A C A A T A
Rhesus monkey	T A A T C A C A C T G
Spider monkey	T A A T C A C A A T C

* Missing (deleted) bases.
Note: The complete gene is 8,474 nucleotides long.
Source: Douglas J. Futuyma, *Evolutionary Biology,* third edition (Sunderland, MA: Sinauer Associates, 1998), p. 104.

TABLE 18.2
Percent Differences among Species in Nucleotide Sequence of an Entire Gene

	Human	Chimpanzee	Gorilla
Human	0.38*		
Chimpanzee	1.56		
Gorilla	1.69	1.82	
Orangutan	3.30	3.42	3.39

* The difference in nucleotide sequence between two human alleles of this gene was found to be 0.38%.
Source: Douglas J. Futuyma, *Evolutionary Biology,* third edition (Sunderland, MA: Sinauer Associates, 1998), p. 730.

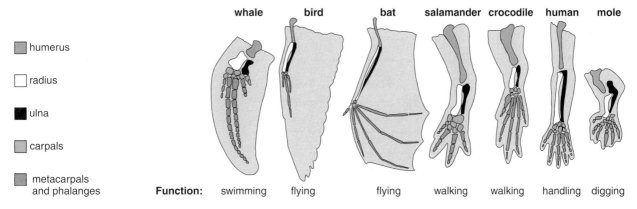

whale   bird   bat   salamander   crocodile   human   mole

humerus
radius
ulna
carpals
metacarpals and phalanges

Function: swimming   flying   flying   walking   walking   handling   digging

FIGURE 18.9

Bones of the forelimbs of seven vertebrates. Follow the color key to homologies; they provide evidence that these animals share an ancestor. Also compare the structure of each forelimb to the function for which it is adapted to support the animal's way of life.

> **CHEMISTRY TIP**
>
> There are chemical limits to variation in amino-acid sequences. Even sequences that vary among species tend to keep chemically similar amino acids in the same positions. For example, it is rare for a polar amino acid to replace a nonpolar one or for an acidic amino acid to replace a basic one.

different organisms have produced much new evidence of evolutionary history. However, chemical homology can be more difficult to identify than anatomical homology. There are only four possible DNA bases that can appear at each position in a DNA sequence. In many cases, only one or two bases in a certain position would code for a functional protein. These limits on chemical variation make chemical homology difficult to distinguish from mere similarity. Comparisons of DNA sequences must be interpreted along with anatomical data and other information before we can decide how species are related.

18.3 The Linnean Classification System

The system we use to classify species is based on the work of the Swedish botanist Carl von Linné (1707–78), who wrote under his Latinized name, Carolus Linnaeus. This system uses homologies to group species into larger and more general categories (Figure 18.10). Species with many similar characteristics are grouped in the same **genus** (plural: genera). For example, domestic dogs, coyotes, and wolves belong to different species but are similar in many ways. Therefore, taxonomists group them in the genus *Canis.*

Similar genera are grouped in the same **family.** Foxes, genus *Vulpes,* belong with *Canis* in Canidae, the dog family. The dog family is related to the cat, bear, and raccoon families, among others. All of these animals have the distinctive anatomy and way of life of flesh eaters. Therefore, they are grouped in the **order** Carnivora (Figure 18.11).

Cats, wolves, and raccoons have differences, but they also have more in common with one another than with rabbits. Rabbits are placed in a different order, Lagomorpha. Like the members of order Carnivora, rabbits

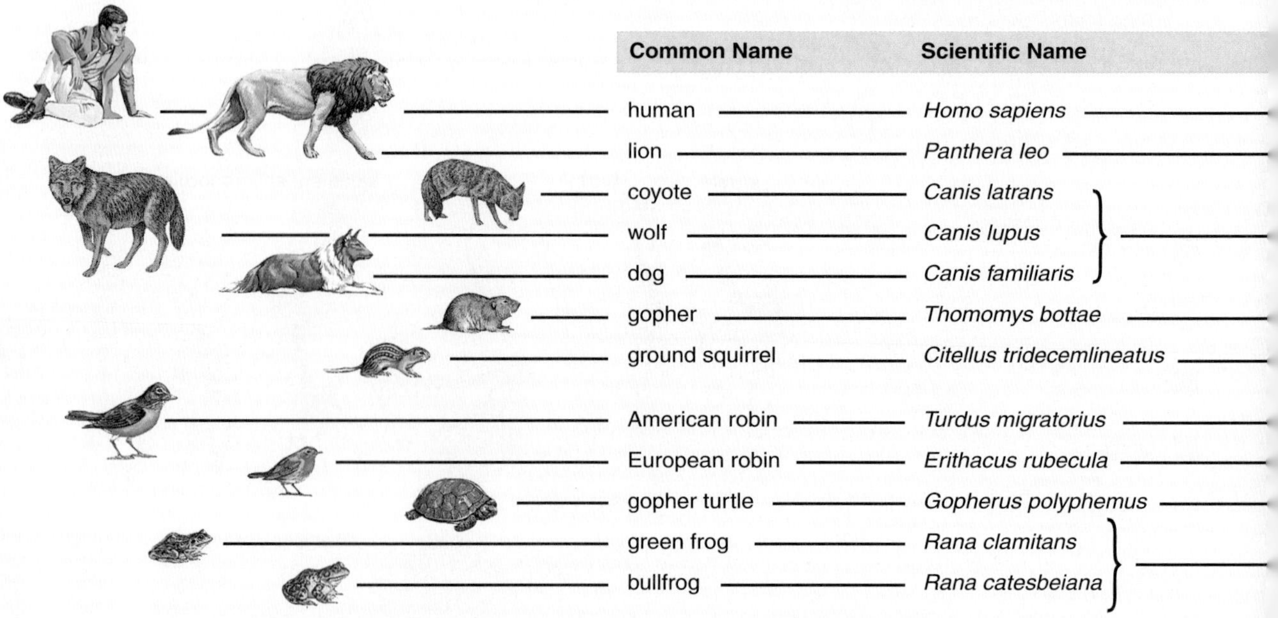

Common Name
Scientific Name
human
Homo sapiens
lion
Panthera leo
coyote
Canis latrans
wolf
Canis lupus
dog
Canis familiaris
gopher
Thomomys bottae
ground squirrel
Citellus tridecemlineatus
American robin
Turdus migratorius
European robin
Erithacus rubecula
gopher turtle
Gopherus polyphemus
green frog
Rana clamitans
bullfrog
Rana catesbeiana

FIGURE 18.10

The classification of 12 animal species. All species are classified in this way.

bear
(*Ursus arctos*)

wolf
(*Canis lupus*)

coyote
(*Canis latrans*)

fox
(*Vulpes fulva*)

weasel
(*Mustela frenata*)

FIGURE 18.11

Some animals in the order Carnivora. Which two look most alike? Which two look most different? In what ways? How do these similarities and differences relate to their classification?

(along with rats, horses, pigs, and animals from other orders) have hair and produce milk. These similarities are the basis for putting these orders together in the **class** Mammalia.

Continuing with this method of grouping, taxonomists place the classes containing birds, snakes, fish, and frogs with Mammalia into the **phylum** (plural: phyla) Chordata. Many botanists (plant biologists) do not use the term *phylum*. Instead, they group organisms at this level into **divisions.**

ETYMOLOGY

phylum or **phulon** = class or race (Latin, Greek)

A **phylum** is a group of classes with a shared ancestor.

Genus	Family	Order	Class	Phylum	Kingdom
Homo	Hominidae	Primates			
Panthera	Felidae				
Canis	Canidae	Carnivora	Mammalia		
Thomomys	Geomyidae	Rodentia		Chordata	Animalia
Citellus	Sciuridae				
Turdus	Turdidae	Passeriformes	Aves		
Erithacus					
Gopherus	Testudinidae	Chelonia	Reptilia		
Rana	Ranidae	Salientia	Amphibia		

Most botanists use the term *division* because the term *phylum* implies a knowledge of ancestries that is often lacking for the major plant groups.

Finally, the chordates, snails, butterflies, and thousands of other organisms are grouped in the **kingdom** Animalia. This kingdom contains all the living things you think of as animals.

As you go from species to kingdom, the organisms that are grouped together share fewer characteristics at each succeeding level. At the species level, individuals are so alike they can interbreed. Organisms at the kingdom level share only a few common characteristics. How many common characteristics can you name for the animals whose classifications are shown in Figure 18.10?

The most important change to this system since Linnaeus's time grew out of the work of Charles Darwin. Linnaeus tried to make sense of the living world by classifying species according to their similarities and differences. Later, when Darwin's ideas became widely accepted, the meaning of classification changed. Biologists began to see classification as a way of describing evolutionary relationships. Today they try to classify organisms so that each group is descended from one ancestral species.

The Linnean system uses **binomial nomenclature,** or a two-word naming system, to identify each species. The first word in each name is the genus name, and the second word describes the particular species. Together these two words form the **scientific name** (or species name). The genus name is always capitalized, but the descriptive word is not. In print, the scientific name appears in italics. (It is underlined when handwritten.) For example, the white-footed mouse is *Peromyscus leucopus*, and the piñon mouse is *Peromyscus truei*. Both are mice, but each is a different species (Figure 18.12). Other species of mice fall within the genus *Peromyscus*. Figure 18.13 shows another example of different species in the same genus— the ponderosa pine *(Pinus ponderosa)* and the loblolly pine *(Pinus taeda)*.

Why use a Latinized binomial system? First, a classification system has to be understood all over the world. For this reason, Latin or Latinized names are used rather than names from a particular modern language (Figure 18.14). A second reason for using the binomial system is its precision. Common names are local and imprecise. For example, if people from different parts of North America are discussing a gopher, they may not

CONNECTIONS
The classification of species demonstrates the unity and diversity of life.

a   b

FIGURE 18.12

Two species of mice in the same genus. Note differences between *Peromyscus leucopus* **(a),** the white-footed mouse, and *Peromyscus truei* **(b),** the piñon mouse.

a

b

FIGURE 18.13

Two species of pines in the same genus. The ponderosa pine, *Pinus ponderosa* **(a),** and the loblolly pine, *Pinus taeda* **(b),** are not members of the same species even though they look similar.

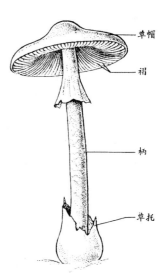

圖 12—6 瓢菌 *Amanita verna* 圖示蕈托
杯狀物位在柄的基部（應藏地下）。這是許多
頭毒蕈的特徵。蕈帽下放射狀蕈褶是蕈的特徵。

FIGURE 18.14

Importance of binomial nomenclature. An illustration of a poisonous mushroom from a Chinese BSCS biology book. Note the scientific name in the caption.

be talking about the same species at all. In California, a gopher is a small burrowing rodent with the scientific name *Thomomys bottae.* In Florida, a gopher is a type of tortoise whose scientific name is *Gopherus polyphemus.* Imagine how confusing communications would be if people used the same common name for different organisms. The use of scientific names avoids this problem.

Try Investigation 18A ◇ Using Cladistics to Construct Evolutionary Trees.

18.4 Three Ways to Classify Species

Taxonomists have always relied heavily on anatomical similarities and differences among organisms. For example, birds are divided into orders partly on the basis of differences in their beaks and claws (Figure 18.15). When microscopes and other tools became available, taxonomists were able to use other kinds of information. For example, they found that the phyla of fungi differ in the cellular details of reproduction. Growing collections of fossils provided clues about early ancestry. Still, taxonomy was fairly subjective. Biologists would examine many organisms and try to judge which characteristics were most useful for classifying them. They had no accepted method for resolving disagreements. This intuitive approach to

a

b

c

d

FIGURE 18.15

Two orders of birds. Hawks **(a)** and eagles **(b)** have sharp grasping claws and flesh-tearing beaks. They are classified in the order Falconiformes, the raptors or hunting birds. Ducks **(c)** and geese **(d)** have webbed feet and broad bills. They are placed in the order Anseriformes, one of several orders of waterbirds.

	mouse	bat	robin	tuna
lungs	✔	✔	✔	✘
4 bony limbs	✔	✔	✔	✘
milk glands	✔	✔	✘	✘
hair	✔	✔	✘	✘
wings	✘	✔	✔	✘
feathers	✘	✘	✔	✘

a

✔ = present   ✘ = absent

b

Number of Shared Characteristics			
	Mouse	Bat	Robin
Bat	5		
Robin	2	3	
Tuna	2	1	2

c

mouse   bat   robin   tuna

class   Mammalia

phylum   Chordata

FIGURE 18.16

A phenetic classification. (a) Many characteristics of the organisms are compared. **(b)** Then the number of characteristics that each pair of organisms shares is calculated. **(c)** Finally, the organisms are grouped according to overall similarity.

classification is still used. It is sometimes called orthodox classification. The grouping of birds into orders is an example.

Two systematic approaches to classification developed in the 1950s. One method compares organisms on the basis of as many characteristics as possible (Figure 18.16). The taxonomist then gives each pair of organisms a similarity score based on the number of traits they share. This method, called **phenetics,** gives equal importance to all characteristics. Species are grouped into genera, families, and so forth according to their overall similarity. All members of a group may not share a specific characteristic, such as the form of their beaks or feet. Phenetics does not consider other evidence, such as fossils that may represent shared ancestors. This system is most useful in classifying large groups of similar organisms, but it is no longer popular.

The other systematic method of classification is simpler and does consider ancestry (Figure 18.17). Known as **cladistics,** it groups species according to ancestry and homologous characteristics not found in other organisms. Cladists assume that each group has an ancestor that other

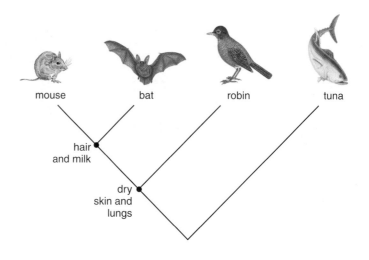

FIGURE 18.17

A cladogram. A cladistic classification of some animals. A small number of characteristics are used to define each group.

Classification of the Giant Panda

Giant pandas live in bamboo forests in the mountains of western China. They form one of the most widely recognized and cherished species. From an evolutionary point of view, the species is also one of the most puzzling. For almost 120 years, biologists have disagreed about the taxonomic classification of the giant panda. Some have placed it in the bear family. Others believe it belongs in the raccoon family with the red panda, a much smaller animal.

Giant pandas look like bears, but they have many traits that are not at all bearlike. They are mainly plant eaters, living mostly on bamboo. The panda has a big, heavy skull (Figure 18.18a) with flattened teeth and jaws that are well developed for grinding. Pandas also have a highly modified wrist bone which they use like a thumb to strip off leaves from bamboo stalks (18.18b).

Giant pandas also differ from bears in other ways. Mountain bears usually hibernate; giant pandas do not. Pandas do not sound like bears either. Instead of growling or roaring, pandas bleat like sheep. The number and type of giant-panda chromosomes (21 pairs) are much more like the red panda's (22 pairs) than a bear's (37 pairs).

Through the study of homologies in DNA, a new classification of bears, raccoons, and pandas has emerged. These techniques are based on what is called the molecular clock hypothesis. According to this hypothesis, as new species develop, their genes become isolated from those of related species. Because of this isolation, differences among the species increase over time. Such differences arise from mutations in each species' DNA that are passed from generation to generation. Some mutations occur in regulatory genes or portions of the DNA that code for particular amino acids. Others occur in noncoding portions of the DNA that have no apparent effect on the organism.

The rate of accumulation of mutations can be estimated. Therefore, the differences between the DNA of two species provide a good indication of how closely related they are. On the basis of this information, the approximate time when two species diverged from a common ancestor can be determined.

The giant panda is now classified in the bear family, Ursidae. Within that family, pandas are distinctive enough to be placed in a subfamily. Many biologists now place the red panda in its own family, Ailuridae. DNA evidence suggests that the raccoon and bear families separated between 35 and 40 million years ago.

a   b

FIGURE 18.18

Giant panda, *Ailuropoda melanoleuca*. (a) Classification of this species remains controversial. Molecular, anatomical, and genetic evidence are helping to clarify the classification of pandas. The panda's thumb **(b)** is actually an enlarged wrist bone that can be pushed forward and used to grasp bamboo.

species do not share. For example, all mammal species have milk glands, but no other organisms do. Therefore, all mammals must be descended from a species that has no other living descendants. The mammals form a clade, or branch, on a diagram of the history of the animals. Cladistics is popular partly because it states exactly which features define each clade.

Check and Challenge

1. What condition must organisms meet in order to be considered members of the same species?

2. What is binomial nomenclature, and what function does it serve?

3. Give two examples of how the number of characteristics shared by members of a group changes as you proceed from species to kingdom.

4. Define the term *homologies* and explain why they are important in classification.

5. Explain how species maintain their distinctness.

6. How would you expect an orthodox classification of the species in Table 18.2 to differ from a cladogram? Explain your answer.

The Kingdoms of Life

18.5 Five Kingdoms

As you move through the classification system from species to kingdoms, each level includes more types of organisms. The more types of organisms that a category includes, the less similar they are. Cell structure, however, separates all organisms into two major groups, the prokaryotes and the eukaryotes (see Section 6.2). Therefore, prokaryotes and eukaryotes are classified in separate kingdoms.

Each kingdom includes thousands or millions of species. What can so many species have in common? In grouping species by kingdom, biologists consider several key questions: Is the organism a prokaryote or a eukaryote? Is it an autotroph or a heterotroph (see Section 2.2)? Does the organism develop from an embryo? Is it unicellular or multicellular? Finally, what is its general structure and function? Most of today's biologists use a form of the five-kingdom classification system first proposed by Robert Whittaker in 1959. Use Figure 18.19 as a guide as you read the description of each kingdom.

CONNECTIONS

Despite the differences between prokaryotes and eukaryotes, the unity of all life can be seen in the fact that all organisms depend on the same basic chemical tools, including enzymes, lipid membranes, the genetic code, and the system of protein synthesis.

Biology Online   BSCSblue.com/check_challenge

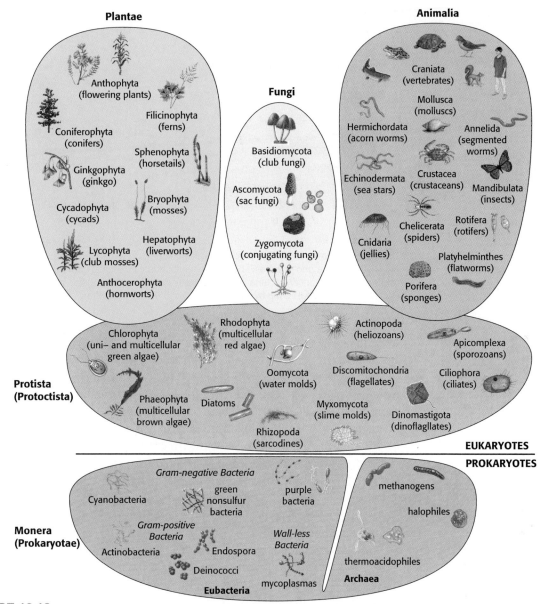

Plantae

Anthophyta
(flowering plants)

Filicinophyta
(ferns)

Coniferophyta
(conifers)

Sphenophyta
(horsetails)

Ginkgophyta
(ginkgo)

Bryophyta
(mosses)

Cycadophyta
(cycads)

Hepatophyta
(liverworts)

Lycophyta
(club mosses)

Anthocerophyta
(hornworts)

Fungi

Basidiomycota
(club fungi)

Ascomycota
(sac fungi)

Zygomycota
(conjugating fungi)

Animalia

Craniata
(vertebrates)

Mollusca
(molluscs)

Hermichordata
(acorn worms)

Annelida
(segmented
worms)

Echinodermata
(sea stars)

Crustacea
(crustaceans)

Mandibulata
(insects)

Cnidaria
(jellies)

Chelicerata
(spiders)

Rotifera
(rotifers)

Platyhelminthes
(flatworms)

Porifera
(sponges)

**Protista
(Protoctista)**

Chlorophyta
(uni– and multicellular
green algae)

Rhodophyta
(multicellular
red algae)

Actinopoda
(heliozoans)

Apicomplexa
(sporozoans)

Oomycota
(water molds)

Discomitochondria
(flagellates)

Ciliophora
(ciliates)

Phaeophyta
(multicellular
brown algae)

Diatoms

Myxomycota
(slime molds)

Dinomastigota
(dinoflagllates)

Rhizopoda
(sarcodines)

EUKARYOTES

PROKARYOTES

Gram-negative Bacteria

green
nonsulfur
bacteria

purple
bacteria

methanogens

Cyanobacteria

halophiles

*Gram-positive
Bacteria*

**Monera
(Prokaryotae)**

Actinobacteria

Endospora

*Wall-less
Bacteria*

Deinococci

thermoacidophiles

mycoplasmas

Archaea

Eubacteria

FIGURE 18.19

An updated form of Whittaker's five-kingdom system. All organisms can be classified in the five kingdoms shown in this diagram. However, increased understanding of the prokaryotes has led some biologists to place Archaea in a separate kingdom. Some taxonomists use the name Protoctista and Prokaryotae instead of Protista and Monera.

Whittaker grouped the prokaryotes in the kingdom **Monera.** Most prokaryotes are unicellular or colonial, though multicellular forms also exist. Reproduction occurs by simple cell division (see Section 6.3). Prokaryotes include heterotrophs, photoautotrophs, and chemoautotrophs (see Section 4.1). Many types change their form of nutrition in response to changes in the environment. Even closely related prokaryotes can differ

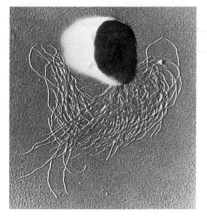
a

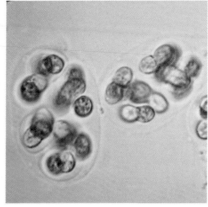
b

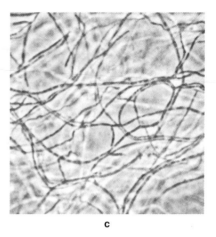
c

FIGURE 18.20

Examples of Monera. (a) *Pyrococcus furiosus* can grow at temperatures up to 103°C.
(b) *Gloeocapsa,* a photosynthetic bacterium, is abundant in aquatic ecosystems.
(c) Diverse decomposer bacteria in the soil break down complex organic compounds
into inorganic materials used by plants. Some, such as the ones shown here, also
produce antibiotics.

more in cell structure and function than do eukaryotes (Figure 18.20).
Because of this variability, bacterial taxonomy relies heavily on comparisons
of DNA sequences and the composition of cell walls and membranes. Even
distantly related bacteria can exchange DNA, however, and their asexual
reproduction makes the usual definition of species useless. Monera is
divided into two main groups, **Eubacteria** and **Archaea.** These groups are
so different that many biologists now treat them as two kingdoms.

The eukaryotes are placed in four kingdoms. The kingdom **Protista**
consists mostly of microscopic unicellular eukaryotes. Some protists, such
as kelp, are multicellular organisms that can be more than 10 m long.
However, these organisms do not differentiate into many types of cells, as
plants and animals do.

Protists descended from bacteria. They include the ancestors of all other
eukaryotes. Protista is far more diverse than the other kingdoms. These
species are grouped in one kingdom for convenience. However, some
biologists favor dividing Protista into several smaller, more homogeneous
kingdoms. Comparisons of DNA and RNA sequences show more diversity in
Protista than among all other eukaryotes combined. The classification of
protists is also complicated because they originated by symbiosis between
different species (see Section 17.7).

Protists vary greatly in structure, reproduction, and lifestyle. They
include producers, consumers, and decomposers. Some protists switch from
one form of nutrition to another in response to environmental conditions.
Protista includes algae (photoautotrophs), **protozoa** (swimming or creeping
heterotrophs), slime molds (funguslike protists), and other organisms
(Figure 18.21). These categories are based on superficial appearance. They
are not taxonomic groups. For example, some "animal-like" swimming
protozoa are photoautotrophs that contain chloroplasts.

Photoautotrophic multicellular eukaryotes that develop from embryos
belong to the kingdom **Plantae.** Plants have cellulose-containing cell

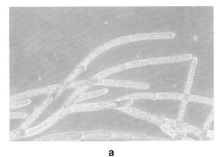

a   b   c

FIGURE 18.21

Examples of Protista. (a) A filamentous alga, genus *Cladophora*; **(b)** a protozoan, or motile protist, *Stentor coeruleus*; **(c)** red raspberry slime mold, *Tubifera ferruginosa*.

walls, and most store food as starch. Their cells contain chloroplasts. The bulk of the world's food and much of its oxygen are derived from plants. Mosses, ferns, conifers, and flowering plants all belong to this kingdom (Figure 18.22).

Heterotrophic multicellular eukaryotes that develop from embryos are placed in the kingdom **Animalia.** Ranging in size from microscopic organisms to giant whales, these organisms are the most diverse in form of the members of all five kingdoms. Most reproduce sexually. The

a

b

c   d

FIGURE 18.22

Members of the kingdom Plantae. (a) A moss, genus *Polytrichium*; **(b)** a fern, *Cystopteris fragills*; **(c)** Douglas fir, *Pseudotsuga menziesli;* **(d)** sand lily, *Leucocrinum montanum.*

CHEMISTRY TIP

The cell walls of archaeans lack muramic acid, which occurs in eubacterial cell walls. Muramic acid consists of a glucose molecule bound to a lactic-acid molecule. Archaean membrane lipids contain branched carbon chains, rather than the usual straight ones. These carbon chains are also linked differently to glycerol molecules.

branched carbon chain

archaean membrane lipid

straight carbon chain

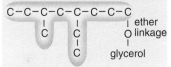

other membrane lipid

a

b

c

d

FIGURE 18.23

Four members of the kingdom Animalia. (a) Sunflower star, *Pycnopodia helianthoides;* **(b)** treehoppers, genus *Thelia;* **(c)** orange-finned anemone fish, *Amphiprion crysopterus;* **(d)** giraffe, *Giraffa camelopardalis.*

Try Investigation 18B ◇ Structural Characteristics of Animals.

CHEMISTRY TIP

Chitin is a tough polymer with a basic subunit consisting of a molecule of glucose and a molecule of acetate connected by a nitrogen atom. It forms the chemical framework for the cell walls of fungi and the exoskeletons of arthropods.

acetate
C – C
|
N H
|
C – C – C – C – C – C
glucose

**chitin monomer
(N-acetylglucosamine)**

arthropods, animals such as insects and crabs that have exoskeletons and jointed legs, may be the majority of all multicellular species. The **vertebrates**—animals with a spinal column, or backbone, such as birds, frogs, fish, and humans—are another large group of animals. Some familiar animals are shown in Figure 18.23. Most members of this kingdom are **motile,** or capable of locomotion, and have senses and nervous systems.

The kingdom **Fungi** includes heterotrophs that absorb small molecules from their surroundings through their outer walls. Most fungi are multicellular (although yeast is an exception) and have cell walls composed of a tough carbohydrate called chitin. They reproduce by forming spores, either sexually or asexually. Mushrooms are the sexual structures of fungi. Most fungi are decomposers. Fungi include yeasts, molds, bracket fungi, and mushrooms (Figure 18.24).

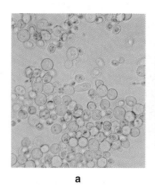

a

b

c

FIGURE 18.24

Examples of the kingdom Fungi. (a) A yeast, genus *Saccharomyces;* **(b)** scarlet waxy cap mushroom, *Hygrophorus coccineus;* **(c)** a rust, order *Uredinales,* a parasite of plants.

Focus On
Lichens

Lichens (Figure 18.25) often resemble mosses or other simple plants, but lichens are not plants at all. In fact, they are not even individual organisms. Lichens are actually fungi and algae (or photosynthetic bacteria) that live together in close association. The nature of their relationship is unclear. The algae provide the fungi with food, but biologists do not know whether the algae receive any benefits from the fungi other than protection from dehydration. Furthermore, the fungi consume some of the algae. For this reason, some biologists argue that the relationship is parasitic—that is, one organism survives directly at the expense of another. The merging of the fungi and algae is so complete, however, that lichens are given binomial species names as if they were single organisms.

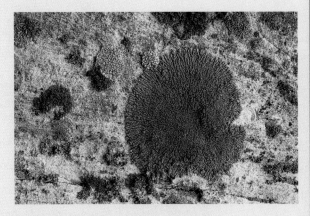

FIGURE 18.25

Lichens colonizing a rock. Thousands of different species of fungi and algae are capable of forming these associations.

18.6 Classification and Change

Remember that *taxonomic classification is not fixed*. It depends on the interpretation of evidence. Biologists do not always agree about where each organism fits. Systems of classification reflect how science works—it is open to change in response to new data. As we gain more information, the relationships among organisms appear more complex. New knowledge often requires changes in the way we group organisms. This is especially true at the kingdom level.

Linnaeus's original system had only two kingdoms, plants and animals. Later, as our understanding of microorganisms improved, they did not seem to fit well into either kingdom. As a result, biologists suggested additional kingdoms.

Electron microscopes confirmed the profound differences in cell structure between prokaryotes and eukaryotes. Taxonomists then added the kingdom Monera for the prokaryotes. Whittaker recognized that fungi are very different from plants and created a separate kingdom for them. During the 1970s and 1980s, Carl Woese of the University of Illinois and others compared ribosomal RNA sequences of many species. Their results suggested that Archaea are as different from other bacteria as bacteria are from eukaryotes. Many biologists responded to these studies by treating Archaea as a separate kingdom. Others compared the DNA sequences of animals and the genes involved in their development. This led to proposals to rewrite the evolutionary history of Animalia. Some of

 CHEMISTRY TIP

Ribonuclease (RNAase) is an enzyme that breaks down RNA. Because this enzyme is found nearly everywhere and is very difficult to inactivate, it is often easier to analyze the DNA sequences of ribosomal RNA genes than the RNA itself.

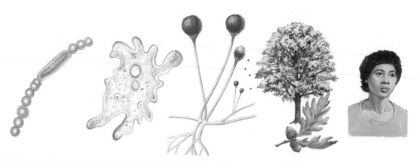

	Anabaena (Cyanobacteria)	Amoeba	Rhizopus (bread mold)	Quercus alba (white oak)	Homo sapiens (human)
kingdom	Monera	Protista	Fungi	Plantae	Animalia
phylum/division	Eubacteria	Sarcodina	Zygomycota	Anthophyta	Chordata
class	Cyanobacteria	Lobosa	Phycomycetes	Dicotyledoneae	Mammalia
order	Oscillatoriales	Amoebina	Mucorales	Fagales	Primata
family	Nostocaceae	Amoebidae	Mucoraceae	Fagaceae	Hominidae
genus	*Anabaena*	*Amoeba*	*Rhizopus*	*Quercus*	*Homo*
species	*circinalis*	*proteus*	*stolonifer*	*alba*	*sapiens*

FIGURE 18.26

Some examples of classifications. Some of these classifications may soon change.

these ideas would drastically change the relationships in Figure 18.19. Figure 18.26 shows the classification of an organism from each kingdom.

Some biologists do not accept the five-kingdom system. Some place multicellular green algae with the plants. Others have proposed new kingdoms for the red algae and other protist phyla.

Classification systems also reflect the aims of the person performing the classification. Grouping organisms as producers, consumers, or decomposers may work quite well for biologists studying ecology. It would not work well for those studying anatomy or evolutionary history.

Classification systems reflect our knowledge of the living world. Orders, kingdoms, and other groupings do not exist in nature. Classification simply helps us think more clearly about the great diversity that surrounds us. The "Brief Survey of Organisms" (Appendix 18A) shows a few of the millions of organisms that account for the diversity of life.

CONNECTIONS

Classification systems reflect the changes that groups of organisms undergo as they evolve.

Check and Challenge

1. Describe the most important difference that separates prokaryotes and eukaryotes.
2. List characteristics used to separate organisms into kingdoms.
3. Describe the characteristics of the five kingdoms, and give an example of an organism from each one.
4. List factors that result in changes in classification systems.

Biology Online   BSCSblue.com/check_challenge

Archaea and Classification

In some ways the Archaea are more like eukaryotes than like bacteria. These similarities include the shapes of their ribosomes and the sequences of their ribosomal RNA.

There are three types of Archaea. One type lives in hot, acidic environments (70° to 90°C and pH 2 or less) such as the hot sulfur springs at Yellowstone National Park. Another group requires a high concentration of salt to survive, such as the Great Salt Lake in Utah. The third type, the methanogens, are the most widely distributed. They live only in anaerobic conditions, including stagnant water, sewage-treatment plants, the ocean bottom, hot springs, and the digestive tracts of animals. Methanogens produce methane gas (CH_4) from hydrogen and carbon dioxide.

Extreme environments may have been more common when life began and the atmosphere was rich in carbon dioxide and hydrogen. This possibility led some scientists to suggest that the Archaea were among the first organisms on Earth. Although the name Archaea (ancient) remains, there is no clear-cut evidence to support this idea. It now appears that the eukaryotes may have originated from Archaea.

At this time, some biologists still place the Archaea in the kingdom Monera with other types of bacteria. Other biologists consider them a separate kingdom. After Carl Woese and his associates first recognized Archaea as a group, they proposed a superkingdom, or domain, level of classification (Figure 18.27). The three domains are Archaea, Eubacteria, and Eukarya (the four eukaryote kingdoms). This classification is based on ribosomal RNA sequences. The sequence of bases in rRNA changes very slowly. Since Woese first published his conclusions in 1977, biologists have compared other genes in the Archaea and other organisms. These studies have complicated the tree of life by showing that members of Archaea and Eubacteria are different in some ways but similar in others.

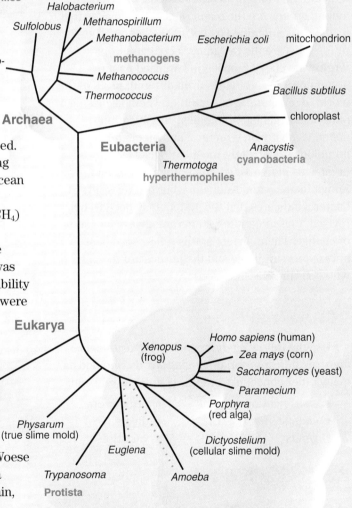

FIGURE 18.27

Carl Woese's classification. Woese's evolutionary diagram of Archaea, Eubacteria, and Eukarya, is based on the percentage of identical ribosomal RNA sequences. The farther apart two groups are, the more their rRNA sequences differ (e.g., the dotted green line shows the distance between *Euglena* and *Amoeba*).

Chapter
HIGHLIGHTS

Summary

Classification systems provide an organized approach to the study of the great diversity of organisms. Taxonomists have devised a system of groups and subgroups to name organisms and indicate their ancestral relationships. Differences among groups are the result of genetic changes that have been maintained by natural selection. These differences are the raw material of evolution. Worldwide use of Linnaeus's Latinized naming system enables biologists from different parts of the world to understand one another when discussing a particular organism. Biological classification depends mostly on structural and biochemical homologies. Currently, most biologists classify all organisms in five kingdoms—Monera, Protista, Fungi, Plantae, and Animalia—but further study of bacteria indicates that the system may need to be revised. Classification systems are always in a state of change. Their current status reflects the best information available and the purpose of the person who does the classification.

Key Concepts

Use the following terms to develop a concept map. Include appropriate linking words, and add as many other concepts from this chapter as you can.

- classification systems
- change
- genus
- prokaryote
- diversity
- variation

Reviewing Ideas

1. Define *species*. What are the problems associated with defining distinct species?
2. Give examples of chemical and structural relationships that indicate related ancestry.

3. How do eukaryotes and prokaryotes differ?
4. Compare the concepts of variation and diversity.

Using Concepts

1. How does Protista differ from the other kingdoms?
2. Is Woese's designation of Archaea as a kingdom separate from Monera more like a phenetic classification or a cladistic classification? Give reasons for your answer.
3. How do advances in technology change classification?

Synthesis

1. Can homologies suggest that many species evolved from a few species? Defend your answer through examples.
2. How does the Linnean system of groups and subgroups express the idea of degree of relatedness?

Extensions

Design and classify an imaginary organism in one of the five kingdoms. Sketch or write about the organism in detail. Give reasons for classifying it as you did.

Web Resources

Visit BSCSblue.com to access
- An online guide to the classification of species
- Taxonomy-related activities
- Guides to the diversity of organisms
- An introduction to microfossils
- Web links related to this chapter

A Brief Survey of Organisms

The variety of organisms seems almost endless. Biologists name and classify living things to reflect current scientific understanding of the biological and evolutionary relatedness among organisms. The classification scheme presented below divides living things into five major groups based on cell structure (prokaryotic or eukaryotic), body structure, and mode of nutrition, as proposed by Robert Whittaker in 1959 and advanced by Lynn Margulis and Karlene V. Schwartz in *Five Kingdoms: An Illustrated Guide to the Phyla of Life on Earth.* Superkingdom Prokarya has a single kingdom, Monera. Superkingdom Eukarya has four kingdoms: Protista, Fungi, Plantae, and Animalia. Another classification scheme, proposed by Carl Woese, is based on ribosomal RNA sequences and divides living things into three major groups: Prokaryotic cells are classified as either Archaea or Bacteria, and all other organisms are Eukarya. Biologists are studying both systems in order to better understand which one best expresses our changing understanding of the natural world. Where possible, common names and approximate size indications are given for the illustrated organisms. Scientific names are used for the organisms that have no common names. Major characteristics are listed for each group. Consult specialized reference works for more detailed information on a specific group of organisms.

Superkingdom Prokarya: prokaryotes; lack nuclei and other membrane-enclosed organelles; reproduction by several types of fission or budding; DNA-level recombination independent of reproduction

▶ **Kingdom Bacteria (Monera): circular DNA without protein; unicellular, multicellular, single or branched filaments; all major types of metabolism; more than 10,000 species described; two major groups (Subkingdoms): Archaea (Archaebacteria) and Eubacteria**

Phylum Archaea (Archaebacteria): distinctive small-subunit ribosomal RNA sequences; may represent the first organisms; widespread in seawater, lakes, soils, muds; many adapted to extreme conditions; methanogens, halophiles, thermoacidophiles

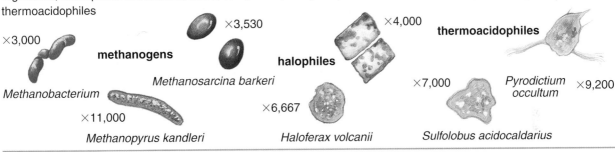

×3,000

×3,530

×4,000

thermoacidophiles

methanogens

halophiles

Methanobacterium

Methanosarcina barkeri

×7,000

Pyrodictium occultum

×9,200

×11,000

×6,667

Methanopyrus kandleri

Haloferax volcanii

Sulfolobus acidocaldarius

Phylum Eubacteria: three major groups based on characteristics of cell wall; includes cyanobacteria, proteobacteria (anaerobic phototrophic bacteria), chemolithotrophs, sulfur-reducing bacteria, spirochetes, aerobic nitrogen fixers, lactic acid bacteria, actinobacteria

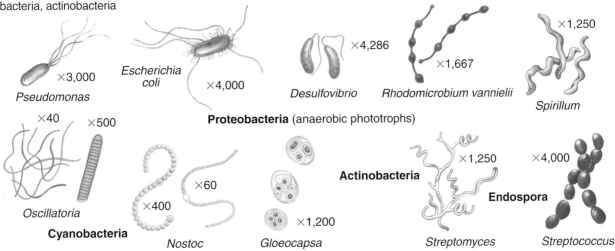

×3,000

Escherichia coli

×4,000

×4,286

×1,667

×1,250

Pseudomonas

Desulfovibrio

Rhodomicrobium vannielii

Spirillum

Proteobacteria (anaerobic phototrophs)

×40

×500

×1,250

×4,000

Actinobacteria

Endospora

×60

Oscillatoria

×400

Cyanobacteria

×1,200

Nostoc

Gloeocapsa

Streptomyces

Streptococcus

Superkingdom Eukarya: membrane-enclosed nuclei and organelles; reproduction by mitosis involving dissolution and reformation of nuclei in offspring cells; DNA organized into chromosomes complexed with protein

▶ **Kingdom Protista:** eukaryotic microorganisms and their descendants, exclusive of plants, animals, and fungi; includes algae, flagellated water molds, slime molds and slime nets, protozoa, and other obscure aquatic organisms; evolved from symbioses between two or more types of prokaryotes; estimates of extant species vary between 65,000 and 250,000

sarcodines (several phyla): locomotion by pseudopods; unicellular, many with intricate skeletal structures; heterotrophic

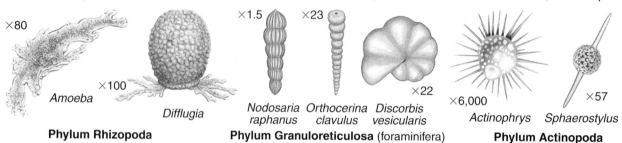

×80

×100

Amoeba

Difflugia

Phylum Rhizopoda

×1.5

×23

×22

Nodosaria raphanus *Orthocerina clavulus* *Discorbis vesicularis*

Phylum Granuloreticulosa (foraminifera)

×6,000

×57

Actinophrys *Sphaerostylus*

Phylum Actinopoda

Phylum Myxomycota (plasmodial slime molds): amoebalike colonies that feed on bacteria by engulfing them with pseudopodia; heterotrophic; sexual reproduction by spores produced in stalked reproductive structures

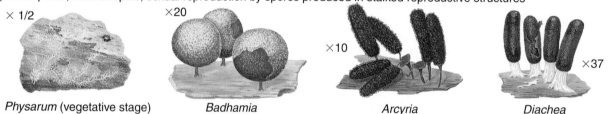

× 1/2

×20

×10

×37

Physarum (vegetative stage)

Badhamia

Arcyria

Diachea

Phylum Ciliophora (ciliates): locomotion by cilia; mostly unicellular; heterotrophic; contain macro- and micronuclei; about 10,000 species described

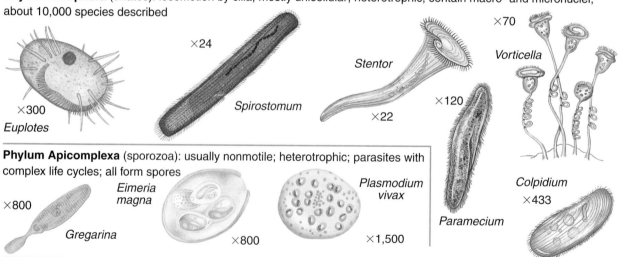

×24

×70

Stentor

Vorticella

×300

Spirostomum

×22

×120

Euplotes

Colpidium

×433

Paramecium

Phylum Apicomplexa (sporozoa): usually nonmotile; heterotrophic; parasites with complex life cycles; all form spores

Eimeria magna

Plasmodium vivax

×800

Gregarina

×800

×1,500

flagellates (several phyla): locomotion by flagella; unicellular or colonial; heterotrophic, autotrophic, or both

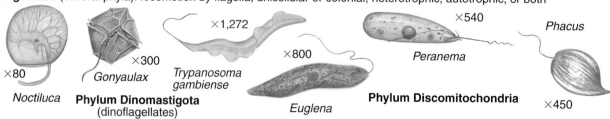

×1,272

×540

Phacus

×300

×800

Peranema

×80

Gonyaulax

Trypanosoma gambiense

Noctiluca **Phylum Dinomastigota** (dinoflagellates)

Euglena

Phylum Discomitochondria

×450

484

Phylum Diatoms (Bacillariophyta); unicellular, colonial, or multicellular; autotrophic; food stored as oil; two-part shells of silica; important as the base of aquatic food chains; about 10,000 living species

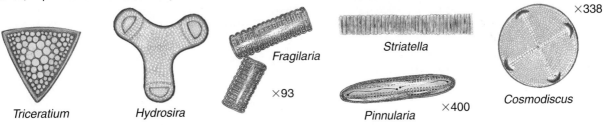

Triceratium

Hydrosira

Fragilaria ×93

Striatella

Pinnularia ×400

Cosmodiscus ×338

Phylum Phaeophyta (brown algae): multicellular, macroscopic (up to 10 m); marine; autotrophic; crucial primary producers, especially in intertidal zones; about 900 species

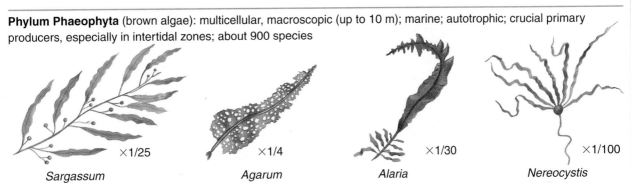

Sargassum ×1/25

Agarum ×1/4

Alaria ×1/30

Nereocystis ×1/100

Phylum Rhodophyta (red algae): multicellular, macroscopic; mostly marine, nonmotile gametes; autotrophic; source of agar; about 4,100 species

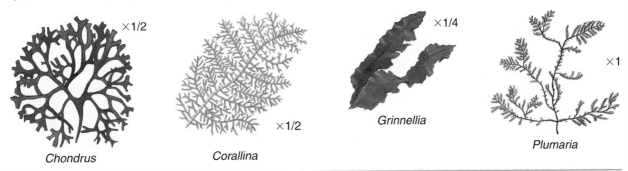

Chondrus ×1/2

Corallina ×1/2

Grinnellia ×1/4

Plumaria ×1

Phylum Chlorophyta (green algae): unicellular, colonial, or multicellular; autotrophic; major component of freshwater plankton; asexual and sexual reproduction; probably ancestral to plants; about 16,000 species described

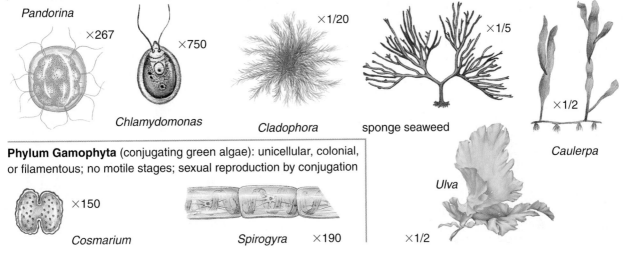

Pandorina ×267

Chlamydomonas ×750

Cladophora ×1/20

sponge seaweed ×1/5

Caulerpa ×1/2

Phylum Gamophyta (conjugating green algae): unicellular, colonial, or filamentous; no motile stages; sexual reproduction by conjugation

Cosmarium ×150

Spirogyra ×190

Ulva ×1/2

485

▶ **Kingdom Fungi:** eukaryotic; nonmotile; heterotrophic (saprotrophic); many parasitic; important decomposers; sources of antibiotics; used in production of cheese, beer, wine; reproduction by spores; three phyla (traditionally called divisions); about 60,000 species described

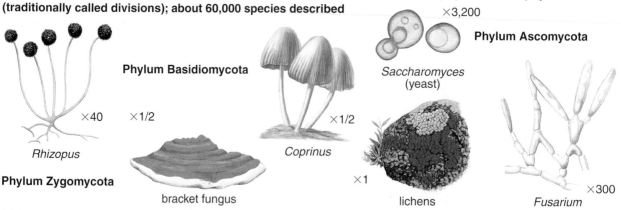

×3,200

Phylum Ascomycota

×40   ×1/2

Saccharomyces
(yeast)

Phylum Basidiomycota

×1/2

Rhizopus

Coprinus

×1

×300

Phylum Zygomycota

bracket fungus

lichens

Fusarium

▶ **Kingdom Plantae:** eukaryotic; multicellular; autotrophic; cellulose-containing cell walls; chloroplasts containing chlorophylls a and b; asexual and sexual reproduction, with alternation of generations; 12 phyla (traditionally called divisions); about 500,000 species described

Phylum Bryophyta (mosses): small; nonvascular; moist habitats; haploid gametophyte dominant; about 10,000 species

Phylum Hepatophyta (liverworts): nonvascular; moist habitats; gametophyte a flat structure (thallus); about 6,000 species

Phylum Anthocerophyta (hornworts): nonvascular; horn-shaped, elongate sporophytes; about 100 species

×1

Polytrichium
×1

scale moss
×1

×2

Sphagnum
(peat moss)

Marchantia

Anthoceros

Phylum Lycophyta (club mosses): vascular; with roots; horizontal stems; small leaves; spores borne on conelike tips; about 1,000 species

Phylum Psilophyta (whisk ferns): vascular; diploid sporophyte dominant; no roots; oldest fossil land plants; 3 species

Phylum Sphenophyta (horsetails): vascular; with roots; jointed stems containing silica; scalelike leaves; spores in conelike structures; about 40 species

×1/3

Lycopodium

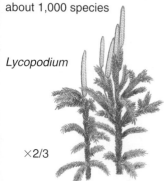

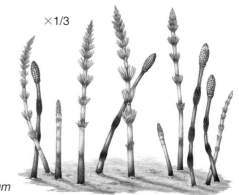

×2/3

×1/2

Psilotum

Equisetum

Phylum Filicinophyta (ferns): vascular; with roots; horizontal underground stems; compound leaves; gametophytes independent; about 12,000 species

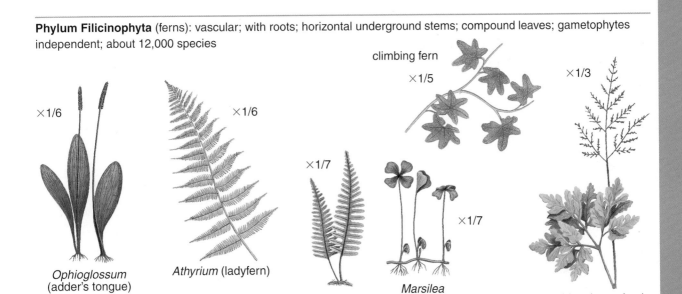

climbing fern
×1/5

×1/3

×1/6

×1/6

×1/7

×1/7

Ophioglossum
(adder's tongue)

Athyrium (ladyfern)

boulder fern

Marsilea
(water fern)

Botrychium (grapefern)

Phylum Cycadophyta (cycads): vascular; naked seeds in cones; about 185 living species

×1/10

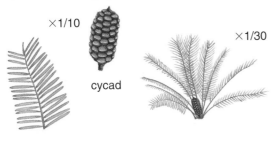

×1/30

cycad

Phylum Ginkgophyta
(ginkgoes): vascular;
naked seeds; flagellated
sperm; fan-shaped
leaves; 1 species

×1/2

Ginkgo

Phylum Coniferophyta (conifers): vascular; naked seeds in cones; needle-shaped leaves; many evergreen; about 700 species

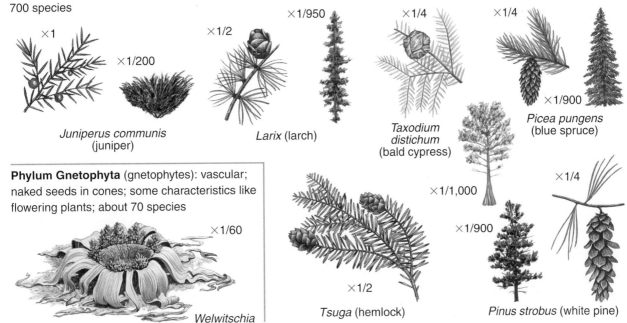

×1

×1/200

×1/2

×1/950

×1/4

×1/4

×1/900

Juniperus communis
(juniper)

Larix (larch)

*Taxodium
distichum*
(bald cypress)

Picea pungens
(blue spruce)

Phylum Gnetophyta (gnetophytes): vascular;
naked seeds in cones; some characteristics like
flowering plants; about 70 species

×1/60

×1/1,000

×1/900

×1/4

×1/2

Welwitschia

Tsuga (hemlock)

Pinus strobus (white pine)

487

Phylum Anthophyta (flowering plants): vascular; seeds enclosed in fruit; sperm in pollen tube; gametophyte reduced, inside sporophyte; about 250,000 species

Class Monocotyledoneae (monocots): flower parts in threes; parallel-veined leaves; embryo with one cotyledon; about 50,000 species

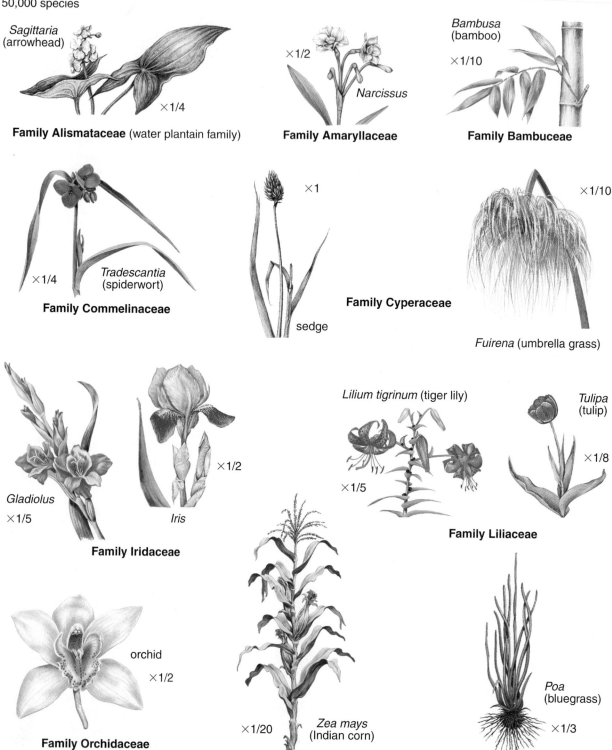

Sagittaria (arrowhead)

×1/4

Family Alismataceae (water plantain family)

×1/2

Narcissus

Family Amaryllaceae

Bambusa (bamboo)

×1/10

Family Bambuceae

×1/4

Tradescantia (spiderwort)

Family Commelinaceae

×1

sedge

Family Cyperaceae

×1/10

Fuirena (umbrella grass)

Gladiolus
×1/5

Iris

×1/2

Family Iridaceae

Lilium tigrinum (tiger lily)

×1/5

Tulipa (tulip)

×1/8

Family Liliaceae

orchid

×1/2

Family Orchidaceae

Zea mays (Indian corn)

×1/20

Poa (bluegrass)

×1/3

Family Poaceae (grass family)

Class Dicotyledoneae (dicots): flower parts in fours or fives; net-veined leaves; embryo with two cotyledons; about 200,000 species

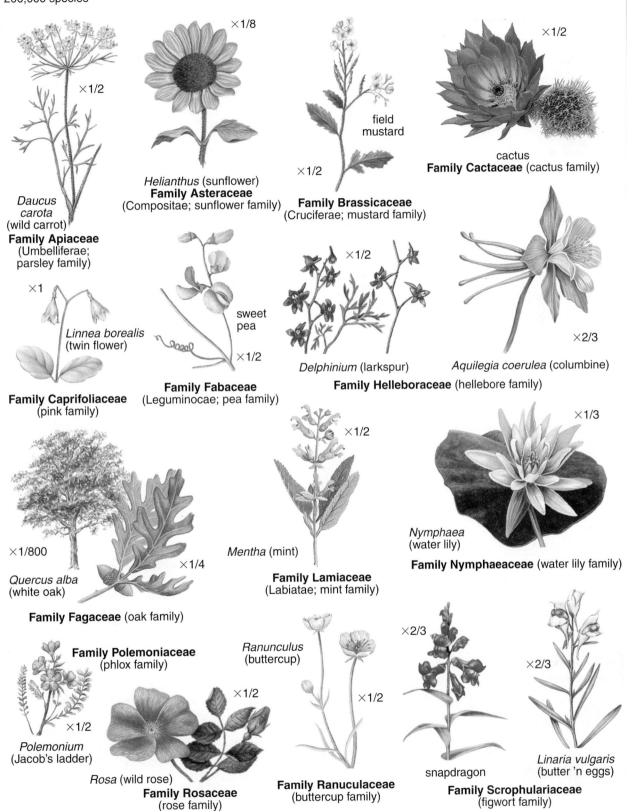

×1/8

×1/2

Daucus carota (wild carrot)
Family Apiaceae (Umbelliferae; parsley family)

Helianthus (sunflower)
Family Asteraceae (Compositae; sunflower family)

field mustard
×1/2
Family Brassicaceae (Cruciferae; mustard family)

×1/2
cactus
Family Cactaceae (cactus family)

×1
Linnea borealis (twin flower)
Family Caprifoliaceae (pink family)

sweet pea
×1/2
Family Fabaceae (Leguminocae; pea family)

×1/2
Delphinium (larkspur)

×2/3
Aquilegia coerulea (columbine)
Family Helleboraceae (hellebore family)

×1/800
Quercus alba (white oak)
×1/4
Family Fagaceae (oak family)

×1/2
Mentha (mint)
Family Lamiaceae (Labiatae; mint family)

×1/3
Nymphaea (water lily)
Family Nymphaeaceae (water lily family)

Family Polemoniaceae (phlox family)
×1/2
Polemonium (Jacob's ladder)

×1/2
Rosa (wild rose)
Family Rosaceae (rose family)

Ranunculus (buttercup)
×1/2
Family Ranuculaceae (buttercup family)

×2/3
snapdragon

×2/3
Linaria vulgaris (butter 'n eggs)
Family Scrophulariaceae (figwort family)

489

▶ **Kingdom Animalia:** eukaryotic; multicellular; heterotrophic; most motile at some stage; reproduction mainly sexual; 33 phyla; more than 1 million species

Phylum Porifera (sponges): sessile; two cell layers; body with pores; asexual and sexual reproduction; about 10,000 species

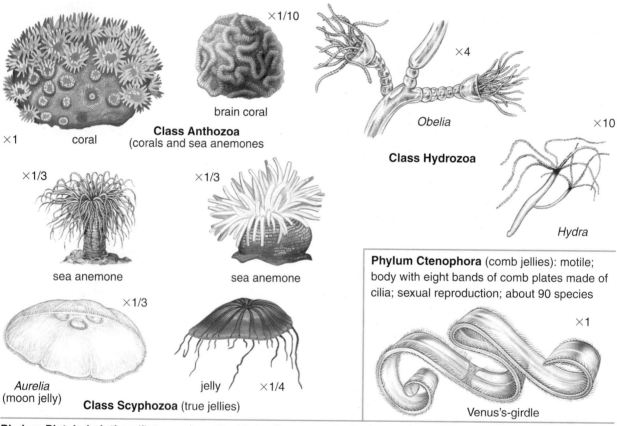

×1/2

×1/4

×1/7

×1

Grantia

finger sponge

bath sponge

sheep's wool sponge

Phylum Cnidaria (stinging tentacled animals): motile stages; radially symmetrical; two cell layers; mouth surrounded by stinging tentacles; saclike gut; nerve network; asexual and sexual reproduction; about 9,700 species

×1/10

×4

brain coral

Obelia

×1

coral

Class Anthozoa
(corals and sea anemones

Class Hydrozoa

×10

×1/3

×1/3

Hydra

sea anemone

sea anemone

Phylum Ctenophora (comb jellies): motile; body with eight bands of comb plates made of cilia; sexual reproduction; about 90 species

×1/3

×1

Aurelia
(moon jelly)

jelly   ×1/4

Class Scyphozoa (true jellies)

Venus's-girdle

Phylum Platyhelminthes (flatworms): motile, bilaterally symmetrical; three cell layers; head; gut with one opening; many parasitic; asexual and sexual reproduction; about 15,000 species

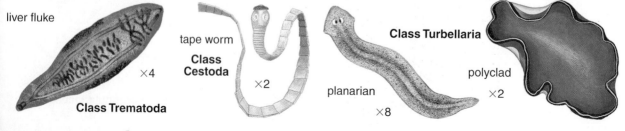

liver fluke

tape worm

Class Turbellaria

Class
Cestoda

polyclad

×4

×2

planarian

×2

Class Trematoda

×8

Phylum Nematoda (roundworms): motile; bilaterally symmetrical; slender, cylindrical body; gut with two openings; many parasitic; more than 80,000 species

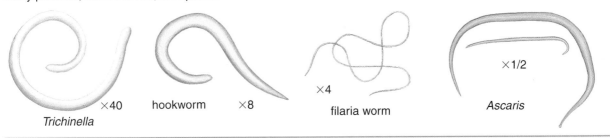

Trichinella ×40

hookworm ×8

filaria worm ×4

×1/2 Ascaris

Phylum Rotifera (rotifers): mostly motile; bilaterally symmetrical; cilia form a wheel around mouth; asexual and sexual reproductive stages; about 2,000 species

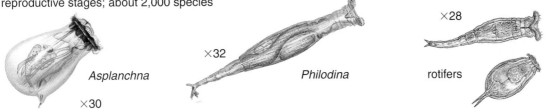

Asplanchna ×30

×32 Philodina

×28 rotifers

Phylum Chelicerata (horseshoe crabs, spiders, scorpions, and ticks; formerly part of Phylum Arthropoda): motile; bilaterally symmetrical; two body parts; jointed appendages; a chitinous exoskeleton; no wings; lack the antennae and mandibles of crustaceans and insects; gas exchange through trachea, lungs, or gill books; about 75,000 species

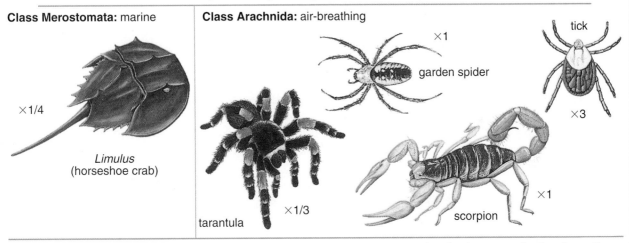

Class Merostomata: marine

×1/4

Limulus
(horseshoe crab)

Class Arachnida: air-breathing

×1 garden spider

tick ×3

tarantula ×1/3

scorpion ×1

Phylum Crustacea (formerly part of Phylum Arthropoda): two body parts; two pairs of antennae on the head; mostly aquatic; gas exchange through gills; about 45,000 species

Class Branchiopoda

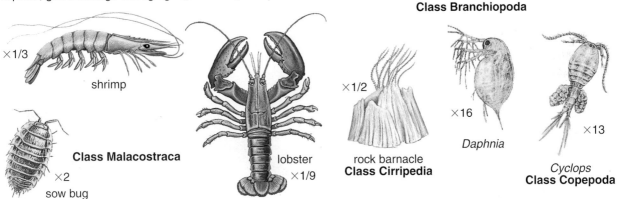

×1/3 shrimp

Class Malacostraca
×2 sow bug

lobster ×1/9

×1/2 rock barnacle
Class Cirripedia

×16 Daphnia

Cyclops ×13
Class Copepoda

Phylum Mandibulata (insects, millipedes and centipedes; formerly part of Phylum Arthropoda): three body parts; one pair of antennae; three pairs of legs; generally one or two pairs of wings; gas exchange through trachea

Class Hexapoda (Insecta): essentially terrestrial; one pair of antennae; generally one or two pairs of wings; usually three pairs of legs; gas exchange through trachea; perhaps 10 million species

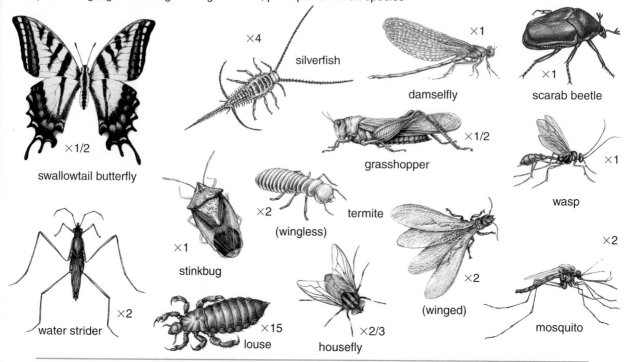

×1/2
swallowtail butterfly

×4
silverfish

×1
damselfly

×1
scarab beetle

×1/2
grasshopper

×1
wasp

×1
stinkbug

×2
termite
(wingless)

×2
(winged)

×2
mosquito

×2
water strider

×15
louse

×2/3
housefly

Class Myriapoda (centipedes and millipedes)

Order Diplopoda (millipedes): many-segmented, cylindrical body; one pair of short antennae; two pairs of legs per segment; gas exchange through trachea; important forest decomposers; about 10,000 species

Order Chilopoda (centipedes): many-segmented, flattened body; one pair of long antennae; one pair of legs per segment; gas exchange through trachea; predators; about 2,500 species

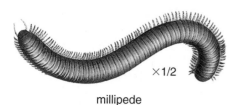

×1/2
millipede

×1
centipede

Phylum Annelida (segmented worms): motile; bilaterally symmetrical; body internally and externally segmented; asexual and sexual reproduction; about 8,700 species

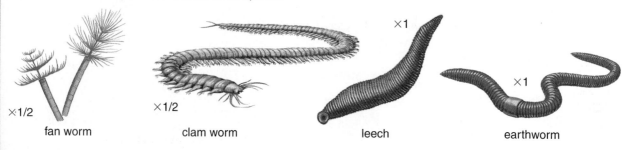

×1/2
fan worm

×1/2
clam worm

×1
leech

×1
earthworm

492

Phylum Mollusca (mollusks): sessile or motile; bilaterally symmetrical; soft-bodied, usually with shell; true body cavity; well-developed digestive, circulatory, and nervous systems; sexual reproduction; about 50,000 species

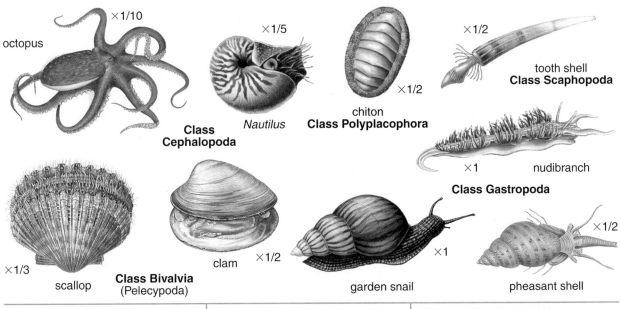

octopus ×1/10

Nautilus ×1/5

Class Cephalopoda

chiton ×1/2
Class Polyplacophora

tooth shell ×1/2
Class Scaphopoda

nudibranch ×1
Class Gastropoda

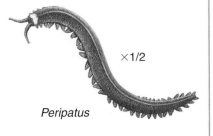

scallop ×1/3

clam ×1/2
Class Bivalvia (Pelecypoda)

garden snail ×1

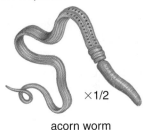

pheasant shell ×1/2

Phylum Onycophora (velvet worms): unsegmented; one pair of short antennae; a link between annelids and arthropods; about 80 species

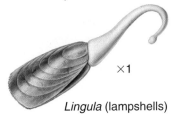

Peripatus ×1/2

Phylum Brachiopoda (lampshells): attached or burrowing; bilaterally symmetrical; dorsal-ventral shells enclosing tentacle-bearing arms; open circulatory system; about 335 species

Lingula (lampshells) ×1

Phylum Hemichordata (acorn worms): unsegmented; bilaterally symmetrical; dorsal nerve cord and pharyngeal gill slits; about 65 species

acorn worm ×1/2

Phylum Echinodermata (echinoderms): no head or brain; unsegmented; adults radially symmetrical; larvae bilaterally symmetrical and chordatelike; tube feet serve for feeding, locomotion, and gas exchange; about 6,000 species

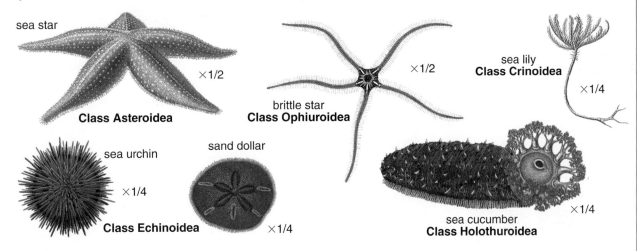

sea star ×1/2
Class Asteroidea

brittle star ×1/2
Class Ophiuroidea

sea lily
Class Crinoidea ×1/4

sea urchin ×1/4
Class Echinoidea

sand dollar ×1/4

sea cucumber ×1/4
Class Holothuroidea

Phylum Urochordata (tunicates; formerly part of Phylum Chordata): notochord disappears marine; adults nonmotile; during development; no brain; about 1,400 species

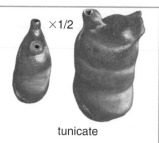

×1/2

tunicate

Phylum Cephalochordata (lancelets; formerly part of Phylum Chordata): found on shallow, sandy sea floors; notochord and nerve cord extend the length of body and persist in adults; no brain; 23 species

×3/4

lancelets

Phylum Craniata (vertebrate animals; formerly part of Phylum Chordata): single, hollow nerve cord becomes the brain and spinal cord; brain lies within a cranium (skull); cartilaginous notochord present, may be replaced by bony vertebral column; gill slits present at some developmental stage, may grow shut or become other structures; about 45,000 species

Class Cyclostomata (jawless fish; lampreys, hagfish): cartilaginous skeleton; no true jaws or paired appendages; two-chambered heart; gas exchange through gills; about 50 species

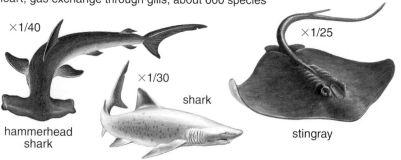

lamprey
×1/10

Pacific hagfish
×1/7

×1/7

Atlantic hagfish

Class Chondrichthyes (cartilaginous fish): cartilaginous skeleton; gill slits visible; jaws; paired fins; two-chambered heart; gas exchange through gills; about 600 species

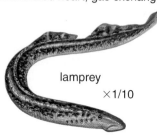

×1/40

×1/30

shark

hammerhead shark

×1/25

stingray

Dinichthys (extinct)

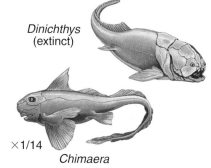

×1/14

Chimaera

Class Osteichthyes (bony fish): bony skeleton; gill slits covered; jaws; paired fins; scales; two-chambered heart; gas exchange through gills; about 25,000 species

×1/8

toadfish

tuna
×1/25

flying fish

×1/8

×1/10

cowfish

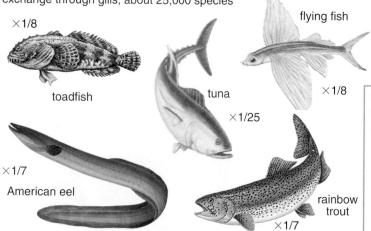

×1/7

American eel

rainbow trout
×1/7

Class Choanichthyes (lungfish): live in freshwater lakes with low dissolved oxygen; can gulp air when lakes dry up

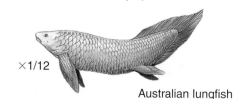

×1/12

Australian lungfish

Class Amphibia (amphibians): bony skeleton; moist, glandular skin; most with two pairs of limbs; three-chambered heart; gas exchange through skin or lungs in adult; gas exchange through gills in larvae; about 200 described species

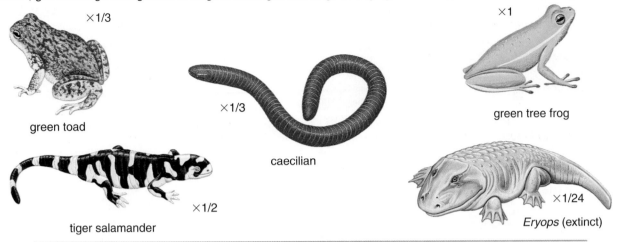

×1/3

green toad

×1/3

caecilian

×1

green tree frog

×1/2

tiger salamander

×1/24

Eryops (extinct)

Class Reptilia (reptiles): bony skeleton; dry, scaly skin; most with two pairs of limbs; leathery-shelled eggs; incompletely divided four-chambered heart; gas exchange through lungs; internal fertilization; about 5,000 species

×1/3

chameleon

×1/4

box turtle

×1/40

alligator

×1/2

six-lined race runner

×1/5

Gila monster

×1/10

rattlesnake

Pteranodon (extinct)

×1/120

×1/6

corn snake

×1/115

Triceratops (extinct)

×1/120

Apatosaurus (extinct)

495

Class Aves (birds): bony skeleton; scales modified as feathers; no teeth; forelimbs modified as wings; hard-shelled eggs; four-chambered heart; endothermic; gas exchange through lungs; nearly 9,000 living species

×1/14
turkey vulture

×1/7
yellow-headed parrot

×1/80
ostrich

×1/8
quetzal

×1/16
barred owl

×1/3
broad-tailed hummingbird

×1/2
tree sparrow

×1/7
mourning dove

×1/3
lazuli bunting

×1/5
American robin

×1/36
albatross

×1/25
flamingo

×1/27
brown pelican

×1/24
king penguin

×1/12
kiwi

×1/8
grebe

×1/16
red jungle fowl

×1/20
Hesperornis (extinct)

×1/14
prairie chicken

×1/14
ring-necked pheasant

Class Mammalia (mammals): bony skeleton; scales modified as hairs; mammary glands in females secrete milk; four-chambered heart; endothermic; gas exchange through lungs; about 4,500 living species

rhinoceros
×1/40

×1/27
aardvark

×1/16
koala

horse
×1/60

×1/35
red kangaroo

×1/4
long-eared bat

×1/13
armadillo

elephant
×1/80

×1/20
porcupine

×1/32
cougar

baboon
×1/22

sea lion
×1/30

platypus
×1/8

gorilla
×1/40

×1/80
manatee

×1/14
cottontail rabbit

×1/10
pika

×1/32
human

blue whale
×1/450

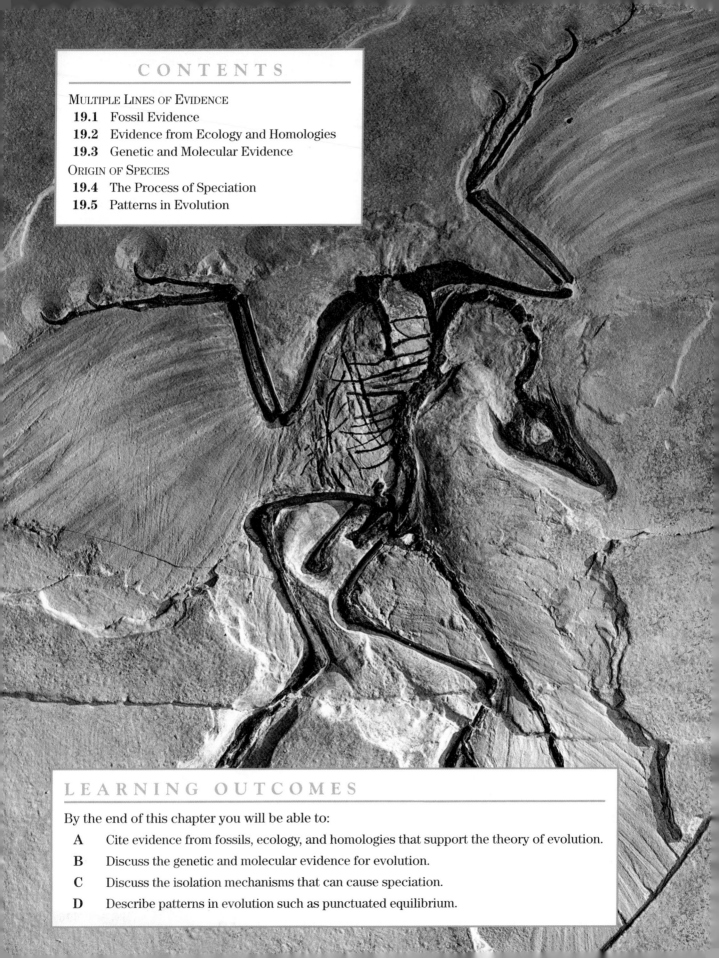

CONTENTS

LEARNING OUTCOMES

By the end of this chapter you will be able to:

A Cite evidence from fossils, ecology, and homologies that support the theory of evolution.

B Discuss the genetic and molecular evidence for evolution.

C Discuss the isolation mechanisms that can cause speciation.

D Describe patterns in evolution such as punctuated equilibrium.

◀ **About the photo**
This photo shows a
casting of *Archaeopteryx*.

CHAPTER

19

Changes in Species

■ *How does evolution account for the appearance of new species and the disappearance of other species?*

■ *Why do scientists need to use different lines of evidence to understand evolution?*

The history of life is a story of ancient organisms and of their modern descendants. Earth is inhabited by an enormous variety of organisms. However, several lines of evidence—including the fossil record, genetic data, comparative morphology, biochemistry, and embryology—support the conclusion that all life had a common ancestor.

How did different species arise? Inherited variation in populations gives some organisms an improved chance to survive and reproduce. Their characteristics are better represented in the next generation because the individuals with these characteristics have more offspring. This process of natural selection is an important part of evolution. Evolution is long-term change in inherited characteristics, a process that Charles Darwin called "descent with modification." Over millions of years, new species evolve from some members of existing species.

This chapter examines the evidence of evolution as it occurred in the past and as it continues to take place. It also describes evidence of different patterns of evolutionary change.

Multiple Lines of Evidence

19.1 Fossil Evidence

The idea of biological evolution did not start or end with Charles Darwin's publication of *The Origin of Species* in 1859. His grandfather Erasmus Darwin, among others, put forward many aspects of evolution long before Charles became a naturalist. Charles Darwin stands out because his book introduced evolution as a testable scientific theory. His book is based on an enormous amount of data, collected during 28 years of study. The idea of evolution is presented in a well-reasoned way and is supported by multiple lines of evidence. Since the publication of Darwin's book in 1859, scientists have expanded, refined, and retested this powerful theory to explain long-term changes in biological species.

You may have heard someone refer to evolution as "just a theory," as though this important idea is somehow different or less well accepted than other major scientific explanations of natural phenomena. Statements like this often show that the speaker does not fully understand what a scientific theory is. Evolution—like the scientific theories of gravity, plate tectonics, atomic structure, and cell structure—is indeed a theory. It is one of the most important

Focus On
Geological Time

The physical appearance of Earth has not always been what it is today. For 4.6 billion years, volcanoes, earthquakes, continental drift, weathering, plate tectonics, chemical change of minerals, and the effects of living organisms have all helped produce our planet's surface features. Modern geology began to develop in the 18th century. At that time, James Hutton studied the chemical and physical processes involved as one type of rock changes into another. Another pioneer in geology, William Smith, was an engineer who cut canals. He observed the layers, or strata, of rock with embedded fossils that he could see in the canal cuts he was making (Figure 19.1). These strata tell a story of geological changes. Charles Lyell, a 19th-century geologist, was a friend of Charles Darwin who also greatly influenced his thinking about long-term changes in the Earth's environments.

Evidence for geological time is based on radioactive decay. For example, the uranium isotope ^{238}U decays to lead-206 (^{206}Pb). The rate of decay is very slow. It takes 4.8 billion years for half of the initial amount of ^{238}U to decay to ^{206}Pb. The ratio of these isotopes in a rock or fossil is used to calculate its age (see Biological Challenges: Using Radioactivity to Date Earth Materials in Section 17.1).

FIGURE 19.1

Rock layers. Layers of exposed rock along Interstate 68 in Maryland. These rock layers and the fossils they contain reveal the history of the area. The deepest layers are the oldest.

ideas in biology. Among professional biologists, evolution, the long-term change in species, is not a controversial idea. Evolution is an area of active research. Scientists are currently exploring the details of how certain species are related, how molecules undergo evolution, and how modern species are changing. To a scientist, calling evolution a theory is a statement of confidence if the theory has been well-tested scientifically and is supported by a large body of evidence.

Evidence of evolution includes physical remains, or fossils, that are the record of ancient organisms. Fossil evidence provides an exciting, tangible view of life in very ancient times. The fossil record in different strata of the earth shows the order of evolutionary change. The study of fossils is a branch of biological science called **paleontology.** To understand evolution, you must first grasp what is meant by "ancient" in the sense of geological time.

Microfossils (Section 17.6) are examples of ancestral species recorded in the fossil record that are related to modern species. Cyanobacteria and diatoms may be the descendents of some of the earliest life forms on Earth. Diatoms are single-celled photosynthetic organisms that live in water (see Unit 1 opener photo on page 20). Their hard glassy shells can be preserved as fossils. The hard parts of other organisms are also the most likely to be preserved. Bones and shells, for instance, can survive decay. Harder minerals replace the original substance of these animal parts, preserving them as fossils. Softer tissue sometimes makes a fossil record if it leaves an impression in soft mud that hardens and is preserved, as seen in the fossil of an ancient plant in Figure 19.2a. Occasionally, ancient insects were completely preserved when they were trapped in tree sap that became amber (Figure 19.2b).

Around 250,000 species of fossil organisms have been found. This huge body of evidence has helped scientists reconstruct much of the history of life. Even so, biologists estimate that fossils of only 1 in 10,000 extinct species have been discovered. The fossil record is incomplete because most dead organisms decay and do not become fossils. The adventure of fossil hunting remains a great lure to both scientists and amateurs.

Fossil evidence supports the theory of evolution in a variety of ways. One of the most obvious ways is that fossils offer physical records of organisms not found on Earth today. Comparisons of fossils in younger, shallower rock

E T Y M O L O G Y

paleo- = ancient (Greek)
onto- = thing or being (Greek)
-logy = study (Greek, Latin)

Paleontology is the study of ancient things such as fossils.

Try Investigation 19A
Geological Time.

CHEMISTRY TIP

Diatom shells consist mostly of silica, SiO_2. Bones and teeth are made mostly of a form of calcium phosphate, $Ca_5H(PO_4)_3$, and mollusk shells consist of calcium carbonate, $CaCO_3$. During fossilization, silica often replaces the softer calcium compounds. The organic compounds in soft tissues are usually consumed by decomposer organisms, so fossils of these tissues are less common.

a

b

FIGURE 19.2

Two kinds of fossils. (a) A leaf fossil impression of *Pecopteris* dates back to the Pennsylvanian Period, approximately 300 million years ago. **(b)** This sample of amber with trapped flies and a cricket is approximately 35 million years old. The dark reddish area around the cricket shows where the insect tried to get free after being caught in the liquid resin.

Biological Challenges

Soft-Tissue Fossils

The deep thick deposits of ancient ice in the Arctic can hold some remarkable surprises. In rare cases, whole organisms have been found in ice, preserved fairly well for thousands of years. Although these fossils are not nearly as old as the fossilized bones of dinosaurs or other very ancient organisms, they are remarkable because some of their soft tissue has been preserved.

One exciting "icebox fossil" is a 20,000-year-old woolly mammoth (Figure 19.3a). It was recovered whole from the ice of Siberia in October 1999. This species, a close relative of the two surviving elephant species, has been extinct for 10,000 years. Mammoths lived during the Pleistocene epoch, the period between 1.8 million and 10,000 years ago. They overlapped in time with early humans. Mammoths and elephants (Figure 19.3b) are two species that share a common ancestor.

This mammoth was not the first to be found, but it is unusual in being removed intact. Called the Jarkov mammoth in honor of the family who found it, this specimen will be the subject of many scientific studies.

A group in Japan hopes to revive this extinct species. Although the project is unlikely to succeed, there are several methods that scientists can attempt. If sperm from the mammoth can be recovered intact, scientists may try to crossbreed a mammoth with an elephant. This hybrid cross would be difficult because the mammoth has 58 chromosomes and the elephant has 56. There is a close genetic homology, however (see Section 18.2). DNA sequences are only about 5% different between the extinct and the modern species.

Another interesting experiment would be to use DNA or a cell nucleus from the mammoth for a cloning experiment. In a few cases, scientists have cloned a mammal, but experiments such as this rarely work. Even if no new mammoths are born, cloning fragments of the mammoth's DNA will enable scientists to study its genes and compare them to those of elephants. This research may shed new light on the evolution of both mammoths and elephants.

a

b

FIGURE 19.3

An extinct animal and its living relative. (a) The woolly mammoth was a large mammal that has been extinct for 10,000 years. **(b)** This African elephant, *Loxodonta africana,* is one of two modern living elephant species; the other is the Indian elephant.

FIGURE 19.4

Some extinct animals. (a) *Eryops,* living approximately 280 million years ago, was one of the first air-breathing animals in a line that would give rise to amphibians. **(b)** The *Diatryma,* living 38 to 2 million years ago, was a flightless bird that may have gone extinct because small mammals ate their eggs. **(c)** The *Syndyoceras* may be related to modern ruminants (cows and deer). **(d)** The *Megatherium* was a ground sloth that went extinct about 11,000 years ago.

deposits and older, deeper rocks also help scientists determine the ancestral relationships of extinct and living species. Figure 19.4 shows a variety of extinct organisms.

Extinctions occur much more often in modern times than they did in the past. This rate change is due mostly to the huge increase in human population. The resulting changes and their impact on the environment have eliminated thousands of species. People have observed some extinctions directly; the passenger pigeon is one such example. This species was one of the most abundant birds in North America in the early 19th century. By 1914, however, hunters had killed enough to bring about this species' extinction.

In some cases, fossils record organisms before and after new species split off from older ones. Fossils are also evidence of the rate of evolutionary change. One of the most intriguing types of fossil evidence is that of intermediate stages between species and their ancestors. In recent years, great excitement has been raised over fossils that seem to bridge the morphological gap between carnivorous dinosaurs, the theropods *Tyrannosaurus rex* and *Velociraptor,* and birds, an early example of which is *Archaeopteryx* (see Figure P.5b). Since the mid-1990s, scientists have uncovered several examples of fossilized "feathered dinosaurs" in China (Figure 19.5). The feathers, wings, and reptilelike teeth and tails of these fossils reveal some of the intermediate stages in the evolution of birds from dinosaurs.

FIGURE 19.5

An example of a transitional fossil.
(a) A model of *Sinornithosaurus millenii,* a turkey-sized feathered dinosaur whose fossilized remains were found in China. **(b)** Despite the dinosaurlike claws, teeth, and tail, this fossil shows the highly advanced shoulder girdle that allowed for flapping arms, a feature almost identical to that of *Archaeopteryx,* the earliest known bird.

a   b

19.2 Evidence from Ecology and Homologies

Evidence of evolution is not limited to remnants of past organisms. Direct observations of modern, living species show how species are related and how natural selection acts on inherited variation to change species. Differences among closely related species often reflect adaptations to different environments. For example, during his visit to the Galápagos Islands in 1835 on the voyage of the *Beagle*, Darwin observed many differences among the organisms on various islands. The giant tortoises of the Galápagos often feed on cactus. On islands where the soft edible leaves of the cacti grow high on tough woody stems, the tortoises have evolved long necks and flared saddleback shells (Figure 19.6a). On islands where cacti are low-growing, the tortoises lack this adaptation (Figure 19.6b). The development of woody cactus stems and flared tortoise shells is an example of **coevolution,** the continuous adaptation of different species to each other.

Evidence of coevolution is often found in the adaptations of predators and their prey. For example, many plants produce toxic compounds that protect them from being eaten by animals. Toxic plants often have a few specialized predators that are adapted to tolerate, store, or excrete these compounds. The sensitivity and distaste of humans for bitter substances probably protect us from many plant toxins. Many symbiotic species (see Section 24.3) have coevolved to the point that neither can survive without the other (see Appendix 12A, "Pollination by Insects Aids Fertilization").

One of the strengths of the theory of evolution is that it can be tested. Artificial selection is one way to drive evolution experimentally. The breeding of domestic species provides evidence of how evolution occurs through natural selection. Darwin was particularly intrigued by the work of English breeders who had produced great variation among domestic pigeons (see Figure 16.10). Artificial selection works in the same manner as natural selection but with deliberate choices by the breeder.

ETYMOLOGY

co- = with (Latin)

Coevolution is evolution of interacting species in response to one another.

CHEMISTRY TIP

Many plant toxins are alkaloids, nitrogen-containing compounds synthesized from amino acids. Because they are bases, alkaloids have a bitter taste. Examples of alkaloids include the deadly poison strychnine ($C_{21}H_{22}N_2O_2$) and the narcotic drug morphine ($C_{17}H_{19}NO_3$).

a

b

FIGURE 19.6

Coevolution in Galápagos tortoises. (a) On some islands, tortoises must feed on cactus with tough lower stems. These tortoises have flared saddleback shells. The elevated anterior portion of the shell allows this tortoise to raise its head high enough to reach the edible cactus leaves. **(b)** On islands where the cacti do not have tall, woody stems, tortoises without saddleback shells can easily reach the edible leaves.

FIGURE 19.7

A controlled experiment in natural selection. The coat color of mice must blend with the background if the mice are to have their best chance of surviving.

CONNECTIONS

Biologist Theodore Dobzhansky said, "Nothing in biology makes sense except in the light of evolution." The theory of evolution ties together and explains adaptation, genetics, and the history and diversity of life.

Observations of the effects of natural and artificial selection provide evidence of evolution, but nature's "experiments" rarely include good controls. Controlled experiments can produce stronger evidence and show how natural selection occurs. For example, the hypothesis that natural selection by moth-eating birds was responsible for changes in the colors of peppered moths (see Figure 16.4) was based on observations of wild moth populations. Around 1930, Lee R. Dice of the University of Michigan conducted an experiment with mice and barn owls that showed how attacks by predators can lead to changes in protective coloration. The mice were either buff-colored or gray, and the experiment used matching colors of soil. Dice alternated the color of soil on the floor of an enclosed room each day. An ample supply of sticks was also added to give the mice a place to hide. Each day four mice of each color were released and exposed to an owl for 15 minutes. Depending on the color of soil used, one set of mice was more easily seen against the background than the other set (Figure 19.7). In 44 trials with each soil type, the owl caught almost twice as many (107 to 65) of the more visible color of mice.

Another class of evidence comes from homologies. Homologies help biologists understand the history of evolutionary changes. Any aspect of an organism—including its anatomical structure, its behavior, and the structure of its DNA, proteins, and other chemical components—can be compared to other species in the search for homologies. As biologists have learned more about the genes involved in the development of plant and animal embryos, they have found that homologous genes are responsible for the formation of the body plan and organs even in very distantly related species (see Figure 10.12).

CHEMISTRY TIP

Chemical analysis helps demonstrate homology. Some plants produce lipids, such as cinnamic acid ($C_9H_8O_2$), that are polymers of the 5-carbon molecule isoprene (C_5H_8). These compounds are responsible for the unique colors, flavors, and odors of many spices and other plant products. Comparisons of the biochemical pathways that produce these substances reveal homologies that help biologists classify these plant species.

isoprene   **cinnamic acid**

19.3 Genetic and Molecular Evidence

Inheritance was poorly understood during Darwin's time. One of Darwin's major problems in defending his theory of evolution by natural selection was explaining how variations are inherited. Since inheritance became more clearly understood in the 20th century, the study of genetics has provided fundamental support for the theory of evolution.

Try Investigation 19B 🔻 **A Model Gene Pool.**

One source of genetic variation is mutation (see Section 15.4). A second source of variation is the recombination of alleles in sexually reproducing eukaryotes. Crossing over, independent assortment, and fertilization produce entirely new combinations of genes. This genetic variation is the raw material of evolution.

One kind of mutation that provides evidence of the history of evolution is gene duplication. Duplication of a gene produces gene families—multiple copies of nearly identical DNA sequences. Some of the copies, called **pseudogenes,** no longer function. They are neither transcribed nor translated. Because pseudogenes are not expressed, they are not subjected to natural selection. Therefore, evolutionary theory predicts pseudogenes accumulate mutations faster than functional genes in the same family. Genetic studies have confirmed this prediction.

Molecular data provide detailed evidence of the degree of relatedness between species (see Tables 18.1 and 18.2). Scientists compare the amino-acid sequences of homologous proteins in different species. They also compare the nucleotide sequences of homologous genes. For example, humans and other vertebrates have two forms (α and β) of a protein called globin. Close similarities in the two forms of globin indicate homology between them. The similarity is evidence that the two globin genes are the result of gene duplication. Together with anatomical evidence, the duplicated genes support the hypothesis that all vertebrates share a common ancestor. Similarities in *Hox* genes (see Section 10.4) that help regulate organ development are also evidence of relatedness. Biologists are now studying *Hox* genes to determine whether mutations in these genes contribute to the formation of new species.

The role of evolution in human disease is a growing focus of medical science. For example, a dramatic and somewhat dangerous evolutionary process in bacteria has been analyzed at the genetic and molecular level. Many strains of bacteria that once were killed by antibiotics are now resistant. A serious health threat is the increase in tuberculosis infections that are resistant to a variety of antibiotics. Widespread use of antibiotics for the past 50 years has produced a selective pressure to favor the previously rare resistant bacteria. In the laboratory, scientists can demonstrate this evolutionary process by gradually increasing the exposure of bacteria to certain antibiotics. This selection pressure results in a new population that is resistant. Consequently, the antibiotic removes most of the competition from the new population. A similar process has led to the development of pesticide resistance in insects.

Mutation is especially common in short repeated nucleotide sequences called microsatellites. They may contribute to evolution by increasing genetic variation. For example, the repeat CTCTT in the pathogenic bacterium *Neisseria gonorrhoeae* is the site of frequent mutations (Figure 19.8). These mutations determine whether the bacteria can infect a person and produce the disease gonorrhea. In a population of *N. gonorrhoeae*, some bacteria cause disease and others survive without causing disease. The microsatellite acts as a genetic switch. It keeps the bacterial population diverse and ensures that at least some bacteria can survive under changing conditions.

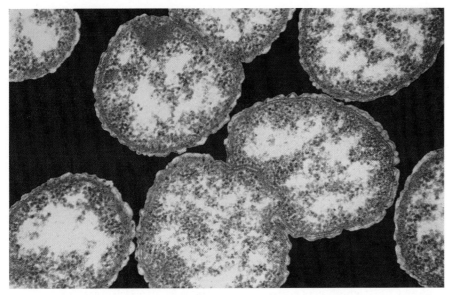

FIGURE 19.8

Mutations in microsatellites. This transmission electron micrograph shows several cells of *Neisseria gonorrhoeae,* the bacterium that causes the sexually transmitted disease gonorrhea (color added, ×30,000). A highly mutable region of microsatellite DNA determines whether the organism will cause disease.

Check and Challenge

1. Compare evidence of evolution from the fossil record to evidence from direct observation of modern organisms.

2. What evidence did the Galápagos Islands provide for evolution by means of natural selection?

3. Compare natural selection and artificial selection.

4. How does evolution play a role in the development of antibiotic-resistant bacteria?

5. Distinguish between mutation and genetic recombination. Explain how they contribute to evolution.

6. What is the significance of a nearly universal genetic code with respect to evolution?

Origin of Species

19.4 The Process of Speciation

Has anyone actually observed **speciation,** the appearance of a new species? Yes. Although most evolutionary change is too slow to see, there are examples in which it has been observed. In addition to changes in bacteria, recent history has witnessed the appearance of several new species of multicellular organisms. Since 1969 a new species of saltbush, *Atriplex robusta,* appears to have developed from hybrids between *Atriplex canescens* (fourwing saltbush) and *Atriplex tridentata* (saltsage). It occurs mostly in the alkaline deserts of western Utah on the shoulders of

FIGURE 19.9

A recent example of speciation.
A new species of saltbush, *Atriplex robusta,* appears to have evolved in Utah since a highway was built in 1969. The highway provided a habitat that allowed two species to come together and hybridize.

highway I-80, which was constructed in 1969 (Figure 19.9). Artificial selection can speed up the speciation process. New species of fruit flies have been created by subjecting existing species to radiation. This treatment increases the rate of mutation and thus increases variation. Another example is a relatively new species of grain, called triticale (Figure 19.10), that is produced by crossing wheat and rye.

Keep in mind that populations evolve, not individual organisms within a population. Genetic changes in populations indicate that evolution is taking place. The frequency of alleles in a gene pool tends to remain stable, especially in a large population that is well adapted to its environment. Such a gene pool is in a state of equilibrium. A variety of factors can, however, change the equilibrium of a gene pool (see Table 16.2).

Speciation in sexually reproducing organisms happens when two populations become so different in their genetic makeup that they can no longer interbreed. Occasionally, new species simply out-compete and replace the species from which they originated. In most cases, however, a

FIGURE 19.10

Triticale *(Triticum × Secale).* This plant is a species of grain that was produced by crossing wheat *(Triticum aestivum)* and rye *(Secale cereale).*

rye   ×   wheat   ⟶   triticale

a   b

FIGURE 19.11

Contribution of a geographic barrier to speciation. The Kaibab squirrel, *Sciurus kaibabensis* **(a),** and the Abert's squirrel, *Sciurus aberti* **(b),** are related species. When they became geographically isolated on opposite rims of the Grand Canyon, they began to develop differences. Note the differences in coloration.

small population that is isolated from the rest of its species develops into a new species. As you saw in the case of the founder effect (see Section 16.5), evolution is much faster in small populations than in large ones. The most common mechanism that separates populations is geographic isolation. Organisms that cannot come into contact with one another cannot interbreed. In animal species, geographic isolation can occur when groups are separated by water, a mountain range, or an impasssable canyon. For example, the Grand Canyon separates and isolates two species of tufted-ear squirrels. These squirrels have already developed some differing characteristics, shown in Figure 19.11.

Ecological isolation occurs when two populations adapt to different habitats. The Traill's flycatcher, found in Canada, was once considered to be one species of bird. New observations show that there are two species in this group. The habitats of the two species are different, and no crossbreeding takes place. While the willow flycatcher lives on brushy slopes, the alder flycatcher lives in alder swamps (Figure 19.12).

Not all forms of isolation are the result of physical location. As Section 12.6 explains, mating behavior and physical characteristics are important in the reproductive success of some organisms. In some cases, the gametes are not chemically compatible. For example, the sperm may not be chemically attracted to the egg. Differences in size of the organisms or in the size or shape of their reproductive organs can also prevent mating. This type of separation is often found in plants.

Behavioral isolation can occur among animal populations. If the mating pattern of a small group of organisms becomes different from that of the main group, then they can become reproductively isolated. Eventually the groups become separate species. Such a change is happening with the leopard frog, shown in Figure 19.13. If one sex from the northern limits of its range and the opposite sex from the southern limits of its range are brought together, the two will not mate. Subtle differences in premating behavior, together with other events, have isolated these groups of frogs. Leopard frogs are still in the process of speciation. Eventually scientists may group some of them in a new species.

a

b

FIGURE 19.12

Contribution of ecological isolation to speciation. The alder flycatcher, *Empidonax alnorum* **(a),** and the willow flycatcher, *Empidonax traillii* **(b),** were once considered the same species until it was noticed that their habitats are completely different and they do not crossbreed.

a   b   c

FIGURE 19.13

Three species of leopard frogs. Leopard frogs formerly were considered all one species, *Rana pipiens.* Today, however, biologists consider some of the populations to be separate species. The three frogs pictured are from Massachusetts **(a)**, Oklahoma **(b)**, and Arizona **(c)**. Northern and southern leopard frogs do not mate with each other. The intermediate forms of the leopard frog can and do mate with each other and with both the northern and southern populations.

CHEMISTRY TIP

Binding of complementary proteins on the surfaces of pollen grains and stigmas (and on egg and sperm cell membranes) helps prevent cross-species fertilization.

CONNECTIONS

Reproduction is part of a species' adaptation. Asexual reproduction enables well-adapted plant populations to expand in existing habitats. Sexual reproduction produces new genotypes that may be better suited to populating a new and different environment.

ETYMOLOGY

poly- = many (Greek)
tetra- = four (Greek)
ploidy = number of chromosome sets (from *diploidy* and *haploidy*)
Polyploid cells contain more than two copies of each chromosome.
Tetraploid cells have four copies of each chromosome.

Some leopard-frog populations are isolated from each other because the frogs mate at slightly different times during the year. This seasonal isolation occurs in plants and animals. In plants, the most common example of seasonal isolation is the flowering of closely related species at different times. The gametes of two species cannot fuse if one species' pollen is not present when the other species' carpels are ready for pollination.

Isolation mechanisms can be classed as those that occur prior to zygote formation and those that occur after zygote formation. Geographic isolation, mismatched mating behavior, and seasonal isolation are all examples of prezygotic mechanisms. In those cases, gametes never meet and fuse. An example of a postzygotic mechanism is failure of the zygote to develop normally. If the parents are too different, even though mating and fertilization have occurred, the zygote may not develop normally. The embryo or offspring may die before birth or sexual maturity, or it may be unable to reproduce. Matings between individuals of related species often produce offspring that cannot produce normal gametes. The lack of fertile offspring makes the hybrid cross a dead end. This helps keep the parents' species separate.

Often a new species of plant forms because of **polyploidy,** duplication of chromosomes. Polyploidy can occur accidentally during cell division. The offspring generally cannot mate successfully with plants having the parental number of chromosomes. They can reproduce asexually or often can mate with other polyploid offspring. In this way, they establish a new species. Polyploidy can also arise from hybrid crosses between species. Most cross-species hybrid plants are sterile. They only reproduce asexually. In time, a chromosome duplication in a hybrid plant can produce a tetraploid ($4n$) new species. More than half of known species of flowering plants are polyploid. In rare cases, animals can be polyploid.

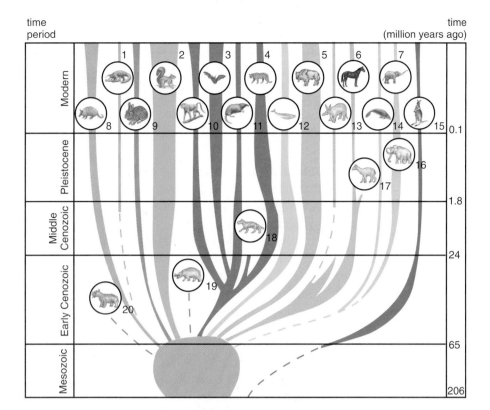

FIGURE 19.14

Adaptive radiation in mammals. As the first mammals dispersed to new areas, they adapted to the new environmental conditions. Gradually, these adaptations produced different species and, eventually, all the different mammals alive today. Representative animals from the orders included are *1,* pangolin; *2,* squirrel; *3,* bat; *4,* lion; *5,* buffalo; *6,* horse; *7,* elephant; *8,* armadillo; *9,* rabbit; *10,* monkey; *11,* mole; *12,* whale; *13,* aardvark; *14,* manatee; *15,* kangaroo; *16,* mastodon; *17,* litoptern; *18,* creodont; *19, Brontotherium; 20, Uintatherium.*

19.5 Patterns in Evolution

The fossil record and direct observations of organisms show that when a population enters an environment with few competing species, it often divides into several smaller populations. These populations avoid competing with each other by adapting to different habitats or by using different resources in the same habitat. These diverse adaptations keep the populations separate. Over time each population can become a new species. This rapid increase in speciation is known as **adaptive radiation.** The adaptations of Darwin's finches to different food sources and nesting sites on the Galápagos Islands provide an example of adaptive radiation. The great numbers of species in groups such as the insects, mammals (Figure 19.14), and the grass and lily families reflect a history of adaptive radiation.

Although evolutionary processes are continuously at work, in some cases, the rate of large-scale change remains very slow for a long period. This condition, **stasis,** may occur because a population or species is well adapted and its environment remains stable. Some evidence for stasis is found in the fossil record. In addition, a few modern species have changed little from their ancient ancestors. One example of a "living fossil" is the lungfish (Figure 19.15).

ETYMOLOGY

stasis = standstill (Greek, Latin)
Species in evolutionary **stasis** change very little over many generations.

FIGURE 19.15

A "living fossil." This Australian lungfish *(Neoceratodus)* has changed little in millions of years.

Biological Challenges

RESEARCH

Mass Extinction

A species may become extinct because its habitat has been destroyed or because its environment has changed. For example, if ocean temperatures fall a few degrees, many species that are otherwise well adapted may perish. Evolutionary changes that affect one species can also affect other species in the same environment.

About two to five families of organisms become extinct per million years, on average. However, at times, Earth's environments have changed so rapidly that a majority of species died out. During these periods of mass extinctions, more than 19 families of organisms may have been destroyed in 1 million years.

One mass extinction occurred about 65 million years ago. At that time, more than half the marine species and many land-dwelling plant and animal species (including nearly all species of dinosaurs) died. The climate was cooling during this period, and the shallow seas were receding from the continental lowlands.

Evidence also indicates that at least two asteroids or comets struck Earth while these extinctions were in progress. Sediments deposited during that time are composed of a thin layer of clay enriched with iridium (Ir), an element that is very rare on Earth but common in meteorites and other debris that occasionally fall to Earth (Figure 19.16). Luis W. Alvarez of the University of California at Berkeley and his colleagues studied the clay. They concluded that it was the fallout from a huge cloud of dust that billowed into the atmosphere when an asteroid hit Earth. The cloud would have blocked sunlight and disturbed the climate for several years.

The evidence for asteroid collisions 65 million years ago is credible, but critics of the impact–collision hypothesis point out that the extinctions, while rapid on a geological time scale, were not that abrupt. Some geologists and paleontologists suggest that changes in climate due to factors other than asteroids are sufficient to account for the mass extinctions.

Mass extinctions provide some advantage to surviving species. Habitats vacated by extinctions become available for the surviving species. Thus, a mass extinction not only removes many species but also helps ensure survival and dispersal of the fortunate few that remain.

FIGURE 19.16

A clue to an ancient mass extinction. A 65-million-year-old layer of iridium-rich sediment (the dark band) is located between older and younger sediments. Iridium, a rare element on Earth, is thought to indicate a collision between a large asteroid and Earth. This collision may have caused a mass extinction. The coin near the center of the photograph shows the scale.

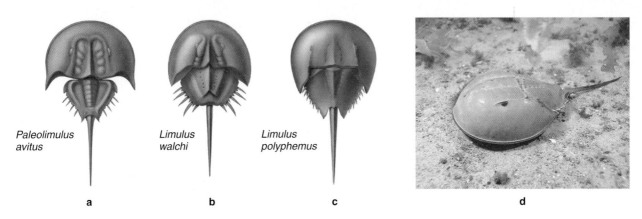

FIGURE 19.17

An example of evolutionary stasis. The modern horseshoe "crab," *Limulus polyphemus*, which is not a true crab, has changed only slightly in appearance from its 250-million-year-old ancestors. A comparison of these species shows how little change has occurred. *Paleolimulus avitus* **(a)** from the Permian Period is about 250 million years old. *Limulus walchi* **(b)** from the Jurassic Period is approximately 180 million years old. *Limulus polyphemus*, the modern species, is shown here as a drawing **(c)** and in its natural habitat in the Gulf of Maine **(d).**

The ancestors of the lungfish underwent fairly rapid evolution about 200 million years ago, but since then the rate has been slow. They live in an isolated, stable environment with little competition. They do not appear to have any special genetic mechanisms that reduce the rate of molecular change. Another example of a living fossil is the horseshoe "crab" (Figure 19.17) found along the eastern Atlantic coast of the United States.

Scientists are now studying two patterns of change. Some evidence indicates cases in which evolutionary change and speciation occurred through accumulation of many gradual and fairly constant changes. New species split off from ancestral species, and both continued to change at similar rates. This pattern is called **gradualism.** Another pattern of

Focus On

Measuring the Rate of Evolution

The fossil record provides evidence of the rate of evolutionary change. Fossils indicate that anatomical structures do not change at a constant rate. Measurement of molecular evolution shows a more constant rate for a particular DNA sequence or protein sequence (see Biological Challenges: Classification of the Giant Panda in Section 18.4). However, the rate of change for the functional part of a protein is generally slower than the change in a less important region of the molecule. For instance, the rate of evolution in the surface parts of the hemoglobin protein is 10 times greater than the rate of change at the site where the protein binds to heme.

A unit of evolutionary change is called a darwin. One darwin is equal to the change in a characteristic being studied by a factor of e in 1 million years. The number e, the basis of natural logarithms, is approximately 2.718. Because it is possible to count the number of changed bases in DNA or amino acids in a protein, scientists can easily use numerical units such as the darwin to measure the rate at which these sequences change. Changes in qualitative traits, such as behavior and some anatomical traits, are more difficult to measure and describe in this way.

The Theory of Punctuated Equilibrium

After 4 years and thousands of generations of bacteria, biologists at Michigan State University think they have observed punctuated equilibrium in progress. The researchers set out to observe how quickly species change over time. The theory of punctuated equilibrium proposes that a species changes very little during most of its existence. Occasionally, mutation or environmental changes suddenly make some individuals in the population significantly better adapted than others. Natural selection then makes a huge difference in survival. The group with the new adaptation competes successfully and flourishes as a new species.

How might one test this model when evolution takes so long to occur? Scientists at Michigan State came up with a simple answer. Bacteria reproduce very quickly (15,000 times faster than humans do). A few years of bacterial growth would be like hundreds of thousands of years of human evolution. That length of time might be enough to see how fast species change.

Biologists grew the common intestinal bacteria *Escherichia coli* under conditions of poor nutrition to induce natural selection. They reasoned that under intense competition for nutrients, any mutant bacteria that were more efficient at absorbing and using nutrients would have a significant survival advantage. The experiment produced no change for thousands of generations of bacteria. Then a bacterium mutated and grew 30% larger than the typical size. Within a few dozen generations, the mutation swept through the population, replacing the ancestral type with the larger form. The population then remained stable for hundreds of generations until a major change again altered the population. Changes evolved quickly, not gradually.

This experimental result helped to establish the theory of punctuated equilibrium. Biologists are now evaluating the theory's other merits as well, such as its answer to a long-standing problem in paleontology. Scientists have discovered fossils of hundreds of thousands of extinct species. In some cases, however, major gaps remain in the fossil record. Punctuated equilibrium can help account for these missing data.

The theory of punctuated equilibrium implies that many transitional fossils are missing because the transitions between species happen so quickly. The right conditions for a significant mutation are rare, and when they happen, they radically change the anatomy of the species. Often this change occurs within a small isolated part of a population. A small population ranging over a restricted territory for only a brief time is not likely to leave many fossils.

This theory also provides a fruitful way of thinking about Earth's history. Its model of stability marked by fitful changes has consequences for many fields—from geology to cosmology. It suggests that major changes in Earth's environment can be rare and rapid rather than gradual. At the very least, it has prompted some unique experimental studies of evolution.

A freshwater snail, *Melanopsis (right)*, lives in environments such as lakes, springs, and streams. Its half-million-year-old fossil ancestors *(left and center)* are found preserved in sediments of the same environments. *Melanopsis* and other freshwater gastropod mollusks exhibit significant evolutionary change in shell form during short intervals of geologic time because snails cannot migrate easily from one body of fresh water to another. In addition, their populations are likely to be reproductively isolated. Over time, local genetic differences lead to evolutionary change. Because the smaller habitats are less likely to leave deposits in the geologic record, fossils of intermediary forms are less likely to be preserved. When such evolutionary change occurs during a brief portion of an organism's evolutionary history, the change is termed punctuated equilibrium.

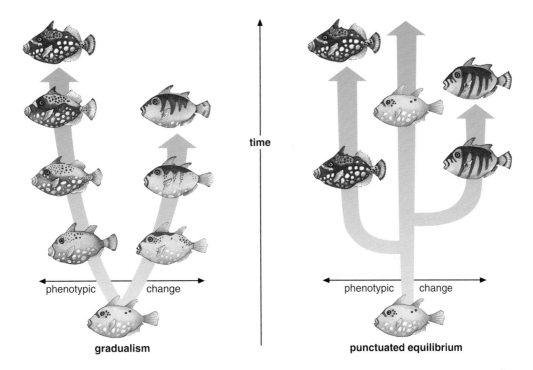

time

phenotypic change

gradualism

phenotypic change

punctuated equilibrium

FIGURE 19.18

Two evolutionary patterns. (a) In gradualism, new species descend from a common ancestor. They steadily become more and more different from the ancestral species as they acquire new adaptations. **(b)** In punctuated equilibrium, a new species changes most at the time it branches from its parent species. It changes little as time goes on but may become extinct (as suggested by the shorter branches) or produce additional species.

evolution, called **punctuated equilibrium,** involves a short period of rapid change just after a population becomes isolated and forms a new species, after which the process slows down and approaches stasis (Figure 19.18). Most species change little during their existence. However, these two patterns most likely represent the extremes in a continuum of evolutionary process. In both cases, the same fundamental principles apply. Selective pressure acts on the raw material of genetic variation to produce heritable, long-term change in living populations.

Check and Challenge

1. Why did Darwin have difficulty explaining variation within a species?
2. What are isolating mechanisms, and how do they operate? Give examples.
3. What is polyploidy? What is its connection to speciation?
4. What is the significance of "living fossils"?
5. Explain the statement "Populations evolve, not individuals within a population."

Chapter HIGHLIGHTS

Summary

The theory of evolution provides a scientific explanation for changes in species and the history of life on Earth. It is a testable theory that is supported by fossil evidence. Additional evidence comes from observation of populations of living organisms and from anatomical, molecular, and genetic information.

Populations, not individual organisms, evolve. Small isolated populations of organisms evolve much more rapidly than large populations. Speciation occurs when variation within a population becomes so great that subgroups no longer interbreed. Isolating mechanisms speed up the process of speciation. Geographic, ecological, behavioral, seasonal, and mechanical isolation all occur and function in natural populations. Other types of isolating mechanisms include prevention of gamete fusion, failure of hybrid zygotes to develop normally and survive, and failure of hybrid offspring to reproduce.

A large increase in the number of species that develop in newly available habitats is called adaptive radiation. Anatomical and molecular data are used to estimate the rates of evolution on different scales. Molecular evolution proceeds at a more constant rate. Parts of proteins that are not functionally important accumulate amino-acid changes at a faster rate than parts that are essential for function. In some very stable environments, small populations with little competition appear to evolve more slowly than other species. This situation can result in "living fossils" that closely resemble very ancient predecessors.

There is evidence for more than one pattern of evolution. Gradualism describes the steady accumulation of small changes over time. Punctuated equilibrium suggests that long periods of evolutionary stability are interspersed with brief periods of rapid change.

Key Concepts

Use the following concepts to develop a concept map. Include additional concepts from the chapter as you need them.

- adaptation
- evidence
- evolution
- extinction
- speciation
- isolating mechanisms
- mutations
- homology
- variations
- genetic recombination

Reviewing Ideas

1. Explain how adaptations in beak structure could help the finches of the Galápagos Islands take advantage of a variety of niches.
2. How does the theory of evolution account for new species?
3. What are some similarities and differences between artificial and natural selection?
4. What reptile characteristics does *Sinornithosaurus millenii* have? what bird characteristics? Why is the fossil significant?
5. What did the University of Michigan mouse-and-owl experiment add to the observations of the peppered moth?
6. What factors influence natural selection when strains of bacteria become resistant to penicillin?
7. Explain the meaning of the sentence "Mutations supply the raw material for evolution."
8. Why do scientists compare differences in amino-acid sequences of proteins from different species as a way to study evolution?
9. Describe examples of evolution in action that have been observed.
10. Compare punctuated equilibrium and gradualism.
11. What is adaptive radiation?

Using Concepts

1. Why are island populations so important in the study of evolution?
2. Explain the connections among variation, adaptation, and natural selection.
3. Explain how radioactive dating methods contribute to the study of evolution.
4. Explain how populations can evolve although the organisms in a population do not.
5. Where would you go to look for a "living fossil"? Explain your answer.
6. Compare advantages and disadvantages of evidence of evolution from the fossil record with evidence from observation of living species.
7. Why are islands good places to find examples of adaptive radiation?

Synthesis

1. Why was the rediscovery of Mendel's work an important step in understanding evolution through natural selection?
2. Relate the forms of isolation presented in this chapter to the problems of maintaining species diversity discussed in Chapter 18. What similarities and differences can you find?
3. Many species are in danger of extinction because human civilization and technology have altered their habitats. Select an urban setting, and describe how successful species have been in coping with the city environment.

Extensions

1. Identify a current area of evolutionary research. Report on some recent discoveries or ideas related to this topic.
2. Use resources in other textbooks, in the library, in a museum, or on the World Wide Web to draw a chart showing the major events in the history of life on Earth. For example, you could make a time line showing major evolutionary periods and indicating examples of species that were common at various times.
3. Select a favorite organism, and make a tree diagram indicating its close relatives and major evolutionary adaptations.
4. Devise a game that demonstrates how a population changes through natural selection.
5. Suppose Charles Darwin had been given a time machine after he published his book *The Origin of Species* in 1859, and the machine was set to bring him into your classroom today. What new evidence of evolution could you report to him that would allow him to revise his book and bring it up to date?

Web Resources

Visit BSCSblue.com to access
- Information and activities for further study related to evolution
- Web links related to this chapter

CONTENTS

LEARNING OUTCOMES

By the end of this chapter you will be able to:

A   Describe how modern humans differ from other primates.

B   Evaluate the techniques used to study evolutionary relationships in humans.

C   Compare early hominids with *Homo erectus* and *Homo sapiens*.

D   Give reasons for the differences in the gene pools of modern human populations.

CHAPTER

20

Human Evolution

■ *How do humans differ from closely related
species?*

■ *What evolutionary processes could have
produced modern humans?*

It is natural to be especially curious about human evolution. Who were
our ancestors? How did they live? What is the future of the human
species?

Although this chapter looks at human evolution in detail, there is no
basic difference between our evolution and the evolution of other species.
Natural selection, changes in allele frequencies, speciation, and all the other
principles that describe the evolution of other species also apply to humans.
Natural selection acts on heritable human traits and produces the changes
that have made modern humans a distinct species.

Recent humans are unusual in having a fast-changing cultural history.
For instance, only a few thousand years ago humans figured out how to
grow plants and domesticate animals. In the last 50 years, we have figured
out how to leave the earth, replicate animals from single cells, and achieve
advances in telecommunications, all of which are nothing less than
phenomenal. Physical changes, however, take much longer. Millions of years
ago, human ancestors had smaller brains and could not speak as we do. This
chapter traces the steps in the evolution of the human species from its early
ancestors to the present.

Common Origin of Primates

20.1 Identifying Primates

Humans are classified within the class Mammalia as members of the order Primates. Primates include monkeys and apes (Figure 20.1), which are closely related to humans in many details of their anatomy and their DNA. All these species share an ancestor that most likely lived in trees and ate insects. Fossil evidence points to a major split in the order Primates about 55 million years ago. At that time, the ancestors of today's monkeys, apes, and humans diverged from other primates, such as lemurs, lorises, and tarsiers (Figure 20.2).

What sets primates apart from other mammals? Most primates live in the trees. Some of the New World monkeys found in Central and South America even have tails that grasp. The fingers and toes of most primates are long and flexible, giving the animals a powerful grasp. Most primates possess thumbs and big toes that are **opposable.** This means that they can be used to grasp branches. Primates also can manipulate objects with their hands by moving the objects forward and inward toward the palm and the other fingers, as shown in Figure 20.3. Primate fingers and toes have nails, rather than claws. The fingers are extremely sensitive, allowing excellent control when manipulating objects (Figure 20.4). These anatomical adaptations help support the way primates live.

Primate shoulder and hip joints allow a wide range of limb motion. This enables primates to jump, run, and scamper and swing through trees. The leg and foot structure of more erect forms such as the apes is adapted to bear much of the animal's weight. The eyes of primates are directed forward

a   b   c

FIGURE 20.1

Some primates. (a) A Japanese macaque *(Macaca fuscata)* sitting beside a natural hot spring in Japan's Shiga Highlands; **(b)** a mountain gorilla *(Gorilla gorilla)* from Africa; and **(c)** an adult orangutan *(Pongo pygmaeus)* in the rain forest of Borneo. Our closest relatives are the tailless great apes—the gorillas, orangutans, chimpanzees, and gibbons.

a

b

c

FIGURE 20.2

Some simpler primates. (a) A black and white ruffed lemur *(Varecia v. variegata)* from Madagascar; **(b)** a slender loris *(Loris tardigradus);* and **(c)** a Philippine tarsier *(Tarsius syrichta).* These small, relatively simple primates are not closely related to humans and the great apes in Figure 20.1. They do have the typical primate traits, however, including grasping hands and binocular vision.

E T Y M O L O G Y

bi- or **bin-** = twice (Latin)

ocul- = eye (Latin)

Animals with **binocular vision** see the same objects with both eyes.

instead of to the side. This results in **binocular vision,** in which both eyes view the same objects simultaneously from slightly different angles. Binocular vision, along with another primate development, the optic chiasma, allows the brain to perceive objects in three dimensions and to judge distances accurately. Most primates also have color vision, something most other mammals lack.

Old World primates (those that evolved in Africa, Asia, and Europe), such as baboons, gorillas, and humans, have the most elaborate color vision

FIGURE 20.3

The primate opposable thumb and grasp. Notice how the thumb and fingers of this batter's hands point in opposite directions.

FIGURE 20.4

Tool use by a nonhuman primate. Baseball is not the only activity for which an opposable thumb is helpful. This chimpanzee *(Pan troglodytes)* is using a twig as a tool to dig inside a tree trunk.

Vision is part of an animal's adaptation. Color-sensitive eye cells are less sensitive to light than the light-detecting cells responsible for black-and-white vision. For this reason, color vision is a more common adaptation in animals that are active in daylight (diurnal) than in nocturnal animals, which are active at night.

FIGURE 20.5

Primate social behavior. One olive baboon *(Papio cynocephalus anubis)* grooming another. Social grooming, which occurs mainly before and after foraging, helps strengthen bonds among troop members. Grooming also has health benefits in the removal of dirt and plant debris from fur as well as removal of dead skin and parasites living on the skin.

omni- = all (Latin)

-vore = devourer (Latin)

Omnivorous animals eat both plant and animal foods.

among mammals. Human eyes have three types of color-detecting cells that absorb red, green, or blue light. The balance of light-absorbing pigments in these cells is well adapted to distinguish the colors of fruits against the background colors of leaves. How might this adaptation have been useful for early tree-living primates?

Primates are among the largest animals. They also have the most advanced brains. Primate brains are larger than the brains of other animals when the size of the brain is compared to the size of the animals' body. Large, complex brains enable primates to learn the complicated behavior that is important for living in social groups. Most primates have well-established patterns of communication. They use elaborate body language (Figure 20.5) and intricate vocal signals. Some primates are **omnivorous.** This means that they eat a wide variety of both plant and animal foods.

Primates also share a common birth pattern. They generally bear only one offspring at a time. This pattern enables adults to provide extended care for the young. Offspring generally need parental help to survive. A longer period of care gives offspring a chance to learn involved behaviors from the parents. These shared traits show the close evolutionary relatedness of primates. Other characteristics distinguish one primate species from another, as shown in Figure 20.6.

galago
125–300 g

tarsier
80–165 g

squirrel monkey
0.75–1 kg

gibbon
4–8 kg

gorilla
70–275 kg

FIGURE 20.6

Primate diversity. Although primates share many characteristics, there is also great diversity within the group. Sizes listed are typical ranges.

What traits distinguish modern humans from other animals, especially other primates? The human brain is particularly large and complex, even compared to those of other primates. Compared to body size, however, chimpanzee brains and human brains are more similar in size. Humans, chimps, and gorillas share a brain structure, the hippocampus, that is critical in memory and learning. This feature is missing in simpler primates, such as lemurs. The human skeleton is especially well adapted for upright, or **bipedal,** walking. Bipedal posture leaves hands even freer to grip things. Human fingers are capable of extremely fine manipulations. Many primates share these features to some degree, but in humans they are more developed and widely used.

The structures of the human vocal chords, mouth, tongue, and ears enable human communication through speech. Language, including speech, is a powerful adaptation for human survival as social animals. The vocal chords of chimpanzees and gorillas are not capable of the huge range of sounds found in human speech. However, these primates do use sounds and gestures as part of their communication systems. Some chimpanzees and gorillas have been taught to communicate with humans through sign language (see Figure 22.21) and computers.

Humans are a distinct species, but they differ from other primates by *degree* rather than by special characteristics. Humans are particularly closely related to chimpanzees and gorillas, as shown by a variety of lines of evidence.

E T Y M O L O G Y

bi- or **bin-** = twice (Latin)

-ped or **-pod** = foot (Latin)

Bipedal animals walk on two feet. Those that walk on four feet are called tetrapods; six-legged insects are hexapods.

One of the most important physiological differences between humans and other primates is their reproductive cycle. All female mammals have a reproductive cycle, called the menstrual cycle in humans. In most primates, the cycle is approximately 4 weeks long. Most mammals mate only during a phase of the cycle called heat, or **estrus.** Humans are unique in lacking an estrus phase. In most primates, estrus lasts about 3 to 5 days when the female is ovulating. In some monkeys and chimpanzees, the female genital region swells during estrus, attracting male sexual attention. Female chimpanzees typically mate only at this time. A female chimpanzee may mate with many of the adult males in the area.

Humans have lost this linking of physiology and behavior associated with ovulation. Women can choose to be sexually active at any time in their cycle. The ovulatory stage of the cycle cannot be detected through any obvious visible sign.

The evolutionary loss of the clear indications of estrus may be associated with the development in humans of relatively permanent male-female bonds. This bonding may have helped establish male cooperation in child rearing and protection. Such behaviors may have led to the development of social groups and communities.

20.2 Comparing Skeletal Evidence

Try Investigation 20A
Interpretation of Fossils.

What scientific evidence reveals the path of human evolution? **Anthropologists,** scientists who study human evolution and culture, rely on many lines of evidence. A major source of evolutionary information is skeletal evidence. The structure of a skeleton tells a lot about how an animal moves, eats, and behaves (Figure 20.7). Fossil skeletons of **hominids**—bipedal, humanlike animals that belong to the same family as modern humans—have been studied for many years. In rare cases, footprints made by hominids have been preserved. These impression fossils are even more fragile than skeletal fossils. Footprints help reveal how an animal moved and how large and heavy it was. What clues can you find in the shapes of the backbones in Figure 20.7? The amount and location of curvature reflect the way an individual moves. Similarly, the form of the femur (thighbone) and especially its relationship to the pelvis suggest whether the individual walked on two legs or on four (see Appendix 20A, "Physical Adaptations").

Anthropologists compare fossil hominid skeletons with each other and with the skeletons of modern humans and other primates. Unfortunately, most bones do not survive as fossils. Few complete fossil skeletons have been found. Anthropologists usually must make their inferences about human ancestry by examining small portions of skulls, a lot of teeth, and occasionally part of a leg or arm bone. Some of the most important features used in comparing fossils include the skull, the pelvis, the backbone, and the femur. Important structures of the skull include the jaws, teeth, and the cranium, the bones that enclose the brain. Table 20.1 summarizes some of the most common skeletal features that anthropologists use to determine the relationship of fossil remains to modern primates.

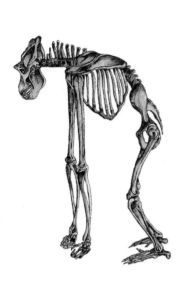
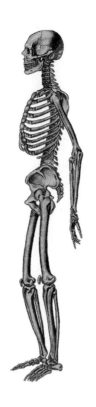

FIGURE 20.7

Skeletal comparison of two primates. Shown above are skeletal proportions and postures of a gorilla *(Gorilla gorilla)* and a human *(Homo sapiens).* What similarities and differences do you see? What does the skeletal evidence suggest about how humans and gorillas eat and move?

TABLE 20.1
Comparison of Ape and Hominid Characteristics

Characteristic	Ape	Hominid
posture	semi-erect or quadrupedal; knuckle-walking common	upright and bipedal
leg and arm length	arms longer than legs; adapted for swinging by arm (brachiation), usually among trees	legs usually longer than arms; adapted for striding
feet	low arches; can grasp with opposable big toe	high arches; non-opposable big toes in line for walking
teeth	prominent; clear gaps between canines and neighboring teeth	smaller; reduced or absent gaps next to canines
skull	angled forward from spinal column; rugged; prominent brow ridges	upright on spinal column; smoother
face	jaws jut out; very heavy; very wide nasal opening; protruding face	distinct chin in modern man *(Homo sapiens);* narrow nasal opening, prominent nasal arch; vertical profile
brain size	280–705 cm³ (living species); chimpanzee average of 400 cm³	400–2,000 cm³ (fossil to present); *H. sapiens* average of 1400 cm³
age at puberty	usually 10–13 years	usually 13 years or earlier
breeding season	estrus at regular intervals	no definite time for modern humans; not known for early hominids

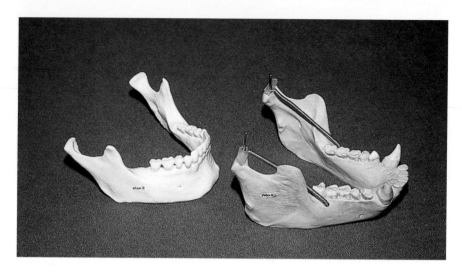

FIGURE 20.8

Teeth of the lower jaw. Differences in tooth size and shape can indicate differences in diet and behavior. Which jaw is human?

The shape of the teeth and their wear patterns (Figure 20.8) give anthropologists an idea of an individual's diet and behavior. If a newly found fossil has teeth similar to those of another fossil, an anthropologist might consider the two to be related. A great deal of difference in these features would lead to the conclusion that the two individuals are not closely related. The form of the skull also is used to classify fossil remains. The size and shape of the skull puts a limit on the possible size of the brain (Figure 20.9). However, brain size alone does not determine intelligence. Whales' brains, for instance, are bigger than those of humans, yet whale communications and other behaviors are simpler. In addition, the skull also is a clue to how much muscle was present. Examine the skulls in Figure 20.9. Which individual probably had more powerful jaws?

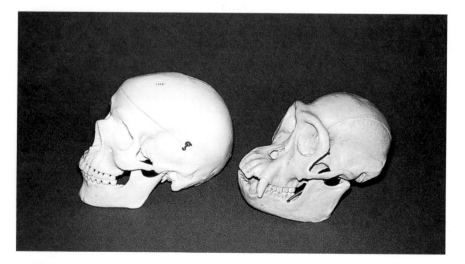

FIGURE 20.9

A human skull and a chimpanzee skull. What differences can you see?

20.3 Comparing Molecular Evidence

Only one species of human, ***Homo sapiens,*** exists today. Earlier species in the genus *Homo* are all extinct. No genetic barriers to mating divide modern human populations into separate species, so all humans are one species. The genetic differences between human groups are too small to make separate species, but they do provide clues to the history of our species. Most anatomical traits, such as facial features, body size, and body shape, are too complicated to investigate genetically. Many genetic and environmental factors affect these characteristics. For example, improved diets throughout much of the world in the past few hundred years generally have led to increased body size.

Molecular data are more useful than anatomical measurements in tracing our evolutionary history. Comparisons of the DNA and protein sequences of different species help determine how closely related they are (Figure 20.10). Such comparisons, in turn, can yield information about human evolution (see Tables 18.1 and 18.2).

Comparisons of DNA and protein sequences indicate that chimpanzees are the closest living relatives to humans. The human and chimpanzee

ETYMOLOGY

homo = man (Latin)

sapiens = wise or tasteful (Latin)

Homo sapiens is the scientific name of the human species.

CHEMISTRY TIP

The greater the percentage of complementary bases in two DNA strands, the more hydrogen bonds hold them together. The percentage of complementary base pairs in experiments such as the one shown in Figure 20.10 can be estimated from the temperature to which the sample must be heated to separate the two strands.

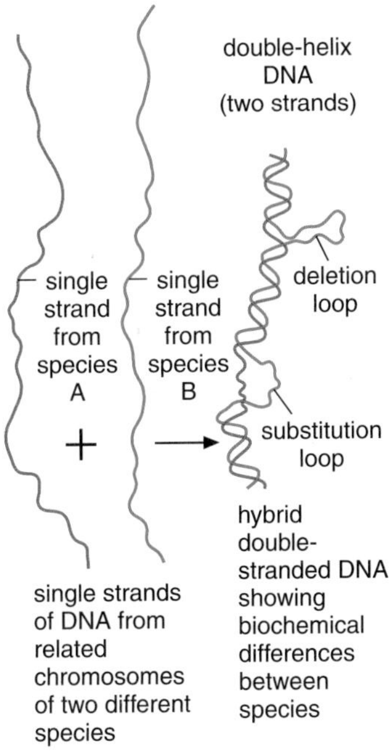

double-helix
DNA
(two strands)

single strand from species A

single strand from species B

deletion loop

substitution loop

hybrid double-stranded DNA showing biochemical differences between species

single strands of DNA from related chromosomes of two different species

FIGURE 20.10

Molecular comparison of two species' DNA. A single DNA strand from one species and a homologous strand from another species are mixed for comparison. Where the base sequences of the two strands are complementary, they form a typical double-helix structure. The more closely they fit together, the more closely they are related. Differences between the two base sequences (substitutions) cause the strands to separate, resulting in a substitution loop. DNA segments that are present in one species but missing (deleted) in the other produce deletion loops.

The Mother of Us All?

Did all modern humans descend from one or a few females who lived in Africa about 200,000 years ago? That idea is one possible interpretation of data collected from research on mitochondrial DNA (mtDNA, shown in Figure 20.11).

Mitochondria have their own DNA. Unlike nuclear DNA, however, it appears that mtDNA does not come from both parents. Sperm cells contain mitochondria that provide energy for locomotion, but these mitochondria apparently do not enter the ovum during fertilization. Therefore, all the mitochondria in your cells came from your mother (see Section 14.2). Because mtDNA is transmitted only through females, it is theoretically possible to trace the origin of modern human mtDNA to our earliest women ancestors. How long ago did they live?

To answer that question, scientists measure variation in mtDNA among modern individuals and how frequently mtDNA mutates to produce those changes. The enzymes responsible for DNA replication in mitochondria lack some of the error-correction functions present in the nucleus. Molecular biologists calculate that mtDNA accumulates mutations 5 to 10 times faster than does nuclear DNA. This characteristic makes mtDNA a good clock for measuring short spans in evolutionary time—spans measured in hundreds of thousands of years rather than many millions of years.

Molecular biologists at the University of California at Berkeley collected mtDNA samples from 21 people who belonged to different population groups from all over the world. The scientists used RFLP analysis (see Section 15.3) to compare the mtDNA samples. The results, combined with the information about how fast mtDNA accumulates mutations, led the researchers to conclude that it would have taken about 200,000 years for modern humans to have accumulated that amount of variation from one ancestral female. Because the mtDNA of Africans was the most diverse, the biologists also concluded that the ancestors of modern humans originated in Africa.

Other biologists contend that mtDNA accumulates mutations more slowly. If this is true, then the variation in modern human mtDNA accumulated over a longer period of time. Some biologists suggest that variation in human mtDNA indicates that our species began about 500,000 years ago. Ongoing research continues to resolve these questions.

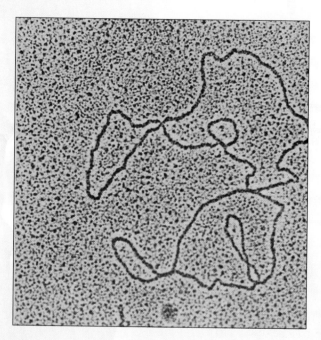

FIGURE 20.11

Electron micrograph of mitochondrial DNA. Replication of mtDNA is not linked to cell division. Each mitochondrion may accumulate many copies of its DNA molecule. The circumference of the DNA loop is 6 μm.

TABLE 20.2
Chimpanzee and Human Protein Differences

Protein	Number of amino acids	Amino-acid differences
hemoglobin	579	1
myoglobin	153	1
cytochrome C	104	0
serum albumin	580	7

genomes are more than 98% identical. Consequently, humans and chimpanzees are more alike genetically than many pairs of species in the same genus (Table 20.2). Molecular data show that gorillas are only slightly more distantly related to humans than chimpanzees are (Table 20.3). The evolutionary distances between chimpanzees, humans, and gorillas may be less than the distance between foxes and dogs and about the same as that between zebras and horses.

Most molecular estimates of relatedness are expressed as percentages of similarity in amino acid or nucleotide sequences. Other estimates are based on comparisons of the chromosome banding patterns of different species (see Section 13.3). Scientists also compare which codons for a particular amino acid are used most often in different species. For example, there are six codons for the amino acid arginine (see Figure 9.4). The codons AGA and CGC are used with about equal frequency in humans. In contrast, the bacterium *Escherichia coli* uses CGC 39 times more often than AGA. The great difference in codon usage between humans and *E. coli* reflects the enormous age of their evolutionary divergence.

Molecular comparisons of living species also help establish the pattern of their evolutionary relationships. Unfortunately, molecular data are not available for fossilized remains except in the rarest of instances. Little protein and DNA survives in fossils. Otherwise, we could perform direct molecular tests of kinship between extinct species and the one living species of *Homo*. In one case, scientists have been able to extract DNA from a fossil of the extinct hominids called Neanderthals. Comparison of this DNA to that of modern humans suggest that Neanderthals were probably not our direct ancestors. Most of the evidence of the course of human evolution, however,

CONNECTIONS
Small differences in genetic information provide the variation on which natural selection acts.

TABLE 20.3
DNA Sequence Comparisons between Closely Related Primates

Species compared	Difference in DNA sequence (%)	Estimated time since divergence
chimpanzee vs. bonobo (pygmy chimpanzee)	0.7	3 million years
human vs. chimpanzee	1.6	7 million years
human vs. gorilla	2.3	10 million years
gorilla vs. chimpanzee	2.3	10 million years

Source: Based on Jared Diamond, *The Third Chimpanzee* (Harperperennial Library 1993).

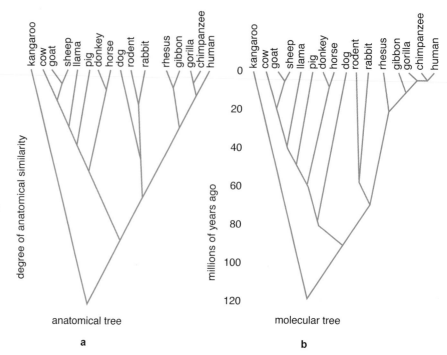

FIGURE 20.12

Evolutionary trees represent how closely organisms are related. This example shows evolutionary trees of some mammals based on anatomy **(a)** and molecular structure **(b).** Anatomical trees are based on physical characteristics. Molecular trees are based on molecular data, such as the amino-acid sequences of proteins or the base sequences of DNA. These two methods generally agree. Notice how closely related the primates are to each other compared with kangaroos or sheep.

degree of anatomical similarity

anatomical tree

a

millions of years ago

molecular tree

b

has come from the study of the physical structure of fossils. Scientists use molecular data together with other lines of evidence to explain evolutionary history and current relationships (Figure 20.12).

20.4 Dating Human Fossils

Scientists use molecular data, fossils, and radiometric age dating to determine when and where our ancestors lived. The age of fossil remains is an important component in the evidence needed to build a picture of evolutionary history. There are several ways to determine the ages of fossils. The first is through the study of the layers of soil and rock and their relationships to each other. The analysis of these layers—a process called stratigraphy—provides a rough picture of Earth's geological history. Typical sequences of common fossils also help determine the relative ages of the rock layers in which they occur. From this information, geologists can arrange newly discovered fossils in order of relative age, depending on the rock layer in which they were found (Figure 20.13).

Another source of information is provided by the groupings of fossil species found in a given stratigraphic layer. These groups of fossils tend to be made up of plants and animals that lived together, then died at approximately the same time. The known dated fossil species help establish the age of an unknown fossil.

A third technique for obtaining the ages of rocks, minerals, and fossils is called radiometric age dating. (see Appendix 1B, "Radioisotopes and Research in Biology," and Biological Challenges: Using Radioactivity to Date Earth Materials in Section 17.1). Isotopes of several elements that decay at known and constant rates can be measured in rocks and fossils. Scientists select an isotope to measure in a particular sample according to the isotope's half-life

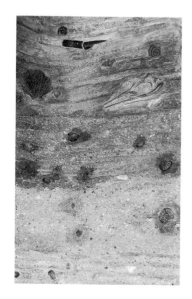

FIGURE 20.13

Stratified rock layers. Fossils of different ages are found in different layers.

Focus On

Pseudogenes

Molecular comparisons of genes give clues to evolutionary connections. Oddly enough, comparisons of nonfunctioning pseudogenes (see Section 19.3) also provide important clues to how closely related two species may be.

Species can share functional genes with similar sequences because of shared ancestry (homology) or, in rare cases, in response to similar selective pressures. Because pseudogenes are not expressed, natural selection does not affect their sequences. Therefore, shared pseudogenes are stronger evidence of relatedness than are shared functional genes. The likelihood of identical mutations is too small to account for the many cases of shared pseudogenes that have been observed. Thus, when two or more species share a genetic mistake, they must have inherited it from the same ancestor.

For example, many animal species have enzymes that enable them to synthesize ascorbic acid, also known as vitamin C. Guinea pigs and primates lack these enzymes, so they require vitamin C in their diets. Molecular analysis revealed that guinea pigs and primates have nonfunctional pseudogenes for the enzymes needed to make ascorbic acid. These species must have inherited the useless pseudogenes from a common ancestor. The pseudogenes, therefore, are evidence of the relatedness of these mammals.

and the estimated age of the sample. Carbon 14 decays relatively rapidly, and it is useful for dating materials up to about 50,000 years old. Other radioactive isotopes (potassium 40 and uranium 235 and 238) that decay much more slowly are useful for dating rocks and fossils that are older than about 50,000 years. Many important changes in human evolution occurred between 50,000 and 400,000 years ago. Hominid fossils from this period must be dated on the basis of stratigraphy radiometric methods and other fossils found with them.

The study of our evolutionary ancestors and relatives is a very active area of research. As new discoveries are made, existing ideas are tested against new data. If the data support the idea, confidence in it is strengthened. If the data do not support the idea, scientists must modify the explanation to make it consistent with more of the data. The basic pattern of our evolution from common ancestors that we share with gorillas and chimpanzees is fairly well understood. Many details, however, remain to be found.

Check and Challenge

1. Describe some characteristics that distinguish primates from other mammals.
2. If a primate had upright posture and was bipedal, what could you infer about its movements and behavior?
3. What are some of the most important skeletal features used to make comparisons between apes and humans?
4. How are biochemical comparisons helpful in determining evolutionary relationships?
5. Why is it important to establish dates for fossils?
6. What species are most closely related evolutionarily to humans?
7. Explain this statement: "The differences between humans and chimpanzees or gorillas are a matter of degree rather than completely separate characteristics."

Human Origins and Populations

20.5 Early Hominids

Many of the key events in the evolution of humans are well established by fossil and genetic evidence. Early hominid species separated from the ancestors of chimpanzees and gorillas at least 4 million years ago. Fossils of skeletons and teeth and impressions of footprints provide evidence that hominids walked upright by this time. How much earlier they may have lived remains to be learned, but genetic evidence indicates it was between 4 and 6 million years ago that humans and chimpanzees last shared a common ancestor.

The earliest hominids lived in Africa. As they evolved, they became taller and their brains more than doubled in size. Eventually human ancestors lived in open land, used tools, moved out of Africa, and evolved physically and culturally into modern *Homo sapiens.* These key points in human evolution are strongly supported by a variety of lines of evidence, such as the morphological evidence in the fossil record, advances in technological innovation (use of tools), and examinations of the molecular relatedness of DNA sequences.

Current understanding of human evolution is being enhanced by active research that builds on accumulated knowledge while raising intriguing questions. Exactly when and where did early populations migrate out of Africa? What was the last ancestor shared by apes and humans? How many distinct species of hominids and *Homo* have lived? Which ones evolved

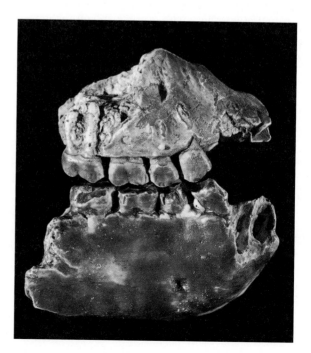

FIGURE 20.14

Remains of *Ramapithecus.* Once thought to be an apelike prehominid, *Ramapithecus* is now considered an ancestor of the orangutan. Fossils attributed to *Ramapithecus* include Yale Peabody Museum 13799, cast of maxillary fragment *(top),* and 13806, cast of mandibular fragment *(bottom).*

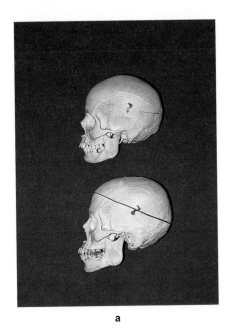

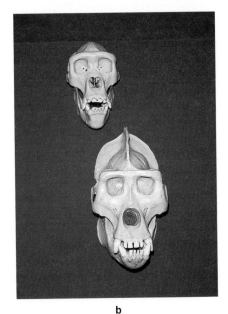

FIGURE 20.15
Sex differences in primate skulls. Variation, the basis for evolutionary change, is present in all species. In most primates, sex differences are a major source of variability as shown in **(a)** a human male skull *(bottom)* and a human female skull *(top),* and **(b)** casts of a male gorilla skull *(bottom)* and a female gorilla skull *(top).* What differences can you see between the male and female skulls?

directly into modern humans, and which have left no living descendants? These fascinating issues are being studied by many scientists today.

Identifying the species that was the last common ancestor of humans and the modern apes is an area of great interest. Fossils have been found of an apelike prehominid known as *Ramapithecus* (Figure 20.14). It lived 8–12 million years ago. *Ramapithecus* was once thought to be a common ancestor of humans and apes. However, examination of more complete fossils has led to the conclusion that *Ramapithecus* was probably related to the ancestor of the orangutan, not the hominids. The exact time that the ape and hominid lineages diverged is unknown but is now thought to have been about 5, and possibly 7, million years ago, well after the time of *Ramapithecus.*

Classification of primate fossils into different species requires great care. For example, there are skeletal differences between the sexes in many species (Figure 20.15). Size differences between the sexes are also common in many animals. Sex differences, individual variation, and fragmentary fossils make identification of species difficult.

The hominid family, Hominidae, includes two genera—*Homo* and the extinct *Australopithecus.* **Australopithecines** (members of *Australopithecus*) differed from humans (members of *Homo*) in having larger cheek teeth and smaller brains. Australopithecines were also shorter than humans and had longer arms given their relative heights. Many fossils of both hominid genera have been found in eastern and southern Africa. This is part of the evidence for the theory that humans originated in Africa. The earliest known, clearly hominid fossil is *Australopithecus anamensis.* This 4-million-year-old fossil was found in Kenya. Other very early hominid fossils include the fossil footprints of hominids that were found in the Laetoli region of Tanzania, also in eastern Africa. These footprints were made about 3.5 million years ago. They were found in 1978 by anthropologist Mary Leakey and her colleagues. The footprints show that the hominids who made them walked upright on two legs and were probably australopithecines.

E T Y M O L O G Y

australo- = southern (Latin)
pithecus = ape (Latin, Greek)
Australopithecus is an apelike hominid first discovered in southern Africa.

Almost half of a complete skeleton of an australopithecine individual nicknamed Lucy (Figure 20.16) was found in Ethiopia in 1974. She is an excellent example of a species known as *Australopithecus afarensis*. She lived about 3.2 million years ago. Like other australopithecines, Lucy was bipedal. A detailed analysis of the structure of the hip, leg, and foot bones of *A. afarensis* and other australopithecines shows that they were capable of walking upright, though maybe not exactly as modern humans do. Like other hominids, her canine teeth are shorter and less sharp than those of other primates but longer than later hominids. Lucy's teeth have a thick coating of enamel, possibly an adaptation for eating large quantities of tough plant material, such as seeds and roots.

Lucy differed from modern humans in several important ways. First of all, she was only 1 m tall, and her brain was only the size of a chimpanzee's. Her skull shows that her face protruded more than a modern human's face. In addition, her arms were relatively longer than ours. Perhaps Lucy or members of closely related species still used their long arms to climb through trees. There is no firm evidence that australopithecines used or made tools. Chimpanzees and other apes have been known to use sticks as simple tools, however, and australopithecines may have done the same. Wooden tools would not have survived as fossils. Although rocks used as tools would have survived, they would not be recognized as tools.

Fossils of several *Australopithecus* species have been found in East and South Africa. For example, *A. africanus* lived in southern Africa slightly

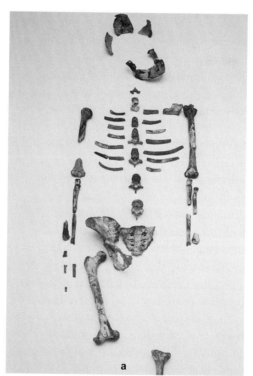

FIGURE 20.16

Australopithecine fossil and footprints. (a) Lucy *(Australopithecus afarensis)* is the most complete skeleton known from the genus *Australopithecus*. She stood slightly more than 1 m tall and lived about 3.2 million years ago. **(b)** Australopithecine footprints are preserved in 3.7-million-year-old volcanic tuff in Laetoli, Tanzania.

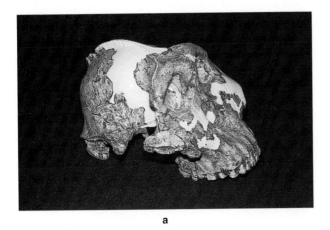

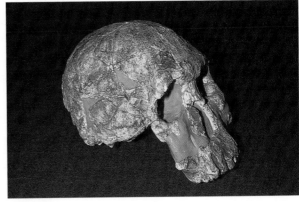

a

b

FIGURE 20.17

Skulls of an early species of *Homo* and its australopithecine neighbor. (a) An *Australopithecus boisei* skull. **(b)** *Homo habilis* skull number 1470, named for its index number at the National Museum of Kenya, East Africa. This skull is at least 1.8 million years old. Fragments of other *H. habilis* skulls have been found, but skull 1470 is the most nearly complete. The white material in **(a)** and the blue material in **(b)** were added during assembly of the fossils.

later than Lucy's species. At some point between 2.5 and 2 million years ago, African hominids underwent adaptive radiation. Also around this time, an important cultural event occurred: Some hominid species began to use crudely made stone tools as early as 2.5 million years ago. By 2 million years ago, at least two, and probably more, species of hominids were present, although no evidence exists for more than three in any one part of Africa. In eastern Africa, the strong, muscular australopithecine *A. boisei* (Figure 20.17a) lived at the same time as a larger-brained hominid called *Homo habilis* (Figure 20.17b). Many anthropologists hypothesize that *Homo habilis* was the first member of our own genus *Homo*. This point is, however, one of the details of human evolution about which there is much debate. In southern Africa, there were two other australopithecine species— *A. robustus* and *A. africanus*—and possibly *H. habilis* as well. By about 1.5 million years ago, *H. habilis* apparently disappeared from Africa and was replaced by even larger-brained hominids.

20.6 The First Humans

What is the oldest species that had generally the same skeletal features as modern humans? The answer continues to change as older fossils are found. Currently, one of the oldest fossils with a roughly modern skeleton is the 1.6-million-year-old remains of an adolescent boy. This fossil, *Homo ergaster*, has been nicknamed Turkana Boy. *H. ergaster* appeared at least 1.9 million years ago. Its size was similar to modern humans. The brain was twice the size of apes' brains but only half the size of the average modern human brain.

A widely distributed direct ancestor of humans is *Homo erectus*. There are many examples of this generally later species. *H. erectus* probably appeared first in Africa. By 1 million years ago, the species was present in southeastern and eastern Asia and survived in that area possibly until as late

Try Investigation 20B
Archeological Interpretation.

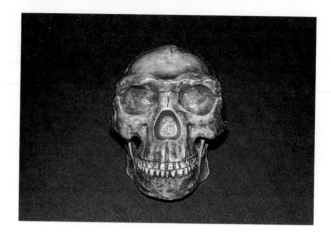

FIGURE 20.18

A reconstruction of a *Homo erectus* skull. The original fossil skull fragments were unearthed near Beijing, China. More *H. erectus* fossils have since been discovered in China, Java, East Africa, Europe, and the Middle East. Compare this skull with those in Figures 20.17, 20.19, and 20.20.

as 300,000 years ago. In that span of time, the fossil record shows that *H. erectus* changed little physically. There were cultural advances, however. For example, *H. erectus* made large, symmetrical stone hand axes. There is also some evidence that some *H. erectus* populations used fire.

Homo erectus resembled later species of *Homo* (except modern humans) in both size and robustness. Larger-brained than *H. habilis* (about 937 cm³ compared to 640 cm³), *H. erectus* had incisors that were nearly as big as those of some earlier hominids. The cheek teeth and face, however, were smaller (Figure 20.18).

To date, the oldest human fossils in Europe have been found in a cave in Spain. They are 780,000 years old. Some scholars have assigned them to the species *Homo heidelbergensis*. This species may also have lived in Africa. It is the most likely direct ancestor of a group known as Neanderthals (Figure 20.19). Anthropologists are still debating whether to classify Neanderthals as a separate species (*Homo neanderthalensis*) or as one of two subspecies of *Homo sapiens* (*H. sapiens neanderthalensis* and our own *H. sapiens sapiens*). The evidence is not yet complete enough to settle this question. Neanderthals were much more robust than modern humans. Stress at the muscle-attachment sites on their bones indicate they were far stronger than we are. Their teeth were larger and more worn. The wear might have resulted from using their teeth as tools. For example, they could have chewed hides to soften them. The wear also could show that they had a diet of tough or gritty foods. Neanderthals buried their dead and perhaps wore jewelry. The interpretation of Neanderthal fossils has been controversial. In some cases, it has been complicated by personal prejudice (see Appendix 20B, "The Old Man from La Chapelle-aux-Saints").

Neanderthals spread out over much of Europe and southwestern Asia. They thrived from about 150,000 to about 28,000 years ago. Early modern *Homo sapiens* appeared about 130,000 years ago, probably in Africa. Modern *H. sapiens* have thinner bones, smaller jaws, a higher skull, and smaller or no brow ridges compared to a Neanderthal. In some places, remains of Neanderthals and modern humans indicate that the two groups lived alongside each other for thousands of years until modern humans completely replaced Neanderthals in a short time. What was their

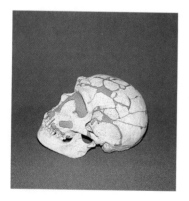

FIGURE 20.19

A fossil Neanderthal skull. The disappearance of these widely distributed archaic people has never been satisfactorily explained. They survived as recently as 28,000 years ago.

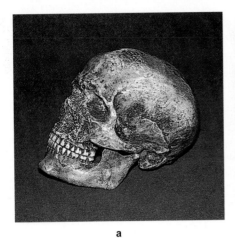
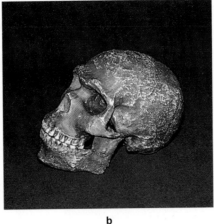

a b

FIGURE 20.20

Two casts of anatomically modern human skulls, possibly representing two somewhat genetically different populations **(a, b)**. These people, like all people today, have been classified as *Homo sapiens*.

relationship? The evidence is still unclear about how much they fought or competed, whether they interbred, and how exactly modern *H. sapiens* replaced the Neanderthals.

A case may be made for the evolution of modern humans from a group of hominids that lived at the same time as Neanderthals. These anatomically modern hominids, formerly called Cro-Magnon people, show a great deal of skeletal variation (Figure 20.20). Some early modern human fossils from Europe have Neanderthal-like skeletal characteristics—large brows and teeth, for example. These traits may have been derived from mating with Neanderthals, or they may already have been present in the population.

There are currently two competing hypotheses about the origin of modern humans. The out-of-Africa theory holds that modern *Homo sapiens* and its ancestors each evolved in turn in Africa. The new species then migrated into Asia and Europe, replacing the more primitive hominids it encountered there. In this view, there was little or no crossbreeding of early *H. sapiens* with Neanderthals or *H. erectus*.

The multiregional theory, in contrast, states that populations of *H. sapiens* evolved from *H. erectus* (and possibly Neanderthals) at approximately similar times in Africa, Europe, and Asia. This version of human evolution involves extensive crossbreeding among *H. erectus*, Neanderthals, and *H. sapiens*. It also includes the idea that neighboring *Homo* populations interbred throughout the period when *H. sapiens* was emerging.

Figure 20.21 shows the major points on which both theories agree. Evidence does not yet permit a clear choice between the two theories. However, each theory leads to different predictions. The out-of-Africa theory implies that fossils that are transitional between *H. sapiens* and earlier hominids should be found only in Africa. It also leads to the prediction that geographic variation in the traits of *H. erectus* fossils will not match the patterns of geographic variation in later *H. sapiens*. At present, there is evidence that seems to support each theory. As additional fossil and molecular evidence appears, one theory may become much stronger. Alternatively, a new theory may account for the evidence more successfully than either the out-of-Africa or the multiregional theory.

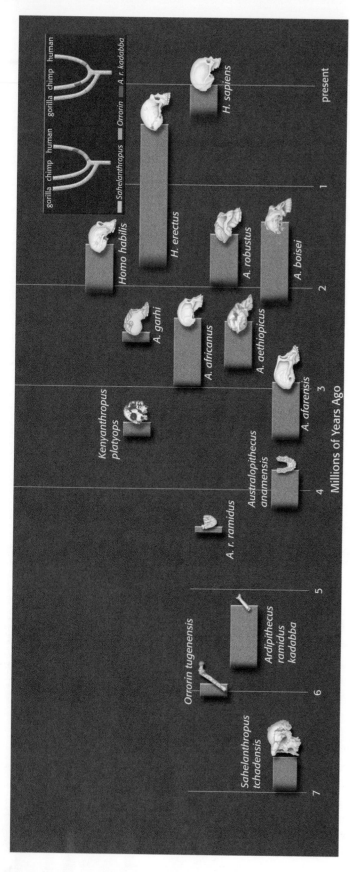

FIGURE 20.21

Hominid evolution. The fossil record of hominids shows that several species of hominids existed at the same time during the later stages of human evolution. New discoveries cause scientists to reconsider which of the early species led to modern humans. The branching diagrams in the upper right of the diagram show different hypotheses about how recently discovered fossils, *Sehelanthropus, Orrorin,* and *Ardipithecus,* might be related to humans. Although scientists are not certain about the evolutionary pathway that led to modern humans, new discoveries will undoubtedly lead to new ideas.

FIGURE 20.22

Prehistoric cave art by early *Homo sapiens.* Several examples of stone sculptures have survived that show great skill by the ancient artists. This example is from Lascaux Cave in France. It is about 15,000 years old.

There is much more evidence of the life of early modern *Homo sapiens* than is available for earlier human ancestors. These people were cave dwellers, who generally were taller than their ancestors. Many were taller than the average height of people today. Their specialized bone and stone tools and art set them apart from Neanderthals. Elaborate stone carvings of humans and animals and paintings in caves show their intelligence and skill (Figure 20.22). These creative representations suggest that language helped their makers think and plan, but exactly when human language began is still unclear (see Appendix 20C, "Cultural Evolution"). New hunting and settlement patterns and the wide use of fire and clothing may have helped modern human populations to grow and range more widely than Neanderthals. By 11,000 years ago, humans had developed agriculture and were no longer dependent solely on hunting and gathering.

20.7 Gene Pools

One intriguing question in human evolution is which populations interbred. Both ancient and modern populations show genetic variation. In the past, geographic and cultural isolation produced differences among the gene pools of human populations. Increased travel and contact with other groups have greatly changed the gene pools of some small human populations in modern times.

The ABO blood groups (see Section 13.7) show how populations change through crossbreeding. The ABO blood types were discovered around 1900. About 1920, scientists began to test the blood types of people from populations all over the world. They found that most Native Americans, Basques of northern Spain, and Australian Aborigines had no I^B alleles. It now is believed that this gene was absent from these populations until crossbreeding introduced it.

Allele frequencies are an important tool in determining genetic relationships between populations. The more similar the allele frequencies of two populations, the more closely related they are. Scientists continue to accumulate more data on blood-group and other allele frequencies. Enough data now exist to draw maps showing the frequencies of alleles in various parts of the world. Study the allele-frequency map of Europe in Figure 20.23. How can you explain the spreading of the I^B allele and the geographic variation in its frequency?

Many genetic disorders are found mainly in specific populations. For example, the gene mutation that leads to sickle-cell anemia is found primarily among people descended from or still living in the following regions: central and coastal Africa, non-Saharan northern Africa, countries on the Mediterranean Sea, and parts of central India. This genetic mutation, when found in heterozygotes (people with one mutated and one normal allele), protects against malaria, which is rampant in these tropical areas. Eye and hair color, hair texture, skin color, and many other traits also vary

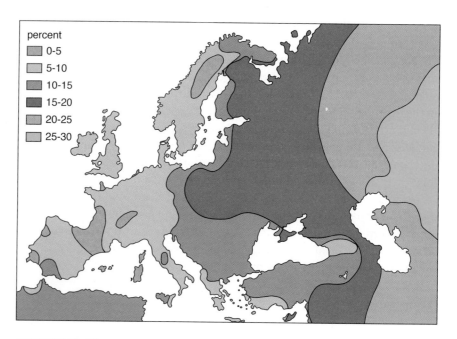

FIGURE 20.23

Allele frequency map. The distribution of the human blood group allele I^B for type B blood in Europe, western Asia, and northern Africa. The frequencies are expressed as percentages of the gene pool of the local population.

among populations. These variations are evidence of long periods of geographic and partial genetic isolation between populations.

Allele frequencies give geneticists new insights into human groups that formerly were called races. In biology, a race is defined as a genetically distinct population that does not interbreed with other populations of its species because it is separated by geographic or behavioral barriers. Is race a useful idea for describing human populations? Easily seen traits such as facial structure; the color of skin, hair, and eyes; hair texture; male facial hair; and others once led to racial classification. These easily seen traits are misleading, however. Research into other inherited traits has shown that human genetic variation is gradual, as you can see in Figure 20.23. A similar map for an allele of another gene would show a very different pattern— alleles of different genes are distributed differently. This evidence shows that the human species is not divided into separate discrete races. There is much more genetic variation within human "races" than among them. All humans share genes that make them *Homo sapiens*. There are fewer genetic differences between any two groups of humans—no matter what their physical differences—than there are between closely related but separate species, such as chimpanzees and humans.

Check and Challenge

1. Explain the evolutionary relationship between modern humans and modern chimpanzees or gorillas.

2. How are australopithecines similar to humans (genus *Homo*)? How are they different?

3. Summarize the major evolutionary changes in the hominids since they last shared a common ancestor with any other primates.

4. What is the approximate geological age of the oldest clearly identified hominid fossils?

5. Compare the out-of-Africa and multiregional theories for the origin of moden humans.

Chapter HIGHLIGHTS

Summary

Humans are primates that evolved from early ancestors shared with the great apes. Humans have particularly well-developed fingers for manipulation, bipedal posture, and unusually large brains. Language is important to human social behavior. Our closest relatives are chimpanzees, which differ from us by less than 2% of DNA sequences. Gorillas are only slightly less related.

Evidence of the course of human evolution includes dated fossils and artifacts, as well as molecular and skeletal studies of living species. The best known early hominid fossils are 4-million-year-old African australopithecines. A well-preserved fossil of *Australopithecus afarensis* shows that these creatures walked upright. They were smaller than modern humans, with longer arms and much smaller brains. Evidence of tool use appears about 2.5 million years ago, near the time of the first of our genus, *Homo*. A very well-preserved early *Homo* species, *H. ergaster*, greatly resembles modern humans. A similar species, *H. erectus*, was more geographically widespread.

Early modern *Homo sapiens* appeared before the Neanderthals disappeared. The contribution of the Neanderthals to the ancestry of modern humans remains unclear.

Geneticists use information from modern humans and other primates to estimate how closely their populations are related. By studying human gene pools, they have found that the racial classifications of humans used in the past are not biologically valid. Although diverse physical appearances exist among populations, our species is not divided into genetically distinct populations.

Key Concepts

Use these terms to develop a concept map of human evolution. Add additional terms as you need them.

- fossils
- molecular comparisons
- hominid
- populations
- skeletal comparisons
- social behavior
- language
- bipedal
- grasping
- *Homo*
- complex brain
- primate
- tools
- evidence
- chimpanzees

Reviewing Ideas

1. How can genetics and the study of DNA help shed light on the origin of humans? What are the limitations of this method?
2. What kinds of information can the study of modern and fossil skeletons provide about the organisms and how they lived?
3. Why is it desirable to have more than one method for establishing the ages of fossils?
4. What species are our nearest relatives evolutionarily? What evidence supports this conclusion?
5. How can anthropologists determine if australopithecines were bipedal?
6. How do modern people differ from earlier populations of *Homo sapiens?*
7. How do modern humans differ from other primate forms?
8. What are five skeletal features used for determining evolutionary relationships among primates?
9. What can an anthropologist infer from the shape of teeth and their wear patterns?
10. Why is stratigraphy important in establishing evolutionary sequences?

11. Where and when did the first known hominids live? What are their taxonomic names?
12. Compare the relative age and characteristics of the first *Homo* species with australopithecines.
13. What were the major differences—both biological and behavioral—between Neanderthals and earlier *Homo* species?
14. Explain why human populations were once classified as races and why genetic evidence no longer supports the use of that term.

Using Concepts

1. The pattern of life on Earth shows both unity and diversity. How are these concepts exhibited among the primates?
2. What may account for the diversity among the australopithecines?
3. Suggest possible reasons why the Neanderthals disappeared.
4. Does the fossil record indicate physical diversity among the first anatomically modern people? How?
5. What types of mistakes can be made in fossil interpretation?
6. What effects can molecular comparisons have on determining evolutionary relationships?
7. Explain what is inaccurate scientifically in the statement "Humans are descended from modern apes."
8. What is the importance of comparing molecular and skeletal evidence in determining relationships between modern primates?

Synthesis

1. How can genetic diversity among populations come about? How does diversity within a population come about?
2. Is anthropology a science? Why or why not?

3. What is the significance of bipedal posture for tool use? What other anatomical features are important for tool use?

Extensions

1. Discovery of new hominid fossils is frequently in the news. Find at least three reports of fossil discoveries from the last 5 years, and report on their significance.
2. A *Homo habilis* fossil indicates that the individual was about 0.91–1.07 m tall. Australopithecines were about 0.9 m tall. *H. erectus* is thought to have been almost 1.7 m or more. What effect might this new discovery have on hypotheses concerning skeletal and brain growth in hominid populations? Research the correlation between brain size and stature in fossil populations. Write an essay addressing this question.
3. Make a poster showing the evolutionary relationships between humans and at least five modern and two fossil species of primates. Include diagrams and charts that show degree of relatedness, time of divergence, and major distinguishing characteristics.
4. Make a time line of human evolution. What other species of animals or plants were prominent during the evolution of hominids into modern humans?

Web Resources

Visit BSCSblue.com to access
- Information and activities related to human evolution and primates
- Web links related to this chapter

CONTENTS

LEARNING OUTCOMES

By the end of this chapter you will be able to:

A   Distinguish between sensory and motor systems.

B   Distinguish between the peripheral and central nervous systems.

C   Describe the cells of the nervous system.

D   Describe how impulses are transmitted and processed.

E   Describe the effects of drugs on the brain.

F   Discuss trends that characterize vertebrate brain evolution.

G   Describe the molecular evolution of the nervous system.

CHAPTER

21

Nervous Systems

■ *How does the runner translate the sound of the starting gun into action?*

■ *What organ systems are involved in the runner's movement?*

Often what we seem to do so naturally requires an amazingly complex set of biological actions. Think of what happens when a runner hears a starter's pistol. First, her ears convert the sound into signals and send them to her brain. Her nervous system then processes the signals, makes decisions, stores information, and begins a complex pattern of actions that allows her to begin running down the track. These actions are accomplished by networks of precisely connected cells. Her nervous system's ability to efficiently interpret sensory information and quickly activate her muscles is a critical factor in determining whether or not she wins the race.

In this chapter, you will examine the organization of the nervous system, the functions of each region of the brain, and the molecules that form the building blocks of the nervous system. The components that make up the human nervous system are very similar to those of simpler animals. Evolution has acted on how these components are arranged. In humans, the arrangement enables a person to react to sensory information, to think, to feel, and to remember.

Organization of the Nervous System

21.1 Sensory Systems

Try Investigation 21A
Sensory Receptors.

Nervous systems make it possible for animals to know and react to the world around them. A change in the environment that causes an organism to respond is called a **stimulus** (plural: stimuli). A stimulus can be either internal (such as hunger) or external (such as sound or odor). Responses can also be internal (an upset stomach) or external (running or speaking). The nervous system interprets stimuli and coordinates other organ systems to respond to the stimuli.

The nervous system gathers and interprets stimuli through **sensory receptors.** These specialized structures are made up of sensory nerve cells and support cells (Figure 21.1). Sensory receptors can be single nerve cells or complex organs, such as eyes and ears, that integrate information from thousands of receptor cells. Different types of sensory receptors transmit information about different types of environmental stimuli, including light, pressure, and chemicals. For example, the taste receptors on your tongue gather stimuli about the chemical composition of the food you eat.

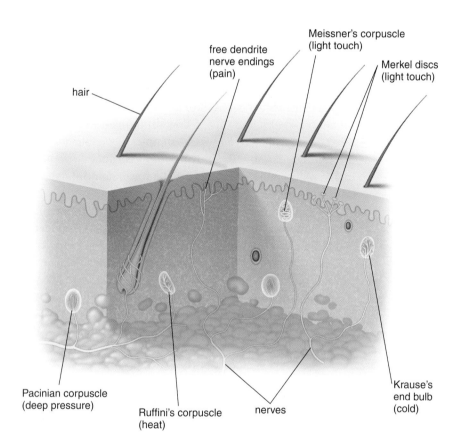

FIGURE 21.1

Sensory receptors found in human skin. Each type of receptor is specialized to detect one type of stimulus, which may include pressure, touch, pain, and high or low temperature. Other types of receptors are found in other organs. Among these are light receptors in the eye and stretch or movement receptors in the muscles and inner ear, which detect body position, muscle tension, and sound.

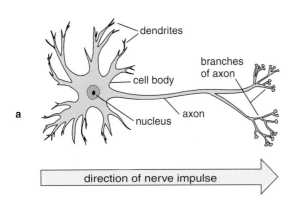

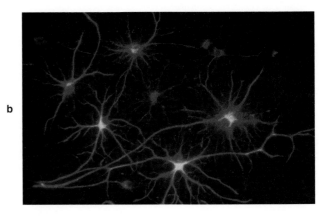

FIGURE 21.2

A typical vertebrate neuron. (a), The dendrites can receive stimuli from sensory receptors such as those in Figure 21.1 or from the axons of other neurons. In response, an electrical impulse travels from the dendrite through the neuron to the tip of its axon. From there, the impulse stimulates another neuron or cell. **(b),** This micrograph clearly shows cell bodies, dendrites, and axons from the human brain.

Sensory nerve cells, called sensory **neurons,** transmit impulses from the sensory receptors to the central nervous system (section 21.4). A typical neuron has three main parts (Figure 21.2). The **cell body** contains the nucleus and other organelles. Two types of fibers extend from the cell body. **Dendrites** receive stimuli and conduct impulses toward the cell body. The **axon** carries impulses away from the cell body. A **nerve** is a bundle of fibers consisting of both axons and dendrites. The structure of neurons is similar in most types of animals, but the shape varies in different animals and even within different parts of one animal's nervous system.

The dendrites of sensory neurons are specialized to collect and transmit information about environmental stimuli. The stimuli, such as heat or light, to which sensory receptors respond are different forms of energy. These receptors translate the energy of specific stimuli into electrical energy, the language of the nervous system. The resulting electrical impulses can travel throughout the nervous system, where interpretation, comparison with other information, and eventually all decisions occur.

For example, when you hear a siren, sensory receptors in your ears vibrate in response to the mechanical energy of sound. The receptors convert this energy to a series of electrical impulses that travel through the axons of sensory neurons to your brain. Groups of neurons in your brain receive these electrical impulses and interpret their pattern as sound. A comparison of this pattern to others stored in your memory leads to an identification of the sound as a siren. Other parts of your brain then signal your muscles to begin the appropriate response—to look around for the source of the sound and to get out of the way. Notice that your ears do not hear the siren. Instead, the sensory receptors in your ears convert the sound of the siren to electrical impulses. Perception—hearing, vision, and other sensory experiences—happen in the brain, not in the sensory organs.

Different animals depend on various types of sensory receptors. For example, some animals have poor vision but a keen sense of smell. Fish in the deep sea have poor vision because light usually is not present. Owls that

Focus On

The Eye

The human eye is sensitive to almost 7 million shades of color and about 10 million variations of light intensity. What makes this sensitivity possible?

The lens focuses light on a thin tissue called the **retina** (Figure 21.3a). The iris (the colored part of the eye) consists of muscles that can contract and relax, changing the size of the opening, called the pupil. This regulates the amount of light that enters the eye. The incoming light stimulates receptor cells, called rods and cones, in the retina (Figure 21.3b). **Rods** are responsible for black-and-white and peripheral vision (the edges of the field of vision). **Cones** are responsible for color vision.

Rods contain rhodopsin, a protein that absorbs light of all visible wavelengths. Cones require brighter light than rods do. There are three types of cones, each containing a different pigment that absorbs red, green, or blue light. Absorbed light energy signals the cell to begin an electrical impulse. Nerve impulses from the rods and cones travel through the optic nerves (one from each eye) to the brain. At the molecular level, light absorption is similar even in distantly related species. For example, some photosynthetic bacteria contain a form of rhodopsin (see Focus On Protein Structure and Function, Section 4.3).

Neurons in the retina encode visual information before it is sent to the brain. For example, both red and green cones absorb yellow light. When they do, they act together to stimulate neurons that carry the perception of yellow light to the brain. However, red and green cones inhibit the transmission of each other's impulses, so we cannot see an object as both red and green.

The ventral position of our eyes enables us to see objects from two slightly different angles. Comparing information from the two eyes allows the brain to perceive depth and estimate distance. This adaptation is typical of predators. Many herbivores, such as horses, have eyes on the sides of their heads. This placement gives them a wide view of their surroundings, which helps them spot predators, but because their eyes see two different scenes, they have more difficulty in judging distance.

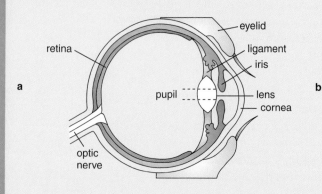

a

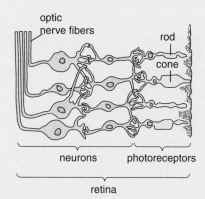

b

FIGURE 21.3

A cross section of the human eye. (a), Light passes first through the transparent cornea and then through the pupil, the adjustable opening in the iris. The light then reaches the lens, which focuses the image sharply on the retina. **(b),** In the retina, rods and cones absorb light, sending impulses to retinal neurons. Connections among these neurons form interacting circuits that process visual information. The output of the circuits eventually goes to the brain through optic nerve fibers, enabling vision of the image.

fly at night have very sensitive nocturnal vision. Bats and whales are especially sensitive to high-pitched sounds; this evolutionary adaptation enables them to use echoes to perceive their surroundings. Migrating birds find their way partly by using their ability to detect the Earth's magnetic field. Our bodies can detect and respond to changes in our blood pressure. Our sensory receptors convert sound, light, electrical, and magnetic energy into electrical impulses that the brain can interpret.

Perhaps the most amazing thing about sensory systems is the ability of the brain to process many kinds of signals simultaneously. Consider the runner, whose fingertips feel the track beneath her and whose muscle sensory systems feed back information on the tension and position of each muscle. At the same time, other sensory systems relay information on heart rate, the concentration of carbon dioxide in the blood, and so forth. Her brain integrates all this information as it awaits the sound of the starter's pistol. The runner's success depends on how effectively her brain tells her muscles that the race has begun.

21.2 Motor Systems

Once the nervous system receives and processes sensory information, the motor system responds. The **motor system** enables an organism to react to stimuli. It is made up of **effectors,** such as muscles and glands, that carry out responses. Motor neurons carry the instructions from the nervous system to the effectors. These neurons have cell bodies within the brain and spinal cord. Motor axons can be extremely long. For example, giraffes have motor axons that extend all the way from cell bodies in the spinal cord to muscles in their lower legs.

Effectors may be under either voluntary or involuntary control. **Voluntary** control involves a conscious decision to act in response to a stimulus. For example, the runner chooses to move when she hears the starter's pistol. **Involuntary** control does not involve a conscious decision. The runner does not choose to speed up her heartbeat, for example, but it is part of her response to the same stimulus.

A complex act such as running a race involves activation and inhibition of many responses. For example, while some motor neurons stimulate the leg muscles, other motor neurons act to increase the heart rate and breathing, ensuring adequate delivery of oxygen and glucose to the muscles. At the same time, other responses, such as directing more blood flow to the digestive system, are inhibited. Even the stimulation of the leg muscles involves some inhibition: For each leg to alternately flex and extend, motor neurons that activate opposing pairs of muscles must be alternately stimulated and inhibited. Complex circuits of neurons that operate without conscious choice make smooth, efficient voluntary movement possible.

21.3 The Peripheral Nervous System

Because the vertebrate nervous system is complex, it is helpful to divide it into parts based on function and structure. The main division is between the central nervous system and the peripheral nervous system. The **central nervous system (CNS)** consists of the brain and spinal cord. The **peripheral nervous system (PNS)** carries information between the CNS and the other organs.

The PNS consists of all nerve tissue except the brain and spinal cord. It contains both sensory and motor neurons. The human PNS includes 12 pairs of cranial nerves that extend from the brain into the head and upper body. There are also 31 pairs of spinal nerves that branch out from the spinal cord

ETYMOLOGY

motor = movement (Latin)
Motor neurons cause movement by stimulating muscles to contract.

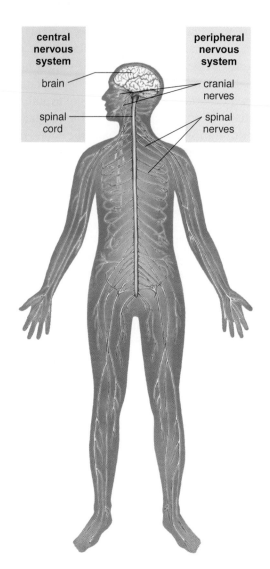

FIGURE 21.4

Human nervous system. The human central nervous system (CNS) and peripheral nervous system (PNS). The PNS consists of bundles of axons, or nerves, that connect the CNS to the rest of the body. These nerves contain sensory neurons that carry impulses from the skin and internal organs to the CNS. Motor neurons in these same nerves carry impulses from the CNS to effector organs. These motor neurons control both voluntary responses, such as skeletal muscle contraction, and involuntary responses, such as changes in breathing, production of digestive juices, and the release of hormones into the bloodstream.

ETYMOLOGY

soma = body (Greek)
auto- = self (Greek)
nom- = law (Greek)

The **somatic** nervous system maintains communication between the CNS and the rest of the body. The **autonomic** nervous system is more or less independent; it does not depend only on instructions from the CNS.

throughout the entire body (Figure 21.4). Sensory neurons of the PNS help maintain homeostasis by coordinating the functions of internal organs and providing the CNS with sensory information. Motor neurons of the PNS enable an organism to respond to its environment by stimulating muscle contraction and the activities of glands.

The motor neurons of the PNS make up two subsystems (Figure 21.5). The **somatic nervous system** is responsible for skeletal muscle contraction, which is usually voluntary. The **autonomic nervous system** is responsible for involuntary responses, such as changes in the activities of glands and the digestive system.

The somatic and autonomic nervous systems often work together. For example, when you are cold, your brain signals your autonomic nervous system to constrict surface blood vessels in your skin. This constriction helps to reduce loss of heat energy. At the same time, your brain signals your somatic nervous system to stimulate your skeletal muscles, causing you to shiver. Together the autonomic nervous system and the glands of the endocrine system regulate many involuntary functions, such as sleep, appetite, and digestion.

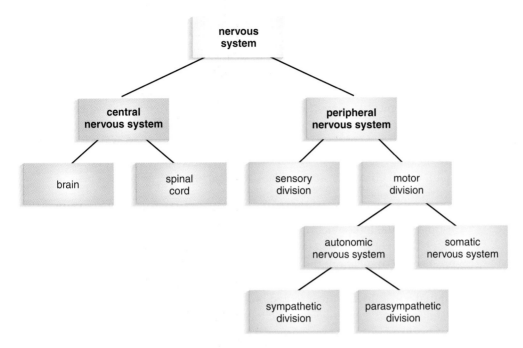

FIGURE 21.5

Divisions of the vertebrate nervous system. The vertebrate nervous system can be divided into various subsystems that perform different functions and have different structures. The main division is between the CNS and the PNS. The autonomic nervous system regulates involuntary responses. Its sympathetic and parasympathetic divisions have opposite effects on effector organs. Although the somatic nervous system controls voluntary contraction of the skeletal muscles, it can also produce involuntary movements of those muscles, such as shivering.

The autonomic nervous system has two systems that have opposite effects. In general, the **sympathetic system** prepares an animal for quick action. For this reason, it is often called the fight-or-flight system. The sympathetic system reduces the activity of the digestive system and stimulates the heart to beat faster. The **parasympathetic system** reduces an animal's readiness for action, slows the heart, and stimulates homeostatic activities such as digestion. Most internal organs are connected to both of these systems. The sympathetic and parasympathetic nerves often have opposite effects on the same organ. If one system stimulates a particular organ, the other inhibits the organ (Table 21.1).

CONNECTIONS

The nervous system is essential for maintaining homeostasis and influencing an organism's behavior. It coordinates the reception of stimuli with appropriate responses.

TABLE 21.1 Some Actions of the Sympathetic and Parasympathetic Nervous Systems		
Organ	Action of Sympathetic Nervous System	Action of Parasympathetic Nervous System
heart	beats faster	beats slower
liver	breaks down glycogen and releases glucose to bloodstream	absorbs glucose from bloodstream and synthesizes glycogen
lungs	expand air passages	constrict air passages
digestive system	reduces production of digestive juices and inhibits muscular contractions that move food through intestines	increases production of digestive juices and stimulates muscular contractions that move food through intestines

21.4 The Central Nervous System

The vertebrate CNS forms the bridge between the sensory and motor functions of the PNS (Figure 21.6). The CNS interprets nerve impulses and works with the glands to coordinate the activities of all body systems. The CNS also can store experiences in memory and learn by establishing patterns of responses based on previous experiences.

The brain and spinal cord are covered by protective membranes called the **meninges.** They are further cushioned from shock by cerebrospinal fluid, which circulates around the CNS. The entire CNS is also encased in protective bone.

The brain of an adult human weighs about 1.5 kg and is composed of an estimated 100 billion neurons. The senses of sight, hearing, taste, smell, and touch are experienced in the part of the brain called the **cerebrum.** Much has been learned about the cerebrum from studies of people whose brains were damaged by injuries. By determining the parts of the brain that were injured and studying the affected body functions of these individuals, scientists have identified specific functions of the brain's areas. Figure 21.7 shows the major functional areas of the human cerebrum.

The cerebrum is divided into left and right hemispheres, each having different functions. The left hemisphere is associated with verbal and analytical skills, such as language, mathematical calculation, and logic. The right hemisphere is responsible for the ability to judge spatial relationships

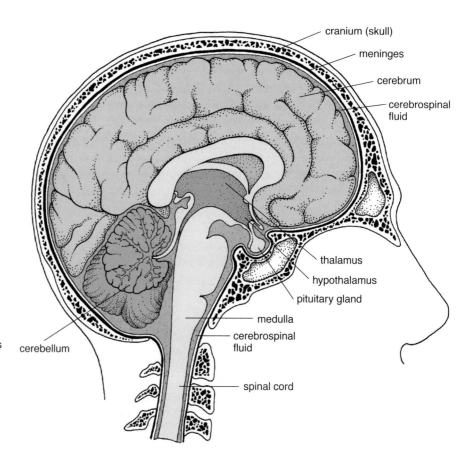

FIGURE 21.6

The human central nervous system, the most complex vertebrate CNS. The CNS consists of the brain and spinal cord, both of which are covered with meninges and bathed in cerebrospinal fluid. The brain is divided into several major regions. The most prominent parts are the cerebellum, which coordinates movement and maintains balance, and the cerebrum, which is responsible for sense perception, voluntary movement, and thought. The medulla regulates basic homeostatic functions, such as the activities of the circulatory and respiratory systems.

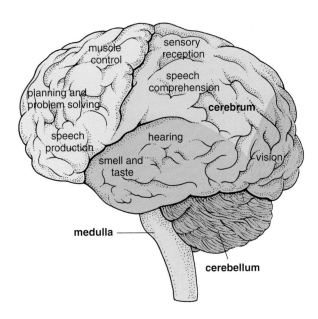

FIGURE 21.7

Major functional areas of the human cerebrum. Specific regions of the brain serve specific functions. Some regions are responsible for certain kinds of sensory information, such as touch, sight, and hearing. Other regions initiate voluntary movement, memory, language skills, and other activities.

and is involved in imaginative or creative thinking. Each hemisphere also controls the sensory and motor functions of the opposite side of the body. Many functions of the right and left hemispheres usually are reversed in left-handed people. The cortex (outer layers) of the human cerebrum has distinct specialized areas that control the sensory and motor functions of each body part (Figure 21.8).

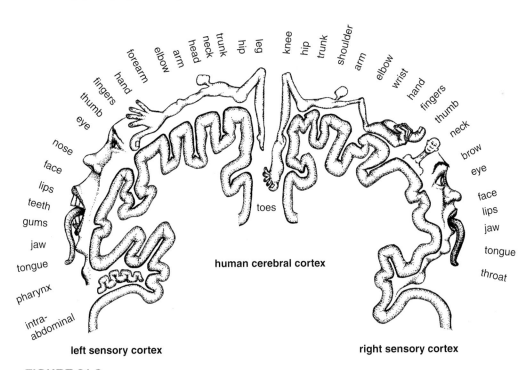

FIGURE 21.8

The sensory and motor control regions of the human cerebral cortex. The cartoon drawings show the body parts served by each part of the cerebrum. Note the distorted sizes of the cartoons; some body parts are served by disproportionately large or small brain regions. How can you explain this?

Pain Is in the Brain

Perhaps the most important sensation that we can feel is pain. Although no one enjoys pain, it serves an important function as an early warning system. It tells you that something is causing an injury and that you need to react. Some people are born without pain receptors or lose the sensation of pain through diseases such as leprosy. Many of these people suffer disfiguring damage because they are not aware it is occurring.

Generally, we think of pain as coming from the site of an injury. Stimulating pain receptors in skin and other organs leads to pain. Like other sensations, however, pain occurs in a specific set of neurons in the brain (Figure 21.9). If these cells are stimulated directly, you feel intense pain even if the receptors themselves are not stimulated.

A lot has been learned about the physiology of pain by examining people with amputations. Some of these people feel pain in the missing limb. This "phantom limb" pain occurs when some event in the brain stimulates the neurons responsible for the sensation of pain in the amputated limb. The sensation is the same as pain caused by direct stimulation of receptors in the missing limb.

Recently it has become possible to observe activation of the pain cells in the brain. In some cases, scientists have determined that the same cells in the brain are activated whether the pain is real or only a hypnotist's illusion. These experiments illustrate that nerves of the PNS and their connections in the spinal cord and brain make up our pain system.

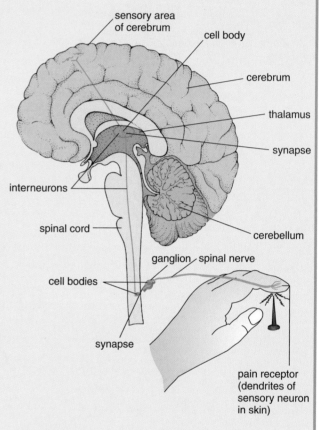

FIGURE 21.9

Pain perception. One of the major pain pathways from receptor to cerebral sensory cortex. Like all sensory information, pain is perceived in the brain, not in the stimulated organ.

The **cerebellum** mainly coordinates contractions of the skeletal muscles. For example, when the cerebellum receives an impulse from the cerebrum to move a leg, it sends a pattern of nerve impulses to each of the muscles involved in the desired movement, ensuring that each of them contracts or relaxes with the right timing and strength to move the leg smoothly. Complex movements, such as walking and running, are under the influence of this part of the brain.

The brain stem, also called the **medulla,** regulates vital involuntary functions, such as heart rate, contractions of blood vessels, and the depth and rate of breathing. The medulla keeps these vital functions operating properly without any conscious intervention.

21.5 Cells of the Nervous System

There are two major classes of cells in vertebrate nervous systems—neurons and glial cells. Neurons can be divided into three general types. You have already learned about sensory and motor neurons. **Interneurons,** the third type of neuron, transmit signals from one neuron to another within the CNS. Interneurons are the most abundant type of neuron in vertebrate nervous systems.

Although **glial cells,** or glia, are generally smaller than neurons, they are 10–50 times as numerous and make up more than half the weight of the nervous system (Figure 21.10a). Unlike most neurons, glial cells retain the capacity to divide in the adult vertebrate. They have various functions. Some support the neurons, filling the spaces between their cell bodies and coating the surfaces of the CNS and blood vessels in the brain. Others provide nutrients and growth factors to neurons and remove wastes. The larger axons of many neurons in vertebrate animals are surrounded by a membrane, or **myelin sheath,** that serves as insulation. Glial cells known as Schwann cells wrap themselves around the axons of neurons in the PNS to form this layer (Figure 21.10b). The gaps between neighboring Schwann cells are known as **nodes of Ranvier.** Scientists continue to discover a large number of roles for different types of glial cells, expanding our knowledge of nervous-system function.

Cell bodies of neurons form aggregations called **ganglia** in the PNS and **nuclei** in the CNS. The CNS also has layered groups of connected neurons called **lamina.** Each of these networks has specific functions. Neurons perform three major functions: They respond to chemical and physical stimuli, they conduct impulses, and they release chemical regulators. Combinations of these three functions enable the nervous system to store

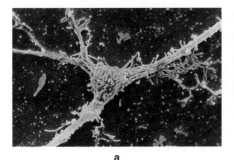
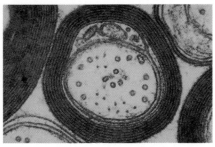

a   b

FIGURE 21.10

Glial cells. There are several types of glial cells in the vertebrate nervous system, and their many functions are still being investigated. **(a),** Glial cells provide physical support for neurons in the spinal cord, ×766. **(b),** Schwann cells wrap themselves around axons in vertebrates, forming an insulating myelin sheath, ×57,000. How do myelinated axons help make the large bodies and brains of many vertebrates possible?

Learning and Memory

The brain has a remarkable ability to store information for years and recall it at will. Unlike information stored in a computer, memory is not stored in specific places. Rather, memory seems to be stored in many places. The same region of the brain may participate in multiple memories.

Psychologists have divided memory into two types. Explicit memories, which are of facts and events, involve regions of the cerebrum called the temporal lobe and hippocampus (Figure 21.11). Implicit memories, which are of how to do something, are stored in the sensory or motor systems of the CNS. Both kinds of memories form by the same chemical and cellular processes.

Diseases can affect each kind of memory. Patients who suffer from amnesia may be able to remember implicit memories but may have great difficulty recalling explicit memories, such as names and places. People who suffer from brain damage may be unable to store new memories but can easily recall events from years earlier. Other brain injuries can lead to forms of amnesia where years of events are completely lost. Sometimes these memories return, indicating that the process of retrieval, not the memories, was injured.

Biologists have begun to understand how short-term memory storage leads to long-term memory retention. Short-term memory lasts only a few minutes, and unless additional changes take place, the memory is lost. You rely on short-term memory to store telephone numbers just before dialing. Short-term memory has a small capacity, and distraction or new information will cause you to lose information stored earlier. If you use the same information often or decide that it is important, then additional molecular changes take place in the brain. An important result of these changes is that interneurons involved in storing the memory become more sensitive to impulses from certain axons that connect to their dendrites.

The results of some experiments suggest that these changes begin with activation of the enzyme protein kinase A. This enzyme activates a protein that binds to specific DNA sequences and promotes transcription of genes that encode proteins of the cytoskeleton. These proteins also make cells more responsive to impulses from other neurons in the memory pathway. These changes strengthen the pattern of impulse transmission from neuron to neuron, forming memory that can last for many years.

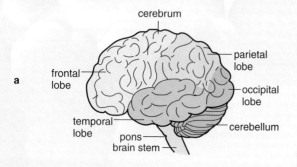

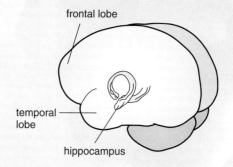

FIGURE 21.11

Brain structures involved in memory and learning. The temporal lobes **(a),** regions located just above the ears, seem to be responsible for storage of long-term memory. The hippocampus **(b),** found tucked in behind the temporal lobes, may be especially important in transferring information from short-term to long-term memory.

your memories, help you analyze situations and make decisions, and regulate your organs and glands.

Check and Challenge

1. Describe the differences between sensory and motor systems.
2. Compare the functions of the central nervous system and the peripheral nervous system.
3. What are the components of the peripheral nervous system?
4. What are the functions of glial cells?

Cellular Communication

21.6 Transmission of Impulses

How are stimuli converted to the energy that the nervous system interprets? Like all cells, neurons are biochemical batteries. A cell's plasma membrane is electrically charged because ions are distributed unequally on the outside and inside of the cell (Figure 21.12).

The unequal distribution is due to three factors (Figure 21.13). First, cells contain many negatively charged molecules, such as phosphates and proteins, that cannot diffuse through the plasma membrane. The two most numerous positively charged ions that are available to balance this negative

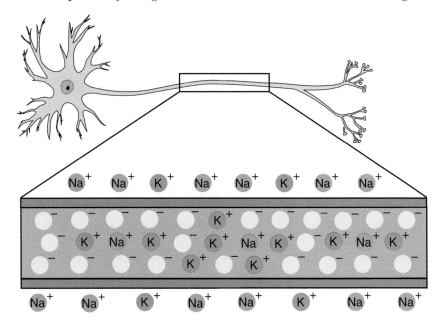

FIGURE 21.12

Electrical charging of the plasma membrane of a neuron. Note that the inside of the cell has a negative charge with respect to the outside. Most of the excess negative charges in the cytoplasm are due to proteins and phosphate ions. Sodium ions (Na^+) are more concentrated outside the cell, and potassium ions (K^+) are more concentrated in the cytoplasm.

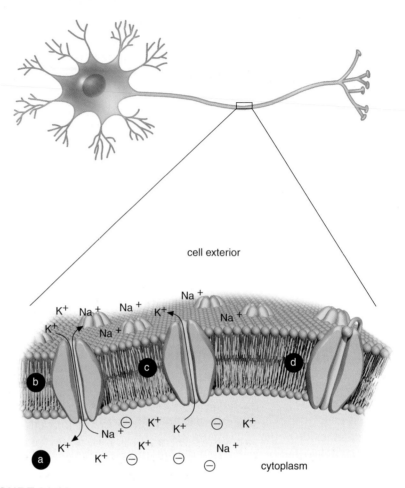

cell exterior

cytoplasm

FIGURE 21.13

How the plasma membrane of a neuron becomes charged. In step **(a),** negative ions trapped inside the cell attract positive ions. In step **(b),** an active-transport protein takes potassium ions (K^+) into the cell and removes sodium ions (Na^+). This process keeps the sodium concentration low and the potassium concentration high in the cytoplasm. These concentrations are reversed outside the cell. In step **(c),** the concentrated potassium ions leak out through potassium channels in the cell membrane, leaving the negative charges in the cytoplasm partly unbalanced. The result is a relatively negative cytoplasm and a relatively positive exterior. In step **(d),** this electrical difference changes the tertiary structure of the sodium-channel proteins, which keeps the sodium-channels closed. They can open in response to a stimulus that depolarizes the membrane.

charge are potassium ions (K^+) and sodium ions (Na^+). Second, an active transport protein in the neuron plasma membrane pumps potassium into the cell and sodium out of the cell. This ensures that potassium is the main positive ion in the cytoplasm. Third, potassium ions easily diffuse back out of the neuron though protein channels in the plasma membrane that facilitate diffusion. This diffusion reduces the ability of the active transport system to concentrate potassium in the cytoplasm. As a result, there are relatively more negative charges in the cytoplasm and more positive charges outside the cell. Sodium ions diffuse back into the cell much more slowly than potassium ions leak out because they are too big to fit into the channels that permit facilitated diffusion of potassium. The slow leakage of sodium into the cell balances some of the negative charges there. However, there is still more negative charge in the cell and more positive charge outside, a difference called an **electric potential.** An electric potential is a form of potential energy. A membrane with a positive side and a negative side is also said to be polarized. The potential of the membrane when it is not transmitting an impulse is called the neuron's **resting potential.**

Stimulation of a neuron results in a sudden change in permeability. Channels in the membrane that are highly selective for sodium ions open, and sodium ions quickly diffuse into the neuron. This rapid influx of positive ions reverses the membrane's electric potential; the membrane becomes depolarized (Figure 21.14b). The change in electric potential causes the

CHEMISTRY TIP

The negatively charged oxygen atoms in water molecules (H_2O) are attracted to the positive ions of potassium (K^+) and sodium (Na^+). As a result, shells of water molecules surround the ions. Sodium ions are surrounded by larger shells than potassium ions, making the sodium ions too big to fit through the potassium channels. Hence, diffusion of the polar sodium ions through the hydrophobic lipid membrane is very slow.

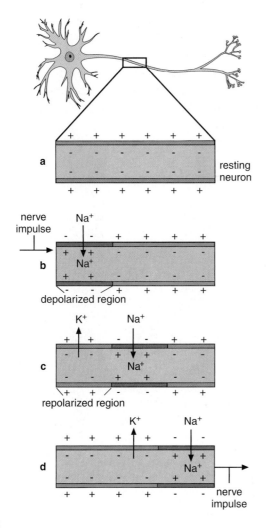

FIGURE 21.14

How an action potential occurs. (a), The plasma membrane of a polarized resting neuron is more positive on the outside than on the inside. **(b),** A stimulus opens the sodium channels in part of the membrane, sodium rushes in, and part of the neuron is depolarized. **(c),** Potassium diffuses out, depolarizing the membrane. The sodium-potassium active-transport system then pumps the ions back to regenerate the resting potential (repolarizing). **(d),** The impulse moves as a wave of such changes in electrical charges.

sodium channels to close and potassium channels to open. Potassium ions diffuse out, restoring the positive charge outside of the membrane or repolarization (Figure 21.14c). The brief depolarization and repolarization, also known as an **action potential,** lasts about 3 milliseconds (msec). The number of action potentials a neuron can experience per second depends on the time required for active transport to restore the resting potential.

Depolarization at one point in a neuron membrane sets up an electric current that spreads rapidly along the axon. This wave of depolarization, caused by the movement of ions across the cell membrane, is called a **nerve impulse.** The nerve impulse moves in one direction along an axon, away from the point of stimulation, because sodium ion channels on the axon behind the area of depolarization temporarily close.

a

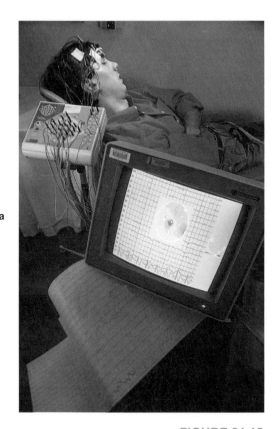

b

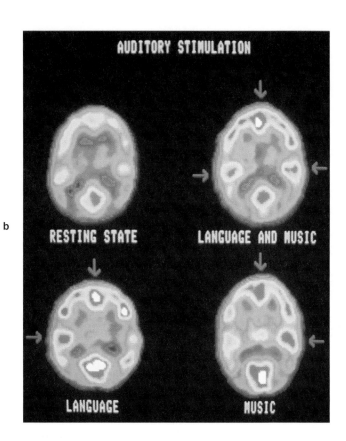

FIGURE 21.15

Making the electrical activity of the nervous system visible. These techniques help reveal the workings of normal and diseased brains. **(a),** An electroencephalograph (EEG) measures the potential that develops across the skull as a result of the simultaneous action potentials of large numbers of neurons in the brain. Wires attached to the skin around the head transmit this electrical activity to a sensitive recording device. **(b),** Positron-emission tomography (PET) involves injecting the subject with safe, weakly radioactive water or glucose. The more active parts of the brain consume more water and glucose. The PET camera forms an image that shows the intensity of the radiation due to radioactive decay in each part of the brain. The more metabolically active brain regions show up as more intense sources of radiation. This PET scan shows a resting state control compared to brain activity changes caused by language only, music only, and both language and music. The red arrows indicate brain activity in the temporal side lobes and frontal lobe of the brain resulting from auditory stimuli.

Neurons obey an **all-or-none law** in transmitting impulses. In other words, a stimulus must be strong enough to cause an impulse to begin traveling down the neuron. The level at which the impulse is triggered is called the **threshold.** However, all of a neuron's impulses are nearly identical. How then can the brain and effectors distinguish between weak and strong stimuli? Usually, strong stimuli cause more neurons to transmit impulses. They also cause individual neurons to transmit more impulses per second. For example, the more sensory neurons in the eye that transmit impulses to the brain and the more often they transmit them, the brighter the light the brain perceives.

In the myelinated neurons of vertebrates, the nerve impulse actually jumps from one node of Ranvier to the next. Jumping speeds up the transmission of the impulse. Many of the neurons that control your muscles are myelinated, allowing for speedy reaction times. Axons with myelin sheaths transmit their impulses at up to about 120 m per second, or about 10 times faster than unmyelinated axons.

Because neural activity is electrical, electronic devices can be used to measure it. Electroencephalographs (EEGs) measure the electrical activity of the part of the brain closest to the electrode (Figure 21.15a). Brain waves, electrical activity in the brain, exhibit identifiable patterns depending on whether a person is awake and active or awake and calm or is in one of the different phases of sleep. Variations in brain wave patterns detected by EEG may indicate abnormal electrical activity, called seizures, which are characteristic of a number of neurological diseases, one of which is epilepsy. Positron-emission tomography (PET) is a useful research tool for identifying the parts of the brain that are most active during various activities (Figure 21.15b).

21.7 Synapses

A **synapse** is a junction between a neuron and another neuron (Figure 21.16). The cell that carries the electrical impulse to a synapse is a **presynaptic** cell. The cell that receives the impulse is a **postsynaptic** cell. The most typical example is a synapse between the tip of the axon of a presynaptic cell and a dendrite of a postsynaptic cell. The axon tip is swollen to form a specialized structure called a synaptic terminal.

Synapses do not conduct electricity, so electrical impulses cannot travel directly across them. Instead, nerve impulses are transmitted by chemical messengers, called **neurotransmitters.** Neurotransmitters are contained in vesicles within the synaptic terminals at the ends of axons. When a nerve impulse arrives at the synaptic terminal, the change in electrical potential causes a calcium channel in the plasma membrane there to open. Calcium ions (Ca^{+2}) flow into the synaptic terminal, stimulating exocytosis (see Section 3.4). The vesicles fuse with the plasma membrane, releasing neurotransmitters into the synaptic space. The neurotransmitters quickly diffuse across the synapse and stimulate the postsynaptic cell's dendrite by binding to a specific receptor protein in the plasma membrane of the dendrite. This stimulation can depolarize part of the plasma membrane,

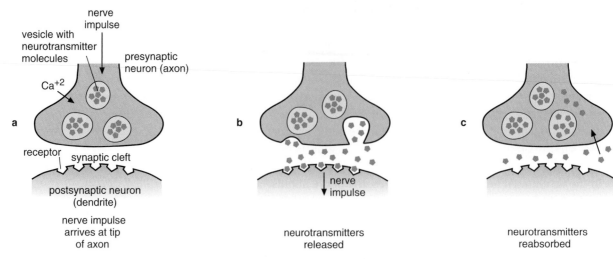

a

b

c

nerve
impulse

vesicle with
neurotransmitter
molecules

presynaptic
neuron (axon)

Ca^{+2}

receptor   synaptic cleft

postsynaptic neuron
(dendrite)

nerve impulse
arrives at tip
of axon

nerve
impulse

neurotransmitters
released

neurotransmitters
reabsorbed

FIGURE 21.16

Nerve-impulse transmission at a synapse. (a), When a nerve impulse arrives at the synaptic terminal of an axon, calcium (Ca^{+2}) channels in the terminal open. **(b),** This stimulates vesicles filled with a neurotransmitter to fuse with the presynaptic membrane, releasing the neurotransmitter into the synaptic space. The neurotransmitter binds to receptors in the plasma membrane of the postsynaptic dendrite, stimulating or inhibiting the formation of an action potential. **(c),** The presynaptic cell then reabsorbs the neurotransmitter.

sometimes enough to trigger an action potential. In this way, a strong stimulus travels rapidly along a connected network of neurons. Neurotransmitters cause muscle cells to contract and gland cells to release secretions such as hormones. After neurotransmitters have acted on the postsynaptic cell, the cell releases them into the synaptic space. They may be broken down or the presynaptic cell may reabsorb them and recycle them into new vesicles.

Various neurotransmitters can either excite or inhibit a postsynaptic cell, increasing or reducing the chance that it will experience an action potential. Many neurons receive both kinds of input from several synapses. These neurons act like simple computers, combining all of these stimuli (inputs) and producing impulses (outputs) whenever they are stimulated enough to pass the threshold potential for initiating an action potential.

Frequent signal transmission across a synapse can strengthen the connection between the two cells in any of the following ways: (1) More neurotransmitter can be released, (2) the shape of the synapse can be altered to allow more efficient transmission, (3) the uptake of the neurotransmitter can be increased or decreased to prolong or shorten the duration of stimulus, and (4) the number of receptors for the neurotransmitter can be altered. Many drugs that are used to treat neurological disorders act at specific types of synapses to alter synaptic transmission.

Several classes of neurotransmitters have been identified, including some amino acids, small peptides, amines, and simple gases (Table 21.2). All classes of neurotransmitters include both excitatory and inhibitory compounds. Some compounds have opposite effects on different

TABLE 21.2
Major Classes of Neurotransmitters

Chemical Class	Examples	Effect on Postsynaptic Cells	Location
amines	acetylcholine $CH_3-C(=O)-O-CH_2-CH_2-N^+-(CH_3)_3$	excites vertebrate skeletal muscle; can excite or inhibit other cells	CNS, PNS, vertebrate neuromuscular synapse
	dopamine (structure shown)	excites most cells	
amino acids	glutamate $R = CH_2-CH_2-COOH$	excites most cells	CNS, invertebrate neuromuscular synapse
peptides	met-enkephalin tyrosine—glycine—glycine—phenylalanine—methionine	inhibits most cells	CNS
gases	carbon monoxide CO	inhibits most cells	CNS, PNS, vertebrate blood-vessel walls
	nitric oxide NO	inhibits most cells	

postsynaptic cells. A number of helpful medicines and harmful drugs have been found to affect the brain by mimicking neurotransmitters. These compounds bind to receptors in postsynaptic plasma membranes because they are structurally similar to the neurotransmitters normally found there.

Studies of neurotransmitters have helped researchers understand some diseases that affect the nervous system. For example, investigators found that people with Alzheimer's disease, which is characterized by a loss of memory and mental abilities, produce less of the neurotransmitter acetylcholine than people without the disease. Parkinson's disease—characterized by slow movements, lack of muscle control, and tremors—is caused by a lack of dopamine. Myasthenia gravis, which causes weak muscles, is due to the production of antibodies by the immune system that block acetylcholine receptors. Attention deficit hyperactivity disorder (ADHD) is likely due to a neurotransmitter imbalance that can be controlled by drugs that alter serotonin levels.

21.8 Integration

The CNS determines responses to specific stimuli through a process called neural integration. Many different sensory inputs can converge on neurons within the CNS. These neurons then perform the equivalent of a set of calculations. They integrate, or combine, the inhibitory and excitatory synaptic inputs that they receive and activate or inhibit particular output

Try Investigation 21B
Reaction Time.

pathways. Such complex input-output relationships within the CNS allow a runner to respond selectively to the sound of the starter's pistol and not the cheers of the crowd.

The simplest input-output relationship is a **reflex arc** (Figure 21.17). Reflex arcs involve both the PNS and the CNS. A **reflex** is an automatic, involuntary response to a stimulus. Because it occurs before the stimulus is received and interpreted by conscious areas of the brain, a reflex arc is a very rapid response. The simplest reflex arc involves a receptor, a sensory neuron, a motor neuron, and the effector to which the motor neuron is connected.

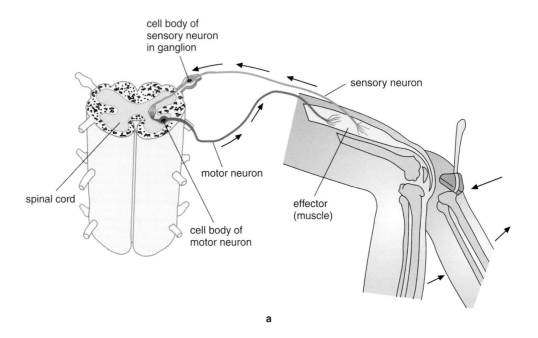

a

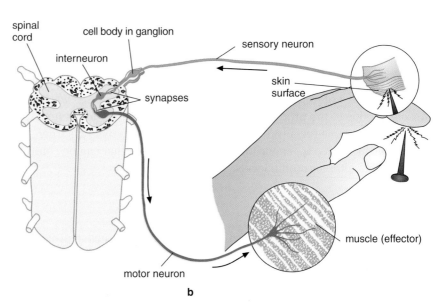

b

FIGURE 21.17

Two examples of reflex arcs. (a), A simple, two-neuron reflex arc involving a sensory neuron and a motor neuron. The sensory neuron stimulates the motor neuron directly. The same sensory neuron may also stimulate sensory regions of the CNS, but the reflex does not depend on conscious perception of the sensation. **(b),** A jab with a pin initiates a three-neuron reflex that results in withdrawal of the hand. Most reflexes involve at least three neurons.

For example, your doctor has probably checked your reflexes by striking your leg just below the knee. The impact of the rubber hammer triggers a receptor on the dendrites of a sensory neuron. The impulse travels along the sensory neuron to a synapse with a motor neuron in your spinal cord and continues to the muscles in your leg, which then contract and jerk your leg and foot upward. The sensory neuron also transmits its information to your brain. This enables you to consciously feel the pressure of the hammer, but your leg is already moving in response before you experience this sensation.

Consider how much more complex neural circuits must be for our runner, whose nervous system must balance a great number of simultaneous stimuli and responses. Our nervous systems achieve this kind of delicate balance for every thought and act we have. Through sets of neural computations, our nervous systems can integrate diverse kinds of information and selectively respond to the most important events.

21.9 Drugs and the Brain

A drug is any substance that has no nutritional value and is introduced into the body to produce some biological effect. Three major categories of drugs that affect the brain include depressants, stimulants, and hallucinogens. Many of these **psychoactive** drugs (which affect psychological processes) alter synaptic transmission by modifying or blocking the synthesis, release, or uptake of neurotransmitters. Many such drugs are addictive—that is, they cause a strong physical and sometimes psychological need for the drug. In addition, people who chronically use psychoactive drugs can develop a tolerance, or reduced response, to them. Because homeostatic mechanisms enable the nervous system to adapt to the drug and its effects, a person who develops a tolerance needs increasing doses of the drug to produce the same effects. Eventually this may lead an addict to consume a lethal overdose.

Depressants, such as alcohol and barbiturates, reduce nerve-impulse transmission in certain parts of the brain. This reduces activity throughout the body. Some depressants, including alcohol, inhibit synaptic transmission. Alcohol may also inhibit the transmission of action potentials by dissolving in plasma membranes and making them leaky.

Caffeine, nicotine, amphetamines, and cocaine are stimulants. Stimulants initially increase alertness and activity but eventually lead to a period of depression. Nicotine has powerful effects on both the CNS and the PNS. It chemically mimics a neurotransmitter and can directly stimulate a number of sensory receptors. Amphetamines and cocaine apparently mimic or enhance the effect of the neurotransmitter norepinephrine, which underlies the fight-or-flight response. Cocaine blocks the reabsorption of the neurotransmitter from the synaptic space. This chemical change prolongs the stimulation of the postsynaptic cell, leading to overactivity, hallucination, and eventually, dangerous nervous exhaustion. Figure 21.18 shows the effect of amphetamines on the web-building ability of a spider.

CHEMISTRY TIP

Ethyl alcohol (C_2H_6O) is less polar than water, so it can dissolve in lipid membranes. This makes the membrane less orderly and more likely to leak sodium (Na^+) or potassium (K^+) ions.

CHEMISTRY TIP

The activity of many psychoactive drugs depends on their structural similarity to neurotransmitters. For example, the drug amphetamine and the neurotransmitter epinephrine (adrenaline) have similar chemical structures.

amphetamine

epinephrine (adrenaline)

a   b

FIGURE 21.18

Effect of amphetamine on spider-web design. Psychoactive drugs can disturb many of the complex behavior patterns of animals. **(a),** A female spider built a web in the characteristic regular form. **(b),** Several hours after feeding on sugar water containing a small dose of amphetamine, the spider built an unusually small, irregular web. Although we cannot directly compare the behavior patterns of spiders and humans, experiments such as this one provide important information about the effects of chemical substances on brain function.

 **CHEMISTRY TIP**

Lithium ions (Li⁺) are chemically similar to sodium ions (Na⁺). Lithium is frequently used to treat severe depression by raising the resting potential of neurons. This change brings them closer to the threshold for initiating an action potential. The effect often makes the patient more active and less depressed.

Hallucinogens are drugs that alter sensory perception, particularly visual and auditory perception. Some hallucinogens, such as lysergic acid diethylamide (LSD) and mescaline, interfere with the normal activity of the neurotransmitters serotonin and dopamine by mimicking their structure and function in the brain. These neurotransmitters affect sleep, mood, attention, and learning. An imbalance in these neurotransmitters can lead to mental illness.

Focus On

Molecular Biology and the Brain

Today molecular biologists are advancing our understanding of normal and abnormal brain function. Their research has shown that the loss of function or the abnormal expression of single genes can cause disease. For example, a particular neurological disease in which motor neurons die and cause paralysis was shown to be caused by mutation in a gene that regulates the synthesis of the enzyme superoxide dismutase. Normally this enzyme catalyzes the breakdown of the harmful superoxide ion (O_2^-). There is increasing evidence that the mutant enzyme also breaks down normal cell components in motor neurons. Other

diseases may be due to mutations in genes that regulate or encode neurotransmitter receptors or other proteins involved in nerve-impulse transmission.

Because human behavior is so complex, it is likely that many genes and environmental factors are involved in conditions such as schizophrenia. Biologists have devised methods such as QTL mapping (see Section 16.6) to locate and identify these genes. A large number of human genes are active only in the brain. It is likely that the complexity of the brain depends on these genes. Identifying these genes and understanding their function may help us unravel the secrets of brain function.

Evolution of Nervous Systems

21.10 Evolutionary Trends

There is tremendous similarity in how nerve cells function throughout the animal kingdom. However, there is also great diversity in how nervous systems are organized. For example, nerve cell bodies in many simple animals are neither uniformly distributed nor concentrated in a brain; they are grouped in ganglia throughout the body. These ganglia allow each part of the nervous system to coordinate its activities without involving the entire system.

The *Hydra* has the simplest type of nervous system (Figure 21.19a). The system consists of nerve cells evenly distributed throughout the animal, without ganglia. Such a system is called a nerve net. When a stimulus is applied to any part of the *Hydra*, the resulting impulse eventually spreads along the nerve net in all directions throughout the organism. For the effect to spread very far, the stimulus must be strong and last a long time. The nerve net carries signals more slowly than does a more specialized CNS.

Unlike the *Hydra*, flatworms such as planarians, members of the phylum Platyhelmenthes (see A Brief Survey of Organisms, p. 483), have a definite anterior end with a small brain (Figure 21.19b). Information travels along two parallel nerve cords on each side of their bodies. Many versions of this basic nervous system are found in the animal kingdom (Figure 21.19c).

Among the vertebrates, the evolutionary trend toward centralized control by an anterior brain is dramatically accelerated. Vertebrates have a highly developed CNS that contains most of the neurons (Figure 21.19d). The vertebrate brain has evolved from a series of three bulges at the anterior end of the spinal cord. These regions—the hindbrain, midbrain, and forebrain (Figure 21.20)—are present in all vertebrates but have become increasingly specialized. Three trends seem to characterize vertebrate brain evolution.

First, the size of the brain relative to the whole body increases in certain evolutionary lines, notably birds and mammals. In the line that includes amphibians, fishes, and reptiles, the ratio of brain size to body size is fairly constant: The brains of a 100-g lizard and a 100-g fish are about the same size.

CONNECTIONS

The various types of nervous systems found in the animal kingdom demonstrate unity of pattern at the molecular level and diversity of type in their behavior as organ systems.

CONNECTIONS

The nervous systems of various animals reflect adaptation to specific environments. For example, in animals that move forward through their environments, sensory organs are concentrated in a head at the anterior end.

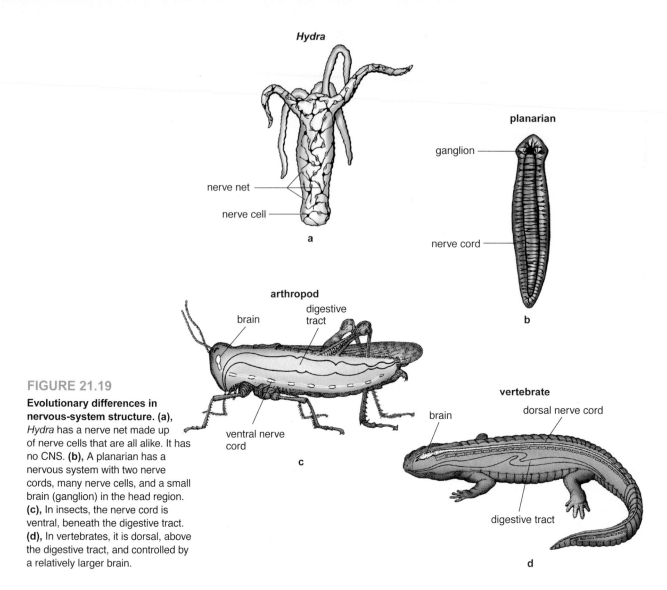

Hydra

nerve net

nerve cell

a

planarian

ganglion

nerve cord

b

arthropod

brain

digestive tract

ventral nerve cord

c

vertebrate

dorsal nerve cord

brain

digestive tract

d

FIGURE 21.19

Evolutionary differences in nervous-system structure. (a), *Hydra* has a nerve net made up of nerve cells that are all alike. It has no CNS. **(b),** A planarian has a nervous system with two nerve cords, many nerve cells, and a small brain (ganglion) in the head region. **(c),** In insects, the nerve cord is ventral, beneath the digestive tract. **(d),** In vertebrates, it is dorsal, above the digestive tract, and controlled by a relatively larger brain.

In birds and mammals, however, the ratio of brain to body size is much greater: A mouse weighing 100 g has a much larger brain than a lizard weighing 100 g.

A second evolutionary trend is increased specialization of function. The three ancestral major brain regions can still be found in modern vertebrates, but they have been subdivided into areas with specific functions. In the hindbrain, the cerebellum became the major structure for coordinating movement. The forebrain is subdivided into the cerebrum, the thalamus, and the hypothalamus. The cerebrum is the part of the brain most important in learning and memory. The thalamus analyzes and transmits sensory input. The hypothalamus regulates many homeostatic mechanisms and serves as a link between the nervous and endocrine systems. As these structures became more complex, the original divisions between the three bulges became blurred.

The third trend in vertebrate brain evolution is the increasing sophistication and complexity of the forebrain. The appearance of more complex behaviors in birds and mammals parallels the evolutionary expansion of the cerebrum. The surface layer, or cortex, of the cerebrum

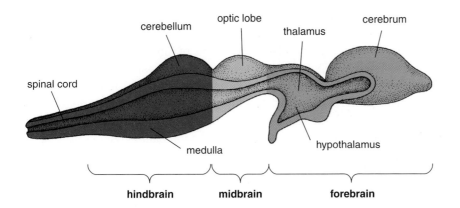

FIGURE 21.20

A generalized vertebrate brain. Notice the characteristic bulges that divide the brain into forebrain, midbrain, and hindbrain. Vertebrate brains vary in size, relative proportion of the forebrain, and number and complexity of synaptic connections.

consists of the cell bodies of neurons. Because of its color, this layer is sometimes called gray matter. Below the gray matter are axons that run to and from the cortex. This area is sometimes called white matter because the myelinated nerve fibers are white.

Among mammals in particular, complex behavior is associated with the relative size of the cerebrum and the presence of folds, or convolutions, that increase the surface area of the brain. Because the cerebrum's cell bodies are in the cortex, the brain's surface area is more important than its size in determining performance. Although the human cerebral cortex is less than 5 mm thick, it covers such a large, highly folded surface area that it makes up more than 80% of the total brain mass. Primates (monkeys, apes, humans, and their relatives) and porpoises have dramatically larger and more complex cerebral cortices than any other vertebrates. Relative to its body size, the cerebral cortex of the porpoise is second in surface area only to that of humans. Figure 21.21 compares the brains of five vertebrates.

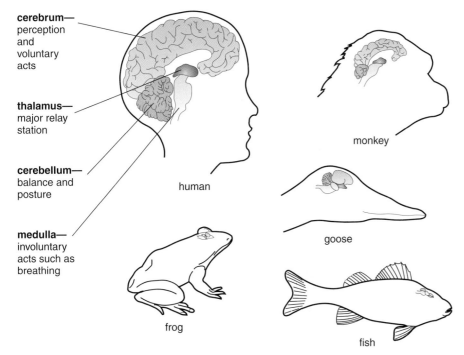

FIGURE 21.21

Five vertebrate brains. Note the increase in brain size and in the surface area of the cerebral cortex that parallels the more complex behavior and greater intelligence of birds and mammals. How can you explain the fact that birds have the largest cerebellums relative to total brain size?

21.11 Molecular Evolution of Nervous Systems

Evolutionary change occurs at every level of biological structure. Examination at the cellular and molecular levels reveals both the conservation of ancestral structures and the slow changes imposed by natural selection. For example, Purkinje cells, the major type of neuron in the cerebellum, have the same basic structure in fish, reptiles, and mammals (Figure 21.22). At the same time, they have become more complex along with the rest of the brain. The Purkinje cells of alligators have more dendritic branches than those of catfish. The mouse Purkinje cell is even more highly branched. Evolution of the Purkinje cell shows how increasing cell complexity has lead to increasingly complex behavior.

Evolutionary changes at the molecular level also help produce the various specializations of diverse species. For example, all neurons have sodium-channel proteins with highly similar primary structures. However, some species produce or are exposed to powerful toxins that block sodium

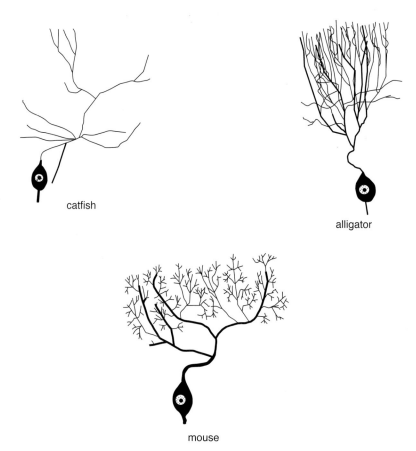

FIGURE 21.22

Evolutionary changes in cell structure. These Purkinje cells are the most numerous type of neurons in the cerebellum. What evolutionary trends do you notice by comparing the Purkinje cells of a fish, a reptile, and a mammal?

channels, preventing nerve conduction. Puffer fish (Figure 21.23) and some other animals accumulate tetrodotoxin, a toxic substance that is made by several types of microorganisms. Animals that use these toxins or are frequently exposed to them (such as shellfish) have become resistant to their harmful effects. The sodium-channel proteins of these tetrodotoxin-resistant animals differ from those of other animals by a single amino acid. This small change is sufficient to confer resistance.

A dramatic example of cellular change that produced an evolutionary specialization is the electric organ found in electric rays, or torpedo fish (Figure 21.24). This organ likely evolved from the neuromuscular junction. The number and density of the synapses are dramatically increased compared with a typical junction between a nerve and a muscle. In addition, the muscle cells are modified to form a large plate. Electrical shocks are delivered by generating a pulse of current in response to an impulse from the nerve. This protective adaptation evolved by the development of a specialized form of neuromuscular junction.

Such evolutionary changes occur over millions of years. For example, the receptors for the neurotransmitter acetylcholine (see Table 21.2) contain 437 amino acids. Over the millions of years since the ancestors of torpedo fish, cows, and humans diverged, only 87 of those 437 amino acids have become different in humans and torpedo fish. Humans are more closely related to cows; our acetylcholine receptors differ from theirs by only 11 amino acids.

The nervous system provides a good example of the evolution of a complex organ system. Evolutionary changes in the sequences and expression of genes lead to changes in the structure and function of cells. These cellular changes, in turn, result in new patterns of nervous-system activity. Populations and species evolve simultaneously on all these levels—molecular, cellular, and organismal. Natural selection shapes these changes in ways that enhance adaptation.

FIGURE 21.23

A spotted puffer fish. This species *(Arothron meleagris),* is one of about 120 species in the family Tetraodontidae that accumulates the nerve poison tetrodotoxin. Symbiotic bacteria in the fish's intestine synthesize the toxin, which the fish stores in its liver and other organs. Tetrodotoxin blocks sodium channels in the neurons of most animals. The sodium channels of puffer fish, however, are insensitive to tetrodotoxin.

 **CHEMISTRY TIP**

The tetrodotoxin molecule ($C_{11}H_{18}N_3O_8$) contains a positively charged amino group ($=NH_2^+$) that is thought to mimic a sodium ion. The amino group can enter the sodium channel, but the molecule is too big to fit through the channel and becomes stuck, blocking the passage of sodium ions.

Check and Challenge

1. How does the relative size of a particular region of the brain in a given animal relate to the functional abilities of that animal? Give an example.
2. Describe the major trends in the evolution of the vertebrate brain.
3. How can the evolution of the vertebrate brain help explain the fact that mammals and birds are capable of more complex learning and behavior than other vertebrates?
4. Do evolutionary trends in the forms of Purkinje cells support or oppose the hypothesis that increasingly complex behavior depends on increases in the number and complexity of connections among neurons? Explain your answer.

FIGURE 21.24

A torpedo ray (the lesser electric ray). This fish *(Narcine entemedor)* has specialized neuromuscular junctions that generate large electric pulses. These pulses help the fish navigate in murky water and stun its prey.

Chapter HIGHLIGHTS

Summary

Neurons, the structural and functional units of an animal's nervous system, are arranged in complex circuits that perform distinct functions. Glial cells support neurons in a variety of ways and sometimes form myelin sheaths around axons. Sensory neurons of the peripheral nervous system (PNS) conduct impulses from the body toward the central nervous system (CNS); motor neurons of the PNS conduct impulses from the CNS toward a muscle or gland. Interneurons form connections between other neurons in the CNS.

Sensory receptors can be grouped in receptor organs, such as in the eye or ear. Receptors convert the energy of a stimulus into electrical impulses. Impulses from sensory receptors are sent to sensory-neuron dendrites. These dendrites receive stimuli and send impulses through the neuron. Axons, with or without myelin, carry impulses away from the neuron's cell body to a synapse. Neurons respond to stimuli, conduct impulses through their axons to synapses with other neurons or effectors, and release neurotransmitters into synapses that stimulate or inhibit other cells. Many drugs mimic neurotransmitters or interfere with their action.

Nerve impulses travel through neurons in the form of moving action potentials. Both diffusion and active transport of sodium and potassium ions through a neuron's plasma membrane are involved in transmitting the impulse and restoring the neuron to its resting state.

In the brain, information in the form of nerve impulses is distributed to specific regions, processed, and compared with stored memories. Decisions to respond are initiated in the brain, and the output is carried to effectors. Reflexes are immediate responses by the motor system to sensory inputs that do not reach the level of conscious thought and may not involve the brain at all.

Evolution has produced animal species that are highly diverse in the complexity and specialization of their nervous systems. In vertebrates, the CNS consists of the spinal cord and the brain. Among vertebrates, and especially among birds and mammals, the size of the brain has increased significantly relative to body size, and the forebrain has enlarged dramatically. These evolutionary changes occurred partly as a result of increasing complexity and specialization at the cellular and molecular levels.

Key Concepts

Use the following concepts to develop a concept map of this chapter. You may add additional terms as you need them.

- neurotransmitter
- cell body
- stimulus
- medulla
- autonomic
- parasympathetic
- drugs
- dendrite
- sodium
- motor neuron
- ion channels
- memory
- nerve
- effectors

Reviewing Ideas

1. Describe the similarities and differences between axons and dendrites.
2. Describe different kinds of sensory receptors.
3. What is the difference between depolarization and a nerve impulse?
4. Describe the function of the threshold in the transmission of nerve impulses.
5. How do neurotransmitters function?
6. What are the main functions of the CNS?
7. Compare the functions of the rods and cones in the human eye.

8. Describe how drugs can modify synaptic activity.
9. Compare the functions of the sympathetic and parasympathetic nervous systems.
10. Describe the differences between a nerve net and the nervous system of mammals.
11. Describe some evolutionary trends in vertebrate brain development.

Using Concepts

1. What are the adaptive advantages of myelin sheaths?
2. Describe the relationships among a receptor, a stimulus, and a motor neuron.
3. Explain why it is impossible for a neuron to continually transmit impulses, one after the other, without resting.
4. What is the role of active transport in transmitting impulses along neurons?
5. Describe specific responses of the sympathetic nervous system that you might experience before a difficult exam. How might those responses affect your performance?
6. What difficulties would a person experience if some neurotransmitters stopped functioning?

Synthesis

1. Stretch receptors in skeletal muscles stimulate sensory neurons. Describe how these neurons could participate in an activity such as walking or riding a bicycle, in which opposing muscles alternately contract and relax.
2. Would you predict that distracting sensory information is likely to interfere with the knee-jerk reflex? Use what you have learned about the nervous system to justify your answer. Describe an experiment that could test your hypothesis.

Extensions

1. Write a short story about the life and times of a sodium ion (Na^+). Describe how it got into an animal, where it might be located, and what some of the body functions are with which it is associated on a daily basis.
2. Visit a local organization or treatment center that deals with people who are trying to overcome physical addictions to alcohol or some psychoactive drug. Report your findings to your class.
3. With the help of your teacher, invite a doctor or counselor into your class to discuss and answer questions about drug abuse.
4. Research some medicines that people take for depression, hyperactivity, or Parkinson's disease. Report to the class on how these drugs act.
5. Prepare a report on epilepsy. Find out what is known about the disease, how the disease affects the nervous system, and how the disease can be treated.

Web Resources

Visit BSCSblue.com to access
- Resources about current research in neuroscience
- Information about alcoholism, other forms of drug addiction, and neurological diseases
- Web links related to this chapter

CONTENTS

LEARNING OUTCOMES

By the end of this chapter you will be able to:

A   Distinguish between innate and learned behavior.

B   Distinguish between the biological and environmental aspects of behavior.

C   Describe methods for studying behavior.

D   Evaluate the roles of populations and communication in social behavior.

CHAPTER

22

Behavior

■ *What triggered the behaviors of these animals?*

■ *Have these animals learned to act this way or were they born with this behavioral ability?*

Why are some activities, such as gathering food and mating, common to nearly all animals, whereas other activities, such as building nests, are limited to certain species? Why do some animals live in groups and others live alone? Questions such as these are the focus of the study of behavior. Behavior is an organism's conduct—the way it acts. It is the most immediate way an organism interacts with its environment.

Behavior is partly the result of natural selection. It is a biological process that is shaped by evolution. Behavior mostly affects the ability of an individual organism to survive and reproduce, but it also plays a part in preserving the species. If the behavior of individuals is successful, they may live to produce many offspring. The number of descendants, in turn, affects the survival of the entire species. Therefore, the behavior of each member of a species helps determine whether or not the species as a whole continues to reproduce and thrive.

This chapter introduces the major concepts in the study of behavior. In addition, you will learn about scientific methods of studying behavior and the social patterns of animal populations. Studying behavior can help us understand why animals, including humans, act the way they do.

Major Elements of Behavior

22.1 Stimulus and Response

Organisms react to changes in their environment through various behaviors (Figure 22.1). For example, a dog begins barking as a stranger approaches it. Ducks fly south when the days become shorter and cooler. Anything that triggers a behavior is called a stimulus. A stranger's approach triggers the dog's barking, and the change in weather triggers the ducks' migration. These are examples of external stimuli. Some internal stimuli also trigger behaviors. For example, thirst usually leads to drinking. An organism's reaction to a stimulus is called a **response.** Thus, the dog's barking, the ducks' migration, and the person's drinking to quench a thirst are all examples of responses.

22.2 Innate and Learned Behaviors

Where do behaviors come from? In general, behaviors are either **innate** or learned. Innate behavior is inborn—influenced by genes (Figure 22.2), not based on experience. Some innate behaviors, such as sucking or crying, occur as soon as an organism is hatched or born (see Appendix 22A, "Innate Behavior"). Others appear later in life. The familiar term *instinct* refers to

ETYMOLOGY

in- = inside (Latin)

-nate = birth, to be born (Latin)

Innate characteristics are inborn; they are not acquired by learning.

CONNECTIONS

Behaviors such as shivering and hibernation, which aid in body-temperature regulation, help organisms maintain homeostasis.

FIGURE 22.1

Stimulus and response. Normally a pill bug *(Armadillidium vulgare)* has its legs on a surface, such as this leaf *(right).* However, when a pill bug is touched, it rolls into a ball *(left).* The touch is the stimulus. The rolling behavior is the response. How is this behavior adaptive?

innate behaviors. In general, an instinct is a response to a specific stimulus that does not require learning or practice. Feeding behaviors of young animals are typical instincts. In many cases, it is difficult to determine whether experience has influenced a behavior that appears to be instinctive.

Some innate behaviors do not change as a result of experience. For example, dogs that have spent their entire lives indoors will attempt to bury a bone by scratching on the carpet as if they were digging. Captive raccoons go through the motions of dunking and washing their food in water even if the food is already clean or wet and even if water is not available. Such patterns of behavior that are characteristic of a given species are called **fixed action patterns.** Some fixed action patterns develop as an animal matures. For example, a baby mouse may flail its leg in the air; later it will use this motion to scratch an itch.

Learned behavior, on the other hand, develops as a result of experience. Most animals can learn to behave in new ways. Learned behavior ranges from simple changes in innate behaviors to the complex series of behaviors you would use to drive a car. Other examples include dogs learning to sit on command and parrots learning to imitate sounds.

There are several forms of learned behavior. **Imprinting** is a type of learning that requires little or no practice but can occur only during a genetically determined time during an animal's development. For example, as soon as a newly hatched duckling or gosling is strong enough to walk, it follows its mother away from the nest (Figure 22.3a). After it has followed its mother for awhile, it will no longer follow any other animal. If goslings hatch away from their mother, they will follow the first moving object they see, even if it is a human being (Figure 22.3b). As they grow, they prefer the artificial

FIGURE 22.2

Innate behavior. In South Georgia, islands east of the Falklands, a mother king penguin *(Aptenodytes patagonicus)* feeds her chick. Both parents regurgitate food (fish, crustaceans, and squid) only for their own chick, which the parents recognize by the chick's distinctive call.

a

b

FIGURE 22.3

Examples of imprinting. (a), Normally young waterfowl, such as these Canada geese *(Branta canadensis),* follow their mother as a result of imprinting. **(b),** When animal behavior scientist Konrad Lorenz hatched goslings away from their natural mother, they followed him instead.

mother to their true mother. German behavior scientist Konrad Lorenz found that, in geese, this imprinting can occur only during the first 2 days of life.

A second type of learning is **habituation.** If an animal is exposed to a stimulus over and over again, it may slowly lose its response, or habituate, to that stimulus. For example, a dog responds to the sound of an airplane passing overhead by turning its head. If airplanes continue to fly overhead, the dog may learn to disregard the sound. Such a reaction can be thought of as learning *not* to respond to certain stimuli.

A more complex type of learning is **conditioning.** Conditioning occurs when one stimulus is associated with another unrelated stimulus. In the early 1900s, Russian physiologist Ivan Pavlov investigated conditioning (Figure 22.4). Pavlov studied the flow of saliva in a dog that was presented with food.

Pavlov first showed that when a bell rang, the dog did not salivate. Then the bell rang again as meat was presented to the dog. The dog responded to the meat by secreting saliva. The acts of ringing the bell and presenting the meat were repeated together many times. Eventually the dog secreted saliva on hearing the bell, even when no food was present. Pavlov reasoned that a new automatic response had been established. The dog had learned to associate the bell's ringing with food and thus salivated in response to the bell, just as it would in response to food.

Habits and fears are often the result of conditioning. A child may not be afraid of a mouse until she or he learns to associate the screams of other people with the mouse. Similarly, deer may learn to associate the sound of gunshots with humans, causing them to avoid people.

Trial-and-error learning, sometimes called trial-and-success learning, is a fourth type of learned behavior. An animal faced with two or more

Try Investigation 22A ▽ **A Lesson in Conditioning.**

CONNECTIONS

Innate behaviors are characteristic of a species. They are inherited traits.

Try Investigation 22B ▽ **Trial-and-Error Learning.**

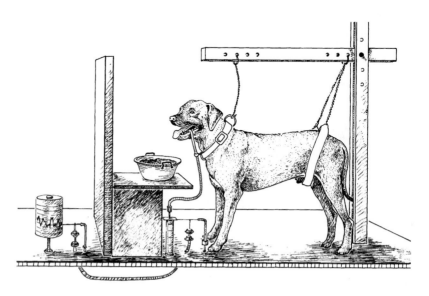

FIGURE 22.4

Pavlov's apparatus. This experiment demonstrated the principle of behavioral conditioning.

Focus On

A Detour Problem: Reason or Trial-and-Error Learning?

Humans are capable of complex learned behavior, such as using abstract concepts to solve problems. Do other animals have this ability? Researchers have studied the problem-solving ability of other animals by using detour problems. In these experiments, an animal must follow an indirect path to get a desired object.

In detour experiments, scientists try to determine whether the animal uses reason or trial-and-error learning to solve the problem. This is a difficult question to answer. Figure 22.5 compares the response of a dog and the response of a squirrel to the same detour problem. Dogs are not good at such tasks—a dog strains at its leash, whines, and runs about wildly. It may even fall asleep. Then it starts again. By luck alone, it might find itself on the same side of the post as the food. In contrast, tree squirrels solve this problem quickly. After first failing to reach the food, the squirrels try an alternative path around the post and to the food. What might explain the difference between their responses?

This type of result can be puzzling unless evolution is taken into account. Although dogs have a larger and more complex brain than squirrels, they live almost exclusively in a horizontal, two-dimensional world. They see food on their level and take the most direct path to it. Squirrels, on the other hand, are adapted to live in trees. In its three-dimensional world, a squirrel moving from tree to tree has a choice. It can climb down the tree, go along the ground, and climb the next tree. It can also try to jump directly from one tree to the next. The first choice would expose the squirrel to ground-dwelling predators. The second choice would be safer. However, the second choice may require the squirrel to start by moving away from the goal (the food). The ancestors of the squirrels that were relatively more adept at solving such problems were better nourished, more likely to survive, and more likely to leave more offspring than squirrels that were not so adept.

Natural selection thus favored the ability to conduct successful detours. Over time, each population of squirrels became composed of individuals that were skilled at approaching goals indirectly. Is this behavior a product of reasoning or trial-and-error learning? Is it an innate behavior?

FIGURE 22.5

A detour problem. The dog strains at the leash to get to the food. The squirrel, however, backs up and goes around the post to reach the food.

a   b   c

FIGURE 22.6

An example of learned behavior. The toad eats a bumblebee **(a)**, which stings the toad's tongue. Although initially appearing interested in a mimic of a bumblebee, a harmless robber fly **(b)**, the toad does not attempt to capture it **(c)**.

responses may eventually learn the one that leads to a reward. Examples of trial-and-error learning include techniques used for training pets and performing animals. In nature, animals learn to eat certain foods and avoid others based on whether the food tastes good or bad.

Although it is useful to describe behaviors as either innate or learned (Figure 22.6), biologists no longer classify behavior as exclusively innate or learned. Most behaviors are a product of both processes. Genes and environment each influence the development and expression of behaviors to varying degrees.

Check and Challenge

1. How does an individual organism's behavior relate to the species as a whole?
2. In what way can behavior be adaptive? Give specific examples not mentioned in this chapter.
3. Distinguish between a stimulus and a response. Give an example of each not mentioned in this chapter.
4. Compare and contrast innate behavior and learned behavior.
5. Describe imprinting, habituation, and conditioning. How are they similar and different?

Biology Online BSCSblue.com/check_challenge

Roots of Behavior

22.3 Biological Aspects of Behavior

Because the brain and the nervous system control behavior, it's clear that biology plays a role in shaping behavior. Genes affect the structure and development of the brain at every level. Consider the gene that codes for the enzyme alcohol dehydrogenase in humans. This enzyme helps break down alcohol. People who have a mutated form of the gene that codes for this enzyme get sick when they drink alcohol. These people are less likely than other people to drink alcohol. In this way, this gene directly influences drinking behavior. In humans, few behaviors are controlled by only one gene. Most human behaviors are influenced by several genes and many environmental factors.

Biologists have found a genetic basis for a variety of behaviors in many organisms. For example, fruit flies tend either to forage widely for food or to stay close to home. This behavior is determined by a single gene. In fact, biologists can convert nonforagers into foragers by altering the gene that codes for this behavior. This gene produces an enzyme that affects the part of the nervous system that controls foraging behavior. Likewise, in honeybees, one gene controls the expulsion of diseased bees from the hive. In crickets, the hybrid offspring of different species show a chirp pattern that is intermediate between the parents' chirp patterns.

Species vary in the degree to which their genes influence behavior. Figure 22.7 compares the influence of genes and environment on the behavior of different groups of organisms. Notice that none of the organisms has a collection of behaviors that is controlled entirely by genes or the environment. In general, a species whose behaviors are more strongly influenced by genes tends to have less flexibility in its behavior than a species whose behaviors are more strongly influenced by environmental factors.

Although most behavior is directed by the nervous system, the glands of the endocrine system also play an important role in regulating behavior. For

CHEMISTRY TIP

The enzyme alcohol dehydrogenase removes a hydrogen atom from ethyl alcohol, converting it to acetaldehyde.

$$CH_3CH_2OH + NAD^+ \longrightarrow CH_3CHO + NADH + H^+$$

CH_3CH_2OH = ethyl alcohol
CH_3CHO = acetaldehyde

Acetaldehyde is responsible for much of the damage alcohol does to the liver and brain.

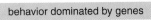

behavior dominated by genes

| insects | fish | reptiles | birds | nonhuman mammals | humans |

behavior dominated by environment

FIGURE 22.7

Genetic and environmental influences on behavior. The behavior of all species is influenced by genes and environment to some degree. The behavior of some animals, such as insects, is closely regulated by genes. Animals such as mammals, which have large, complex brains, are more capable of learning and have greater behavioral flexibility.

example, vertebrate adrenal glands, located just above the kidneys, release epinephrine (also called adrenaline) into the bloodstream when an organism is under stress, such as when it is being pursued by a predator. This hormone increases the activity of the heart and skeletal muscles, stimulating a rapid and vigorous response, such as quickly running or flying away. The effects of genes and learning on behavior are expressed through the endocrine and nervous systems.

22.4 Environmental and Cultural Aspects of Behavior

Although genes are responsible for the biological basis of behavior, behavior is also influenced by experience of the environment, which may include culture, individual experience, and other external stimuli, including exposure to chemicals, light, and temperature. Behavior is part of an organism's adaptation to its environment. For example, you do not put on a heavy coat unless you are cold or anticipate being cold in the near future.

Clothing is an excellent example of the effect of experience on human behavior. People choose clothing in part as an adaptation to the local climate. Clothing choices also reflect cultural and individual differences (Figure 22.8a). In addition, some clothing—such as work attire and uniforms (Figure 22.8b) and the traditional garments worn by judges, brides (Figure 22.8c), and clowns—carries symbolic value.

External and internal stimuli can influence the same behavior. For example, the courting and reproductive behavior of birds and fishes depends on the levels of certain hormones. Hormone levels, in turn, may be affected by external factors, such as temperature or the length of day.

a

b

c

FIGURE 22.8

Wearing particular styles of clothing, a culturally influenced behavior. (a), This man's kilt, historically conventional dress in Scotland and still worn today on special occasions, might not be considered acceptable in certain other cultures. **(b),** The shirt, tie, and slacks that this man is wearing conforms to the current global standard of acceptable dress for many types of jobs. This type of business dress is essentially a uniform. **(c),** Some clothing, such as this couple's wedding gown and tuxedo, is a symbol of the wearer's special status.

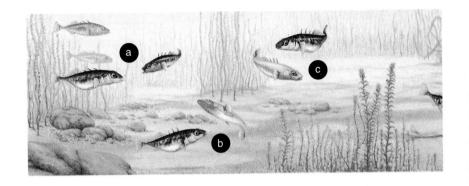

FIGURE 22.9

Interaction of internal and external stimuli. At breeding time, a male three-spined stickleback fish *(Gasterosteus aculeatus)* has a red belly. The red belly is caused by internal stimuli—hormonal changes linked to reproduction. **(a),** Males respond defensively to other red-bellied males. **(b),** Females tend to follow the red-bellied males. **(c),** The male attracts a female with dancelike, zigzag swimming motions and leads her to the nest he has built. **(d),** The male guides the female into the nest. **(e),** The male taps the female near her tail. She responds by laying eggs. **(f),** The male then bites the female. She leaves the nest, and the male enters the nest and fertilizes the eggs. He remains there, guarding the nest and eggs.

The courting and reproductive behavior of the three-spined stickleback fish illustrates the interaction of internal and external stimuli. In early spring, reproductive hormones cause changes in the appearance and behavior of the stickleback. The male moves to the warm, shallow edge of the pond to select a breeding area, or territory. The male's belly becomes red, a stimulus that affects the behavior of females and other males (Figure 22.9).

Male sticklebacks defend their territory against other red-bellied males of the same species. They will even defend their territory against a wooden dummy fish of any shape, as long as the dummy has a red belly (Figure 22.10). If a realistic model of a male stickleback without a red underside is presented, however, the male stickleback will not attack it. The red underside is the stimulus that provokes the male's defensive behavior. This example shows how genes, hormones, and environmental factors all interact to produce behavior.

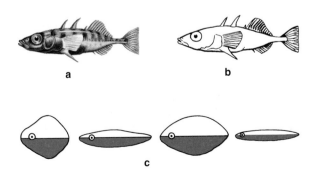

FIGURE 22.10

Behavior provoked by a specific stimulus. Aggression in the male stickleback fish **(a),** is triggered by an external stimulus—a red belly. When presented with a realistic model of a stickleback fish that has no red belly **(b),** a male stickleback does not respond defensively. Unrealistic models that have a red underside **(c),** elicit a strong defensive response.

Methods of Studying Behavior

22.5 Tools for Measuring Behavior

Scientists use several techniques to measure behavior. These include observing behaviors in nature and in the laboratory, asking people questions about their behaviors, and using genetic methods such as analyzing family trees and searching for QTLs (see Section 16.6).

Some behaviors can be measured simply by observation. Activity levels in humans can be measured with a strap-on device that records how often a person moves. Violence can be quantified by how often a person has been arrested for assault. In the laboratory, animal behavior can be studied by how active an animal is at night or during the day, how often an animal crosses a line in its cage, or how often it eats or drinks (Figure 22.11).

Self-reporting, people's descriptions of their own behaviors, is another way to study behavior. Self-reports can include descriptions of thoughts, feelings, and activities. Researchers sometimes ask people to keep diaries of the behaviors being studied. Self-reports often are used to arrive at numerical measurements of personality characteristics.

Studies of human behavior rely primarily on observations and self-reporting. It is not practical or ethical to perform certain experiments on humans. Nonhuman animal studies allow more control over behavior and other variables. These experiments can help identify genes that affect animal behaviors. Researchers then can search for homologous human genes with similar effects. This method is based on our evolutionary relationships with other animal species. Thus, if a certain gene is important in the behavior of other mammals, it is reasonable to expect that the homologous gene has a similar effect on human behavior.

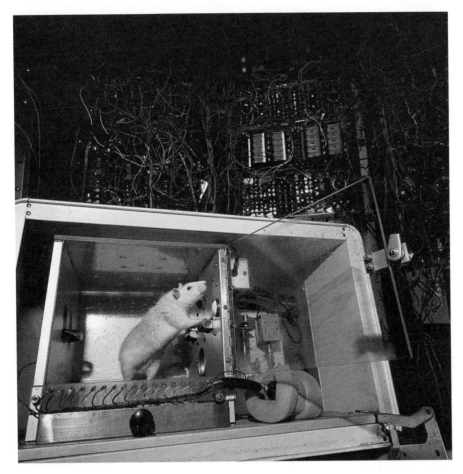

FIGURE 22.11

One type of animal behavior experiment. This rat in a Skinner box must press a bar for water when it is thirsty. Recording devices with sensors in the bottom of the cage allow researchers to track the animal's movements and can determine both when and how many times in a day the animal needs to drink.

22.6 Genetic Methods

Scientists cannot study human behavior by performing genetic crosses. Consequently, two types of genetic methods are used in the study of human behavior—**heritability** studies and molecular-genetic studies. Heritability is an estimate of the importance of genetic factors in explaining how a trait or behavior varies within a specific population. Heritability studies often compare indentical twins or other family members. It is important to note that estimates of heritability apply only to a particular population studied in a specific environment.

An obvious place to look for genetic effects is within families because family studies can provide information about genetic influences on a behavior and how they are inherited. Because family members also share environmental influences, it is difficult to identify genetic contributions to traits that run in families. For example, if you assume that all family similarities are genetic, you will wrongly conclude that differences in language and wealth have a genetic basis.

Studying identical twins, especially those who are raised apart, is one way to separate genetic and environmental influences. Identical twins are

CONNECTIONS

Natural selection of behaviors that are favored by the environment can result in adaptation.

FIGURE 22.12

Identical twins who were reared apart. Separated at birth, Mark Newman *(left)* and Gerald Levey *(right)* were reunited after 31 years of not even knowing about each other. (Also shown here is Dr. Nancy Segal, former codirector of the University of Minnesota project on Twins Reared Apart.)

Identical twins have identical DNA and therefore identical genes.

genetically identical. Fraternal twins, in contrast, share only half their genes. Because identical twins differ from fraternal twins only in the amount of DNA they share, greater similarity in a particular behavior in identical twins than in fraternal twins is evidence for a genetic influence on that behavior. If identical and fraternal twins are about equally similar in a particular behavior, this is evidence for a shared environmental influence, rather than a genetic one.

Studying twins who have been adopted and raised separately makes it easier to separate genetic and environmental influences on behavior (Figure 22.12). Similar behaviors in biological parents and their adopted children is evidence of a genetic effect. In contrast, similarity between adoptive parents and children probably reflects environmental effects.

Most of the methods in behavioral genetic research, such as family studies and twin studies, estimate genetic and environmental effects using statistics (Figure 22.13). **Correlation** is a statistic that indicates how closely two measurements are related. The value of a correlation can range from −1.0 to +1.0.

A correlation of +1.0 between a particular allele and a certain behavior means that they always occur together. If individuals that carry the allele never demonstrate the behavior, the correlation between them is −1.0. A value between −1.0 and +1.0 indicates that there is a less-than-perfect relationship between them. A correlation of 0 indicates that the behavior and the allele are unrelated. Correlations do not prove causation. In this example,

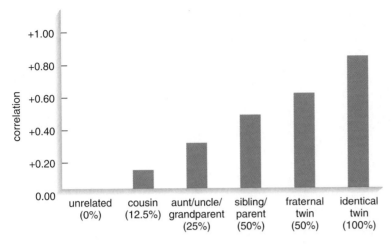

FIGURE 22.13

The use of correlation statistics in family and twin studies. The percentages indicate how similar two relatives' genes are. If genes are important in influencing a behavioral trait, then the correlation between the behavior of relatives is expected to increase as genetic relatedness increases. For example, if a person has a certain behavior, then the person's identical twin (correlation +0.85) is very likely to have that behavior as well. In contrast, aunts, uncles, and grandparents (correlation +0.30) are less likely to have the behavior.

the allele may not cause the behavior. The allele and the behavior may be common in the same individuals for some other reason. For example, many members of a family may have an allele for brown eyes and an affection for dogs. This does not mean, however, that the allele causes them to like dogs.

Molecular genetic methods, on the other hand, involve the direct study of DNA. These methods are used to find and locate specific genes that are linked with certain behaviors. Molecular-genetic methods take several forms. Linkage analysis (see Section 13.8) is used to map genes by comparing the inheritance of a trait with an unknown genetic basis to the inheritance of genes at known chromosomal locations. If a trait and a known marker gene are transmitted together within families, the trait and the marker are said to be linked. This linkage indicates that a gene that affects the trait is located close to the marker gene on the same chromosome.

Linkage analysis has located genes for many different traits, including one related to schizophrenia. Schizophrenia is a mental illness that usually begins in early adulthood and often includes hallucinations and paranoia. Adoption studies show that the risk for schizophrenia is greater among the biological children of schizophrenic parents. Twin studies show that identical twins are more likely to share the trait than fraternal twins. These clues and the results of linkage analysis led scientists to discover a gene on chromosome 6 that affects the chance of developing schizophrenia.

Linkage analysis can tell us approximately where a gene lies. To actually identify the gene, however, is more difficult. Scientists use labeled DNA probes and comparisons of DNA sequences to identify the gene and its function. They use this information to better understand the trait and how it is affected by different stimuli.

FIGURE 22.14

Sky diving, an example of novelty-seeking behavior. Novelty seeking, like other behaviors scientists have studied using genetic methods, appears to have a genetic component.

In one example of this method, scientists studied novelty-seeking behavior. Novelty seeking is the tendency to look for and enjoy new, and sometimes risky, experiences (Figure 22.14). Through twin studies, scientists found that novelty-seeking behavior is better correlated between identical twins than fraternal twins. These findings suggested that genes influence the behavioral trait. Molecular studies then showed that novelty-seeking behavior is correlated with an allele of a specific gene.

In many cases, scientists hypothesize that a known gene influences a behavioral trait. They can then test that hypothesis using an **association study,** which compares the frequency of an allele in populations that do and do not have the behavioral trait. If a particular allele is observed more often in the group with the trait than in the group without the trait, then that allele may influence the trait. For example, people who are lactose intolerant avoid consuming dairy products because they lack the enzyme beta-galactosidase, which is necessary for digestion of lactose, or milk sugar. This food-choice behavior is most common in populations whose ancestors historically did not raise domestic cattle or consume milk.

You might be surprised to learn that when scientists study behavioral traits using genetic methods, they almost always find a genetic effect on the behavior. Every inherited physical trait has a genetic component, so we should expect to find genetic effects on behavior as well.

Check and Challenge

1. What types of information can be inferred by studying the traits of identical twins?
2. How can studying the behavior of nonhuman animals help us understand human behavior better?
3. Explain what correlation values of −1.0, 0, and +1.0 mean.
4. Explain how knowing the location of a certain gene can help scientists locate a gene for a specific behavioral trait.
5. What is an association study?

Social Behavior

22.7 Populations and Behavior

Many animal populations are considered a group only because they share a common habitat. They are not organized in communities or other social groups as humans are. For example, all the earthworms in a garden form a population, though they do not interact as a group as humans do.

Other animals live in complex social groups organized by genetically determined behavior patterns. For example, this organization occurs among the social insects—bees, ants, wasps, and their relatives. These insects live in colonies in which each individual's role depends on its genotype. Some ants, for example, are genetically programmed to care for herds of insects called aphids. The ants place the aphids to feed on succulent plant leaves and then "milk" the aphids to obtain food for the colony (Figure 22.15).

FIGURE 22.15

Genetically determined behavior in a social insect. These little black ants *(Monomorium minimum)* are tending milking aphids by carrying them to their food plants and protecting them from predators. The aphids provide the ant colony with a sugary fluid that the ants use as food.

Try Investigation 22C A Field Study of Animal Behavior.

Other genetically influenced interactions between animals include the recognition of prey and predators; the physical distribution of individuals; types of communication; and the distribution of food, territory, and mates. The environment influences these behaviors, too. Learning, planning, or reasoning can change such behaviors, especially in humans.

The presence of leaders, followers, and specialists in a population is evidence of organized and usually cooperative behavior. Populations that exhibit such patterns are called **animal societies.** Societies may be loosely structured or highly organized. For example, in a stable city neighborhood, there is usually a top dog that dominates the other dogs. Each dog in the neighborhood has its own social standing. There is no constant, serious fighting, although skirmishes do occur when an underdog tries to improve its social position. You can easily observe the nonviolent behavioral acts that acknowledge which dog has the higher place in the society. If a new dog is introduced, the neighborhood dogs quickly challenge it. The new dog submits immediately and accepts a less dominant position, bluffs the challenger into submitting, or fights. The outcome of the fight determines the new dog's social position. Such societies, also seen in wolves, are called **dominance hierarchies.**

Many insect societies are highly organized **caste systems** in which the population's survival depends on each type of individual doing its job. Genetic factors heavily influence whether an individual will develop a body type specialized for food gathering, defense, reproduction, or other tasks. In many cases, closely related members of different castes can look as different from each other as members of different species.

Focus On

Caste Systems

Social insects, such as termites and bees, have a caste system. Body types that are adapted for certain jobs indicate the division of labor within the insect society. The honeybee society has three castes: queen, drone, and worker (Figure 22.16). A drone supplies sperm, and the queen supplies eggs. The workers carry out all other activities in the hive.

Both diet and heredity determine body type. The workers and queens develop from female larvae. The drones develop from male larvae. Adult workers feed a substance called royal jelly to all the female larvae for the first 3 days of development. After that, only the few larvae that will become queens are fed royal jelly. When the queens emerge, they fight until a lone survivor becomes the hive's queen.

Neither the queen nor the drones gather or produce food. Their only role in the hive is to reproduce. To do so, a drone must catch and mate with a flying queen. Drones with more highly developed sensory organs, longer wings, and longer antennae are the most successful in mating.

The workers gather food, produce wax, rear and protect offspring, and carry on all the work of the hive. Although they are genetically female, they do not produce eggs. Their body specializations increase the work efficiency of the hive and thus the chance of the society's survival.

FIGURE 22.16

The three castes of honeybees. Think about how the differences you see among the queen, the worker, and the drone cause each individual to be specifically suited to its job in the colony.

Living in societies can benefit individual organisms in many ways. Social organization aids defense from predators. Musk oxen form a defensive circle when threatened by wolves. In this group, they are less vulnerable than an individual facing these predators alone (Figure 22.17). Similarly, the members of a prairie dog town cooperate by barking to warn each other of approaching predators. Thus, each individual in the social organization benefits from association with the group. Large numbers of prey grouped together also may distract or even frighten away predators.

Social behavior also can benefit reproduction. Animals that live in groups can synchronize their reproductive behavior so that the young are born at the most favorable time of year. For example, a bird's mating calls and displays can start endocrine changes that are required for reproduction in fellow colony members. Due to higher levels of social stimulation, large gull colonies produce more young per nest than do small gull colonies.

Other benefits of social behavior include increased care of the young through social networks and cooperation in obtaining food. Social behavior

Sociobiology: An Evolutionary Theory of Behavior

Have you ever heard the phrase "Nice guys finish last"? Can you imagine that kindness would be something that affects your reproductive success? A theory called sociobiology attempts to explain social traits such as kindness as evolutionary adaptations that increase the reproductive success of generous individuals. Sociobiologists point out that sacrifices by individuals can help their relatives, who share many of the same alleles, to survive and pass on those alleles. In fact, animal behavior that appears generous or self-sacrificing almost always occurs among close relatives.

Social insects such as ants will sacrifice their lives for the good of the colony. How can this behavior benefit the individual's reproductive success? If the death of a single ant helps the colony, then the ant has increased the chance of survival of its genetically similar nestmates. Its behavior enhances the survival of the alleles that it shares with the rest of the colony.

Ant societies, however, are very different from human cultures. Unlike ants, most of our behavior is learned. Some sociobiologists try to apply evolutionary ideas to both innate and learned behavior. They suggest that learned and cultural behaviors change in a way that resembles natural selection, as more successful, new ideas, customs, and technologies replace older, less useful ones. For example, messengers who traveled on foot were first replaced by horseback riders and later by the telegraph, the telephone, and the Internet. Of course, none of these changes resulted from genetic mutations, but many social behaviors do change in ways that are similar to evolution.

Other biologists and psychologists question this reasoning. They argue that there is not enough evidence to support the theory that genes and natural selection shape human social behavior. Genes may contribute to particular personality types and mental attributes, but it is unscientific to extend that correlation to cultural traits as a whole.

The application of sociobiological ideas to human societies has not yet been subjected to strict tests. While it is easy to imagine how evolution might affect certain behaviors, experiments to test these ideas are difficult to design. The lack of experimental evidence weakens the value of sociobiological explanations as scientific hypotheses.

Critics of sociobiological theory also argue that such ideas are dangerous because they imply that social injustices are biologically inevitable. They worry that sociobiology could be used to justify poverty, for example, on the grounds that social inequalities are the result of genetic differences in ability. Differences in social status could then be seen as simply part of the process of natural selection. Without more evidence to support the sociobiological hypothesis, many scientists remain hesitant to accept many of its predictions.

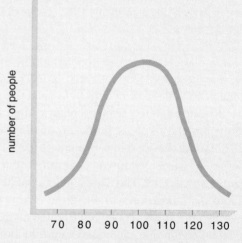

Data that have led to a popular misinterpretation of sociobiology. The graph shows the distribution of IQ (intelligence quotient) scores in a human population. Claims that genetic factors are mainly responsible for variation in human intelligence (and therefore for differences in income and social class) have been discredited. Even so, critics of sociobiology charge that this theory provides a basis for discrimination.

FIGURE 22.17

One benefit of social behavior. As long as these musk oxen *(Ovibos moschatus)* remain in a tight circle, wolves cannot attack them. Young oxen that have not yet grown horns are protected in the center of the circle.

also increases opportunities for reproduction, division of labor, and sharing of useful information.

If social living is so advantageous, why do animals of some species live alone? Living alone also offers advantages. For example, species that avoid predators by camouflage are more likely to survive if spread out in their environment. Large predators such as bears and tigers come together only to mate. These animals live alone because they require large territories to support sufficient prey. Solitary living reduces competition within these species. A species' lifestyle is closely linked to its ecological role and environment.

CONNECTIONS

A social or solitary lifestyle is part of a species' overall adaptation that has evolved to maximize its chance of survival.

22.8 Animal Communication

Communication is one way in which an animal can influence the behavior of another. Imagine that you see a dog chewing a bone. You approach the dog and reach out as if to take the bone away. As a result of your behavior, you would probably notice several changes in the dog's behavior and appearance. The hair on its back might rise, its lips might curl back to expose its teeth, and it might growl. The dog is communicating with you. You would soon learn the meaning of the message if you tried to take the bone away from the dog.

Many animals communicate via sound. For example, a hunting wolf gives a short bark and howl when closing in for the kill. A human may shout a warning when someone is about to get hit by a baseball. All these sounds tell

FIGURE 22.18

Visual communication in pronghorns. Pronghorns' *(Antilocapra americana)* rump patch of white hairs usually lies flat. When muscles cause these hairs to stand up, nearby pronghorns are warned of an intruder or another danger.

something specific to the listener. The listener's reaction depends on the type of sound.

Animals also communicate by sight, smell, touch, and taste. The pronghorns of the western United States (Figure 22.18) give each other a visual alarm by flashing the white hair of their rump patch when they spot a predator. Some visual signals are part of a display. For example, courtship displays are common in birds (Figure 22.19). Such displays can establish cooperation and species recognition between males and females and help ensure mating or care of the young.

Many animals communicate by chemical cues. For example, female silkworm moths make a chemical, called a **pheromone,** that alters the behavior of members of its species. The female's pheromone acts as a sexual lure that the male moths smell with their antennae. Small amounts of

FIGURE 22.19

A courtship display in a bird. Male prairie chickens *(Tympanuchus cupido)* inflate colored neck sacs and produce loud, booming calls during their mating season.

pheromone are carried downwind from the female. When a few molecules reach the male's antennae, he is stimulated to find the female and mate. Natural selection favors males with receptors sensitive enough to detect the chemical at great distances. Males with less well-developed sensory systems fail to locate a female and thus are reproductively eliminated from the population.

Honeybees communicate the location of food to other colony members through dance. A honeybee that finds nectar returns to the hive, gives the nectar to the other bees, and begins a dance. A round dance tells the others to search for food nearby (Figure 22.20a). A figure eight or waggle dance means the food is far away (Figure 22.20b). The orientation of this dance describes the nectar's location in relation to the Sun (Figure 22.20c, d, and e).

Sound and touch help wild chimpanzees communicate. When they play, they use a panting laughter. When young chimpanzees are frightened or lonely, they whimper. When chimpanzees meet after a separation, they greet each other with sound and hugging or other contact, much like humans. They also use sounds as warnings or threats.

Physical contact among chimpanzees seems to extend their vocal communication. They pat each other on the back, embrace, and hold hands. Grooming is also an important social activity (see Figure 20.5). Many

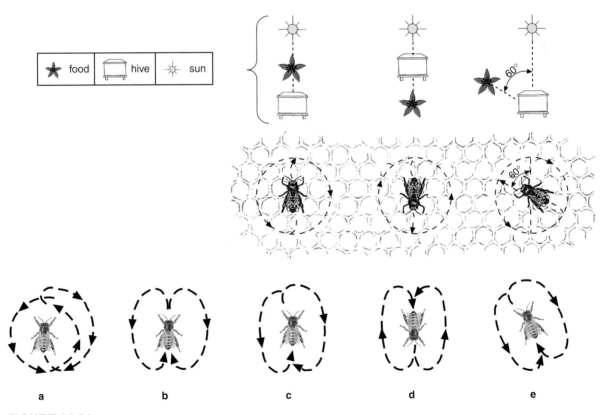

FIGURE 22.20

Communication by movement in honeybees. (a), The round dance in honeybees *(Apis mellifera)* indicates that food is nearby. **(b),** The direction of the waggle dance apparently tells observing bees where the food is in relation to the Sun. For example, in **(c),** the food is in a straight line from the hive to the Sun; in **(d),** the food is on the opposite side of the hive from the Sun; and in *e* the food is 60 degrees to the left of the Sun.

Teaching Language to Nonhuman Primates

People have long been fascinated by the similarities between humans and our close relatives, the chimpanzees and gorillas. In fact, many researchers have even tried to teach language to these animals. Despite painstaking efforts by researchers, chimps and gorillas have never learned humanlike speech.

Chimps have dexterous hands and are intelligent social animals. In the 1960s, researchers decided to try to teach them American Sign Language instead of speech (Figure 22.21). Allen and Beatrice Gardner pioneered these efforts. They spent 51 months working with a young chimp named Washoe. Washoe learned over 30 signs, but she was never able to make new sentences. Most of her signing was simple imitation of her instructors. Several other apes have been trained to associate meaning with symbols. Since the 1970s, scientists also have used computers to try to teach language to nonhuman primates.

Language experts agree that, like Washoe, the animals in these studies simply learned to imitate their trainers. Some subjects are able to associate symbols with objects or behaviors. Nevertheless, they do not generate new sentences using grammatical rules the way humans—even children as young as 2 1/2 years old—can. Dr. Herb Terrace of Columbia University says, "If a child did exactly what the best chimpanzee did, the child would be thought of as disturbed." Despite criticism, some scientists continue to try to teach nonhuman primates human language, and many believe that they are making progress in this area.

Dr. Sue Savage-Rumbaugh of Georgia State University thinks that too much emphasis has been placed on trying to teach human grammar to chimps. Instead, she has focused her research on determining whether nonhuman primates understand the signs and symbols that they learn to use. She suggests that the chimps in past experiments simply imitated, rather than learned, the signs because the experimenters did not use the signs symbolically. Savage-Rumbaugh maintains that her subjects do grasp the meaning of the symbols they use. In addition, she claims that they do not simply imitate their trainers but use the symbols to communicate real messages. In this controversial scientific arena, critics continue to question Savage-Rumbaugh's conclusions.

Herb Terrace, who worked with Washoe and other chimps for more than 10 years, now says that trying to teach nonhuman primates human language is fruitless. He believes that scientists should be pursuing different questions. "How do chimpanzees think without language, how do they remember without language? Those are much more important questions than trying to reproduce a few tidbits of language from a chimpanzee trying to get rewards," he says.

FIGURE 22.21

Language or imitation? Joyce Butler shows this chimpanzee (*Pan troglodytes*) named Nim the sign for "drink," and he copies her sign language.

chimpanzee behaviors are similar to those that we associate as expressive of human emotions. However, we cannot be sure that these animal behaviors express love, anger, or friendship. There is no way to know for certain what an animal experiences. Careless interpretation of animal behavior in terms of human motives and experiences is called **anthropomorphism.**

Humans live in highly organized societies that depend on complex communication systems. Human communication involves a wide variety of written and vocal symbols and other nonverbal signals, such as facial expressions. We can write poetry and novels for others to read. We can compose, paint, sculpt, and build things to convey thoughts and feelings. We smile at one another when we are happy and cry when we are sad. Humans often engage in physical contact, such as holding hands and hugging, to communicate. We build communication tools, such as telephones and the Internet, that extend our senses.

Our most important communication tool is language (see Section 20.6). Unlike other forms of animal communication, human language has a complex grammatical structure, or set of rules, that allows us to make entirely new statements that others can understand. Language enables humans to disseminate information among individuals and to future generations. Language development begins early in young children as they acquire their native language through exposure to other speakers.

Uses of language also extend beyond communication. Humans use language to think about things that we do not perceive in the present, such as the past, the future, and abstract ideas and emotions, such as love, atoms, our personalities, and death. The human ability to think about ourselves, the future, and the unknown has enabled us to develop the rich culture and technology that form a civilized world within the biosphere in which other species live.

Check and Challenge

1. How do dominance hierarchies help organize a population?
2. Explain how a caste system differs from a dominance hierarchy.
3. How is social behavior advantageous for individuals? for populations?
4. Describe how an individual animal can influence the behavior of another.
5. Why is communication necessary for a cooperative society?
6. How do displays promote cooperation?
7. How are pheromones an example of natural selection?

Chapter
HIGHLIGHTS

Summary

Behavior is the way an organism acts. It is an organism's reaction, or response, to stimuli. Both internal and external stimuli play a role in triggering certain behaviors. Internal stimuli include metabolic factors, such as hormone levels. External stimuli include any factors in the organism's external environment, such as light, gravity, and temperature. An organism's behavior is adapted to the demands of the environment in which it lives. Behavior is partly the result of evolution and, as with other characteristics of an organism, is affected by both heredity and the environment.

There are two general types of behavior, innate and learned. Both forms of behavior are influenced by stimuli. Innate behavior is inborn, or genetically influenced. It is not determined by experience, although it may be stimulated by environmental stimuli. Learned behaviors are the responses an organism develops as a result of environmental experiences. Generally speaking, innate behavior changes slowly over generations, while learned behavior may change over the course of an organism's lifetime as it gains experience.

Societies exhibit cooperative behavior. Different societies may have different social structures, such as dominance hierarchies or caste systems. Cooperative behavior can provide the members of the society with advantages, such as protection and increased reproductive success.

Social animals must be able to communicate with each other. Animals communicate through chemical, visual, acoustical, and physical interactions. Human language is one form of communication. Language enables humans to exchange information in many different forms.

Key Concepts

Develop a concept map of this chapter using the following words from the chapter and linking words of your choice.

- behavior
- innate
- heredity
- environment
- fixed action patterns
- conditioning
- learned

Reviewing Ideas

1. What are some environmental stimuli that might bring about responses in organisms?
2. How is behavior related to the survival of individual organisms? the survival of a population?
3. What type of information might an association study tell you about a behavior that an identical twin study could not?
4. Explain how societies differ from populations.
5. How does the form of a social group affect the life of its members?
6. If two behaviors have a correlation of 0.9, what does this statistic tell you about the behaviors? Can you determine from the statistic whether one behavior causes the other?
7. How does innate behavior affect the formation of societies?
8. How is learning related to social behavior?
9. Female African electric fish emit distinctive electric pulses. During breeding season, males of the same species respond with their own distinctive electric pulses. How might this behavior benefit the species?

Using Concepts

1. Suppose you want to investigate whether or not the type of web a spider spins is genetically determined. What kind of study might help you find out this information?
2. How would you determine if a dominance hierarchy existed in one of your classes? What evidence would you seek?
3. How would you determine if walking is an innate or learned behavior?
4. Explain how individual survival and species survival in humans may correspond in some behavioral situations yet conflict in others.
5. How would you plan an experiment to determine whether imprinting occurs in baby chicks?
6. A female greylag goose uses her bill to retrieve an egg that has rolled from her nest. Any round object will provoke this behavior. Identify the stimulus and response involved in this behavior. Which of the following term(s) would you use to describe this behavior—innate, learned, imprinting, trial-and-error, fixed action pattern? Explain your choice(s).
7. Ants spread scent trails from the anthill to food. How would you find out whether the ants require the scent trails to find their way back to the anthill?
8. Compare the scent trails of ants with the dancing of honeybees. Are the two functions similar? Are *both* forms of communication? Explain your answer.
9. What evidence supports the idea that societies are a product of evolution?

Synthesis

1. Which is more efficient, a society organized by conscious cooperation or by innate behavior? Defend your reasoning with examples.

2. John Maynard Smith has been called one of the most important evolutionary biologists since Darwin. He once said, "The one thing that really separates us from other animals is our ability to talk." Use the information in this chapter and elsewhere in your study of biology to either support or refute this statement.

Extensions

1. Do plants respond to internal and external stimuli? Give examples. Would you call these plant responses "behavior"? Defend your position.
2. Go to a park, shopping mall, or bank. Observe the ways that people behave toward one another in various circumstances. Write a report detailing your observations.
3. Go to the zoo. Pick a group of animals—for example, gorillas. Spend several hours observing their behavior, and write a report about what you see.

Web Resources

Visit BSCSblue.com to access
- Additional resources about behavior
- Web links related to this chapter

CONTENTS

LEARNING OUTCOMES

By the end of this chapter you will be able to:

A Describe immune system functions and distinguish between specific and nonspecific defenses.

B Summarize the functions of lymphocytes, B cells, and T cells.

C Describe the structural features of antibodies and explain how antibody diversity is achieved.

D Explain the workings and regulation of the immune system.

E Relate malfunctions of the immune system to disease.

CHAPTER

23

Immune Systems

- *How does this immune system cell protect the organism?*

- *How can immune systems distinguish invaders from an animal's own cells?*

Immune systems protect organisms from pathogens in their environment. We have all had firsthand experience with the immune system. For example, you probably had chicken pox as a child. Because your immune system has specialized cells that "remember" the first chicken-pox infection and defend your body against this virus, you do not worry about getting chicken pox again.

In recent years, research has begun to answer many fundamental questions about how this system functions. You might wonder why your immune system can protect you against getting chicken pox again but not against colds. You might also be curious about how your immune system can make you allergic to certain foods or other substances. Why do transplant recipients often reject their new organs? How does the immune system recognize that microorganisms or another person's cells are foreign? This chapter examines how immune systems protect organisms against foreign substances, the molecular and cellular components of the immune response, and some diseases that result from immune system disorders.

Protection against Infection

23.1 Immune System Functions

Pathogens, disease-causing viruses and organisms such as parasitic bacteria and fungi, invade other organisms and cause disease. **Infection,** the invasion of an organism by pathogens, is the primary health threat to multicellular organisms. Diseases such as hepatitis and pneumonia result when invading pathogens damage cells and tissues so much that they can no longer function properly. All organisms have some kind of defense against pathogens. Without healthy defenses, we would be constantly ill or dead. Our immune systems protect us against both pathogens and hazardous foreign substances.

Some defenses prevent foreign invaders from ever entering an organism. Other defenses remove them from the organism, and still others fight them if they remain inside. The two basic types of resistance to infection are nonspecific and specific defenses. Nonspecific defenses, such as tough epidermal coverings, protect us from most foreign substances. In contrast, specific defenses, such as cells and enzymes that attack a particular type of pathogen, become active only when their target is present.

Nonspecific immunity can limit certain infections, but it cannot eliminate them. These defenses provide temporary protection until specific immunity to the invading pathogen develops. In humans, specific immunity develops in 4–14 days. As you will see later in this chapter, this process builds on the nonspecific response. Thus, we need both arms of the immune system to fully control and recover from infections.

Any foreign material that causes an immune response is called an **antigen.** Most antigens are proteins. Because all cells and organisms contain many different types of proteins, they also contain many antigens. Normally the immune system discriminates well between an organism's own molecules and foreign antigens. A malfunctioning immune system, however, sometimes attacks an organism's own molecules as if they were foreign, destroying cells and tissues.

23.2 Nonspecific Defenses

Nonspecific defenses provide the first line of defense against foreign substances. Nonspecific defenses include physical and biochemical barriers. In animals, for example, the skin and mucous membranes provide a physical barrier against most foreign materials. In plants, bark and cuticle serve a similar function. Plants also produce defensive compounds called phytoalexins that bind to pathogens. The low pH of the stomach and the antibacterial enzymes in saliva, tears, and sweat are biochemical barriers to many microorganisms. In addition, fever is a nonspecific defense that raises body temperature above the optimum for the reproduction of many viruses and other pathogens.

In animals, **phagocytic cells,** or phagocytes, play a critical role in clearance of common bacterial infections and are the first line of defense against most

FIGURE 23.1

Phagocytosis by a macrophage. In this colored scanning electron micrograph (×7,500), the yellowish white macrophage is engulfing a red blood cell in the liver. Such macrophages, found mostly in the venous blood vessels of the liver, destroy cellular debris, bacteria, foreign bodies, and old red blood cells.

microbes. These cells consume both foreign materials and damaged or dead cells (Figure 23.1). Phagocytic cells are found in the blood and lymphatic systems, as well as in the skin, lungs, liver, brain, and most other tissues. Phagocytes have cell-surface molecules that recognize some components of bacterial surfaces. One type of phagocyte, the macrophage (see Section 7.9), can engulf and destroy a wide variety of molecules, particles, and organisms. This phagocytic process is also important in activating specific defenses.

Phagocytosis is an important part of a nonspecific defense called **inflammation.** Inflammation is a general response to tissue damage (such as cuts) and to antigens in foreign materials (such as splinters and infectious organisms). Capillaries in the damaged tissue leak cells and fluid into the inflamed tissue, causing redness, swelling, and warmth. Millions of phagocytic cells help remove the foreign material. The pus that may form consists of phagocytic cells, dead pathogens, and cell debris. Inflammation is the primary nonspecific protective mechanism in vertebrates.

Another type of white blood cell, or leukocyte (see Section 7.9), called a natural killer cell is the first line of defense against tumor and virus-infected cells. Natural killer cells accumulate in secondary lymph tissue (Figure 23.5) and act quickly to destroy cells recognized as abnormal. Their action also stimulates other immune responses. This is a critical first step in fighting viral infections and some cancers.

ETYMOLOGY

phago- = to eat (Greek)
-cyte = cell (Greek)
Phagocytes recognize pathogens and destroy them.

ETYMOLOGY

inflamm- = to set on fire (Latin)
-ation = action or process (Latin)
Inflammation is redness, swelling, and warmth produced by phagocytosis during an infection.

23.3 Specific Defenses and Adaptive Immunity

Nonspecific defenses are not sufficient to protect organisms against the large variety of pathogens in the biosphere. Specific defenses, or **adaptive immunity,** are essential for the successful fight against microbes. These are provided by two types of **lymphocytes,** called **B cells** and **T cells** (Figure 23.2). T cells develop in bone marrow and mature in the thymus (hence, the name "T cell"), the organ that lies above the heart. In mammals, B cells develop in the bone marrow of the long bones. When B cells encounter foreign particles or antigens, they produce large amounts of a protein, called an **antibody,** that binds to that specific antigen. This process is called the **antibody-mediated immune response.** The antibodies mark the antigen for destruction and removal by the nonspecific defenses. T cells, however, do not secrete antibodies. They secrete specific chemicals that help initiate and maintain the immune response against the invading pathogen. T cells are part of the **cell-mediated immune response,** in which they recognize antigens bound to other cells. Both the cell-mediated response and the antibody-mediated response contribute to the immune system's defense against pathogens.

Specific immunity not only removes pathogens but also provides long-term resistance to pathogens it has previously destroyed. When the immune

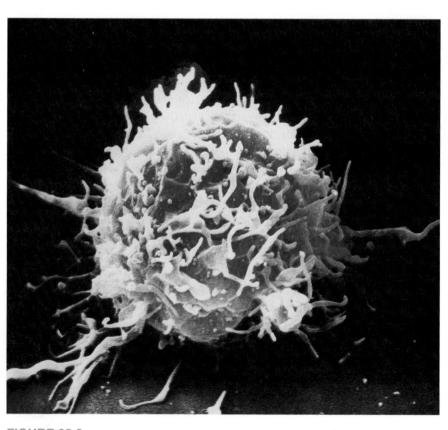

FIGURE 23.2

Colored scanning electron micrograph of a T cell. Notice the characteristic long microvilli that extend from the surface of this normal T cell, × 5,400.

The Cow-Pock — or — the Wonderful Effects of the New Inoculation! — Vide the Publications of y' Anti-Vaccine Society

FIGURE 23.3

Vaccination, as imagined by a cartoonist of Jenner's day. The cartoon illustrates the fears that vaccination against smallpox aroused. The fear of developing animal characteristics arose because of Jenner's use of vaccine derived from cowpox sores.

system encounters antigens it has dealt with before, it responds quickly and efficiently to them. Knowledge of this long-term resistance, or **immune memory,** led scientists to discover that immunity could be produced artificially with **vaccines,** suspensions of dead or weakened organisms that induce specific immunity without actually causing disease. Injecting or swallowing vaccines to produce immunity is called immunization.

Perhaps the most successful story of immunization is the elimination of the disease smallpox. In ancient China, children developed immunity to smallpox after they inhaled powder made from the scabs of smallpox sores. In 1798, the English physician Edward Jenner observed that milkmaids who had recovered from a mild disease called cowpox were protected against smallpox. Jenner inoculated a boy with pus isolated from cowpox sores and subsequently demonstrated that the boy was protected against smallpox. Jenner called the process **vaccination,** and by 1800, at least 100,000 people had been vaccinated against smallpox (Figure 23.3). Smallpox has been eliminated worldwide; the last two deaths from smallpox occurred in 1978.

ETYMOLOGY

vacca- = cow (Latin)

-ation = action or process (Latin)

Jenner's inoculations were named **vaccinations** because they were developed from the cowpox virus.

Jenner's experiments probably have been responsible for saving more lives than any other experiments in history. In the 19th century, Louis Pasteur developed vaccination procedures for rabies and other diseases. The task of developing effective vaccines is challenging. Today scientists continue to work on developing vaccines for infectious diseases such as acquired immunodeficiency syndrome (AIDS).

Everyone has had practical experience with immunity. You have probably been vaccinated against the polio virus. This immunization provides long-term, specific immunity against polio, a neuromuscular disease that often results in paralysis. The polio vaccine, however, provides no protection against other viruses, such as the one responsible for chicken pox. The effect of the polio vaccine is limited because the antigens on the polio virus stimulate only those lymphocytes that specifically recognize this virus. Similarly, because more than 100 viruses can cause the common cold, immunity to one type of cold virus does not provide resistance to another type. In addition, many pathogens—such as flu viruses and the bacteria responsible for meningitis (an infection of the brain), gonorrhea, and Lyme disease—can evade the immune system by changing their antigenic surface proteins. If you became immune to a specific flu virus last year, a slight alteration of surface proteins allows the virus to evade your immune system and cause disease again this year. Medical science has not yet found a way to

CONNECTIONS

The ability of some pathogens to change their surface antigens in response to their hosts' immune responses is an example of coevolution.

Focus On

Evading the Immune System

Trypanosomes (Figure 23.4a) are microscopic protists that spend part of their life cycle in the blood of humans and other mammals. In humans, they cause African sleeping sickness, a fatal disease of the nervous system characterized by fever, fatigue, tremors, and weight loss. Trypanosomes enter the human bloodstream via tsetse fly bites (Figure 23.4b). The key to the trypanosome's success is its ability to circumvent the human immune system by changing its surface proteins.

With the initial infection, the host's immune system produces antibodies that bind to antigens on the trypanosome's surface and signal cells of the immune system to destroy the organism. During this time, however, some of the trypanosomes change their surface proteins and are not immediately attacked. Eventually the immune system produces antibodies against these new antigens. Meanwhile, some trypanosomes again change their proteins, staying one jump ahead of the immune system. The result is a persistent infection of the host's bloodstream.

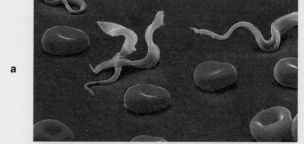

a

b

FIGURE 23.4

Immune system evaders. Trypanosomes *(Trypanosoma brucei)* cause African sleeping sickness, and their carrier, the tsetse fly *(Glossina palpalis)*. **(a),** These protist parasites, ×4,600, live part of their life cycle in the human bloodstream. **(b),** They are transmitted by the bite of the tsetse fly, found in the humid areas of central Africa.

prevent the flu virus from infecting us repeatedly. While immunological research has provided us with many useful vaccines and therapies, there is still much to learn.

Check and Challenge

1. What is the difference between infection and disease?
2. What nonspecific protective mechanisms do humans have?
3. Define the term *antigen* and provide several examples.
4. What are antibodies?
5. Distinguish between antibody-mediated immunity and cell-mediated immunity.

Components of the Specific Immune Response

23.4 Lymphocytes

There are approximately 2 trillion (2×10^{12}) lymphocytes in a mature human—about as many cells as there are in the liver or brain. Lymphocytes move as single cells in the blood and in the lymphatic system (see Section 7.9). They also develop and collect in specialized organs, called **lymphoid organs.** Figure 23.5 shows the structures of the human immune system.

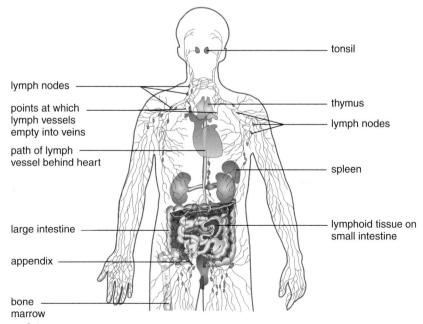

tonsil

lymph nodes

points at which lymph vessels empty into veins

path of lymph vessel behind heart

thymus

lymph nodes

spleen

large intestine

appendix

bone marrow

lymphoid tissue on small intestine

FIGURE 23.5

The human immune system. The bone marrow and thymus *(shown in blue)* are primary lymphoid organs responsible for the development of B cells and T cells, respectively. Secondary lymphoid organs *(shown in reddish colors),* where mature lymphocytes take up residence, include the lymph nodes, spleen, appendix, and tonsils.

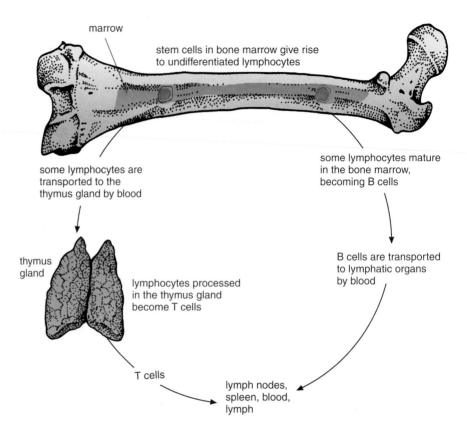

FIGURE 23.6

The processing of lymphocytes.
Lymphocytes develop from stem
cells in bone marrow. Some
lymphocytes continue development
in the marrow and become B cells.
Other lymphocytes are transported
to the thymus gland, where they
mature into T cells. Both types of
lymphocytes are found in lymphoid
organs and in the blood.

In a fetus, lymphocytes, like all other blood cells, are produced in
the liver. As a child develops, the bone marrow gradually takes over
lymphocyte production and continues to produce lymphocytes throughout
life (Figure 23.6). In the bone marrow, undifferentiated cells called stem
cells differentiate into all the types of blood cells (Figure 23.7). Bone

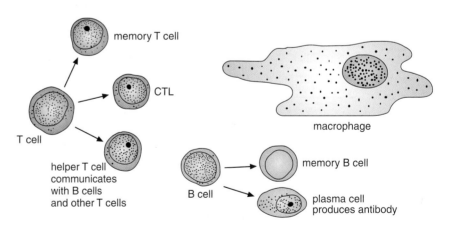

FIGURE 23.7

Major cells of the immune system. T cells are of two main types. Cytotoxic T lymphocytes
(CTLs) attack infected cells and invading pathogens by puncturing their plasma membranes.
Helper T (T_H) cells secrete cytokines, signaling proteins that enhance the immune response.
B cells, when stimulated by an antigen, differentiate and become plasma cells that secrete
antibodies. Both B cells and T cells also give rise to long-lived memory cells. Macrophages
ingest pathogens and display their antigens to T cells, triggering the specific immune response.

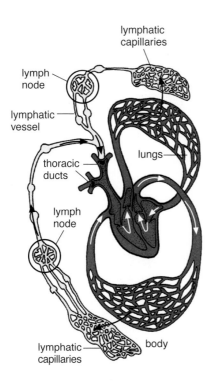

lymphatic
capillaries

lymph
node

lymphatic
vessel

thoracic
ducts

lungs

lymph
node

lymphatic
capillaries

body

FIGURE 23.8

Lymphatic circulation. Lymphocytes and antibody molecules circulate through the lymphatic vessels *(yellow)* and return to the blood in the circulatory system *(red)* via the thoracic ducts.

marrow is a primary lymphoid organ—B cells produced there are mature and functional when they leave the marrow. T cells are produced in the marrow and then travel to the thymus, where they continue to divide and mature. Thus, the thymus is another primary lymphoid organ.

When mature T cells and B cells enter the circulation, many of them lodge in secondary lymphoid tissues, such as the spleen, lymph nodes, tonsils, appendix, and lymphoid tissues beneath the skin and in the lining of the intestinal tract. Interaction of T cells and B cells with antigens usually occurs in these tissues. Swollen glands are really enlarged lymph nodes in which lymphocytes have multiplied to fight infection.

Unlike red blood cells, lymphocytes travel between the blood and the lymph, entering the lymphatic circulation primarily in the lymph nodes. After passing through numerous lymph nodes, lymphocytes return to the blood by way of the thoracic duct (Figure 23.8). T cells and B cells constantly migrate throughout the body seeking out and destroying foreign cells or substances that they encounter. This behavior is often called **immune surveillance.**

23.5 B Cells and the Antibody-Mediated Immune Response

Although many antigens that induce an immune response are proteins, almost any large foreign molecule can be an antigen. The immune system is capable of discriminating among millions of different antigens, some of which are only slightly different from each other.

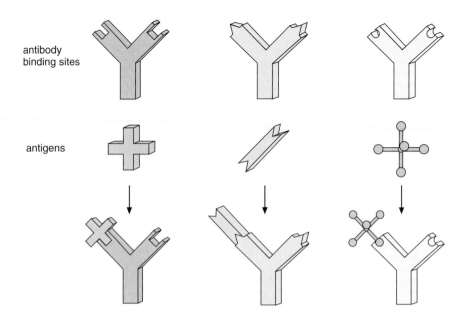

antibody binding sites

antigens

FIGURE 23.9

Antigen-antibody binding.
Antibodies are produced to fight particular antigens and bind specifically to them. Antibodies and antigens have complementary structures that produce tight binding.

The primary role of B cells is to make antibody molecules that react with a specific antigen (Figure 23.9). B cells express antibody molecules on their outer surface. These antibodies are called B-cell receptors. Each B cell is covered by thousands of identical B-cell receptors. Therefore, each B cell can bind to only one specific antigen. There are many thousands of B cells, however, each with a different receptor. Once the protruding ends of a B-cell receptor bind to a specific antigen, the cell begins to divide rapidly. Its offspring cells differentiate into two types—**plasma cells** and **memory B cells.** Plasma cells are like tiny factories that produce antibodies identical to the B-cell receptor that bound to the original antigen. The antibodies bind to the same type of antigen molecules wherever they are in the body, marking them for destruction. While plasma cells live only a few days, memory cells continue to circulate for long periods of time—up to a lifetime. The next time the immune system encounters a particular pathogen, memory cells can immediately begin the large-scale production of antibodies. Thus, memory cells account for immunity to subsequent infections by the same pathogen.

23.6 T Cells and the Cell-Mediated Immune Response

Unlike B cells, T cells do not secrete antibodies. Two kinds of T cells— **cytotoxic T lymphocytes (CTLs)** and **helper T (T_H) cells**—play a critical role in the cell-mediated immune response. T_H cells help immune responses develop by secreting signaling proteins called **cytokines,** which stimulate B cells and other T cells. For example, some cytokines are critical for B-cell development and antibody production. T_H cells may also be required for CTL generation. CTLs are essential for limiting viral infections. These cells recognize virus-infected cells and kill them by puncturing their plasma membranes.

E T Y M O L O G Y

cyto- = cell (Greek)

-kine = related to regulation (Greek)

-toxic = poison (Latin)

A **cytokine** is a molecule produced by one cell that regulates activities in another cell. **Cytotoxic** T lymphocytes inject poisons into their target cells.

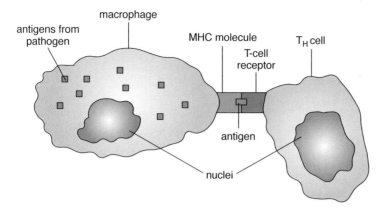

FIGURE 23.10

Recognition of an antigen by T cells. A macrophage processes a pathogen into fragments containing antigens. These fragments are displayed on the macrophage surface, bound to MHC proteins. The T$_H$ cells recognize and respond to the MHC-antigen combination.

T cells also have proteins, called T-cell receptors, on their surface. These proteins have some similarities in structure to antibodies. However, they are not secreted by the T cells as antibodies are from B cells. In addition, T cells recognize antigens in a different way than B cells. T cells are activated when they recognize pieces of antigen proteins that are bound to specialized cell surface proteins called **major histocompatibility complex (MHC)** molecules. MHC proteins occur on the surface of all vertebrate cells. When a phagocytic cell such as a **macrophage** digests a pathogen, the phagocyte displays fragments of the pathogen's proteins bound to MHC proteins on its surface. T cells recognize this complex of protein and MHC but not either one of them alone (Figure 23.10). Like phagocytes, cells that are infected by viruses also display viral protein fragments on their MHC proteins.

23.7 Clonal Selection in the Immune Response

During embryonic development and throughout life, developing lymphocytes become committed to producing antibody or receptor molecules that bind specifically to a single antigen. This commitment occurs without any antigens present. Once a B or T lymphocyte becomes committed to producing a specific receptor or antibody, all of its descendants will have the same commitment. Each clone of lymphocytes binds a different antigen from other clones. When an antigen appears, antigen binding stimulates specific clones to divide, giving rise to many identical descendants (Figure 23.11). This process is known as **clonal selection** because only those clones that can bind to the antigen are stimulated to divide. An analogy to help understand clonal selection is a department store that carries shirts of various sizes, styles, and colors. The store orders and stocks more of the shirts that customers demand, just as the immune system produces more of the lymphocytes it needs to remove specific antigens.

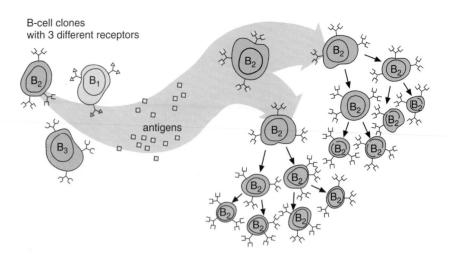

B-cell clones
with 3 different receptors

antigens

FIGURE 23.11

Clonal selection. Millions of different B-cell clones with different B-cell receptors are present before an antigen is first encountered. The antigen binds to the best-fitting receptor and thus selects the clone that will be stimulated to divide, forming thousands of identical descendants.

Our genes determine which antibodies and T-cell receptors the immune system will make and, thus, which antigens the system will recognize. Even without any antigen exposure, there is a large library of preexisting antibody and T-cell receptor types. From this library, an antigen "selects" which clones will be stimulated. The millions of different preexisting lymphocyte clones, with receptors capable of binding to almost any antigen, provide broad protection against a host of pathogens.

Clonal selection also helps explain other characteristics of the immune response. For example, if you were immunized against the measles virus, you did not become immune the day you received the vaccine. It takes 7–14 days after the first exposure to a new antigen to reach maximum antibody levels. This is the **primary immune response** (Figure 23.12). Prior to antigen exposure, the B cells, with their surface receptors, divide only

FIGURE 23.12

Response of the immune system to antigens. In the primary response, the concentration of antibodies in the blood rises to a peak 7–14 days after the initial exposure. In the secondary response, subsequent exposure to the same antigen triggers a greater response with a much higher concentration of antibodies.

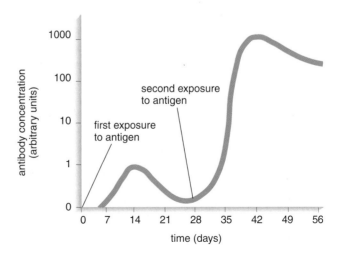

occasionally. Exposure to a specific antigen stimulates the clones capable of binding to that antigen to divide, producing plasma cells and antibodies—a process that takes several days. Because most foreign materials contain many different antigens, numerous B-cell clones will be stimulated to produce antibodies.

During the primary immune response, many of the stimulated B cells produce long-lived memory cells, which retain the original specific antigen-binding capability. They greatly increase the number of lymphocytes capable of binding to a specific antigen. A second exposure to the same antigen stimulates the memory cells to divide and produce more plasma cells. These cells produce high levels of antibodies rapidly, within 2 to 6 days. This rapid response of memory cells to a familiar antigen is called the **secondary immune response** (see Figure 23.12).

23.8 Antibody Structure

An effective immune defense must respond selectively to specific antigens. The binding of antibodies to specific antigens is critical to this selectiveness. Therefore, the immune response requires antibodies with structures that are highly selective in their binding properties. Figure 23.13 shows that each antibody molecule consists of four polypeptide chains. There are two identical, smaller light chains and two identical, larger heavy chains. One light chain and one heavy chain combine to make half of an antibody molecule. The two identical halves then combine to make a complete antibody molecule. In each half, one end of the heavy chain and one end of the light chain together provide one antigen binding site. Thus, each antibody molecule has a total of two antigen binding sites (see Appendix 23A, "Antibody Classes").

Try Investigation 23A ◇ Antigen-Antibody Binding.

CHEMISTRY TIP

Disulfide bonds (—S—S—) link the four protein strands of an antibody molecule to one another (Figure 23.13). These bonds form between the sulfur atoms of two cysteines that are close to one another. Cysteine is an amino acid that has a sulfhydryl group (—SH) on its side chain. The formation of disulfide bonds is important in determining protein structure and function.

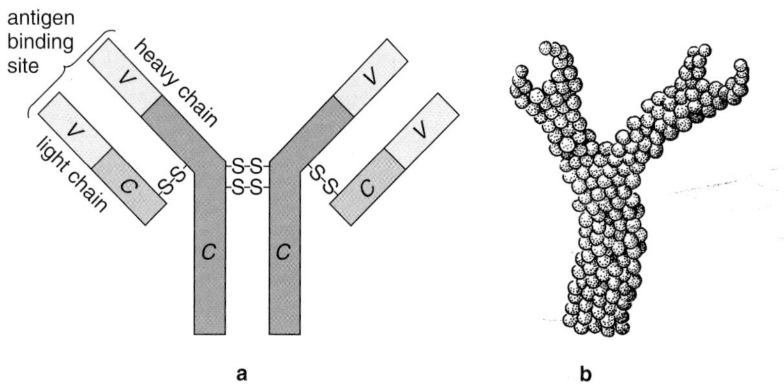

FIGURE 23.13

Two models of antibody structure. (a), This diagram shows the variable (V) and constant (C) regions of the light and heavy polypeptide chains and the identical halves of the molecule held together by disulfide bonds (—S—S—). **(b),** A drawing based on a computer-generated model gives a sense of the three-dimensional shape of an antibody molecule. The variable regions of one light and one heavy chain combine to form an antigen-binding site; each antibody molecule has two antigen-binding sites.

CONNECTIONS

In biology, structure and function are interdependent. The structure of an antibody suits it for its function: to bind specifically to one of millions of different antigens.

Antibody molecules have two important actions. First, they bind to specific antigens, which means that antibody molecules have unique antigen-binding sites. Second, after antigen binding, antibody molecules activate specific defenses, such as phagocytes, that destroy the antigen. Thus, antibody molecules are unique in structure, yet similar in their function of eliminating antigens.

Among all antibody molecules, the sequences of amino acids vary greatly at the end that forms the antigen-binding sites. This highly variable region (V in Figure 23.13) explains how so many different antigens can be bound. At the other end, the sequences of amino acids are very similar among all antibody molecules. This part of the molecule is called the constant region (C in Figure 23.13).

As you saw in Figure 23.9, antigens and their antibodies have complementary structures. Antigen-antibody binding is highly specific, and in many cases, quite strong. Surprisingly, antigen-antibody binding often involves weak electrostatic forces, such as hydrogen bonds and hydrophobic attractions. However, many weak attractions add up to a strong molecular interaction.

23.9 Generation of Antibody Diversity

Try Investigation 23B ◇ Antibody Diversity.

As in all proteins, the sequence of amino acids in an antibody determines its shape and function. In turn, genes determine the sequence of amino acids. However, humans have only about 35,000 genes. This is not enough to account for the millions or more antibodies that each person carries.

The basis for generating antibody diversity is similar to creating a diversity of lock combinations. If the numbers (genes) 0 through 9 are used singly, ten combinations (antibodies) are possible. Using the ten numbers two at a time provides 10×10, or 100, different combinations (00 through 99). In the same way, ten numbers used three at a time yields $10 \times 10 \times 10$, or 1,000 different combinations (000 through 999), and so on.

What if there were 100 genes that coded for the light antibody chains and 100 genes that coded for the heavy antibody chains? If any light chain could associate with any heavy chain, there would be 10,000 (100×100) possible different combinations using only 200 genes. Figure 23.14 shows the combinations possible with only three different light chains and three different heavy chains. In addition, different parts of the variable regions of the heavy and light chains of antibodies are coded by more than one gene fragment. This provides even greater possibilities for splicing together different combinations of gene fragments to make antibodies that have different specific structures.

In this way, the immune system produces large numbers of different antibodies with a relatively small number of genes (see Appendix 23B, "Generating Antibody Diversity"). Similarly, but with a different set of genes, T cells make a large number of receptors that recognize specific antigens.

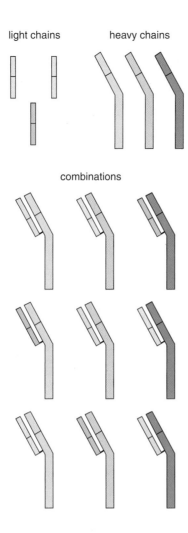

light chains   heavy chains

combinations

FIGURE 23.14

Generating antibody diversity. Three different light chains can join individually with three different heavy chains to form nine different combinations for each half of an antibody molecule. Two identical halves join to form a complete antibody molecule.

Check and Challenge

1. How do lymphocytes circulate throughout the body?
2. How do antigens and antibodies bind together?
3. Why does the secondary immune response display a faster response with more antibodies than the primary response?
4. What does clonal selection mean?
5. Where are antibody molecules found? Describe their structure.
6. How is it possible for lymphocytes to make such a large number of different antibodies from a limited amount of genetic information?

Susumu Tonegawa

For a long time, scientists thought that genes in an individual were stable and that alterations were relatively rare. However, research by Dr. Susumu Tonegawa (Figure 23.15) produced a new picture. In B and T lymphocytes, specific parts of antibody genes are cut, reshuffled, and assembled in ways that make it possible for the cell to produce millions of different proteins.

Tonegawa was born in Japan, where he earned his bachelor's degree. In 1969, he received his Ph.D. in biology from the University of California at San Diego. Then he went to the Basel Institute in Switzerland, where he began investigating the immune system.

The explanation for how the immune system produces millions of different antibodies was discovered by Tonegawa and one of his Basel colleagues, Nobumichi Hozumi. Using restriction enzymes (see Section 15.3) and DNA hybridization (see Section 10.7), Tonegawa and Hozumi mapped the location of the DNA coding for the variable (V) and constant (C) regions of light chains in mice. They found that, in embryonic B cells, gene segments for the V and C regions were far apart, separated by a third segment, called J for joining. In mature B cells, however, the V and C regions were close together. These findings provided a clue that "cutting and pasting" of the DNA might occur. In later experiments, they found that genes for heavy chains contain still another type of gene segment, called D for diversity.

Tonegawa's studies showed that antibody genes are expressed in two stages. First, the V, J, and D gene segments move closer together when enzymes remove the intron segments that separate the coding regions. Second, the resulting DNA chain is then transcribed into RNA, along with RNA representing the C portion of the gene. This RNA is translated into an antibody protein. The combinations are random, and only cells producing antibodies to invading antigens are selected to multiply. Because of the randomness of combinations, billions of different antibodies are possible.

FIGURE 23.15

Susumu Tonegawa. His research on the immune system showed how a large number of different antibodies could be produced from a much smaller number of genes.

The Immune System in Action

23.10 Eliminating Invading Pathogens

The immune system acts in various ways to eliminate antigens. For example, during a viral infection, an infected cell generates pieces of viral antigens and displays them on its surface along with the MHC molecules. This serves as a red flag, alerting CTLs to the presence of an invader. CTLs kill the infected cell and limit the spread of infection. Cancer cells may also display specific antigens along with MHC molecules. CTLs can sometimes recognize these antigens and destroy the tumor cells (Figure 23.16).

FIGURE 23.16

Cytotoxic T lymphocytes in action. These small cytotoxic T lymphocytes, CTLs, *(shown in yellow)* are attacking a cancer cell *(shown in pink)* in this scanning electron micrograph. The CTLs induce the cancer cell to initiate a genetically encoded self-destruct sequence called programmed cell death, or apoptosis. Pink vesicles, or apoptotic bodies, emerging from the cancer cell indicate that this form of cell death is occurring.

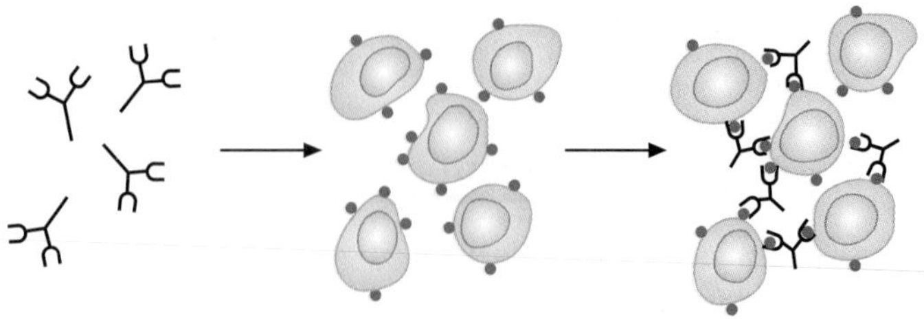

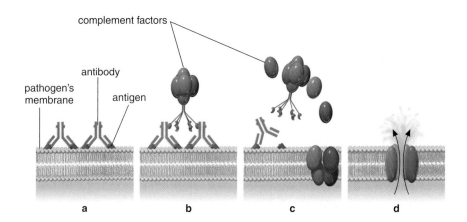

antibodies

foreign cells with
antigens on surface

antibodies form
bridges between antigens

FIGURE 23.17

Antibodies in action. Antibodies can form bridges between foreign cells or antigens, making clusters for phagocytes to engulf.

When antibody molecules encounter antigens during an antibody-mediated immune response, several outcomes are possible. If the antigen is small, such as a soluble toxic protein, the antibody may coat it by binding to it directly. If the antigen is part of a microorganism, antibody binding has little direct effect on the microorganism. However, antibody-coated antigens are easier than noncoated antigens for macrophages to engulf and destroy. Because antibody molecules have two binding sites, bridges can form between antigens. Macrophages and other phagocytes then can engulf the antibody-antigen clusters (Figure 23.17).

complement factors

antibody

pathogen's
membrane

antigen

a   b   c   d

FIGURE 23.18

Activation of the complement system. When two antibody molecules bind to adjacent sites on a pathogen **(a),** they can activate the first factor in the complement system **(b),** which in turn activates other complement factors. Activated complement factors aggregate within the plasma membrane of the pathogen **(c),** causing the membrane to break **(d).** As a result, the pathogen dies.

In addition, antibodies bound to antigens can activate the **complement system,** a complex group of plasma proteins that can destroy or eliminate antigens (Figure 23.18). Antibody-antigen complexes activate the first complement factors. These factors in turn activate other complement factors. Activated complement factors eliminate antigens in two ways. First, some of the activated complement proteins can attract phagocytes and greatly enhance phagocytosis and destruction of the antigen. Second, activated complement molecules can aggregate within the plasma membrane of an invading pathogen. This action eventually forms holes that cause the pathogen's cells to burst. Some complement factors can be activated directly by antigens. That way certain antigens are destroyed early in an infection before antibodies are formed. Such activated complement factors also stimulate inflammation.

CHEMISTRY TIP

The complement system is an example of a regulatory cascade, in which each enzyme in a series increases the activity of the next. (See Figure 7.24 for another example of a cascade.) A cascade can generate a large response quickly.

23.11 Regulation of the Immune System

A healthy immune system depends on the coordination of many individual tasks. These tasks include turning on needed responses, producing the components necessary for the response, eliminating the invader, turning off responses that are no longer needed, and regulating the many immune responses that may be occurring at the same time. This complex defense system is a network of specialized cells that constantly interact with one another to distinguish self—the organism's own cells and molecules—from nonself—foreign cells and molecules.

T cells determine that an antigen is nonself by comparing the antigen bound to MHC molecules with proteins the system recognizes as self. In humans, these molecules are called human leukocyte antigens **(HLA).** Except for identical twins, each individual has a nearly unique combination of HLA proteins on the surfaces of all cells except red blood cells.

Your T cells do not react against your HLA. This lack of reaction to self is called **tolerance.** Your T cells, however, are not tolerant to HLA proteins from another individual. Therefore, when surgeons transplant tissues or organs between individuals, the recipient's CTLs begin a cell-mediated immune response against the transplanted cells. If the two individuals happen to have several HLA proteins in common, the immune response is weaker, and there is a good chance the transplant will succeed. A good donor-recipient match means that several HLA proteins of the two individuals are identical. The recipient's T cells recognize the shared proteins as self and do not react against them. If there are major differences between the HLA sets, however, the recipient will reject the transplant. In those cases, patients may be given medications that suppress the immune reaction. However, these medications need to be given in appropriate doses. If the immune system is suppressed too much, patients may be unable to defend themselves against infection.

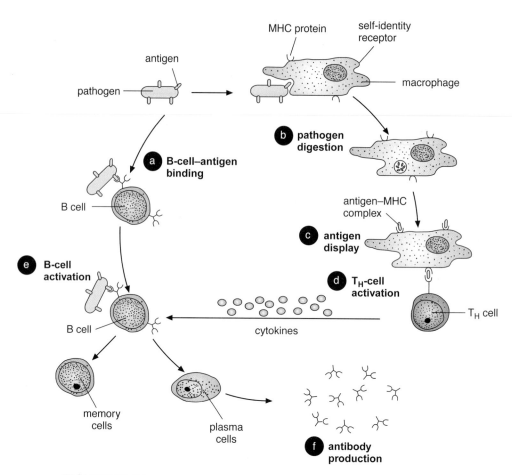

FIGURE 23.19

Production of antibodies in the specific immune response. Pathogenic cells may bind to a B-cell surface receptor **(a)**, or be engulfed by a macrophage **(b)**. The macrophage digests the pathogen and displays its antigens along with MHC molecules on the macrophage's plasma membrane **(c)**. Helper T cells (T$_H$) bind to the antigens displayed on the macrophages and B cells and become activated **(d)**. Activated T$_H$ cells begin to produce cytokines. Some cytokines stimulate microorganism-bound B cells to produce a clone of identical plasma cells **(e)**, which secrete thousands of antibodies to the pathogen into the bloodstream **(f)**. Other cytokines stimulate T$_H$ cells to divide and produce still more cytokines.

Most antigens to which you are exposed are not cells from another human. Microorganisms, which carry many antigens, are the main reason immune responses begin. A specific immune response to a pathogen can begin when receptors on a B cell bind to a pathogen or when a macrophage consumes a pathogen and displays the pathogen's antigens on its own surface MHC proteins (Figure 23.19). In both cases, helper T cells respond to the pathogen's antigens by releasing cytokines that activate a variety of immune cells. Their most important effect is to stimulate B cells bound to pathogens to multiply, producing plasma cells that release antibodies into the bloodstream. Helper T cells are at the center of this system. They

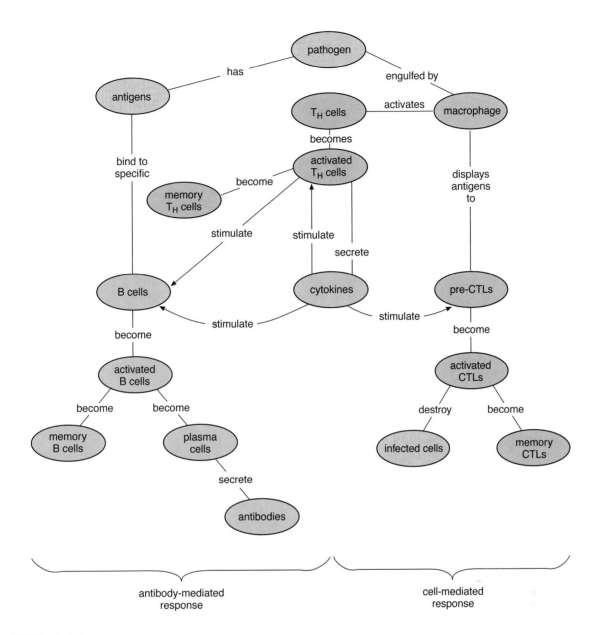

FIGURE 23.20

Interactions in the specific immune response. Note the central role of T_H cells in connecting the antibody-mediated immune response with the cell-mediated immune response.

produce the cytokines that stimulate both the antibody-mediated immune response and the cell-mediated response (Figure 23.20).

Once there are sufficient antibody molecules and T cells to eliminate the antigens, the immune system must shut down its response. This control is achieved in several ways, such as death of activated lymphocytes after several days, production of cytokines that regulate the function of other activated cells, and expression of inhibitory molecules on B-cell and T-cell surfaces.

23.12 Malfunctions and Disease

There are many reasons for malfunctions in the immune system. How many can you think of? One malfunction you may have experienced is getting sick during times of stress. This phenomenon may be explained by the chemical connections between the brain and the immune system. For example, hormones made by the adrenal gland in response to stress can inhibit the function of T cells. In turn, substances released by the brain and the pituitary gland into the bloodstream control the function of the adrenal gland. Thus, being stressed out can suppress the immune system, causing susceptibility to infection.

A variety of external factors can also suppress the immune system. Certain medications, such as cortisone, and illegal drugs, such as angel dust (PCP), suppress the function of lymphocytes. Certain infectious agents, such as HIV, can also suppress the immune response. Although your body can mount good primary and secondary immune responses against this virus, the long-term effect is immunodeficiency, or weakening and eventual destruction of the immune response. For example, HIV selectively infects and destroys lymphocytes, especially T_H cells. HIV can destroy a previously normal immune system, leading to acquired immunodeficiency syndrome (AIDS). In such severe cases of immunodeficiency, normally minor infections can cause life-threatening diseases.

A hypersensitive immune system also can result in disease. Hypersensitive responses, such as allergies, may be either cell-mediated or antibody-mediated. Contact with poison ivy, for example, can cause a hypersensitive cell-mediated response. A surface lipid in poison ivy *(Toxicodendron radicans)* called urushiol can combine with the cell-surface proteins of your skin cells and change their structure. Your T_H cells then recognize these altered proteins as foreign. The T_H cells signal CTLs and phagocytes to attack the affected skin cells. The destruction of these cells leads to inflammation, a rash, and blisters. The same type of allergic reaction is sometimes caused by metals released from jewelry, such as rings and earrings.

Hypersensitivities can also cause allergies and are responsible for hay fever, asthma, and bee-sting reactions. Antigens in pollen, animal dander, food protein, or insect venom may bind to antibodies attached to mast cells, immune cells that remain in connective tissue, rather than circulating in the blood. If the antigen is reintroduced (for example, during another pollen season), antigen-antibody binding causes mast cells to dump their contents into the surrounding tissues (Figure 23.21). Mast cells contain many biologically active substances, including histamine. When released, histamine causes the sneezing, itchiness, and teary eyes associated with hay fever. Histamine also causes smooth muscles to contract and blood vessels to swell. This reaction causes the airways in the lungs to constrict, resulting in the severe breathing difficulty of an asthma attack. Antihistamines can often reduce the effects of such hypersensitivities.

Another malfunction of the immune system is **autoimmunity.** As the name implies, the immune system no longer tolerates self molecules, and it produces an immune response against them. Autoantibodies or CTLs bind

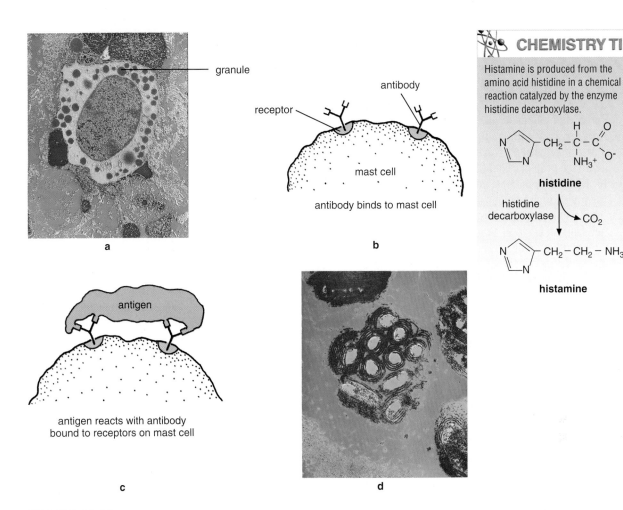

granule

receptor

antibody

antibody binds to mast cell

mast cell

a

b

antigen

antigen reacts with antibody
bound to receptors on mast cell

c

d

FIGURE 23.21

Mast cells, immune system cells involved in allergic reactions. **(a),** This colored transmission electron micrograph, ×2760, of a mast cell *(central light green body)* shows numerous cytoplasmic granules surrounding the cell's large oval nucleus *(red and dark green)*. The large, fuzzy, reddish spots are fat globules. **(b),** Antibodies attach themselves to the surface of the mast cell. **(c),** When an antigen binds to these antibody molecules, the cell will release the contents of the granules. **(d),** Four granules *(red structures)* are visible in this colored transmission electron micrograph of a mast cell, ×72,000. These granules contain substances such as histamine, heparin, serotonin, and proteins, shown here in characteristic "scrolls" that form spirals. Release of these substances is responsible for setting off inflammation and allergic responses.

to self molecules and interfere with or destroy them. Diseases such as rheumatoid arthritis, multiple sclerosis, and diabetes have an autoimmune component. Although the exact mechanisms are not fully understood, it appears that a combination of genetic susceptibility and unknown environmental agents may cause autoimmune diseases.

Often the immune systems of individuals with cancer do not function properly for several reasons. For example, in leukemia, lymphocytes become cancerous and increase greatly in number. The cancerous lymphocytes are nonfunctional (as lymphocytes) and often crowd out the normal lymphocytes in the blood and bone marrow. This change leads to a general decrease in immune functions. Cancer cells often break off a tumor and float in the bloodstream to

Focus On

AIDS

Human immunodeficiency virus, or HIV, causes AIDS. HIV primarily affects T_H cells, but it also may infect macrophages and nerve cells. These cells all have a cell-surface protein called CD4 that is the attachment site for HIV. Inside the host cell, a viral enzyme called reverse transcriptase copies viral RNA into DNA, which then may be incorporated into the host genome. Each infected cell can serve as a factory for HIV production. Through budding or death and rupture, each infected cell releases hundreds of viruses, each able to infect other cells. Infection of T_H cells produces a gradual decrease in T_H cells and a weakening of the immune system. This leads to increased susceptibility to infections and cancers. Although HIV can damage organs directly, infections account for up to 90% of the deaths from AIDS.

Individuals are infected with HIV through contact with internal body fluids such as blood and semen. Hemophiliacs and others requiring blood transfusions have been infected by contaminated blood supplies. Health care workers exposed to HIV through accidental cuts or injections have contracted AIDS. Wearing gloves is a standard precautionary procedure that health care professionals must practice (Figure 23.22). Transmission of the virus also occurs when contaminated needles are used to inject intravenous drugs. Infected mothers can transmit HIV to their babies during pregnancy or in their breast milk.

AIDS can be transmitted sexually by both heterosexuals and homosexuals. The only ways to avoid HIV infection are to abstain from drug injection and sexual activity or to limit sexual activity to a monogamous relationship in which both partners have repeatedly tested negative for HIV. High-quality condoms used in combination with gels containing nonoxynol-9 reduce, but do not completely eliminate, the risk of HIV infection. There is no evidence that AIDS can be transmitted through casual contact, such as shaking hands or swimming in the same pool.

Immediately after infection, the incubation period for HIV is 2 weeks to 3 months. Anti-HIV antibodies can be detected in the blood at this time. However, an individual infected with HIV may not develop AIDS symptoms for several months or even up to 12 years. Until very recently, AIDS was invariably fatal. Some new therapies—such as the drug combination called the protease inhibitor cocktail—have provided hope and at least temporary relief from disease progression. However, these medicines are very expensive and need to be taken several times a day.

Tragically, the majority of the AIDS cases in Africa and Asia still remain fatal within a few years of diagnosis. There are two reasons for this. First, effective treatment is not always available and often begins too late. Second, failure to treat the infection early and aggressively gives the virus population time to accumulate mutations. This increased genetic diversity reduces the effectiveness of antiviral drugs.

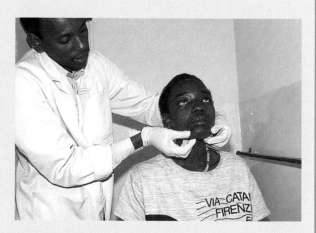

FIGURE 22.22

Medical examination of a potential AIDS patient. A doctor in Nairobi, Kenya, wears gloves for protection while examining a woman who is HIV positive.

other sites, where they develop into secondary tumors (see Focus On Cancer, Section 8.9). Cancers often spread in this way to the bone marrow, where they replace the normal cells and interfere with immune functions.

Disorders of the immune system are the subject of much current medical research. Scientists are finding that autoimmunity and other malfunctions of the immune system are involved in a wide range of diseases. A well-ordered immune system is critical to health.

Check and Challenge

1. How does the immune system eliminate viruses?
2. What is the role of the complement system in the immune response?
3. What are HLA molecules? Why are they important?
4. What role do cytokines play in the immune response?
5. What are the roles of the various types of T cells?
6. What are two ways the immune system can be weakened?
7. Give an example of a hypersensitivity reaction and explain how it occurs.

Chapter
HIGHLIGHTS

Summary

Both nonspecific and specific types of immunity provide protection against hazards from the environment, especially infectious agents. Nonspecific protection against foreign materials includes physical and biochemical barriers. Phagocytosis and inflammation are also nonspecific defense mechanisms. Specific immunity is carried out by lymphocytes and antibodies. The immune response distinguishes between self and nonself. Using a limited amount of genetic information, lymphocytes produce an extremely diverse family of antigen-binding molecules known as B-cell receptors, antibodies, and T-cell receptors.

When foreign material such as a microorganism is detected, the immune system usually eliminates the infectious organism before it can cause disease. If disease does occur and recovery follows, the immune system then provides long-lasting protection against the original foreign material. The immune system sometimes malfunctions, causing diseases and other adverse outcomes, such as allergies, rather than providing protection. Certain pathogens, such as HIV, can adversely affect the cells of the immune system, resulting in a nonfunctioning system.

Key Concepts

Use the concepts below to build a concept map, linking to as many other concepts from the chapter as you can.

- lymphocyte
- nonself
- antigen
- immune response
- nonspecific
- B cells

Reviewing Ideas

1. What is inflammation, and how does it differ from specific immunity?
2. What function is served by lymphoid organs?
3. Give some examples of nonself.
4. What is the evidence that immunity is specific and long-lasting?
5. How does the structure of an antibody promote its function?
6. Explain the differences between the primary immune response and the secondary immune response.
7. Briefly describe the role of each type of cell involved in a successful immune response.
8. What causes autoimmunity?
9. What is the relationship between stress and the immune system?
10. What is the cause of poison ivy symptoms?
11. Why are you more likely to get sick during final exams than summer vacation?

Using Concepts

1. How can a small amount of genetic material give rise to a large number of different antibodies?
2. Explain how vaccination is like an artificial infection.
3. How would you make a vaccine against the flu virus?
4. What would happen if a person were born without a thymus?
5. What is the importance of immune tolerance? What happens if it is lost?

Biology Online BSCSblue.com/vocabulary_puzzlemaker

6. A person is to receive a transplanted kidney. What can be done to increase the chances for a successful transplant in a prospective kidney transplant recipient?

7. The bacterium meningococcus type B causes a serious infection of the brain. Some immunologists think that the membranes of these bacterial cells include structures that are similar to those in the membranes of human brain cells. How might the similarity provide a selective advantage for meningococcus type B?

8. Viruses evolve by changing their surface antigens. Explain how this impacts vaccine development.

Synthesis

1. Humans are born before their immune systems are mature. In what way might that provide a selective advantage?

2. Throughout the 1950s, 1960s, and early 1970s, biologists accumulated evidence that supported the hypothesis that "one gene produces one polypeptide." In other words, one segment of DNA encodes only one polypeptide. How has knowledge about the production of antibodies caused scientists to revise this hypothesis?

Extensions

1. Investigate the mechanisms by which the immune system can produce allergies. Write an essay hypothesizing in what way such a reaction might be helpful.

2. Draw a cartoon of how the cells and molecules of the immune system interact to protect against foreign material.

Web Resources

Visit BSCSblue.com to access
- The basics of immunology
- Detailed discussions of antigen-antibody interactions
- Information about complement factors, viral replication, and the action of poison ivy
- Web links related to this chapter

UNIT

6

Interactions and Interdependence

■ *What environmental factors affect the organisms in this mountain pond and the surrounding landscape?*

■ *What evidence do you see that this ecosystem is changing?*

Organisms have adapted to live in almost every environment on Earth, from the top of the highest mountain to the bottom of the deepest ocean trench. There are algae that grow only in melting snow, fish that survive in lightless caves, and bacteria that thrive in underground petroleum deposits. Other bacteria live only in hot springs, where the water is nearly boiling.

These extreme adaptations demonstrate the importance of the nonliving part of the environment. Organisms also form part of one another's environment. For example, fish must adapt to one another. In order to survive, each species must evolve adaptations that enable it to successfully prey on other fish while avoiding being eaten itself. All the fish in this lake also depend on autotrophs, such as algae and plants, to fix the carbon that feeds the entire ecosystem.

What determines which species are present in an ecosystem? How do the species in an ecosystem interact? What happens to an ecosystem when its environment changes? These are some of the questions that you will explore in this unit.

CONTENTS

LEARNING OUTCOMES

By the end of this chapter you will be able to:

A Relate abiotic factors to the character of ecosystems.

B Describe the relationship between autotrophs and heterotrophs in determining the flow of energy and nutrients in the ecosystem; state the 10% rule of energy pyramids and explain its significance.

C Explain the competitive-exclusion principle and the role of adaptation in reducing competition between species with similar niches.

D Explain how carbon, nitrogen, and water are recycled in an ecosystem.

E Explain how the principle of limiting factors applies to ecosystems and the populations they include.

F Compare and contrast exponential growth and logistic growth.

◀ **About the photo**
This photo shows the
landscape in Owens
Valley, located in the
eastern Sierra Nevada
Mountains.

CHAPTER

24

Ecosystem Structure and Function

■ *Based on the vegetation you see, what inferences can you make about the climate here?*

■ *How have these plants adapted to this environment?*

All life on Earth is interconnected. Populations of organisms constantly interact among themselves and with other populations. They also interact with the nonliving environment. The actions of the organisms affect their environment, and the environment influences the organisms.

These interactions help determine the distribution of organisms throughout the world. Living things must adapt to their environment, or they will not survive. In this way, the environment changes organisms, but organisms also change the environment. The environment and the organisms that live in it complement one another. This complementarity is a product of evolution through natural selection.

In this chapter, you will learn how the environment influences the types of organisms that live in it and how organisms adapt to their environment. You will also explore how all organisms depend on each other and how flow of energy and cycling of matter support the diversity of life on Earth.

The Structure of Ecosystems

24.1 Abiotic Factors

The biosphere is composed of all Earth's ecosystems. It is a patchwork of habitats that differ in abiotic factors such as range of temperature and amount of rainfall and sunlight. Climate and other physical conditions limit the geographic range of many species (Figure 24.1). For example, some bacteria are adapted to live only in hot springs. Their enzymes have evolved to function best at high temperatures. On the other hand, certain Antarctic fish have "antifreeze" molecules in their blood that allow them to tolerate the low temperatures of their environment. Temperature limits the distribution of these organisms.

The availability of water is another abiotic factor that affects the distribution of organisms. Water is essential for life, but often there is too little or too much water for a particular organism. Special adaptations enable some organisms to tolerate extremes of water availability. Ocean fish must resist becoming dehydrated in hypertonic seawater, and freshwater fish must keep their tissues from being diluted (see Section 3.4). Ocean fish depend on active transport to preserve their water balance by removing salts from their tissues. Freshwater fish excrete excess water. Land organisms are adapted to conserve both water and salts. Desert organisms are especially vulnerable to water loss. Cacti, for example, have thick cuticles that minimize the loss of water. Small desert mammals avoid the hot, dry air of the day and become active mostly at night.

FIGURE 24.1

Typical arctic vegetation. What abiotic factors affect these plants and lichens? Why are these plants, and not others, the dominant vegetation here?

Another abiotic factor that influences the distribution of organisms is sunlight, the primary source of energy in most ecosystems. The amount of light available for photosynthesis is usually not an important limiting factor in the productivity of ecosystems on land. In water, however, the intensity and quality of light for photosynthesis decreases with depth. As a result, most photosynthesis in aquatic environments occurs near the surface of the water.

Other abiotic factors that affect land organisms include the physical structure, pH, and mineral composition of soil. These factors limit the distribution of many plant species and the animals that feed on them. For example, many forest soils contain only small amounts of the nitrogen compounds and the phosphate that plants need to synthesize DNA, amino acids, and other essential compounds. Two adaptations that enable forest plants to survive on nutrient-poor soils include symbiotic relationships with fungi, which help to absorb phosphate, and dense networks of shallow roots that absorb nutrients from decaying organic matter in the surface soil before these nutrients wash deep into the soil.

Wind cools land organisms by carrying off their heat energy, and it increases loss of water by speeding up evaporation. Wind can also affect the form and structure of plants. Figure 24.2 shows the effects of wind on the growth of tree limbs.

Fires, floods, avalanches, and other abiotic catastrophes may remove organisms from an area temporarily. Sooner or later, however, the area is repopulated by survivors or by organisms from outside. For example, many of the shrubs in coastal California, where fires are common, survive by storing food in fire-resistant roots, enabling them to resprout after a fire.

CHEMISTRY TIP

Most plants can absorb and use soil nitrogen in the form of nitrate (NO_3^-) or ammonium (NH_4^+) ions. Nitrate is the preferred form; high concentrations of ammonium ions can be toxic to plants. Most mineral soil particles are negatively charged, however, so they bind positive ions, such as ammonium, more strongly than negative ions, such as nitrate. Rainwater easily washes nitrate deep into soils, out of reach of roots.

FIGURE 24.2
Fir trees on Mt. Hood in Oregon. Most of the branches of these trees point in one direction. Think about the different ways that wind could cause this kind of growth.

Plants and Salt Tolerance

A major stress on plants in coastal areas and deserts is the high level of salt (NaCl) in the soil. In such areas, plants face two problems: obtaining enough water from the soil and dealing with potentially toxic levels of sodium (Na^+), chloride (Cl^-), and carbonate (CO_3^{-2}).

Some crop plants, such as beets and tomatoes, are more salt-tolerant than others, such as onions and peas. Some plants can live where salt levels in the soil are extremely high. One plant, saltbush, can tolerate high salt levels because it has specialized cells that concentrate the sodium and chloride it absorbs. Saltbush accumulates salt in epidermal bladders on its surface (Figure 24.3). This keeps internal salt levels low. Salt from leaf tissues is transferred through a small stalk into each bladder cell. As a leaf ages, the salt concentration in its bladders increases. Eventually the bladders burst, releasing salt to the outside of the leaf.

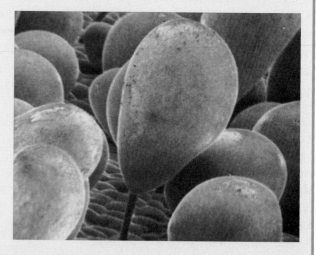

FIGURE 24.3

Adaptation to salty soils. Scanning electron micrograph of salt bladders on the leaf of the saltbush, *Atriplex spongiosa*, ×450. This unusual adaptation enables saltbush to survive in harsh desert environments.

Organisms affect the abiotic environment in various ways. For example, plant roots, lichens, and microorganisms help break down rocks into soil. Dead material and animal wastes contribute organic matter and various nutrients to soil. Roots and organic matter hold soil particles together, reducing erosion. Transpiration (see Section 3.6) from forests increases humidity, which helps form clouds and promotes rainfall.

Various organisms survive and successfully reproduce nearly everywhere in the biosphere. Many tolerate extreme physical conditions. Usually, this involves a trade-off. Adaptation to one environment generally means exclusion from another. For example, desert saltbush cannot compete with other shrub species in moist soils; tall, light-requiring trees, such as oaks, cannot survive in the shade; lake trout would be pickled in seawater; and ocean fish, such as tuna, would drown in a freshwater lake. Thus, the diversity of abiotic environments contributes to the great variety of species discussed in Chapter 18.

24.2 Energy in Food Webs

Try Investigation 24A Producers in an Ecosystem.

All organisms require energy. Autotrophs, or producers (see Section 2.2), obtain energy from photosynthesis or inorganic chemical reactions (see Section 4.8). Consumers, or heterotrophs, acquire their energy second-, third-, or fourth-hand through food webs. Ultimately, photosynthesis determines the amount of energy available for most ecosystems.

How does energy pass through an ecosystem? The nutritional relationships among the producers and consumers in an ecosystem form its

trophic structure (Figure 24.4). This pattern of relationships determines the flow of energy and nutrients in the ecosystem. Producers, such as photosynthetic plants and algae make up the trophic level that supports the system. The other levels include consumers that depend either directly or indirectly on the producers for energy and nutrients. **Herbivores** (animals that consume plants or algae) are the primary consumers. Deer, grasshoppers, and garden snails are examples of primary consumers. At the next trophic level are secondary consumers—**carnivores** (meat eaters, such as wolves and most fish) that eat herbivores. These carnivores may in turn be eaten by other carnivores that are tertiary consumers. For example, hawks are tertiary consumers that prey on birds that eat insects. Some ecosystems have even higher levels of consumers. Finally, decomposers, such as fungi and bacteria, consume organic wastes and dead organisms from all trophic levels.

The main producers in land ecosystems are plants. In most aquatic environments, the primary producers are algae (photosynthetic protists and cyanobacteria). What are some primary and secondary consumers in terrestrial and aquatic ecosystems?

In most ecosystems, bacteria and fungi are the primary decomposers. They secrete acids and enzymes that digest complex organic compounds, such as cellulose. Then they absorb the smaller digested products. Earthworms and scavengers, such as cockroaches, are two examples of animal decomposers.

ETYMOLOGY

troph- = feed or nourish (Greek)

herb- = plant (Latin)

carni- = flesh (Latin)

omni- = all (Latin)

-vore = devourer (Latin)

The **trophic** structure of a community is a description of feeding or nutritional relationships among organisms in the community. It includes plant-eating **herbivores,** meat-eating **carnivores,** and **omnivores** that consume both animals and plants.

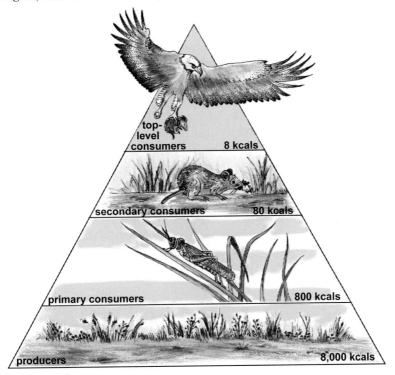

top-level consumers 8 kcals

secondary consumers 80 kcals

primary consumers 800 kcals

producers 8,000 kcals

FIGURE 24.4

Ecosystem trophic structure. An energy pyramid, one way of representing the trophic structure of an ecosystem. In this example, the top level is a tertiary consumer. In this idealized energy pyramid, 10% of the energy at each trophic level is available at the next higher level. In actual ecosystems, the productivity varies somewhat depending on the organisms involved in the food chains.

Food webs can be very complex. Several species of consumers may feed on the same species, and one species may feed on several species. For example, many fish species consume squid. Most of these species eat other fish too. A consumer may prey on both primary and secondary consumers as well as on decomposers. Robins, for example, consume both plant-eating insects and decomposers, such as earthworms. Omnivores, such as humans, eat a variety of producers as well as consumers from different levels.

The diagram of trophic structure in Figure 24.4 is called an **energy pyramid.** The area of each level roughly represents the quantity of energy it receives from the level below. The primary producers are the foundation. In many ecosystems, each level receives about one-tenth of the energy of the level below it. This concept is often referred to as the **10% rule.** Typically, organisms use most of their energy in metabolism or movement or release it as heat energy.

Just as available energy declines at higher trophic levels, so does **biomass,** the total quantity of living matter at each trophic level. For example, the total mass of all the wolves in a forest is much less than the mass of rabbits and other prey or the mass of trees and other plants. **Productivity**—the rate at which new biomass forms—is also highest among primary producers and declines at higher trophic levels. An ecosystem can sustain far fewer top-level carnivores than low-level consumers and producers. Only about one-thousandth of the chemical energy synthesized by the producers can flow all the way through a three-consumer food chain to a top predator, such as a hawk.

The energy pyramid also has implications for humans. Meat and other animal products are concentrated sources of protein and energy. However, animal foods provide only a small part of the energy fixed by photosynthesis. For this reason, they form a large part of the human diet mainly in places where food is relatively plentiful, such as the United States. Animal foods are also important in dry or rocky places such as the Middle East. These regions are better suited to raising grasses and other animal feeds, rather than grains, fruits, and vegetables that require more water.

24.3 Relationships in Ecosystems

Try Investigation 24B 🔹 Relationships between a Plant and an Animal.

CONNECTIONS
Predation contributes to evolution by favoring predators that are well adapted to capture and consume prey and by favoring prey that have effective defenses against predators.

The most obvious relationship between two species in an ecosystem is **predation,** in which a predator species consumes a prey species. Predation moves energy and nutrients through food webs. When two species consume the same prey or other resource, they compete with each other. Competition is an important force in natural selection. The predator that catches more prey or digests it more thoroughly, the fungus that absorbs nutrients more efficiently, and the plant that overshadows its neighbors and receives a greater share of sunlight all have advantages that make them more likely to survive and reproduce.

The particular combination of resources that a species is adapted to exploit is called its **niche.** For example, a niche is available in many forests for an animal that can live in the treetops and eat fruit. In some forests, various species of birds fill this niche. In others, it is filled by monkeys. The more closely two species resemble one another in their patterns of resource use, the more they compete with each other. The **competitive exclusion principle** states that two species cannot occupy the same niche in the same ecosystem for long. The species that competes more successfully for food, space, or other resources eventually displaces its competitors. Adaptive radiation helps reduce competition (see Section 19.5). Adaptations that enable competing species to divide or share resources also reduce competition in species with similar niches. For example, hawks and owls hunt for many of the same species, but hawks hunt mostly by day and owls hunt mostly at night.

Three types of interactions between species are grouped under the term *symbiosis* (Figure 24.5). All involve relationships in which species live together, usually in physical contact. **Mutualism** (Figure 24.5a) is the most familiar of the three. Both species benefit in a mutualistic relationship. For example, plants of the legume (bean) family form mutualistic relationships with bacteria of the genus *Rhizobium*. The bacteria live in nodules that form on the roots, where the plant provides them with carbohydrates formed during photosynthesis. The bacteria in turn convert nitrogen gas (N_2) to nitrogen compounds that the plant can use to synthesize amino acids.

Parasitism (Figure 24.5b) is a second type of symbiosis in which one organism gains at the expense of the other. A parasite lives in or on a host that provides it with nutrients or other resources. The host is the loser in this relationship. All disease-causing microorganisms are parasites. Mistletoe is a parasitic plant that does not root in soil. Instead, it grows into the xylem tissue of a tree and draws water and dissolved soil-mineral nutrients from its host. Viruses are parasites too, although they are not true organisms.

b

a

c

FIGURE 24.5

Three types of symbiosis. (a) This kudu, *Tragelaphus strepsiceros,* and these oxpecker birds, *Buphagus africanus,* have a mutualistic relationship. The birds eat parasitic insects that attack the kudu. **(b)** This tomato hornworm, *Manduca quinquemaculata,* is covered with the parasitic young (pupae) of a braconid wasp (family Braconidae). **(c)** Some species of birds, such as these mistle thrushes, *Turdus viscivorus,* have a commensal relationship with the trees in which they nest. If the birds do not eat part of the tree (such as its seeds) or animals that prey on the tree (such as leaf-eating insects), then their nest may have little or no effect on the tree.

com- = together (Latin)

mensa = table (Latin)

A **commensal** organism receives a benefit from another organism without helping or injuring it. The two organisms "eat at the same table" but do not consume each other's resources.

In **commensalism** (Figure 24.5c), the third type of symbiosis, one organism benefits and the other is unaffected. True commensalism is rare. For example, algae and barnacles that attach to whales benefit by being carried to new environments. They slightly harm the whales that carry them by increasing drag and turbulence, which increases the amount of energy the whales must use to swim. The arctic fox and polar bear are closer to true commensals. The foxes follow the bears, eating the scraps that the bears leave behind. However, the bears occasionally eat the foxes, so their relationship also includes predation.

Close contact between symbiotic organisms can lead to an occasional exchange of DNA. For example, many parasites have acquired genes from their hosts. Chloroplasts and mitochondria may be thought of as symbionts that have lost their individuality as organisms (see Section 17.7). Many of their proteins and other components are products of nuclear genes. Some of these genes are thought to have migrated from these organelles to the nucleus early in the history of this symbiosis.

Check and Challenge

1. How do abiotic factors affect the adaptations of organisms? Give an example not mentioned in this chapter.

2. Explain why organisms that are able to resist the stresses of desert life are rare in milder environments.

3. Would you expect most ecosystems to contain more species of primary consumers or secondary consumers? Explain your answer.

4. What relationship is represented by drawing each level of an energy pyramid smaller than the one below it?

5. Give an example of a parasitic relationship and a mutualistic relationship not described in the text. Explain how these relationships are parasitic and mutualistic.

Ecosystem Dynamics

24.4 Nutrient Cycles

CHEMISTRY TIP

The carbon, oxygen, and nitrogen cycles are global oxidation-reduction systems.

Chemical elements in ecosystems are limited, and those essential for life must be recycled. Organisms absorb and release carbon, oxygen, and nitrogen as gases (CO_2, O_2, and N_2). As a result, these elements move through large-scale global cycles. Solid elements, such as phosphorus (P) and sulfur (S), are less mobile. They do not move far from where they originate. The soil is the main abiotic reservoir for those elements.

Three of the most important nutrient cycles are the carbon, nitrogen, and water cycles. In the carbon cycle, carbon fixation and cell respiration move

carbon through the biosphere. Photosynthetic autotrophs absorb carbon dioxide from the atmosphere and reduce it to form organic compounds. Some of the carbon remains in the bodies of producers and consumers. Cell respiration by producers, consumers, and decomposers returns the rest to the air as carbon dioxide. When fossil fuels such as oil, coal, and gas are burned, the carbon stored in these materials is released into the air as carbon dioxide. Figure 24.6 summarizes the carbon cycle.

All photosynthetic autotrophs can fix carbon, but most organisms cannot use nitrogen gas (N_2). Only certain prokaryotic soil bacteria can convert the nitrogen in the atmosphere to ammonia (NH_3), a process called nitrogen fixation. Other bacteria convert the ammonia to nitrite (NO_2^-) or nitrate (NO_3^-). This process is called nitrification. Plants can absorb either ammonia or nitrate from soil or water and use it to synthesize nitrogen-containing compounds, such as amino acids and nucleotides. Consumers and decomposers, on the other hand, must obtain their nitrogen from other organisms. Finally, denitrifying bacteria convert nitrate to nitrogen gas, completing the cycle. Figure 24.7 summarizes the nitrogen cycle.

Figure 24.8 shows the water cycle in an ecosystem. Plants absorb water from the soil. Land animals and other consumers and decomposers absorb water from their food, or they drink it directly. Aquatic organisms are constantly bathed in water. Water returns to the atmosphere through transpiration, cell respiration, and evaporation, mostly from the oceans. Eventually much of the atmospheric water vapor condenses and returns to the system as rain or snow.

CHEMISTRY TIP

The bacterial enzyme nitrogenase catalyzes the fixation reaction in which nitrogen gas is reduced. The activity of this enzyme is easily measured because the triple bond in the N_2 molecule is similar to the one in the gas acetylene (C_2H_2). When nitrogen-fixing bacteria are provided with acetylene, they reduce it to ethylene gas (C_2H_4), which can be measured chemically.

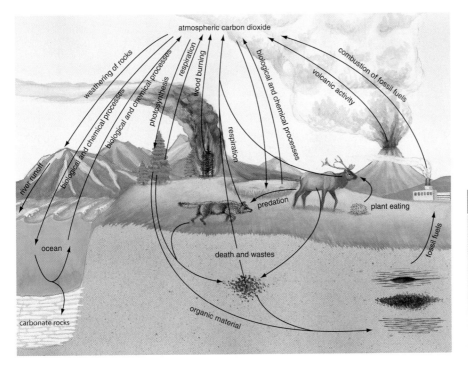

FIGURE 24.6

The carbon cycle. Both living organisms and the physical environment contribute to the carbon cycle. The influence of humans is shown here by fuel combustion at the factory.

CONNECTIONS

The law of conservation of matter limits the supply of nutrient elements to an ecosystem. The flow of any element into an ecosystem is equal to the sum of the quantity it extracts from abiotic sources and the quantity it receives from other ecosystems minus the quantity it releases to the environment.

FIGURE 24.7

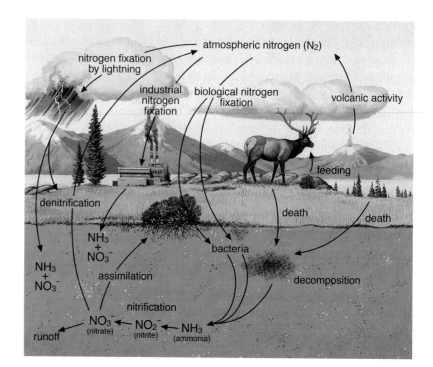

The nitrogen cycle. Oxidation of nitrogen gas by lightning adds significant quantities of nitrate to soil and water. In much of the world, nitrogen fixation in fertilizer factories is a major source of nitrate and ammonia. The unintended oxidation of atmospheric nitrogen gas that occurs when fossil fuels are burned also contributes large quantities of nitrate.

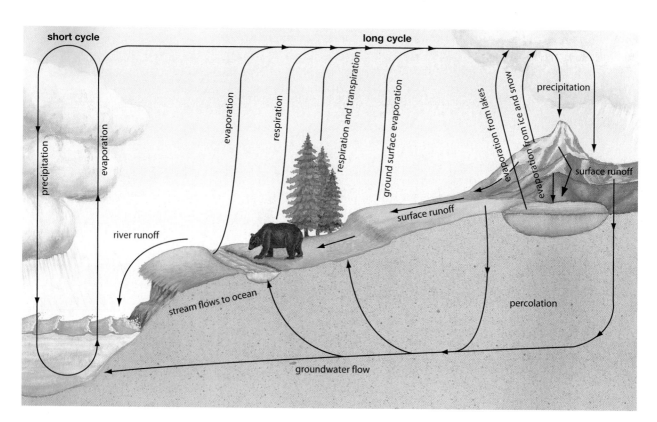

FIGURE 24.8

The water cycle. Water is the most plentiful substance in the tissues of organisms. Most evaporation occurs from the surface of the oceans. Rainwater that percolates (drips) into soil can remain there as groundwater or join streams and rivers that carry it to the sea. The decomposition of water in photosynthesis balances its synthesis in cell respiration.

24.5 Limiting Factors

In Section 4.5, you learned about the principle of limiting factors, the idea that the supply of light, water, or other needs can limit the rate of a process such as photosynthesis. This principle also applies to ecosystems and the populations they include. For example, in one series of experiments, scientists added a soluble iron compound to the surface water of a patch of ocean. Soon after this, they observed a sharp increase in the productivity of the ecosystem in that area. The experimenters interpreted this result as evidence that the supply of iron limits the productivity of ocean ecosystems. The algae that are the producers in this ecosystem grow and reproduce rapidly when they receive plenty of this nutrient.

Various factors limit the productivity of other ecosystems. As in the ocean, limiting factors affect terrestrial ecosystem productivity mostly by acting on producers. The supply of water, for example, limits plant growth in deserts. After a heavy downpour, desert plants grow rapidly for a short time (Figure 24.9). Populations of consumers and decomposers rise and fall quickly in response to this brief, explosive plant growth. In grasslands, the limiting factor is often the supply of soil nitrogen (nitrate or ammonium ions). In arid regions, both water and nitrogen may limit grassland productivity.

The productivity and population size of various species in the same ecosystem can be limited by different factors. For example, water, soil pH, or mineral nutrients can limit the growth and number of various plants. At the same time, the population size of a species of bird in the same ecosystem may be limited by the productivity of the insect species that it eats or the size of the population of the trees in which it nests. In turn, the insect population size may be limited by temperature.

a

b

FIGURE 24.9

Water as a limiting factor in a desert ecosystem (Death Valley, California). (a) Most of the time, the water supply severely limits the productivity of the desert ecosystem. **(b)** Desert gold *(Geraea canescens)* blooms following an abundant rainfall. After brief but heavy rains, the desert fills with this kind of vegetation, mostly short-lived plants that germinate, flower, produce seeds, and die following the short rainy period.

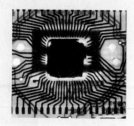
Robots for Ecology

Many ecosystems contain an enormous number of species. For example, tropical rain forests probably contain millions of species. For an ecologist, identifying all the species in an ecosystem and determining how they interact with each other can be an overwhelming task. The job is even more difficult in the ocean, where organisms and processes near the surface affect those at deeper levels and the whole system spreads over millions of square kilometers of moving water.

For many years, scientists have used instruments on floating buoys scattered over the ocean to record changing air and water temperatures, winds, and other data. During the 1960s and 1970s, many of these sensors were connected to battery-powered radio transmitters. This enabled scientists to collect data without having to visit each buoy. Now a variety of new tools is dramatically increasing our knowledge of the ocean.

In 1990, scientists began placing a fleet of robotic data collectors in the oceans (Figure 24.10). These probes automatically sink and ascend in the upper 2,000 m of water on a programmed schedule, measuring temperature and salt concentration. These factors affect the distribution and growth of many species and reveal the presence of currents. As the robots drift with the currents, they periodically come to the surface and transmit their data to satellites.

Recent advances in microelectronics, extended-life batteries, and materials that resist corrosive seawater and survive the crushing pressure of the ocean depths are making more ambitious robots possible. Off the coast of New Jersey, the Long-term Ecosystem Observatory known as LEO-15 sits under 15 m of water. LEO-15 carries sensors that record chemical, physical, and biological activities in the surrounding water. Small, mobile, self-guiding robots called AUVs (autonomous underwater vehicles) travel through the surrounding water surveying the environment. From time to time, they dock at LEO-15 and download their data.

LEO-15 and other underwater robotic observatories have detected the upwelling of cold, nutrient-rich water from near the seafloor to the surface. This commonly occurs near the coast in summer when winds blow warmer surface water out to sea. The increased supply of mineral nutrients in the surface water often increases the growth of algae, leading to algal blooms or red tides. Data collected by robotic ocean observatories have helped scientists understand the factors that determine whether or not upwelling will produce an algal bloom.

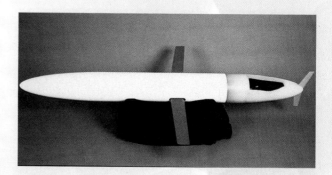

FIGURE 24.10

An ocean robotic data collector. This newly developed six-foot-long glider is a remote-controlled, self-propelled vehicle that sails across the ocean at various depths, using underwater currents and its own changing buoyancy for propulsion. This glider, called Spray, uses satellite relays to report its position to shore. Spray was developed jointly at Scripps Institution of Oceanography in California and Woods Hole Oceanographic Institution in Massachusetts.

24.6 Population Dynamics

Several environmental factors usually limit the population of each species in an ecosystem. What would happen if those limitations were removed? Imagine a single bacterial cell in a large container of nutrient medium. If the cell and its descendants divide every 30 minutes, how many cells would you expect to find in the container in 24 hours? Under ideal conditions, such a culture could produce more than 280 trillion bacteria. How can that be? The number of cells doubles every 30 minutes, producing 2 cells in the first half hour, $2^2 = 4$ in the second, $2^3 = 8$ in the third, and so on. Because the number of organisms doubles every half hour, their population size is equal to 2^n, where the exponent n is the number of half hours that have passed. You can use a calculator to check the number of cells produced in 48 half hours. This is an example of **exponential growth** (Figure 24.11a).

Why hasn't exponential growth resulted in enormous numbers of organisms that completely overwhelm Earth's resources? In reality, populations rarely grow exponentially. As a population grows, its **population density**—the number of individuals per unit of land area or water volume—increases as well. Competition within the population increases as nutrients and other resources are used up. Their toxic metabolic wastes may accumulate in the environment. Predators and parasites may become common. All of these factors reduce reproduction and increase the death rate. This slows population growth. A population that develops in a new environment may begin to grow exponentially, but it soon slows to a linear growth pattern and eventually approaches a stable maximum. This type of pattern is called **logistic growth.** Figure 24.11b shows the typical S-shaped logistic-growth curve.

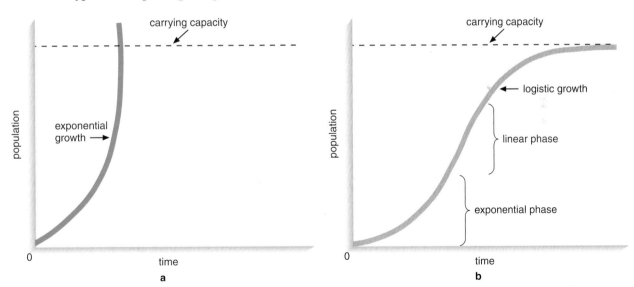

FIGURE 24.11

Two types of population-growth curves. The exponential curve **(a)** represents a population's maximum growth rate under ideal conditions. Usually, population growth follows a logistic curve **(b)**, it slows down as competition and other factors limit reproduction and survival in an increasingly dense population. Eventually reproduction and death reach similar rates, and the population stabilizes.

Notice that the logistic curve approaches a stable maximum. This maximum is interpreted as the **carrying capacity,** the largest population of a species that the environment can support. Carrying capacities are not fixed. They can vary as a result of changes in food supply or other requirements. For example, the invention of artificial fertilizers enabled farmers to increase the population size of crop species. The increased food supply led to an increase in Earth's carrying capacity for humans. What factors other than food might affect carrying capacity?

Populations do not always approach their carrying capacities as smoothly as Figure 24.11b suggests. In species that reproduce rapidly, exponential growth can lead to a brief period when the population exceeds its carrying capacity. This is followed by a rapid decline in reproduction and a sharp increase in the death rate (Figure 24.12a), events that cause the population to crash. Such a pattern is referred to as a **boom-and-bust cycle.**

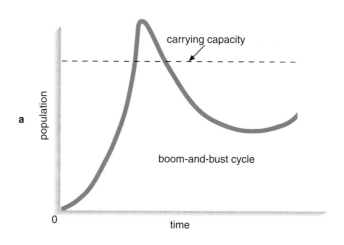

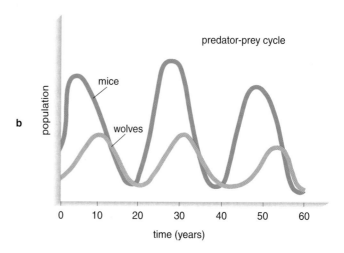

FIGURE 24.12

Changes in populations near carrying capacity. Rapidly growing populations can exceed their environment's carrying capacity, leading to a disastrous boom-and-bust cycle **(a).** Cyclic variation in prey population is followed by a similar variation in the predator population in a predator-prey cycle **(b).**

Variation in the population size of one important species can ripple through an entire ecosystem. For example, year-to-year variation in weather may affect the growth of grasses and their production of seeds. This leads to a cycle of rising and falling numbers of mice that eat the seeds. The populations of owls and wolves that eat the mice rise and fall in response to the changes in the mouse population. The cycles of these predator populations lag behind the cycle of the mouse population in a **predator-prey cycle** (Figure 24.12b).

Populations are always changing. This is typical of ecosystems, which are complex, constantly fluctuating systems. Because ecosystems are so complex, discovering the causes of these changes is often extremely difficult.

Check and Challenge

1. How does carbon cycle through the biosphere?
2. How can abiotic factors in an ecosystem affect the water cycle? Give an example.
3. What is the connection between the idea of limiting factors and the idea of carrying capacity?
4. Can the different limiting factors, such as water and light, influence one another? Explain your answer and give an example.

Chapter HIGHLIGHTS

Summary

Organisms are adapted to the abiotic factors that shape their ecosystem. Temperature, light energy, and water are the most important factors in determining the character of an ecosystem. The organisms' success in surviving to reproduce depends on their tolerance of the prevailing abiotic factors.

Ecosystems consist of complex networks of relationships among species. These relationships include food webs, mutualism, commensalism, predation, and parasitism.

Energy flows and chemicals cycle through ecosystems. Producers capture light energy. Only about 10% of that incoming energy passes from one trophic level to the next. Therefore, just a small percentage of the light energy is available for consumers at higher levels. Water and compounds of nitrogen, carbon, and other essential elements cycle through an ecosystem. The chemicals may change form during the cycle, but ultimately they are returned to the atmosphere or the soil to begin the cycle again.

Various biotic and abiotic factors can limit the size and growth rate of a population. Population size can vary in a cyclic way in response to these factors. Rapidly growing populations can briefly exceed their environment's carrying capacity until these limiting factors produce a population crash.

Key Concepts

Construct two concept maps, one for the cycling of matter and one for population growth. Try to find a way to link the two together.

Reviewing Ideas

1. Describe how abiotic factors affect biotic factors in an ecosystem. Give examples.
2. Where are the autotrophs found in an energy pyramid? the heterotrophs?
3. What are decomposers, and how do they affect the recycling of materials in an ecosystem?
4. Compare the three types of symbiosis and give an example of each.
5. Why are nitrogen-fixing bacteria important in ecosystems?
6. Explain how climate affects the adaptations of terrestrial animals to ecosystems. Give examples.
7. What is the 10% rule, and how does it affect productivity? What are the implications of this rule for human populations?
8. What is a limiting factor?
9. Why are infectious diseases often common in crowded cities and refugee camps?
10. Describe an example of how a predator-prey cycle might develop. (Do not use the example in Section 24.6.)

Using Concepts

1. Describe some relationships among species in a particular ecosystem.
2. Some biologists are trying to transfer the genes needed for nitrogen fixation to crop plants. Use the concept of limiting factors to predict how this would affect the nitrogen cycle.
3. Explain why different ecosystems have different plants and animals.
4. Why is energy said to *flow through* an ecosystem, whereas chemicals *cycle?*
5. Explain how trade-offs are involved in an organism's ability to tolerate extreme environments. Give some examples.
6. Nonnative species that people bring into new environments sometimes grow and reproduce rapidly, endangering or displacing native species. Use your knowledge of limiting factors and population growth to explain how this happens.
7. Use the concept of limiting factors to explain the control of household insect pests.

Synthesis

1. Explain how evolution through natural selection affects the plants and animals in different abiotic environments.
2. What might be the major limiting factor for the world's human population? Justify your answer.
3. Each species can survive in a certain range of conditions. How must this statement be modified in applying it to humans?

Extensions

1. Much of the tropical rain forest in South American countries is being clear-cut for farming and other uses. Some of the farm families have no other source of income. The resulting loss of habitat is causing many rain forest species to become endangered or extinct. What are the likely results if the rain forests disappear? What are the likely results of a complete stop to rain forest clearing? What is the best policy for use or protection of the rain forests? Choose a position and write an essay defending your position.

2. Sketch a food web for a specific environment. Be sure to include as many decomposers, producers, and different-level consumers as you can. If you would like to draw the organisms of the ecosystem, library reference books have photos you can copy.
3. Describe an imaginary new ecosystem with particular limiting factors. Create and sketch an organism that has specific adaptations to that environment.

Web Resources

Visit BSCSblue.com to access
- An interactive mathematical model of population growth
- Web links related to this chapter

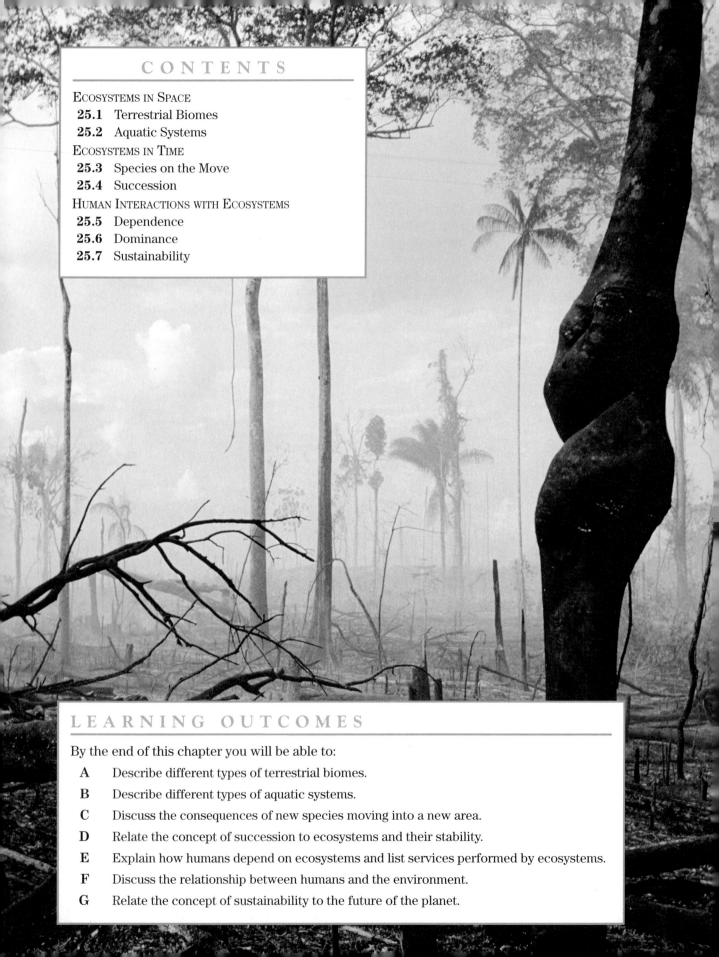

CONTENTS

LEARNING OUTCOMES

By the end of this chapter you will be able to:

A   Describe different types of terrestrial biomes.

B   Describe different types of aquatic systems.

C   Discuss the consequences of new species moving into a new area.

D   Relate the concept of succession to ecosystems and their stability.

E   Explain how humans depend on ecosystems and list services performed by ecosystems.

F   Discuss the relationship between humans and the environment.

G   Relate the concept of sustainability to the future of the planet.

CHAPTER

25

Change in Ecosystems

■ *How are forests important to humans?*

■ *What will happen to this site in the next several years?*

Earth is covered with ecosystems, ranging from simple fields of wheat or corn to great complex oceans. There are many kinds of ecosystems, and if we could observe the history of any landscape, we would see that the ecosystems on that site have changed and continue to change over time. Species come and go as events—sometimes the species themselves—disturb the equilibrium of an ecosystem.

The one species with the greatest impact on ecosystems is us—humankind. Now more than 6 billion in number, humans interact with ecosystems in two basic ways: through dependence—our absolute reliance on the resources in ecosystems—and through dominance—our exploitation of ecosystems, in which we change them (often for the worse). Our future depends on whether or not we learn to live sustainably on Earth. To do this, we will need to take care of the natural world so that it can continue in time and space, in all of its diversity and complexity.

This last chapter in the book deals with ecological change—change in ecosystems in space and time, change in populations and species, and especially, change brought about by human behavior.

Ecosystems in Space

25.1 Terrestrial Biomes

ETYMOLOGY

bio- = life (Greek)

-ome = group (Latin, Greek)

The word **biome** describes a particular geographical/climatic region and the large community of plants and animals living there.

The 25% of Earth's surface that is above water consists of regions with different physical environments and vegetation. These areas are called **biomes.** Biomes are groups of similar ecosystems, usually defined by the most conspicuous types of vegetation—those plants that are the largest or most numerous and convert the most energy from the Sun. The main climatic factors affecting the growth of vegetation, and thus the geographic distribution of biomes, are the temperature and precipitation at different altitudes and latitudes. In terms of temperature, increasing altitude up a mountain is similar to increasing latitude, or distance from the equator. Where these climatic factors are similar, they will support the same kind of biome even if they are on different continents, although the actual species may be quite different. The boundaries of biomes are usually indistinct because changes in climate from place to place are gradual. Within biomes, differences

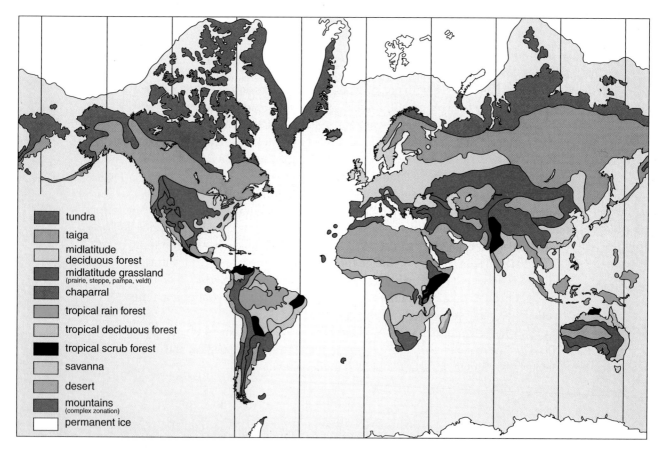

tundra

taiga

midlatitude deciduous forest

midlatitude grassland
(prairie, steppe, pampa, veldt)

chaparral

tropical rain forest

tropical deciduous forest

tropical scrub forest

savanna

desert

mountains
(complex zonation)

permanent ice

FIGURE 25.1

Major biomes of Earth. Many parts of Earth have not been thoroughly studied for accurate biome classification. Even where observations are plentiful, ecologists sometimes disagree about the correct interpretation.

in soil, topography, and local disturbances result in a patchwork of interacting ecosystems, referred to as a landscape. Figure 25.1 shows the distribution of major biomes. Refer to the figure as you read about several biomes.

One type of biome is the **tropical rain forest.** This biome (Figure 25.2a) occurs on lowlands near the equator, where sunlight is most intense and rainfall (250 cm or more per year) and evaporation load the air with moisture. Tropical rain forests are found throughout Central America, along the Amazon basin of Brazil, on the Ivory Coast of Africa, and in central Africa and eastern Madagascar. Tropical rain forests characterize the western coast of southern India, the island of Ceylon, Thailand, and Indonesia.

Very little light reaches the forest ground because the tall, dense canopy of foliage blocks the light. The low light intensity on the forest floor limits the growth of vegetation there (Figure 25.2b). These forests are the most complex of all communities and support more species of plants and animals than any other biome. Rain-forest soils are often poor in organic matter and plant nutrients (such as nitrogen and phosphorus); the nutrients are all tied up in the rich biomass. Deforestation has removed more than half of Earth's original rain forests through conversion to grazing lands or crops.

The **savanna** is a tropical or subtropical grassland with scattered trees and woodlands (Figure 25.3). These grasslands are usually in areas with low or highly seasonal precipitation (30 cm or less), generally in the interior of continents. Savanna vegetation characterizes the high plateau of Brazil, much of Central Africa, western Madagascar, and a narrow belt of northern Australia.

a

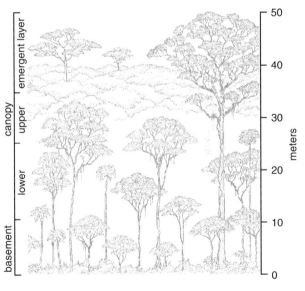

b

FIGURE 25.2

Tropical rain forests. (a) This rain forest on the Big Island of Hawaii includes tree ferns, *Cibotium glaucom,* and ohia lehua trees, *Metrosideros polymorpha.* **(b)** A tropical rain forest is made up of layered canopies. Because virtually no sunlight reaches the rain forest floor, the basement region, few organisms are able to live on the ground. Many more are found in the trees, in the canopy formed by leafy branches competing for sunlight. Note that the emergent layer consists of the tallest trees that have broken through the canopy and receive unrestricted light.

FIGURE 25.3

A savanna biome. Giraffe, *Giraffa camelopardalis,* and gemsbok, *Oryx gazella,* share a savanna watering hole in Kalahari Gemsbok National Park, South Africa.

The savanna has three distinct seasons: cool and dry, hot and dry, and warm and wet. Savanna soils are richer in plant nutrients than those of the rain forest and are often capable of supporting agriculture or grazing.

The savanna is home to some of the world's largest herbivores. These include elephants, giraffes, rhinoceros, and kangaroos, as well as many smaller species of grazers, such as antelope. Because of poaching and human settlement, many of these herbivore species and their predators (lions, cheetahs, leopards, and hyenas) are experiencing serious population declines.

Deserts often have severely hot and dry climates, in which annual precipitation is less than 25 cm and temperatures may reach 54°C (Figure 25.4a). The hot, dry desert is found in northern Africa (the Sahara) and much of the Arabian Peninsula, as well as across central Australia. North American examples are the Chihuahuan Desert of southern New Mexico and northern Mexico, the Sonoran Desert of Arizona, and the Yuman and Colorado Deserts of southern Arizona and southern California.

There are also cool deserts, where winter temperatures are very cold (Figure 25.4b). Cool deserts are typified by the Great Basin Desert of Nevada and Utah and the Painted Desert of the Four Corners region (Utah, Colorado, Arizona, and New Mexico). Both types are characterized by a lack of precipitation.

The driest deserts have almost no **perennial** vegetation (plants that live more than 1 year). In less arid areas, the predominant producers are usually

a

b

FIGURE 25.4

Two kinds of deserts. (a) The hot Sonoran Desert of Arizona; **(b)** a cool desert scene from northeastern Nevada. Plants of the hot deserts tend to be succulents and cacti that have leaves reduced to spines which minimize moisture loss and stems that are green to conduct photosynthesis. These plants leaf out and flower in response to rain, and the stems have evolved to store water. Plants of the cool deserts tend to be widely and evenly spaced in their environment. Able to withstand cold, snowy winters, these plants are low-growing, compact shrubs with small leaves.

scattered shrubs, which are interspersed with cacti or other **succulents** (plants having thick leaves with large cells able to store water). Growth and reproduction in the desert are stimulated by brief heavy rains, which result in spectacular blooms of plants. Desert plants have physiological and structural adaptations to extreme drought and temperatures. For example, many of them have small leaves and thick cuticles that reduce water loss (Figure 25.5). The special adaptations of CAM (crassulacean-acid metabolism) plants are discussed in Section 4.6.

Seed-eating herbivores, such as ants, sparrows, and rodents, are common in deserts. Lizards and snakes are important predators. Like the desert plants, many of these animals are adapted to living with extreme temperatures and little available water. Many desert animals live in burrows and are active only at night. The kangaroo rat, which lives in the American Southwest, has a remarkable adaptation of surviving without drinking any water (Figure 25.6). What adaptations do you think might enable the kangaroo rat to survive in this manner?

Midlatitude coastal areas near cool ocean currents experience mild, rainy winters and long, hot, dry summers. These areas include **chaparral,** or brushland communities (Figure 25.7). This biome is characterized by dense, spiny shrubs with tough leaves, often coated with a thick cuticle that inhibits moisture loss but increases the plants' flammability. Chaparral vegetation occurs along the coastlines of California, Chile, southwestern Africa, southwestern Australia, and the Mediterranean Sea. Plants from these regions are unrelated, but they resemble one another in their appearance and adaptations.

The organisms of the chaparral are adapted to and maintained by periodic fires. Many of the shrubs have large, deep root systems that store carbohydrates and mineral nutrients. These reserves enable the plants to regenerate quickly after fires. Other chaparral species produce seeds that germinate only after a hot fire. Deer, fruit-eating birds, rodents, lizards, and snakes are characteristic chaparral animals.

FIGURE 25.5

A plant adapted to desert conditions. The pleated structure of the trunk of the saguaro cactus, *Cereus giganteus,* allows the plant to expand or contract as it gains or loses water.

FIGURE 25.6

An animal adapted to desert life. The Ord's kangaroo rat, *Dipodomys ordii,* of the Mojave, Colorado, and Yuman Deserts, can survive without drinking any water.

FIGURE 25.7

A chaparral landscape in California, showing firebreaks on ridges. These dense stands of spiny shrubs have evergreen leaves that burn easily. Many of these plants grow back quickly from their roots after periodic fires destroy invading trees.

CONNECTIONS

The development of similar adaptations in unrelated species is an example of convergent evolution.

FIGURE 25.8

Adaptations to predation. The white-tailed jackrabbit, *Lepus townsendii,* a grassland herbivore that survives by being swift. Coyotes, *Canis latrans,* are their primary predators.

Temperate grasslands, such as the prairies of North America, are similar to savannas, except that prairie trees are found only near streams. These grasslands typically receive 25–75 cm of annual precipitation, have low winter temperatures, and develop rich, deep soils. Drought and periodic fires keep woody shrubs and trees from invading the environment.

Grasslands expanded in areas when climates became warmer and drier after the last ice age about 10,000 years ago. Grazing animals, such as the bison and pronghorns of North America, also increased in population. Because of the openness of the grasslands, speed and endurance are important adaptations that enable many animals to escape predators (Figure 25.8). Small mammals such as prairie dogs seek shelter below ground. Three distinct types of grasslands are native to North America: the tall-grass prairie of the northern Midwest, the short-grass prairie of the area east of the Rocky Mountains to about Kansas, and the mixed-grass prairie of the central Great Plains (Figure 25.9). Because of their rich soils

b

a

c

FIGURE 25.9

Three major types of prairies. North American temperate grasslands. **(a)** The tall-grass prairie of the northern Midwest is characterized by a predominance of big bluestem along with other grasses and herbaceous plants. **(b)** Pawnee National Grassland (northeast of Denver, Colorado) is covered by short-grass prairie and patches of yucca, *Yucca flamentosa.* **(c)** The mixed-grass prairie stretches across portions of South Dakota.

and favorable climate, these prairies have been converted almost entirely to agricultural use. Cattle ranching and raising crops have made the American prairies one of the world's largest food-producing regions.

Temperate forests can include both deciduous and coniferous trees. **Temperate deciduous forests** develop throughout the midlatitude regions, where there is enough moisture to support large trees (more than 75 cm per year). Temperatures range from −30°C in the winter to +35°C in the summer, with a 5- to 6-month growing season. Rain and snowfall are abundant, humidity tends to be high, growth and decomposition rates are high, and the soil is fertile. Although they are more open and not as tall as tropical forests, temperate forests also have several layers of producers, usually short herbs, intermediate shrubs, and tall trees. These forests are found in eastern North America, western and central Europe, eastern Asia, and on the eastern coast of Australia extending to Tasmania.

Temperate forests support diverse animal species (Figure 25.10). A large variety of organisms, such as beetles, sow bugs, and leafhoppers, live in the soil and leaf litter or feed on the leaves of trees and shrubs. The forest is home for many species of birds and small mammals. In some forests, larger predators such as wolves and mountain lions can still be found. Like the grasslands, temperate forests have soil and climate suitable for agriculture. The temperate forests of the Northern Hemisphere have been heavily cleared for this purpose.

The northern coniferous forest, or **taiga,** extends across northern North America, Europe, and Asia north to the tree line (beyond which no trees can grow). Taiga is also found at high elevations, such as the Rocky Mountains (Figure 25.11). It is characterized by harsh winters and short, cool summers. Snow is the major form of precipitation, but this water is not available to plants until the spring thaw.

a

b

FIGURE 25.10

Animals of the temperate forest. (a) The pileated woodpecker, *Dryocopus pileatus,* lives in a wide range from the eastern to northwestern United States. (b) The red fox, *Vulpes vulpes,* lives in the same forests.

FIGURE 25.11

Taiga, or conifer forest, in the Rocky Mountains. The northern coniferous forests and high-altitude coniferous forests of all continents make up the taiga biome, in which ponds, lakes, and bogs are abundant.

FIGURE 25.12

Flowering plants of the tundra. These bearberry *(Arctostaphylos alpina)* and lichens in Denali National Park, Alaska, tend to grow in low, spreading mats. Think about environmental conditions that lead to this form of growth.

 **CHEMISTRY TIP**

Oxygen diffuses from air to plant roots. In wet soils, the presence of water greatly slows the process of diffusion, allowing only roots near the surface of the soil to obtain enough oxygen to survive.

The taiga is dominated by a dense canopy of conifers that absorbs most of the sunlight. Only mosses, lichens, and a few shrubs can grow in the dim light near the ground. Most photosynthesis, therefore, takes place in the upper parts of the trees. Conifers, including spruce, fir, pine, and hemlock, grow in a variety of combinations with deciduous trees, such as willow, alder, and aspen. Taiga animals include squirrels, jays, moose, elk, hares, beavers, and porcupines. Some examples of predators living in the taiga are grizzly bears, wolves, lynxes, wolverines, and owls. What adaptations to the cold might these predators have?

Beyond the tree line or at very high altitudes is the **tundra,** where plants are compact, shrubby, and matlike (Figure 25.12). In the Arctic tundra, permafrost, or continuously frozen ground, prevents roots from penetrating very far into the soil. A permafrost layer in the soil blocks drainage, keeping the soil soggy. Only the shallow roots of small plants can obtain enough oxygen to survive there. Spring comes in mid-May; only plants remaining near the sun-warmed ground can grow, flower, and produce seeds before the advent of fall in August. During the brief, warm Arctic summer, daylight lasts for almost 24 hours, allowing almost constant growth. The tundra may receive less precipitation than many deserts, yet the combination of permafrost, low temperatures, and low evaporation keeps the soil saturated. These conditions also restrict the species that grow there. Dwarf perennial shrubs, sedges, grasses, mosses, and lichens are the dominant producers.

Alpine tundra occurs above the tree line on high mountains, for example, above 3,800 m on Pikes Peak in the Colorado Rocky Mountains. Alpine tundra vegetation is found on Mount Rainier in Washington State and on mountain tops in Utah, as well as on the European Alps. Because of the similar conditions, many of the same plant and animal species are found in both Arctic and alpine tundra. Alpine tundra occurs at all latitudes if the elevation is high enough (Figure 25.13). However, summer days are shorter there than in the Arctic tundra because these high mountains are found at

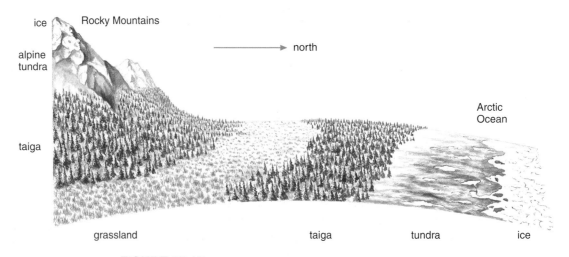

FIGURE 25.13

Effects of altitude and latitude on types of vegetation and animal life. Taiga replaces grassland and temperate forests in the colder environments found at both high altitudes and high latitudes. Where conditions are even colder, tundra replaces the taiga.

a b

FIGURE 25.14

Animals of the arctic tundra.
(a) The willow ptarmigan, *Lagopus lagopus,* shows its winter white plumage. **(b)** In Denali, Alaska, a caribou bull, *Rangifer tarandus-caribou,* stands amid autumn vegetation.

lower latitudes than the early polar Arctic tundra. Brief and intense periods of productivity, plant growth, and reproductive development occur daily in summer, but freezes may happen any night. Long winters, with heavy snowfall and gale winds, keep the growing season very short.

Many animals of the tundra withstand the cold by living in burrows, having a large body size, or possessing good body insulation that retains heat (Figure 25.14). Other species, especially birds, leave the area during the harshest part of the winter. Large animals of the tundra include musk oxen and caribou in North America and reindeer in Europe and Asia. Smaller animals include lemmings, marmots, pikas, white foxes, and snowy owls.

25.2 Aquatic Systems

Most of the earth's biosphere is aquatic. Life, which began in water billions of years ago (see Section 17.4), could not exist without water. Although aquatic ecosystems are not called biomes, they can be divided into two major types, freshwater ecosystems (rivers and ponds) and marine ecosystems (oceans and estuaries, where freshwater streams meet the ocean).

Rivers and streams change significantly from their source, or headwaters, to the point where they empty into an ocean or a lake (Figure 25.15). At the source, the water is usually cold, clear, and rapidly moving; has a very low salt content; and contains few nutrients. Downstream, the water becomes more turbid, or cloudy, and moves more slowly. As a result, the nutrient and salt concentrations increase.

The physical and chemical changes of a river flowing downstream are reflected in the communities of its organisms. Upstream, the communities are relatively simple and include fish such as trout that thrive in water with high levels of dissolved oxygen and low temperatures. Downstream, as pools form, the communities become more complex. As oxygen levels decrease and the temperature increases, species such as sunfish, catfish, and bass replace the trout.

> **Try Investigation 25A** ◇ **Producers in an Aquatic Ecosystem.**

> **CHEMISTRY TIP**
>
> The turbulence and limited depth of shallow streams allow the atmosphere to keep the water saturated with oxygen. Downstream, where the water is deeper and moves more slowly, organisms consume more of the oxygen in the water and sediments. These factors reduce the concentration of oxygen.

Rivers and streams may end in lakes or ponds, standing bodies of freshwater ranging from a few square meters to thousands of square kilometers (Figure 25.16). As a river or stream enters a lake, the speed of the current decreases and suspended particles begin to settle to the bottom. The amount of material that remains suspended in the water and the amount of light scattered by these particles determine the depth to which photosynthesis can occur.

Near the surface of the water where sunlight penetrates, **phytoplankton,** cyanobacteria and photosynthetic protists, act as producers. Nutrients such as phosphorous and nitrogen are the limiting factors in the growth of phytoplankton populations. **Zooplankton** (Figure 25.17), protists and small animals, feed on phytoplankton. Zooplankton often migrate between the surface and deeper water to feed and to escape predators. Which direction would they go in the day? in the night?

Many small fish, aquatic insects, and mollusks feed on plankton. These animals are, in turn, fed on by larger fish, snakes, frogs, toads, and salamanders. Larger predators include raccoons, otters, and birds such as kingfishers and ospreys.

Marine, or ocean, ecosystems can be described in several ways (Figure 25.18). For example, the distribution of marine organisms depends on the penetration of light, which varies with the depth and clarity of the water. The **photic zone** is the shallow top layer of the ocean where enough light penetrates for photosynthesis to occur. Almost all the energy that sustains marine communities comes from the photosynthesis of phytoplankton in this zone. Below the photic zone is the **aphotic zone.** Light is insufficient there for photosynthesis. Most heterotrophs that live in this zone obtain energy by consuming living or dead organic material

FIGURE 25.16

A natural pond in Cedarville, Maryland. It is easy to see the close relationship between the forest and aquatic ecosystems.

FIGURE 25.17

A diverse group of zooplankton. These small organisms feed on phytoplankton and are themselves used as a food source by larger animals.

produced in the photic zone. Others depend on chemoautotrophic bacteria associated with hydrothermal vents (see Section 4.9).

In addition to the amount of light, marine communities are also described according to depth. The shore area between the high-tide and low-tide marks is called the **intertidal zone.** Barnacles, mussels, sponges, clams, algae, and worms are some of the organisms that inhabit this zone (Figure 25.19a). Farther out is the **neritic zone**—the shallow water over the

> ## CONNECTIONS
> Decomposers are vital to every biome and aquatic ecosystem. Without their activities, organic matter would accumulate and nutrients would cease to recycle.

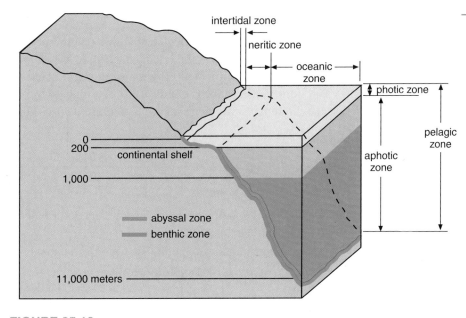

FIGURE 25.18

Marine ecological zones. Marine ecosystems are three-dimensional. Zones depend on depth, light penetration, and distance from the shoreline.

a

b

c

FIGURE 25.19

Some marine animals. (a) Intertidal zone animals include the ochre sea star and the blood star. **(b)** The spotted moray eel, *Gymnothorax moringa,* lives in the benthic (sea floor) zone. **(c)** The deep-sea hatchet fish *(left), Argyropelecus hemigymnus,* and the deep-sea viper *(right), Chauliodus sloani,* live in the dark, cold abyssal zone.

continental shelf. The major fisheries of the world are found there. Past the shelf is the open ocean, or **oceanic zone.**

In all these zones, the open water not associated with the seafloor is the **pelagic zone.** The seafloor is the **benthic zone.** The benthic community consists of bacteria, sponges, worms, sea stars, crustaceans, fish, and other organisms (Figure 25.19b). The area of the benthic zone where light does not penetrate is the **abyssal zone.** The organisms living there are adapted to continuous darkness, cold temperatures, scarce food, and extreme pressure (Figure 25.19c).

Check and Challenge

1. How does the chemical composition of rock and soil help determine the type of biome that will develop in an area?

2. Describe how climate determines which biome is found in an area.

3. What biomes are most suitable for agricultural use, and how has this use changed them?

4. What are the most probable limiting factors in each marine zone? Explain your answer.

Biology Online   BSCSblue.com/check_challenge

Ecosystems in Time

25.3 Species on the Move

Organisms can disperse (in other words, move away from their current site or normal range). Seeds fall from plants or are carried by wind or animals (Figure 25.20a). Winged animals have great dispersal ability. Many aquatic animals have planktonic larvae—immature forms that are members of the zooplankton—that move with the currents, sometimes for great distances (Figure 25.20b). Competition puts pressure on organisms to disperse to new areas. Of course, the new areas might already be occupied by competing organisms of the same or other species, and the seed or individual may be doomed to failure.

Sometimes the dispersing organisms encounter an unoccupied habitat and can successfully settle in, or **colonize,** the new area. This is what happened as the glaciers that once covered much of North America retreated about 7,000 to 10,000 years ago. Plants and animals gradually moved northward as climatic conditions changed. Some species of trees, such as American beech and eastern hemlock, are still extending their range northward.

In recent years, humans have become major agents of dispersal of plants and animals, some by accident and some intentionally. Often the new organism (called an **exotic** species) simply perishes. But sometimes it takes well to the new area and spreads out, establishing itself firmly. Exotic

a

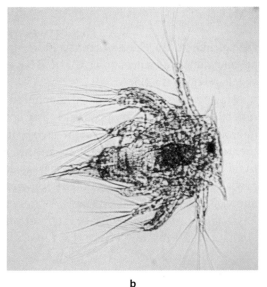
b

FIGURE 25.20

Adaptations that favor dispersal. (a) The seed of the milkweed, *Asclepias,* has tufts of fine hairs that favor dispersal on the winds. **(b)** Planktonic larvae such as this barnacle, *Balanus,* in its nauplius larval stage, ride the currents to new areas.

a

b

c

FIGURE 25.21

Exotic species causing problems in the new ecosystems they have invaded. (a) Native vegetation at Horn Lake, Michigan, is choked by kudzu, a fast-growing vine in the pea family (Fabaceae). Kudzu was first introduced into the U.S. from Japan at the 1872 Centennial Exposition; once considered ornamental, it was declared a weed in 1972. **(b)** The South American aquatic rodent nutria, *Myocaster coypus,* was introduced into Louisiana in 1937. Following its escape from captivity, the nutria has become a pest in swamplands. **(c)** The fire ant, *Solenopis,* with its destructive jaws and painful sting, is spreading north and west into the United States from the Southeast.

species like kudzu, the zebra mussel, fire ants, and African bees have become major nuisances to people and have had serious impacts on native species (Figure 25.21).

The great diversity of species adapted to different ecosystems is a reflection of evolutionary processes. The communities of plant, animal, and microbial species of an ecosystem have probably been together for thousands of years in such habitats and are well adapted to each other and to the abiotic environment of the area. Small-scale changes, such as predator-prey ratios (see Section 24.6) generally cycle over time and do not alter the nature of a healthy, stable ecosystem. However, if conditions of the physical environment change or if new species move into established communities, the entire ecosystem can change dramatically.

25.4 Succession

Try Investigation 25B 🔹 **Ecosystem Diversity within a Biome.**

How stable are ecosystems? They are constantly subjected to disturbances such as fires, storms, species invasions, and changing climate. When the environment changes, species come and go, and the community changes. The process by which one type of community replaces another is

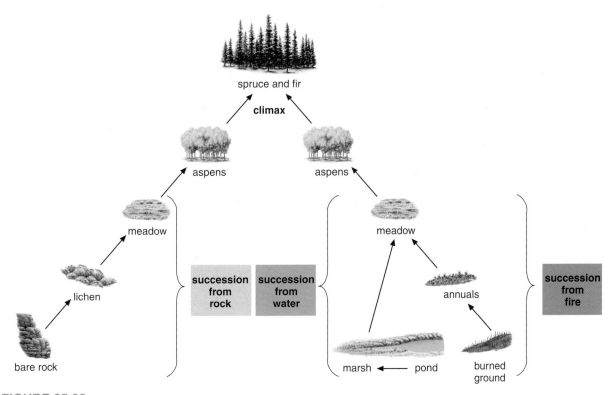

FIGURE 25.22

Possible successional sequences in the taiga biome. Abiotic factors in this region, such as soil and climate, favor development of a coniferous forest.

succession. Succession can begin even on a bare, lifeless surface such as a rock. Lichens grow first, breaking down the stone into small pockets of soil. Mosses then may colonize the area, eventually followed by ferns and the seedlings of flowering plants. In time, organic matter accumulates, enriching the soil, and a mature forest results (Figure 25.22).

Lakes and ponds are also subject to succession. Shoreline plants such as cattails and rushes, submerged plants such as pondweeds, and other organisms all contribute to the formation of rich organic sediment (Figure 25.23). As this process continues, the lake gradually fills with vegetation. Eventually the area where the lake once stood becomes a forest. Succession that begins on bare rock, on glacial deposits, or in a lake bed is called **primary succession.** More often, succession takes place on a disturbed site where soil is already present, such as an abandoned farm field. This process is called **secondary succession.**

The early stages of succession are generally dominated by plants that disperse easily, grow rapidly, and have adaptations that favor colonization. The earliest colonizing plants are small **annuals** (plants such as crabgrass that live for only 1 year) that grow close to the ground and produce many seeds. Typical colonizers can tolerate a wide range of stresses, such as poor soil, high temperatures, and low precipitation. In time, these species are replaced by larger perennials. In later stages of succession, the dominant plants are tree species that live longer and can endure increased shade and competition. The types and numbers of species of plants and animals change

CHEMISTRY TIP

Lichens and plants can break down rocks by producing acids that slowly dissolve the rock minerals. The most important acid is carbonic acid (H_2CO_3), formed from carbon dioxide (CO_2) produced by cell respiration and H_2O. Plant roots, bacteria, fungi, and lichens also release organic acids into soil.

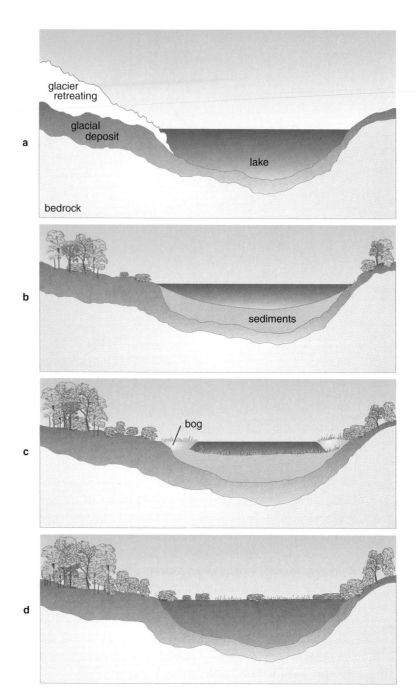

FIGURE 25.23

An example of lake succession.
(a) Meltwater from a retreating
glacier forms a lake. **(b)** Plants
colonize the area, beginning a
successional sequence. **(c)** Plants
and other organisms present at
each stage add sediment to the
lake. **(d)** This alters the environment,
paving the way for the organisms of
the next successional stage.

throughout all the stages of succession. The course of succession depends
on which species colonize an area in the early stages and on environmental
disturbances that cause further changes. The process of succession is driven
mostly by the organisms themselves. The organisms present at each stage
change the ecosystem in ways that pave the way for later species.

The recovery of the Mount St. Helens area in Washington State after
the volcanic eruption in 1980 is an interesting example of succession
(Figure 25.24). Some scientists predicted an orderly progression beginning

FIGURE 25.24

Another example of succession.
An area of Mount St. Helens
colonized by beardtongue, genus
Penstemon. This photograph shows
the succession that followed the
1980 eruption.

with pioneer plants and eventually ending in a mature forest. Others thought recovery would begin as a matter of chance—a seed would arrive, perhaps blown by the wind, take root, and then play a role in determining the types of plants that could grow nearby. To the surprise of these two scientific groups, however, recovery is occurring both ways. Alpine lupine, a typical colonizer, was among the first plants to grow in areas that were nearly sterile. Dead lupine stalks and dead leaves collected sand and dust particles, allowing perennials and bushes to root in an orderly succession. At the same time, however, individual trees, mostly alder, quickly became established in the middle of desolate areas, where they were the first plants to appear. Because alder is symbiotic with nitrogen-fixing bacteria (see Section 4.9), it grows well in soils that are nutrient poor. Is the recovery of Mount St. Helens a primary or secondary succession?

As succession progresses, interactions such as predation and competition become more complex. This opens up new niches and increases the diversity of species that live in the community. Eventually the web of interaction may become so intricate that no additional species can fit into the community unless others disappear from the area. Such a community is considered to be a **climax community;** it has reached a stable equilibrium. Unlike other stages in the succession process, a climax community is not replaced by another community. For example, when a spruce tree dies, the space it leaves behind is too shaded by neighboring spruces for trees such as aspens, which need bright sunlight. A young spruce, tolerant of shade, will probably take the place of the old one, thus maintaining the biological equilibrium of the community.

Biomes are climax communities that reflect the climate, soils, and topography of the area. However, most landscapes are a patchwork of communities in different stages of succession because of disturbances. Fires, floods, hurricanes, and even unseasonable temperatures can interrupt the succession process and create gaps that prevent the community from reaching a state of equilibrium. Where disturbances have recently occurred, good colonizer species are common. These may be weed species of annuals (plants that only live 1 year and produce numerous, small, easily dispersed seeds). Elsewhere, the species that are the most competitive may dominate the community. Species diversity, therefore, may be greatest when disturbances are fairly frequent but not too severe. Hence, many different successional community stages make up the landscape.

Even climax communities undergo variations when the climate changes, causing some scientists to question the basic concept of the climax. Following the retreat of the glaciers, the north temperate-zone forests have shifted from coniferous (taiga) to deciduous. These changes in conditions and species took place over hundreds or thousands of years. What happens if climate changes more rapidly, say, in a few decades? How quickly can forests adapt? During the 21st century, we may learn the answers to these questions because of the human-induced impacts of global climate change.

Check and Challenge

1. What adaptations do organisms have for dispersal?
2. What role does dispersal play in ecosystem changes?
3. What is succession? Distinguish between primary and secondary succession and give examples.
4. Explain the concept of the climax community and give reasons why it may not always occur.
5. Describe the impacts of humans on species dispersal and succession in natural ecosystems.

Human Interactions with Ecosystems

25.5 Dependence

Ecosystems on land and sea provide natural resources that are essential for human existence. Forests and woodlands are sources of wood, fuel, paper, and wild game. Grasslands are grazed by domestic animals that give us food, leather, and wool. Coastal oceans are harvested for fish and shellfish, providing 20% of the total human consumption of protein (Figure 25.25). People don't just harvest materials from the land and oceans; they actively change habitats in certain ecosystems to increase the yield of specific resources. For example, many forests and grasslands have been converted to crop lands, which provide most of our food and beverages, oils, and raw materials for fabrics. The production and processing of all of these goods employ more people than any other economic activity on Earth.

Biology Online BSCSblue.com/check_challenge

FIGURE 25.25

Human dependence on the oceans. The crew of a fishing boat hauls in a net full of fish. These fish are ecosystem goods vital to commerce and nutrition.

Thus, the world economy is highly dependent on the extraction and processing of resources from these ecosystems. In fact, one important measure of a country's wealth is the natural resources available to meet its people's needs and to generate economic profit.

The goods that ecosystems provide for human communities are only part of the picture, however. The energy flow and nutrient cycling that normally occur in ecosystems also support human life. The following are a few of the most significant services that ecosystems perform:

1. **Ecosystems modify climate.** Plants transpire water and absorb solar radiation. As a result, vegetated ecosystems such as forests and wetlands moderate temperature and help maintain a stable climate.

2. **Ecosystems prevent erosion and build soil.** Plant roots and leaf litter bind the soil and prevent rainfall from breaking up the soil and washing it away (Figure 25.26). Most of the soil that supports farming gained much of its fertility when the land was covered with forests or grasslands.

3. **Ecosystems break down wastes.** Domestic and agricultural wastes are released untreated or partially treated into many ecosystems, especially wetlands. These ecosystems decompose organic wastes, transforming many toxic substances into harmless products. In doing so, they purify the water and the air.

4. **Ecosystems store carbon and maintain the carbon cycle** (see Section 24.4). Energy flow maintains the global carbon cycle and has stored over 500 billion metric tons of carbon in forest organisms and much more than that in the organic matter of soils. Without this service, we would have severe and rapid global warming.

Ecosystems perform many other functions, such as controlling pests, maintaining nutrient cycles and the water cycle, providing recreation, and

FIGURE 25.26

An aerial view of the Betsiboka River Delta, Madagascar. Deforestation, removal of the native forest for cultivation and pastureland, during the last 50 years has led to massive annual soil losses. In this photograph, the effects of water erosion are seen as brown streaks leading to the blue deeper water over a 4,000-square-km land surface area. The trees that were removed are no longer available to remove carbon dioxide from the atmosphere. Consequently, this environmental change adds to global warming.

sustaining biodiversity. These services are provided free of charge year after year. Ecosystem goods and services are renewable, but only if the ecosystems continue to function. For example, when people remove trees over vast areas, as they have on most continents, the climate changes, soil erosion increases, and less wood is available for building and fuel. On the other hand, if we harvest trees from forests on a sustainable basis—by cutting them no faster than they can grow—we keep all of the services performed by the forests intact while we enjoy the economic benefits of the wood.

Many of the goods and services provided by ecosystems can be viewed as **common pool resources,** or just **commons.** A common pool resource is shared by many but controlled by no one person. There are many kinds of commons: the coastal or open-ocean fisheries; federal grasslands grazed by privately owned livestock; groundwater drawn by private owners and towns; forests and woodlands in the developing world; and the entire atmosphere, which is polluted by both large industries and individuals. There is often little or no regulation of access to the commons, so it is exploited by competing parties, all of whom profit by using the shared resources. For example, if a local fishery resource is open to fishing by all, there is a strong tendency for each party to take as many fish as possible. Everyone acting in his or her own self-interest leads to the decline of the resource and eventually the ruin of the commons—something biologist Garrett Hardin called "the tragedy of the commons." To solve this problem, access to the commons is often regulated so that it can be managed sustainably.

In 1997, a team of scientists and economists produced a report entitled *The Value of the World's Ecosystem Services and Natural Capital.* They identified 17 major goods and services provided by the world's ecosystems. The research team also calculated the value these goods and services contribute to human welfare. In one sense, their value is infinite since human survival is absolutely dependent on them. The researchers made the best approximation they could, however, and came up with a total value of $33 trillion for a year's worth of ecosystem goods and services. This figure is almost double the world's economic production.

Given the great value of ecosystems, it would seem wise to protect them and take their value into consideration when exploiting the land and sea to meet our needs. Unfortunately, it has proven too easy to overlook the long-term value of natural systems; the result has been destruction and degradation of many of those systems by pollution and irresponsible resource use—the tragedy of the commons.

25.6 Dominance

In 2003, the human population passed the 6.3 billion mark (Figure 25.27). It is increasing at a rate of about 80 million people per year. With so many of us on Earth, humans have become a major force in changing the natural landscape. Every person has needs that must be met from the environment. If we increase to 8 billion or more by 2050, as many think we will, are we in

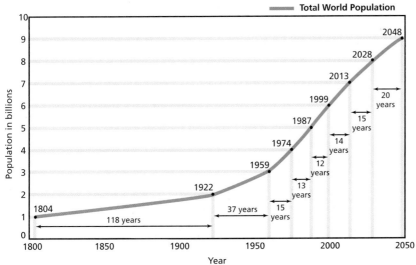

Source: United Nations (1995b); U.S. Census Bureau, International Programs Center, International Data Base and unpublished tables.

FIGURE 25.27

Growth of the human population since 1800. World population is now more than 6 billion and is projected to increase to at least 9 billion by midcentury. The largest gains are expected to occur in sub-Saharan Africa and the Near East. Moderate gains are expected in North Africa, North and South America, Asia, and the Pacific. Most populations of countries in Europe and the former Soviet Union are expected to decline.

danger of exceeding Earth's carrying capacity (see Section 24.6) for humans? What would a world with 8 billion people be like? In fact, how are we affecting the biosphere now?

Human activities affect almost every ecosystem. Our impacts are both local and global. The most important local effects involve land use. No activity is more disruptive to natural ecosystems than agriculture. We now use 11% of the land surface for agricultural crops. A further 7% has been converted to pasture, and 4% is covered by cities, towns, and roads. Thus, 22% of the land surface is now devoted exclusively to human use. There may be ecosystems of a sort on much of this land, but these are managed ecosystems that, for the most part, do not contain the typical or naturally found organisms of the region. This destruction of natural habitat is the major driving force behind the loss of biodiversity worldwide. Although this devastation of ecosystems happens at the local level, the effects are global: The resulting loss of species is unprecedented in human and possibly world history.

The effects of air pollution range from local to global. Locally air pollution harms human health. Many cities in eastern Europe lack air-pollution controls; infant mortality due to respiratory disease is 20 times greater there than it is in North America. Furthermore, acid rain affects larger regions. Burning fossil fuels in power plants (Figure 25.28) and motor vehicles adds

CHEMISTRY TIP

Various sulfur oxides are produced when sulfur-containing fossil fuels are oxidized during combustion. Sulfur dioxide (SO_2), for example, combines with water vapor in the atmosphere, forming sulfuric acid.

$$SO_2 + H_2O \longrightarrow H_2SO_4$$

Atmospheric nitrogen (N_2) oxidizes in furnaces and automobile engines. As a result, various nitrogen oxides (NO_x) are produced and released into the atmosphere. There they react with water to form nitric acid.

$$NO_x + H_2O \longrightarrow HNO_3$$

Both nitric and sulfuric acids are very strong and damaging acids.

FIGURE 25.28

Major sources of acid rain. Acid rain originates from the burning of fossil fuels, such as oil and coal, by hundreds of power plants and millions of motor vehicles.

The Atmosphere in Danger

Between 20 and 40 km above the earth's surface, high in the atmosphere, is a layer of ozone (O_3) that protects living things on Earth from ultraviolet (UV) radiation. This layer normally absorbs over 99% of damaging UV rays. The small fraction that still gets through is responsible for sunburn and many skin cancers. Unlike the ozone in the lower stratosphere (just above 16 km), this ozone does not originate from air pollution. It comes from the interaction of UV with atmospheric oxygen (O_2).

In 1974, chemists Mario Molina and Sherwood Rowland published a research paper on a group of compounds of the elements carbon, chlorine, and fluorine called chlorofluorocarbons (CFCs—known commercially as Freon). The report suggested that CFCs, commonly used as refrigerants and spray-can propellants, could lead to a breakdown in the ozone shield and an increase in skin cancers. They reasoned that CFCs would drift up into the stratosphere, break down, and release chlorine, as follows:

$$CFCl_3 \text{ (Freon 11)} + \text{UV radiation} \rightarrow Cl\bullet + CFCl_2$$

The dot symbol following Cl in this equation indicates that the chlorine atom has an unpaired electron and is a highly reactive **free radical.** The chlorine would then react with ozone to form chlorine monoxide and oxygen:

$$Cl\bullet + O_3 \rightarrow ClO\bullet + O_2$$

Two molecules of chlorine monoxide can subsequently react to release more chlorine and oxygen:

$$ClO\bullet + ClO\bullet \rightarrow 2\,Cl\bullet + O_2$$

Notice that the last reaction regenerates the chlorine radical produced in the first reaction. The chlorine radical can thus act as a **catalyst,** breaking down more ozone. It is very stable and can destroy thousands of ozone molecules before it leaves the stratosphere.

In 1985, scientists reported an **ozone hole** over Antarctica that started in 1970. Ozone levels there were less than half their usual concentration (Figure 25.29). The hole now covers more than 25.9 million square km every spring over Antarctica. With summer conditions, the hole breaks up and the ozone-depleted air spreads out, increasing UV radiation in Australia and New Zealand.

World-wide use of CFCs has been phased out with an agreement called the Montreal Protocol. Upon awarding Molina and Rowland the Nobel Prize in 1995, the Nobel committee said that the "researchers contributed to our salvation from a global environmental problem that could have catastrophic consequences." CFC concentrations in the stratosphere have peaked, and the ozone shield will likely be back to normal some time during the mid–21st century.

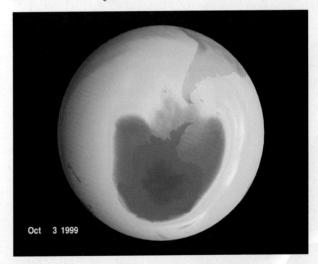

Oct 3 1999

Ozone Level (Dobson Units)

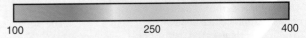

100   250   400

FIGURE 25.29

The Antarctic ozone hole. Satellites map ozone concentrations in the stratosphere using a special spectrometer, the total ozone mapping spectrometer (TOMS).

large amounts of sulfur and nitrogen oxides to the atmosphere. There these oxides react with water and change into sulfuric and nitric acids, which fall to Earth as acid rain, often hundreds of miles downwind of the sources. Acid rain has been responsible for the destruction of life in thousands of lakes in Canada, Scandinavia, and the United States. Acid rain also damages forest trees, mainly by leaching minerals from the soils. Consequently, forests in North America and Europe affected by acid rain have experienced a severe decline in growth.

Two forms of air pollution have a global impact on ecosystems. CFC use has led to a serious decline in the ozone layer of the stratosphere. Additionally, most scientists are convinced that global climate change represents our most serious environmental problem. Human activities have been adding a number of gases to the atmosphere ever since the beginning of the Industrial Revolution in the 1800s. Some of these gases are greenhouse gases, which means that they contribute to the greenhouse effect. Greenhouse gases trap infrared radiation in the atmosphere and cause the atmospheric temperature to rise—thereby simulating a greenhouse. The most important of these is carbon dioxide, which has increased from 280 parts per million (ppm) in the atmosphere in 1850 to 373 ppm in 2002, and it is still increasing (Figure 25.30). The added carbon dioxide comes from burning fossil fuels and vegetation cut from newly cleared land, currently at a rate of 6.5 to 8 billion metric tons of carbon per year.

The greenhouse effect is actually beneficial. Without it, Earth's average surface temperature would be 21°C cooler, and life would be impossible. By continually adding greenhouse gases to the atmosphere, however, we risk a permanent change in the climate as the gases trap more and more heat energy. Atmospheric scientists calculate that, if we continue on our current course of burning fossil fuels, in 100 years the average global temperature will be at least 2°C higher, and patterns of rainfall and temperature will

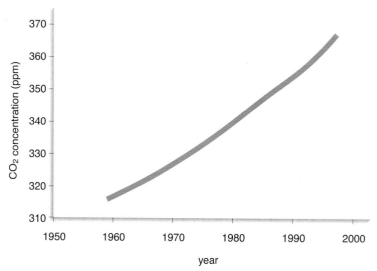

FIGURE 25.30

Atmospheric carbon dioxide concentrations since 1958 in Mauna Loa, Hawaii. The data show the long-term increase in carbon dioxide concentration.

Source: Dave Keeling and Tim Whorf (Scripps Institution of Oceanography)

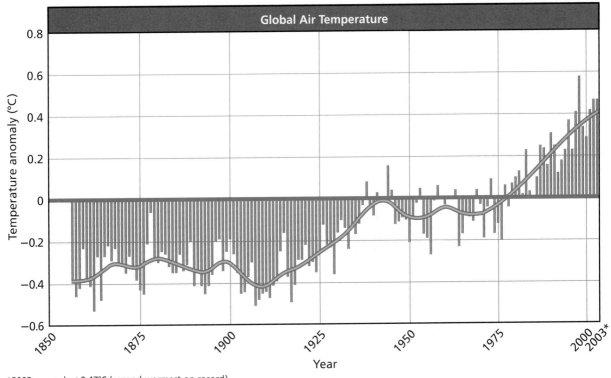

*2003 anomaly +0.47°C (second warmest on record)

FIGURE 25.31

Global temperatures. The graph shows average global air temperature anomalies since 1850.

change dramatically. Since adding heat energy to the oceans causes seawater to expand and glaciers to melt, scientists also predict that sea level will rise at least 50 cm. Although these figures are only estimates, the climate is already showing signs of change (Figure 25.31), and sea level is already rising. This evidence supports the prediction of global warming.

The possible consequences of climate change and sea-level rise are severe. Agriculture and ecosystems are adapted to current climatic conditions and can adapt to changes only if they occur gradually over hundreds or thousands of years. Our climate, however, is likely to change over a few decades. A rising sea level will also threaten many small island nations and will have a serious impact on coastal cities, which are home to half the world's population.

What can be done about this problem? At present, the industrial countries are so dependent on fossil fuels that it is hard to see how we could change our current dependency. However, most scientists and many politicians agree that we must begin to cut back on our use of oil, coal, and gas; use energy more efficiently; and make more use of renewable energy sources, such as wind and solar energy. At a conference held in Kyoto, Japan, in 1997, many nations agreed to begin to reduce their use of fossil fuels.

THEORY

Organism Earth: The Gaia Hypothesis

Most people divide objects into living and nonliving things. In the 1970s, biologists James Lovelock and Lynn Margulis challenged this way of thinking about the earth with their Gaia hypothesis. They proposed that our planet is a single complex living organism. As changes in the abiotic environment affect organisms, the biosphere as a whole responds in ways that help maintain its overall homeostasis. For example, as humans clear rain forests and produce air pollution, they alter the atmosphere. Changes to the atmosphere and climate in turn affect the survival of organisms, including humans. The abiotic and biological parts of the world evolve together as a larger unit. Similarly, global warming due to greenhouse gas emissions is supposed to have increased the activity of marine bacteria that release sulfur compounds into the atmosphere. These substances act as seeds for the formation of clouds. As the figure at right suggests, the additional clouds in turn reflect more sunlight back into space, cooling the earth slightly.

Most scientists doubt that organisms and their environments are so bound together that the entire Earth qualifies as an organism. What would that mean? Margulis paints a completely new picture of evolution. In the usual description of evolution, competition for limited resources drives natural selection. Margulis suggests that, more often, natural selection and speciation favor successful symbiotic cooperation. For example, her endosymbiont hypothesis (see Section 17.7), which is now widely accepted, explains the origin of eukaryotic cells as the development of a successful symbiosis among several types of prokaryotes. In much the same way, the ability of ruminants (such as cattle and deer) to digest cell walls that other mammals cannot and the ability of legumes (such as pea and bean plants) to compete effectively with grasses in nitrogen-poor soils are due to symbioses—with cellulose-digesting bacteria in the first case and nitrogen-fixing bacteria (see Section 4.9) in the other.

Symbiosis may have been important in the origin of eukaryotic cells, but how far should biologists extend this idea? Lovelock and Margulis claim that cooperation is the main driving force of evolution. The Gaia hypothesis is an interesting idea that has led to a great deal of debate and some experimental research. However, this hypothesis has failed to attract many scientific supporters for two reasons. First, the Gaia hypothesis leads to few specific, testable predictions; second, few data have been found to support it. The idea that the Earth itself is an evolving life-form is appealing and thought-provoking, but it is not a strong or widely supported theory. The Gaia hypothesis may help inspire efforts to conserve our planet's ecosystems even if it is not successful as a scientific hypothesis.

A single living system? The Gaia hypothesis suggests that increased cloud formation over the oceans compensates for global warming tendencies by reflecting more sunlight away from the Earth.

25.7 Sustainability

The human species has been remarkably successful in spreading over Earth and using the goods and services provided by Earth's ecosystems. However, much of what we are now doing is unsustainable; that is, it cannot continue at its present rate without serious and undesirable consequences. Human civilization must form a long-term relationship with the natural world that is sustainable—where we can continue to use ecosystem goods and services without ruining them for future generations. Many people believe that we can do this only if we begin to take more seriously our responsibilities as caretakers of those ecosystems. Our responsibilities cannot stop there. We also must be concerned about the people on Earth. The living conditions of many of the world's people are terrible. Over 1 billion live in abject poverty, unable to meet their most basic needs for food, shelter, and health. At least hundreds of millions are malnourished or chronically hungry. As we enter a new millennium, your generation is faced with this enormous challenge. Will you use your time and talents to make things better? The future depends on it.

Check and Challenge

1. What are some of the most important goods and services provided to humans by natural ecosystems? How much are these worth?

2. What is "the tragedy of the commons," and how can it be avoided? Give an example.

3. Provide examples of local, regional, and global impacts of human activities on natural systems.

4. Explain how two chemists helped to bring about a major global treaty concerning the atmosphere.

Chapter
HIGHLIGHTS

Summary

Biomes are groupings of similar terrestrial ecosystems. They are determined by the prevailing climate and soil and are characterized by the dominant energy-producing vegetation. Both latitude and altitude affect the climate and, hence, the biomes as well. Similar groupings of aquatic ecosystems depend on the water's depth, salt concentration, and currents. Deeper aquatic systems, such as lakes and oceans, are divided into zones occupied by various types of plants and animals.

Most organisms are adapted to disperse to new locations. When colonizing new areas through dispersal, they often bring about major changes in the ecosystems they invade. Invasions, physical processes, and human interventions frequently disturb ecosystems. These disturbances often promote the process of succession, in which a sequence of organisms occupies a given site over time, leading to a climax community. A given landscape may contain ecosystems at different successional stages at any one time, leading to greater species diversity.

Humans are completely dependent on natural ecosystems. We derive many goods from them, such as wood, fish, leather, wool, paper, and meat. People engaged in harvesting and processing natural products provide a major part of the world economy. Some of these are overexploited, leading to "the tragedy of the commons." We also depend on natural ecosystems for the services they provide. Ecosystems modify climate; prevent erosion; build soil; break down wastes; maintain the cycles of nutrient elements such as carbon, nitrogen, and phosphorus; control pests; provide recreation; and perform many other vital services. These goods and services were recently estimated to be worth $33 trillion per year.

The more than 6 billion people now living on Earth have an impact on almost all ecosystems. Habitat destruction has led to the loss of biodiversity; local and regional air pollution is harmful to human

health and natural systems; burning fossil fuels and forests has led to a buildup of atmospheric CO_2. In addition to CO_2, other gases linked to human activities are threatening to permanently change the climate. Evidence of global warming is clearly visible. Unless we learn to live sustainably, we will leave a degraded Earth to our descendants.

Key Concepts

1. Construct a concept map that links the biomes and aquatic ecosystems described in this chapter with the factors that influence and distinguish them.
2. Build a concept map of terms from this chapter that shows how humans depend on and dominate natural ecosystems. Add other terms as needed. (Hint: Use the distinction between goods and services and the ways we interact with natural resources.)

Reviewing Ideas

1. What are the major factors controlling the distribution of terrestrial biomes? How do they differ in the seven biomes described in the chapter?
2. Describe the most important characteristics of each of the following biomes: tropical rain forest, temperate deciduous forest, savanna, and taiga.
3. Distinguish among streams, lakes, and the oceans. What characterizes each aquatic system?
4. What are phytoplankton and zooplankton? What limits them, and how are they linked?
5. Define the following marine zones: benthic, intertidal, aphotic, and pelagic.
6. What are some means of dispersal used by organisms? How have humans affected the dispersal of exotic species?

Biology Online BSCSblue.com/vocabulary_puzzlemaker

7. What is succession? Distinguish between primary and secondary succession.
8. Describe a typical primary succession on newly created land up until it reaches a climax stage.
9. What are some of the more important natural resources we depend on?
10. What natural services vital to human existence do ecosystems perform?
11. What are commons? How does human use of a commons lead to "the tragedy of the commons"?
12. Describe local, regional, and global examples of impacts of human activities on natural systems.
13. Explain the link between chlorofluorocarbons and the depletion of the ozone layer.

Using Concepts

1. Why do similar biomes on different continents have different plant and animal species?
2. What are the limiting factors for each of the following: a desert, the tundra, a tropical rain forest, and the open ocean?
3. Describe the biome of your community. How is it similar to or different from the biomes described in this chapter?
4. For a biome where a temperate deciduous forest is the climax stage, describe several disturbances that could lead to greater biodiversity in the landscape. Explain your answer.
5. Describe how a federally owned grassland might be regulated so that use by many ranchers to graze their cattle would be sustainable.
6. Describe a situation in which all of the microorganisms in terrestrial ecosystems could be killed by pollutants. What would be the consequences to humans?
7. What common pool resources do you share with those around you? Is your community exploiting or conserving these commons?

Synthesis

1. Explain how evolution through natural selection affects the plants and animals in different biomes.
2. Phytoplankton in the photic zone of the oceans undergo seasonal increases and decreases. Explain how temperature, light, nutrients, and grazing by zooplankton could help cause these seasonal changes.
3. Select three biomes, and describe how each contributes to human needs.

Extensions

1. Write a treaty among the nations of the world that would lead to a stable concentration of greenhouse gases in the atmosphere. How would your agreements affect the wealthier and poorer nations?
2. In the United States, national forests are managed by the USDA Forest Service. Use the web to investigate how well the Forest Service carries out its responsibilities and the management concepts it employs.
3. Research and evaluate the sustainability of such practices as controlling pests with chemical sprays, using or preventing forest and grassland fires, and hunting.

Web Resources

Visit BSCSblue.com to access
- Links to information related to biomes, succession, use of natural resources, and human impacts on ecosystems
- Web links related to this chapter

INVESTIGATIONS

Introduction to Laboratory Work

The laboratory is a scientist's workshop—the place where ideas are tested. In the laboratory portion of this course, you will see evidence that supports major biological concepts. To pursue your investigations effectively, you need to learn certain basic techniques, including safe laboratory practices, record keeping, report writing, and measurement. The information on the following pages will help you learn these skills and techniques.

Laboratory Safety

The laboratory can be either safe or dangerous. The difference depends on your knowledge of and adherence to safe laboratory practices. It is important that you read the information here and learn how to recognize and avoid potentially hazardous situations. Basic rules for working safely in the laboratory include the following:

1. Be prepared. Study the assigned investigation before you come to class. Be prepared to ask questions about the procedures you do not understand before you begin to work.
2. Be organized. Arrange investigation materials in an orderly fashion.
3. Maintain a clean, open work area, free of everything except those materials necessary for the assigned investigation. Store books, backpacks, and purses out of the way. Keep laboratory materials away from the edge of the work surface.
4. Tie back long hair, and remove dangling jewelry. Roll up long sleeves, and tuck long neckties into your shirt. Do not wear loose-fitting sleeves or open-toed shoes in the laboratory.
5. Wear a lab apron and safety goggles whenever working with chemicals, hot liquids, lab burners, hot plates, or apparatus that could break or shatter. Wear protective gloves when working with preserved specimens or toxic or corrosive chemicals or when otherwise directed.
6. Never wear contact lenses while conducting any experiment that uses chemicals. If you must wear them (by a physician's order), inform your teacher prior to conducting any experiment involving chemicals.
7. Never use direct or reflected sunlight to illuminate your microscope or any other optical device. Direct or reflected sunlight can cause serious damage to your retina.
8. Keep your hands away from the sharp or pointed ends of equipment such as scalpels, dissecting needles, and scissors.
9. Observe all cautions in the procedural steps of the investigation. CAUTION, WARNING, and DANGER are signal words used in the text and on labeled chemicals or reagents that tell you about the potential for injury. They remind you to observe specific practices. *Always read and follow these statements.*
10. Become familiar with caution symbols, identified in Figure A.

safety goggles
Safety goggles are for eye protection. Wear goggles whenever you see this symbol. If you wear glasses, be sure the goggles fit comfortably over them. In case of splashes into the eye, flush the eye (including under the lid) at an eyewash station for 15 to 20 minutes. If you wear contact lenses, remove them *immediately* and flush the eye as directed. Call your teacher.

lab apron
A lab apron is intended to protect your clothing. Whenever your see this symbol, put on your lab apron and tie it securely behind you. If you spill any substance on your clothing, call your teacher.

gloves
Wear gloves when you see this symbol or whenever your teacher directs you to do so. Wear them when using *any* chemical or reagent solution. Do not wear your gloves for an extended period of time.

sharp object
Sharp objects can cause injury, either as a cut or a puncture. Handle all sharp objects with caution, and use them only as your teacher instructs you. *Do not* use them for any purpose other than the intended one. If you do get a cut or puncture call your teacher and get first aid.

irritant
An irritant is any substance that, on contact, can cause reddening of living tissue. Wear safety goggles, lab apron, and protective gloves when handling any irritating chemical. In case of contact, flush the affected area with soap and water for a least 15 minutes and call your teacher. Remove contaminated clothing.

reactive
These chemicals are capable of reacting with any other substance, including water, and can cause a violent reaction. *Do not* mix a reactive chemical with any other substance, including water, unless directed to do so by your teacher. Wear your safety goggles, lab apron, and protective gloves.

corrosive
A corrosive substance injures or destroys body tissue on contact by direct chemical action. When handling any corrosive substance, wear safety goggles, lab apron, and protective gloves. In case of contact with a corrosive material, *immediately* flush the affected area with water and call your teacher.

flammable
A flammable substance is any material capable of igniting under certain conditions. Do not bring flammable materials into contact with open flames or near heat sources unless instructed to do so by your teacher. Remember that flammable liquids give off vapors that can be ignited by a nearby heat source. Should a fire occur, *do not* attempt to extinguish it yourself. Call your teacher. Wear safety goggles, lab apron, and protective gloves whenever handling a flammable substance.

poison
Poisons can cause injury by direct action within a body system through direct contact (skin), inhalation, ingestion, or penetration. *Always* wear safety goggles, lab apron, and protective gloves when handling any material with this label. Before handling any poison, inform your teacher if you have preexisting injuries to your skin. In case of contact, call your teacher *immediately.*

biohazard
Any biological substance that can cause infection through exposure is a biohazard. Before handling any material so labeled, review your teacher's specific instructions. *Do not* handle in any manner other than as instructed. Wear safety goggles, lab apron, and protective gloves. Any contact with a biohazard should be reported to your teacher immediately.

NO FOOD OR DRINKS SHOULD BE PRESENT IN THE LAB AT ANY TIME.

FIGURE A Caution symbols.

11. Never put anything into your mouth, and never touch or taste substances in the laboratory unless specifically instructed to by your teacher.

12. Never smell substances in the laboratory without specific instructions. Even then, do not inhale fumes directly; wave the air above the substance toward your nose and sniff carefully.

13. Never eat, drink, chew gum, or apply cosmetics in the laboratory. Do not store food or beverages in the lab area.

14. Know the location of all safety equipment, and learn how to use each piece of equipment.

15. If you witness an unsafe incident, an accident, or a chemical spill, report it to your teacher immediately.

16. Use materials only from containers labeled with the name of the chemical and the precautions to be used. Become familiar with the safety precautions for each chemical by reading the label before use.

17. When diluting acid with water, *always add acid to water.*

18. Never return unused chemicals to the stock bottles. Do not put any object into a chemical bottle except the dropper with which it may be equipped.

19. Clean up thoroughly. Dispose of chemicals, and wash used glassware and instruments according to your teacher's instructions. Clean tables and sinks. Put away all equipment and supplies. Make sure all water, gas jets, burners, and electrical appliances are turned off. Return all laboratory equipment and supplies to their proper places.

20. Wash your hands thoroughly after handling any living organisms or hazardous materials and before leaving the laboratory.

21. Never perform unauthorized experiments. Do only those experiments assigned by your teacher.

22. Never work alone in the laboratory, and never work without a teacher's supervision.

23. Approach laboratory work with maturity. Never run, push, or engage in horseplay or practical jokes of any type in the laboratory. Use laboratory materials and equipment only as directed.

In addition to observing these general safety precautions, you need to know about some specific categories of safety. Before you do any laboratory work, familiarize yourself with the following precautions:

Heat

1. Use only the source of heat specified in the investigation.

2. Never allow flammable materials, such as alcohol, near a flame or any other source of ignition.

3. When heating a substance in a test tube, point the mouth of the tube away from other students and yourself.

4. Never leave a lighted lab burner, hot plate, or any other hot object unattended.

5. Never reach over an exposed flame or other heat source.

6. Use tongs, test-tube clamps, insulated gloves, or pot holders to handle hot equipment.

Glassware

1. Never use cracked or chipped glassware.
2. Use caution and proper equipment when handling hot glassware; remember that hot glass looks the same as cool glass.
3. Make sure glassware is clean before you use it and clean when you store it.
4. When putting glass tubing into a rubber stopper, use a lubricant such as glycerine or petroleum jelly on both the stopper and the glass tubing. When putting glass tubing into or removing it from a rubber stopper, protect your hands with heavy cloth. Never force or twist the tubing.
5. Sweep up broken glassware immediately (never pick it up with your fingers), and discard it in a special labeled container for broken glass.

Electrical Equipment and Other Apparatus

1. Before you begin any work, always be sure you learn how to use each piece of apparatus safely and correctly to obtain accurate scientific information.
2. Never use equipment with frayed insulation or with loose or broken wires.
3. Make sure the area under and around electrical equipment is dry and free of flammable materials. Never touch electrical equipment with wet hands.
4. Turn off all power switches before plugging an appliance into an outlet. Never jerk wires from outlets or pull appliance plugs out by the wire.

Living and Preserved Specimens

1. Properly mount and support specimens for dissection. Do not cut a specimen while holding it in your hand.
2. Wash down your work surface with a disinfectant solution both before and after using living microorganisms.
3. Always wash your hands with soap and water after working with live or preserved specimens.
4. Care for animals humanely. General rules are
   a. Always carefully follow your teacher's instructions concerning the care of laboratory animals.
   b. Provide a suitable escape-proof container in a location where the animal will not be constantly disturbed.
   c. Keep the container clean. Cages of small birds and mammals should be cleaned daily. Provide proper ventilation, light, and temperature.
   d. Provide water at all times.
   e. Feed the animal regularly, depending on its needs.
   f. Treat laboratory animals gently and with kindness in all situations.
   g. If you are responsible for the regular care of any animals, be sure to make arrangements for weekends, holidays, and vacations.
   h. When animals must be disposed of or released, your teacher will provide a suitable method.
5. Many plants or plant parts are poisonous. Work only with the plants specified by your teacher. Never put any plant or plant parts in your mouth.

6. Handle plants carefully and gently. Most plants must have light, soil, and water, although requirements differ.

Accident Procedures

1. Report *all* incidents, accidents, injuries, breakages, and spills, no matter how minor, to your teacher.
2. If a chemical spills on your skin or clothing, wash it off immediately with plenty of water and notify your teacher.
3. If a chemical gets into your eyes or on your face, wash immediately at the eyewash station with plenty of water. Wash for at least 15 minutes, flushing the eyes—including under each eyelid. Have a classmate notify your teacher.
4. If a chemical spills on the floor or work surface, do not clean it up yourself. Notify your teacher immediately.
5. If a thermometer breaks, do not touch the broken pieces with your bare hands. Notify your teacher immediately.
6. Smother small fires with a wet towel. Use a blanket or the safety shower to extinguish clothing fires. Always notify your teacher.
7. Report all cuts and abrasions (no matter how small) received in the laboratory to your teacher.

Chemical Safety

All chemicals are hazardous in some way. A hazardous chemical is defined as a substance that is likely to cause injury. Chemicals can be placed in four hazard categories: flammable, corrosive, toxic, and reactive.

In the laboratory investigations for this course, every effort is made to minimize the use of dangerous materials. However, many "less hazardous" chemicals can cause injury if not handled properly. The following information will help you become aware of the types of chemical hazards that exist and of how you can reduce the risk of injury when using chemicals. Be sure also to review the basic safety rules described previously before you work with any chemical.

Flammable/Combustible Substances. Flammable/combustible substances are solids, liquids, or gases that will sustain burning. The process of burning involves three interrelated components—fuel (any substance capable of burning), oxidizer (often air or a specific chemical), and ignition source (a spark, flame, or heat). The three components are represented in Figure B. For burning to occur, all three components (sides) of the fire triangle must be present. To control a fire hazard, you must remove, or otherwise make inaccessible, at least one side of the fire triangle. Flammable chemicals should not be used in the presence of ignition sources such as lab burners, hot plates, and sparks from electrical equipment or static electricity. Containers of flammables should be closed when not in use. Sufficient ventilation in the laboratory will help keep the concentration of flammable vapors to a minimum. Wearing safety goggles, lab aprons, and protective gloves are important precautionary measures when using flammable/combustible materials.

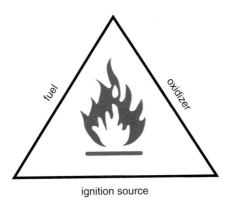

FIGURE B
The fire triangle.

Toxic Substances. Most of the chemicals in a laboratory are toxic, or poisonous to life. The degree of toxicity depends on the properties of the specific substance, its concentration, the type of exposure, and other variables. The effects of a toxic substance can range from minor discomfort to serious illness or death. Exposure to toxic substances can occur through ingestion, skin contact, and inhaling vapors. Wearing a lab apron, safety goggles, and protective gloves are important precautions when using toxic chemicals. A clean work area, prompt spill cleanup, and good ventilation also are important.

Corrosive Substances. Corrosive chemicals are solids, liquids, or gases that by direct chemical action either destroy living tissue or cause permanent change in the tissue. Corrosive substances can destroy eye and respiratory-tract tissues, causing impaired sight or permanent blindness, severe disfigurement, permanent severe breathing difficulties, and even death. Lab aprons, safety goggles, and protective gloves should be worn when handling corrosive chemicals to prevent contact with the skin or eyes. Splashes on the skin or in the eye should be washed off immediately while a classmate notifies your teacher.

Reactive Substances. Reactive chemicals promote violent reactions under certain conditions. A chemical may explode spontaneously or when mechanically disturbed. Reactive chemicals also include those that react rapidly when mixed with another chemical, releasing a large amount of energy. Keep chemicals separate from each other unless they are being combined according to specific instructions in an investigation. Heed any other cautions your teacher may give you. Always wear your lab apron, safety goggles, and protective gloves when handling reactive chemicals.

Record Keeping

Science deals with verifiable observations. No one—not even the original observer—can trust the accuracy of a confusing, indefinite, or incomplete observation. Scientific record keeping requires clear and accurate records made *at the time of observation.*

The best method of keeping records is to jot them down in a logbook. It should be a hardcover book, permanently bound (not loose-leaf).

Keep records in diary form, recording the date first. Keep observations of two or more investigations separate. Data recorded in words should be notes that are brief but to the point. Complete sentences are not necessary, but single words are seldom descriptive enough to represent accurately what you have observed.

You may choose to sketch your observations. A drawing often records an observation more easily, completely, and accurately than words can. Your sketches need not be works of art. Their success depends on your ability to observe, not on your artistic talent. Keep the drawings simple, use a hard pencil, and include clearly written labels.

Data recorded numerically as counts of measurements should include the units in which the measurements are made. Often numerical data are most easily recorded in a table.

Do not record your data on other papers to copy into the logbook later. Doing so might increase neatness, but it will decrease accuracy. Your logbook is *your* record, regardless of the blots and stains that are a normal circumstance of field and laboratory work.

You will do much of your laboratory work as a member of a team. Your logbook, therefore, will sometimes contain data contributed by other members of your team. Keep track of the source of observations by circling (or recording in a different color) the data reported by others.

Writing Laboratory Reports

Discoveries become a part of science only when they are made known to others. Communication, therefore, is an important part of science. In writing, scientists must express themselves so clearly that another person can repeat their procedures exactly. The reader must know what material was used (in biology this includes the species of organism) and must understand every detail of the work. Scientific reports frequently are written in a standard form as follows:

1. Title
2. Introduction: a statement of how the problem arose, often including a summary of past work
3. Materials and equipment: usually a list of all equipment, chemicals, specimens, and other materials
4. Procedure: a complete and exact account of the methods used in gathering data
5. Results: data obtained from the procedure, often in the form of tables and graphs
6. Discussion: a section that demonstrates the relationship between the data and the purpose of the work
7. Conclusion: a summary of the meaning of the results, often suggesting further work that might be done

8. References: published scientific reports and papers that you have mentioned specifically

Your teacher will tell you what form your laboratory reports should take for this course. Part of your work may include written answers to the Analysis questions at the end of each investigation. In any event, the material in your logbook is the basis for your reports.

Measurement

Measurement in science is made using the Systeme International d'Unités (international system of units), more commonly referred to as SI. A modification of the older metric system, it was used first in France and now is the common system of measurement throughout the world.

Among the basic units of SI measurement are the meter (length), the kilogram (mass), the kelvin (temperature), and the second (time). All other SI units are derived from these four. Units of temperature that you will use in this course are degrees Celsius, which are equal to kelvins.

SI units for volume are based on meters cubed. In addition, you will use units based on the liter for measuring volumes of liquids. Although not officially part of SI, liter measure is used commonly. The liter is a metric measurement ($1 L = 0.001 m^3$). It is equivalent to 1,000 cubic centimeters (cc).

Below are some of the SI units derived from the basic units for length, mass, volume, and temperature. Note, especially, those units described as most common to your laboratory work.

Length
1 kilometer (km) = 1,000 meters
1 hectometer (hm) = 100 meters
1 dekameter (dkm) = 10 meters
1 meter (m)
1 decimeter (dm) = 0.1 meter
1 centimeter (cm) = 0.01 meter
1 millimeter (mm) = 0.001 meter
1 micrometer (μm) = 0.000001 meter
1 nanometer (nm) = 0.000000001 meter

Measurements under microscopes often are made in micrometers (also called microns). Still smaller measurements, as for wavelengths of light used by plants in photosynthesis, are made in nanometers. The unit of length you will use most frequently in the laboratory is centimeter (cm).

Units of area are derived from units of length by multiplication. One hectometer squared is a measure often used for ecological studies; it is commonly called a hectare and equals 10,000 m^2. Measurements of area made in the laboratory most frequently will be in centimeters squared (cm^2).

Units of volume are also derived from units of length by multiplication. One meter cubed (m^3) is the standard, but it is too large for practical use in the laboratory. Centimeters cubed (cm^3) are the more common units you will see. For the conditions you will encounter in these laboratory investigations, 1 cm³ measures a volume equal to 1 mL.

Mass

1 kilogram (kg) = 1,000 grams
1 hectogram (hg) = 100 grams
1 dekagram (dkg) = 10 grams
1 gram (g)
1 decigram (dg) = 0.1 gram
1 centigram (cg) = 0.01 gram
1 milligram (mg) = 0.001 gram
1 microgram (μg) = 0.000001 gram
1 nanogram (ng) = 0.000000001 gram

Measurements of mass in your biology laboratory usually will be made in kilograms, grams, centigrams, and milligrams.

Volume

1 kiloliter (kL) = 1,000 liters
1 hectoliter (hL) = 100 liters
1 dekaliter (dkL) = 10 liters
1 liter (L)
1 centiliter (cL) = 0.01 liter
1 milliliter (mL) = 0.001 liter

Your measurements in the laboratory will usually be made in milliliters and liters.

Temperature

On the Celsius scale, 0°C is commonly known as the freezing point of water and 100°C is the boiling point of water. (Atmospheric pressure affects both of these temperatures.) Figure C illustrates the Celsius scale alongside the Fahrenheit scale, which is still used in the United States. On the Fahrenheit scale, 32°F is the freezing point of water and 212°F is the boiling point of water. The figure is useful in converting from one scale to the other.

Concentration

Another type of measurement you will encounter in this course is molarity (labeled with the letter M). Molarity measures the concentration of a dissolved substance in a solution. A high molarity indicates a high concentration. The solutions you will use in the investigations frequently will be identified by their molarity.

If you wish to learn more about SI measure, write to the U.S. Department of Commerce, National Institute of Standards and Technology (NIST), Washington, DC 20234.

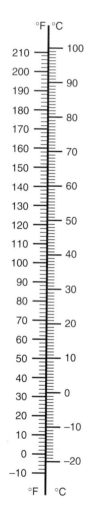

FIGURE C
Temperature-conversion scale.

Investigations for the Prologue
Biology and the Molecular Perspective

Biological Challenges in the prologue describes bioethics as the "application of ethics to biological issues." Ethical analysis is a process of critical inquiry. There are various models for conducting such inquiries; this investigation introduces you to one such model, which focuses on goals, rights, and duties. You will apply this model to a case study that involves the use of biosynthetic human growth hormone (GH).

Materials (per person)
paper and pen

Procedure

1. Read the following case study:
   Sharon has just completed ninth grade at Deerfield Junior High School in Iowa. She attended summer basketball camp for the past 3 years and already is one of the best players in the state. Deerfield's girl's basketball team has won two district championships, and Sharon—a forward—has been the leading scorer. It is clear she has great potential both as a high school and college player.

   Sharon and her coaches, however, are concerned about her height. She is only 5 feet 7 inches (170 cm) tall. Sharon's coaches want to develop her skills through high school and would like her to attain as much height as possible. She is not likely to gain much more height naturally. Sharon and her parents have discussed her problem with an endocrinologist and asked that he provide her with biosynthetic GH. Thus far he has refused, saying that GH is used only to treat children who are abnormally short. The biotech company that sells the hormone recommends that it be prescribed only to children who are among the shortest 3% in the population. Furthermore, a deficiency of GH is not the only reason for being short of stature. The doctor maintains that Sharon is not handicapped and should not risk the possible side effects of GH when she does not need the hormone for a legitimate medical reason. Although the consequences of receiving extra GH are not fully understood, there is reason for concern. Mice that received extra GH attained twice their normal size and appeared to be otherwise healthy. Pigs, on the other hand, did not fare as well. Most pig embryos that received genes for GH did not survive, and of those that did, many were arthritic, had vision problems, and displayed deficient immune systems. The potential side effects for humans, which are irreversible, include the development of diabetes and heart problems and elongation of the facial bones, hands, and feet.

   Sharon has told the doctor that she can get GH on the black market and will do so if he does not prescribe it for her. She prefers to have the hormone administered under the doctor's supervision because the hormone will be pure and he can monitor her progress. What should the doctor do?

2. Read the following information about goals, rights, and duties.

   Goals: One way to judge the morality of an action is by looking at what it intends to accomplish. If this view is the basis for determining whether an action is morally correct, then a "good" outcome may be judged morally correct no matter how the outcome is achieved. Assume, for example, that a given physician's primary goal is the preservation of life. She might then refuse to disconnect a respirator that is keeping a terminally ill patient alive, even if the patient or the patient's family wishes her to do so. In her view, going against the rights of the patient is justified by her goal of the preservation of life.

   Rights: Moral arguments based on rights are familiar to all of us. Our Constitution, for example, guarantees the right to free speech, the right to religious freedom, and the right to trial by a jury of one's peers. Most physicians agree that a patient has the right to know all of the relevant information about a given treatment. Consider a situation in which the physician knows that a patient will refuse lifesaving treatment if the patient knows all of the potential side effects. Should the physician violate the patient's *right* to the information to further the physician's *goal* of the preservation of life?

Duties: Some moral arguments are based on the obligation, or duty, to act in a certain way. For example, we generally have a duty to tell the truth, keep a promise, or help a friend. The justification for a duty often is based on the achievement of a worthy goal or on the basis of someone's right. Duties, therefore, can be derived from goals or rights, but they can also be in conflict with goals or rights. Suppose, for example, a dying man asks a physician not to prolong his life. Does the physician have a *duty* to respect the man's *right* to die? Or does the physician have a *duty* to pursue his own *goal* of the preservation of life?

3. Work in teams of three or four to discuss the case study of Sharon and her desire to use GH to increase her height. List the goals, rights, and duties for Sharon, the doctor, Sharon's parents, and the basketball coaches in this situation. When you have completed your discussion, your teacher will compile a list of the responses from each group so you can discuss the Analysis questions as a class.

Analysis

1. Where are the major conflicts in the goals, rights, and duties for Sharon, the doctor, Sharon's parents, and the basketball coaches?
2. Should the doctor prescribe GH for Sharon?
3. Is there any additional scientific information that would affect your ethical analysis?
4. Can you think of reasons for short stature besides a lack of GH?
5. What are the most important justifications for the position you have chosen?
6. Two U.S. companies now produce GH for distribution through prescription and under a physician's supervision. What are the goals, rights, and duties of these companies with respect to production and distribution of GH?
7. Is the recommendation that GH be administered to the shortest 3% of the population appropriate? Why or why not?

8. Who should determine the population eligible for receiving GH?
9. Is Sharon handicapped?
10. Assuming that GH is completely safe, would you take it?

Investigation PB ◆ Scientific Observation

Careful observation is important in any science. The results you obtain from an experiment must be replicable. Detailed observations and notes not only help identify possible areas of error, but also allow others to duplicate, and thus verify, your work. This investigation is an exercise in observation.

Sometimes careful observation is needed to distinguish one species from another, or even one sex from another within the same species. In this investigation, you will examine a mixed population of male and female fruit flies and attempt to separate them into two groups based on their sex (Figure PB.1). It is not expected that you will know which group corresponds to males and which group corresponds to females. The challenge is to develop observational criteria that will allow not only you but others to accurately distinguish between the sexes of fruit flies.

Materials (per team of 3)
hand lens
metric ruler
small paintbrush
mixture of male and female fruit flies (containing an equal number of each sex)

Procedure

1. Your team will be given a petri dish containing an equal number of male and female fruit flies. As you observe individual flies, use the paintbrush to gently move them about the dish.

FIGURE PB.1

What observations best describe the sex differences in fruit flies?

2. Use the hand lens and ruler to measure the length of each fly to the nearest 0.5 mm, and record it in your logbook. Observe each fly carefully, and record all of your observations in your logbook. Describe the shapes and colors of the major body parts (or sketch them).

3. When you have recorded as many observations as you can, discuss with your lab partners which observations allow you to separate the flies into two equal groups.

4. Recombine the two groups of flies, and use your agreed-upon criteria to re-sort the flies. If your criteria do not sort the flies into two equal groups, refine your observations so that you can clearly sort each fly into one of the two groups.

5. When you and your lab partners are satisfied with your sexing criteria, your teacher will give you instructions for another part to this exercise. It will be a realistic test of how careful you were in your observations and note taking.

Analysis

1. What were the different ways you found to distinguish male from female fruit flies?

2. What proved to be the most helpful information in distinguishing the sexes of fruit flies?

3. What percentage of your class could reproducibly sort their flies into two groups?

4. What percentage of your class could use another team's criteria to accurately sort flies by sex?

5. What steps could be taken to improve both percentages?

6. People often confuse observations with inferences. Observations are collected on the scene, using your senses. Inferences are ideas or conclusions based on what you observe or already know. Based on this distinction, which of the following statements are observations and which are inferences?
   • The size of individual flies in one group is larger than in the other group.
   • The wings of all the flies look similar.
   • The female flies are slightly larger than the male flies.
   • The two groups can be distinguished by the shapes of their abdomens.

• The male flies have bristles (sex combs) on their front legs.
• Both groups of flies have red eyes.
• One group has a striped abdomen.
• The female's abdomen becomes expanded prior to egg laying.

7. Look at your notes and label any inferences that you included.

Investigation PC ◆ The Compound Microscope

The human eye cannot distinguish objects much smaller than 0.1 mm. The microscope is a tool that extends vision and allows observation of much smaller objects. The most commonly used compound microscope (Figure PC.1) is monocular (one eyepiece). Light reaches the eye after passing through the objects to be examined.

In this investigation, you will learn how to use and care for a microscope.

Materials (per team of 2)

3 coverslips
dropping pipette
3 microscope slides
compound microscope
newspaper
scissors
transparent metric ruler
prepared slide of colored threads

Procedure

PART A   Care of the Microscope

1. The microscope is a precision instrument that requires proper care. Always carry the microscope with both hands—one hand under its base, the other on its arm.

2. When setting the microscope on a table, keep it away from the edge. If a lamp is attached to the microscope, keep its wire out of the way. Keep everything not needed for microscope studies off your lab table.

3. Avoid tilting the microscope when using temporary slides made with water.

4. The lenses of the microscope cost almost as much as all the other parts put together. Never

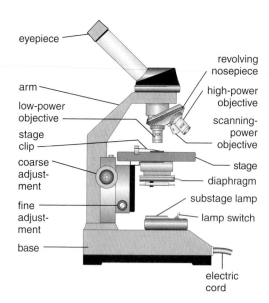

FIGURE PC.1

Parts of a compound microscope.

clean lenses with anything other than the lens paper designed for this task.

5. Before putting away the microscope, *always* return it to the low-power setting. The high-power objective reaches too near the stage to be left in place safely.

PART B   Setting Up the Microscope

6. Rotate the low-power objective into place if it is not already there. When you change from one objective to another, you will hear a click as the objective sets into position.

7. Move the mirror so that even illumination is obtained through the opening in the stage, or

TABLE PC.1
Microscopic Observations

Object being viewed	Observations and comments
Letter *o*	
Letter *c*	
Etc.	
Millimeter ruler	

turn on the substage lamp. Most microscopes are equipped with a diaphragm for regulating light. Some materials are best viewed in dim light; others, in bright light. Remember that direct sunlight can damage eyes. If you use natural light as your light source, do not reflect direct sunlight through the diaphragm.

8. Make sure that the lenses are dry and free of fingerprints and debris. Wipe lenses with lens paper only.

PART C   Using the Microscope

9. In your logbook, prepare a table like Table PC.1.

10. Cut a lowercase *o* from a piece of newspaper. Place it right side up on a clean slide. With a dropping pipette, place one drop of water on the letter. This type of slide is called a wet mount.

 CAUTION: Scissors are sharp. Handle with care.

11. Wait until the paper is soaked before adding a coverslip. Hold the coverslip at about a 45% angle to the slide, and slowly lower it. Figure PC.2 shows these first steps.

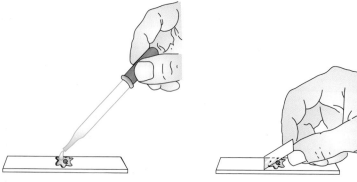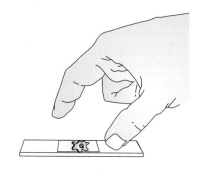

a

b

c

FIGURE PC.2

Preparing a wet mount with a microscope slide and coverslip.

12. Place the slide on the microscope stage, and clamp it down. Move the slide so the letter is in the middle of the hole in the stage. Use the coarse-adjustment knob to lower the low-power objective to the lowest position.

13. Look through the eyepiece, and use the coarse-adjustment knob to raise the objective slowly until the letter *o* is in view. Use the fine-adjustment knob to sharpen the focus. Position the diaphragm for the best light. Compare the way the letter looks through the microscope with the way it looks to the naked eye.

14. To determine how greatly magnified the view is, multiply the number inscribed on the eyepiece by the number on the objective being used. For example, eyepiece (10×) × objective (10×) = total magnification (100×).

15. Follow the same procedure with a lowercase *c*. In your logbook, describe how the letter looks when viewed through a microscope.

16. Make a wet mount of the letter *e* or the letter *r*. Describe how the letter looks when viewed through the microscope. What new information (not revealed by the letter *c*) is revealed by the *e* or *r?*

17. Look through the eyepiece at the letter as you use your thumbs and forefingers to move the slide *away* from you. Which way does your view of the letter move? Move the slide to the right. In which direction does the image move?

18. Make a pencil sketch of the letter as you see it under the microscope. Label the changes in image and movement that occur under the microscope.

19. Make a wet mount of two different-colored hairs, one light and one dark. Cross one hair over the other. Position the slide so that the hairs cross in the center of the field. Sketch the hairs under low power; then go to Part D.

PART D   Using High Power

20. With the crossed hairs centered under low power, adjust the diaphragm for the best light.

21. Turn the high-power objective into viewing position. Do *not* change the focus.

22. Sharpen the focus with the *fine-adjustment knob only. Do not focus under high power with the coarse-adjustment knob.*

23. Readjust the diaphragm to get the best light. If you are not successful in finding the object under high power the first time, return to step 20 and repeat the whole procedure carefully.

24. Using the fine-adjustment knob, focus on the hairs at the point where they cross. Can you see both hairs sharply at the same focus level? How can you use the fine-adjustment knob to determine which hair crosses over the other? Sketch the hairs under high power.

25. Remove the wet mount of the hairs, and replace it with the prepared slide of the colored threads. The prepared slide contains three colored threads that overlap in a specific order.

26. Focus the threads under low power, and adjust the diaphragm for best light.

27. Turn the high-power objective into viewing position. Do *not* change the focus.

28. Sharpen the focus with the fine-adjustment knob only.

29. Readjust the diaphragm to get the best light. If you are not successful in finding the threads under high power, return to step 26 and repeat the procedure.

30. Using the fine-adjustment knob, focus on an area where the threads overlap. Use the fine-adjustment knob to determine the order in which the colored threads lie on the slide.

PART E   Measuring with a Microscope

31. Because objects examined with a microscope usually are small, biologists use units of length smaller than centimeters or millimeters for microscopic measurement. One such unit is the micrometer, which is 1/1,000 of a millimeter. The symbol for micrometer is μm, the Greek letter μ (called mu) followed by m.

32. You can estimate the size of a microscopic object by comparing it with the size of the circular field of view. To determine the size of the field, place a plastic metric ruler on the stage. Use the low-power objective to obtain a clear image of the divisions on the ruler. Carefully move the ruler until its marked edge passes through the exact center of the field of view. Count the number of divisions that you can see in the field of view. The marks on the

ruler will appear quite wide; 1 mm is the distance from the center of one mark to the center of the next. Record the diameter, in millimeters, of the low-power field of your microscope.

33. Remove the plastic ruler, and replace it with the wet mount of the letter *e*. (If the wet mount has dried, lift the coverslip and add water.) Using low power, compare the height of the letter with the diameter of the field of view. Estimate as accurately as possible the actual height of the letter in millimeters.

Analysis

1. Summarize the differences between an image viewed through a microscope and the same image viewed with the naked eye.
2. When viewing an object through the high-power objective, not all of the object may be in focus. Explain your answer.
3. What was the order of the overlapping colored threads in step 30?
4. What is the relationship between magnification and the diameter of the field of view?
5. What is the diameter in micrometers of the low-power field of view of your microscope?
6. Calculate the diameter in micrometers of your high-power field. Use the following equations:

$$\frac{\text{magnification number of high-power objective}}{\text{magnification number of low-power objective}} = A$$

$$\frac{\text{diameter of low-power field of view}}{A} = \frac{\text{diameter of high-power field of view}}{}$$

For example, if the magnification of your low-power objective is $12\times$ and that of your high-power objective is $48\times$, $A = 4$. If the diameter of the low-power field of view is 1,600 μm, the diameter of the high-power field of view is 1,600 μm ÷ 4, or 400 μm.

7. Use your sketch of the hairs under high power and the diameter of your high-power field calculated above to estimate the diameter of your human hair.

Investigation PD ◆ Developing Concept Maps

As you study biology, you will be exposed to many new ideas and processes. At times the amount of information and the number of new words can seem overwhelming. There is, however, a method you can use to manage this new information. Concept maps are tools that can help you organize and review ideas in a way that emphasizes the relationships among ideas. Ideas are much easier to learn and remember once you understand how they are related to one another.

Concept maps are constructed using ideas, objects, processes, and actions as the *concept words*. Other words that explain the relationship between two concepts are the *linking words*. For example, look at the simple concept map in Figure PD.1 of the information about AIDS presented in Section P.2.

The words that appear in the boxes are the concepts. *Notice that the concepts become increasingly specific as you travel down the map.* The relationships between the concepts are shown by the connecting lines. The words that appear with the lines are the linking words that describe the relationships among the concepts.

Examine a second concept map of the AIDS discussion in Figure PD.2. Notice that this map makes different connections between concepts.

There is no single, correct concept map for a body of information. Each map may be different and may emphasize different concepts. In the second map, notice that some of the connecting lines do not follow the downward trend. These lines are cross-

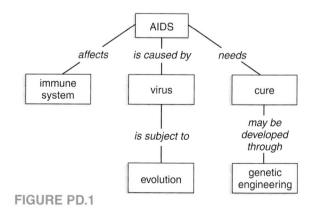

FIGURE PD.1

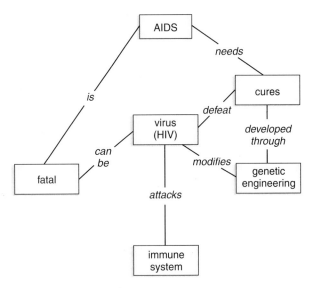

FIGURE PD.2

linkages. They describe additional, and perhaps more complex, relationships. You can follow a few simple steps that will help you build your own concept maps.

1. Identify and list the major concepts you want to map.
2. Decide which idea is the main concept. Group similar remaining concepts together, and rank them with each group from the more general to the most specific.
3. Choose linking words that identify the relationships between the concepts. Be sure that the linking words are not concepts themselves.
4. Begin constructing your map by branching one or two concepts from your major concept. Add the remaining more specific concepts as you progress. Be sure to look for opportunities to establish cross-linkages.

Materials (per team of 3)
pictures of laboratory equipment
3 pairs of scissors

PART A   Laboratory Equipment

Procedure
You will work in cooperative teams during this investigation, which means you must share your information to build a complete concept map. Each of you has received only part of a set of pictures. You will need to build a map of your set of pictures and then work with your teammates to construct a team map for all the pictures.

1. Cut out each individual picture of equipment from the sheet you receive.
2. Group the pieces of equipment that are somehow similar.
3. Subdivide the groups according to more specific characteristics. For example, the microscope is larger than the slides, and the slides are larger than the coverslips.
4. Using *laboratory equipment* as your main concept, arrange the pictures on a sheet of paper in the groups you have determined.
5. Link the pictures with linking words. For example, laboratory equipment *may be* glass or *may be* metal.
6. Try to establish some cross-linkages in your map. For example, glass items *can be used* with metal items.
7. Once you have completed your own map, work with your teammates to construct a team map of all the items, still using *laboratory equipment* as your main concept. Your teacher will assign you a job to do during the construction of the team map: The checker will make sure all concepts have been included and determine whether all team members agree on the structure of the map. The recorder will draw the map and make changes and corrections suggested by the team. The arranger will make sure linking words are accurate and that cross-linkages have been identified.
8. Once you have completed your map, your teacher will randomly select one member of your team to explain your map to the class. Any teammate may be called on, so all members must be able to explain the team's work.

Analysis
1. Compare your map with your classmates' maps. How is your concept map different?
2. Which concept map is easiest for you to understand?

3. What did members of your team do that worked well? What did not work well? What problems did you encounter? How did you solve those problems?

4. What would you do differently next time you work in a group?

PART B Evolution

9. Working with your teammates, use the concepts and linking words listed here to complete the concept map for *evolution* in Figure PD.3, based on Sections P.5 and P.6. (Linking words can be used more than once.) Add other concepts to the map if you wish.

Concept Words	Linking Words
evolution	is a
natural selection	developed by
observations	made
predictions	based on
acquired characteristics	lead to
Darwin	accounts for
Lamarck	about
theory	does not account for
diversity	
variation	

Analysis

1. Compare the concept maps you drew in Parts A and B. Do their characteristics differ? If so, how?

2. Compare your own concept maps for Part B with the maps of other teams. How are the maps different? How are they the same?

PART C Science and Pseudoscience

10. Construct a concept map dealing with science and pseudoscience based on Section P.8 in the text.

11. Use the following concepts and linking words. You may add additional concepts and linking words as you need them.

Concept Words	Linking Words
theory	may be
hypothesis	based on
data	may include/includes
observation	such as

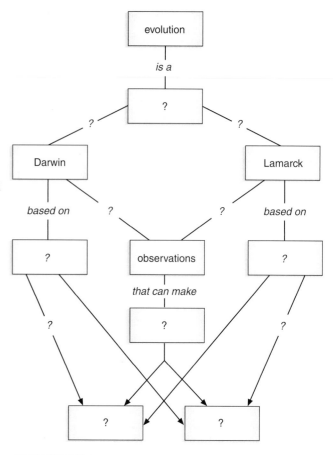

FIGURE PD.3

experimentation	from
faith	affects
science	results in
pseudoscience	

Analysis

1. Compare your concept map for Part C with the maps of other teams. How are they similar? How are they different?

Investigations for Chapter 1
The Chemistry of Life

Investigation 1A ◆ Organisms and pH

Individual organisms and cells must maintain a relatively stable internal environment, but many factors can affect the stability—for example, the relative concentrations of hydrogen ions (H^+) and hydroxide ions (OH^-). The biochemical activities of living tissues frequently affect pH, yet life depends on maintaining a pH range that is normal for each tissue or system. Using a pH meter or wide-range pH paper, you can compare the responses of several materials to the addition of an acid and a base.

Hypothesis: Before you begin, study the investigation and develop a hypothesis that addresses the question "How do organisms survive and function despite metabolic activities that tend to shift pH toward either acidic or basic ends of the scale?"

Materials (per team of 4)
4 pairs of safety goggles
4 lab aprons
50-mL beaker or small jar
50-mL graduated cylinder
3 colored pencils
pH meter or wide-range pH paper
forceps
tap water
HCl ($0.1M$) in dropping bottle
NaOH ($0.1M$) in dropping bottle
sodium phosphate pH 7 buffer solution
liver homogenate
potato homogenate
egg white (diluted 1:5 with water)
warm gelatin suspension (2%)

Procedure

1. In your logbook, prepare a table similar to Table 1A.1, or tape in the table provided by your teacher.
2. Pour 25 mL of tap water into a 50-mL beaker.
3. Record the initial pH by using a pH meter, or use forceps to dip small strips of pH paper into the water and compare the color change to a standard color chart.
4. Add $0.1M$ HCl a drop at a time. Gently swirl the mixture after each drop. Determine the pH after 5 drops have been added. Repeat this procedure until 30 drops have been used. Record the pH measurements in your table.

CAUTION: 0.1 M HCl is a mild *irritant*. Avoid skin/eye contact; do not ingest. If contact occurs, flush affected area with water for 15 minutes; rinse mouth with water; call your teacher.

5. Rinse the beaker thoroughly, and pour into it another 25 mL of tap water. Record the initial pH of the water, and add $0.1M$ NaOH drop by drop, recording the pH changes in exactly the same way as for the $0.1M$ HCl.

TABLE 1A.1 Testing pH														
	Tests with 0.1 *M* HCl							Tests with 0.1 *M* NaOH						
Solution Tested	pH after addition of							pH after addition of						
	0	5	10	15	20	25	30 drops	0	5	10	15	20	25	30 drops
Tap water														
Liver														
Potato														
Egg white														
Gelatin														
Buffer														

CAUTION: 0.1M NaOH is a mild *irritant*. Avoid skin/eye contact; do not ingest. If contact occurs, flush affected area with water for 15 minutes; rinse mouth with water; call your teacher.

6. Using the biological material assigned by your teacher, repeat steps 2–5. Record the data in your table.
7. Test the buffer solution (a nonliving chemical solution) using the same method outlined in steps 2–5. Record the data in your table.
8. Wash your hands thoroughly before leaving the laboratory.

Analysis

1. Summarize the effects of HCl and NaOH on tap water.
2. What was the total pH change for the 30 drops of HCl added to the biological material? for the 30 drops of NaOH added? How do these data compare with the changes in tap water?
3. In your logbook, prepare a simple graph of pH versus the number of drops of acid and base solutions added to tap water. Plot two lines—a solid line for changes with acid and a dashed line for changes with base. Using different colored solid and dashed lines, add the results for your biological material. Compare your graph to the graphs of teams who used a different biological material. What patterns do the graphs indicate for biological materials?
4. How do biological materials respond to changes in pH?
5. Use different colored solid and dashed lines to plot the reaction of the buffer solution on the same graph. How does the buffer system respond to the HCl and NaOH?
6. Is the pH response of the buffer system more like that of water or of the biological material?
7. How does the reaction of the buffer solution serve as a model for the response of biological materials to pH changes?
8. Would buffers aid or hinder the maintenance of a relatively stable environment within a living cell in a changing external environment?
9. What does the model suggest about a mechanism for regulating pH in an organism?

Investigation 1B ◆ Compounds of Living Things

The compounds your body needs for energy and growth are carbohydrates, proteins, fats, vitamins, and other nutrients. These compounds are present in the plants and animals you use as food. In this investigation, you will observe tests for specific compounds and then use those tests to determine which compounds are found in ordinary foods.

Materials (per team of 4)
4 pairs of safety goggles
4 lab aprons
4 pairs of plastic gloves
250-mL beaker
10-mL graduated cylinder
6 18-mm × 150-mm test tubes
test-tube clamp
test-tube rack
hot plate
Benedict's solution in dropping bottle
Biuret solution in dropping bottle
indophenol solution in dropping bottle
Lugol's iodine solution in dropping bottle
silver nitrate solution (1%) in dropping bottle
isopropyl alcohol (99%) in screw-top jar
brown wrapping paper
3 foods: apple, egg white, liver, onion, orange, or potato

SAFETY Put on your safety goggles, lab apron, and gloves. Tie back long hair.

Procedure

PART A   Test Demonstration

1. In your logbook, prepare a table similar to Table 1B.1.
2. Reagents are chemical solutions that scientists use to detect the presence of certain compounds. Observe the six reagent tests your teacher performs. In your table, describe the results of each test.

TABLE 1B.1
Reagent Tests of Known Food Substances

Food substance	Reagent test	Results
Gelatin	Biuret solution	
Glucose	Benedict's solution	
Starch	Lugol's iodine solution	
Vitamin C	indophenol solution (0.1%)	
Sodium chloride	silver nitrate solution (1%)	
Butter or vegetable oil	brown paper	

PART B   Compounds in Food

3. In your logbook, prepare a table similar to Table 1B.2. Record the presence (+) or absence (–) of each chemical substance in the foods you test.

WARNING: The reagents you will use in this procedure may be corrosive, poisonous, and/or irritants, and they may damage clothing. Avoid skin and eye contact; do not ingest. If contact occurs, flush the area with water for 15 minutes; rinse mouth with water; call your teacher immediately.

4. Predict the substances you will find in each sample your teacher assigns to you. Then test the samples as your teacher demonstrated or as described in steps 5–10. Record the result of each test in your logbook, using a + or –.

5. Protein test: Place 5 mL of the assigned food in a test tube. Add 10 drops of Biuret solution.

6. Glucose test: Add 3 mL of Benedict's solution to 5 mL of the assigned food. Place the test tube in a beaker of boiling water, and heat for 5 minutes.

7. Starch test: Add 5 drops of Lugol's iodine solution to 5 mL of the assigned food.

8. Vitamin C test: Add 8 drops of indophenol to 5 mL of the assigned food.

9. Chloride test: Add 5 drops of silver nitrate solution to 5 mL of the assigned food.

10. Fat test: Rub the assigned food on a piece of brown wrapping paper. Hold the paper up to the light. When food contains only a small amount of fat, the fat may not be detected by this method. If no fat has been detected, place the assigned food in 10 mL of a fat solvent such as isopropyl alcohol. Allow the food to dissolve in the solvent for about 5 minutes. Pour the solvent on brown paper. The spot should dry in about 10 minutes. Check the paper.

WARNING: Isopropyl alcohol is *flammable* and is a *poison*. Do not expose the liquid or its vapors to heat, sparks, open flame, or other ignition sources. Do not ingest; avoid skin/eye contact. If contact occurs, flush affected area with water for 15 minutes; rinse mouth with water. If a spill occurs, flood spill area with water; *then* call your teacher.

11. Wash your hands thoroughly before leaving the laboratory.

TABLE 1B.2
Analysis of Compounds in Common Food

Substance		Protein	Glucose	Starch	Vitamin C	Chloride	Lipid
Egg	Prediction						
	Test results						
Potato	Prediction						
	Test results						
Etc.	Prediction						
	Test results						

Analysis

1. How did your predictions compare with the test results?
2. Which of your predictions was totally correct?
3. Which foods contained all the compounds for which you tested?
4. On the basis of your tests, which food could be used as source of protein? glucose? starch? vitamin C? fat?
5. How might the original colors of the test materials affect the results?

Investigations for Chapter 2
Energy, Life, and the Biosphere

Investigation 2A ◆ Are Corn Seeds Alive?

Observe a corn seed carefully. Can you tell whether the seed is alive? How could you determine if it is? If a seed is alive, it is said to be viable—capable of growing and developing. There are two ways to investigate whether or not a seed is viable. One way is to perform a tetrazolium test. Tetrazolium is a colorless chemical that turns pink or red in the presence of hydrogen, which is released by all living organisms as they carry on their daily chemical activities. The second way is to do a germination test. You will do both these tests using corn seeds from two different batches, labeled I and II. Keep the two groups separate.

Materials (per team of 4)
4 pairs of safety goggles
4 lab aprons
4 pairs of plastic gloves
10-mL graduated cylinder
2 jars
2 petri-dish halves
forceps
glass-marking pencil
2 paper towels
2 rubber bands
2 scalpels
wax paper
20 mL tetrazolium reagent (1%)
soaked corn seeds (50 type I and 50 type II)

SAFETY Put on your safety goggles, lab apron, and gloves.

Procedure

PART A Tetrazolium Test

1. Mark the outside bottom of one petri-dish half *type I* and the outside bottom of the other petri-dish half *type II*.
2. Obtain 25 corn seeds of type I. Using the scalpel, cut each kernel lengthwise down the middle, as shown in Figure 2A.1a. The seeds may have been treated with a pesticide. Do not handle them without gloves. You should be able to see the miniature plant (the embryo) inside the seed after it is cut.

CAUTION: Scalpels are sharp; handle with care.

3. Discard one half of each seed, and place the other half in the team's petri dish with the cut surface down.
4. Cover the seed halves with 10 mL of tetrazolium reagent.

WARNING: Tetrazolium is a contact *irritant* and *poison.* Avoid skin/eye contact; do not ingest. If contact occurs, flush affected area with water for 15 minutes; rinse mouth with water; call your teacher immediately.

5. Repeat steps 2–4 with 25 seeds of type II.
6. After 25 minutes, use gloves and forceps to remove the seeds. Examine the cut surface of each seed for a color change. A red or pink color indicates a living substance.
7. Copy Table 2A.1 in your logbook, and record your results there.

PART B Germination Test

8. Place a wet paper towel on a sheet of wax paper (see Figure 2A.1b).
9. Place 25 type I corn seeds in five rows of five across the wet paper towel. Place the seeds so the pointed end is toward the top of the towel.

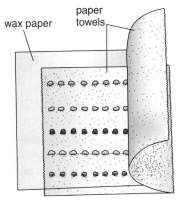
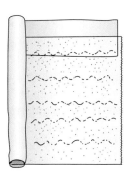

a   b

FIGURE 2A.1

(a), Procedure for cutting corn seed for tetrazolium test. **(b),** Technique for preparing seeds for germination in a moist paper roll.

10. Cover the seeds with another wet paper towel.
11. Fold the left side of the wax paper over the edge of the wet paper towels. Roll the wax paper, wet paper towels, and corn seeds from the left side toward the right side of the paper. Make the roll tight.
12. Secure the roll with a rubber band so it will stay rolled up, and place it upright in a jar of water (see Figure 2A.1b).
13. Repeat steps 8–12 with 25 type II corn seeds.
14. Label each jar either *type I or type II.* Place the jars in a dark area designated by your teacher.
15. Wash your hands thoroughly before leaving the laboratory.
16. After 3 days, determine the total number of seeds in each jar that have begun to germinate (grow). Record the data in the table.

Analysis

1. For each treatment done per team and per class, calculate the percentage viability as follows and enter the percentages in the table.

$$\frac{\text{number of seeds showing viability}}{\text{total number of seeds per treatment}} \times 100 = \%\text{ viability}$$

2. In Part A, if a color change indicates activity in living things, what can you conclude about the type I and type II seeds?
3. What evidence have you found that the type I seeds are alive? not alive?
4. What evidence have you found that the type II seeds are alive? not alive?

TABLE 2A.1
Results of Tetrazolium Test and Germination Test on Corn Seeds

	Tetrazolium test				Germination test			
	Type I seeds		Type II seeds		Type I seeds		Type II seeds	
	Team	Class	Team	Class	Team	Class	Team	Class
Number of seeds used								
Number of seeds viable or germinated								
Percentage of seeds viable or germinated								

5. Discuss in class the results of both experiments. Does the information provided by your teacher change any of your conclusions? Explain your answer.

6. How does the percentage of viability as determined from the tetrazolium test compare with the percentage from the germination test?

7. Are the two percentages the same? If not, can you suggest a reason for the difference?

8. Is it possible to tell if seeds are alive by just looking at them? Why or why not?

9. What is the advantage of combining data from several teams in the class?

Investigation 2B ◆ Food Energy

All foods contain energy, but the amount varies greatly from one food to another. You can use a calorimeter (Figure 2B.1) to measure the amount of energy, in calories, in some foods. A calorie is the amount of heat required to raise the temperature of 1 g (1 mL) of water 1°C. Calorie values of food on diet charts are given in kilocalories (1,000 calories), or kcals. (The kilocalorie is also referred to as the Calorie.) Your teacher will provide tables listing caloric values for common foods.

Using a thermometer, you can measure the change in temperature (ΔT) of a known volume of water. The water changes temperature by absorbing the heat given off by the burning of a known mass of food. Based on ΔT, you can calculate the amount of energy in the food.

Materials (per team of 3)
3 pairs of safety goggles
3 lab aprons
100-mL graduated cylinder
250-mL flask
nonmercury thermometer
balance

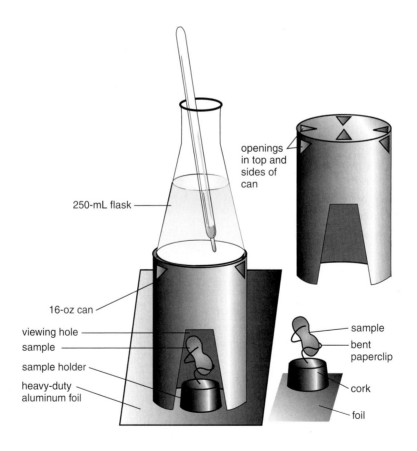

FIGURE 2B.1
The calorimeter setup.

16-oz. can with cutout air and viewing holes
cork with sample holder
kitchen matches
20-cm × 30-cm piece of heavy-duty aluminum foil
2 pot holders
small container of water
3 whole peanuts
3 walnut halves

 SAFETY **Put on your safety goggles and lab apron. Tie back long hair, and roll up long, loose sleeves.**

Procedure

1. Decide who in your team will be the experimenter, who will be the recorder, and who will be the safety monitor to assure correct safety procedures are followed.
2. Copy Table 2B.1 in your logbook, or tape in the table your teacher provides.
3. Using the balance, determine the mass to the nearest 0.1 g of each peanut and each walnut half. Record the masses in the table.
4. Obtain a 250-mL flask, a can, a cork with sample holder, and a piece of heavy-duty aluminum foil. Use equipment to make a calorimeter like the one shown in Figure 2B.1. Practice assembling and disassembling the equipment.
5. With the calorimeter disassembled, measure 100 mL of tap water and pour it into the flask.
6. Set the thermometer in the flask.
7. Measure the temperature of the water, and record it in the table.
8. Place a peanut in the wire holder anchored in the cork. Then place the cork on the piece of aluminum foil.
9. Carefully set fire to the peanut. This may require several matches. Discard burned matches in the container of water.

 WARNING: Matches are *flammable* solids. In case of burns, *immediately* place burned area under *cold* running water; *then* call your teacher.

10. Place the can over the burning sample with the viewing hole facing you. Place the flask of water on top of the can.
11. Take temperature readings as soon as the sample has burned out and then at 30-second intervals until the water temperature begins to decrease. (The temperature will continue to rise after the sample has burned out as the water absorbs heat from the can.)
12. Allow the calorimeter to cool about 2 minutes before disassembling.

TABLE 2B.1
Energy Content of Nut Samples

	Mass of sample	Temperature of water, °C			Food energy		
		Before burning	After burning	Change	calories	kcal	kcal per gram
Walnut sample 1							
Walnut sample 2							
Walnut sample 3							
Average							
Peanut sample 1							
Peanut sample 2							
Peanut sample 3							
Average							

WARNING: Use pot holders to handle hot flasks. Boiling water will scald, causing second-degree burns. Do not touch the flask or allow boiling water to contact your skin. Avoid vigorous boiling. If a burn occurs, *immediately* place burned area under *cold* running water; *then* call your teacher.

13. Repeat steps 5–12 until you have data for three samples each of a peanut and walnut half. Change the water in the flask each time.

14. Wash your hands thoroughly before leaving the laboratory.

Analysis

1. Prepare graphs of the data. Determine the average change in temperature for each sample. Calculate the number of kilocalories produced per gram. To do this, multiply the increase in water temperature (average change) by 100 (the number of milliliters of water used). This step will give you the number of calories. To convert to kilocalories, divide by 1,000 calories/kilocalorie. To calculate kilocalories produced per gram of food, divide this number by the number of grams of food burned. Enter all data in the table.

2. How do your data (adjusted for 100 g) compare with the values for 100 g of the same or similar food listed by your teacher? (The kilocalories listed in most diet charts are per 100 g, per ounce, per cup, or per serving. To compare your results, you may need to convert to common units.)

3. How do you account for any differences?

4. If the same amount of food you tested were completely burned in the cells of the human body, would you expect the energy release to be greater or less than your results? greater or less than published charts of the same caloric content of foods?

5. Which of the two foods tested seems to be the better energy source?

6. Why might some foods with fewer kilocalories be better energy sources than other foods with more kilocalories?

7. What was the original source of energy in all the foods tested?

Investigation 2C ◆ Enzyme Activity

Enzymes are biological catalysts (usually proteins) that speed up the rates of chemical reactions that take place within cells. In this investigation, you will study several factors that affect the activity of enzymes. The specific enzyme you will use is catalase, which is present in most cells and found in high concentrations in liver and blood cells. You will use liver homogenate as the source of catalase. Catalase promotes the decomposition of hydrogen peroxide (H_2O_2) in the following reaction:

$$2H_2O_2 \xrightarrow{\text{catalase}} 2H_2O + O_2$$

Hydrogen peroxide is formed as a by-product of chemical reactions in cells. It is toxic and soon would kill cells if not immediately removed or broken down. (Hydrogen peroxide is also used as an antiseptic. It is not a good antiseptic for open wounds, however, as it quickly is broken down by the enzyme catalase, which is present in human cells.)

Materials (per team of 3)
3 pairs of safety goggles
3 lab aprons
50-mL beaker
2 250-mL beakers
10-mL graduated cylinder
50-mL graduated cylinder
reaction chamber
6 18-mm × 150-mm test tubes
forceps
square or rectangular pan
test-tube rack
nonmercury thermometer
filter-paper disks
ice
water bath at 37°C
buffer solutions: pH 5, pH 6, pH 7, pH 8
catalase solution
fresh H_2O_2 (3%)

 SAFETY Put on your safety goggles and lab apron. Tie back long hair.

Procedure

In all experiments, make certain that your reaction chamber is scrupulously clean. Catalase is a potent enzyme. If the chamber is not washed thoroughly, the enzyme will adhere to the sides and make subsequent tests inaccurate. Measure all substances carefully. Results depend on comparisons between experiments, so the amounts measured must be equal or your comparisons will be valueless. Before you do the experiments, read through the instructions completely. Make sure that you have all required materials on hand, that you understand the sequence of steps, and that each member of your team knows his or her assigned function.

PART A   The Time Course of Enzyme Activity

1. Prepare two tables in your logbook similar to Table 2C.1. One will record your team's data while the other will record data averaged for the entire class.
2. Obtain a small amount of catalase solution in a 50-mL beaker.
3. Obtain a reaction chamber and a number of filter-paper disks.
4. Place four catalase-soaked filter-paper disks on *one* interior sidewall of the reaction chamber. (They will stick to the sidewall.) Prepare a disk for use in the reaction chamber by holding it by its edge with a pair of forceps and dipping it into the catalase solution for a few seconds. Drain excess solution from the disk by holding it against the side of the beaker before you transfer it to the reaction chamber.
5. Stand the reaction chamber upright, and carefully add 10 mL of 3% hydrogen peroxide (H_2O_2) solution. *Do not allow the peroxide to touch the filter-paper disks.*

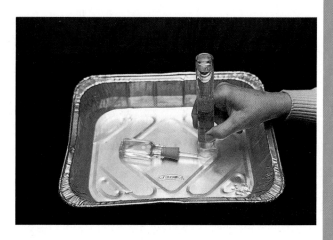

FIGURE 2C.1

Apparatus for measuring oxygen production in a reaction between catalase and hydrogen peroxide.

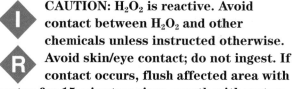

CAUTION: H_2O_2 is reactive. Avoid contact between H_2O_2 and other chemicals unless instructed otherwise. Avoid skin/eye contact; do not ingest. If contact occurs, flush affected area with water for 15 minutes; rinse mouth with water; call your teacher immediately.

6. Tightly stopper the chamber.
7. Fill a pan almost full with water.
8. Lay the 50-mL graduated cylinder on its side in the pan so that it fills with water completely. If any air bubbles are present, carefully work these out by tilting the cylinder slightly. Turn the cylinder upside down into an upright position, keeping its mouth underwater at all times.
9. Making certain the side with the disks is at the top, carefully place the reaction chamber and its contents on its side in the pan of water.
10. Move the graduated cylinder into a position (Figure 2C.1) where its mouth lies directly over

---

TABLE 2C.1
Catalase Activity under Various Conditions

Experiment	mL O_2 evolved/30 sec																			
Full concentration	1	2	3	4	5	6	7	8	9	10	11	12	13	14	15	16	17	18	19	20
3/4 concentration																				
Etc.																				

the tip of the dropping pipet that extends from the reaction chamber. One member of the team should hold it in this position for the duration of the experiment.

11. Rotate the reaction chamber 180 degrees on its side so that the hydrogen peroxide solution comes into contact with the soaked disks.

12. Measure the gas levels in the graduated cylinder at 30-second intervals for 10 minutes. Record the levels in your data table.

13. Pool your team's data with the other teams in the class. Record class average values of gas levels for each 30-second interval in your second data table.

PART B   The Effect of Enzyme Concentration on Enzyme Activity

14. Test 3/4, 1/2, and 1/4 concentrations of enzyme solution using the procedures for Part A with the following changes:
    - 3/4 concentration: Use three catalase-soaked disks instead of 4.
    - 1/2 concentration: Use two catalase-soaked disks and a 10-mL graduated cylinder.
    - 1/4 concentration: Use one catalase-soaked disk and a 10-mL graduated cylinder.

15. Record all data in your data table.

16. Pool data over the entire class, and record average values in your second data table.

PART C   The Effect of Temperature on Enzyme Activity

17. Add 10 mL of 3% H_2O_2 to each of two test tubes. Place one tube in a beaker of ice water and the other in a beaker with water maintained at 37°C.

18. When the temperature of the contents of the chilled H_2O_2 reaches approximately 10°C, repeat Part A with the following changes:
    - In step 5, use 10 mL of chilled 3% H_2O_2.
    - In step 7, add ice to the pan to chill the water to approximately 10°C.

19. When the temperature of the warmed H_2O_2 reaches approximately 37°C, repeat Part A with the following changes:
    - In step 5, use 10 mL of warmed 3% H_2O_2.
    - In step 7, fill the pan with water warmed to approximately 37°C.

20. Record the data in your data table.

21. Pool data over the entire class, and record average values in your second data table.

PART D   The Effect of pH on Enzyme Activity

22. Label four test tubes as follows: *pH 5*, *pH 6*, *pH 7*, and *pH 8*. Add to each of these 8 mL of 3% H_2O_2.

23. Add 4 mL of pH 5 buffer solution to the *pH 5* test tube, shaking well to ensure mixing. To each of the other three test tubes, add 4 mL of pH 6, pH 7, and pH 8 buffer solutions, shaking each test tube well.

24. Repeat the procedure of Part A for each pH value, substituting the buffered 3% H_2O_2 solutions in step 5.

25. Record the results in your data table.

26. Pool data over the entire class, and record average values in your second data table.

27. Wash your hands thoroughly before leaving the laboratory.

Analysis

1. Why is it a good idea to pool data from the entire class? Use the class averaged data to answer the following questions.

2. In your logbook, plot the data from Part A on the graph. Label the horizontal axis *Time (sec)* and the vertical axis *mL O_2 evolved*. Does the action of the catalase change through time? Explain your answer.

3. Plot the data from Part B on the grid used for Part A, and label the enzyme concentrations on the graph. Based on these data, how does enzyme activity vary with concentration?

4. Copy the graph for Part A, and plot the data from Part C on it. Based on these data, how does temperature affect enzyme action?

5. Plot the results for all four runs of Part D on a third graph. How does pH affect the activity of enzymes?

6. What is a buffer? Would the results of Parts A, B, and C have been different if buffers also had been used in those experiments? If so, how?

7. Summarize the general conditions necessary for effective enzyme action. Are these conditions the same for each enzyme? Why or why not?

8. How would you design an experiment to show how much faster H_2O_2 decomposes in the presence of catalase than it does without the enzyme?

9. Explain why the enzyme catalase was still active even though the liver cells from which you obtained the enzyme were no longer living.

Investigation 2D ◆ Starch Digestion

How do you obtain the energy you need for your activities? Starch makes up a large part of the food of many organisms and is a major source of energy for them. Human saliva contains an enzyme, amylase, which begins the breakdown of starch into sugar molecules that can be absorbed into the bloodstream and taken up by cells. In this investigation, you will explore the action of amylase on starch and identify some sources of this enzyme.

To determine if the results you observe are caused by amylase or by some other factor, such as elapsed time, you will set up controls. In an experiment, all the factors that might cause the observed results are referred to as variables. When scientists test a particular variable, they must control all the other variables so they know what causes the results. One way to do this is to set up additional tests, called controls, to rule out the other variables. For example, when starch and amylase are dissolved in water, a change in the starch is observed. To be sure this change is caused by the amylase rather than by the water, you could set up one test with starch and amylase and another with starch and water. The starch and water become the control for the test of the effect of amylase on starch.

Scientists used special chemical solutions called indicators to detect the presence of certain substances. For example, Lugol's iodine solution is used to detect the presence of starch.

Materials (per team of 4)
4 pairs of safety goggles
4 lab aprons
7 dropping pipettes
6 18-mm × 150-mm test tubes
10-mL graduated cylinder
spot plate
test-tube rack
6 rubber stoppers for test tubes
glass-marking pencil
funnel
scissors
mortar and pestle
cheesecloth (several layers)
pinch of sand
10 mL 5% starch suspension
10 mL 0.1% starch suspension
10 mL 1% amylase solution
25 mL distilled water
Lugol's iodine solution in dropping bottle
7 glucose test strips
5 g fresh spinach

SAFETY Put on your safety goggles and lab apron. Tie back long hair.

PART A Designing a Controlled Experiment
Participate in the discussion/demonstration that your teacher leads, keeping notes in your logbook.

PART B Starch Digestion in Animals
Half of the teams will conduct this part of the activity.

Day 1
1. Read through the procedure for day 1, and prepare a logbook in which to record predictions, observations, and results of any tests.
2. Place 5 mL of starch suspension and 5 mL of amylase solution into a test tube labeled *starch/amylase*.
3. What factors (variables) might affect your results in step 2? Set up controls for those variables in additional test tubes. Label the controls appropriately, and record the contents of each control in your data table.
4. Stopper the tubes, and store them in a dark place for 24 hours.
5. Based on the introduction to this activity, predict what substances will be present in each tube after 24 hours, recording your predictions in your data table.
6. Wash your hands before leaving the laboratory.

Day 2

7. Observe the tubes that you set up yesterday, and record your observations in the table.

8. What indicator tests should you use to determine what has occurred in each tube? Perform those tests, recording the results in your data table and in the class data table that your teacher has set up.
   Use dropping pipettes to remove samples, a spot plate for the tests, and the test indicators you observed in Part A.

 WARNING: Lugol's iodine solution is a poison if ingested, is a *strong irritant*, and can stain clothing. Avoid skin/eye contact; do not ingest. If contact occurs, flush affected area with water for 15 minutes; rinse mouth with water; call your teacher immediately.

9. Wash your hands thoroughly before leaving the laboratory.

PART C Starch Digestion in Plants
Half of the teams will conduct this part of the activity.

Day 1

1. Using scissors, cut up about 5 g of fresh spinach leaves into small pieces.

2. Put the pieces in a mortar with 10 mL of distilled water and a pinch of sand. Grind them thoroughly with a pestle.

3. Line the funnel with several layers of cheesecloth, and set it in a test tube labeled *filtrate.* Filter the crushed plant material through the funnel. Save the resulting filtrate (plant juice).

4. Prepare a data table in which to record predictions, observations, and results of any tests.

5. Using the indicators you observed in Part A, can you tell what substances probably are in the filtrate? Perform appropriate tests to find out. Record your results in your data table.
   Use a dropping pipette to remove samples, a spot plate for the tests, and the test indicators you observed in Part A.

 WARNING: Lugol's iodine solution is a poison if ingested, is a *strong irritant*, and can stain clothing. Avoid skin/eye contact; do not ingest. If contact occurs, flush affected area with water for 15 minutes; rinse mouth with water; call your teacher immediately.

6. Into a test tube labeled *filtrate/starch*, place 5 mL of filtrate and 5 mL of starch suspension, swirling to mix the solutions.

7. What factors (variables) might affect your results in step 6? Set up controls for those variables in additional test tubes. Label the controls appropriately, and record the contents of each control in your data table.

8. Stopper the tubes, and store them in a dark place for 24 hours.

9. Based on the results from step 5, predict what substances will be present in each tube after 24 hours. Record your predictions in your data table.

10. Wash your hands thoroughly before leaving the laboratory.

Day 2

11. Using the procedures of step 5, test the contents of the tubes that you set up in steps 6 and 7. Record the results in your data table and in the class data table that your teacher has set up.

12. Wash your hands before leaving the laboratory.

Analysis

1. From the results of Part B, what can you conclude about the effect of amylase on starch?

2. From the results of Part C, what can you conclude about a plant's ability to digest starch?

3. If you had not controlled the variables in this experiment, what conclusions could you have drawn?

4. Of what value to an organism is the ability to digest starch?

5. Animals have special glands for the production of enzymes that break large food molecules into small ones. What, if any, evidence from the results of this experiment indicates that plant cells possess similar enzymes?

6. Describe the processes responsible for food breakdown in the major compartments of the digestive system and the role of each.

7. Why do doctors give glucose rather than starch as intravenous injections to patients whose digestive systems are not functioning properly?

Investigations for Chapter 3
Exchanging Materials with the Environment

Investigation 3A ◆ Cells and Movement of Materials

To survive, all organisms need to balance their internal environment despite constantly changing conditions. Multicellular organisms have complex systems of organs that maintain this balance, but a single cell must rely on a solitary structure. The contents of a cell are surrounded by a thin membrane—the plasma membrane. Anything entering or leaving the cell must pass through this membrane. Certain materials can pass into the cell, whereas others cannot. Similarly, some materials pass out of the cell, while others that are vital to the cell's existence remain inside.

In this investigation, you will explore the movement of substances through membranes—both living plasma membranes and nonliving material that models plasma membranes. A nonliving material—called dialysis tubing—separates larger molecules from smaller molecules. The size of the pores in the tubing determines which molecules may pass through it. A balloon also serves as a model of a plasma membrane, allowing some molecules to pass through easily while retaining others.

Materials (per team of 2)
2 pairs of safety goggles
2 lab aprons
2 250-mL beakers
5 coverslips
4 dropping pipettes
5 microscope slides
2 test tubes
glass-marking pencil
compound microscope

2 15-cm pieces of dialysis tubing
3 balloons
hot plate
paper towel
4 10-cm pieces of string
test-tube clamp
test-tube rack
glucose test strips
brilliant cresyl blue solution in dropping bottle (5%)
distilled water
mineral oil
white vinegar
vanilla extract
15 mL glucose solution
salt solution (5%)
15 mL soluble-starch solution
Lugol's iodine solution in dropping bottle
yeast suspension
elodea leaf *(Anacharis)*

 SAFETY Put on your safety goggles and lab apron. Tie back long hair.

PART A Diffusion through Dialysis Tubing

Procedure
1. Twist one end of a piece of dialysis tubing. Fold the twisted end over, and tie it tightly with a piece of string. Prepare the other piece the same way.
2. Pour soluble-starch solution to within 4 cm of the top of one piece of tubing. Twist and tie the end as in step 1. Rinse the tubing under running water to remove any starch from the outside.
3. Place the tubing in a beaker of water labeled *A* (Figure 3A.1). Add enough Lugol's iodine solution to give the water a distinct yellowish color.

 WARNING: Lugol's iodine solution is a poison if ingested, is a *strong irritant*, and can stain clothing. Avoid skin/eye contact; do not ingest. If contact occurs, flush affected area with water for 15 minutes; rinse mouth with water; call your teacher immediately.

Lugol's iodine solution

dialysis tubing containing starch solution

water in beaker

FIGURE 3A.1
Setup for diffusion.

4. Repeat step 2 with the second piece of dialysis tubing, using glucose solution instead of the soluble starch. Place this tubing in a beaker of water labeled *B*.

5. Allow the pieces of tubing to stand for about 20 minutes. Dip a glucose test strip into the water in beaker B. Record the color on the strip.

6. Observe the tubing in beaker A. Record any changes, including color, that you see in either the tubing or the water in the beaker.

7. Let beakers A and B stand overnight. Record any changes observed the next day.

8. Wash your hands thoroughly before leaving the laboratory.

Analysis

1. On the basis of the results from steps 3 and 6, what must have happened to the iodine molecules in beaker A?

2. On the basis of the chemical test for glucose (step 5), what must have happened to the glucose molecules in beaker B?

3. From the evidence obtained after the beakers stood overnight, what other substances passed through the membrane in beaker B?

4. Which substance did not pass through a membrane? Explain your answer.

5. Physicists can show that all the molecules of a given substance are about the same size but that molecules of different substances are different in size. Measurements show that iodine

molecules and water molecules are very small, glucose molecules are considerably larger, and starch molecules are very large. On the basis of this information, suggest a hypothesis to account for your observations.

6. What assumption did you make about the structure of the dialysis tubing?

PART B   Diffusion through a Balloon

Procedure

9. Blow up a balloon, and tie off the end.

10. Use a dropping pipette to place one drop of water on the surface of the balloon. Note how the water drop behaves as it interacts with the balloon surface. Use a paper towel to wipe the water drop off the surface of the balloon.

11. Use a dropping pipette to place one drop of mineral oil on the surface of the balloon. Note how the oil drop behaves as it interacts with the balloon surface.

12. Use a dropping pipette to add about 2 mL of white vinegar to the inside of an uninflated balloon. (White vinegar contains water and acetic acid.) Blow up the balloon, and tie off the end.

13. Consider your observations from step 10, and predict whether you will be able to smell the vinegar inside the balloon. Take the balloon to an area of the room away from open containers or pipettes that contain vinegar, and conduct the smell test. Record your observations in your logbook.

14. Use a dropping pipette to place about 2 mL of vanilla extract inside a second uninflated balloon. Blow up the balloon, and tie off the end.

15. Consider your observations from step 11, and predict whether you will be able to smell the vanilla extract inside the balloon. Take the balloon to an area of the room away from open containers or pipettes that contain vanilla extract, and conduct the smell test. Record your observations in your logbook.

Analysis

1. On the basis of your smell test, did the vinegar molecules pass through the balloon?

2. On the basis of your smell test, did the vanilla extract molecules pass through the balloon?
3. Although a balloon membrane is much thicker than a plasma membrane of a cell, both can be described as nonpolar membranes. Nonpolar membranes repel charged particles and polar molecules but let nonpolar (fat soluble) molecules pass through. Use this information to suggest a hypothesis that explains your observations.
4. Why can polar water molecules pass through a cell's plasma membrane?

PART C   Osmosis and the Living Cell

Procedure

16. Place a leaf from the growing tip of an elodea plant in a drop of water on a clean slide. Add a coverslip and examine under low power. Position the slide so that the cells along one edge of the leaf are near the center.
17. Switch to high power, and focus sharply on a few cells near the edge of a leaf. Place a small piece of absorbent paper at the edge of the coverslip opposite the side of the leaf you are observing. (Remember, directions are reversed when you look through a microscope.) Have your lab partner place several drops of glucose solution at the coverslip edge nearest the part of the leaf being observed. Use the fine adjustment to adjust the focus while the water is being replaced. Continue observing the cells until you see changes in them.
18. Make simple sketches showing cells both before and after the glucose solution was added.
19. Remove the glucose solution, and replace it with distilled water. Use a new piece of absorbent paper, and allow 2 or 3 drops of distilled water to flow across the slide into the paper to make sure that most of the glucose solution is washed away. Make observations while this is being done.
20. Exchange places with your lab partner. Repeat steps 17–19.
21. Repeat steps 17–19 with salt solution in place of the glucose solution.

Analysis

1. Did water move into or out of the cells while the leaf was surrounded by the glucose solution? by the salt solution? What evidence do you have to support your answer.
2. In which direction did water move through the plasma membrane when the cell was surrounded by distilled water?
3. What do you think would happen to elodea cells if they were left in the glucose solution for several hours? Could elodea from a freshwater lake be expected to survive if transplanted into the ocean? (Assume that the salt concentration of the ocean is about the same as the salt solution used in this experiment.)
4. An effective way to kill plants is to pour salt on the ground around them. Using principles discovered in this investigation, explain why the plants die.
5. Bacteria cause food to spoil and meat to rot. Explain why salted pork, strawberry preserves, and sweet pickles do not spoil even though they are exposed to bacteria. Name other foods preserved in the same manner.

PART D   Membranes in Living and Dead Cells

Procedure

22. Place one drop of yeast suspension on a slide, add a coverslip, and observe the yeast cells under low power and then high power. Describe what you see, and sketch two or three cells to show their general appearance.
23. Place about 1 mL of yeast suspension in each of two small test tubes. Label one tube *boiled* and the other *unboiled*. Heat one of the test tubes in a beaker of boiling water until the contents have boiled for at least 2 minutes. This action will kill the yeast cells. Allow the test tube and its contents to cool for a few minutes.

WARNING: Use test-tube clamps to hold hot test tubes. Boiling water will scald, causing second-degree burns. Do not touch the beaker or allow boiling water to contact your skin. Avoid vigorous boiling. If a burn occurs, *immediately* place the burned area under cold running water; *then* call your teacher.

24. Add 5 drops of brilliant cresyl blue solution to the boiled yeast suspension and 5 drops to the unheated yeast suspension.

> **CAUTION: Brilliant cresyl blue solution is a mild *irritant*. Avoid skin/eye contact; do not ingest. If contact occurs, flush affected area with water for 15 minutes; rinse mouth with water; call your teacher.**

25. Label one microscope slide *boiled* and another *unboiled*. Prepare a slide from each test tube and examine under high power. Record any differences between the yeast cells in the two suspensions.
26. Wash your hands thoroughly before leaving the laboratory.

Analysis

1. What effect does heat seem to have on the yeast plasma membrane?
2. In a preparation of unheated yeast solution and brilliant cresyl blue, a few blue yeast cells are usually visible. What assumption can you make concerning these cells?
3. Which passes more easily through membranes of living cells, brilliant cresyl blue molecules or water molecules? Develop a hypothesis to account for your observation and answer.

Investigation 3B ◆ Diffusion and Cell Size

Does diffusion proceed rapidly enough to supply a cell efficiently with some of its materials? The same question can be asked about removing cell wastes. In this investigation, you will discover how the rate of diffusion and the size of a cell are related.

Materials (per team of 2)
2 pairs of safety goggles
2 lab aprons
2 pairs of plastic gloves
250-mL beaker
metric ruler
paper towel
plastic knife
plastic spoon
3 cm × 3 cm × 6 cm cube of phenolphthalein agar
150 mL HCl (0.1%)

> **SAFETY** Put on your safety goggles, lab apron, and gloves.

Procedure

1. Using a plastic knife, trim the agar block to make three cubes—3 cm, 2 cm, and 1 cm on a side.
2. Place the cubes in the beaker, and add 0.1% HCl until the cubes are submerged. Record the time. Use the plastic spoon to turn the cubes frequently for the next 10 minutes.

> **CAUTION: 0.1% HCl is a mild *irritant*. Avoid skin/eye contact; do not ingest. If contact occurs, flush affected area with water for 15 minutes; rinse mouth with water; call your teacher.**

3. Prepare a table similar to Table 3B.1, and do the calculations necessary to complete it. The ratio of surface area to volume is calculated as follows:

$$\text{ratio of surface area to volume} = \frac{\text{surface area}}{\text{volume}}$$

This ratio also may be written as "surface area:volume." The ratio should be expressed in its simplest form (for example, 3:1 rather than 24:8).

4. Wear gloves and use the plastic spoon to remove the agar cubes from the HCl after 10 minutes. Blot them dry. Avoid handling the cubes until they are blotted dry. Use the plastic knife to slice each cube in half. Rinse and dry the knife between cuts. Record your observations of the sliced surface. Measure the depth of diffusion of the HCl in each of the three cubes.

Analysis

1. List the agar cubes in order of size, from largest to smallest. List them in order of the ratios of surface area to volume, from the largest to the smallest ratio. How do the lists compare?
2. Calculate the ratio of surface area to volume for a cube 0.01 cm on a side.
3. Which has the greater surface area, a cube 3 cm on a side or a microscopic cube the size of an onion-skin cell? (Assume the cell to be 0.01 cm

TABLE 3B.1
Comparison of Agar Cubes

Cube dimension	Surface area (cm²)	Volume (cm³)	Simplest ratio
3 cm			
2 cm			
1 cm			
0.01 cm			

on a side.) Which has the greater surface area *in proportion* to its volume?

4. What evidence is there that HCl diffuses into an agar cube? What evidence is there that the rate of diffusion is about the same for each cube? Explain your answer.

5. What happens to the ratio of surface area to volume of cubes as they increase in size?

6. Most cells and microorganisms measure less than 0.01 cm on a side. What is the relationship between rate of diffusion and cell size?

7. Propose a hypothesis to explain one reason why large organisms have developed from *more* cells rather than *larger* cells.

Investigation 3C • The Kidney and Homeostasis

The cells of the human body are surrounded by liquid that is remarkably constant in its properties. The continuous regulation of the many dissolved compounds and ions in this internal environment is referred to as homeostasis.

The kidneys play an important role in homeostasis by regulating blood composition and by regulating the levels of many important chemicals and ions. The production of urine and its elimination from the body are critical functions of the kidneys and the urinary system (Figure 3C.1).

Materials (per team of 3)
pencil and paper

PART A   Blood versus Urine

Procedure
The relationship of structure and function in the kidney is illustrated in Figure 3.21 in Chapter 3. Use this illustration and the data in Table 3C.1 to answer the Analysis questions.

Analysis
1. What do the data for water indicate?
2. Protein molecules are not normally found in the urine. Explain why.
3. The information for glucose is similar to that for protein. Explain these data.
4. Based on what the sodium data indicate, what do you think may happen to the sodium content in the urine of a person who increases his or her intake of sodium chloride?

TABLE 3C.1
Comparison of Substances in Blood and Urine

	% in blood as it enters kidney	% in urine as it leaves kidney
Water	91.5	96.0
Protein	7.0	0.0
Glucose	0.1	0.0
Sodium	0.33	0.29
Potassium	0.02	0.24
Urea	0.03	2.70

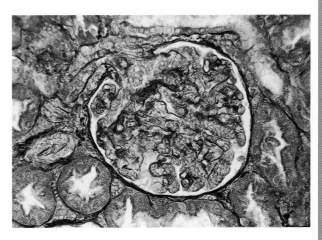

FIGURE 3C.1

Micrograph of the glomerulus of a nephron (×260). The glomerulus is a capillary network through which fluid and other materials pass from the blood into the tubules of the nephron.

TABLE 3C.2
Proportions of Substances at Four Points along the Nephron*

	In blood entering glomerulus	In tubule from glomerulus	In urine leaving nephron	In blood leaving nephron†
Water	100	30	1	99
Protein	100	0	0	100
Glucose	100	20	0	100
Sodium	100	30	1	99
Potassium	100	23	12	88
Urea	100	50	90	10

*The numbers in this table represent proportions, not actual numbers of molecules or ions. For example, for every 100 molecules of water in the blood, 30 will be found in the tubule.

†The numbers in this column were obtained by subtracting the proportionate number of molecules of the substance in the urine from the proportionate number of molecules of the substance originally in the blood (100).

5. How do the data for potassium differ from those for sodium?
6. How would you interpret the data for urea?
7. Summarize the processes that take place between blood and urine, and identify the structures where they occur.

PART B   Filtration, Reabsorption, and Secretion

Procedure
The micropuncture method was used in a second study of the six materials listed in Table 3C.1. Under a microscope, a very fine pipet was used to withdraw samples of fluid at four points along the nephron (see Figures 3.21, 3.22, and 3.23). Study Table 3C.2, which shows the data that were collected using this technique. Use the data to answer the Analysis questions.

Analysis
1. Which function, secretion or reabsorption, involves the movement of a greater amount of water in the kidney? Explain your answer.
2. Proteins are involved in which of the three kidney functions?
3. Compare the protein data with the glucose data. What is the difference? Explain the difference.
4. In some samples, glucose is found in the urine. What might cause this condition?

5. Why are excess glucose molecules in the blood excreted?
6. The data tell us that the concentration of sodium in the blood is greater than in the urine, yet most of the sodium ions in the urine move back into the blood. What process makes this movement possible?
7. Urea is a by-product of amino-acid metabolism. Next to water, urea is the most abundant material found in urine. If urea were allowed to accumulate in the blood, what might happen?
8. Homeostasis is the maintenance of a relatively stable internal environment in an organism. Summarize how the kidney functions as a homeostatic organ.

Investigations for Chapter 4
Autotrophy: Collecting Energy from the Nonliving Environment

Investigation 4A ◆ Photosynthesis

Section 4.2 of your text developed the following equation to represent the materials and products of photosynthesis:

$$3CO_2 + 3H_2O \xrightarrow{\text{light energy}} C_3H_6O_3 + 3O_2$$

carbon dioxide   water   chlorophyll   3-carbon sugar   oxygen gas

This equation raises several questions, which can serve as a basis for experiments that will help you understand the process of photosynthesis.

A. Does a green plant use carbon dioxide in the light?

B. Is light necessary in order for this reaction to take place?

C. Are the materials in the equation involved in any plant process other than photosynthesis?

D. Do plants release the oxygen produced in photosynthesis?

In this investigation, you are asked to design experiments to answer the questions above. You will need to consider several factors before starting work.

- What type of plant could best serve your purpose, a water plant or a land plant?
- What factor affecting photosynthesis could best be used to start and stop the process?
- What type of detector can be used to show that photosynthesis has or has not occurred?
- How can you identify the substances that are produced or given off during photosynthesis?
- What type of controls are necessary?

Materials (per team of 2)
2 lab aprons
2 pairs of safety goggles
18-mm × 150-mm test tubes (number used will vary)
wrapped drinking straws
rubber stoppers for test tubes
carbonated water
bromthymol blue solution
elodea *(Anacharis)*

SAFETY **Put on your safety goggles and lab apron. Tie back long hair.**

Procedure

PART A   Use of Carbon Dioxide in Light

1. Add enough bromthymol blue solution to a test tube to give a light blue color, and using a drinking straw, gently bubble your breath through it until you see a color change. Discard the straw after use.

CAUTION: Do not suck any liquid through the straw.

2. Add a few drops of carbonated water to a small amount of bromthymol blue in a test tube, and observe any color change. What do carbonated water and your breath have in common that might be responsible for the similar result? What action would be necessary to restore the original color of the bromthymol blue?

3. Using elodea, bromthymol blue solution, and test tubes, set up an experiment to answer Question A. (Hint: Bromthymol blue solution is not poisonous to elodea.)

4. Using Table 4A.1 as a guide, prepare a table listing the test tubes in your experiment. Show what you added to each tube, what change you expected in the bromthymol blue solution, and what change actually occurred. Explain the change. Fill in the first three columns of the table on the day the experiments are set up, and fill in the last two columns the next day.

5. Wash your hands before leaving the laboratory.

PART B   Light and Photosynthesis

6. Using the same types of material as in Part A, set up an experiment to answer question B. Use as many plants and test tubes as necessary to be sure of your answer.

7. Prepare and complete a data table as in Part A.

Test tube	Material added (procedure)	Expected indicator change (hypothesis)	Actual indicator change (data)	What the change shows (interpretation)
TABLE 4A.1 **Form for Data Table**				
1	Bromthymol blue solution, elodea, CO_2, and light	Yellow bromthymol blue solution will turn blue.		

PART C Carbon Dioxide in Other Plant Processes

8. Using the same types of materials as in Part A, set up an experiment to answer question C for carbon dioxide.
9. Prepare and complete a data table as in Part A.

PART D Oxygen and Photosynthesis

10. Your teacher or a selected group of students will set up a demonstration experiment to answer question D. What observations in Parts A, B, or C indicate that the elodea in light was giving off a gas? How might some of this gas be collected and tested to determine its identity?

Analysis

1. Do you have evidence from Part A that light alone does not change the color of the bromthymol blue solution? Explain your answer.
2. What test tubes show that light is necessary for a plant to carry on photosynthesis?
3. How is carbon dioxide used by a plant that is not carrying on photosynthesis? What test tubes show carbon dioxide's role? What biological process accounts for your findings?
4. Determine where your expected changes disagree with the actual changes. Are the differences, if any, due to experimental error or a wrong hypothesis? Explain your answer.

Investigation 4B ◆ Rate of Photosynthesis

There are several ways to measure the rate of photosynthesis. In this investigation, you will use elodea and pH paper. The rate of photosynthesis is determined indirectly by measuring the amount of carbon dioxide removed from water by elodea. Carbon dioxide is added by bubbling breath into the water, which absorbs the carbon dioxide. The carbon dioxide combines with water to produce carbonic acid ($H_2O + CO_2 \rightleftharpoons H_2CO_3$), a reversible reaction. As elodea uses carbon dioxide in photosynthesis, less carbonic acid is present, and the pH of the water increases.

Hypothesis: After reading the procedure, develop hypotheses that predict the effects of light intensity and color on the rate of photosynthesis.

Materials (per team of 4)
4 pairs of safety goggles
2-L beaker
100-mL beaker
250-mL flask
2 1-mL pipettes
2 25-mm × 200-mm test tubes
forceps
2 wrapped drinking straws
lamp with 100-watt flood-lamp bulb
narrow-range pH paper
nonmercury thermometer

TABLE 4B.1
Rate of Photosynthesis

	Experimental condition																	
	30 cm			10 cm			50 cm			Red			Blue			Green		
Reading	pH1	pH2	°C	pH1	pH2	°C	pH1	pH2	°C	pH1	pH2	°C	pH1	pH2	°C	pH1	pH2	°C
1																		
2																		
3																		
4																		
5																		
6																		

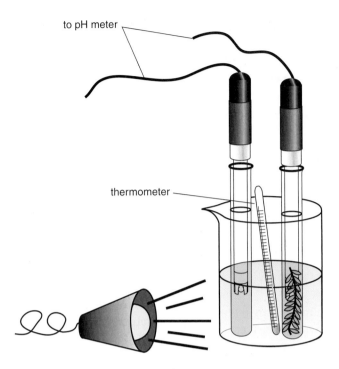

to pH meter

thermometer

FIGURE 4B.1
Experimental setup.

red, blue, and green cellophane
distilled water
ice water
tap water at 25°C
2 15-cm sprigs of young elodea

 SAFETY Put on your safety goggles.

Procedure

PART A   Equipment Assembly

1. In your logbook, prepare a table similar to Table 4B.1, or tape in the table your teacher provides.
2. Put 125 mL of distilled water in the flask. Blow through a small straw into the water for 2 minutes. Discard the straw after use as your teacher directs. What is the purpose of blowing into the water in the flask?

CAUTION: Be careful not to suck any liquid through the straw.

3. Place two sprigs of elodea, cut end up, into one of the test tubes (experimental tube).
4. Fill both test tubes three-fourths full with the water you blew into. What is the purpose of the second test tube?
5. Stand the test tubes in a 2-L beaker; add 25°C tap water to the beaker until it is about two-thirds full (Figure 4B.1).
6. Insert a thermometer into the water in the 2-L beaker. Determine the temperature throughout the experiment. (Use the small beaker to add ice water and/or to remove water from the 2-L beaker.)
7. Let the entire assembly stand for about 5 minutes to permit the temperature to become uniform throughout the system. Note the initial pH in the test tubes, and record both readings in the data table. Use the 1-mL pipettes to transfer a drop of water from halfway down each test tube to a piece of narrow-range pH paper, and read the pH from the comparison chart on the strip.

PART B   Basic Photosynthetic Rate

8. Place the lamp with the 100-watt bulb 30 cm from the beaker, and illuminate the two test tubes.
9. Take pH and temperature readings every 5 minutes for a total of 30 minutes. Record the readings in the data table.
10. Wash your hands thoroughly before leaving the laboratory.

PART C   Effects of Light Intensity on Photosynthetic Rate

11. Repeat steps 2–9 with the light source first 10 cm and then 50 cm away from the test-tube assembly.

PART D   Effects of Light Color on Photosynthetic Rate

12. Repeat Steps 2–9 using first red, then blue, and then green cellophane over the beaker.
13. Wash your hands thoroughly before leaving the laboratory.

Analysis

1. What chemical change occurred in the water you blew into?
2. What happens during photosynthesis that causes the pH to increase in the test tubes?
3. What two environmental factors are being controlled by the test tube without the elodea?
4. How much did the pH change during the 30-minute period for each condition tested?
5. How would you use any change in pH in the test tube without elodea to correct the data for the experimental test tube?
6. Prepare graphs showing pH every 5 minutes under the various conditions being tested. What variables are on the horizontal and vertical axes of the graphs?
7. Use the change in pH that occurs at the three light intensities you tested to determine the effect of light intensity on the rate of photosynthesis. Do your data support your hypothesis?
8. Use the change in pH with the three colors (red, blue, and green) to determine the effect of light color on the rate of photosynthesis. Do your data support the hypothesis you constructed?

Investigation 4C ◆ Chemoautotrophs

All life on Earth can be placed into categories according to the organism's sources of energy and carbon. Phototrophs obtain their energy from sunlight, while chemotrophs obtain energy by oxidizing various organic or inorganic molecules. Autotrophs use carbon dioxide as their source of carbon, while heterotrophs get carbon from a variety of organic compounds.

A Winogradsky column is a self-contained microbial ecosystem. Mud and water from a freshwater or marine source supply the microorganisms. Energy is derived from sunlight and organic molecules. Careful observation of a Winogradsky column over a period of weeks or months reveals a great deal of microbial diversity, including examples of all four basic life strategies described above.

Materials (per team of 3)
3 pairs of safety goggles
3 lab aprons
3 pairs of plastic gloves
1-L graduated cylinder
mud from freshwater or marine source
water from freshwater or marine source
16 g chalk
8 g newspaper
8 g sodium sulfate
clear plastic wrap
rubber band
balance
light source
scissors
stirring rod
hammer
bucket

SAFETY Put on your safety goggles, lab apron, and gloves. Tie back long hair.

Procedure

1. Cut or shred newspaper into small pieces.
2. Weigh out 8 g of shredded newspaper, and add it to the bottom of the 1-L graduated cylinder.
3. Add approximately equal amounts of mud and water to the bucket, and use the stirring rod to mix them together.
4. Pour the mud/water mixture into the graduated cylinder until the column is about half full.
5. Use the stirring rod to mix the mud and water, removing any bubbles.
6. Use a hammer to crush a few pieces of chalk.
7. Weigh out 16 g of crushed chalk.
8. Add the crushed chalk to the column, and mix using the stirring rod.
9. Weigh out 8 g sodium sulfate.
10. Add the sodium sulfate to the column, and mix using the stirring rod.
11. Let the mixture settle in the column for about 5 minutes.
12. Add more of the mud/water mixture as described in steps 4–5 until the volume of packed mud occupies approximately one-third of the column volume.
13. If necessary, add more water to fill the column to the top.
14. Cover the top of the column with clear plastic wrap and secure with a rubber band.
15. Place the column near a window that receives full sunlight.
16. Position a light source to illuminate the side of the column facing away from the window. Do not move or disturb the column once it is in position.
17. During the first week following construction of the column, make daily observations of it in your logbook. Describe the appearance of the column. Include any layers that you see, and note their colors. Also record any odors you can detect.
18. Make weekly observations until your teacher instructs you to stop.

Analysis

1. Describe how the appearance of the Winogradsky column changes over time.
2. Which layers in the column are the most aerobic? most anaerobic?
3. Why was newspaper added to the column?
4. Why was sodium sulfate added to the column?
5. Explain how sulfur is recycled in the Winogradsky column.

Investigations for Chapter 5
Cell Respiration: Releasing Chemical Energy

Investigation 5A ◆ How Does Oxygen Affect Cells?

Oxygen is very important in the release of energy from food. Most organisms, including plants and humans, cannot live without a constant supply of oxygen. Some organisms, however, can get energy from food without oxygen. Additionally, there are a few kinds of microorganisms that use oxygen if it is available but still can get energy from food if oxygen is not available. These microorganisms that are both aerobic and anaerobic are called facultative anaerobes. The bacterium *Aerobacter aerogenes* is an example of a facultative anaerobe.

This investigation provides data from an experiment with *Aerobacter aerogenes*. The bacteria were allowed to grow in test tubes containing distilled water to which a few salts and various concentrations of glucose were added. Some of the test tubes were sealed so that no air was available to the cells. Other tubes had a stream of air bubbling through the growth solution. You will work with and interpret the data and develop a hypothesis to explain the findings.

Materials (per student)
graph paper
pencil

Procedure

1. Graph the data shown in Table 5A.1. Label the vertical axis *millions of cells per mL* and the horizontal axis *glucose (mg/100 mL)*. Plot the data from series A (test tubes without air). Label the line *growth without air*.
2. On the same graph, plot the data from series B (test tubes with air). Label the second line *growth with air*.
3. Use your graphs to help you answer the Analysis questions.

Analysis

1. What are the two obvious differences between the graph curves for series A and B?
2. Look at Table 5A.1, and compare test tubes 4A and 4B. How many times greater was the growth when air was present?
3. Compare the other test tubes in series A and B at the various glucose concentrations. How much greater is the growth in air for each pair of test tubes from 1A and 1B through 5A and 5B?
4. Notice that the number of bacteria is not given for test tubes 6B, 7B, 8B, and 9B. How many bacteria would you predict in test tube 6B? in 7B? These numbers were omitted from the table because the bacteria were too numerous to count.

TABLE 5A.1
Effects of Different Glucose Concentrations on Growth of Bacteria

Concentration of glucose (mg/100 mL of H_2O)	Number of bacteria at maximum growth (millions per mL)			
	Series A		Series B	
	Tube	Tubes without air	Tube	Tubes with air
18	1A	50	1B	200
36	2A	90	2B	500
54	3A	170	3B	800
72	4A	220	4B	1100
162	5A	450	5B	2100
288	6A	650	6B	
360	7A	675	7B	
432	8A	675	8B	
540	9A	670	9B	

5. After each test tube reached maximum growth, the solution was tested for the presence of glucose. In all the test tubes from 1A to 6A and from 1B to 9B, there was no glucose. (Test tubes 7A, 8A, and 9A contained some glucose even after maximum growth had been reached.) Compare test tubes 4A and 4B. How many bacteria were produced per milligram of glucose in each case?
6. Develop a hypothesis that explains the numbers you calculated for question 5. Why are there many more bacteria per milligram of glucose in the B test tubes than in the A test tubes?
7. Each milligram of glucose has the same amount of energy available to do work. The series B test tubes produced more bacteria per milligram of glucose than did the series A test tubes. Assuming that each bacterium produced requires a certain amount of energy, which test tube should contain some products of glucose that still contain some "unused" energy?
8. In additional tests, it was determined that alcohol accumulates in the series A test tubes. How does this information relate to your answer concerning "unused" energy in question 7?

Investigation 5B ◆ Rates of Respiration

Precise measurement of the rate of respiration requires elaborate equipment. Reasonably accurate measurements can be obtained by placing living material in a closed system and measuring the amount of oxygen consumed within the system or the amount of carbon dioxide produced over a period of time. In this investigation, you will use a simple volumeter to measure the amount of oxygen taken up by dormant and germinating seeds at different temperatures.

Hypothesis: After reading the procedure, develop a hypothesis to explain the movement of the colored drops in the capillary tubes.

Materials (per team of 3)
3 pairs of safety goggles
3 lab aprons
volumeter jar with spacers and screw-on lid

3 volumeters
100-mL graduated cylinder
Pasteur pipette
2 jars
glass or plastic beads or washed gravel
nonmercury thermometer
nonabsorbent cotton
81/2" × 11" piece of cardboard
81/2" × 11" piece of white paper
forceps
ring stand with support ring
glass-marking pencil
paper clips
paper towels
masking tape
rubber band
wax paper
soda lime packets
colored water
85 Alaska pea seeds
water bath
water at room temperature (about 22°C)
water at 37°C
water at 10°C
ice

Procedure

In this investigation, you will use three volumeters inserted into a jar with spacers and a screw-on lid (Figure 5B.1). Each volumeter consists of a 25-mm × 200-mm test tube, a 30-cm glass capillary tube calibrated at 1-cm intervals, and a stopper assembly. The stopper assembly includes a two-hole rubber stopper, two glass tubes with latex tubing attached, and a 5-mL syringe.

All the test tubes must contain equal volumes of material to ensure that an equal volume of air is present in each tube. A small drop of colored water is inserted into each capillary tube at its outer end. If the volume of gas changes in the tube, the drop of colored water moves, and the direction of water movement depends on whether the volume of gas in the system increases or decreases.

When measuring respiration with the volumeter, consider not only that oxygen enters the living material (and thus leaves the test-tube environment),

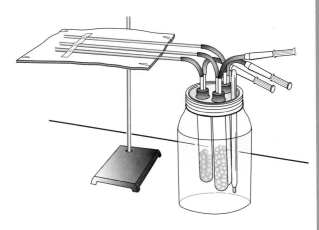

FIGURE 5B.1
Volumeter setup.

but also that carbon dioxide leaves the living material (and enters the test-tube environment). To measure the oxygen uptake by the respiring material, a substance is added to absorb the carbon dioxide as it evolves, so it is not added to the volume of gas in the tube.

Day 1
1. Germinate 45 pea seeds according to the procedure used in Investigation 2A, Part B. Label the jars with your team, class, experiment, and date. As the seeds begin to germinate, what biochemical process increases?

 CAUTION: Wash your hands immediately after handling the seeds. They may be treated with a fungicide.

Day 2
2. Select 40 pea seeds from the germination jar. Discard the 5 extra seeds. Determine the volume of the 40 soaked seeds by adding them to a measured volume of water in a graduated cylinder and reading the volume of displaced water. Record the volume of the seeds. Return the 40 seeds to the germination jar.
3. Repeat step 2 using 40 dry ungerminated pea seeds.
4. Wash your hands thoroughly after handling the seeds.

SAFETY Put on your safety goggles and lab apron. Tie back long hair.

5. The volume in the test tubes of the germinated and dry pea seeds must be equal. Measure the volumes of both types of seeds again, and add beads or gravel to the dry seeds until their volume is equal to that of the germinated seeds. What change occurred in the volume of the seeds after 48 hours in the germination jar?

6. Measure an amount of beads or gravel to equal the volume determined for the germinated seeds.

7. Fill the volumeter jar about two-thirds full with room-temperature water. Screw on the lid. Why is water added to the volumeter jar?

8. Clip a piece of white paper to the cardboard, and place it on the ring-stand support ring.

9. Remove the stopper assemblies from each of the three test tubes. Add the germinated peas to one test tube; add the dry peas (and beads or gravel) to another. In the third test tube, place the equal volume of beads or gravel you measured in step 6. This third tube is a thermobarometer, which is used to determine any changes in the system. What two variables will the thermobarometer help you measure?

10. Place a 2-cm plug of *dry* nonabsorbent cotton about 1 cm above the seeds or beads in each test tube (Figure 5B.2). Use forceps to place a small packet of soda lime wrapped in gauze on top of the plug.

C CAUTION: Soda lime is a *corrosive* solid. Do not touch; do not ingest. If it gets on skin or clothing or in the mouth, rinse thoroughly with water; if in eyes, wash gently but thoroughly for 15 minutes. Call your teacher.

11. Gently but firmly press a stopper assembly into each test tube. Insert the test tubes through the lid into the volumeter jar.

12. Insert the thermometer into the jar through the thermometer hole in the lid. Record the

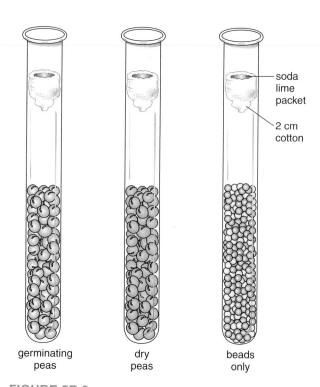

soda lime packet

2 cm cotton

germinating peas

dry peas

beads only

FIGURE 5B.2

Volumeter tubes after preparation.

temperature of the water in the volumeter jar, and maintain this temperature throughout the experiment.

13. With a Pasteur pipette, insert a small drop of colored water into the calibrated end of each *dry* capillary tube. Rotate each tube until the drop is correctly positioned. The drop of colored water in the thermobarometer should be positioned at about the middle of the calibrations, and the other drops should be positioned at the outermost calibration.

14. Carefully attach each capillary tube (by its *un*calibrated end) to the longer glass tube in a stopper assembly. Support the tubes in a level horizontal position on the paper prepared in step 8.

15. Use the syringe to reposition the drop in the capillary tube, if necessary. Tape the tubes in place on the paper.

16. Allow the apparatus to stand for about 5 minutes to permit the temperature to become uniform throughout the system.

TABLE 5B.1
Distance Colored Drop Moves in a Closed System

Time	Thermobarometer readings (mm)	Germinating-pea readings (mm)		Dry-pea readings (mm)	
		Uncorrected	Corrected	Uncorrected	Corrected

17. Prepare a table in your logbook similar to Table 5B.1. On the paper beneath the capillary tubes, mark the position of one end of each drop.

18. Record the position of each drop every 5 minutes for 20 minutes. If respiration is rapid, you may need to reposition the colored drop as you did in step 13 or 14. If you do this, be sure to add both measurements of the distance moved by the drop to calculate the total change during the experiment.

19. Disassemble the volumeter, and dry out the capillary tubes.

20. Reassemble the volumeter using 37°C water, and place the jar containing the volumeter tubes in a water bath that also contains 37°C water. Ideally, when the volumeter is placed in the water bath, the water level should reach the same level as the water in the jar.

21. Position a drop of colored water in each capillary tube, attach each capillary tube to its longer glass tube, and record drop positions as described in steps 13–18.

22. Repeat steps 19–21 using water at 10°C. Add ice as necessary to lower the water temperature.

23. Wash your hands thoroughly before leaving the laboratory.

Analysis

1. What is the effect of moisture on the respiration rate of germinating pea seeds?

2. Would adding more water to the soaked seeds result in an increased rate of respiration? Explain your answer.

3. What if the carbon dioxide absorbent (soda lime) were not used? Use the equation
$C_6H_{12}O_6 + 6O_2 \rightarrow 6H_2O + 6CO_2$
to calculate how much, if any, the volume within the volumeter would change if the carbon dioxide were not removed. Do you think the six water molecules that are released per molecule of sugar should be considered? Why or why not?

4. Use any changes in the thermobarometer to determine the corrected distance moved by the drops in the volumeters containing the pea seeds.

5. Calculate the total volume of oxygen used by the pea seeds. Each 1-cm mark on the capillary tube equals 0.063 mL of oxygen by volume. For each temperature, determine the rate of respiration for the pea seeds by calculating oxygen consumed per minute.

6. Does the rate of respiration for pea seeds change with temperature?

7. How does the rate of respiration of the ungerminated seeds compare to the germinated seeds? What is the significance of this difference as far as the seeds' ability to survive in nature is concerned?

8. Make a linear graph comparing the rates of respiration of pea seeds at the three different temperatures.

Investigations for Chapter 6
Cell Structures and Their Functions

Investigation 6A ◆ Cell Structure

In this investigation, you will examine some cells from unicellular and multicellular organisms. By comparing selected cell preparations, you will be able to identify different cell structures and propose explanations for their functions.

Materials (per team of 2)
2 pairs of safety goggles
2 lab aprons
coverslip

dropping pipet
microscope slide
fine-pointed forceps
scalpel
paper towel
Lugol's iodine solution in dropping bottle
salt solution in dropping bottle (5%)
onion piece
prepared slide of Gram-stained bacterial cells
prepared slide of human and frog blood
prepared slide of plant and animal cells
prepared slide of paramecia
compound microscope

 SAFETY Put on your safety goggles and lab apron. Tie back long hair.

Procedure

1. Separate one layer from an onion quarter, and hold it so that the concave (curved inward) surface faces you. Snap it backward (Figure 6A.1a) to separate the transparent, paper-thin layer of cells from the outer curve of the scale.
2. Use forceps to peel off a small section of the thin layer, and lay it flat on a microscope slide.

Discard the rest of the onion piece. Trim the piece with a scalpel if necessary, and smooth any wrinkles.

 CAUTION: Scalpel blades are sharp; handle with care.

3. Add 1 or 2 drops of Lugol's iodine solution and a coverslip.

 WARNING: Lugol's iodine solution is a *poison* if ingested, is a *strong irritant*, and can stain clothing. Avoid skin/eye contact; do not ingest. If contact occurs, flush affected area with water for 15 minutes; rinse mouth with water; call your teacher immediately.

4. Examine the slide first with low power and then with high power.
5. Sketch a few cells as they appear under high power. How many dimensions do the cells appear to have when viewed through high power? Sketch a single cell as it would appear if you could see three dimensions. Using the procedure from Investigation PC, estimate and record cell size.
6. Identify the cell wall, nucleus, and cytosol. Sketch each in a diagram. You may be able to see a nucleolus within a nucleus. If so, sketch it also. Add a drop of salt solution (Figure 6A.1b)

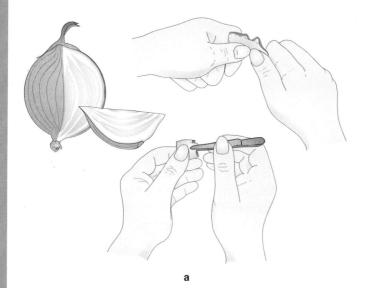

a

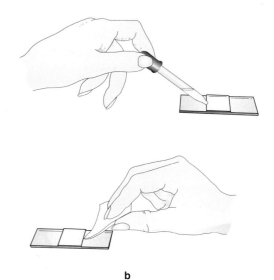

b

FIGURE 6A.1

(a), Removing the thin layer of cells from a section of onion. **(b),** Adding sugar or salt solution under a coverslip.

at one edge of the coverslip. The paper will act as a wick, pulling the salt solution across the slide and into contact with the cells. Continue observing the cells until the cytosol appears to pull away from the cell walls. The boundary of the cytosol is the plasma membrane.

7. Obtain a prepared slide of Gram-stained bacteria. The slide has three sections, each of which contains a different species of bacteria. Examine each section with the low-power objective and then the high-power. Sketch in your logbook a few bacteria of each species under high power. Note the color and shape of each species. Estimate and record the cell sizes.

8. Obtain a prepared slide containing human and frog blood. Examine each blood sample under low power, then high. Find a field where the cells are separate and distinct, and sketch a few cells of each type under high power. Estimate and record the cell sizes.

9. Obtain a prepared slide containing plant and animal cells. Examine the plant and animal cells under low power, then high. Find a field where the cells are separate and distinct, and sketch a few cells of each type under high power. Estimate and record the cell sizes.

10. Obtain a prepared slide of paramecia. Examine the cells under low power, then high. Find a field where the cells are separate and distinct, and sketch a few under high power. Estimate and record cell sizes. Paramecia are mobile organisms. Try to identify cell structures that play a role in locomotion.

11. Wash your hands before leaving the laboratory.

Analysis

1. Are the plasma membranes of plant cells difficult to see? Explain your answer.
2. Did the Lugol's iodine solution aid in your observation of cells? Why do biologists use stains to study cells?
3. What is the significance of the observation that some bacteria stain purple with Gram's stain and others stain red?
4. What differences did you observe between the human and frog blood samples? Do these observations suggest that humans and frogs have different ways of maintaining their blood cells?
5. Construct a summary table comparing the common organelles of plant and animal cells. What parts of the plant cell were not present in the animal cell? What are the functions of these plant-specific cell structures?
6. What cell structures are used by the paramecia for locomotion? Are there similar structures in any human cells, and if so, what might be their functions?

Investigation 6B ◆ From One Cell to Many

Single-celled organisms come in many different shapes and sizes. Despite their varied appearance, these organisms all must carry out certain essential life processes. To that end, they have evolved specific cell structures to meet each of those needs. Among multicellular organisms, some protists are colonial, where each member of the colony is nearly identical to the next. A true multicellular organism, such as a plant or an animal, requires that cells exhibit a division of labor.

In this investigation, with the aim of identifying the life processes that all cells share, you first will observe microorganisms found in pond water. Next, you will examine prepared slides of various mammalian tissue types. Your task is to observe them carefully and, based on their appearance, form a hypothesis as to which life process that cell type contributes.

Materials (per team of 2)
2 pairs of safety goggles
2 lab aprons
coverslip
dropping pipet
microscope slide
compound microscope
key to pond-water organisms
pond-water culture
Detain™ in dropping bottle (slowing agent)
prepared slide A
prepared slide B
prepared slide C
prepared slide D
prepared slide E

| SAFETY | **Put on your safety goggles and lab apron. Tie back long hair.** |

Procedure

1. Place 1 or 2 drops of the pond-water culture on a microscope slide, add a coverslip, and examine under low power with your microscope.

2. Try making observations through the high-power lens, keeping in mind that most of the organisms are transparent or almost transparent. Decrease the amount of light by adjusting the diaphragm. Many of the organisms move rapidly and are hard to find under high power. The coverslip will slow some of them, but you can also add a drop of Detain™, a slowing agent.

3. Use the key to pond-water organisms to identify some of the organisms. Roundworms, daphnia, cyclops, rotifers, and immature forms (larvae) of insects are among the multicellular creatures that could be in your pond water. In your logbook, list as many life processes shared by the microorganisms as you can. For each process in your list, describe a tissue or cell type in a human that is involved in performing that function.

4. The five prepared slides are all from mammals and are associated with the functions of reproduction, movement, protection, food storage, and sensory communication. Obtain prepared slide A, and observe first under low power, then high. Sketch a few cells under high power.

5. Obtain prepared slides B through E in turn, and observe each under low power, then high. For each slide, sketch a few distinct cells under high power.

6. Use your observations to match each of the five prepared slides with one of the five functions: reproduction, movement, protection, food storage, and sensory communication.

Analysis

1. How many different organisms did you observe in the pond water? Describe some of the differences among them.

2. What limits the size of single-celled organisms?

3. At the genetic level, how does a lung cell differ from a brain cell?

4. Which prepared slides correspond to which functions? Explain your answers.

5. List other functions carried out by single-celled organisms and the cell types that correspond to them in multicellular organisms.

Investigations for Chapter 7
Transport Systems

Investigation 7A ◆ Water Movement in Plants

The normal pathway of moving water in a living plant is first into the roots, then through the stem, and finally into the leaves. Environmental influences, the chemical properties of water, and structures in the plant are involved in the movement. This laboratory investigation deals with the following questions:

1. What plant structures—roots, stems, or leaves—are most important in the movement of water?

2. What are the types of cells that transport water in a plant? What are their characteristics?

3. Is all the water that is delivered to the leaves used, or is some lost?

4. How would you describe the source and direction of the force that moves water upward in a plant, against the force of gravity?

Hypothesis: Before beginning the investigation, read the procedure for Part A. Write a prediction that describes what you think will happen to each test tube.

Materials (per team of 3)
3 pairs of safety goggles
6 18-mm × 150-mm test tubes
test-tube rack
glass-marking pencil
heavy-duty aluminum foil
scalpel
petroleum jelly
cotton swab
compound microscope
prepared slide of woody-stem cross section
prepared slide of leaf cross section

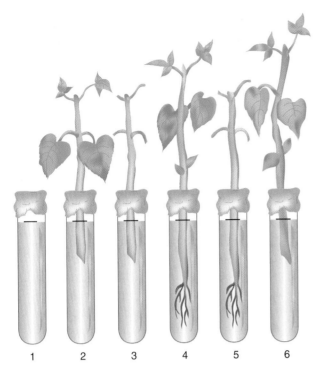

FIGURE 7A.1
Setup of test tubes and plants for the water-uptake experiment.

prepared slide of leaf epidermis
6 bean or sunflower seedlings
radish or grass seedlings in a petri dish

 SAFETY **Put on your safety goggles.**

PART A   Measuring Water Uptake

Procedure

1. Fill six test tubes with water to within 2 cm of the top, and cover the tops with aluminum foil. Label the test tubes *1* through *6*. Treat each test tube as follows, using Figure 7A.1 as a guide.
   Tube 1: Mark the water line.
   Tube 2: With a scalpel, remove the roots of one bean or sunflower seedling 6 cm below the cotyledons. Pierce the aluminum foil with a pencil point, and ease the plant through the hole into the water. Mark the water line after the plant is in place.

 CAUTION: Scalpels are sharp; handle with care.

Tube 3: Remove all leaf blades from another seedling, leaving only the leaf stems. Then repeat the procedure for test tube 2.
Tube 4: Remove the aluminum foil from the test tube. Sink an intact plant's roots into the water. Mark the water line. Mold the aluminum foil to the rim of the test tube so that it seals the tube and supports the plant stem.
Tube 5: Remove the leaf blades from another seedling. Only the leaf stems should remain. Then repeat the procedure described for test tube 4 for this leafless plant.
Tube 6: With a cotton swab, brush petroleum jelly on the upper and lower surfaces of a seedling's leaves. Repeat the procedure described for test tube 2.

2. Allow the rack of treated plants to stand in indirect light overnight. After 24 hours, observe the test tubes for any changes. Record your observations and any measurements you take.

Analysis

1. What is the purpose of the first test tube?
2. Identify variables in this experiment and how they were controlled.
3. Do the results of the investigation support or disprove your prediction? Explain your answer.
4. Based on your results, what do you predict would happen if
   a. you used seedlings that had twice the number of leaves as those you actually used for test tubes 2, 4, and 6? Explain your answer.
   b. the seedling in test tube 5 had twice the number of roots as the one you actually used? Explain your answer.
   c. you put petroleum jelly on the stem cut of the plant in test tube 6, as well as on the leaves? Explain your answer.
5. Based on your observations alone, which of the following statements is most likely correct? Give the reasons for your choice.
   a. Water is pushed upward in a plant by a force created in the lower parts of the plant.
   b. Water is pulled upward in a plant by a force created in the upper parts of the plant.

6. Account for your observations and results in terms of any changes that occur in the experimental setup.

7. Do the results of this experiment support or disprove the cohesion-tension hypothesis? Explain your answer.

8. Using your experimental results and your reading, describe the path of a water molecule through a seedling. Where does it begin? Where does it end up?

9. Describe an experiment that would trace the path of water molecules through a plant. How would you track the water molecules? How would you set up the experiment? How would you control the variables? (Hint: See Appendix 1B, "Radioisotopes and Research in Biology.")

PART B   Transport Structures in Plants

Procedure

3. Study the cross section of a leaf first with low power of the microscope and then with high power. Identify the following structures using Figure 7A.2 as a reference:

   Upper epidermis: thick-walled, flat cells without chloroplasts, covered with a waxy layer

   Palisade layer: tightly packed and column-shaped cells containing many chloroplasts

   Spongy layer: rounded cells containing chloroplasts, with air spaces between them

   Veins: transport tissue composed of thick-walled xylem and phloem cells

   Lower epidermis: thick-walled, flat cells without chloroplasts but with openings at intervals

   Stomates: openings in the lower epidermis

   Guard cells: two small cells surrounding each stomate and containing chloroplasts

4. Observe a slide showing the lower epidermis of a leaf under high power. Locate the guard cells enclosing a small slit, the stomate. Sketch the stomate and guard cells.

5. Study a cross section of a woody stem under low power. Identify the regions of the pith, vascular bundles, cambium, and bark. Turn to high power, and observe a vascular bundle in more detail. Identify the xylem, heavy-walled cells toward the center area of the stem, and the phloem, small thin-walled cells toward the outside area of the stem. Xylem and phloem cells are separated by a layer of living cells called cambium.

6. Carefully place a radish or grass seedling in a drop of water on a microscope slide, and observe with low power. Notice the extent of absorptive area provided by the root hairs. The young root's darker, denser center region

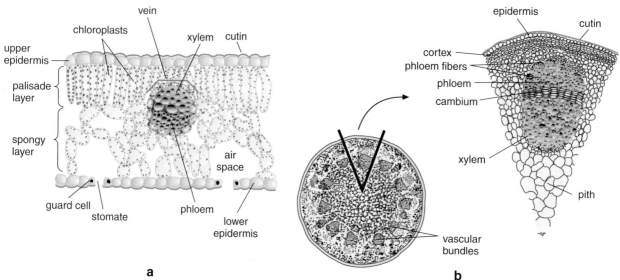

FIGURE 7A.2

Cross section of a leaf (a) and a young woody stem **(b).** Note that in a leaf only the chloroplasts have the pigments we see as a green color. The many chloroplasts in the palisade layer of cells give a leaf an overall green appearance.

includes developing vascular bundles, also composed of xylem, phloem, and cambium cells.

7. Wash your hands before leaving the laboratory.

Analysis

1. Which adaptations of the leaf for the prevention of water loss did you observe? Describe them. In what ways may water loss by leaves help the plant? In what ways might it harm the plant?

2. Describe the adaptations to absorb water from the soil that you observed.

3. How does water enter the root hairs and epidermal cells?

4. What structure in the stem connects the water-conducting vessels of the roots and the veins of the leaf? Describe the structure.

5. Illustrate a pathway showing the route followed by water molecules as they move from the soil, through a plant, and into the atmosphere. Include the specific structures of the root, stem, and leaf that are involved. Use the information and observations you gathered during this investigation to create your illustrations.

Investigation 7B ◆ Exercise and Pulse Rate

You probably have experienced the sensation of your heart beating strongly inside your chest when you participated in a physical activity such as running, aerobic exercise, or athletic competition. The heartbeat rate increases in response to signals from the nervous and endocrine systems, which are monitoring the entire body. Pulse rate, as measured manually at the wrist, is a measure of how fast the heart is beating.

Read the Analysis questions, and develop a hypothesis about how exercise (mild or vigorous) affects heart rate and recovery time. Recovery time is the length of time it takes for the heartbeat to return to the resting rate. Design an experiment to test your hypothesis using the materials listed below.

Materials (per team of 2)
stopwatch or clock with a second hand

Safety
Keep in mind that you must have plenty of room to safely perform your pulse-raising activity; avoid bumping into other persons or objects. Make sure your activity area is free of hazards.

Procedure
Measure pulse at the wrist to carry out the experiment that you designed and that your teacher has approved (Figure 7B.1). In your logbook, record the pulse rate you measure. Organize your data in a table, and construct the graphs at the end of the experiment to display the data.

Analysis

1. What was the control in your experiment?

2. What label did you put on the y-axis of your graph? on the x-axis?

3. Are there any differences in heartbeat rates and recovery times for the different types of exercise among the students in your class? Explain your answer.

4. Describe the effect of varying degrees of exercise—mild, moderate, and strenuous—on heartbeat.

5. Are there any differences in heartbeat rates and recovery times between males and females in your class? Explain your answer.

6. Note any other conclusions you can make based on your data.

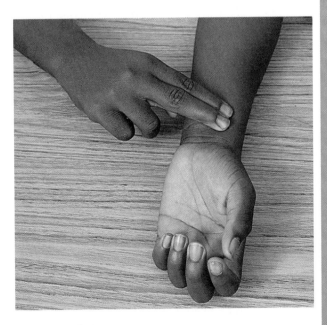

FIGURE 7B.1

Taking a pulse at the wrist.

Investigations for Chapter 8
The Cell Cycle

Investigation 8A ◆ DNA Replication

DNA replication is the process by which exact copies are made of the DNA in prokaryotes and in the chromosomes of eukaryotes. During replication, the genetic code contained within a sequence of nucleotide bases in DNA is preserved. How does replication take place? This investigation will give you an opportunity to observe some of the basic steps involved in the process of replication.

Materials (per team of 2)
pop-it beads: 40 black, 40 white, 32 red, 32 green
string or twist ties
4 tags with string

Procedure

PART A   Building the DNA Molecule
You will build a double-stranded segment of DNA using colored pop-it beads and the following key for the nucleotide bases:

    black = adenine (A)   white = thymine (T)
    green = guanine (G)   red = cytosine (C)

1. Construct the first DNA strand by linking the colored pop-it beads to represent the following sequence of nucleotide bases:
   A A A G G T C T C C T C T A A T T G G T C T C C T T A G G T C T C C T T
2. Attach a tag to the AAA end of the strand, and label the strand roman numeral *I* by marking the tag.
3. Now construct the complementary strand of DNA that would pair with strand I. Remember that thymine (T) bonds with adenine (A) and that guanine (G) bonds with cytosine (C).
4. Attach a tag to the TTT end of the strand, and label the strand *II* by marking the tag.
5. Place strand II beside strand I, and check to make certain that you have constructed the proper sequence of nucleotide bases in strand II. Green pop-it beads should be opposite red pop-it beads, and black pop-it beads should be opposite white pop-it beads. Make any

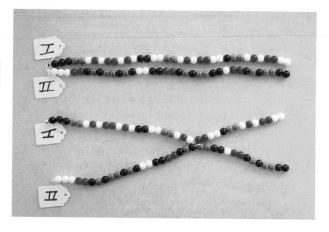

FIGURE 8A.1
Top, **an overhand knot with string holds the ends of the two strands together.** *Bottom,* the knot at the 22nd pop-it bead represents the temporary end point of DNA replication.

necessary corrections in strand II in order to have the proper sequence of base pairs.
6. Tie a simple overhand knot with a short piece of string, or use a twist tie to join the first pop-it beads in strand I and strand II, as shown in Figure 8A.1 (top).
7. Repeat step 6 for the pop-it beads on the other end of strands I and II.

PART B   DNA Replication
8. Place your double-stranded DNA molecule on your table so that both strands are in horizontal straight lines and the AAA end of strand I is on your left. (Do not unfasten the ends.)
9. Beginning at the AAA end of strand I, count 22 pop-it beads from left to right. Tie the strands together between the 22nd and 23rd pairs of beads, as shown in Figure 8A.1 (bottom).
10. Untie the string holding the two strands together at the AAA end of strand I. Separate strand I and strand II so they form a Y.
11. DNA replication on strand I begins with the action of DNA polymerase at the AAA end and proceeds toward the replication fork (the point at which the nucleotide bases are joined at the 22nd base pair). Construct the new complementary strand for strand I beginning at the AAA end and working toward the replication fork. When you have finished, tie another

overhand knot at the AAA end of strand I to join it to its new complementary strand.

12. Replication of strand II begins with the action of DNA polymerase at the replication fork and works outward toward the TTT end. (This is the lagging strand of DNA.) Build the complementary DNA strand for strand II, proceeding from right to left; then tie the two strands together at the TTT end.

13. Now untie the overhand knots that join strands I and II at the 22nd base pair and at the right-hand end of the original double strand.

14. Continue the replication of strand I from left to right until you complete the new complementary strand. Tie another overhand knot to join strand I and its complement at the right-hand end of the molecule.

15. Continue the replication of strand II from the right-hand end to the left. When you reach the 22nd base pair, use another knot to join the two segments of the newly formed strand (the complement of strand II).

16. Tie another overhand knot in the right-hand end of strand II to join it to its new complementary strand.

Analysis

1. Compare the two new double-stranded molecules you have just completed. How are they similar to the original DNA molecule containing strands I and II? How are they different?

2. In your own words, describe a replication fork.

3. Describe the differences in the way strands I and II are replicated.

4. Describe how DNA replication makes it possible to produce two identical cells from one parent.

Investigation 8B ◆ Mitotic Cell Division

To study mitosis, you will examine groups of cells that have been preserved and then stained. Their nuclear structures are visible with the compound microscope. Some of the cells you will see are at very early stages of mitosis, some will be at later stages, and others may be in interphase—a term used for G1, S, and G2, collectively. From slides such as these, biologists have been able to trace the steps a cell goes through during mitotic division. It is very difficult to tell from slides just which stages come first and which come later. Keep this in mind as you try to reconstruct the process for yourself.

Materials (per team of 2)
compound microscope
modeling clay
prepared slide of animal embryo cells (*Ascaris* or whitefish)
prepared slide of onion root-tip cells

Procedure

1. Place the slide of root-tip cells on the microscope stage, and examine it under low power. Scan the entire section. Observe that cells far from the tip and cells right at the tip are not actively dividing. Locate the region of active mitosis between these two regions.

2. Change to the high-power objective. As you observe the cells, focus up and down slowly with the fine-adjustment knob to bring different structures into sharp focus. Find cells at various stages of mitosis. When the slide was prepared, the cells were killed at different stages of a continuous process. The cells can be compared to scrambled single frames of film. Figure out how you would piece the frames of the film together. Refer to Figure 8.11 for help. Make sketches from the slide of cells in interphase, prophase, metaphase, anaphase, and telophase as described in Section 8.6. Identify each by stage.

3. Examine a slide of developing *Ascaris*. Find a cell in which the chromosomes are long and threadlike. Try to count the number of individual chromosomes.

4. Find a cell in which the chromosomes are at the equator of the spindle. Compare the poles of this spindle with those of the spindles in the dividing plant cells you studied in steps 1 and 2.

5. Find a cell in which the chromosomes are separating and the cell is beginning to pinch together in the middle. Compare this method of cytoplasmic division with the method you observed in plants.

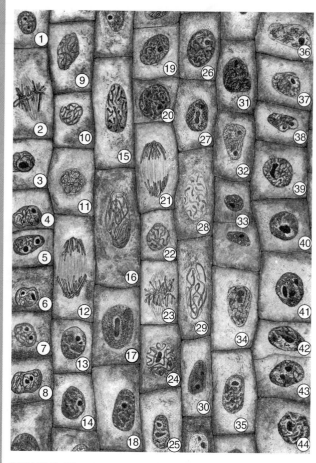

FIGURE 8B.1

Drawing of dividing cells in an onion root as seen through a compound microscope. Arrange cells 2, 7, 9, 11, 12, 13, 16, and 33 in the correct order, demonstrating stages in mitosis.

6. With clay and four large sketches of a plant cell, model each stage of mitosis, from prophase to telophase, for a plant cell with three pairs of chromosomes. Make each pair of chromosomes a different length or color.

Analysis

1. How is the process of mitosis in plant and animal cells similar?
2. How does mitosis in plant and animal cells differ?
3. Refer to Figure 8B.1, and study cells numbered 2, 7, 9, 11, 12, 13, 16, and 33. Rewrite the order of these cells to reflect the sequence of stages you would see if only one cell were undergoing mitosis.

4. Compare the number and types of chromosomes in the two new nuclei of the clay model with the number and types from the original parent nucleus.
5. If mitosis occurs in a cell but cell division does not occur, what is the result?

Investigations for Chapter 9
Expressing Genetic Information

Investigation 9A ◆ Transcription

DNA is the molecule in which all of the genetic information for the cell is stored. The information is encoded as a triplet code in which each sequence of three nucleotide bases codes for a specific piece of information. The DNA is contained in the nucleus, but the cellular processes take place in the cytosol. How does the information from the DNA get into the cytosol where it can be used? This investigation will help you understand that process.

Materials (per team of 2)
pop-it beads: 40 black, 40 white, 32 green, 32 red, 20 pink
string
tags

Procedure

PART A   DNA Transcription
You will build a double-stranded segment of DNA and a single-stranded segment of messenger RNA (mRNA) using colored pop-it beads and the following color code for the nucleotide bases:

    black = adenine (A)   white = thymine (T)
    green = guanine (G)   red = cytosine (C)
    pink = uracil (U)

1. Construct DNA strand I, as you did in Investigation 8A, by linking the colored pop-it beads together to represent the following sequence of nucleotide bases:
A A A G G T C T C C T C T A A T T G G T C T C C T T A G G T C T C C T T
2. Attach a tag labeled roman numeral *I* to the AAA end of strand I.

3. Now construct the complementary strand of DNA that would pair with strand I. Remember that thymine (T) bonds with adenine (A) and that guanine (G) bonds with cytosine (C).

4. Attach a tag labeled *II* to the TTT end of the complementary strand.

5. Place strand II beside strand I, and check to make certain that you have constructed the proper sequence of nucleotide bases in strand II. Green pop-it beads should be opposite red pop-it beads, and black pop-it beads should be opposite white pop-it beads. Make any corrections necessary to have the proper sequence of base pairs.

6. Position the strands on your work surface so that the AAA end of strand I is to your left. Using two pieces of string, tie together nucleotide bases 22 and the right end of the double strands with an overhand knot, as you did in Investigation 8A.

7. Open the left-hand side of the DNA molecule to form the replication fork. Assume that a molecule of RNA polymerase has just attached to the left-hand end of strand I and that a molecule of DNA polymerase has just attached to strand II at the replication fork.

8. Strand I will produce a single-strand molecule of mRNA instead of a new double-stranded molecule of DNA. RNA is produced by the process of transcription from strand I. The rules for forming mRNA are the same as for DNA except that uracil (pink) is used in place of thymine (white). Beginning at the AAA end of strand I, construct the mRNA molecule according to the sequence of bases in strand I. Continue moving toward the replication fork.

9. Untie the overhand knot at base pair 22. Continue the mRNA transcription on strand I, moving from left to right toward the right-hand end of the molecule. When you have finished, attach a tag labeled *mRNA* at the left-hand end of the mRNA molecule.

PART B DNA Replication

10. Complete the DNA replication of strand II by moving to the right-hand end of the molecule

and working back to the left. Join the two segments of the complementary strand to complete the new DNA molecule. Tie together both ends of the double-stranded DNA molecule with string, and set it aside.

11. Check the sequence of nucleotide bases on the mRNA molecule against strand I of the DNA. Remember that uracil (pink) is substituted for thymine (white) in RNA. The mRNA molecule is now ready to move from the nucleus into the cytosol, where its message will be translated. Save your mRNA molecule and the double-stranded DNA molecule to use in Investigation 9B.

Analysis

1. Compare the sequence of nucleotide bases in the mRNA molecule with the sequence of nucleotide bases in strand II. How are they the same? How are they different?

2. Compare the mRNA molecule with the DNA molecule you have set aside for use in Investigation 9B. How are they the same? How are they different?

3. Why do you think mRNA can leave the nucleus and DNA cannot?

Investigation 9B ◆ Translation

You have seen how DNA is replicated and how messenger RNA is formed from a DNA template. The mRNA can leave the nucleus and move into the cytosol, where the message it carries from the DNA can be translated into a sequence of amino acids in the formation of a protein. In this investigation, you will study another type of RNA, transfer RNA (tRNA), and the important role it plays in translation.

Materials (per team of 2)
DNA molecule from Investigation 9A
masking tape
modeling clay
mRNA molecule from Investigation 9A
pop-it beads: 7 black, 10 red, 6 green, 13 pink
tags

Procedure

Throughout this investigation, refer to Figure 9B.1, which shows what amino acids are coded for by each DNA triplet. The DNA triplets are in black type, and the mRNA complementary codons are in blue type.

The tRNA anticodons pair with the mRNA codons in the same way that mRNA complementary codons pair with DNA codons except for the substitution of uracil for thymine in tRNA. If you have difficulties decoding the tRNA codons, ask your teacher for help. Use the same color code for the pop-it beads that you used in Investigation 9A:

    black = adenine (A)   white = thymine (T)
    green = guanine (G)   red = cytosine (C)
    pink = uracil (U)

1. Place the double-stranded DNA molecule and the mRNA molecule across your work surface so that the mRNA lies next to strand I of the DNA and their complementary codons are side-by-side. Beginning at the AAA end, record the letters representing the first three nucleotide bases (the first codon) in strand I of your DNA molecule. Next, record the letters representing the first three bases (the first complementary codon) in your mRNA molecule. In the same way, record the letters representing the bases of the remaining 11 codons in the DNA and mRNA molecules.

2. Use Figure 9B.1 to determine which amino acid is coded for by each mRNA codon. (The DNA triplet and the mRNA complementary codon are adjacent in the figure.) List the appropriate amino acid opposite each mRNA codon.

3. Write the name of the first amino acid on a tag. Make a tag for each of the 11 other amino acids indicated by the codons in the mRNA strand.

4. Determine the anticodon sequence for the tRNA that would pair with the mRNA codon. Build the tRNA anticodon from pop-it beads using the code given above. Tie the tag showing the name of the appropriate amino acid near the middle pop-it bead, as shown in Figure 9B.2 (left). This model represents the tRNA anticodon with its attached amino acid that will take part in protein synthesis.

5. Repeat step 4 to build tRNAs for each of the remaining 11 triplet codons in the mRNA.

First base	Second base				Third base
	A or U	G or C	T or A	C or G	
A or U	AAA *UUU* ⎤ phenylalanine AAG *UUC* ⎦ AAT *UUA* ⎤ AAC *UUG* ⎦ leucine	AGA *UCU* ⎤ AGG *UCC* ⎥ serine AGT *UCA* ⎥ AGC *UCG* ⎦	ATA *UAU* ⎤ tyrosine ATG *UAC* ⎦ ATT *UAA* ⎤ stop ATC *UAG* ⎦	ACA *UGU* ⎤ cysteine ACG *UGC* ⎦ ACT *UGA* ⎤ stop ACC *UGG* ⎦ tryptophan	A or U G or C T or A C or G
G or C	GAA *CUU* ⎤ GAG *CUC* ⎥ leucine GAT *CUA* ⎥ GAC *CUG* ⎦	GGA *CCU* ⎤ GGG *CCC* ⎥ proline GGT *CCA* ⎥ GGC *CCG* ⎦	GTA *CAU* ⎤ histidine GTG *CAC* ⎦ GTT *CAA* ⎤ GTC *CAG* ⎦ glutamine	GCA *CGU* ⎤ GCG *CGC* ⎥ arginine GCT *CGA* ⎥ GCC *CGG* ⎦	A or U G or C T or A C or G
T or A	TAA *AUU* ⎤ TAG *AUC* ⎥ isoleucine TAT *AUA* ⎦ TAC *AUG* ⎥ methionine	TGA *ACU* ⎤ TGG *ACC* ⎥ threonine TGT *ACA* ⎥ TGC *ACG* ⎦	TTA *AAU* ⎤ asparagine TTG *AAC* ⎦ TTT *AAA* ⎤ lysine TTC *AAG* ⎦	TCA *AGU* ⎤ serine TCG *AGC* ⎦ TCT *AGA* ⎤ arginine TCC *AGG* ⎦	A or U G or C T or A C or G
C or G	CAA *GUU* ⎤ CAG *GUC* ⎥ valine CAT *GUA* ⎥ CAC *GUG* ⎦	CGA *GCU* ⎤ CGG *GCC* ⎥ alanine CGT *GCA* ⎥ CGC *GCG* ⎦	CTA *GAU* ⎤ aspartic CTG *GAC* ⎦ acid CTT *GAA* ⎤ glutamic CTC *GAG* ⎦ acid	CCA *GGU* ⎤ CCG *GGC* ⎥ glycine CCT *GGA* ⎥ CCC *GGG* ⎦	A or U G or C T or A C or G

FIGURE 9B.1

The genetic code. The DNA codons appear in black type; the complementary mRNA codons are in blue. *A* = adenine, *C* = cytosine, *G* = guanine, *T* = thymine, *U* = uracil, *stop* = chain termination or "nonsense" codon.

FIGURE 9B.4

Two amino acids joined to form a dipeptide on the A site *(top)*; mRNA and tRNA transposed to the P site and a third tRNA on the A site *(bottom)*.

FIGURE 9B.2

Amino-acid tag attached to the middle pop-it bead of a tRNA *(left)*; the two parts of a ribosome as simulated with modeling clay *(right)*.

6. To represent the large subunit of a ribosome, use one color of modeling clay to make a large oval the length of six pop-it beads. To represent the small subunit of a ribosome, make a narrower oval of a different color of modeling clay, also the length of six pop-it beads. Push the long sides of the two ovals together so their edges are joined and the small subunit lies on top of the large subunit, as shown in Figure 9B.2 (right). These two ovals represent a functional ribosome. When you finish you should have the following:

   1 double-stranded DNA molecule
   1 single-stranded mRNA molecule
   12 tRNA molecules with amino acids attached
   1 ribosome

7. Mark a small piece of masking tape *A site* and another piece *P site*. Press the A-site tape into the small ribosome subunit near the right side and the P-site tape near the left side.

8. Place the first codon of the mRNA molecule on the A site of the small subunit of the ribosome, and lightly press the three pop-it beads of the codon into the clay.

9. Select the tRNA anticodon that will pair with the mRNA codon. Place the tRNA on the larger subunit of the ribosome so that the amino acid points away from the mRNA molecule. Figure 9B.3 (top) shows these positions.

10. Move the mRNA molecule to the left so that the first codon is on the P site and the second codon is on the A site. Move the first tRNA anticodon to keep it paired with its mRNA codon.

11. Select the correct tRNA molecule to pair with the second mRNA codon that is now on the A site of the ribosome. Press both the mRNA and the tRNA molecules lightly into the clay.

12. Remove the tag representing the first amino acid from its tRNA anticodon, and tape it to the tag representing the second amino acid, as shown in Figure 9B.4 (top). You have just formed a peptide bond.

FIGURE 9B.3

Messenger RNA and tRNA positioned on the ribosome to line up the first *(top)* and second *(bottom)* amino acids.

13. Move the mRNA molecule to the left by one codon so that the first codon with its tRNA anticodon is no longer on the ribosome. Put the first tRNA anticodon off to the side.

14. Select the appropriate tRNA anticodon to pair with the third mRNA codon, which is now on the A site of the ribosome, as shown in Figure 9B.4 (bottom). Attach the dipeptide formed in step 12 to this third amino acid.

15. Continue moving the mRNA molecule across the ribosome and pairing the tRNA anticodons with the mRNA codons. After each tRNA anticodon with its amino acid has arrived at the ribosome, remove the amino acid and attach it to the growing chain of amino acids.

16. When all 12 amino acids have been joined, remove the chain of amino acids from the tRNA and the ribosome. This model represents part of a protein molecule.

Analysis

1. Describe similarities and differences between the first DNA triplet and the first tRNA anticodon that you built.

2. What role does the ribosome play in protein synthesis?

3. What role does tRNA play in protein synthesis?

4. What role does mRNA play in protein synthesis?

5. Write a short paragraph that summarizes the roles of transcription and translation in protein synthesis.

Investigations for Chapter 10
Animal Growth and Development

Investigation 10A ◆ Development in Polychaete Worms

The marine polychaete worm *Chaetopterus variopedatus* lives in the sand near the low-tide level inside a leathery, U-shaped tube (Figure 10A.1). The sperm or eggs are visible inside the parapodia (leglike extensions from the body) near the posterior end of the worm. The sperm appear ivory white and give the parapodia a smooth, white appearance. The ovaries are in yellow coils and contain eggs that give

the female parapodia a grainy appearance. You will observe the worm, remove eggs and sperm, fertilize the eggs, and observe the development of the embryos. The initial stages of development that you will observe look the same for nearly all sexually reproducing organisms.

Materials (per team of 2)
3 microscope slides
3 coverslips
dropping pipette
3 finger bowls
compound microscope
forceps
dissecting scissors
2 dissecting needles
cheesecloth
paper towels
seawater
male and female parapodia of *Chaetopterus*

> **SAFETY** Spilled water can cause slippery footing and falls. Wipe up any spilled water immediately. Handle dissecting tools with care; immediately report any injuries to your teacher.

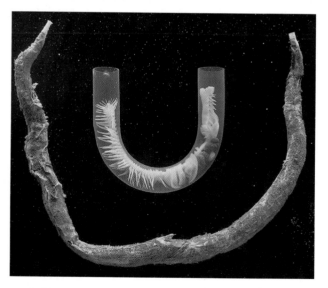

FIGURE 10A.1

Polychaete worm in a glass tube *(top)* and the worm's natural tube *(bottom)*.

Procedure

1. Rinse a small folded-over piece of cheesecloth in fresh water; then rinse it in seawater. Place the wet cheesecloth in the bottom of a finger bowl.

2. Using forceps and dissecting scissors, remove one parapodium with eggs from the female worm and place it on the wet cheesecloth in the finger bowl. Use the pipette to remove any eggs that spilled out of the parapodium, and squeeze the seawater and eggs out of the pipette onto the cheesecloth. Cut open the parapodium, and with the dissecting needle and forceps, tear the parapodium apart and release the eggs.

    CAUTION: Dissecting tools are sharp; handle with care.

3. Lift the cheesecloth, and pour about 1 cm of seawater into the finger bowl. Pick up the cheesecloth by the four corners; gently dip it up and down in the water, and move it around slowly. The movement should allow most of the eggs to filter into the seawater while sand and other debris stay behind. The eggs are tiny yellow or yellow orange dots. Note the time in your logbook. Discard the cheesecloth in a container designated by your teacher.

4. *You must wait 15 minutes from the time you put the eggs in seawater before you add sperm.* While you wait, you can obtain and study the sperm. Add 10 mL of fresh seawater to another finger bowl. Remove a parapodium with sperm from the male worm, and quickly place the parapodium into the fresh seawater. With a dropping pipette, pick up any sperm that were released in the bowl with the worm. The sperm will appear as a small cloud in the water. Add these sperm to those already in the 10 mL of water.

5. Place one drop of the sperm and seawater mixture onto a clean microscope slide, and add a coverslip. Examine the sperm using the low-power objective of a microscope. They will appear as small dots moving around. Switch to the high-power objective, and observe the sperm.

6. If you do not have many moving sperm, take another sample from the 10 mL of seawater. If moving sperm still are not evident, repeat steps 4 and 5.

7. After the eggs have been in the seawater for 15 minutes, add one drop of the sperm and seawater mixture to the eggs. The eggs should all be fertilized within 30 minutes. After 30 minutes, remove the eggs from the finger bowl and place them in another finger bowl with fresh seawater. The development of the embryos will take place in this bowl and can be observed after about 24 hours.

8. While you are waiting for 30 minutes, obtain fertilized eggs from bowls designated by your teacher and examine them under the microscope. Look for embryos at the 4-, 8-, and 16-cell stages, and note any other developmental stages you observe.

9. Wash your hands before leaving the laboratory.

Analysis

1. Examine photographs of human sperm in Figures 10.2 and 12.17. How do the sperm of the polychaete worm compare with the picture of human sperm? Describe any similarities or differences you observe.

2. Describe the development process you observed in the polychaete worm embryos.

3. The polychaete worm larvae usually develop within 24 hours after fertilization. Based on what you observed and your knowledge of cleavage and cell division, approximately how many cell divisions have occurred between the one-cell stage and the swimming larvae?

4. Describe any similarities between the development of the polychaete worm embryos and the development of human embryos.

Investigations for Chapter 11
Plant Growth and Development

Investigation 11A ◆ Seeds and Seedlings

A seed is a packaged plant—a complete set of instructions for growing a plant such as a maple tree or a geranium. The seed contains everything needed to produce a young plant. How does the seed change to a plant? What are the functions of the structures of the seed and the growing plant? What changes can you observe? In this investigation, you will have the opportunity to find out.

Hypothesis: After reading the procedures, construct a hypothesis that predicts how petri dishes C and D will differ.

Materials (per team of 2)
2 pairs of safety goggles
2 lab aprons
10× hand lens
scalpel
petri dish with starch agar and germinating corn grains—dish C
petri dish with starch agar and boiled corn grains—dish D
Lugol's iodine solution in dropping bottle
soaked germinating corn grains
soaked bean seeds
bean seeds germinated 1, 2, 3, and 10 days

 SAFETY **Put on your safety goggles and lab apron.**

PART A   The Seed

Procedure
1. Examine the external features of a bean seed. Notice that the seed is covered by a tough, leathery coat. Look along the concave inner edge of the seed, and find a scar. This scar marks the place where the seed was attached to the pod.
2. Remove the seed coat, and examine the two fleshy halves, the cotyledons, which are part of the embryo.
3. Using a scalpel, cut a small sliver from one of the cotyledons. Test the sliver with a drop of Lugol's iodine solution. Record the results in your logbook.

 CAUTION: Scalpels are sharp; handle with care.

 WARNING: Lugol's iodine solution is a _poison_ if ingested, is a _strong irritant_, and can stain clothing. Avoid skin/eye contact; do not ingest. If contact occurs, immediately flush affected area with water for 15 minutes; rinse mouth with water. If a spill occurs, call your teacher; _then_ flood spill area with water.

4. Separate the two cotyledons, and find the little plant attached to one end of one of the cotyledons. Use a hand lens to examine the plant. You will see that this part of the embryo has two miniature leaves and a root. The small leaves and a tiny tip make up the epicotyl (_epi_ = above: above the cotyledon) of the embryo. The root portion is the hypocotyl (_hypo_ = below: below the cotyledon).

Analysis
1. Based on your observations, what do you think is the function of the seed coat? Explain your answer.
2. What do you think is the function of a connection between a parent plant and a developing seed?
3. Based on your observations, what do you conclude is the primary function of the cotyledons? Explain your answer.
4. What was the original source of the matter that makes up the cotyledons?

PART B   The Seedling

Procedure
5. Examine bean seedlings that are 1, 2, and 3 days old.
6. Compare the 10-day-old seedling with the 3-day-old seedling.

Analysis
1. What part of the plant becomes established first?
2. Where are the first true leaves of the 3-day-old seedling?

3. What has happened to the cotyledons in the 10-day-old seedling?
4. Where is the seed coat in this plant?
5. Which part or parts of the embryo developed into the stem?
6. How are the first two tiny true leaves arranged on the stem?

PART C   Corn Seeds

Procedure

7. Cut a soaked, germinated corn grain lengthwise with the scalpel.
8. Test the cut surfaces with a few drops of Lugol's iodine solution. Record your observations and conclusion.
9. The starch agar and plain agar in petri dishes A and B were tested with Lugol's iodine solution as a demonstration. Observe the dishes. Record your observations and conclusion in your logbook.
10. On petri dish C, two or three corn grains have started to germinate on starch agar. Each grain was cut lengthwise, and the cut surfaces were placed on the starch agar for about 2 days. Petri dish D contains starch agar and boiled corn grains.
11. Cover the surface of the starch agar in petri dishes C and D with Lugol's iodine solution.
12. After a few seconds, pour off the excess.
13. Wash your hands thoroughly before leaving the laboratory.

Analysis

1. When you tested the cut surface of corn grains, what food was present?
2. What other nutrients might be present in the corn that were not demonstrated by the test of the cut surfaces?
3. What difference did you observe in the test of the agars in petri dishes A and B?
4. What difference did you observe when you tested petri dishes C and D? Suggest hypotheses that might account for what you observed.

5. What food substance would you expect to find in the areas where the germinating corn grains were?

Investigation 11B ◆ Tropisms

This investigation allows you to observe tropisms. One-half of each team will conduct either Part A or Part B of the investigation. Observe all your team's work so you can discuss all the results as a class.

Hypothesis: After reading the procedures, construct two hypotheses—one predicting how the germinating corn plants will be affected and the other predicting the patterns of growth of the radish seeds.

Materials (per team of 4; one-half of each team performing Part A or Part B)

PART A
petri dish
scissors
glass-marking pencil
nonabsorbent cotton
heavy blotting paper
transparent tape
modeling clay
4 soaked corn grains

PART B
4 flowerpots, about 8 cm in diameter
4 cardboard boxes, at least 5 cm higher than the flowerpots
red, blue, and clear cellophane
scissors
transparent tape
40 radish seeds
soil

PART A   Orientation of Shoots and Roots in Germinating Corn

Procedure

1. Place four soaked corn grains in the bottom half of a petri dish. Arrange them cotyledon side down, as shown in Figure 11B.1.

2. Fill the spaces between the corn grains with wads of nonabsorbent cotton to a depth slightly greater than the thickness of the grains.

3. Cut a piece of blotting paper slightly larger than the bottom of the petri dish, wet it thoroughly, and fit it snugly over the grains and the cotton.

 CAUTION: Scissors are sharp; handle with care.

4. Hold the dish on its edge, and observe the grains through the bottom. If they do not stay in place, pack them with more cotton.

5. When the grains are secure in the dish, seal the two halves of the petri dish together with tape.

6. Rotate the dish until one of the grains is at the top. With the glass-marking pencil, write an *a* on the petri dish beside the topmost grain. Then proceeding clockwise, label the other grains *b*, *c*, and *d*. Also label the petri dish with a team symbol.

7. Use modeling clay to support the dish on edge, as shown in Figure 11B.1, and place it in dim light.

8. When the grains begin to germinate, make sketches every day for 5 days, showing the direction in which the root and the shoot grow from each grain.

9. Wash your hands thoroughly before leaving the laboratory.

Analysis

1. From which end of the corn grains did the roots grow? From which end of the grains did the shoots grow?

2. Did the roots eventually turn toward one direction? If so, what direction?

3. Did the shoots eventually turn toward one direction? If so, what direction?

4. To what stimulus did the roots and shoots seem to be responding?

5. In each case, were the responses positive (toward the stimulus) or negative (away from the stimulus)?

6. Why was it important to have the seeds oriented in four different directions?

PART B   Orientation of Radish Seedlings

Procedure

10. Turn the four cardboard boxes upside down. Number them 1 to 4. Label each box with your team symbol.

11. Cut a rectangular hole in one side of boxes 1, 2, and 3. (Use the dimensions shown in Figure 11B.2.) Do not cut a hole in box 4.

12. Tape a piece of red cellophane over the hole in box 1, blue cellophane over the hole in box 2, and clear cellophane over the hole in box 3.

13. Number four flowerpots 1 to 4. Label each with your team symbol. Fill the pots to 1 cm below the top with soil.

14. In each pot, plant 10 radish seeds about 0.5 cm deep and 2 cm apart. Press the soil down firmly over the seeds, and water them gently. Place the pots in a location that receives strong light but not direct sunlight.

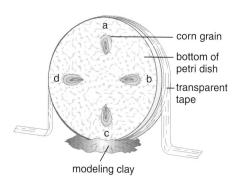

FIGURE 11B.1
Petri dish setup.

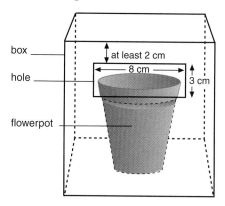

FIGURE 11B.2
Flowerpot setup.

15. Cover each pot with the box labeled with its number. Turn the boxes so the sides with holes face the light.
16. Once each day remove the boxes and water the soil. Do not move the pots; replace the boxes in their original positions.
17. When most of the radish seedlings have been above the ground for 2 or 3 days, record the height (length) of each seedling and calculate an average seedling height for each pot. Record the direction of stem growth in each pot—upright, curved slightly, or curved greatly. If curved, record in what direction with respect to the hole in the box.
18. Wash your hands thoroughly before leaving the laboratory.

Analysis

1. Construct a data table to organize your observations. Include box number, average seedling height, direction of curving, and amount of curving.
2. In which flowerpot were the radish stems most nearly upright?
3. In which pot were the radish stems most curved? In what direction were they curved? Were the stems curved in any of the other pots? Which ones? In what direction did they curve?
4. To what stimulus do you think the radish stems responded?
5. What effect, if any, did the red and blue cellophane have on the direction of the radish stem growth?
6. Speculate about possible biological mechanisms that could account for your observations.

Investigations for Chapter 12
Reproduction

Investigation 12A ◆ A Model of Meiosis

Many biological events are easier to understand when they are modeled. In this investigation, you will use a model to simulate the events of meiosis.

Materials (per team of 2)
modeling clay, red and blue, or red and blue pop-it beads
4 2-cm pieces of pipe cleaner
large piece of paper

PART A   Basic Meiosis

Procedure

1. Use the clay to form two blue and two red chromatids, each 6 cm long and about as thick as a pencil. Alternatively, use pop-it beads.
2. Place the pairs of similar chromatids side by side. Use pipe cleaners to represent centromeres. Press a piece of pipe cleaner across the centers of the two red 6-cm chromatids (made of clay or beads). This represents a chromosome that has replicated itself at the start of meiosis. Do the same for the blue replicated chromosomes. (Figure 12A.1 shows an example using clay.)
3. Form four more chromatids, two of each color, 10 cm long. Again, press a piece of pipe cleaner across the centers of the two pairs of red and blue chromatids.
4. On a sheet of paper, draw a spindle large enough to contain the chromosomes you have made. Assume that the spindle and chromatids have been formed and the nuclear membrane has disappeared.

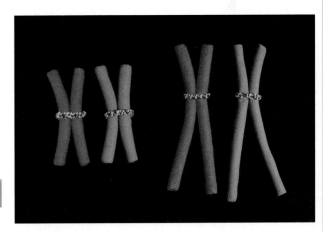

FIGURE 12A.1
Chromosome models.

5. Pair the two 6-cm chromosomes so that the centromeres touch. Pair the two 10-cm chromosomes. Assume that the red chromosome of each pair came from the female parent. Its matching chromosome, the blue one, came from the male parent.

6. Arrange the two chromosome pairs along the equator (middle) of the spindle so that the red chromosomes are on one side and the blue on the other.

7. Holding on to the centromeres, pull the chromosomes of each matching pair toward opposite poles of the spindle. Once the chromosomes have been moved to the two poles, you have modeled the first meiotic division.

8. Draw two more spindles on the paper. These new spindles should be centered on each pole of the first meiotic division. Both spindles should be perpendicular to the first spindle. Your model cells are now ready for the second division of meiosis.

9. Place the chromosomes from each pole along the equator of each of the two new spindles. Unfasten the centromere of each chromosome. Grasp each chromatid at the point where the centromere was attached. Pull the chromatids to opposite poles of their spindles. Try to move each spindle's chromatids simultaneously, as occurs in a living cell. Draw a circle around each group of chromosomes.

Analysis
1. How many cells were there at the start of meiosis? How many cells are formed at the end of meiosis?
2. How many chromosomes were in the cell at the beginning of meiosis? How many chromosomes were in each of the cells formed by meiosis?
3. What types of cells does meiosis produce?
4. How many of your cells at the end of meiosis had only red chromosomes in them? How many had only blue chromosomes in them?

PART B Effects of Chromosome Position on Sorting
10. A real cell is three-dimensional. Although the red chromosomes (from the female) may be on one side of the equator and the blue (from the male) may be on the other when they line up on the spindle, there is an equal chance that one red and one blue chromosome will be on the same side. Attach the chromatids as they were at the beginning of the investigation. Go back to step 6, and arrange the chromosomes so that one blue and one red chromosome are on each side of the equator. Complete meiosis I and II.

Analysis
1. How do these gametes compare with those you made earlier?
2. What difference does this change in position make in terms of genetic variation in the offspring?
3. How many different types of gametes could be made if there were three sets of chromosomes instead of just two?

PART C Effects of Crossing-Over
11. Reassemble your chromosome models. To show crossing-over (see Figure 13.19), exchange a small part of the clay from a chromatid making up one chromosome with an equal part from a chromatid of its homologous pair. The colors make the exchange visible throughout the rest of the investigation.
12. Place your chromosome pairs along the equator of the spindle as in step 6, and complete meiosis I and II.

Analysis
1. How many different types of gametes did you form? Did you form any gametes different from those formed by others in your class?
2. In general, how do you think crossing-over affects the number of different types of gametes that are formed?
3. In crossing-over, what actually is exchanged between the chromatids?
4. What are some of the advantages of using a model to visualize a process?
5. How did this model improve your understanding of the process of meiosis?
6. What are some disadvantages of this model?

Investigation 12B ◆ The Yeast Life Cycle

Baker's yeast *(Saccharomyces cerevisiae)* is a unicellular organism that reproduces both sexually and asexually. Because the cells have a characteristic shape at each stage, it is possible to distinguish all the major stages of the life cycle under the microscope.

Yeast cells may be either haploid or diploid. Haploid cells occur in two mating types (sexes): mating type **a** (HAR) and mating type **α** (HBT). When **a** and **α** cells come in contact, they secrete hormonelike substances called mating pheromones, which cause them to develop into gametes. The **a** and **α** gametes pair and then fuse, forming a diploid zygote. The fusion of **a** and **α** gametes is similar to fertilization in animals except that both parents contribute cytoplasm and nuclei. Yeast zygotes reproduce asexually by budding. When cultured on a solid growth medium, a yeast zygote may grow into a visible colony that contains up to 100 million cells.

Diploid yeast do not mate, but in times of stress, such as when they have an unbalanced food supply, the diploid cells may sporulate, or form spores. The spores remain together, looking like ball bearings, in a transparent saclike structure called an ascus.

In this investigation, you will start with two haploid yeast strains of opposite mating types, mate them to form a diploid strain, and try to complete an entire yeast life cycle.

Materials (per team of 2)
microscope slide
coverslip
dropping pipette
glass-marking pencil
compound microscope
container of clean, flat toothpicks
6 self-sealing plastic bags labeled *waste* (1 for each day)
2 YED medium agar plates
MV medium agar plate
unknown medium agar plate
agar slant cultures of HAR and HBT yeast strains

Procedure

Day 0

1. Prepare fresh cultures of both mating types. Colonies of the HAR strain of mating type **a** are red; those of the HBT strain of mating type **α** are cream-colored. Touch the flat end of a clean toothpick to the HAR strain; then gently drag it across the surface of a YED medium agar plate to make a streak about 1 cm long and 1 cm from the edge (Figure 12B.1). Discard the toothpick in the self-sealing plastic waste bag, being careful not to touch anything with it. Use a glass-marking pencil to label the bottom of the plate near the streak with an **α.** *Use a new clean toothpick from the container for each streak you do. Be careful not to touch the ends of the toothpicks to anything except yeast or the sterile agar. Discard used toothpicks in the plastic waste bag. Keep the lid on the agar plate except when transferring yeast.* Label this plate *I*, and add the date and your names. Incubate upside down for 1 day, or 2 days if your room is very cool.

2. Wash your hands thoroughly before leaving the laboratory.

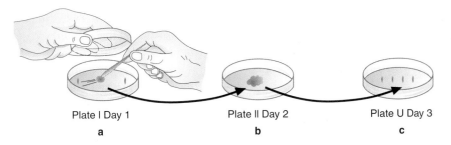

Plate I Day 1
a

Plate II Day 2
b

Plate U Day 3
c

FIGURE 12B.1

Procedure for making a mating mixture from HAR and HBT yeast strains (a), for subculturing the mating mixture (b), and for inoculating the unknown medium plate (c).

Day 1

3. Use a microscope to examine some yeast cells of either mating type. To prepare a slide, touch a toothpick to the streak of either mating type **a** *or* mating type **α** in plate I, mix it with a small drop of water on the slide, and place a coverslip over the drop. Discard the toothpick in the waste bag. Examine the cells with the high-power objective. In your logbook, sketch the cells.

4. Use a clean toothpick to transfer a small amount of mating type **a** from the streak in plate I to the middle of the agar. Use another clean toothpick to transfer an equal amount of mating type **α** to the same place. Being careful not to tear the agar surface, thoroughly mix these two dots of yeast to make a mating mixture. Discard used toothpicks in the waste bag. Invert and incubate plate I at room temperature for 3 or 4 hours; then refrigerate until the next lab period. (Or refrigerate immediately, and then incubate at room temperature for 3 or 4 hours before the next step.)

5. Wash your hands thoroughly before leaving the laboratory.

Day 2

6. Remove plate I from the refrigerator. What color is the mating mixture colony? Use the microscope to examine the mating mixture, as in step 3. Discard used toothpicks in the waste bag. Sketch what you see and compare with your earlier drawings. Describe any differences in the types of cells you see. When haploid cells of opposite mating types (**a** and **α**) are mixed, they develop into pear-shaped haploid gametes. Do you see any gametes? A diploid zygote forms when two gametes fuse. Growing diploid cells are slightly larger and more oval than haploid cells. Do you see any evidence that gametes may be fusing into zygotes? Compared to day 1, are there more or fewer diploid cells?

7. Make a subculture by transferring some of the mating mixture with a clean toothpick to an MV agar plate. Discard used toothpicks in the waste bag. Label this plate *II*. Invert and incubate at least overnight, but not more than 2 nights.

8. Wash your hands thoroughly before leaving the laboratory.

Day 3

9. Use the microscope to examine the freshly grown subculture in plate II. Discard used toothpicks in the waste bag. What types of cells are present? Sketch each type. If any of the types seen in step 6 have disappeared, explain what happened to them.

10. On a plate of unknown medium, make several thick streaks of the freshly grown subculture. Discard used toothpicks in the waste bag. Label this plate *U*. Invert and incubate at room temperature at least 4 days.

11. Wash your hands thoroughly before leaving the laboratory.

Day 7

12. Use the microscope to examine yeast from plate U. Discard used toothpicks in the waste bag. You may need to use the fine adjustment on the microscope to distinguish cells at different levels. What cell types are present today that were not present before? Sketch these cell types, and compare them with the cell types you saw at other stages.

13. Refer to the introduction. How do you think the unknown medium differs from the YED medium? If the cells in the sacs are more frequently found in groups of four, do you think they were formed by meiosis or mitosis? Explain your answer. Are the cells in the sacs haploid or diploid? Explain your answer. What part of the life cycle seen on day 1 do these cells most resemble?

14. Transfer some yeast from plate U to a fresh YED medium agar plate. Discard used toothpicks in the waste bag. Label this plate *III*, invert and incubate at room temperature for about 5 hours, and then refrigerate until the next lab period. (Or refrigerate immediately and then incubate at room temperature for 5 hours before the next step.)

15. Wash your hands thoroughly before leaving the laboratory.

Day 8

16. Use the microscope to examine the growth from plate III. Discard used toothpicks in the waste bag. What life-cycle stages are present? Sketch the cells, and compare them with the stages you observed before. What evidence is there that a new life cycle has started?

17. Discard used culture plates as directed by your teacher, and wash your hands thoroughly before leaving the laboratory.

Analysis

1. In this investigation, you have observed the major events of a sexual life cycle. You could readily observe these events in yeast because it is a unicellular organism. In plants and animals, including humans, similar cellular events occur, but they are difficult to see. Although the changes were occurring too slowly for you to see, your sketches provide a record of the sequences. Think of them as pauses in a tape of a continually changing process. Notice in particular how the cycle repeats. On one page, draw a life-cycle diagram showing the different shapes of cells you observed in the order in which they appeared. Indicate where you first saw each type and when it disappeared, if it did.

2. Compare your sketches with the life-cycle diagram in Figure 12B.2, and try to identify each of the forms you saw.

3. For each of the cell forms you observed, indicate whether it was haploid or diploid.

4. Mark the points in your diagram where cells changed from haploid to diploid and from diploid to haploid.

5. Why do you think the two different mating types are not called female and male?

6. Can you think of a good argument for calling any particular point in the cycle the beginning or the end? Why or why not?

7. What would you expect to happen if you allowed the yeast in step 14 to grow for another day and then put them on the unknown medium again?

8. Table 12B.1 summarizes the similarities between stages in the yeast life cycle and the events in

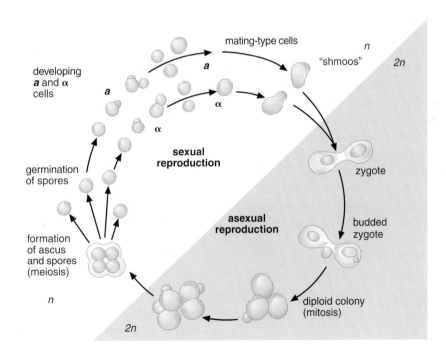

FIGURE 12B.2
Yeast life cycle, showing both asexual and sexual reproduction.

TABLE 12B.1
Comparison of the Yeast Life Cycle and Sexual Reproduction in Animals

Yeast	Animals
Mating pheromones	Sex hormones
Mating-type **a** and **α** gametes	Gametes (ova and sperm)
Fusion	Fertilization
Zygote	Zygote
Asexually reproducing diploid cells	Diploid body cells
Meiosis and formation of spores	Meiosis and formation of gametes

sexual reproduction in animals. For each change you observed in the yeast life cycle, indicate the step in the human sexual reproduction cycle that is most similar.

Investigation 12C ◆ Reproduction in Mosses and Flowering Plants

Although their reproductive organs differ as much as the environments in which they live and reproduce, the basic principles of sexual reproduction are the same in a moss, a flower, a bee, and a human. In this investigation, you will learn how the structures of a moss and a flower serve reproductive functions in their respective environments.

Mosses form mats on logs and on the forest floor, growing best in damp, shaded environments. Sporophytes, or diploid spore-producing structures, grow out of the tops of haploid gamete-producing structures, called gametophytes. Sporophytes often look like brownish hairs growing out of the mat of moss. Mosses cannot reproduce unless they are moist. Flowering plants, on the other hand, are found in many different environments and climates. They need water to live but not to reproduce.

Materials (per team of 2)
3 microscope slides
3 coverslips
dissecting needle
scalpel
forceps
dropping pipette
petri dish
cotton

compound microscope
dissecting microscope or 10× hand lens
modeling clay
prepared slide of filamentous stage of moss
prepared slide of moss male and female reproductive organs
sucrose solution (15%)
moss plant with sporophyte
fresh moss
gladiolus flower
other simple flowers
fresh bean or pea pods

Procedure

PART A   Moss

1. Examine a moss plant with a sporophyte attached. The sporophyte consists of a smooth, brownish stalk terminated by a small capsule. Separate the two generations by pulling the sporophyte stalk out of the leafy gametophyte shoot.

2. Using a dissecting needle, break open the sporophyte capsule into a drop of water on a slide. Add a coverslip and examine under the low power of a compound microscope. What are the structures you observe? How are these structures dispersed in nature? How are they adapted for life on land?

 CAUTION: Needles are sharp; handle with care.

3. Most moss spores germinate on damp soil and produce a filamentous stage that looks like a branching green alga. Examine a prepared slide of this stage.

4. The filamentous stage gives rise to the leafy shoot of the gametophyte. Using forceps, carefully remove a leafy shoot from the fresh moss. How does this shoot obtain water for growth?

5. The reproductive organs of the gametophyte are at the upper end of the leafy shoot. Examine a prepared slide of these organs under the low power of a compound microscope. The male sex organs are saclike structures that produce large numbers of sperm cells. The female sex organs are flask-shaped and have long, twisted necks. An egg forms within the base of the female organ. How does a sperm reach the egg? Would you expect to find moss plants growing where there was little or no water? Explain your answer. The union of the egg and sperm results in a cell called the zygote. Where is the zygote formed? What grows from the zygote?

PART B Flowers

6. Examine the outside parts of a gladiolus flower. The outermost whorl of floral parts may be green and leaflike. These green sepals protected the flower bud when it was young. In some flowers, such as lilies, the sepals look like an outer whorl of petals. Petals are usually large and colorful and lie just inside the sepals. Both sepals and petals are attached to the enlarged end of a branch. These parts of the flower are not directly involved in sexual reproduction. What functions might petals have?

7. Strip away the sepals and petals to examine the reproductive structures. Around a central stalklike body are five to ten delicate stalks, each ending in a small sac, or anther. These are the male reproductive organs, or stamens. Thousands of pollen grains are produced in the anther. The number of stamens varies according to the type of flower. How many stamens are present in the flower you are using? How may pollen be carried from the anthers to the female part of the flower?

8. If the anthers are mature, shake some of the pollen into a drop of sucrose solution on a clean slide. Add a coverslip and examine with the low power of a compound microscope. What is the appearance of the pollen? How is the pollen adapted for dispersal?

9. Make another pollen preparation on a clean coverslip. Use modeling clay to make a 5-mm high chamber, slightly smaller than the coverslip, on a clean slide. Add a small drop of water to the chamber, and invert the pollen preparation over it. Examine after 15 minutes and again at the end of the lab period. What, if any, changes have occurred? (If no changes have occurred, store the slide in a covered petri dish containing a piece of cotton moistened with water, and examine it the next day.)

10. The central stalk surrounded by the stamens is the female reproductive organ, or carpel. It is composed of a large basal part, the ovary, above which is an elongated part, the style, ending in a stigma. How is the stigma adapted to trap the pollen grains and to provide a place for them to grow?

11. Use a scalpel to cut the ovary lengthwise. Using a hand lens or dissecting microscope, look at the cut surface. How many ovules can you see? Each ovule contains one egg. To what stage of the moss life cycle is the ovule comparable? Where is the pollen grain deposited? How does the sperm in the pollen grain reach the egg? To what stage of the moss life cycle is a pollen grain comparable?

 CAUTION: Scalpels are sharp; handle with care.

12. The union of egg and sperm causes extensive changes in the female reproductive parts. Fertilization of the egg stimulates the growth of the ovary and the enclosed ovules. Carefully examine a fresh bean or pea pod. Open the pod to find the seeds. Which of the female reproductive structures is the pod of a bean or pea? What is the origin of a seed? If you plant ripe bean or pea seeds and water them, what will they produce? What can you conclude develops within a seed as a result of fertilization?

13. If time permits, examine other types of flowers. Compare the numbers of various parts and the ways the parts are arranged with respect to each other.

14. Wash your hands thoroughly before leaving the laboratory.

Analysis

1. In alternation of generations in a moss, which is the predominant independent generation? Which is the less conspicuous generation?
2. Compare the life cycle of a moss (with alternation of generations) with your life cycle (with no alternation of generations).
3. Would you expect more variation in flowering plants or in those that reproduce asexually? Explain your answer.
4. Compare and contrast the sporophyte and gametophyte stages of a moss to those of a flowering plant.
5. Do flowering plants demonstrate more or less adaptation to a land environment than mosses? Explain your answer.

Investigations for Chapter 13
Patterns of Inheritance

Investigation 13A ◆ Probability

The probability of a chance event can be calculated mathematically using the following formula:

$$\text{probability} = \frac{\text{number of events of choice}}{\text{number of possible events}}$$

What is the probability that you will draw a spade from a shuffled deck of cards like that shown in Figure 13.13? There are 52 cards in the deck (52 possible events). Of these, 13 cards are spades (13 events of choice). Therefore, the probability of choosing a spade from this deck is 13/52 (or 1/4, or 0.25, or 25%). To determine the probability that you will draw the ace of diamonds, you again have 52 possible events, but this time there is only 1 event of choice. The probability is 1/52, or approximately 2%. In this investigation, you will determine the probability for the results of a coin toss.

Materials (per team of 2)
2 pennies (1 shiny, 1 dull)
cardboard box

Procedure

1. Work in teams of two. One person will be student A and the other will be student B.
2. Student A will prepare a score sheet with two columns—one labeled *H* (heads) and the other *T* (tails). Student B will toss a penny ten times. Toss it into a cardboard box to prevent the coin from rolling away.
3. Student A will use a slash mark (/) to indicate the results of each toss. Tally the tosses in the appropriate column on the score sheet. After ten tosses, draw a line across the two columns and pass the sheet to student B. Student A then will make ten tosses, and student B will tally the results.
4. Continue reversing roles until the results of 100 (ten series of ten) tosses have been tallied.
5. Prepare a score sheet with four columns labeled *H/H, Dull H/Shiny T, Dull T/Shiny H,* and *T/T* (H = heads, T = tails). Obtain two pennies—one dull and one shiny. Toss both pennies together 20 times while your partner tallies each result in the appropriate column of the score sheet.
6. Reverse roles once so that you have a total of 40 tosses.

Analysis

1. How many heads are probable in a series of ten tosses? How many did you actually observe in the first ten tosses?
2. Deviation is a measure of the difference between the expected and observed results. It is not the difference itself. It is the ratio of the sum of the differences between expected and observed results to the total number of observations. Thus,

$$\text{deviation} = \frac{\begin{array}{c}\text{difference between heads expected} \\ \text{and heads observed} \\ + \\ \text{difference between tails expected} \\ \text{and tails observed}\end{array}}{\text{number of tosses}}$$

Calculate the deviation for each of the ten sets of tosses.

3. Calculate the deviation for your team's total (100 tosses).

4. Add the data of all teams in your class. Calculate the class deviation.

5. If your school has more than one biology class, combine the data of all classes. Calculate the deviation for all classes.

6. How does increasing the number of tosses affect the average size of the deviation? These results demonstrate an important principle of probability. State what it is.

7. On the chalkboard, record the data for tossing two pennies together. Add each column of the chart. In how many columns do data concerning heads of a dull penny appear?

8. In what fraction of the total number of tosses did heads of dull pennies occur?

9. In how many columns do data concerning heads of a shiny penny occur?

10. In what fraction of the total number of tosses did heads of the shiny pennies occur?

11. In how many columns do heads of both dull and shiny pennies appear?

12. In what fraction of the total number of tosses did heads of both the pennies appear at the same time?

13. To which of the following is this fraction closest: the sum, the difference, or the product of the two fractions for heads of one penny at a time?

14. Your answer suggests a second important principle of probability that concerns the relationship between the probabilities of separate events and the probability of a combination of events. State this relationship.

15. When you toss two coins together, there are only three possibilities—*H/H*, *T/T*, or *H/T*. These three combinations will occur 100% of the time. The rules of probability predict that *H/H* and *T/T* each will occur 25% of the time. What is the expected probability for the combination of heads on one coin and tails on the other?

16. When you toss a dull penny and a shiny penny together, what is the probability that heads will occur on the dull penny? What is the probability that tails will occur on the shiny penny? Calculate the probability that the dull penny will be heads and the shiny penny will be tails if you toss the two pennies together. Compare this answer to the answer in question 15. How do you account for the different answers? Are there other ways than *Dull H/Shiny T* to get the *H/T* combination?

17. How many different ways can you get the *H/T* combination on two coins tossed together? What is the probability of each of those different ways occurring? Is the probability of getting heads and tails in any combination of pennies closest to the sum, the difference, or the product of the probabilities for getting heads and tails in each of the different ways?

18. Your answer suggests a third important principle of probability that concerns the relationship between (1) the probability of either one of two mutually exclusive events occurring and (2) the individual probabilities of those events. State this relationship.

Investigation 13B ◆ Seedling Phenotypes

Albinism is a rare condition found in both plants and animals. The cells of albino animals or plants lack certain pigments. Albino plants, for example, have no chlorophyll (Figure 13B.1). In this investigation, you will observe the influence of both heredity and environment on a plant's ability to produce chlorophyll.

FIGURE 13B.1

Corn seedlings. Growing among the green plants, the albino corn seedlings lack the pigment chlorophyll.

Materials (per team of 2)
petri dish
filter or blotting paper
light-proof box or aluminum foil
about 50 tobacco seeds

Procedure

1. Evenly sprinkle tobacco seeds over moistened filter paper in the bottom of a petri dish. The seeds should be separated by at least twice their length.

    CAUTION: Wash hands after handling tobacco seeds. They may be treated with a fungicide.

2. Put the cover over the dish. Put the dish in a lighted area for about 4 days. Record observations daily in your logbook

3. After 4 or 5 days, wrap the dish in foil or cover it with a light-proof box. Put the setup in a dark place where it will not be disturbed for 3 or 4 more days. Be sure the paper is moist at all times, and add water if necessary. Do not expose the seeds to light when checking them.

4. After the seeds have germinated, usually in about a week to 10 days, remove the light-proof cover and examine the seedlings. Observe especially the color of the tiny leaves. Record your observations.

5. Replace the cover on the petri dish, and put it in a well-lighted place for a few days. Be sure the paper is moist at all times. Observe the seedlings each day as they are exposed to the light.

6. After being in the light for a few days, count the number of plants of each color. In your logbook, record your observations and the numbers of each type of plant.

7. Wash your hands thoroughly before leaving the laboratory.

Analysis

1. What was the color of the tobacco plants while they were growing in the dark?

2. Did light have the same effect on all of the plants? Explain your answer.

3. The seeds used in this investigation came from a specially bred tobacco plant. Do you suppose the parent plants were both green, one green and one albino, or both albino? (Hint: Consider the role of chlorophyll in the life of a plant.) Explain your answer.

4. Does light have any effect on a tobacco plant's ability to produce chlorophyll? Explain your answer.

TABLE 13B.1
Mendel's Results from Crossing Pea Plants with Single Contrasting Traits

P_1 cross	F_1 plants	F_1 plants (self-pollinated)	F_2 plants	Actual ratio
Round × wrinkled seeds	all round	round × round	5,474 round 1,850 wrinkled 7,324 total	2.96:1
Yellow × green seed (cotyledons)	all yellow	yellow × yellow	6,022 yellow 2,001 green 8,023 total	3.01:1
Green x yellow pods	all green	green × green	428 green 152 yellow 580 total	2.82:1
Long x short stems	all long	long × long	787 long 277 short 1,064 total	2.84:1

5. Is light the only factor required for a tobacco plant to produce chlorophyll? Explain your answer.

6. Which seedlings showed the influence of heredity on chlorophyll development? Explain your answer.

7. What was the approximate ratio of green to albino plants that appeared when seedlings were grown in the light for several days? Compare this ratio with Mendel's data for crossing garden peas shown in Table 13B.1.

Investigation 13C ◆ A Dihybrid Cross

Sexually reproducing organisms have haploid and diploid stages. In flowering plants and animals, only the gametes are haploid, and traits (the phenotype) can be observed only in the diploid stage. When Mendel made dihybrid crosses to study the inheritance of two different traits, such as seed shape and seed color, he could observe the traits only in the diploid cells of the parents and their offspring. He had to use probability to calculate the most likely genotypes of the gametes.

In the yeast *Saccharomyces cerevisiae*, however, phenotypes of some traits can be seen in the haploid cell colonies. In Investigation 12B, for example, you could observe the color trait red or cream in both haploid and diploid stages of the yeast life cycle. Not all yeast traits are visible, however, so geneticists use the methods Beadle and Tatum devised for studying mutations; that is, they isolate mutants that cannot make some essential substance. The inability of the mutant strain to make the substance and the ability of the normal strain to make it are two different forms of a trait.

In the yeast dihybrid cross (the outcomes of which are listed in Table 13C.1), you will follow two forms of each of two traits: red versus cream color and tryptophan-dependent (requiring this amino acid to grow) versus tryptophan-independent (not requiring this amino acid). In Investigation 12B, a red strain of yeast of one mating type is crossed with a cream-colored strain of the other type. The diploid strain is cream-colored. If there is a single gene for this trait, there must be one allele that determines cream color in the haploid strain and another allele that determines red color. In the diploid strain then, there must be one of each of these alleles. Which form of the color trait is dominant?

By observing the color of a colony that a haploid strain forms, you can predict with certainty which allele it carries. In the case of a cream-colored diploid, however, you cannot be sure of the genotype. It could carry either one allele for cream color and one for red or two alleles for cream color. The trait defined by tryptophan-dependence or independence works in much the same way. A tryptophan-dependent haploid must carry the allele for tryptophan-dependence (a defective form of the gene); a tryptophan-independent haploid must carry the allele for tryptophan-independence (the functional form of the gene), but a tryptophan-independent diploid could be either homozygous or heterozygous for the functional allele.

Using symbols for the traits, the allele for the dominant cream form is R and the recessive red

TABLE 13C.1
Growth of Yeast on Two Media

Haploid	Diploid	On COMP medium	On MIN medium
RT	RR TT RR Tt Rr TT Rr Tt	growth, cream	growth, cream
Rt	RR tt Rr tt	growth, cream	no growth, (cream?)
rT	rr TT rr Tt	growth, red	growth, red
rt	rr tt	growth, red	no growth, (red?)

form is *r*; the normal tryptophan-independent form of that gene is *T* and the tryptophan-dependent form is *t*. Whereas the color trait is visible, determining tryptophan dependence or independence requires a simple growth test.

Table 13C.1 shows how all possible combinations of these two traits can be determined by testing the yeast on two types of growth medium: a nutritionally complete medium (COMP) and a medium lacking tryptophan (MIN).

Materials (per team of 2)
container of clean flat toothpicks
glass-marking pencil
3 self-sealing plastic bags labeled *waste* (1 for each day)
biohazard bag
complete growth medium agar plate (COMP)
minimal + adenine growth medium agar plate (MIN)
COMP plate with 24-hour cultures of yeast strains
HAO, HAR, HAT, HART, HBO, HBR, HBT, and HBRT

Procedure

PART A   Predicting the Gametes of the Parents
The diploid parents of this dihybrid cross would have been *rr TT* and *RR tt*. To predict the F$_2$ offspring, it is necessary to predict the different types and relative numbers of gametes that can be produced. The simplest way to make this prediction is to diagram how the chromosomes separate at meiosis. Figure 13C.1a shows the gametes predicted from the pure-breeding diploid red parent *(rr TT)*.

Two pairs of chromosomes are represented, one carrying the gene for color (in this case red, *r*) and the other carrying the gene for the ability to make tryptophan (in this case tryptophan-independent, *T*). At the first meiotic division (M$_I$), chromosome pairs segregate (separate). At the second meiotic division (M$_{II}$), the two chromatids of each chromosome segregate into the nuclei of the gametes. In this case, segregation can occur in only one way because only one allele of each gene is represented. All the gametes from a pure-breeding diploid are the same *(rT)*. The probability of this parent producing a gamete with the genotype of *rT* is 1/1, or 100%.

1. In your logbook, copy the diagram in Figure 13C.1b (or use the diagram your teacher provides) for the cream-colored, tryptophan-dependent parent *(RR tt)*. Fill in the missing symbols. Each chromatid should have one letter.

2. Use the symbols to describe for each parent strain the genotypes and phenotypes of all the possible gametes and the relative probabilities of their occurrence.

3. The haploid gametes produced from these two pure-breeding parents mate to form the diploid zygotes of the F$_1$ generation. Use the symbols to describe the genotype and phenotype of the diploid zygotes that could be formed from the fusion of these gametes. At this point, your diagrams should show that they all will have the dihybrid genotype *Rr Tt*.

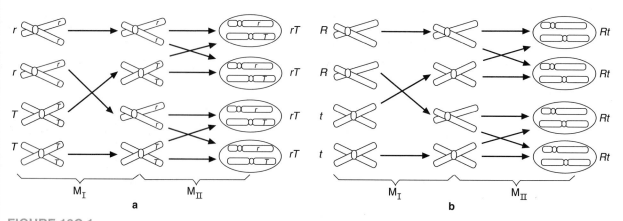

FIGURE 13C.1
Probabilities of gametes that could be formed by yeast organisms that are homozygous for two traits.

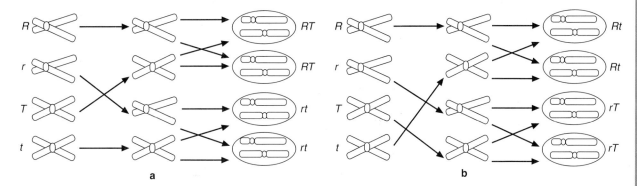

FIGURE 13C.2

Gametes that could be formed by yeast organisms heterozygous for two traits.

PART B   Predicting the Gametes of the F_1 Diploids

Because there are two equally probable ways that the alleles of the two genes can separate at M_I, the chromosome diagram for the segregation of the R, r, T, and t alleles is more complicated. These are shown in Figure 13C.2. Since these two patterns of segregation are equally probable, the two diagrams together illustrate the relative numbers of all possible genotypes.

4. Copy the diagrams in Figure 13C.2 into your logbook, or use the diagrams your teacher provides. Fill in the missing symbols. Each chromatid should have a symbol.
5. What is the total number of different gametes represented in the diagram?
6. How many different genotypes are represented? Give their symbols.
7. How many times is each genotype represented?
8. What is the probability of occurrence of each genotype among the total number of gametes shown? Compare this prediction with the gametes in Figure 13.15.
9. Draw a checkerboard diagram for all the possible crosses among these gametes. This diagram should illustrate the 16 different combinations predicted. Instead of male and female gametes, use mating-type **a** and mating-type **α** haploid strains. In each square of the diagram, construct the diploid genotype that would result from the fusion of the

corresponding gametes. How many different genotypes are there? How many different phenotypes? Does this diagram predict a 9:3:3:1 ratio? What specific phenotypes would be represented in these ratios?

PART C   Testing the Predicted Phenotypes

Day 0

10. Make two templates for setting up crosses. Copy the pattern shown in Figure 13C.3 onto a piece of paper twice so that each fits the bottom of a petri dish. (Or use the templates your teacher provides.)
11. Tape one template to the bottom of a plate of complete growth medium (COMP) so that you can read it through the agar.

mating-type **a** strains

		HAO RT	HAR Rt	HAT rT	HART rt
mating-type **a** strains	HBO RT				
	HBR Rt				
	HBT rT				
	HBRT rt				

FIGURE 13C.3

Template for dihybrid yeast cross.

12. Transfer a small sample of each strain onto the agar directly over its corresponding label. To do this, touch the flat end of a clean toothpick to strain HAO on the plate your teacher provides. Gently drag the toothpick along the box labeled *HAO* on the COMP agar plate to make a streak about 1 cm long. Discard the toothpick in the self-closing waste bag, being careful not to touch anything with it. Repeat this procedure for all eight strains, using a new toothpick for each transfer. Be careful not to touch the ends of the toothpicks to anything except yeast or the sterile agar. Discard used toothpicks in the self-closing plastic waste bag. Keep the lid on the plate at all times except when transferring yeast.

13. Invert and incubate the plate at room temperature.

14. Wash your hands before leaving the laboratory.

Day 1

15. On the same COMP plate, make a mating mixture for each of the mating-type **a** strains with each of the mating-type **α** strains. To do this, use the flat end of a clean toothpick to transfer a dot of the HAO strain of freshly grown cells to each of the boxes below it on the template. Discard the toothpick in the waste bag. Repeat this procedure for HAR, HAT, and HART, using a new toothpick for each strain. Using the same procedure, transfer a dot of the freshly grown cells of each HB strain to each of the boxes to the right of the strain on the template. Place these dots side-by-side but not touching. Use a clean toothpick to mix each pair of spots together. Be sure to use a clean toothpick each time you mix pairs together. Discard all toothpicks in the waste bag.

16. Invert and incubate the plate at room temperature.

17. Wash your hands before leaving the laboratory.

Day 2

18. Test each mating mixture and parent strain for its ability to grow on MIN agar. To keep track of the tests, tape a copy of the template to the bottom of the MIN plate. Use the flat end of a clean toothpick to transfer a small amount of each strain and mixture from the COMP plate to the corresponding position on the MIN plate. Be sure to use a clean toothpick each time you change strains and mixtures. Discard all toothpicks in the waste bag.

19. Invert and incubate both plates until the next day.

20. Wash your hands before leaving the laboratory.

Day 3

21. Copy the score sheet in Figure 13C.4 into your logbook. (Or tape in the copy your teacher provides.) Record the color and growth phenotypes of each parent haploid strain and F_2 diploid on the score sheet for both plates.

22. Tabulate the different phenotypes observed among the F_2 diploids and the number of times each one occurred among the 16 crosses.

23. Compare the F_2 phenotypes with the predictions you made in Part A. Explain how your actual results either support or contradict your predictions.

24. Discard all plates in the biohazard bag.

25. Wash your hands before leaving the laboratory.

Analysis

Because yeast exhibit most of the same traits in the haploid stage (gametes) and the diploid stage, you knew the precise genotypes of the gametes (haploid strains) that you mated to produce the F_2 diploids. This removed the element of chance at this step. In Part B, however, you had to deal with the role of chance to predict the numbers and genotypes of the gametes from the F_1 diploids. Throughout the entire

mating-type **a** haploid parents

	HAO	HAR	HAT	HART
HBO	HAO x HBO	HAR x HBO	HAT x HBO	HART x HBO
HBR	HAO x HBR	HAR x HBR	HAT x HBR	HART x HBR
HBT	HAO x HBT	HAR x HBT	HAT x HBT	HART x HBT
HBRT	HAO x HBRT	HAR x HBRT	HAT x HBRT	HART x HBRT

mating-type **α** haploid parents

F_2 diploids

FIGURE 13C.4

Score sheet for dihybrid yeast cross.

process, beginning with the parental cross (P) and going through to the F$_2$ offspring, there are some steps at which chance plays a role, so the results can only be expressed as a probability. In other steps, chance is not a factor, so you can predict the outcome exactly.

1. List the steps in which chance is a factor.
2. List the steps in which chance is not a factor.
3. Explain why the outcome of some steps involves chance, whereas the outcome of others does not.

Investigations for Chapter 14
Other Forms of Inheritance

Investigation 14A ◆ Jumping Genes

Our understanding of genetics has been greatly aided by studying organisms that display one or more unusual traits due to the impact of altered or mutated genes. Indeed, some of the fundamental behaviors of genes were discovered through careful observation and experimentation, even before the structure of DNA was worked out. In this investigation, you will use your powers of observation and critical thinking to describe mechanisms by which mutations change in the appearance of corn kernels. Remember that each kernel on an ear of corn represents a separate fertilization. This means that corn kernels on an ear can be thought of as siblings.

Materials (per team of 2)
paper and pencils

PART A   Analysis of Corn Ears

Procedure
1. In this variety of corn, purple kernels are the dominant phenotype. The purple pigment is produced through the activity of an enzyme encoded by a gene at the C locus. Colorless (white) kernels result from a mutation in the *C* gene. The dominant allele is designated *C* while the recessive allele is *c*.
2. Examine the two ears of corn pictured in Figure 14A.1, and count how many kernels of

a

b

FIGURE 14A.1
Corn displaying a mutation at the C locus. (Yellow rather than white kernels are shown here.)

each color are on each ear. Count at least 150 kernels per ear of corn. Record the data in your logbook.

Analysis
1. What was the ratio of purple to white kernels on the ear of corn labeled *a?*
2. Using this data, what were the likely genotypes of the parent plants?
3. What was the ratio of purple to white kernels on the ear of corn labeled *b?*
4. Using this data, what were the likely genotypes of the parent plants?
5. Did either of the ears have kernels that were difficult to score as either purple or white?

PART B   Analysis of Individual Corn Kernels

Procedure
3. Barbara McClintock studied the color patterns of kernels in Indian corn in the 1940s. She determined that the speckled appearance of some kernels could be explained by the presence of unstable mutations in the corn genome. Today we know that the mutations originally studied by McClintock are due to the insertion of a foreign DNA sequence, called a transposon or jumping gene, into the coding region of a gene.

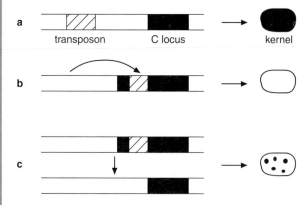

FIGURE 14A.2

Mutation of the C locus.

4. Examine Figure 14A.2, and answer Analysis questions 1 and 2.

5. Figure 14A.2 illustrates how movement of the transposon results in speckled corn kernels (Figure 14A.1b). Using the diagram in Figure 14A.3a as a model, draw in your logbook corresponding diagrams for the corn kernels depicted in b, c, and d. Answer Analysis questions 3–7.

Analysis

1. What happens when a transposon inserts into the C locus?

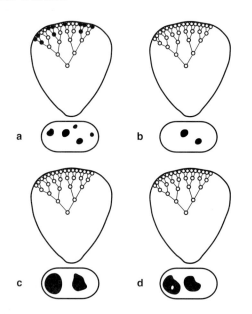

FIGURE 14A.3

How timing and frequency of transposon movement affects kernel development.

2. What causes the appearance of the purple spots on the speckled corn kernels?

3. Why is the transposon in the C locus referred to as unstable?

4. Why does the kernel in Figure 14A.3b display fewer spots than the one in Figure 14A.3a?

5. Why are the spots on the kernel in Figure 14A.3c larger than those in a and b?

6. How can you account for the presence of the white spot in the dark sector of the kernel shown in Figure 14A.3d?

7. Does insertion of a transposon always stop expression of the affected gene?

8. What is the evolutionary significance of transposons?

Investigations for Chapter 15
Advances in Molecular Genetics

Investigation 15A ◆ Determining Mutation Frequency in Bacteria

In this investigation, you will attempt to determine the frequency of formation of white mutants from cultures of the red-pigmented bacterium *Serratia marcescens*, strain D1 (Figure 15A.1). Many factors contribute to mutation frequency (the proportion of mutants that exists in a culture at any given moment). Two factors are especially important in increasing the proportion of mutants as the population grows. The first is the reproduction of the mutants themselves, and the second is the *mutation rate* of the culture. The mutation rate is the probability that a mutation will occur in a generation of cells.

Consider two generations of 100 bacterial cells with a mutation rate of 1/10. (This rate is convenient for demonstration purposes but is unrealistically high.) Since the proportion of mutants is 1:10, of the original 100 cells, 90 are wild types and 10 are mutants. The mutants will divide to produce 20 cells of their own type. Among the wild type cells, however, 90% (or a theoretical 81 cells) will divide to produce 162 wild-type cells, while 10% (or 9 cells) will mutate and divide to produce 18 new mutant cells. These 18 new mutant cells plus the 20 from the reproduction of the old mutants yield 38 mutants in

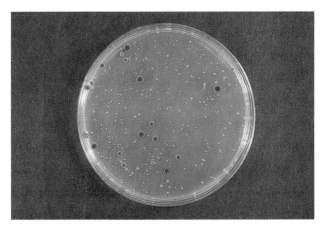

FIGURE 15A.1

Growth on agar of the bacterium *Serratia marcescens*, strain D1.

this generation of 200 cells. This same reasoning carried into the second generation yields a proportion of mutants in the population of about 108:400 as compared to 1:10 in our original population.

Of course, if this increase in the population of mutants were to continue, nearly the entire culture would soon be mutants. This condition is prevented, however, by *back mutations* (the reversion of the mutant genes to their original state). Back mutations too have their own characteristic mutation rates. When enough mutants have accumulated, back mutations begin; the forward and backward mutations should reach an equilibrium and just balance each other. At this point, the culture should reach and maintain a *mutation equilibrium* (a constant proportion of mutant cells).

The procedure for this investigation involves spreading the bacteria over the surface of a sterile agar plate, incubating the bacteria overnight, and analyzing the bacterial colonies that result. Your starting culture of *S. marcescens* is too concentrated to analyze directly, so you will prepare a series of tenfold dilutions (called serial dilutions) and spread them on separate agar plates. Ideally, one of the dilutions will result in about 200–400 colonies growing on the plate. Remember, we assume that each colony results from a single bacterium.

Materials (per team of 3)

3 lab aprons
3 pairs of safety goggles
3 pairs of plastic gloves
4 *S. marcescens* agar plates
2 mL starter culture of red *S. marcescens*, strain D1
glass-marking pencil
6 tubes of sterile water, 9 mL each
bacteria spreader
5 sterile graduated pipettes
rubber pipet bulb
250-mL beaker
50 mL isopropyl alcohol (70%)
incubator

SAFETY **Put on your safety goggles, lab, apron, and gloves. Tie back long hair.**

Procedure

1. Use a sterile pipette to prepare your first tenfold dilution of the starter culture of *S. marcescens*. Pipet 1 mL of the bacterial culture into one of the test tubes containing 9 mL of sterile water. Mix the dilution by gently tapping the tube with your finger, and label it *1:10* or *10^1*.

2. Use a sterile pipette to prepare a 1:100 dilution by adding 1 mL of the 1:10 dilution to a second tube containing 9 mL of sterile water. Mix the dilution by gently tapping the tube with your finger, and label it *1:100* or *10^2*.

3. Repeat step 2 four more times to prepare dilutions of 1:1,000; 1:10,000; 1:100,000; 1:1,000,000 (or 10^3, 10^4, 10^5, 10^6).

4. Obtain four *S. marcescens* agar plates, and label them *10^3*, *10^4*, *10^5*, and *10^6*.

5. Use a sterile pipette to add 0.1 mL (100 μl) of the 10^3 dilution to the surface of the *S. marcescens* agar plate labeled *10^3*.

6. Sterilize the bacterial spreader by dipping it into a beaker of 70% isopropyl alcohol. Let it air dry (about 10 seconds) before spreading cells. Spread the cells evenly over the surface of the plate. Be careful not to press too hard, or you will break the agar surface.

7. Repeat steps 5 and 6 to spread the 10^4, 10^5, and 10^6 dilutions on the appropriately labeled plates.
8. Invert the four plates, and incubate them for 24 hours in a 278°C incubator.
9. Following the incubation, select a plate containing 200 to 400 colonies for counting.
10. Count the total number of colonies on your plate; note the number that are red and the number that are white. Also note colonies of other colors.

Analysis

1. What was the viable count of red and white *S. marcescens* per milliliter in the original starter culture? Remember that each plate was inoculated with 0.1 mL rather than 1 mL. Thus, if 200 colonies were on the plate inoculated with the 10^6 dilution, the number of viable organisms per milliliter would be 200×10^7, or 2,000,000,000 per milliliter.
2. What percentage of the total number of colonies represented white mutants? Pool the results of the class, and calculate the class average. This number will represent the mutation frequency for *S. marcescens*, strain D1.
3. Was the loss in pigmentation by *S. marcescens* accompanied by a change in morphology of the bacterial colonies? If such changes were noticed, how would they complicate the proper identification of the white colonies as mutants of the red forms?
4. Did you notice any single colonies with both red and white segments? How might such an occurrence be explained?
5. Can you suggest a way to allow your white mutants to regain their red pigment?

Investigations for Chapter 16
Population Genetics

Investigation 16A ◆ Sickle-Cell Disease

What is the significance of the variations we see from one individual to another? Does each variation provide some special advantage? Biologists assume that most variations within a species have survival value. Could this assumption include variations that cause diseases? For example, geneticists have described hundreds of variations in the structure of the hemoglobin molecule inside human red blood cells. Some of these variations cause disease, and others do not. Could a variation in hemoglobin structure that causes disease have survival value to a population? If so, under what circumstances?

Sickle-cell disease is a disorder in which the abnormal hemoglobin molecules inside red blood cells combine with each other when the oxygen supply is low, causing the cells to lose their flexibility and assume an abnormal sickle shape (see Figure 16.5). The sickle cells clump together and block small blood vessels, stopping the flow of nutrients and oxygen. Vital organs may be damaged, and the individual may die. In this investigation, you will examine the conditions that influence the inheritance of sickle-cell disease.

Materials (per team of 3)
paper and pencils

PART A

Procedure
Read the following information about sickle-cell disease, and answer the Analysis questions.

Sickle-cell disease is due to the homozygous presence of the sickle-cell allele (symbol: Hb^S). Heterozygous ($Hb^S Hb^A$) individuals have sickle-cell trait and can be identified by a blood test. (Hb^A represents the allele for normal hemoglobin.) About 0.25% of African-Americans are homozygous ($Hb^S Hb^S$) and have sickle-cell disease. In certain parts of Africa, about 4% of black Africans have sickle-cell disease.

The Hardy-Weinberg model (see Section 16.3) enables us to predict when the allele frequencies in a population will remain constant. Because people with sickle-cell disease frequently die in childhood, however, the frequency of the allele causing this disease should not remain constant. Each death removes a pair of the sickle-cell-causing alleles from the population.

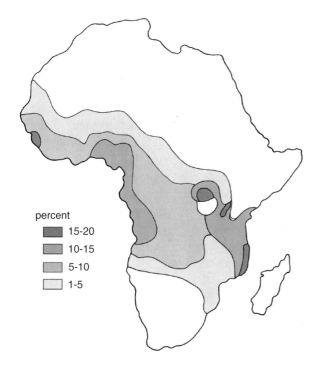

FIGURE 16A.1

Frequency of the allele that causes sickle-cell disease, plotted as percentages of the population gene pool in parts of Africa.

percent
- 15-20
- 10-15
- 5-10
- 1-5

Figure 16A.1 shows the locations in Africa where the allele frequency of the Hb^S form is highest. These locations are also where a fatal form of malaria is found. Studies reveal that people homozygous for the normal allele $(Hb^A Hb^A)$ often die of malaria. However, people with the heterozygous genotype do not contract the fatal form of malaria.

Analysis

1. What are the three possible genotypes involving the sickle-cell allele and its normal allele?
2. What are the phenotypes that would be associated with each genotype?
3. The Hardy-Weinberg model was described in Section 16.3. The model is summarized by this equation: $p^2 + 2pq + q^2 = 1$. Based on this model, what percentage of the African-American population has the genotype $Hb^S Hb^A$?
4. In some areas of Africa, the frequency of the Hb^S allele is very high. What percentage of the black African population in these areas has the genotype $Hb^S Hb^A$?

5. The relatively high frequency of the Hb^S allele has tended to remain constant in these areas of Africa. Under usual conditions, would you expect the frequency of a harmful allele to remain constant? Explain your answer.
6. Propose three possible explanations for the observation that the frequency of the Hb^S and Hb^A alleles remains high in these areas, despite its lethal effect in the homozygous condition.
7. What are the advantages and disadvantages of having each of the three genotypes you determined in question 1? Which genotype(s) would tend to survive?
8. Assuming that $Hb^A Hb^A$ individuals die from malaria, what would be the allele frequencies of Hb^S and Hb^A in a population made up of the surviving genotype(s)?
9. The allele frequencies in the African population under discussion are not 0.5 for each form. Considering that fact, do you think the assumption in question 8 (that an individual with $Hb^A Hb^A$ will die from malaria) is reasonable? Which genotype do you think is less likely to be fatal? Explain your answer.
10. How can you explain the lower frequency of the allele causing sickle-cell disease among people of African descent who now live in America?

PART B

Procedure

Read the following information about the differing chemical structures of normal hemoglobin and sickle-cell hemoglobin, and answer the Analysis questions.

The sickle-shaped red blood cell is caused by a mutation that affects the hemoglobin molecule in the cells. Hemoglobin is a protein and therefore is made of amino acids. Scientists have been able to compare the chemical structure of normal hemoglobin, called hemoglobin A, with the hemoglobin found in sickle cells, called hemoglobin S. They found that both types of hemoglobin molecules contain 560 amino acids of 19 different types but differ in the substitution of valine for glutamate at a specific position in hemoglobin S (see Section 16.4).

Analysis

1. What are the nucleotide codes for the two amino acids that are different in hemoglobin A and hemoglobin S? (Consult the genetic code in Figure 9.4.) What is the simplest error in coding that could have occurred to cause the mutation?

2. Write a few paragraphs to summarize sickle-cell disease. Emphasize the ideas of mutation, selection, survival value, and evolution.

Investigations for Chapter 17
The Origin of Life

Investigation 17A ◆ Molecular Evolution in the Test Tube

When most people think of biological evolution, they think of living organisms and the formation of new species in the wild. Biologists have recognized since the 1960s, however, that the most essential features of biological evolution—replication, variation, and selection—apply equally to molecules. It is possible, for example, to study nucleic-acid molecules reproducing in the test tube (in vitro) and to watch them evolve new properties.

In this investigation, you will simulate an experiment in RNA evolution. The bacteriophage Qβ, which normally infects *Escherichia coli*, has a genome that consists of a single RNA molecule 4,000 nucleotides long. During the infection cycle, the RNA must enter the bacterial cell. To accomplish this, three of the four genes encoded by the RNA's 4,000 nucleotides specify proteins that enable the RNA to enter the bacterial cell and the "progeny" RNAs to spread to new bacteria. The fourth gene encodes viral *replicase*, the protein enzyme that uses the viral RNA as a template on which to assemble monomers into new copies of the RNA. The replicase enzyme initiates copying of the RNA by binding to a small subset of bases within it, called the origin of replication. These few bases alone are all that any Qβ RNA molecule needs in order to be copied by the replicase. Any molecule with an intact sequence at the origin of replication will be copied, and any molecule in which this sequence is either lost or

significantly mutated will be either not copied at all or copied at an altered rate.

This natural system can be streamlined and simplified to study evolution in a test tube. This in vitro system consists of the Qβ RNA molecule plus the materials needed for RNA synthesis (the A, U, C, and G nucleotide building blocks plus the replicase enzyme). This RNA system also has a built-in mutation feature that ensures that the RNA progeny molecules exhibit molecular variation (base-sequence changes). Such mutations come about because the replicase enzyme sometimes makes mistakes. On average, for each Qβ RNA molecule copied, there are one or two random base changes (mutations). Furthermore, the replicase occasionally produces molecules shortened by random amounts. Such shortened RNA molecules cannot infect bacteria but can still be copied in the test tube, provided that they retain the sequence for origin of replication. This system therefore incorporates two of the three features essential for evolution: replication and variation.

The third essential feature of evolution, selection, can be introduced by applying some form of selective pressure to the system. This is accomplished by limiting the time available for the RNA molecules to be copied. Such a constraint confers a selective advantage on RNA molecules that can be copied more quickly. In this scenario, speed of replication becomes a "phenotype" of the molecules and a test of their "fitness" in the test-tube environment.

Materials (per person)
paper and pencils

Procedure

1. The investigator begins the experiment by adding Qβ RNA to a test tube containing replicase enzyme and nucleotide monomers. Replication reactions proceed for just 15 minutes. Then a random sample of the progeny RNAs is transferred from the first tube into a second tube containing a fresh supply of replicase and nucleotides (but no RNA other than that transferred). The replication process

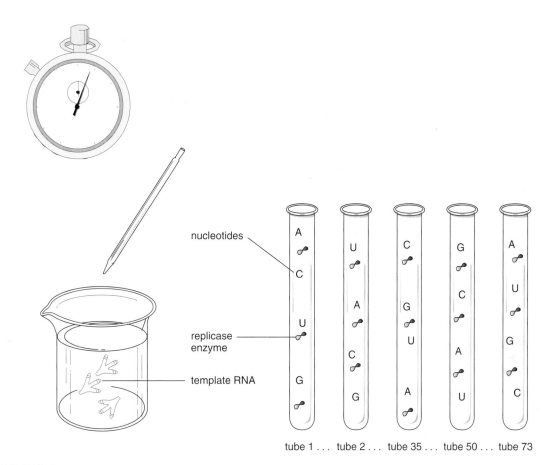

nucleotides

replicase enzyme

template RNA

A	U	C	G	A
C	A	G	C	U
U	C	U	A	G
G	G	A	U	C
	G	A	U	

tube 1 . . . tube 2 . . . tube 35 . . . tube 50 . . . tube 73

FIGURE 17A.1

Experimental setup for in vitro replication of RNA.

again proceeds for 15 minutes, and a sample from this second tube is transferred to a third tube containing more fresh raw materials. This serial-transfer process is repeated 72 times (Figure 17A.1). During the experiment, the investigator monitors the total amount of RNA that accumulates in each tube, as well as the size and nucleotide composition of each "generation" of RNA molecules.

2. Use what you know about the experimental setup and about evolution to formulate a hypothesis and make some general predications about how the starting RNA population changes during the course of the experiment. Your hypothesis should make predications about changes that might occur in three phenotypic traits of the RNA molecules: (1) the speed of

replication; (2) the length of the molecules; (3) the ability of the RNA molecules to infect *E. coli* bacteria.

3. Examine the data in Figure 17A.2, and if necessary, revise your initial hypothesis.

Analysis

1. What can you conclude from the data in Figure 17A.2 about the total number of molecules produced in each generation? What does this imply about the average speed of replication of the molecules?

2. What was the reason for the change in the speed of RNA replication?

3. Two major changes occurred in the RNA molecules during the course of the experiment that account for their altered replication speed:

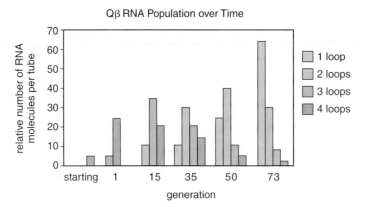

Qβ RNA Population over Time

FIGURE 17A.2
Simulated results from the in vitro evolution experiment.

Copying errors that replaced one base with another were made randomly by the "sloppy" replicase. The second reason for altered replication speed is apparent from the general trend in the size of the molecules over time. How would you describe this trend? What was happening to the longer molecules?

4. Imposing a time limit on the population for replication effectively made the copying process a competition—a race. This competition was certainly not "intentional" on the part of the molecules, but it was nevertheless inevitable simply because of the way the experimental environment operated. Why did each generation of RNA replicate faster than the preceding one?

5. If you were to test the RNAs of successive test-tube generations for their ability to carry out natural infection-replication cycles in bacteria, what would you expect to find?

6. After going from a length of 4,000 bases to approximately 700 bases, the molecules became no shorter. Suggest a reason that the size of the RNA stabilized around 700 bases.

7. A change in environmental conditions will select for new traits in molecules, just as it selects for new phenotypic traits in populations of organisms evolving in nature. Suppose a chemical inhibitor of replication was added to the test-tube system. If the amount of inhibitor was enough to slow down but not completely prevent replication, predict what would happen to the speed of RNA replication over time.

Investigation 17B ◆ Coacervates

Under certain conditions, the proteins, carbohydrates, and other materials in a solution may group together into organized droplets called coacervates. Because coacervates have some properties that resemble those of living things, droplets like them might have been an important step in the origin of life. In this investigation, you can produce coacervates and study the conditions under which they form. You also can compare the appearance of coacervates with the appearance of the one-celled organisms, amoebas.

Materials (per team of 2)
2 pairs of safety goggles
2 lab aprons
3 coverslips
dropping pipette
10-mL graduated cylinder
3 microscope slides
test tube
rubber stopper for test tube
compound microscope
wide-range pH test paper
gelatin suspension (1%)
gum arabic suspension (1%)

HCl (0.1*M*)
amoebas, living or prepared slides

SAFETY Put on your safety goggles and lab apron. Tie back long hair.

Procedure

1. Mix together 5 mL of the gelatin suspension and 3 mL of the gum arabic suspension in a test tube. Gelatin is a protein. Gum arabic is a carbohydrate and is related to sugars and starches. Measure and record in your logbook the pH of this mixture.

2. Place a drop of the gelatin–gum arabic mixture on a slide, and observe it under the low power of the microscope.

3. Slowly add dilute hydrochloric acid (0.1*M*) to the test tube, one drop at a time. After each drop, mix well and then wait a few seconds to see if the mixture becomes uniformly cloudy. If the liquid in the test tube remains clear, add another drop of acid. Continue adding acid one drop at a time until the mixture remains uniformly cloudy.

    CAUTION: 0.1*M* HCl is a *mild irritant*. Avoid skin/eye contact; do not ingest. If contact occurs, flush affected area with water for 15 minutes; rinse mouth with water; call your teacher.

4. When the material turns cloudy, take another pH reading. At this point, carefully observe a drop of the liquid under the microscope. Look for coacervates, structures resembling those in Figure 17.16. If you cannot see them, try adjusting the light and using high power. Add another drop of acid to the test tube, mix, and observe again. If you still do not observe coacervates, repeat the procedure from the beginning, for you may have added the acid too rapidly. When you are successful, record your observations and make sketches of the coacervate droplets.

5. Examine a wet-mount preparation of living amoebas or a prepared slide, and compare their structure and organization with the larger coacervates.

6. When you have finished your observations of the coacervates, add more dilute acid to the test tube, a drop at a time. Mix after adding each drop, and measure the pH after every third drop. Continue until the solution becomes clear again. Examine a drop under the microscope, and measure the new pH.

7. Wash your hands before leaving the laboratory.

Analysis

1. How do the materials you used to make coacervates compare with those that might have been present in the ancient oceans?

2. In what pH range did the coacervate droplets form?

3. Did the pH change as expected as a result of adding more acid to the solution between coacervate formation and clearing?

4. When dilute hydrochloric acid was added beyond a certain point, the coacervates disappeared. What might you add to the test tube to make the coacervates reappear?

5. How might the coacervate droplets be made more visible under the microscope?

6. How might coacervates have contributed to the formation of the first cells?

7. Do coacervates display the necessary criteria to be described as living? Explain your answer.

Investigations for Chapter 18
Diversity and Variation

Investigation 18A ◆ Using Cladistics to Construct Evolutionary Trees

The world's biodiversity may seem overwhelming, but patterns emerge when you examine traits that organisms have in common. Biologists have devised a method of organizing living things into smaller and smaller nested groups based on the presence of newly evolved traits. This method of analysis is called cladistics. The groups represent collections of organisms that belong to various taxonomic categories within Linnaeus's updated taxonomic hierarchy. This hierarchy consists of seven or eight taxonomic categories, including domains (sometimes called empires or superkingdoms), kingdoms, phyla (animals) or divisions (bacteria, plants, and fungi), classes, orders, families, genera, and species.

In this investigation, you will use cladistic analysis to reconstruct the evolutionary history of different groups of organisms. The aim is to use the appearance of newly evolved traits to reconstruct common ancestry. These relationships can be used to sort the organisms to create a natural classification system that is based on evolutionary relatedness rather than an artificial classification that is based on overall superficial resemblances.

Materials (per person)
"IS-Clad Method of Cladistics" handout
"Mammals" handout
"Mammalian Cladogram" handout

Procedure

PART A   Plants
1. Examine the data table in the "IS-Clad Method of Cladistics" handout. Based on the information in the data table, write an appropriate column heading in the upper left portion of the table. Also supply a title for the data table.
2. Using the information in the data table, complete the diagram on your handout. Draw in the remaining plants, and identify their shared traits. Note: The moss doesn't share features with the other plants and should not be included in the diagram. It is called an outgroup and is included in the cladistic analysis for comparison purposes.
3. Use your completed diagram to fill in the cladogram at the bottom of the handout. Look at the innermost circle on the diagram. This circle identifies the characteristic that the fewest types of plants have in common. Fill in the missing name and image of the plant sharing

this trait at the top left side of the cladogram. Follow the lines downward, and fill in the trait in the open circle labeled node A.
4. Look at the next outer circle on the diagram. This circle identifies a trait that a larger group of plants share. Fill in the missing name and image of the plant, and write the trait in node B.
5. Repeat steps 3 and 4 to identify the trait that four plants have in common. Write this trait in the bottom node.

PART B   Mammals
6. Examine the "Mammals" handout. Using the information in the data table, complete the diagram on your handout. Draw in the remaining mammals, and identify their shared traits. Remember that one type of mammal represents the outgroup and is not represented in the diagram. As before, the innermost circle on the diagram identifies the characteristic that the fewest types of mammals have in common. The next outer circle on the diagram identifies a trait that a larger group of mammals share. Repeat the process until you have accounted for all the mammal types.
7. Use your completed diagram to fill in the "Mammalian Cladogram" handout.

Analysis
1. How can cladistic analysis be used in classifying organisms and reconstructing their evolutionary history?
2. What type of information does cladistics use?
3. Why is it important to know what traits are possessed by a common ancestor?
4. How might the discovery of an extinct group that had never been seen before affect the construction of a cladogram?
5. How would the addition of a maple tree affect the plant cladogram in Part A?

Investigation 18B ◆ Structural Characteristics of Animals

In this investigation, you will observe selected external characteristics of animals, and you will use these characteristics to classify the animals. Careful

TABLE 18B.1
Classification Chart for Invertebrates

		Name of animal						
Exoskeleton[1]	present							
	absent							
Body symmetry	radial[2]							
	bilateral[3]							
	part bilateral, part spiral							
Jointed walking legs	3 pairs present							
	4 pairs present							
	more than 4 pairs present							
	absent							
Body segmentation[4]	present							
	absent							
Tentacles[5]	more than 4 present							
	4 or fewer present							
	absent							
Antennae[6]	2 or more pairs present							
	1 pair present							
	absent							
		Phylum	Phylum	Phylum	Phylum	Phylum	Phylum	Phylum

[1]Exoskeleton: a skeleton on the outer surface of an animal, enclosing the animal.
[2]Radial symmetry: body parts arranged in a circular manner around a central part or region, as in a bicycle wheel.
[3]Bilateral symmetry: matching body parts along the right and left sides of a line running from one end of the animal to the other, as in the body of a bus.
[4]Body segmentation: a structural pattern in which the body is divided into a series of more or less similar sections, the boundaries of which are usually indicated by grooves encircling the body.
[5]Tentacles: slender, flexible structures that can be lengthened or shortened; usually attached near the mouth.
[6]Antennae: slender structures that can be waved about but cannot change length; usually attached to the head.

TABLE 18B.2
Classification Chart for Vertebrates

		Name of animal						
Skin structures	hair present							
	feathers present							
	scales present							
	none of above present							
Appendages	wings present							
	legs present							
	fins present							
	none of above present							
Skeleton	bony[1]							
	cartilaginous[2]							
Teeth	present							
	absent							
		Class	Class	Class	Class	Class	Class	Class

[1]Bony skeleton: a skeleton in which most of the parts are hard and relatively rigid because of the hard mineral matter they contain.
[2]Cartilaginous skeleton: a skeleton in which all the parts are tough but flexible because they are composed of cartilage, a substance that does not contain significant deposits of hard minerals.

observation and note taking are required. Two questions may help guide your observations: How are these animals similar? How are these animals different?

This investigation will be a field trip. Your teacher will describe the objectives, schedule, and place or places to be visited. Make careful observations of the animals you see, and attempt to work out a classification scheme based on your observations.

Materials (per student)

charts of animal characteristics to be observed and recorded

Procedure

1. Before the field trip, study Figure 18.11 and the illustrations of Kingdom Animalia in "A Brief Summary of Organisms." Make a mental note of types of characteristics that could help in organizing and recording your data.
2. On the field trip, observe an animal carefully before you record any data. Note those features that can be used to distinguish it from other animals. Also note the more general characteristics that can be used to group the animal with others that appear to be related to it.
3. Record your observations on charts similar to Tables 18B.1 and 18B.2, or use other charts as directed by your teacher. Try to identify the phylum or class to which the animal belongs.
4. Make additional notes beyond the observations you record on the charts. Save these notes for discussion after the field trip.
5. Repeat steps 2–4 for each animal to be observed. If you have questions, try to ask them before making entries in the charts.

Analysis

1. Which of the external characteristics you observed were of the greatest value in grouping animals?
2. What features were of greatest value in distinguishing one species from another?
3. Which of the animals you observed seemed most closely related to one another?

4. Which of the animals you observed seemed least closely related to one another?
5. On the basis of your observations, write a short paragraph that distinguishes animals from organisms in other kingdoms.
6. Based on the animals you observed, what changes, if any, would you make in the classification charts you used?
7. What advantages and disadvantages did you find in studying living animals rather than preserved specimens or illustrations in magazines and books?
8. What types of characteristics that you were unable to observe would have proved helpful in distinguishing between species of some of the animals you saw?
9. What roles did type of food and characteristics of feeding play in your classification of similar and different groups of animals?

Investigations for Chapter 19
Changes in Species

Investigation 19A ◆ Geological Time

Sedimentary rocks such as shale are found in areas that were once sea or lake beds. The fine layers seen in sedimentary rocks were produced by the compression of many layers of sediment that were deposited over thousands or millions of years (Figure 19A.1).

FIGURE 19A.1

Sedimentary rock strata from the Green River Formation shale deposit in Utah.

Assuming that geological forces have not fractured or folded the deposits, scientists can study the layers in sedimentary rocks to estimate the age of the deposit. Strata lying close to the surface are presumed to be deposited more recently than strata lying farther down in the deposit. This type of reasoning leads to relative dating, where samples or fossils can be placed easily in their chronological order. Other types of dating, such as counting tree rings, produce a numerical result and are referred to as absolute dating. Perhaps the most accurate absolute-dating method is radioisotopic dating. This technique measures the abundance of certain radioactive isotopes in a specimen. Knowing the rate of radioactive decay allows the age of the sample to be calculated.

In this investigation, you will examine a shale specimen to estimate its age and, by extrapolation, that of the entire deposit. The Green River Formation is a shale deposit averaging 600 m thick and covering parts of Colorado, Wyoming, and Utah. The strata in the Green River Formation are very thin layers that were deposited in annual sedimentation cycles called varves. Each varve consists of a pair of layers, one light and the other dark. The light layer is thick, coarse-grained, and rich in calcium carbonate; while the dark layer is thin, fine-grained, and rich in organic material.

Materials (per team of 2)
compound microscope
piece of shale
metric ruler
calculator (optional)

Procedure
1. Examine your piece of shale, and use a ruler to estimate its thickness. Record the data in millimeters (mm) in your logbook.
2. Estimate the total number of dark bands running through your sample, and record the prediction in your logbook.
3. Examine your specimen under the microscope, using low power. Be sure that the varved layers are facing toward you. Place the ruler on the surface of the shale so that the ruler markings are parallel to the varve layers. Record in your logbook the number of varves in a typical 1-mm section. Count the dark bands only.

4. Determine the total number of varves for your specimen by multiplying the thickness of your specimen (in millimeters) by the number of varves per millimeter determined in step 3. Record the result in your logbook.
5. When instructed by your teacher, report your varve count per millimeter. The class data will be used to construct a histogram.

Analysis
1. How thick is an average year's deposit?
2. Why is it impossible that one layer would cross over into other layers?
3. How long would it take for 1 m of sediment to be deposited?
4. If the average thickness of the Green River Formation is 600 m, how long was this lake (actually a series of three different lakes) in existence before drying up?
5. What type of evidence would persuade scientists that these strata formed at the bottom of ancient lakes?
6. What causes the formation of varves in this sedimentary rock?
7. How does this investigation illustrate the principles of relative and absolute dating?
8. Since dark bands contain organic material, what is the explanation for the occasional unusually thick dark bands?
9. Use the class data to determine the average value for varves per millimeter.

Investigation 19B ◆ A Model Gene Pool

In 1908, Godfrey Hardy, an English mathematician, wrote to the editor of *Science* with regard to some remarks of a Mr. Yule.

Mr. Yule is reported to have suggested, as a criticism of the Mendelian position, that if brachydactyly is dominant, "in the course of time one would expect, in the absence of counteracting factors, to get three brachydactylous persons to one normal."

It is not difficult to prove, however, that such an expectation would be quite groundless [using] . . . a little mathematics of the multiplication-table type.

The letter addresses an early and common criticism of Gregor Mendel's work on inheritance. Many scientists thought that Mendel's explanations of dominance and recessiveness suggested that recessive traits ultimately would be eliminated from the population and only dominant traits would remain. This investigation allows you to explore the validity of that assumption, using brachydactyly as an example.

Brachydactyly is a dominant disorder of the hands in which the fingers are shortened because of shortening of the bones. The actual frequency of occurrence is about 1 in 1 million births. You will use smaller numbers in setting up a model of a gene pool from which to randomly select pairs of alleles that represent individuals of a new generation.

Materials (per team of 3)
2 containers with lids, one labeled *male* and one labeled *female*
38 white beans
82 red beans

WARNING: Do not eat the beans. Wash your hands immediately after handling the beans or seeds. They may be treated with a fungicide.

Procedure

1. In this activity, you will set up a model of a human population by using red beans to represent the allele *(B)* for the dominant trait—brachydactyly—and white beans to represent the allele *(b)* for the recessive trait—normal hands. Before you begin, review the Hardy-Weinberg model in Section 16.3. List the five assumptions made by the Hardy-Weinberg

model for a population that maintains a genetic equilibrium.

2. Place 19 white beans in the container labeled *male* and 19 white beans in the container labeled *female*. Add 41 red beans to each of the containers. Place the lid on each container, and shake the beans. What do the beans in each container represent?

3. To represent fertilization and the resulting possible allele combinations that make up the genotype of a new individual (F_1 generation), you will select one bean from each container. What genotypes are possible?

4. In your logbook, prepare a table similar to Table 19B.1.

5. Remove the lid, and without looking, select one bean from each container. Use tally marks to record the results of this first selection in the table. Return the beans to their respective containers, cover the containers, and shake them again. Make a total of 60 selections for the F_1 generation, returning the beans to the containers each time. In the table, total the number of each genotype in the F_1 generation. Why should you return the beans to their respective containers after each selection?

6. Examine the data in your table for the F_1 generation. Each pair of beans represents an individual of the F_1 generation having a certain genotype. Assume that half the bean pairs of one genotype represent the males of the population and that half the bean pairs of the genotype represent the females. For example, if you have 18 pairs of red-red *(BB)* combinations, 9 pairs are males and 9 pairs are females. Determine how many of each type *(BB, Bb, bb)* should be

TABLE 19B.1
Genotypes for Two Generations

| | F_1 generation | | | | F_2 generation | |
Genotype	Tally marks	Total	Male	Female	Tally marks	Total
BB						
Bb						
bb						

male and how many female. Record the numbers in your table.

7. Use the information in your table to determine how many red beans *(B)* and how many white beans *(b)* should be in each container to represent the parents of the next generation. Correct the bean counts in each container. Mix the beans thoroughly.

8. Make 60 more selections from the containers of beans, as you did in step 5. This will be the F_2 generation. Tally and record the totals for each genotype.

9. Wash your hands before leaving the laboratory.

Analysis

1. What is the total population of individuals in the F_1 generation?

2. In genetics, *frequency* refers to the probability that a particular event will occur in a population, or

$$\frac{\text{number of individuals of one genotype}}{\text{number of individuals of all genotypes}} = \text{frequency}$$

Determine the frequency of homozygous recessive individuals *(bb)* for the F_1 generation.

3. According to the Hardy-Weinberg model, the frequency of homozygous recessive individuals *(bb)* in a population is expressed as q^2. Therefore, the frequency of the *b* allele = $\sqrt{q^2}$, or q. What is the frequency of the *b* allele for the F_1 population?

4. The Hardy-Weinberg model states that the frequencies for two alleles of a trait add up to 1, or 100%. The frequency of the allele for the dominant trait (in this case, *B*) is represented by p. Therefore, $p + q = 1$, or 100% of alleles for the trait in the population. To calculate the frequency of the allele for the dominant trait, use the formula $p = 1 - q$ (remember, $p = B$ and $q = b$).

5. Prepare a population gene-analysis table like the one in Figure 19B.1 to show the expected frequency of each genotype in the next generation (F_2). Cross two heterozygous individuals, and show the frequencies you have

allele frequencies
(male gametes)

allele frequencies (female gametes)		0.6*S*	0.4*s*
	0.6*S*	0.36*SS*	0.24*Ss*
	0.4*s*	0.24*Ss*	0.16*ss*

FIGURE 19B.1

Population gene analysis with data showing frequencies in a population of two alleles, *S* and *s*.

determined for each allele. Multiply to show the frequency of each F_2 genotype.

6. Review question 2. To determine the number of individuals of a particular genotype, you can multiply the frequency of the genotype by the number of individuals of all genotypes in the population. Predict the number of each genotype *(BB, Bb, bb)* that would occur in an F_2 generation of 60 individuals.

7. How do the results of your bean selections for the F_2 generation compare with your predictions from question 6?

8. Based on this model, respond to Mr. Yule's 1908 critique of Mendelian genetics.

9. Suppose the *BB* and *Bb* genotypes represented a lethal condition in the population you have been studying. Can the Hardy-Weinberg model still be applied in this situation? Explain your answer.

10. Hardy made the following statement in his 1908 paper: "There is not the slightest foundation for the idea that a dominant character should show a tendency to spread over a whole population, or that a recessive should tend to die out." Do your results support this statement?

11. What was the primary mistake in Mr. Yule's suggestion?

Investigations for Chapter 20
Human Evolution

Investigation 20A ◆ Interpretation of Fossils

How do anthropologists learn about evolution? Fossil remains form a record of the evolution of early humans, hominids, and other primates. Even

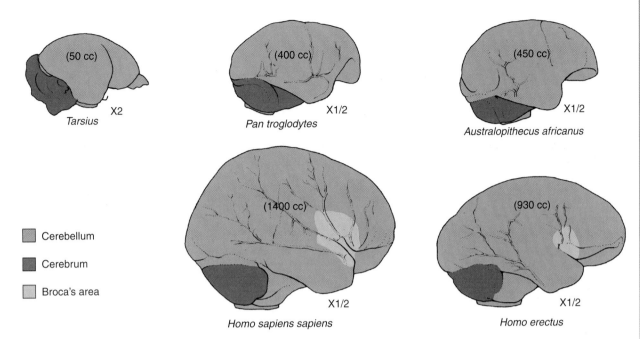

FIGURE 20A.1
Cranial casts of various primates.

Legend
■ Cerebellum
■ Cerebrum
□ Broca's area

though humans and other primates share many similar features, humans did not evolve from apes. Although primate brains have increased in both size and complexity during the course of evolution, the relationships between modern humans and other primates and fossil hominids cannot be determined only by an examination of braincase, or cranial, casts. Anthropologists often compare the skulls, jaws and teeth, pelvises, and femurs of fossil hominids and apes with those of modern primates. This investigation simulates some of the comparisons anthropologists make when studying fossil hominid remains. The activities include examination of several hominid cranial casts and

comparison of some skeletal measurements for a human, an early hominid, and a gorilla.

Materials (per team of 2)
paper and pencils
metric ruler
protractor
graph paper

PART A   Comparison of Cranial Casts

Procedure

1. Examine the five cranial casts shown in Figure 20A.1. What is the volume of each? Which species do you predict to have the largest brain volume compared with body weight?
2. Using the information in Table 20A.1 and a sheet of graph paper, plot the ratio of brain volume to body mass for each of the five species.
3. In Figure 20A.1, which species have Broca's area? On the *Homo sapiens* cranial cast, you can see that Broca's area is an enlargement of part of the cerebellum. Broca's area and two other areas of the brain are important in language and speech. Broca's area sends signals to a part of the brain that controls the muscles

TABLE 20A.1
Body Weight of Selected Species

Species	Average body mass (grams)
Tarsius (tarsier)	900
Australopithecus	22,700
Homo erectus	41,300
Pan troglodytes (chimpanzee)	45,360
Homo sapiens	63,500

TABLE 20A.2
Ratios of Brain Mass to Body Mass

Mammal	Ratio of brain mass to body mass
Tree shrew	1:40
Macaque	1:170
Blue whale	1:10,000
Human	1:45
Squirrel monkey	1:12
House mouse	1:40
Elephant	1:600
Porpoise	1:38
Gorilla	1:200

of the face, jaw, tongue, and upper part of the throat. If a person is injured in Broca's area, normal speech is impossible. What might the presence of Broca's area indicate?

4. Convert the ratios of brain mass to body mass shown in Table 20A.2 to decimal fractions. Draw a bar graph of the fractions. On your graph, the x-axis should represent the five species; the y-axis, the ratio of brain mass to body mass.

Analysis

1. Based on your graph, what could you infer is the relationship between evolution in primates and the ratio of brain mass to body mass?

2. What major portion of the brain has enlarged most noticeably during the course of primate evolution?

3. Do you think the cranial cast of *Australopithecus* indicates that this hominid could have had a Broca's area in its brain? Explain your answer.

4. Does the presence or absence of Broca's area alone determine the language capabilities of a hominid? Explain your answer.

5. How does your bar graph affect your answer to question 1? Are brain size and ratios of brain mass to body mass reliable indicators of the course of primate evolution? Why or why not?

PART A   Skeletal Comparisons

Procedure

5. Examine the three primate skulls in Figure 20A.2 and the drawings of lower jaws and the pelvises in Figure 20A.3. Imagine you are an anthropologist and the fossils have been placed before you for identification. Complete this hypothesis in your logbook: "If the skulls, jaws, and pelvises are significantly different, then . . . "

6. In your logbook, prepare a table similar to Table 20A.3, or tape in the table your teacher provides. Your task is to determine which skull, jaw, and pelvis belong to a human, which belong to an

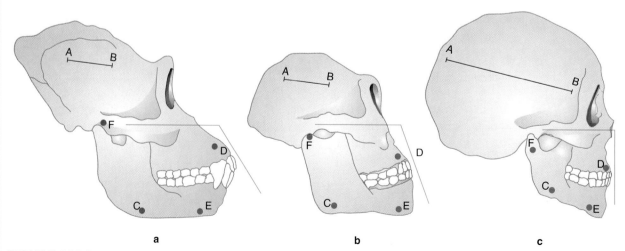

a   b   c

FIGURE 20A.2
A variety of primate skulls.

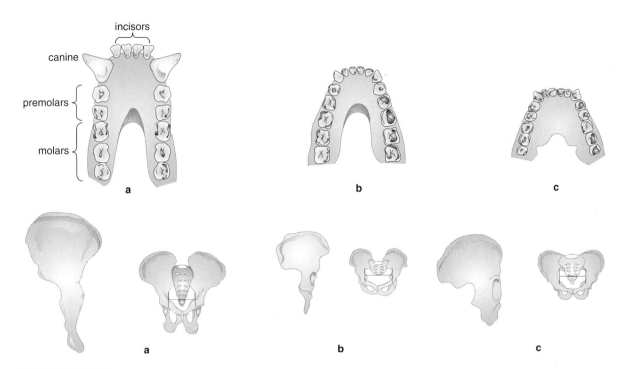

incisors

canine

premolars

molars

a

b

c

a

b

c

FIGURE 20A.3
Primate lower jaws and pelvises.

early hominid, and which belong to a gorilla. Record all observations and measurements in your table.

7. **Cranial volume:** The straight line drawn on each skull represents the brain volume for each primate. Measure in centimeters the distance from point A to point B for each skull. Multiply by 175 to approximate the cranial volume in cubic centimeters.

8. **Facial area:** Measure in centimeters from points C to D and from E to F on each skull. Multiply the measurements of each skull to determine the approximate area of the lower face. What might you infer about how the size of the facial area has changed through primate evolution?

9. **Facial projection:** Use a protractor to determine the angles created by the colored lines. What can you infer from these measurements?

10. **Brow ridge:** This is the bony ridge above the eye sockets. Record the presence or absence of this feature for each skull. Also note the relative sizes.

11. **Teeth:** Study the drawings of the jaws that are shown in Figure 20A.3. Record the number of teeth and the number of each type of tooth. Also look at the relative sizes of the different types of teeth.

12. **Pelvis:** Study the drawings of the pelvises in Figure 20A.3. Notice their relative sizes and whether the flange, or lower portion, projects to the rear. Measure in millimeters the diameter of the pelvic opening at the widest point.

13. Read these additional data.

    a. A larger cranial volume is characteristic of humans.

    b. A smaller lower facial area is characteristic of humans.

    c. A facial projection of about 90 degree is characteristic of humans.

    d. Modern humans have lost most of the brow ridge.

    e. All primates have the same number of teeth and the same number of each type.

TABLE 20A.3
Comparison of Primate Characteristics

Characteristic	Fossil A	Fossil B	Fossil C
Cranial volume			
Facial area			
Facial projection			
Brow ridge			
Number of teeth (lower jaw)			
Number of each type of teeth (lower jaw)			
Relative size of teeth (lower jaw)			
Relative size of pelvis			
Direction of flange			
Width of pelvic opening			

f. Humans and other hominids have smaller canine teeth than do gorillas.

g. A smaller pelvis, with a broad blade, a flange extending to the rear, and a wider opening, is characteristic of primates that walk on two feet.

h. As brain size increased, the width of the pelvic opening increased to accommodate the birth of offspring with a larger head-to-body ratio.

Analysis

1. Compare the data you have assembled for each specimen. Which ones are hominid? Which characteristics are similar in all three primates? Which characteristics are similar in a and b?

2. Compare your data with the additional data. On the basis of your observations and measurements and the additional data, which fossil remains would you say are human, which are gorilla, and which are early hominid?

3. What might an anthropologist infer from the size of the pelvic opening?

4. Write a few paragraphs discussing the methods anthropologists use to determine human ancestry. Which of the characteristics you examined were most helpful? Which were least helpful? List any other additional observations or measurements that could be made for these specimens. Would having a more complete skeleton be helpful? Why?

Investigation 20B ◆ Archaeological Interpretation

Because we cannot travel through time to see how people worked and lived thousands of years ago, we can never be sure that we understand the details of earlier cultures. Archaeologists search for clues among the remains of ancient peoples and civilizations. When they have found, dated, and studied the evidence, they formulate hypotheses to explain their findings. What emerges is a picture of life in a particular place hundreds or thousands of years ago. It is impossible to create a complete, detailed picture, but many reasonable and logical conclusions can be drawn. Some findings, however, can be interpreted in a variety of ways, and archaeologists may disagree about which interpretation is correct.

This investigation requires analysis of data found in archaeological digs. In nearly every case, several interpretations are possible. Think of as many interpretations as you can. Remember that your interpretations should account for all, not just some, of the existing data. In the final section of this investigation, you will predict how archaeologists far in the future might interpret evidence of today's cultures and lifestyles.

Materials (per team of 3)
paper and pencils

PART A A Native American Cemetery in Newfoundland

Procedure
Read the following paragraphs, and answer the Analysis questions.

A scientist investigated a Native American burial site located near Port aux Choix, a small village on the west coast of Newfoundland (Figure 20B.1a). The burial ground appeared to have been used between 4,000 and 5,000 years ago. The graves were located in strips of very fine sand and were often covered with boulders or slabs of rock. One of the sand strips was almost 1.6 km long, varying from about 9 m to 22 m in width. It lay about 6 m above and parallel to the present high-water mark of the ocean shore. The estimated ages of the skeletons found at the site are in Table 20B.1. What do these data suggest about the lives of the Native Americans?

The teeth of many of the adults were worn; often the inner nerve was exposed. Some skeletons also had teeth missing, but there was no evidence of tooth decay.

Figures 20B.1b–e show some artifacts that were found in the graves. A great number of woodworking tools were also found.

Analysis

Answer the following questions. Give reasonable answers that are supported by the evidence found at the site or from your own knowledge and past experience.

1. What type of area might the Newfoundland burial site have been at the time the Native Americans lived there (for example, forest, mountains, beach)? Explain your anwer.
2. List some possible reasons that the sandy strip was chosen as a burial site.
3. Boulders or slabs of rock covered many graves. Why might that practice have been common?
4. From the data on teeth, what can you infer about Native American diet and lifestyle?
5. What do you think the objects shown in Figure 20B.1b were used for? What purpose might the holes have served?
6. What might the objects in Figure 20B.1c and d have been used for?
7. What modern implement do the objects in Figure 20B.1e resemble? What might their function have been?
8. Considering the location of the site and its objects, what might the Native Americans have built from wood?

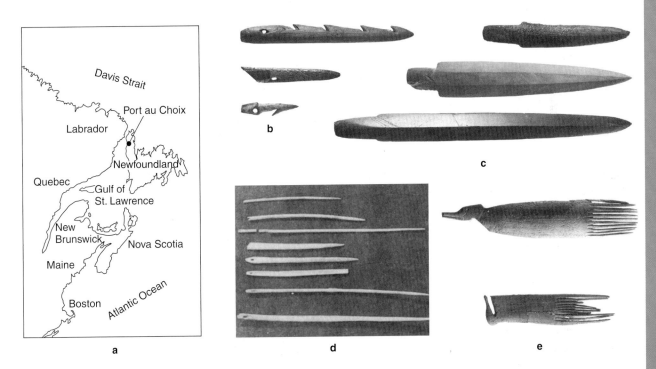

FIGURE 20B.1

(a), **Location of Port au Choix. (b–e), Objects found at Port au Choix excavation.**

TABLE 20B.1
Port au Choix Skeletal Ages

Age at time of death	Number of skeletons
Newborn infants	12
Under 2 years	15
2–6 years	2
6–18 years	15
18–21 years	1
21–50 years	36
50+ years	7

PART B Ancient People of Greece

Procedure

Read the following paragraphs, and answer the Analysis questions.

Stone Age people inhabited Greece long before the dawn of Classical Greek civilization. Until recently, little was known about the culture of these people who lived between 5,000 and 20,000 years ago. Excavations in and around a particular cave on the east coast of Greece began in 1967. Since then, archaeologists have gathered much evidence of the changes that occurred in the culture and lifestyles of Stone Age Greeks living there.

The oldest remains (20,000 years old) in the cave primarily are the bones of a single species of horse and some tools made from flint (rock that forms sharp edges when it breaks). What might explain the presence of the horse bones mixed with flint tools?

Newer remains (10,000 years old) include bones of red deer, bison, horses, and a species of wild goat, as well as remains of wild plants such as vetch and lentils (pealike plants), shells of land snails, marine mollusks, and some small fish bones. These newer remains date from the time when a great ice age had come to an end. In later finds, from about 9,250 years ago, very large fish bones were found—the fish might have weighed 100 kg or more. At about the same time, tools made of obsidian, a type of volcanic glass, were found. The nearest source of obsidian is 150 km away from the excavation site and across a body of water. The site also yielded the oldest complete human skeleton found in Greece. A male of about 25 years of age was buried in a shallow grave

covered with stones. Certain bone abnormalities indicated that he may have suffered from malaria, a tropical disease spread by mosquitoes.

Remains dated about 8,000 years old are dominated by the bones of goats and sheep found in and around the cave. These bones were very different from the goat and sheep bones found away from the cave. Evidence of wheat and barley seeds also was found. Among the tools found were axes and millstones. Only a few of the human graves at the site contained objects such as tools or jewelry. A 40-year-old woman, who probably died about 6,500 years ago, was buried with some bone tools and some obsidian blades. One infant was found near a marble vessel and a broken clay pot.

Analysis

1. In what ways did the diet of the cave's inhabitants change over the centuries? Explain your answer.
2. What does the evidence from about 9,250 years ago suggest about a change in the people's lifestyle?
3. How do you think the ancient people obtained the obsidian?
4. From the 8,000-year-old remains, what can you infer about the cave inhabitants' lifestyle?
5. What might be the significance of finding items buried along with people?

PART C Excavating the Present

Procedure

Imagine that you are an archaeologist living in the year 4000. A catastrophe has destroyed the records of the past. You have only the remains that have survived the last 2,000 years as clues to what life was like in the 1990s. Imagine, furthermore, that you are excavating a site that was a school 2,000 years ago— your own school.

Analysis

1. What objects do you think you would find at the site? Make drawings of or list descriptions of the objects. Exchange them with your teammates.
2. For each object, suggest several interpretations. (Remember, you know nothing about the time

when the objects were left—the only clues are the objects themselves.)

3. Some materials can last thousand of years. Other materials decay rapidly. What important things—of value to you or to society—would be absent in the archaeological dig that once was your school? What ideas about our society might you miss because the clues were not preserved?

4. Consider questions 1 through 3 again. This time, imagine you are excavating a different site, such as a mall, a business park, a residence, a farm, or an industrial area. How would your findings and interpretations differ?

5. Suppose you wanted to build a time capsule to tell people in the distant future about our lives today. How would you construct your capsule? What would you put inside it? Explain your answer.

Investigations for Chapter 21
Nervous Systems

Investigation 21A ◆ Sensory Receptors

How do you know and learn about the world in which you live? Information is received through receptor cells that function as part of the nervous system. Specialized receptors make possible the senses of touch, sight, hearing, smell, and taste. In this investigation, you will test some touch receptors.

Materials (per team of 2)
nonmercury thermometer
2 round toothpicks
2 10d nails with blunt points
pen with water-soluble ink
paper towels
large container of hot water
large container of lukewarm water
large container of ice water

Procedure

PART A   Skin and Temperature Sensations
1. Check the hot water with a thermometer to make sure it is more than comfortably warm (45° to 50°C), but not hot enough to burn your hand.

CAUTION: Hot water can cause serious burns. Do not let the water temperature exceed 50°C.

2. Put one of your hands in the hot water and the other in ice water. Leave them there 1 minute.

3. After 1 minute, remove your hands from the hot water and ice water. Immediately sink both hands in the lukewarm water.

4. Record in your logbook how the lukewarm water felt to each hand.

PART B   Sensory Receptors in Skin
5. Work in pairs for this part of the investigation. Determine who will be the experimenter and who will be the subject. Place one of the 10d nails in the ice-water container; place the other nail in the hot-water container.

6. **Experimenter:** Make a 5 cm × 7 cm grid of small points of ink on the inside of the subject's wrist with the points 5 mm apart (Figure 21A.1). Make a larger copy of the grid in your logbook. Make it large enough to mark an H (hot) and a C (cold) by each grid point. If the sensation is felt by the subject, circle the letter. If it is not felt, put a line through the letter. This will allow you to test each grid point with a hot and cold probe and keep track of which points have been tested. **Subject:** Sit so your arm is resting comfortably on top of the desk with the wrist up. Look away from the experimenter.

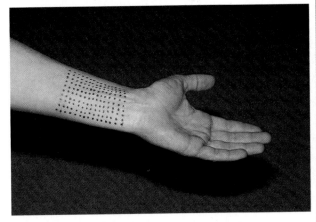

FIGURE 21A.1
Grid on wrist.

7. **Experimenter:** Test hot and cold receptors on the subject's wrist by touching the blunted point of a nail taken from the hot or cold water on each grid spot. Dry the nail quickly with the paper towel before touching the grid point. Alternate the hot and cold nails randomly.

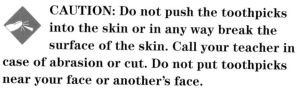 **CAUTION: Do not push the nails into the skin or in any way break the surface of the skin with the nails. Call your teacher in case of abrasion or cut. Do not put nails near your face or another's face.**

Subject: Tell the experimenter whether you have a hot or cold sensation.

Experimenter: Record the results of each touch by marking the appropriate letter in your logbook.

8. **Reverse roles with your partner, and repeat the experiment.**

PART C   Distance between Receptors

9. **Subject:** Close your eyes, and turn your head away from the experimenter.

Experimenter: Gently touch the end of the subject's index finger with the points of two toothpicks. Start with the points relatively far apart (1 cm), and then move them slowly together for each subsequent contact.

CAUTION: Do not push the toothpicks into the skin or in any way break the surface of the skin. Call your teacher in case of abrasion or cut. Do not put toothpicks near your face or another's face.

Subject: Tell the experimenter whether you sense one or two toothpick points.

Experimenter: If the subject senses two points, lift the toothpicks and move them slowly together for the next test. Continue testing the subject until only one toothpick point is sensed.

10. **Experimenter:** Trace the outline of the subject's finger in the logbook. Use two pencil dots to represent the closest points where the subject sensed both toothpick points.

11. Measure the distance, in millimeters, between the dots on the diagram. Measure the width of the finger in millimeters. Express these two measurements as the ratio *width of finger*

(mm)/*minimum distance* (mm). Reduce the fraction to a small whole number.

12. Repeat steps 9 through 11, using the back of the hand. Start the touching technique with the points at least 6 cm apart. Record your results as a ratio and reduce as in step 11.

13. Switch roles and repeat the experiment.

Analysis

1. List the ratios you obtained for the fingertip and the back of the hand on the chalkboard under the correct heading *male* or *female*. How similar are the ratios for the people in your class? Are there any differences between the ratios for males and females?

2. Feeling the forehead is a common way to find out whether body temperature is above normal. On the basis of what you learned in Part A of this investigation, explain how a person with a slightly raised temperature of 40°C could feel cool to the touch.

3. Are the sensitive areas that are stimulated by cold the same sensitive areas that are simulated by heat? Explain your answer.

4. If a person heard a crackling fire and people talking about heat and then suddenly was touched with a piece of ice on the back of the neck, what sensation might he or she first experience? Explain your answer.

5. How does the distance between touch receptors on the fingers and those on the back of the hand compare? Explain the significance of the difference.

6. Develop a hypothesis that could account for the fact that when two toothpicks touch the skin, sometimes only one is felt.

Investigation 21B ◆ Reaction Time

Various sensory organs receive stimuli from the environment and send signals to the brain, where they are interpreted. Messages sent to muscles and glands cause specific reactions to occur. How fast can you react to sound stimuli? How fast can you react to visual stimuli? Are your reactions any faster if you use both sound and visual stimuli? Are your fingers faster than your arms in reacting to stimuli?

This investigation will give you a chance to answer those questions.

Materials (per team of 2)
meterstick
metric ruler

Procedure

PART A Finger Muscles

1. Copy Table 21B.1 in your logbook to help organize your data collection.
2. Work in pairs for this experiment. Determine who will be the experimenter and who will be the subject.
   Subject: Sit down with an arm resting on the desk so that your hand extends past the edge of the desk. Your thumb and forefinger should be parallel to the ground and 4 cm apart. Use the metric ruler to keep the fingers about 4 cm apart. If the distance between the fingers varies too much, the results will be affected.
3. **Experimenter:** Stand and hold a meterstick vertically between the thumb and forefinger of the subject so that the lowest number on the meterstick is between the subject's thumb and forefinger.
4. **Experimenter:** Without warning, drop the stick.
5. **Subject:** Try to catch the meterstick with just your thumb and forefinger.
6. **Experimenter:** Look at the meterstick, and note the number of centimeters the stick dropped before the subject caught it. Record the distance in your logbook, but do not enter it in your table at this time.
7. Repeat the test four times, recording each distance the stick dropped. Determine the

average for the five trials. Record the average in the data table under the heading *"Sight only."*

8. Repeat the investigation with the subject's eyes closed. The experimenter will snap his or her fingers or otherwise signal aloud when the stick is released. Calculate the average for five trials, and enter the figure in the data table under *"Sound only."*
9. Repeat the investigation with the subject's eyes open. The experimenter will use the same signal used in step 8 when the stick is released. Calculate the average for five trials, and enter the figure in the data table under *"Sight and sound."*

PART B Arm Muscles

10. **Subject:** Stand an arm's length away from a classroom or hall wall, and place the palm of your hand, fingers up, flat against the wall. Lean slightly backward or move slightly backward so the palm of your hand is 4 cm away from the wall.
11. **Experimenter:** Stand and hold a meterstick against the wall so that the base of the subject's hand is even with the lowest number on the meterstick.
12. **Experimenter:** Without warning, drop the stick.
13. **Subject:** Try to catch the stick by pinning it to the wall with the flat of your hand.
14. **Experimenter:** Look at the meterstick, and find the number of centimeters the stick dropped before the subject caught it (measure from the base of the hand). Record the distance in your logbook, but do not enter it in your table at this time.
15. Repeat the test four times, recording each distance the stick dropped. Determine the average for the five trials. Record the average in the data table under the heading *"Sight only."*
16. Repeat the investigation with the subject's eyes closed. The experimenter will snap his or her fingers or otherwise signal aloud when the stick is released. Calculate the average for five trials, and enter the figure in the data table under *"Sound only."*
17. Repeat the investigation with the subject's eyes open. The experimenter will use the same signal used in step 16 when the stick is released. Calculate the average for five trials, and enter

TABLE 21B.1
Summary of Reaction Times

Method	Average distance meterstick travels (cm)		
	Sight only	Sound only	Sight and sound
Finger muscles			
Arm muscles			
Difference			

the figure in the data table under *"Sight and sound."*

18. Calculate the differences in the distances dropped by the meterstick for the two methods of catching it. Record the differences.

Analysis

1. In the sight-only test, what parts of your body are involved in catching the meterstick with your fingers? Which parts are involved when you catch the meterstick with your hand? Explain your answer.
2. In the sight-only test, what receptors are involved in both methods of catching the stick?
3. How do the results of the sound-only test compare with the results of the sight-only test?
4. Are there any differences between the reaction times of the finger muscles and those of the arm muscles? Explain this difference.
5. Compare the results of the sight-and-sound test with the sound-only and sight-only tests.
6. What can you conclude about your reactions to different stimuli?
7. What can you conclude about your reactions with different muscle groups?

Investigations for Chapter 22
Behavior

Investigation 22A ◆ A Lesson in Conditioning

Conditioning occurs when the patterns of innate behavior or reflexes change. In this investigation, you will determine if it is possible to condition a classmate to jerk his or her hand away when you make a noise.

Materials (per team of 2)
small rubber ball
scissors
cardboard box (book-box size)
noisemaker, such as an electric buzzer or a metal "cricket"

Procedure

1. Cut a hole near the bottom of one side of a cardboard box, similar to the one shown in

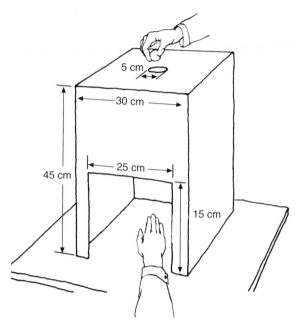

FIGURE 22A.1

Screen made from a cardboard box. Dimensions are approximate.

Figure 22A.1. In the center of the top of the box, cut a small circular hole through which the ball can easily pass.

 CAUTION: Scissors are sharp. Handle with care.

2. Determine which person will be the experimenter and which person will be the subject. *The subject should not read any further directions.* In your logbook, prepare a table similar to Table 22A.1. Be sure you have room to record how the subject reacted and what combination of stimuli you used.
3. Place the subject in front of the box with his or her hand inserted through the hole, directly underneath the circular hole on the top. Make sure that the subject's hand is mostly out of his or her sight.
4. Inform the subject that he or she will be attempting to keep the hand from being hit by the ball as it drops. Be sure the subject understands that the hand must remain in position at all times unless he or she removes it to keep from being hit by the ball.

TABLE 22A.1
Results of Conditioning Trials

Trial number	Type of stimulus			Subject's reaction	
	Noise	Dropped ball	Both	Hand in	Hand out
1					
2					
Etc.					

5. Stand behind the box with your noisemaker hidden from the subject's view. Now drop the ball through the hole *at the same time* you make a noise with your noisemaker. Do this several times in a row. The subject probably will not be hit more than once or twice. Record the subject's reaction and the stimulus used for each trial.

6. On the next trial, instead of dropping the ball, just make the noise. Record your results.

7. Randomly change the order of the stimuli used. Try some trials with both, several with just the noise alone, and several with just the ball. Record the results.

8. Wash your hands before leaving the laboratory.

Analysis

1. Compare your results with those from the rest of the class. Do you observe a pattern in the subjects' reactions?

2. Are there any differences in the results? If so, can you suggest why?

3. Do you think another stimulus besides noise would work the same way? Why or why not?

Investigation 22B ◆ Trial-and-Error Learning

This investigation will allow you to experience how trial-and-error learning operates. You will do something differently from the way you are used to doing it and will examine the learning process that occurs.

Materials (per team of 2)
small mirror
stopwatch or clock with a second hand
20 star diagrams
4 pieces of graph paper
4 to 6 books (enough to make the screen in Figure 22B.2)

Procedure

1. Decide who will begin as the experimenter and who will be the subject. You will change roles later.

2. Divide the 20 star diagrams into two sets of 10. Number each set from 1 through 10. On each diagram, choose one of the star points and label it *start* (Figure 22B.1). Do this for all 20 figures.

3. Construct a screen using the books. Make two piles of books, and lay the last book across the top of the two piles (Figure 22B.2). Leave enough room for your arm to fit through the space between the two piles of books.

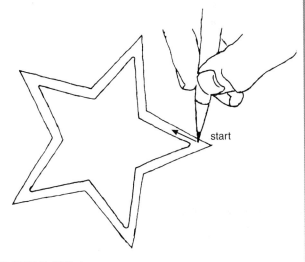

FIGURE 22B.1
Star diagram with start position labeled.

FIGURE 22B.2
Outlining a star reflected in a mirror.

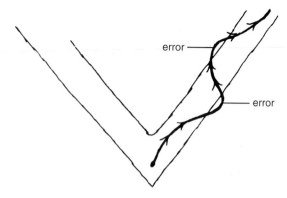

FIGURE 22B.3
Illustration of errors.

4. In your logbook, prepare a table with three headings: *Trial number*, *Number of errors*, and *Time in seconds*.

5. The subject will sit with his or her arm between the piles of books and will outline the star on the diagram without looking directly at the paper. The experimenter will hold a mirror in which the subject can see the reflected star diagram. The experimenter will time how long it takes the subject to outline the star.

6. The subject should begin at the point on the diagram labeled *start* and draw a line all the way around the star, trying to stay within the lines. The subject should look *only* at the reflection of the star diagram in the mirror held by the experimenter.

7. The experimenter should record in seconds how long it takes the subject to complete the outline of the star.

8. Repeat this procedure for the other nine figures. For each diagram, record the time to completion.

9. Change roles with your partner. Repeat steps 6 through 8.

10. When you and your partner have completed all 20 trials, count the number of errors in each trial. An error is counted every time the subject's pencil line went outside the lines of the star diagram and returned. See Figure 22B.3 for an illustration of errors.

11. After you have counted the errors for each trial, graph the results for the ten trials you completed as the subject. Your partner will do the same for the other ten trials. Label the horizontal axis *time* and the vertical axis *number of errors*.

Analysis

1. Examine the number of errors for each of your ten trials, and compare those with the time it took you to complete the outline. What, if any, evidence indicates that learning occurred?

2. Compare your graph with those of your partner and with the other teams. Is there any pattern in the data? Explain your answer.

3. Summarize the results of this investigation in terms of trial-and-error learning. Use your data to support your conclusions.

Investigation 22C ◆ A Field Study of Animal Behavior

The study of an organism's behaviors in an ecosystem is called ethology. These behaviors include its responses to the abiotic environment, to other species, and to other members of its own species. You can learn a great deal about an organism by observing it under natural conditions, where its behavior is likely to be typical of the species. In this investigation, you will develop a systematic field study of an organism in its natural habitat and then relate its behaviors to its basic needs. Be careful not to attribute human values and emotions to the organism.

Procedure

1. Select a nondomesticated animal that is available for observation. Some of the following animals would make good subjects: insects and spiders; birds such as robins, blue jays, pigeons, sparrows, ducks, and geese; mammals such as deer, squirrels, chipmunks, muskrats, and raccoons; snails, slugs, and fish.
2. Most of these organisms live in fields, ponds, parks, or forests. Even an aquarium, though not the same as a natural habitat, provides a suitable environment for observation. The best times to observe are just before sunset and just after sunrise. Many animals feed at these times. You do not need to study the same individual each time because the behaviors you observe should be typical of the species.
3. Some behaviors on which to focus include orientation to external stimuli, such as wind, sun, and moisture; communication with other members of the species or with other species; feeding; courtship and mating; interaction with other members of the species or with members from other species, including protective measures for itself or its young; reactions to the presence of other species, including humans.
4. Once you have selected an organism to study, devise a procedure for your study and submit it to your teacher for approval. If possible, the organism should be observed several times. Your plan should include:
   a. The scientific and common names of your organism.
   b. Where and when you will make your observations.
   c. The question you plan to investigate.
   d. A hypothesis related to your question.

5. Conduct your observations. Your results will more accurately reflect natural behaviors if your subject is not aware of your presence.
6. Record your observations accurately and comprehensively in your logbook. You will use your data to prepare your report.

Analysis

Write a report relating the behaviors you observed to the organism's basic needs. Remember not to attribute human motives to the animal. Your report should include:

a. Title, your name, and date.
b. The procedure approved by your teacher.
c. The actual procedure you followed if it differed from your proposal.
d. Your questions and hypothesis.
e. The data you collected: Organize your notes by quantifying as much data as possible (how many, how much, how often, and so on) using tables, charts, or graphs.
f. Conclusions: Interpretations or explanations of the behaviors observed in light of the basic needs of the animal.
g. Evaluation of your hypothesis: Did the data support or refute your hypothesis?
h. Recommendations for further study: What would you suggest if someone else were going to do the same study, or what would you do differently if you did the same study again?

Investigations for Chapter 23
Immune Systems

Investigation 23A ◆ Antigen-Antibody Binding

One of the body's most important defense mechanisms against infection is the production of antibodies, or immunoglobulins. These proteins circulate in the bloodstream, where they make up a part of the gamma globulin (IgG) fraction of blood plasma (see Appendix 23A, "Antibody Classes"). The production of antibodies can be stimulated by the antigens of an infecting agent such as a bacterium or virus. An antibody binds to an antigen in a reaction that is highly specific—each type of antibody binds to a particular type of antigen and no other.

There are many laboratory procedures designed to detect the presence of antibodies and the interactions between antibodies and antigens. One such test is called ELISA—enzyme-linked immuno-sorbent assay—an immunological technique used to detect and quantify specific serum antibodies. (Serum is blood plasma without the clotting factors.)

In ELISA, serum to be tested is allowed to react with specific antigens. Serum antibodies that combine with the antigens are detected by treating the test system with a conjugate—another antibody linked to an enzyme. This second antibody binds to the antigen-antibody complex that formed earlier. The enzyme serves as a marker. When a substrate for the enzyme is added, a reaction between the substrate and the conjugate is indicated by a color change. (If no serum antibodies are present to bind with the conjugate, no color change will occur during the time of observation.) These reactions are diagrammed in Figure 23A.1. ELISA is used routinely in the screening of blood donors for antibody to HIV.

In this investigation, you will perform ELISA to observe a specific antigen-antibody reaction and to determine the amount of antibody present. You will use a microwell plate—a piece of plastic molded to form many small wells, somewhat like a miniature egg carton. Once applied, the antigen is absorbed onto the surface of the wells and can react with serum antibodies added later.

The test will span 3 days and will include the following steps:

Day 1, application of antigens, blocking unbound sites

Day 2, addition of primary antibodies

Day 3, addition of secondary antibodies with conjugate, addition of substrate, checking for color

Materials (per team of 5)

5 pairs of safety goggles

5 lab aprons

15 pairs of disposable gloves

paper towels

white index card

glass-marking pencil

incubator

3 sheets of parafilm

96-well polyvinyl microwell plate

13 1-mL transfer pipettes

20 1.5-mL microfuge tubes

distilled water

Shared Materials

antigen 1 (bovine serum albumin; BSA) at antigen station 1

antigen 2 (bovine serum transferrin) at antigen station 1

10× PBS-Tween at washing station 7

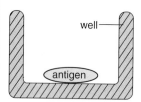

Step 1: Antigen is added and binds to the well.

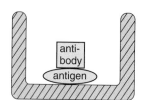

Step 2: Antbody is added and binds to the antigen.

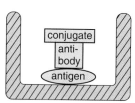

Step 3: Conjugate is added and binds to the antibody-antigen complex.

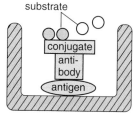

Step 4: Substrate is added for the enzyme in the conjugate. A change from colorless to color indicates antigen-antibody reaction.

FIGURE 23A.1

Steps in ELISA.

10× phosphate buffer saline (PBS) at washing
  station 7
1 mg gelatin mixed with distilled water at station 2
1 mL of antibody 1 at station 3
1 mL of antibody 2 at station 4
1 bottle of antibody 3 at station 5
substrate (ABTS) at station 6

> **SAFETY** **Put on your safety goggles, lab
> apron, and gloves. Tie back long hair.**

Procedure

Day 1

1. In your logbook, prepare a grid like the diagram
   in Figure 23A.2. The numbers identify the rows
   in the microwell plate, and the letters identify
   the wells. Along the sides and top of the grid,
   record what you add to each well as you follow
   the procedures. (On day 3, you will record
   ELISA results in each square of the grid.)

    **CAUTION: The chemicals you will be
   working with are *irritants*. Avoid
   skin/eye contact; do not ingest. Flush
   spills and splashes with water for 15 minutes.
   Call your teacher.**

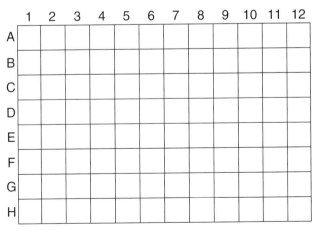

FIGURE 23A.2
Diagram of microwell plate for recording data.

2. With a clean 1-mL transfer pipette, add 1 drop of
   antigen 1 (from station 1) to all 12 wells in rows
   A–D (top of plate).
3. With a clean 1-mL transfer pipette, add 1 drop of
   antigen 2 (from station 1) to all 12 wells in row
   E (bottom of plate).
4. Write your names on a piece of parafilm, and
   cover the plate with the parafilm. Place the plate
   in the 37°C incubator for 30 minutes.
5. Move your plate from the incubator to washing
   station 7, and remove the parafilm. Spray PBS-
   Tween across the plate, and recover it with the
   same parafilm. Gently shake the plate back and
   forth, but be careful not to spill the contents.
6. Remove the parafilm, and empty the plate into
   the sink. Spray PBS (at washing station 7)
   across the plate, and recover with the same
   parafilm. Gently shake the plate back and forth,
   but be careful not to spill the contents. Empty
   the contents into the sink.
7. With a clean 1-mL transfer pipette, add 1 drop of
   gelatin (from station 2) to every well (to block
   the unbound sites). Use the same piece of
   parafilm to recover your plate, and place it in
   the refrigerator.
8. Wash your hands before leaving the laboratory.

Day 2

> **SAFETY** **Put on your safety goggles, lab
> apron, and gloves. Tie back long hair.**

9. Remove your plate from the refrigerator, and
   place it in the incubator for 5 minutes.
10. Take your plate to washing station 7, and
    remove the parafilm. Spray PBS across the
    plate, and recover the plate with the same
    parafilm. Gently shake the plate back and forth,
    but be careful not to spill the contents.
11. Go to station 3, and get 10 1.5-mL microfuge
    tubes. Label five tubes *BSA 1*, *BSA 2*, *BSA 3*,
    BSA 4, and *BSA 5*.

12. With a clean 1-mL transfer pipette, add 1 mL of antibody 1 to the tube labeled *BSA 1*. From tube BSA 1, use the transfer pipette to remove 500 μL (0.5 mL), and place this in tube BSA 2. With a clean transfer pipette, add 500 μL of distilled water to tube BSA 2.

From tube BSA 2, remove 500 μL, and place this in tube BSA 3. With a clean transfer pipette, add 500 μL of distilled water. What is your dilution in tube BSA 3?

From tube BSA 3, remove 500 μL, and place this in tube BSA 4. With a clean transfer pipette, add 500 μL of distilled water.

From tube BSA 4, remove 500 μL, and place this in tube BSA 5. With a clean transfer pipette, add 500 μL of distilled water.

13. Add BSA 1 to all the wells in column 2.
Add BSA 2 to all the wells in column 3.
Add BSA 3 to all the wells in column 4.
Add BSA 4 to all the wells in column 5.
Add BSA 5 to all the wells in column 6.

14. Go to station 4, and get 10 1.5-mL microfuge tubes. Label five tubes *T* (transferrin) *1*, *T2*, *T3*, *T4*, and *T5*.

15. With a clean transfer pipette, add 1 mL of antibody 2 to tube T1. Remove 500 μL, and place it in tube T2. With a clean transfer pipette, add 500 μL of distilled water to tube T2.

From tube T2, remove 500 μL, and place it in tube T3. With a clean transfer pipette, add 500 μL of distilled water to tube T3.

From tube T3, remove 500 μL, and place it in tube T4. With a clean transfer pipette, add 500 μL of distilled water to tube T4.

From tube T4, remove 500 μL, and place it in tube T5. With a clean transfer pipette, add 500 μL of distilled water to tube T5.

16. Add T1 to all wells in column 8.
Add T2 to all wells in column 9.
Add T3 to all wells in column 10.
Add T4 to all wells in column 11.
Add T5 to all wells in column 12.

17. Get a new piece of parafilm, and write your names on it. Place it on the plate. Put your plate in the incubator for 30 minutes. Then place your plate in the refrigerator overnight.

18. Wash your hands thoroughly before leaving the laboratory.

Day 3

> **SAFETY** **Put on your safety goggles, lab apron, and gloves. Tie back long hair.**

19. Remove your plate from the refrigerator, and take it to washing station 7. Remove the parafilm, and spray PBS-Tween across the plate; recover it with the same parafilm. Shake the plate gently back and forth, but be careful not to spill the contents of the plate.

20. Remove the parafilm, and empty the plate into the sink. Spray PBS across the plate, and recover it with the parafilm. Gently shake the plate back and forth, but be careful not to spill the contents of the plate. Remove the parafilm, and empty the plate into the sink.

21. Go to station 5 and with a clean 1-mL transfer pipette, add 1 drop of antibody 3 to each well. Cover the plate with a new piece of parafilm, and place the plate in the incubator for 30 minutes.

22. Take the plate to washing station 7, and remove the parafilm. Spray PBS-Tween across the plate, and recover it with the parafilm. Gently shake the plate back and forth, but be careful not to spill the contents.

23. Remove the parafilm, and empty the plate into the sink. Spray PBS across the plate, and recover it with the parafilm. Gently shake the plate back and forth, but be careful not to spill the contents. Empty the plate into the sink.

24. Go to station 6 and with a clean 1-mL transfer pipette, add 1 drop of substrate (ABTS) to each well.

25. In your logbook or on a 3″ × 5″ card, use a plus sign (+) to record which wells have a color change. Put three plus signs for the strongly changed blocks, two plus signs for the moderately changed blocks, and one plus sign for the lightest blocks.

26. Wash your hands thoroughly before leaving the laboratory.

Analysis

1. Antigen 1 is bovine serum albumin (BSA)—a protein from cattle blood. Antigen 2 is bovine serum transferrin, another protein from cattle blood. What effect would these proteins have if they were injected into a different animal?

2. If the starting dilution of tube BSA 1 is 1:1,000, what are the final dilutions for tubes BSA 2, 3, 4, and 5? If the starting dilution of tube T1 is 1:1,000, are the end dilutions of tubes T2, T3, T4, and T5 the same as for the BSA tubes? Explain your answer.

3. Rabbit antibody 1 is from a rabbit that was previously injected with antigen from cattle (BSA). What should the rabbit serum contain as a result of that injection?

4. Rabbit antibody 2 is from a rabbit that was injected previously with antigen from cattle (transferrin). What should the rabbit serum contain as a result of that injection? How are these antibodies different?

5. Predict the ELISA results by completing the following hypotheses:
   a. If rabbit antiserum BSA is added to wells containing BSA, then . . .
   b. If rabbit antiserum BSA is added to wells containing transferrin, then . . .
   c. If rabbit antiserium BSA is added to wells containing nothing, then . . .
   d. If rabbit antiserum transferrin is added to wells containing BSA, then . . .
   e. If rabbit antiserum transferrin is added to wells containing transferrin, then . . .
   f. If rabbit antiserum transferrin is added to wells containing nothing, then . . .

6. The secondary antibody is goat antirabbit with a conjugate HRP (horseradish peroxidase). How is this antibody made?

7. Why is antibody added to the wells?

8. Based on your results, which well demonstrated an antigen-antibody reaction? Did these results confirm your hypotheses in question 5?

9. What is the highest dilution that gave a positive result? How does the serial dilution enable you to determine how much antibody is present?

10. Identify the controls used in this investigation, and explain the specific purpose of each.

11. Based on the steps in this investigation, describe how ELISA can be used to detect antibodies to HIV.

12. Does a positive ELISA test for the antibodies against HIV indicate that the individual has AIDS? Explain your answer.

13. ELISA also is the basis for the pregnancy-test kits that can be purchased at a pharmacy. How might ELISA work in a pregnancy test? (Hint: What substances might be present in the blood or urine of a woman only during pregnancy?)

Investigation 23B ◆ Antibody Diversity

We live in a microbe-filled world. There are numerous pathogens, or antigens, that have the potential to make us sick or even kill us. As you have read in Chapter 23, the immune system generally keeps these infectious agents in check through an organized set of responses.

One of the mechanisms the immune system uses to control infection and disease is the production of antibody proteins in response to the presence of antigens. Immunologists estimate that humans have the potential to produce approximately 10^6 specific antibody proteins.

As Sections 9.1 and 9.2 explain, proteins are gene products. Therefore, the human genome (the complete complement of an organism's genes) must contain enough genetic information to code for the production of 10^6 antibody proteins in addition to all the other information an organism needs. But consider that there are approximately 3×10^9 base pairs in the human genome. Assuming that each gene is about 3×10^4 base pairs long on average, there are about 10^5 genes in the human genome. Clearly, this number is not large enough to account for one gene devoted to the production of only one antibody. This investigation will help you explore the genetic rearrangement mechanism, discovered by Tonegawa and others, by which the human genome encodes information for such an incredible amount of antibody diversity (see Biological Challenges: Susumu Tonegawa in Section 23.9).

Materials (per team of 2)

pop-it beads: 1 red, 1 pink, 1 orange, 1 yellow, 1 blue, 1 green, 1 white, 1 black, 13 lilac

PART A

Procedure

1. Assume that your body has been invaded by 12 different pathogens and must produce a different antibody light chain against each one. The complete gene for a light chain includes three classes of smaller genes: variable (V), joining (J), and constant (C).
2. Working as a team, use your pop-it beads to construct a DNA sequence that will code for the production of an antibody light chain. Each colored pop-it bead represents a gene segment. Snap the beads together in the sequence indicated in Figure 23B.1, using the following key: R = red, P = pink, O = orange, Y = yellow, Bl = blue, G = green, W = white, B = black, L = lilac.
   The gene segments are labeled J_1–J_3 for the joining region genes. There is only one C gene. The lilac pop-it beads represent introns. If necessary, review Section 9.4 on exons and introns.
3. Each complete light-chain gene is coded for by *one* V gene, *one* J gene, and the C gene. Construct a gene sequence that codes for a light chain by combining V_1 (red), J_1 (blue), and C (black).
4. Return V_1, J_1, and C to the original DNA strand. Now combine V_2, J_2, and C.

Analysis

1. How do the chains constructed in steps 3 and 4 differ?
2. Is there enough genetic information present in these two sequences (V_1–J_1–C and V_2–J_2–C) to produce ten different light chains? How many chains are possible?
3. How many light chains could you produce using V_1 and each of the J genes?

PART B

Procedure

5. Assume that any V gene can combine with any J gene (and C) to produce a complete gene coding for an antibody light chain. Use the pop-it beads to construct complete genes that will code for ten distinct light chains. Record the sequences. You have already completed two:
   R Bl   B V_1 J_1 C
   P G   B V_2 J_2 C

Analysis

1. How many complete light chains could you produce using the genetic information in the original strand?
2. If there were approximately 250 V genes, 5 J genes, and 1 C gene for the antibody light chain of the kappa type, how many different light chains could this information code for?
3. Using the symbols V, J, and C, write an equation for the number of light chains that can be produced by any given segment of DNA.
4. A gene may be defined as a sequence of nucleotides that codes for a functional product, such as tRNA, an enzyme, a structural protein, or a pigment. Does the model for production of

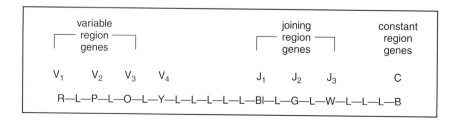

FIGURE 23B.1
Pop-it-bead sequence.

antibodies support that definition? Explain your answer.

5. The classical definition of a gene is "a hereditary unit that occupies a specific position (locus) within the genome or chromosome."* Does the model for production of antibodies support that definition? Why or why not?

*R. C. King and W. Stansfield, *A Dictionary of Genetics*. (New York: Oxford University Press, 1985.)

Investigations for Chapter 24
Ecosystem Structure and Function

Investigation 24A ◆ Producers in an Ecosystem

In this investigation, you will study the producers in a small piece of the biosphere. It need not be a very large piece—only large enough to contain several types of producers that show interrelationships. Different schools have different opportunities for outdoor studies. Therefore, the study procedures used by you and your class will have to be designed to fit the piece of biosphere for your school. Read through the lab carefully to get an overview of what you might do.

Selecting a Study Area. A forest is a complex environment, providing opportunities to collect abundant data, but it is difficult to picture as a whole. A prairie is almost as complex but is somewhat easier to study. Cultivated areas, such as cornfields and pastures, are relatively simple to study. They are as important as forests and prairies since they cover a large portion of the land in the world.

Suitable areas also can be found in cities. Many schools have lawns with trees and shrubs. Here there are many fewer types of organisms than outside the city, but you can be more thorough in your study. You also can study vacant lots and spaces between buildings. Even such small places as cracks in the pavement, gutters, and the areas around trees often contain a surprising number of organisms.

Organizing the Work. Teamwork will be necessary for this study. Plan the work carefully, divide responsibilities, and take extensive notes.

Organizing the Data. It is convenient to divide producers into three groups for study: trees, shrubs, and saplings; herbaceous plants; and seedlings. In addition, organic litter, such as fallen leaves, or ground cover, such as lichen and mosses, may be present. Trees may be classified as either deciduous or coniferous. Canopy trees are at the top of the forest and receive direct sunlight. The trees below this top layer form the subcanopy. Shrubs are low woody plants between 0.5 and 3 m tall. A sapling is a young tree that is 2.5 to 4 cm in diameter and about 1.3 m high. Herbaceous plants are nonwoody plants that die down to the ground in the autumn. These include grasses, grains, and small flowering plants. A seedling is a very young tree with a stem less than 1 cm in diameter.

Materials (per team of 8)
The materials required depend on the methods used, but the following list is minimal for the field.
strong twine, over 100 m long
4 stakes
compass, magnetic
metric ruler
graph paper
2 metersticks
hoop with area of 1 m^2
plastic bags
rubber bands or twist ties
can, 5-cm diameter
8 pairs of plastic gloves

Procedure
1. Choose your site.
   - In a forest, study areas should be approximately 10 m^2 and broken down into smaller areas, as shown in Figure 24A.1a.
   - In unforested areas, study areas should be 2 to 4 m on a side without internal divisions.
   - In vacant city lots, cultivated fields, and pastures, use smaller areas, about 1 m^2. (An area the size of a hula hoop works well.)
2. Each team should develop a hypothesis concerning the type of ecosystem of its site and what vegetation it expects to find.
3. Within the chosen site, measure a study area of appropriate size. Drive stakes into the ground at

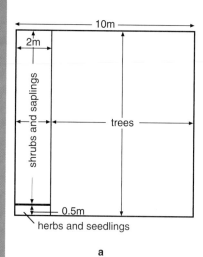
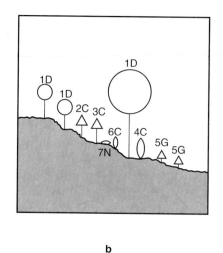
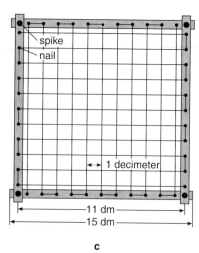

FIGURE 24A.1

(a), Study plot for a forest; (b), a profile showing layers of vegetation; (c), area for study of small plants.

the corner points, and connect the stakes with twine. The study area should be as square as possible. You may want to subdivide the area to make counting organisms easier.

4. Stand back and look at your study area. Walk around the outer boundary. Use symbols like the ones shown in Figure 24A.1b to sketch a profile of the ecosystem as you see it.

5. Indicate with numbers the various plant types that are recognizably different. For example, in Figure 24A.1b, the first plant observed was a deciduous (D) canopy tree (1D). Three of these were in the study area. The second species observed (labeled 2) was a conifer (C) subcanopy species. Species 3 also was a subcanopy conifer species, but different from species 2.

6. Once you have made a profile of your study area, you are ready for a closer look at its vegetation. So that the class as a whole gains a detailed picture of the vegetation, different teams can gather data on trees, shrubs, and saplings; herbaceous plants and seedlings; and litter.

7. Abundance refers to the number of individuals of one kind within a given area. Count the number of each type of producer in your study area. For example, count the number of each type of tree in the forest canopy and subcanopy

layers in the main study area. Count shrubs and saplings in one of the smaller subdivided areas. Count the herbaceous plants and seedlings in an even smaller area. Special problems may arise. In a lawn, for example, there is no need to count blades of grass. However, a count of the weeds might be worthwhile, especially if comparisons will be made between well-trodden areas and protected ones. A frame (Figure 24A.1c) may be useful for this work.

8. Make a table to record your counts. Across the top, list the types of plants in your subdivided area. For example, if you are counting shrubs and seedlings, you may list evergreen shrubs, rosebushes, raspberry bushes, and so on. Under these headings, record your count of each type of plant. You may wish to gather a sample or characteristic part of the plant, such as a leaf, and place it in a plastic bag. Do not collect the whole plant, and be sure to wear plastic gloves. Some plants are rare or endangered species. Be sure to record where you collected the sample.

CAUTION: Do not collect any plants you cannot identify as harmless.

9. Teams that study organic litter and ground cover should collect samples in plastic bags (wear gloves). Take samples from the ground around trees, shrubs, and small plants in a small

subdivided area. Secure the bag with a rubber band or twist tie, and mark it to indicate where it was gathered. You will study these samples later in the laboratory. At the study site, examine a 15-cm² sample of litter in place. Carefully pick the sample apart. Record the layers and composition of each layer. Also try to determine its physical condition. These observations should include the state of decay, moisture level, odor, amount of shade at the location, and nearness to large plants.

Analysis

1. Calculate the number of each type of plant to determine its abundance. If several teams had study areas near one another, average the data for plants of the same type. List all of the plants on the study site in order of their abundance.

2. To determine density, divide the total number of individuals of each type by the area of the study site in square meters.

$$\text{density} = \frac{\text{total number of each species}}{\text{total sample area (in m}^2)}$$

3. List the plants in order of their densities. The teams analyzing the litter and ground cover also should calculate the abundance and density of each species in the samples from the study area.

4. With your classmates, make a summary of the important features of the plants in the ecosystems you studied. Include such factors as profile, abundance, density, and litter. Record this information in your logbook.

5. Ecosystems are usually named for the most obvious or important species of plants. For example, a deciduous forest might be called a beech-maple forest if those trees are dominant. After you have analyzed the data, determine the one or two species that appear to be the most abundant in the area you studied. What would you name this ecosystem? How does the data support or refute your hypothesis concerning the type of vegetation you expected to find?

6. How many layers of vegetation did you find in the entire study area? List the important species in each of these layers.

7. Some of the plants are in a layer because they are limited genetically to that maximum height. Others are young individuals that will some day grow higher. Review your data to determine how many of these young individuals were among the plants in each layer.

Investigation 24B ◆ Relationships between a Plant and an Animal

Plants and animals interact in a variety of ways. By setting up closed systems with an aquatic plant and an aquatic animal, you can study an interaction related to the carbon cycle. Carbon dioxide dissolves in water and forms a weak acid, which lowers the pH. This drop in pH indicates an increase in the concentration of carbon dioxide. Conversely, an increase in pH indicates a decrease in the concentration of carbon dioxide. Thus, you can measure any changes that occur in the system by measuring indirectly the concentration of carbon dioxide using a pH meter or wide-range pH paper.

Before beginning the experiment, read through the procedure and develop hypotheses that predict the changes that will occur in each test tube in the light and in the dark.

Materials (per team of 4)
4 25-mm × 200-mm test tubes
glass-marking pencil
test-tube rack
aluminum foil
light source
pH meter or wide-range pH paper
forceps
200 mL dechlorinated water
2 15-cm pieces of elodea
2 1- to 1 1/2-cm freshwater snails

Procedure

Day 1
1. Record all data in your logbook, or use the table your teacher provides.
2. Label the test tubes *1* through *4*, and place them in the test-tube rack.

3. Pour dechlorinated water into each test tube to approximately 4 cm from the top. (Tap water becomes dechlorinated by standing for 24 hours—the chlorine escapes into the air.)

4. Add nothing more to test tube 1. To test tube 2, add a snail and a leafy stem of elodea. To test tube 3, add only a snail. To test tube 4, add only a leafy stem of elodea.

5. Determine the initial pH of each tube by using a pH meter or by dipping small strips of pH paper held with forceps into the test tubes and comparing the color change to a standard color chart.

6. When the readings have stabilized, record the current pH readings for test tubes 1 and 2 from the numerical data displayed, or record your readings from the pH paper and color chart.

7. To determine the pH of test tubes 3 and 4, remove the pH probe from test tube 2, rinse with dechlorinated water, and insert into test tube 3. When the readings have stabilized, record the current pH for test tube 3. Then remove the pH probe from test tube 3, rinse, and insert into test tube 4; record the current pH for test tube 4. Rinse the pH probe and replace in test tube 2. Alternatively, measure the initial pH values with pH paper as described.

8. Seal the top of each test tube with a double layer of aluminum foil. Press the foil tightly to the sides of the test tube. Place the test tubes in strong artificial light.

9. Repeat steps 6 and 7 every 10 minutes until the end of the period, taking care to work quickly and replace the foil covers. Record all readings in the data table.

10. Seal the test tubes, and leave in the light overnight.

11. Wash your hands thoroughly before leaving the laboratory.

Day 2

12. Observe the test tubes, and take pH readings as in steps 6 and 7. In the data table, record the pH readings and the condition of the organisms in each test tube.

13. Seal the test tubes, and wrap each one in aluminum foil to keep it dark.

14. Wash your hands thoroughly before leaving the laboratory.

Day 3

15. Repeat steps 12 and 13, but this time remove the aluminum foil and place the test tubes in the light.

16. Wash your hands before leaving the laboratory.

Day 4

17. Observe the test tubes again, record pH data, and stop the experiment.

18. Wash your hands thoroughly before leaving the laboratory.

Analysis

1. Construct graphs from your data, using a different color for each test tube.

2. Use information about photosynthesis, cellular respiration, and the carbon cycle to write a paragraph that explains the data.

3. What was the purpose of test tube 1?

4. Did the pH change in test tube 1? If so, how might you explain this change?

5. What effect would a pH change in test tube 1 have on the data for the other three test tubes?

6. What results might you expect if all the test tubes were kept in total darkness for the duration of the experiment?

7. Do the experimental data support your hypothesis? Explain your answer. If not, try to devise a general hypothesis that is consistent with all of your observations.

8. Suppose a snail were kept by itself in a sealed test tube for a week. Predict what is likely to happen, and explain your prediction.

Investigations for Chapter 25
Change in Ecosystems

Investigation 25A ◆ Producers in an Aquatic Ecosystem

Photosynthetic organisms such as algae and cyanobacteria are producers essential to an aquatic ecosystem. The numbers and diversity of these organisms provide a measure of the health of the ecosystem. How clean is the water in a lake, pond, or stream near your school? Technical tests can determine the exact identities and amounts of water pollutants, but even a simple survey of algae and cyanobacteria can be used as an index of water quality. Typically, clean water has a wide variety of algal and cyanobacterial species, with no species being especially dominant. Polluted water has fewer species present and greater numbers of each.

In this investigation, you will learn first to identify key genera of algae and cyanobacteria. Then you will test water samples known to be clean, moderately polluted, and polluted. Finally, you will use what you have learned to analyze a water sample from your area.

Materials

5 microscope slides
5 coverslips
dropping pipette
1-L beaker
plankton net
compound microscope
protist/cyanobacteria key
2-cm² piece of graph paper with 1-mm squares
5 flat toothpicks
Detain™ in a squeeze bottle
water sample with algae and cyanobacteria
water samples—1–3
water sample—from a local source

Procedure

PART A Field-of-View Calculations

1. Place 2-cm² piece of graph paper on a microscope slide, and add a coverslip. Examine the slide with the low power (100×) of your microscope. Align the squares so that one edge of a square just touches the bottom edge of the field of view. Count the number of squares from the bottom to the top of the field (Figure 25A.1a). The number of squares equals the diameter of the field of view in millimeters. Record the results and the magnification.

2. Repeat step 1 with the high power (400×) of your microscope. (Do not use the oil immersion

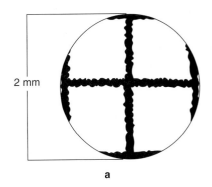

a

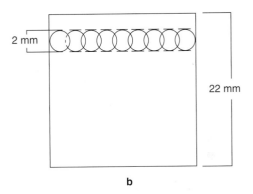

b

FIGURE 25A.1

Measuring microscope field of view.

lens.) The field of view will be less than 1 mm in diameter.

3. You will count individual organisms on a slide made from your water samples. To do so, you will move the slide across the microscope stage, the width of one field of view at a time, as illustrated in Figure 25A.1b. The area of the strip of fields equals the diameter of the field of view multiplied by the width of the coverslip (22 mm). Practice moving the slide one field of view at a time, going from one side of the coverslip to the other.

PART B Algae Identification

4. Place a drop of Detain™ (protist slowing agent) on a clean microscope slide. Using a dropping pipette, add one drop (0.1 mL) of the sample with algae and cyanobacteria, and use a toothpick to mix the two drops. Carefully add a coverslip.

5. Practice locating and identifying different organisms using the protist/cyanobacteria key.

Locate organisms under low power, and then use high power to make a positive identification.

6. Move the slide one field of view at a time, going from one side of the coverslip to the other. In each strip, count the number of individuals for each genus you can identify. Ask your teacher for help if you cannot identify some of the organisms.

7. Wash your hands thoroughly before leaving the laboratory.

PART C Identification of Water Samples

8. The various genera of algae display differing sensitivities to pollution. This means that the microbial diversity in a water sample is a reflection of the water's quality. You will examine water samples that are clean, intermediate, and polluted and determine which genera are associated with clean water and which with polluted water. This information will assist you in analyzing the quality of water obtained from a local aquatic ecosystem.

9. For each of the three water samples, you will count the number of organisms in each genus in three strips of fields of view. You will record in your logbook the genus and number of each type of algae observed in the water sample.

10. Put one drop of Detain™ and one drop (0.1 mL) of sample 1 on a slide. Mix with a clean toothpick. Add a coverslip. In your logbook, note the sample number.

11. Count and record in your logbook the number of individuals from each genus present in a strip of fields of view. Count *units*. (Colonies or filaments are counted as one unit.) Large filaments and colonies only partly lying in the strip should be counted as fractions. For example, if a filament of *Stigeoclonium* is only half in your field of view, record 0.5 instead of 1.

12. Repeat steps 10 and 11 two times.

13. Repeat steps 10–12 using samples 2 and 3.

14. Use your data and the fact that microbial diversity increases with water quality to identify the water samples as clean, intermediate, or polluted. Before proceeding to Part D, check with your teacher to determine if you correctly identified the quality of the three water samples.

PART D Investigation of Local Water Source

15. Using water from a local lake, pond, or stream, filter 1 L through a plankton net to trap the organisms. Rinse the organisms from the net into the 1-L beaker, and add clean water to a total of 100 mL. Repeat steps 10 and 11 to determine the relative water quality of your sample.

16. Wash your hands thoroughly before leaving the laboratory.

Analysis

1. Which algal genera can tolerate the highest levels of pollution?

2. Which algal genera are the most sensitive to pollution?

3. If your local water source has an intermediate or high pollution level, what might be the source of the pollution?

4. How can algae and cyanobacteria be helpful in polluted water?

5. How can algae and cyanobacteria be harmful in polluted water?

Investigation 25B ◆ Ecosystem Diversity within a Biome

The biosphere includes many different biomes, such as deserts, grasslands, and tropical rain forests. The purpose of this investigation is to study species diversity within a single type of biome. Species diversity is an expression of the community structure of an ecosystem. A biome with high species diversity will contain many equally or nearly equally abundant species. Biomes with low species diversity are characterized by few different species, even if they are plentiful.

The distribution of species within a biome varies depending on which area of the biome you examine. For example, grasses are more common in some areas, and shrubs or trees are common in others. The distribution of animals and other species around tall grasses will be different from that among the trees or scrub. The biome you study will depend on your location. First you will observe the plants and animals in your biome. Then you will choose a question that can be addressed by sampling the species diversity from different locations within the biome.

Materials (per team of 6)
6 pairs of safety goggles
2 metal coat hangers or pieces of stiff wire
4 wooden stakes
meterstick or metric tape
trowel or small-bladed garden shovel
white-enameled pan or large sheet of white paper
6 pairs of forceps
wire screen of approximately 1-mm mesh curved into a bowl shape
2 30-cm pieces of PVC pipe, 1/2–1 inch diameter
50-cm wooden dowel that fits inside the pipe
corks
field guides

Procedure

PART A Field-Study Preparations

1. As a class, select an area to study, such as a local park, forest, or grassland. Ideally, choose an area that is convenient to get to but is seldom disturbed by people. The more natural your biome, the greater the species diversity will be. Cultivated areas will exhibit fewer species but are easier to study. If studying natural biomes is inconvenient, suitable areas can be found in urban locations. Many schools have lawns and shrubs. You also can study vacant lots between buildings. These areas will contain fewer species but should allow you to be more thorough in your study.

2. Make and record observations of the plants and animals in your study area. Notice how the various plant species are distributed within the study area. Consider the types of relationships that exist between the plants and animals within the biome. Choose a question to study that

relates to the distribution of species within the biome. For example, you might compare the distribution of species in different locations—under trees versus open areas or on top of a hill versus the bottom. Determine what type of data you will need to collect to answer the question. With your teacher, decide whether the entire class will study the same question or whether different teams will study different questions.

To study plants, use quadrats—square or rectangular samples of the study areas. For trees, a quadrat of 10 m × 10 m is common; for shrubs, 4 m × 4 m is recommended. Where there are few trees or woody species, use quadrats made of stiff wire that are 0.5 m on each side. Note the size of the area inside the quadrat, and use this same size throughout your study.

To study organisms in the soil, take soil cores of equal length from the appropriate locations in your study area. Push a length of pipe into soft soil approximately 25 cm deep. Gently twist the pipe in the ground, and pull it up with the soil core inside. Plug up the ends of the pipe with corks, and bring the pipes back to the lab for study. In the lab, carefully push the soil core out of the pipe with a wooden dowel. Count the numbers of earthworms, pill bugs, and other larger organisms you see, noting their depth in the core.

3. Select a team leader to assign tasks. For example, different team members can be responsible for collecting data on plants, insects, animals, and organisms in the soil. Don't forget to include the physical aspects of the area, such as temperature, light intensity, and soil moisture. Prepare forms on which you can conveniently collect your data. Once you return to the classroom, paste the data forms into your logbook.

CAUTION: Never enter an area unless you have permission from the property owner. Be careful and aware that some plants and animals can be poisonous, harmful, or rare and endangered— make sure you know what you are touching.

4. Decide on a method for recording species-diversity data. Since it is not practical to identify the species of each organism observed and counted, you may want to use general terms, such as grasses, beetles, and spiders. When turning over logs and rocks to look for animals, be sure to return the sheltering objects to their original position once you have your data. Note the types, numbers, and activities of any animals you find.

PART B   Conducting the Field Study

5. If your team is studying a small area, toss the quadrat on the ground at the first selected site. If you are studying a larger area, use stakes and a metric tape to define a quadrat.

6. Record all the different types of plants within your quadrat. If you don't know the names of the plants, describe them well enough so that you will be able to identify them in your other quadrats.

7. Estimate the cover of each plant species. This is the percentage of ground within the quadrat covered by the leaves and stems of that species. Suppose there are three different types of plants, as in Figure 25B.1. Estimate to the nearest 10% the amount of cover for each type of plant as well as the amount of bare area. The greater the amount of cover for a plant species, the greater its importance within the plant community.

8. To increase accuracy, have at least two members of your team estimate the cover of each plant species. Record both estimates and the average value in your logbook. Note that if some taller plants overshadow smaller plants, your estimated cover may exceed 100%.

9. To properly address your question, be sure to analyze at least three quadrats in appropriate locations of your study area. Note any trends in species diversity that relate to the question posed by your team.

10. Examine a sample of the surface soil and/or organic litter. Insert a trowel or small garden shovel into the ground to a depth of approximately 10 cm, and remove a cubic

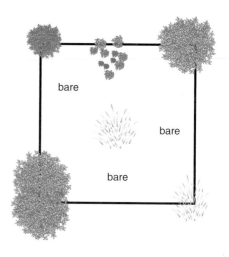

FIGURE 25B.1
Vegetation coverage in a quadrat.

sample of approximately 10 cm × 10 cm × 10 cm. Using the wire screen that has been shaped into a bowl, sift the contents onto a white-enameled pan or a large sheet of white paper. Pick through the remaining litter with forceps. Organisms present probably will include a variety of insects (especially beetles), sow or pill bugs, millipedes and centipedes, spiders or other arachnids, worms, and fungi.

 SAFETY Be sure to wear safety goggles when working with the soil samples.

11. Organize your team's data in your logbook, and discuss whether it answers the question posed by the team at the beginning of the investigation.

Analysis

1. How did the species diversity change as a function of location within your study area?

2. What can you conclude from your data as to the question posed by your team?

3. What producers are found in your study area?

4. Which producer is dominant in your area? What effect does this species have on the rest of the study area?

5. Are the producers evenly divided in your study area? Can you explain the pattern of growth in terms of the question posed by your team?

6. Are there layers of producers? If so, what relationships can you find among the layers?

7. Does your study area produce all its own food, or is food carried in from beyond its boundaries? What evidence do you have for your answer?

8. What consumers are found in your study area?

9. Which consumers are herbivores, and which are carnivores? What evidence supports your answer?

10. What relationship can you find between the numbers of a particular herbivore and the numbers of a carnivore that eats it?

11. What is the evidence that one type of organism affects another in ways other than those involving food relationships?

12. What biome type best characterizes the area you studied? Explain your answer.

APPENDICES

Appendix 1A
The Periodic Table of the Elements

The periodic table of the elements (Figure 1A.1) shows the name and symbol for each element, as well as its atomic number and atomic mass. Atomic number is the number of positive charges, or protons, for an element. Atomic mass is the mass of one atom of the element, as compared to a standard. The accepted standard is a form (isotope) of carbon with a mass of 12.0000. One atomic mass unit is then defined as 1/12 the mass of the standard. All atomic masses are measured as some number of atomic mass units (amu). For example, hydrogen has an atomic mass of 1 amu, oxygen's atomic mass is 16 amu, and nitrogen's atomic mass is 14 amu.

The periodic table is organized to provide information about the characteristics of elements. For example, elements in group 1, at the left of the table, have one electron in the outer energy levels of their atoms. They are highly reactive and tend to lose one electron and become positively charged ions—each forms a +1 ion. Atoms of elements in group 2 tend to lose two electrons. Atoms of elements at the right of the table, such as oxygen and chlorine, tend to gain electrons, becoming negatively charged ions (see Figure 1.7). The behavior of elements such as carbon, nitrogen, and phosphorus is more difficult to predict. Those elements react differently depending on the other elements present.

Appendix 1B
Radioisotopes and Research in Biology

Radioactive tracers and labels are valuable research tools in biology. Use of radioactive isotopes made possible many important discoveries about cells and their functions.

Radioactive isotopes are elements that have unstable atomic nuclei. For example, the common form of the element carbon is carbon 12 (with six protons and six neutrons). Carbon 14 (with two extra neutrons) is a radioactive isotope. Over time, the nuclei of unstable radioactive isotopes give off subatomic particles and energy as they break down to a more stable atomic form. The decay of radioactive isotopes occurs at a constant rate known as the half-life. The decay of carbon 14 is very slow. It takes 5,730 years for half of the carbon 14 in a sample to decay to nitrogen 14. This steady and predictable rate of decay can be used in radioactive dating of fossils and minerals (see Biological Challenges: Using Radioactivity to Date Earth Materials in Section 17.1). Other radioactive isotopes decay much more quickly and reach a stable condition in days or weeks.

Radioactive isotopes can be introduced into living systems and cells as molecular tracers or labels. Special decay-counting instruments, such as a Geiger counter, can detect the energy given off by the isotopes as they decay (Figure 1B.1). Thus, the pathway of the isotopes in chemical reactions can be followed. For example, Melvin Calvin and his coworkers introduced carbon dioxide containing radioactive carbon 14 into photosynthetic algae. Then they traced the path of carbon 14 as the algae produced sugars in photosynthesis. Radioactive nucleotides can also be used to study the synthesis and fate of DNA and RNA in cells.

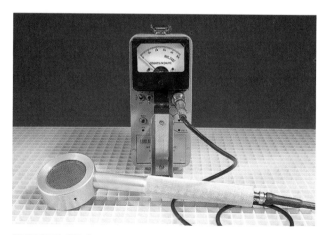

FIGURE 1B.1

A portable instrument for detecting and measuring radiation.
The gas-filled probe of the Geiger-Mueller counter converts radioactive decay into an electrical current. The current travels as a series of pulses into a meter that displays the level of radioactive decay in a small area. This device was developed in the 1920s by Hans Geiger and W. Mueller.

PERIODIC TABLE OF THE ELEMENTS

FIGURE 1A.1

Periodic table of the elements. The atomic masses shown are averages of the most common isotopes of each element. Isotopes vary widely in abundance.

Appendix 4A
ATP Synthesis in Chloroplasts and Mitochondria

In 1961, British biochemist Peter Mitchell proposed a hypothesis to explain the synthesis of ATP in chloroplasts and mitochondria. The process links chemical reactions and transport. Mitchell called the process chemiosmosis, from the Greek *osmos*, to push. He won the Nobel Prize in 1978 for his work.

The chemiosmotic model of ATP synthesis is an example of the relationship between structure and function. Most ATP is made by enzymes called ATP synthetases that are inserted in membranes. The enzymes use a concentration gradient of protons as an energy source to make ATP. The Mitchell hypothesis proposes that the energy to establish the proton gradient comes from the flow of electrons through electron transport systems in membranes. As electrons flow through the thylakoids of chloroplasts or the cristae of mitochondria, they release energy that is used to pump, or expel, protons (H^+ ions) to one side of the membrane.

A proton gradient results, with a higher concentration of protons in the solution on one side of the membrane than on the other side. The relatively high concentration of protons is highly organized and can store energy. ATP synthetase uses that stored energy to form ATP from ADP and phosphate as the protons diffuse down the concentration gradient and through the enzyme (Figure 4A.1). Thus, the energy from electron flow is used to concentrate protons and ultimately to synthesize ATP. In chloroplasts, light supplies the energy to drive the electron flow. In mitochondria, the original energy for electron flow comes from the sugars, fats, or proteins used in respiration.

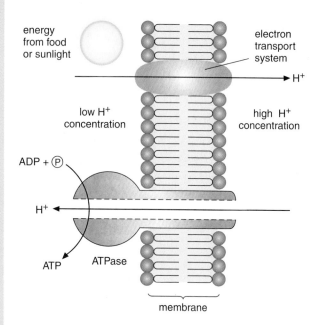

FIGURE 4A.1

Chemiosmotic energy transfer. Food or sunlight supplies energy to generate a proton gradient by way of an electron transport system. ATP forms as protons diffuse down a concentration gradient through the membrane enzyme ATP synthetase (ATPase).

Appendix 6A
Preparing Cells for Study

Cells are very small and complex. It is difficult to directly observe their structure and discover their composition. It is even more difficult to find out how their various components function. For more than 300 years, scientists have worked to develop a variety of techniques to study cells. The strength and limitations of each technique have largely determined our concept of the structure and function of the cell.

Except for water, most cell contents do not interfere with the passage of light through the cell. Therefore, untreated cells are almost invisible when viewed through an ordinary light microscope. One way to make them more visible is to stain them.

The use of stains for preparing tissue samples was a direct outgrowth of the demand for dyes in the 19th-century textile industry. Some of the dyes were found to stain tissues and unexpectedly showed an affinity for particular parts of the cell, such as the nucleus or membrane. This differential staining allowed the internal structure of the cell to be seen clearly. Today many of the organic dyes available bind specifically to particular cellular components (Figure 6A.1). For example, methylene blue binds to proteins, and fuchsin binds to DNA. The chemical basis for the affinity of some dyes, however, is not known.

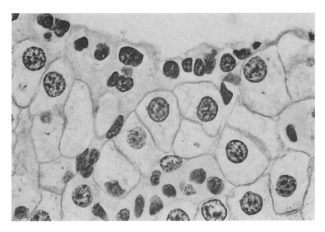

FIGURE 6A.1

Visualizing nuclei. These animal cells were exposed to the Feulgen staining procedure, which dyes DNA a reddish purple color, ×380.

Techniques for observing the components of cells that are even more specific than organic dyes are also available. For example, certain enzymes can be located in cells by exposing them to substrates that produce a visible product. Antibodies against specific macromolecules can be covalently bonded to a fluorescent dye and then used to mark the distribution of that macromolecule in the cell.

Before they can be stained, most tissues must be fixed. Fixation makes the cells permeable to stains and stabilizes their macromolecules. Some of the earliest fixation techniques involved brief immersion in acid or organic solvents, such as methanol. Current procedures usually include exposure of the cell to compounds that form covalent bonds with free amino groups of proteins, thereby linking adjacent molecules.

Most tissues are too thick for their cells to be examined without preparation and must first be cut into thin slices or sections. The tissues are cut with a microtome, a machine that operates somewhat like a meat slicer. Then the sectioned tissue is placed on a glass microscope slide and stained.

Tissues are generally too soft to be cut directly into thin sections. After fixation, they are embedded in a liquid wax or plastic resin that surrounds and permeates the cell. This preparation hardens to a solid block, and then a microtome kept in a cold chamber sections it. Frozen sections produced in this way represent a more original form of the tissue, but they are difficult to prepare, and the presence of ice crystals can destroy many details.

Appendix 12A
Pollination by Insects Aids Fertilization

Although the flowers of many plants have both male and female organs, most species have some means of cross-pollination. This adaptation provides for more genetic variety than self-pollination and thus has been selected for as flowering plants have evolved.

Insects are often the agents that carry pollen from one plant to another. The fossil record indicates that flowering plants and insects probably evolved together; evidence for this relationship is their parallel successes once both groups appeared on land. The most striking characteristics of flowers—their colors, scents, and unusual shapes—are adaptations that attract insects. Pollination by insects makes fertilization more likely. In some cases, the adaptations of plant and insect are intertwined. A single species of insect may be totally or mainly responsible for the pollination of a single plant species. One of the most extraordinary examples of such a relationship is that of the yucca plant and the pronuba moth of the southwestern United States (Figure 12A.1).

a   b

FIGURE 12A.1

Pollination by insects. Examples of coevolution of a plant and an animal. **(a)** A yucca, *Yucca glauca,* only produces flowers for about 6 weeks of the year. **(b)** The timing of the reproductive cycle of the pronuba moth, *Tegeticula yuccasella,* exactly coincides with yucca flowering.

When the yucca flowers open up, several small pronuba moths, apparently attracted by the fragrance, fly into the blossoms and mate there. The female moth collects pollen from the anthers of that flower and rolls it into a ball. She flies to another yucca flower, bores into its ovary, and deposits a fertilized egg in the opening she has made. Afterward, she carries the ball of pollen to the stigma above the ovary, which will serve as a hatching place for the egg. She presses the pollen into the sticky stigma, increasing the chances of pollen-tube formation. This behavior is repeated until several eggs are in the ovary and several balls of pollen dot the stigma of the same blossom. Somewhat later, the moth eggs hatch into larvae that feed on a few, but by no means all, of the seeds that have begun to mature within the ovary of the yucca blossom.

Both the yucca and the pronuba moth benefit from this remarkable relationship. Neither organism can complete its full life cycle without the other. In an area where pronuba moths were accidentally destroyed, yucca blossoms were not pollinated and, of course, no seeds developed. This circumstance illustrates the interdependence of these two organisms. Other insects that are attracted to a yucca blossom may derive benefit for themselves but are unlikely to pollinate the flowers.

Appendix 13A
The Chi-Square Test

Suppose that in studying a test cross between two types of tomato plants, a scientist expects half the offspring to have green leaves and half to have yellow. In the actual experiment, however, 671 out of 1,240 seedlings turn out to have green leaves, and 569 out of the 1,240 have yellow leaves. Is this a minor difference due to chance, or is it a relatively significant deviation from the expected numbers?

Scientists have devised a method to determine whether or not a deviation from expected experimental results is large enough to be significant. This method is called the chi-square test. It consists of two steps: (1) calculating the chi-square (χ^2) value for the data in question and (2) finding how often a chi-square value of that size is likely to result from chance alone.

The chi-square value is a measure of the deviation between observed and expected experimental results. In mathematical shorthand, this value can be expressed as follows:

$$\chi^2 = \Sigma \; \frac{(\text{observed number} - \text{expected number})^2}{\text{expected number}}$$

where χ is the Greek letter chi and Σ is the Greek letter sigma, standing for sum and meaning here "the sum of all." This formula is used to find the chi-square value for the experiment just described. In that experiment, 620 green-leaved plants and 620 yellow-leaved plants were expected, but the numbers observed were 671 and 569, respectively. Therefore,

$$\chi^2 = \frac{(671 - 620)^2}{620} + \frac{(569 - 620)^2}{620}$$

$$= \frac{(51)^2}{620} + \frac{(-51)^2}{620} = \frac{2{,}601 + 2{,}601}{620}$$

$$= \frac{5{,}202}{620} = 8.4$$

Thus, the value of chi-square for this experiment is 8.4. What does that value mean? How do you know if this value is significant?

Mathematicians have compiled tables that provide a basis for judging how likely any chi-square value is to have arisen by chance alone. Table 13A.1 shows that a chi-square value of 8.4 for just two classes (green-leaved plants and yellow-leaved plants) is larger than any of the values listed. This value means that the probability that this deviation is the result of chance alone is less than 1 in 100. By custom, when the probability is as low as (or less than) 5 in 100, the deviation is said to result from other causes. Such a deviation is said to be statistically significant—that is, the results differ significantly from those that were expected. In the case of the green- and yellow-leaved plants, for example, subsequent studies showed that few of the yellow-leaved plants germinated and survived; they were less hardy than the green-leaved ones.

TABLE 13A.1
Values for Chi-Square Test

Number of Classes	χ² Values							
2	0.0002	0.004	0.455	1.074	1.642	2.706	3.841	6.635
3	0.020	0.103	1.386	2.408	3.219	4.605	5.991	9.210
4	0.115	0.352	2.366	3.665	4.642	6.251	7.815	11.345
Number of times per hundred that chance alone would produce a deviation as large or larger	99	95	50	30	20	10	5	1

How is the chi-square test applied to experimental results involving more than two classes of objects or events? Suppose that a scientist crossed pink-flowered four-o'clocks and obtained 236 offspring. From Mendel's laws, the offspring are expected to be red-, pink-, and white-flowered in the ratio of 1:2:1. For the 236 offspring, this ratio would mean 59 red, 118 pink, and 59 white. But the actual results are 66 red, 115 pink, and 55 white. The chi-square test is applied as follows:

$$\chi^2 = \frac{(66-59)^2}{59} + \frac{(115-118)^2}{118} + \frac{(55-59)^2}{59}$$

$$= \frac{49}{59} + \frac{9}{118} + \frac{16}{59} = \frac{98+9+32}{118}$$

$$= \frac{139}{118} = 1.18$$

By consulting Table 13A.1, you will see that a chi-square value of 1.18 for three classes is quite small. Between 50 and 95 times in 100, a value this large or larger might be produced by chance alone. The experimenter therefore concluded that the observed results agree with those predicted.

Appendix 13B
Gene Mapping

Linked genes are inherited together unless recombination occurs between their chromosome sites. In 1911, A. H. Sturtevant, one of Thomas Morgan's students, suggested that recombination frequencies are proportional to the distance separating two genes on a chromosome. He defined a *map unit* as a 1% recombination frequency. With this standard of measurement, the mutant genes for black body and vestial wings in *Drosophila*, for example, are 17 map units apart; thus, they have a recombination frequency of nearly 17%. By determining recombination frequencies in many different crosses, Sturtevant and his coworkers determined the linkage groups and linear arrangement of the known *Drosophila* genes.

In human genetics, determining linkage and map distance of genes has been more difficult. Two genes must be on the same chromosome and close enough to reveal their linkage, at least one parent must be heterozygous for both genes, how the genes were inherited from the grandparent generation must be known, and recombination of the genes must be obvious in offspring.

These criteria may be met more easily for X-linked genes than for autosomal genes because the phenotype of a grandfather and his grandsons is determined by single copies of these genes. If a mother has a mutant allele for both color blindness *(c)* and hemophilia *(h)* on one X chromosome and normal alleles for these genes *(C, H)* on the other, her sons would be either color-blind and hemophiliac or normal for both traits. The percentage of times that either a color-blind, nonhemophiliac son or a normal-sighted, hemophiliac son is born reveals the map distance between these two genes. Recombination has been found to occur between these genes 10% of the time. Therefore, the eggs this mother produces would be 45% *CH*, 45% *ch*, 5% *Ch*, and 5% *cH*. Given the low frequencies of these mutant alleles in the population and the few children produced in most families, you can see why the task of mapping human genes has proceeded slowly.

How have technological advances accelerated the mapping of human chromosomes? In 1960, biologists learned to fuse cells from different organisms. The nuclei of these hybrid cells contained the chromosomes from both cells. Human skin cells and mouse cells fused, and as the hybrid cells divided in culture, many of the human chromosomes were lost. Often one chromosome or only a piece of a human chromosome was left in a stabilized cell line. Any human protein produced by such hybrid cells had to come from a gene on the remaining chromosome. With this technique, many genes were mapped to specific chromosomes and roughly located along the chromosomes.

Recombinant DNA technology has provided new tools for mapping genes. Restriction fragment length polymorphisms (RFLPs) are used as reference points, or genetic markers, along the chromosomes (see Section 15.3). When individuals of a family with a genetic disorder also share an unusual RFLP, that RFLP may serve as a marker of the disorder. The frequency with which the disorder and the marker are inherited together indicates the distance between the two locations. The closer a disease-causing gene and a linked RFLP are, the more often they will be inherited together and the more reliable the marker.

Humans have 3 billion DNA nucleotides distributed over their 23 pairs of chromosomes. A distance of approximately 1.5 million nucleotides between markers (approximately one map unit) means that the markers will recombine only 1% of the time. If a gene is only 10,000 nucleotides long, there may be many genes between two RFLP sites. However, the sites still can be linked closely enough to allow 99% accuracy in discriminating between two RFLP sites.

Researchers are working to develop a complete human gene map. The DNA from large families (three generations with many children) is cut up with restriction enzymes and analyzed for RFLP polymorphisms. This DNA is being shared with scientists worldwide; their findings are collected in computerized data banks. Potential benefits from this effort include the ability to locate and study the genes responsible for the more than 4,000 recognized genetic disorders and the combinations of genes that appear to cause illnesses such as hypertension, heart disease, and certain cancers. The identification, isolation, and cloning of human genes will enable scientists to study how genes function at the molecular level. Determining the products of these genes may lead to new methods of preventing and treating genetic disorders.

The Human Genome Project (HGP) has developed a detailed human gene map. However, even a complete human gene map will not yield immediate understanding of how the DNA sequence produces a human being. For example, the map alone cannot explain how nerve cells become connected in the complex structure of the brain— nor can it provide explanations for all the ways that individuals differ.

What use should be made of the new genetic information? Among the most important issues are personal freedom, privacy, and societal rights versus individual rights of access to genetic information. For example, could such information be used to deny employment? What would be the impact of discovering that, genetically, humans are less equal than we now suppose? These and other questions must be considered in any policy decisions about the human gene map.

Appendix 20A
Physical Adaptations

In becoming a stable bipedal animal, human ancestors underwent evolutionary changes in the shape and proportions of the foot, leg, and pelvic bones and in the muscles of the legs and buttocks. Bonobos (pygmy chimpanzees) often walk upright like humans. A chimpanzee can walk rather comfortably on its hind legs but not for long. It also can run surprisingly fast. Nonetheless, the chimpanzee's body is not adapted for efficient bipedal walking and running. Its feet are better adapted for gripping than for upright walking (Figure 20A.1a), as its big toes stick out to the side like thumbs instead of forward to provide spring in the stride. Its legs are short, and the small size and

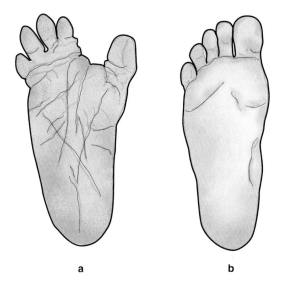

FIGURE 20A.1

Primate anatomical adaptations. (a) The structure and function of a chimpanzee foot reflect adaptation to tree climbing and grasping with the feet. This structure makes prolonged bipedalism difficult. **(b)** A human foot, however, is adapted for bipedal walking.

location of the buttock muscles limit propulsion. The muscles that the chimpanzee can use for bipedalism are attached to the bones in a way that supplies inefficient leverage for a vigorous stride. The chimpanzee walks upright in a waddle or rolling gait because it must shift its body weight at each step.

A more efficient method of walking erect is to have the legs straighter and closer together. This arrangement is possible in humans because of marked evolutionary changes in the shape and proportion of bones in the leg, the foot (Figure 20A.1b), and particularly the pelvis. Over time, the two large pelvic blades have twisted and flattened. The modern human pelvis helps place the torso more vertically over the legs and provides better anchorage and leverage for the three sets of buttock muscles used in walking.

Appendix 20B
The Old Man from La Chapelle-aux-Saints

Fossils of Neanderthals from southwestern France were discovered in the first decade of the 20th century. Among them was the skeleton of an old man

from a cave near the village of La Chapelle-aux-Saints. This fossil was selected for reconstruction as a typical Neanderthal specimen. The task of rebuilding the old man fell to a French paleontologist, Marcellin Boule.

Although there were abundant and well-preserved materials, Boule proceeded to commit an astonishing number of errors that were not recognized and corrected for decades. The full reconstructions of Neanderthal individuals, in turn, were even more misleading (Figure 20B.1). Boule misinterpreted the bones and made the individual appear much like an ape. He arranged the foot bones so that the big toe diverged from the other toes like an opposable thumb, thus forcing the man to walk on the outside of his feet. The knee joint could not be entirely extended, resulting in a bent-kneed gait. The spine lacked the curves that allow modern humans to stand upright, and the head was thrust so far forward that it appeared as if Boule's Neanderthal would have to use his arms to keep from falling on his face. In addition, Boule completely ignored the

FIGURE 20B.1

Inaccurate Neanderthal anatomy. This museum model of a Neanderthal is based on an incorrect skeletal reconstruction.

large cranial capacity (1,600 cm³) of the old man. Instead, Boule stated that there was not much room for the frontal portion of the brain, which was then incorrectly thought to be the center of higher intelligence. Boule ranked the Neanderthal's brainpower somewhere between that of apes and modern humans, but closer to the apes. The hunched shoulders and apelike posture of the reconstructions were quite incorrect; worse still were their dull expressions. Actually, Neanderthals were probably quite intelligent.

The discovery of additional Neanderthal fossils slowly led to a contradiction of Boule's reconstruction of the old man from La Chapelle-aux-Saints. What had Boule done wrong? He had overlooked the effects of arthritis when he reconstructed the skeleton and allowed his biases to influence his interpretation. That oversight led him to incorrectly assume that all Neanderthal people walked with a stooped gait and bent knees.

Appendix 20C
Cultural Evolution

Normally the brains of mammalian fetuses grow rapidly, but the growth slows down and ceases shortly after birth. A prolonged growth period of the human skull, longer than for any other primate, results in a larger human brain. The extended developmental period lengthens the time parents must care for their offspring and subsequently extends the child's learning period. This learning is the basis of human culture—the transmission of accumulated knowledge through time by means of written or spoken language.

The first, very long stage in human cultural evolution probably began with hunters and gatherers of the African grasslands who later spread to other parts of the Old World. They made tools, divided labor, and probably organized communal activities. The second and much more recent stage came with the agricultural revolution in Eurasia and the Americas about 10,000 to 15,000 years ago. Along with agriculture came permanent settlements and eventually the first cities. The third major stage in human cultural evolution was the Industrial Revolution, which began in the 18th century. Since then new technology has grown exponentially. Through all of this recent cultural evolution, humans have not changed biologically in any significant way. We are probably no more intelligent than earlier modern humans. Knowing how to build microcomputers or produce recombinant DNA is not stored in the human genes but is, instead, a product of hundreds of generations of human experience and the capacity for intelligence—which is inherited.

The evolutionary changes in the human brain have had enormous consequences. Cultural evolution, including the development and use of language and tools, has made *Homo sapiens* a species that can intentionally change its own environment. We do not have to wait to adapt to a new environment. Instead, we use artificial "adaptations"—clothing, irrigation, heated or cooled buildings, for example—or otherwise change the environment to suit our needs. Humans are the most numerous and widespread of all large animals. Everywhere we go we bring change, faster than many other species can adapt to it. Of the many upheavals in the history of Earth, the emergence of humans is the latest and perhaps one of the greatest. What do you think the next great stage in human cultural evolution will be?

Appendix 22A
Innate Behavior

There are many different types of innate behavior. One primitive behavior is called a taxis (plural: *taxes*). A taxis is a change in the direction of movement caused by some stimulation in the environment. For example, shining a light head-on to a planarian will cause it to change direction. Random movements are not taxes. The movement must be in response to a definite stimulus. Taxes are innate behavior because they usually are fixed and unchanging for the species—the organism always reacts in the same way to the same stimulus.

Another form of innate behavior is a simple reflex. The human knee-jerk reflex is an example. When the tendon below the knee is struck, the leg responds with a jerk. Simple reflex behavior is innate and characteristic of the species. In general, innate reflexes protect an organism from harm or help it maintain normal conditions. Other reflexes involving more complicated nervous-system responses can be innate or learned.

Many automatic human behavioral responses are learned reflexes. For example, normal posture is maintained by constant muscle adjustment. This adjustment is stimulated by slight changes in the environment. In any normal movement, such as taking a step, coordination of muscles also depends on a series of reflex acts. These acts involve many muscles, nerves, and sensory receptors in different parts of the body.

The most complex patterns of innate behavior are instincts. Instinct is an inherited form of behavior that usually involves a whole series of actions. For example, the nest-building instinct in birds includes all of the activities of searching for good nesting sites and building materials, bringing the materials to the site, and making a particular type of nest. The migratory behavior of certain fish and birds is another example.

Sometimes a species can be recognized by observing an animal's behavior pattern. An expert can determine the species of a spider by examining its web (Figure 22A.1). Clues to the evolution of certain spiders may be gathered by analyzing the behavior patterns recorded in their webs. Similarities in webs may show genetic relationships between groups of spiders.

FIGURE 22A.1

The distinctive web of the "writing spider." Also called the black-and-yellow spider, *Argiope aurantia* is an orb-weaver, spinning a web on which it scribbles a zigzag pattern that is characteristic for each species of the genus *Argiope*.

Appendix 23A
Antibody Classes

The constant regions of the heavy chains of antibody molecules—that is, the ends that do not bind to antigens—determine antibody class. There are five major classes of antibodies, or immunoglobulins (Igs): IgM, IgD, IgG, IgA, and IgE. The antibody class determines certain biological activities of the molecules. For instance, five IgM molecules can join to form a complex with 10 identical antigen-binding sites. IgM is the predominant antibody of the primary immune response. The predominant antibody of the secondary immune response is IgG. It is the most abundant antibody in circulation, and, because it is smaller than IgM, it can pass from a mother to a fetus. IgA molecules often form doublets in body secretions such as breast milk and saliva. IgE appears to be important in fighting parasitic infections and is responsible for allergic reactions such as asthma. Although IgD is present on young lymphocytes, its function is unknown. The structures and functions of the immunoglobulins are summarized in Table 23A.1.

The constant regions of antibody light chains determine the light chain type. The differences between the two light chain types, kappa and lambda, have no known biological connections.

TABLE 23A.1
Antibody Classes

Ig Class	Structure	Where Found	Function
M	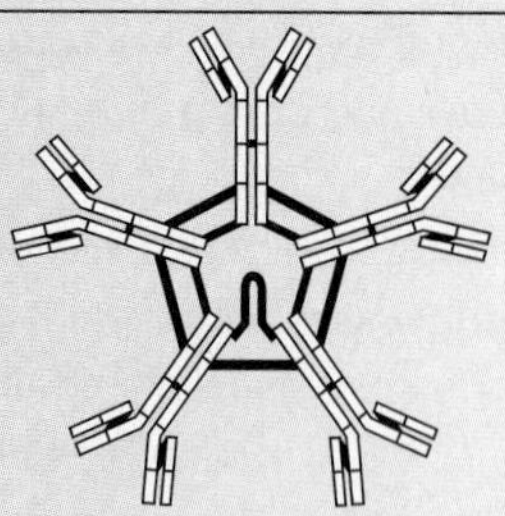	blood	protects blood; works in primary immune response
A		mucous secretions	makes secretions more protective; forms the first line of defense against infection
D	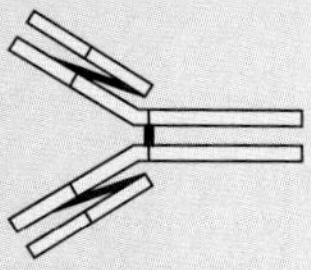	on young lymphocytes	may have a role in differentiation
G	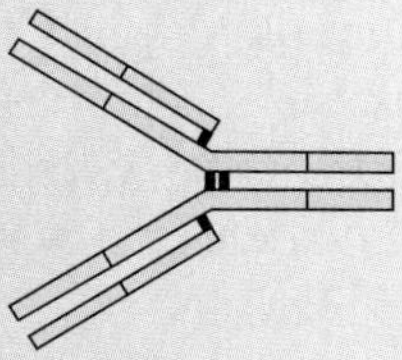	blood and extracellular fluid	works in secondary immune response
E	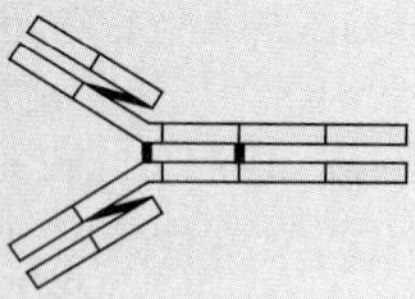	blood and extracellular fluid	causes allergic reactions; combats parasitic infections

Appendix 23B
Generating Antibody Diversity

The enormous diversity in antibodies likely evolved in vertebrates as a protective mechanism against numerous environmental antigens and cancer cells. The discovery of the mechanisms that generate antibody diversity was among the first research benefits of recombinant DNA technology.

Large numbers of different antibody molecules can be made using different combinations of a limited number of light and heavy chains. Even greater numbers of different antibody molecules are made by combining several smaller genes to make a large gene that codes for one polypeptide chain. Three different genes code for each light chain—C for the constant region, V for the variable region, and J for the joining region. Four different genes code for each heavy chain—C, V, J, and D (diversity). Figure 23B.1 shows how V, J, and C genes combine to produce a larger gene that codes for one light chain of the kappa (κ) type. Frequent somatic mutations in the V region also increase antibody diversity.

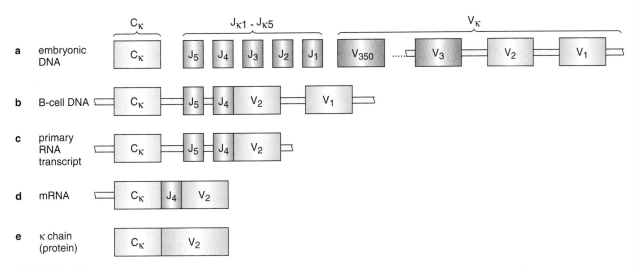

FIGURE 23B.1

Kappa (κ) antibody light chain gene recombination. (a) In the zygote and embryo, kappa genes include several V and J gene segments (V_1–V_{350} and J_1–J_5) and one C gene (C_K). **(b)** In differentiating B cells, a V gene recombines with a J gene to form a VJ combination. The DNA between the V and J genes is deleted. In this example, gene V_2 has combined with gene J_4. **(c)** The B cell transcribes the segment of DNA that includes the VJ combination and the C gene into a primary RNA transcript that also includes introns and additional J segments (J_5). **(d)** Introns and extra J segments are removed from the RNA to form mRNA. **(e)** This mRNA is then translated into kappa light chains. This example is only one of thousands of possible rearrangements.

Thousands of combinations are possible. For example, if there were three sets of 10 genes coding for the light chain polypeptides, recombination of genes would produce $10 \times 10 \times 10$, or 1,000, different polypeptides. Although the exact number of genes in each gene family (kappa, lambda, and heavy chain) is not known, there appear to be enough genes to recombine and ultimately code for millions of different antibody molecules. For example, in mice there are approximately 350 V genes, 5 J genes, and 1 C gene that can recombine to form one light chain of the kappa type.

Exactly when and how this genetic recombination takes place in the B cells is not known. Presumably, every possible combination of genes is produced. The combination of genes in a particular B cell is the final step in determining what binding site that cell and all of its descendants will express.

GLOSSARY/GLOSARIO

Numbers in parentheses indicate the page where a glossary term is introduced and defined in the text.
Los números en paréntesis indican la página del texto donde se introduce y se define un término del glosario.

A multilingual science glossary at BSCSblue.com includes Arabic, Bengali, Chinese, English, Haitian Creole, Hmong, Korean, Portuguese, Russian, Spanish, Tagalog, Urdu, and Vietnamese.

English		Español

abiotic: referring to a physical or nonliving component of an ecosystem   (56)

abscisic acid: a plant growth regulator that protects a plant in an unfavorable environment by promoting dormancy in buds and seeds and closing stomates   (304)

abyssal zone: the deepest portion of the ocean floor, where light does not penetrate, temperatures are cold, and pressure is extreme   (662)

acidic: having a pH of less than 7, reflecting more dissolved hydrogen ions than hydroxide ions   (32)

actin: a protein in muscle fiber that, together with myosin, is responsible for muscle contraction and relaxation (197)

action potential: a rapid change in the membrane potential of an axon that is brought about by the opening and closing of sodium and potassium channels and that transmits a signal   (562)

activation: the increase in cell respiration and protein synthesis that occurs in a newly formed zygote   (262)

activation energy: the energy necessary to start a chemical reaction   (29)

active site: the portion of an enzyme that attaches to the substrate through weak chemical bonds   (61)

active transport: the movement of a substance through a biological membrane against a concentration gradient (84)

adaptation: a characteristic that improves an organism's chance of surviving to reproduce   (10)

adaptive immunity: immune defenses that respond to specific antigens   (606)

adaptive radiation: the development of numerous species from a common ancestor in a diverse environment   (511)

adhesion: the attractive force between different substances   (189)

abiótico: componente físico o inanimado de un ecosistema   (56)

ácido abscísico: regulador del crecimiento de las plantas que protege la planta en un medio desfavorable al promover un letargo en los brotes y semillas y mantener los estomas cerrados   (304)

zona abisal: la parte más profunda del fondo oceánico, donde no penetra la luz, las temperaturas son bajas y la presión es extrema   (662)

ácido: sustancia que tiene un pH menor que 7, más iones hidrógeno disueltos que iones hidróxido   (32)

actina: proteína en la fibra muscular que, junto con la miocina, es responsable de la contracción y relajación musculares   (197)

potencial de acción: cambio rápido en el potencial de membrana de un axón que sucede como consecuencia de un abrir y cerrar de los canales de sodio y potasio y que transmite una señal   (562)

activación: incremento en la respiración celular y la síntesis de proteínas que ocurre en un cigoto recientemente formado   (262)

energía de activación: energía necesaria para empezar una reacción química   (29)

sitio activo: parte de una enzima que se une al sustrato a través de enlaces químicos débiles   (61)

transporte activo: movimiento de una sustancia a través de una membrana biológica en contra de un gradiente de concentración   (84)

adaptación: característica que mejora la oportunidad de un organismo de sobrevivir para reproducirse   (10)

inmunidad adaptiva: defensas inmunológicas que responden a antígenos específicos   (606)

radiación adaptiva: el desarrollo de numerosas especies de un antepasado común en un ambiente diverso (511)

adhesión: fuerza atractiva entre diferentes sustancias (189)

ADP: adenosine diphosphate; the compound that remains when a phosphate group is removed from ATP, releasing energy (66)

aerobic: occurring or living in the presence of free or dissolved oxygen (130)

aldosterone: a hormone secreted by the adrenal gland; helps regulate sodium and potassium concentrations and water balance (95)

allele: one of two or more possible forms of a gene (346)

all-or-none law: the law stating that neurons can either respond to stimuli by producing a full-scale nerve impulse or not respond at all (562)

alternation of generations: life cycle that includes both asexual and sexual forms; characteristic of plants, algae, and other life-forms (324)

alveoli: air sacs in a lung (91)

amino acid: an organic compound composed of a central carbon atom to which are bonded a hydrogen atom, an amino group ($-NH_2$), an acid group ($-COOH$), and one of a variety of other atoms or groups of atoms; the building block of polypeptides and proteins (38)

ammonia: a toxic nitrogen compound (NH_3) excreted by some aquatic organisms; thought to have been present in Earth's early atmosphere (93)

amnion: a sac or membrane filled with fluid and enclosing the embryo of a reptile, bird, or mammal (273)

anaerobic: occurring or living in conditions without free or dissolved oxygen (130)

analogy: similarity in form or function that is not a result of evolution from a common ancestor but is evidence of convergent evolution (similar environmental pressures) (466)

anaphase: the stage in cell division in which chromosomes separate and move toward opposite ends of the cell (224)

androgen: family of male sex hormones, including testosterone (337)

aneuploid: having a number of chromosomes that is not an exact multiple of the haploid number for that species (223)

Animalia: the animal kingdom (477)

animal society: an organized population showing cooperative behavior (592)

annual: a plant that completes its life cycle in 1 year or growing season (665)

anther: the enlarged end of a stamen in a flower, inside which pollen grains containing male gametes form (326)

ADP: difosfato de adenosina; el compuesto que queda cuando un grupo fosfato se elimina del ATP, liberando energía (66)

aeróbico: que ocurre o vive en presencia de oxígeno libre o disuelto (130)

aldosterona: hormona secretada por la glándula adrenal que ayuda a regular las concentraciones de sodio y potasio y el balance de agua (95)

alelo: una de dos o más formas posibles de un gene (346)

ley del todo o nada: ley que establece que las neuronas pueden responder estímulos al producir un impulso nervioso de escala total o no responder en absoluto (562)

alteración de generaciones: ciclo vital que incluye ambas formas sexuales y asexuales; características de plantas, algas y otras formas de vida (324)

alvéolos: sacos de aire en un pulmón (91)

aminoácido: compuesto orgánico compuesto de un átomo de carbón central a los cuales se enlaza un átomo de hidrógeno, un grupo amino ($-NH_2$) y un grupo ácido ($-COOH$), uno de una variedad de diferentes átomos o grupos de átomos, el bloque constructor de polipéptidos y proteínas (38)

amoniaco: compuesto tóxico de nitrógeno (NH_3) excretado por algunos organismos acuáticos, que se piensa ha estado presente en la atmósfera temprana de la Tierra (93)

amnión: saco, o membrana, lleno de fluido y que encierra a un embrión de reptil, ave o mamífero (273)

anaeróbico: que ocurre o vive en condiciones sin oxígeno libre o disuelto (130)

analogía: similitud en la forma o función que no es resultado de la evolución de un antepasado común pero es evidencia de evolución convergente (presiones ambientales similares) (466)

anafase: fase en la división celular en la cual los cromosomas se separan y mueven hacia los extremos opuestos de la célula (224)

andrógeno: familia de hormonas sexuales masculinas incluyendo a la testosterona (337)

aneuploide: que tiene un número de cromosomas que no es un múltiplo exacto del número haploide para esa especie (223)

Animalia: el reino animal (477)

sociedad animal: población organizada que muestra un comportamiento cooperativo (592)

anual: planta que completa su ciclo vital en un año o estación de crecimiento (665)

antera: extremo alargado de un estambre en una flor, dentro del cual los granos de polen contienen la forma de los gametos masculinos (326)

anthropologist: a scientist who studies human evolution, variation, and both past and present cultures and behavior (524)

anthropomorphism: attributing human characteristics to nonhumans (599)

antibody: a blood protein produced in response to an antigen, with which it combines specifically (359)

antibody-mediated immune response: immune response in which B cells produce antibodies that bind with and mark antigens for destruction by nonspecific defenses; initiated by helper T cells (606)

anticodon: three-nucleotide sequence in a transfer-RNA molecule that is complementary to, and base pairs with, a specific codon in messenger RNA (236)

antidiuretic hormone (ADH): a hormone released from the posterior lobe of the pituitary (an endocrine gland in the brain); enhances conservation of water by the kidneys (96)

antigen: any material, usually a protein, that is recognized as foreign and elicits an immune response (604)

aphotic zone: the part of the ocean beneath the photic zone; does not have enough light for photosynthesis (660)

apical meristem: embryonic plant tissue in the tips of roots and shoots that supplies cells for growth in length (293)

Archaea: domain of bacterial species proposed by Carl Woese on the basis of RNA analysis (476)

artery: a vessel that transports blood away from the heart (194)

arthropod: any organism of the phylum Arthropoda; includes organisms such as insects, arachnids, and crustaceans that have jointed exoskeletons (192)

artificial selection: the selective breeding of organisms to encourage the occurrence of desirable traits (429)

association study: study in which the frequency of an allele is compared in populations that do and do not have a certain trait (591)

atom: the smallest particle of an element that retains the properties of that element (25)

ATP: adenosine triphosphate; a compound that has three phosphate groups and is used by cells to store energy and to fuel many metabolic processes (64)

ATP synthetase: an enzyme complex in the inner membrane of a mitochondrion and the thylakoid membrane of a chloroplast that catalyzes the formation of ATP (109)

antropólogo: científico que estudia la evolución humana, la variación y las culturas y comportamientos del pasado y del presente (524)

antropomorfismo: que atribuye características humanas a lo no humano (599)

anticuerpo: proteína de la sangre producida en respuesta a un antígeno, con el cual se combina específicamente (359)

respuesta inmunológica mediada por anticuerpos: respuesta inmunológica en la cual las células B producen anticuerpos que se enlazan y marcan a los antígenos para destruirlos mediante defensas no específicas, iniciadas por células T ayudantes (606)

anticodón: secuencia de tres nucleótidos en una molécula de RNA de transferencia que es complementaria a, y sus bases forman pares con, un codón específico del RNA mensajero (236)

hormona antidiurética (ADH): hormona liberada desde el lóbulo posterior de la pituitaria (una glándula endocrina del encéfalo); aumenta la conservación del agua por los riñones (96)

antígeno: cualquier material, usualmente una proteína, que es reconocido como foráneo y que desencadena una respuesta inmunológica (604)

zona afótica: parte del océano debajo de la zona fótica; no tiene suficiente luz para la fotosíntesis (660)

meristema apical: tejido vegetal embrionario en las puntas de las raíces y brotes que sustenta a las células para el crecimiento en longitud (293)

Arcae: dominio de especies bacterianas propuesto por Carl Woese sobre la base del análisis del RNA (476)

arteria: vaso que transporta a la sangre lejos del corazón (194)

artrópodo: cualquier organismo del filo Artrópoda; incluye organismos tales como insectos, arácnidos y crustáceos que tienen exoesqueletos articulados (192)

selección artificial: cruce selectivo de organismos para promover la ocurrencia de caracteres deseables (429)

estudio de asociación: estudio en el cual la frecuencia de un alelo se compara en poblaciones que tienen y no tienen cierto carácter (591)

átomo: la partícula más pequeña de un elemento que mantiene las propiedades de ese elemento (25)

ATP: trifosfato de adenosina; un compuesto que tiene tres grupos fosfato y lo usan las células para almacenar energía y proveer de energía a muchas reacciones metabólicas (64)

sintetasa de ATP: complejo enzimático en la membrana interna de una mitocondria y la membrana tilacoide de un cloroplasto que cataliza la formación de ATP (109)

atrium (plural: *atria*): a chamber of the heart that receives blood from the veins   (194)

australopithecine: any of the earliest known hominids that walked erect and had humanlike teeth but whose skull, jaw, and brain capacity were more apelike; of the genus *Australopithecus*   (533)

autoimmunity: a disorder in which antibodies are produced against some of the body's own cells   (624)

autonomic nervous system: a division of the nervous system that controls involuntary functions such as blood pressure, body temperature, and others necessary to maintain homeostasis   (552)

autotroph: an organism that forms its own food molecules (carbon compounds) from abiotic materials   (55)

auxins: a group of plant growth regulators that promote growth by enlarging or lengthening cells, rather than by increasing the number of cells   (302)

axon: a structure that extends from a neuron and conducts impulses away from the cell body   (549)

aurícula: cavidad del corazón que recibe sangre de las venas   (194)

australopiteco: cualquiera de los primeros homínidos conocidos que caminó erecto y tuvo dientes parecidos a los del hombre pero cuyo cráneo, maxilar y capacidad cerebral eran más parecidos a los del simio; del género *Australopithecus*   (533)

autoinmunidad: trastorno en el cual se producen anticuerpos en contra de algunas células del propio cuerpo   (624)

sistema nervioso autónomo: división del sistema nervioso que controla las funciones involuntarias como la presión sanguínea, temperatura corporal y otras necesarias para mantener la homeostasis   (552)

autótrofo: organismo que forma su propias moléculas alimenticias (compuestos de carbono) de materiales abióticos   (55)

auxinas: grupo de reguladores del crecimiento vegetal que promueve el crecimiento al aumentar o alargar las células, más que por aumentar el número de células (302)

axón: estructura que se extiende desde una neurona y conduce impulsos lejos del cuerpo celular   (549)

bacteriophage: a virus that infects bacteria; also called a phage   (254)

basic: alkaline; having a pH greater than 7, reflecting more dissolved hydroxide ions than hydrogen ions   (32)

B cell: a lymphocyte that matures in the bone marrow and later produces and secretes antibodies into the bloodstream   (606)

benthic zone: the bottom surface of the ocean   (662)

bile: a secretion of the liver stored in the gallbladder and released through a duct to the small intestine; breaks large fat droplets into smaller ones that enzymes can act on more efficiently   (72)

binocular vision: vision that uses both eyes at once; produces an appearance of solidity or depth because of the slightly different angle from which each eye views an object   (521)

binomial nomenclature: the two-word naming system used in taxonomy, consisting of the Latin genus and species names of an organism   (470)

biodiversity: the number and variety of living organisms in an area at a specific time   (461)

biomass: the dry weight of organic matter composing a group of organisms in a particular habitat   (638)

bacteriófago: virus que infecta a las bacterias; también llamado un fago   (254)

básico: alcalino; que tiene un pH mayor de 7, que refleja más iones hidróxido disueltos que iones hidrógeno   (32)

célula B: linfocito que madura en la médula ósea y el cual más tarde produce y secreta anticuerpos en el torrente sanguíneo   (606)

zona béntica: la superficie del fondo del océano   (662)

bilis: secreción del hígado almacenada en la vesícula biliar y liberada a través de un ducto al intestino delgado; rompe las grandes gotas de grasa en más pequeñas sobre las cuales las enzimas pueden actuar con mayor eficacia   (72)

visión binocular: visión que usa ambos ojos a la vez; produce una apariencia de solidez o profundidad debido al ángulo ligeramente diferente desde el cual cada ojo vislumbra un objeto   (521)

nomenclatura binomial: el sistema de nombres de dos palabras usado en taxonomía y que consiste en los nombres del género y especie en latín de un organismo (470)

biodiversidad: número y variedad de organismos vivos en un área en un tiempo específico   (461)

biomasa: peso en seco de materia orgánica que compone un grupo de organismos en un hábitat particular (638)

biome: the kind of biological community typically associated with a particular physical environment; often named for its plant cover   (652)

biosphere: the outer portion of Earth—air, water, and soil—where life is found   (56)

biotic: relating to a living component of an ecosystem   (56)

bipedal: capable of walking upright on two legs   (523)

blastocyst: the mammalian embryonic stage that corresponds to the blastula of other animals   (272)

blastula: an animal embryo after the cleavage stage, consisting of a hollow, fluid-filled ball of cells   (266)

body plan: the general form of an organism's body structure, including its pattern of symmetry, germ layers, and body cavities   (267)

boom-and-bust cycle: a pattern of population growth in which exponential growth leads to a period when the population exceeds its carrying capacity, causing the population to decrease rapidly or crash   (646)

bundle sheath: a tightly packed layer of cells that surrounds a vascular bundle in the leaves of C_4 plants, plants such as crabgrass and corn in which carbon dioxide is fixed twice   (118)

bioma: tipo de comunidad biológica típicamente asociada con un ambiente físico particular; a menudo nombrado por su cubierta vegetal   (652)

biosfera: la porción exterior de la Tierra—aire, agua y suelo—donde se encuentra la vida   (56)

biótico: relativo a un componente vivo de un ecosistema   (56)

bípedo: capaz de caminar erguido en dos patas   (523)

blastocito: estado embrionario de los mamíferos que corresponde a la blástula de otros animales   (272)

blástula: embrión animal después del estado de ruptura, consistente de una bola de células huecas, llenas de fluido   (266)

plan corporal: forma general de la estructura corporal de un organismo, incluyendo su patrón de simetría, capas germinales y cavidades corporales   (267)

ciclo de auge y caída: patrón de crecimiento de población en el cual el crecimiento exponencial conduce a un periodo en el que la población excede su capacidad de transporte, ocasionando que la población disminuya rápidamente o se colapse   (646)

vaina fascicular: capa de células empacadas fuertemente que rodea a un haz vascular en las hojas de las plantas C_4, las plantas tales como el pasto garranchuelo y el maíz en el cual el dióxido de carbono se fija dos veces   (118)

Calvin cycle: the cycle that incorporates carbon dioxide in sugars during photosynthesis; uses chemical energy previously converted from light energy   (106)

CAM: crassulacean-acid metabolism; an adaptation for photosynthesis in arid conditions in which carbon dioxide entering open stomates at night is converted into organic acids that release carbon dioxide during the day when the stomates are closed   (120)

cambium: a layer of meristematic tissue, a region of self-perpetuating cells, in the stems and roots of plants that thickens these organs by producing new xylem and phloem cells   (187)

cancer: a group of diseases that involve abnormal, uncontrolled growth and division of cells   (229)

capillary: a microscopic blood vessel penetrating the tissues and consisting of a single layer of cells that allows exchange between the blood and tissue fluids   (73)

carbohydrate: an organic compound made of carbon, hydrogen, and oxygen, with the hydrogen and oxygen atoms in a 2:1 ratio; examples are sugars, starches, and cellulose   (34)

ciclo de Calvin: el ciclo que incorpora dióxido de carbono en azúcares durante la fotosíntesis, usa energía química convertida previamente de la energía luminosa   (106)

CAM: metabolismo del ácido de crasuláceas; una adaptación para la fotosíntesis en condiciones áridas en las cuales el dióxido de carbono que entra a los estomas abiertos de noche se convierte en ácidos orgánicos que liberan dióxido de carbono durante el día cuando los estomas se cierran   (120)

cámbium: capa de tejido meristemático, una región de células autoperpetuantes, en los tallos y raíces de las plantas que engrosa estos órganos al producir nuevas células de xilema y floema   (187)

cáncer: grupo de enfermedades que involucran un crecimiento anormal, descontrolado y la división de células   (229)

capilar: vaso sanguíneo microscópico que penetra los tejidos y consiste en una sola capa de células que permite el intercambio entre la sangre y los fluidos tisulares   (73)

carbohidrato: compuesto orgánico hecho de carbono, hidrógeno y oxígeno, con los átomos de hidrógeno y oxígeno en una razón de 2:1; algunos ejemplos son los azúcares, almidones y celulosa   (34)

cardiac cycle: the sequence of events in one complete contraction and relaxation of the atria and ventricles of the heart (196)

carnivore: any organism that consumes animals; a meat eater (637)

carpel: the female reproductive organ of a flower; one or more carpels fuse to form the stigma, style, and ovary (326)

carrying capacity: the maximum population size that can be supported by the available resources of a given area at a given time (646)

caste system: in insects, a highly organized society in which each member has a specific biologically determined task (592)

catalyst: a chemical that promotes a reaction between other chemicals by reducing the energy required to activate the reaction; may take part in the reaction but emerges in its original form (61)

cell: the basic living unit (28)

cell body: the enlarged portion of a neuron that contains the nucleus (549)

cell cycle: an ordered sequence of events in the life of a dividing eukaryotic cell, composed of mitosis (M) and interphase growth and DNA synthesis phases (G1, S, and G2) (214)

cell-cycle arrest: an abrupt halt in the cell cycle when proteins detect mistakes or damage in DNA that needs to be repaired (229)

cell-mediated immune response: immune response in which highly specialized lymphocytes circulate in the blood and lymphoid organs and attack and destroy cells that carry specific surface antigens (606)

cell respiration: the series of chemical reactions by which a living cell breaks down carbohydrates and obtains energy from them (55)

cell theory: the theory that organisms are composed of cells and their products and that these cells are all derived from preexisting cells (156)

cell wall: a stiff covering around the plasma membrane of certain cells, as in plants, many algae, and some prokaryotes; in plants, the cell wall is constructed partly of cellulose (163)

central nervous system (CNS): the brain and spinal cord in vertebrates (551)

centriole: structure in animal cells and some others composed of cylinders of nine triplet microtubules in a ring; helps organize microtubule assembly during cell division (169)

ciclo cardiaco: la secuencia de eventos en una contracción y relajación completas de las aurículas y ventrículos del corazón (196)

carnívoro: cualquier organismo que consume animales; uno que se alimenta de carne (637)

carpelo: órgano reproductor femenino de una flor, uno más carpelos se fusionan para formar el estigma, estilo y ovario (326)

capacidad de carga: tamaño máximo de población que puede ser sostenido por los recursos disponibles de un área dada a un tiempo dado (646)

sistema de castas: dentro de los insectos, una sociedad altamente organizada en la cual cada miembro tiene una tarea específica biológicamente determinada (592)

catalizador: compuesto químico que promueve una reacción entre otros compuestos químicos al reducir la energía requerida para activar la reacción; puede tomar parte en la reacción pero se regenera en su forma original (61)

célula: la unidad básica de un ser vivo (28)

cuerpo celular: porción aumentada de una neurona que contiene al núcleo (549)

ciclo celular: secuencia ordenada de eventos en la vida de una célula eucariótica en división, compuesto de mitosis (M), crecimiento de interfase y fases de síntesis del DNA (G1, S y G2) (214)

atrofia del ciclo celular: paro abrupto del ciclo celular cuando las proteínas detectan errores o daño en el DNA que necesita ser reparado (229)

respuesta inmunológica mediada por células: la respuesta inmunológica en la cual los linfocitos altamente especializados circulan en la sangre y en los órganos linfoides y atacan y destruyen las células que transportan antígenos superficiales específicos (606)

respiración celular: serie de reacciones químicas por las cuales una célula viviente rompe los carbohidratos y obtiene energía de ellos (55)

teoría celular: teoría que dice que los organismos se componen de células y sus productos que estas células se derivan de células preexistentes (156)

pared celular: cubierta dura alrededor de la membrana plasmática de ciertas células, como en las plantas, muchas algas y algunos procariotas; en las plantas, la pared celular se construye parcialmente de celulosa (163)

sistema nervioso central (CNS): el cerebro y la médula espinal en los vertebrados (551)

centriolo: la estructura en las células animales y algunas otras compuesta por cilindros de nueve microtúbulos de triplete en un anillo; ayuda a organizar el ensamblado de microtúbulos durante la división celular (169)

centromere: the specialized region of a chromosome that holds two replicated chromosomal strands together and that attaches to the spindle in mitosis   (222)

cerebellum: the part of the vertebrate brain that is associated with regulating muscular coordination, balance, and similar functions   (556)

cerebrum: the largest portion of the brain in humans and many other animals; controls the higher mental functions such as learning   (554)

CFTR: cystic fibrosis transporter; the name of the gene found to be responsible for causing cystic fibrosis, a respiratory disorder that kills children; the gene encodes a transmembrane protein   (411)

chaparral: a coastal scrubland biome of dense, spiny evergreen shrubs found at midlatitudes; characterized by mild, rainy winters and long, hot, dry summers   (655)

chemical bond: the attraction between two atoms resulting from the sharing or transfer of outer electrons   (28)

chemical energy: energy stored in the structure of molecules   (54)

chemical reaction: change in chemical bonds that produces one or more new substances   (28)

chemoautotroph: an organism that derives energy from the oxidation of inorganic compounds, such as hydrogen sulfide (H_2S)   (102)

chemoheterotroph: an organism that must consume organic compounds for energy and carbon   (103)

chemosynthesis: a biochemical pathway that uses energy from the oxidation of inorganic substances to drive the formation of organic molecules   (55)

chlorophyll: the green pigments of plants and many microorganisms; converts light energy (via changes involving electrons) to chemical energy that is used in biological reactions   (105)

chloroplast: an organelle found in plants and photosynthetic protists; the site of photosynthesis   (104)

chorion: an embryonic membrane that surrounds all the other embryonic membranes in reptiles, birds, and mammals   (273)

chromatin: in eukaryotes, the chromosomal material (DNA and associated proteins) as it ordinarily appears in a cell's nucleus, with individual chromosomes indistinct   (220)

chromosome segregation: separation of the sister chromatids during mitosis in which each new nucleus receives one copy of each chromosome   (223)

centrómero: región especializada de un cromosoma que sostiene dos cadenas cromosómicas replicadas juntas y que se une al huso en la mitosis   (222)

cerebelo: parte del cerebro de los vertebrados que está asociada con la coordinación muscular reguladora, balance y funciones similares   (556)

corteza cerebral: la porción más grande del cerebro en los humanos y muchos otros animales; controla la funciones mentales más complejas tales como el aprendizaje   (554)

CFTR: transportador de fibrosis quística; el nombre del gene que se ha hallado responsable de causar fibrosis quística, un trastorno respiratorio que mata a niños: el gene codifica la formación de una proteína transmembránica   (411)

chaparral: bioma costero de matorrales con arbustos densos, espinosos y de hoja perenne, que se halla en latitudes medias; se caracteriza por inviernos moderados y lluviosos y veranos secos, calurosos y largos   (655)

enlace químico: atracción entre dos átomos que resulta de la transferencia o del compartir electrones externos   (28)

energía química: energía almacenada en la estructura de las moléculas   (54)

reacción química: cambio en los enlaces químicos que produce una o más sustancias nuevas   (28)

quimioautótrofo: organismo que obtiene energía de la oxidación de compuestos inorgánicos, tales como sulfuro de hidrógeno (H_2S)   (102)

quimioheterótrofo: organismo que debe consumir compuestos orgánicos para obtener energía y carbono   (103)

quimiosíntesis: ruta bioquímica que usa la energía de la oxidación de sustancias inorgánicas para impulsar la formación de moléculas orgánicas   (55)

clorofila: los pigmentos verdes de plantas y muchos microorganismos; convierte la energía luminosa (vía cambios que involucran electrones) a energía química que se usa en reacciones biológicas   (105)

cloroplasto: organelo que se halla en plantas y protistas fotosintéticos; el sitio de fotosíntesis   (104)

corión: membrana embrionaria que rodea a todas las demás membranas embrionarias en reptiles, aves y mamíferos   (273)

cromatina: en eucariotes, el material cromosómico (DNA y proteínas asociadas) como aparece comúnmente en un núcleo celular, con cromosomas individuales indistintos   (220)

segregación cromosómica: la separación de la cromátide hermana durante la mitosis, en la cual cada nuevo núcleo recibe una copia de cada cromosoma   (223)

cilia: short, hairlike cell appendages specialized for locomotion and formed from a core of nine outer doublet microtubules and two inner single microtubules (170)

cladistics: a systematic method of classification that relies on shared characteristics not found in other organisms (472)

class: in taxonomy, a group of related orders (469)

cleavage: the process of cell division in animal cells characterized by pinching of the plasma membrane; also, the rapid cell divisions without growth that take place during early embryonic development (266)

climax community: a stable, self-perpetuating community established by succession and sequential development and considered semipermanent; persists until interrupted or destroyed by environmental change (667)

clonal selection: a process in the immune system through which only cells that produce antibodies capable of binding to specific antigens are selected to divide (613)

clone: a lineage of genetically identical individuals produced asexually (318)

closed circulatory system: a type of internal transport in which blood is confined to vessels (193)

coagulate: clot (207)

codominant: the condition in which both alleles in a heterozygous organism are expressed (359)

codon: the basic unit of the genetic code; a sequence of three adjacent nucleotides in DNA or messenger RNA that encodes an amino acid (236)

coenzyme A: a coenzyme, a small molecule required for enzymatic activity, present in all cells; necessary for cell respiration and fatty-acid metabolism (138)

coevolution: the evolution of two species interacting with each other and reciprocally influencing each other's adaptations (504)

cohesion: binding together of molecules of the same substance (189)

colonize: occupy a new habitat (663)

colony: a distinct group of microorganisms growing together (174)

commensalism: a symbiotic relationship in which one species benefits and the other neither benefits nor is harmed by the relationship (640)

common pool resource: a resource that is shared by many people and is not controlled by any one person (670)

cilios: apéndices celulares cortos filamentosos especializados en la locomoción y formados a partir de un centro de nueve microtúbulos dobles externos y dos microtúbulos simples internos (170)

cladística: método sistemático de clasificación que se basa en características compartidas no encontradas en otros organismos (472)

clase: en taxonomía, un grupo de órdenes relacionados (469)

ruptura: proceso de división celular en las células animales caracterizado por una punción de la membrana plasmática; también, las divisiones celulares rápidas sin crecimiento que se llevan a cabo durante las etapas iniciales del desarrollo embrionario (266)

comunidad en clímax: comunidad autoperpetuable y estable, establecida por sucesión y desarrollo secuencial y considerada semipermanente; persiste hasta ser interrumpida o destruida por un cambio ambiental (667)

selección clonal: proceso en el sistema inmunológico a través del cual sólo las células que producen anticuerpos capaces de unirse a antígenos específicos son seleccionadas para dividirse (613)

clon: linaje de individuos idénticos genéticamente producidos asexualmente (318)

sistema circulatorio cerrado: tipo de transporte interno en el cual la sangre está confinada a los vasos (193)

coagular: cuajar (207)

codominante: condición en la cual se expresan ambos alelos en un organismo heterocigoto (359)

codón: la unidad básica del código genético; una secuencia de tres nucleótidos adyacentes en el DNA o el RNA mensajero que codifica para un aminoácido (236)

coenzima A: coenzima, una molécula pequeña requerida para la actividad enzimática, presente en todas las células; necesaria para la respiración celular y el metabolismo de los ácidos grasos (138)

coevolución: la evolución de dos especies que interactúan una con otra e influyen sobre las adaptaciones de cada una (504)

cohesión: la unión de moléculas de la misma sustancia (189)

colonizar: ocupar un nuevo hábitat (663)

colonia: grupo distinto de microorganismos que crecen juntos (174)

comensalismo: relación simbiótica en la cual una de las especies se beneficia y las otra ni se beneficia ni se perjudica (640)

recurso de propiedad comunitaria: recurso que es compartido por mucha gente y que no es controlado por nadie (670)

commons: a natural resource (such as a pasture, a forest, the ocean, or the air) that is shared by many but controlled by no one person   (670)

competitive exclusion principle: the principle that when the populations of two species compete for the same limited resource at the same time, one population will use the resource more efficiently, leading to local extinction of the other species   (639)

complement system: a group of blood proteins that can destroy pathogens   (621)

concentration gradient: a difference in the concentration of a substance over a distance   (81)

conditioning: training that leads an animal to associate a response with a stimulus   (580)

cone: a color receptor in the retina of the eye   (550)

conjugation: a process of exchanging genetic information in which chromosomes pass between two bacteria by means of a tube of cytoplasm that temporarily connects them   (324)

consumer: a heterotroph; an organism that feeds on other organisms or on their organic wastes   (56)

corpus luteum: the structure that forms from the tissues of a ruptured ovarian follicle; secretes female hormones (335)

correlation: a statistic that indicates how closely two variables are related   (588)

cotyledon: the single (monocot) or double (dicot) seed leaf of a flowering-plant embryo   (293)

covalent bond: a chemical bond formed by two atoms sharing a pair of electrons   (30)

crossing-over: during prophase I of meiosis, the breakage and exchange of corresponding segments of homologous chromosomes at one or more sites along their length, resulting in genetic recombination   (323)

cuticle: the waxy outer layer covering the surfaces of most land-dwelling plants, animals, and fungi   (91)

cyclins: a group of proteins whose function is to regulate the progression of a cell through the cell cycle and whose concentrations rise and fall throughout the cell cycle   (226)

cytochrome: an electron-carrying pigment in electron transport systems; cytochrome c is the most abundant (139)

cytokine: a molecule produced by one cell that regulates activities in another cell   (612)

cytokinesis: the division of the cytoplasm of a cell after nuclear division   (216)

bien comunitario: recurso natural (como un pastizal, un bosque, el océano, o el aire) que es compartido por muchos pero no es controlado por nadie   (670)

principio de exclusión competitiva: principio que dice que cuando dos especies compiten por la misma fuente limitada al mismo tiempo, una población usará el recurso con mayor eficacia, causando la extinción local de la otra especie   (639)

sistema complemento: grupo de proteínas de la sangre que puede destruir patógenos   (621)

gradiente de concentración: diferencia en la concentración de una sustancia a lo largo de una distancia   (81)

condicionamiento: el entrenamiento que hace que un animal asocie una respuesta con un estímulo   (580)

cono: receptor de color en la retina del ojo   (550)

conjugación: proceso de intercambio de información genética en el cual los cromosomas se pasan entre dos bacterias por medio de un tubo de citoplasma que las conecta temporalmente   (324)

consumidor: heterótrofo; un organismo que se alimenta de otros organismos o de sus desechos orgánicos   (56)

cuerpo lúteo: estructura que se forma de los tejidos de un folículo ovárico roto; secreta hormonas femeninas (335)

correlación: una estadística que indica el grado de relación entre dos variables   (588)

cotiledón: el cotiledón simple (monocotiledónea) o el doble (dicotiledónea) de un embrión vegetal en floración   (293)

enlace covalente: enlace químico formado por dos átomos que comparten un par de electrones   (30)

entrecruzamiento: durante la profase I de la meiosis, la ruptura e intercambio de los segmentos correspondientes de los cromosomas homólogos en uno o más sitios a lo largo de su longitud, que dan lugar a la recombinación genética   (323)

cutícula: capa externa cerosa que cubre las superficies de la mayoría de las plantas terrestres, animales y hongos (91)

ciclinas: grupo de proteínas cuya función es regular la progresión de una célula a través del ciclo celular y cuyas concentraciones se elevan y caen a lo largo del ciclo celular   (226)

citocromo: pigmento transportador de electrones en sistemas de transporte de electrones; el citocromo c es el más abundante   (139)

citoquina: molécula producida por una célula que regula actividades en otra célula   (612)

citoquinesis: la división del citoplasma de una célula después de la división nuclear   (216)

cytokinins: a group of plant growth regulators that promote cell division, stem and root growth, chlorophyll synthesis, and chloroplast development   (303)

cytoplasm: the entire contents of the cell, except the nucleus, bounded by the plasma membrane   (78)

cytoskeleton: a network of microtubules, microfilaments, and intermediate filaments that run throughout the cytoplasm of eukaryotic cells and serve a variety of mechanical and transport functions   (166)

cytosol: the gelatinlike portion of the cytoplasm that bathes the organelles of the cell   (166)

cytotoxic T lymphocyte (CTL): lymphocyte that recognizes and destroys infected cells   (612)

citoquininas: grupo de reguladores del crecimiento vegetal que promueve la división celular, el crecimiento del tallo y la raíz, la síntesis de clorofilas y el desarrollo de cloroplastos   (303)

citoplasma: los contenidos enteros de la célula, excepto el núcleo, unidos por la membrana plasmática   (78)

citoesqueleto: red de microtúbulos, microfilamentos y filamentos intermedios que se distribuyen a través del citoplasma de las células eucariotas y que sirven para una variedad de funciones mecánicas y de transporte (166)

citosol: porción de tipo gelatinoso del citoplasma que baña a los organelos de la célula   (166)

linfocito citotóxico T (CTL): linfocito que reconoce y destruye células infectadas   (612)

data (singular: *datum*): observations and experimental evidence bearing on a question or problem   (6)

decomposer: an organism that lives on decaying organic material, from which it obtains energy and nutrients   (56)

decomposition: the process of breaking substances down into smaller chemical units   (62)

dendrite: a structure that extends from a neuron and transmits electrical impulses toward the cell body   (549)

descent with modification: Darwin's first theory stating that newer life-forms in the fossil record are modified descendants of older species   (13)

desert: a biome characterized by lack of precipitation and by extreme temperature variation; may be hot or cold (654)

determination: the process in which a cell commits to a particular pathway of differentiation   (281)

differentiation: process in which cells become specialized for a specific structure and function via selective gene expression   (264)

diffusion: the movement of a substance down its concentration gradient from a more concentrated area to a less concentrated area   (81)

digestion: the process by which food breaks down into molecules that an organism can absorb or use   (67)

dihybrid: cross between individuals differing in two alleles (354)

diploid: a cell containing two sets of homologous chromosomes ($2n$)   (319)

datos: observaciones y evidencia experimental que apoyan una pregunta o problema   (6)

descomponedor: organismo que vive de material orgánico en descomposición, del cual obtiene energía y nutrientes   (56)

descomposición: el proceso de romper sustancias en unidades químicas más pequeñas   (62)

dendrita: estructura que se extiende de una neurona y transmite impulsos eléctricos hacia el cuerpo celular (549)

descendencia con modificación: la primera teoría de Darwin que establece que nuevas formas de vida en el registro fósil son descendientes modificados de especies más antiguas   (13)

desierto: bioma caracterizado por la escasez de precipitación y por la variación extrema de temperatura; generalmente, caluroso durante el día y frío por la noche   (654)

determinación: proceso en el cual la célula se compromete a una ruta particular de diferenciación (281)

diferenciación: proceso en el cual las células se vuelven especializadas para una estructura y función específicas mediante la expresión selectiva de genes   (264)

difusión: el movimiento de una sustancia debido a su gradiente de concentración de un área más concentrada a otra área menos concentrada   (81)

digestión: proceso por el cual la comida se descompone en moléculas que un organismo puede absorber o usar (67)

dihíbrido: cruce entre individuos que difieren en dos alelos   (354)

diploide: una célula que contiene dos juegos de cromosomas homólogos ($2n$)   (319)

disaccharide: a double sugar composed of two chemically bonded simple sugars (35)

division: in taxonomy, a group of similar classes; equal in the taxonomic hierarchy to a phylum (469)

DNA: deoxyribonucleic acid; the hereditary material of most organisms; a nucleic acid composed of deoxyribose sugar, phosphate groups, and four nitrogen-containing bases (41)

DNA polymerase: an enzyme that catalyzes the synthesis of a new DNA strand using one of the original strands as a template (218)

DNA-RNA hybridization: pairing of DNA molecules with RNA molecules by hydrogen bonds between complementary base pairs (278)

dominance hierarchy: a linear ranking of animals within the same species in which position dictates characteristic social behaviors (592)

dominant: referring to an allele that masks the presence of another allele of the same gene in a heterozygous organism (352)

disacárido: azúcar doble compuesto de dos azúcares simples unidos químicamente (35)

división: en taxonomía, un grupo similar de clases, igual en la jerarquía taxonómica a un filo (469)

DNA: ácido desoxirribonucleico; el material hereditario de la mayoría de los organismos, un ácido nucleico compuesto de un azúcar desoxirribosa, grupos fosfato y cuatro bases que contienen nitrógeno (41)

polimerasa del DNA: enzima que cataliza la síntesis de una cadena nueva de DNA usando una de las cadenas originales como una plantilla (218)

hibridación DNA-RNA: el apareamiento de moléculas de DNA con moléculas de RNA por puentes de hidrógeno entre pares de bases complementarias (278)

jerarquía de dominancia: una organización lineal dentro de las mismas especies en la cual la posición dicta comportamientos sociales característicos (592)

dominante: referente a un alelo que enmascara la presencia de otro alelo del mismo gene en un organismo heterocigoto (352)

ecosystem: a biological community and its abiotic environment (56)

ectoderm: the outer layer of cells in the gastrula stage of an animal embryo (266)

effector: a muscle or gland activated by nerve impulses or hormones (551)

ejaculation: expulsion of semen, sperm-bearing fluid, from the urethra (337)

electric potential: energy created by a difference in charge distribution (for example, as on two sides of a membrane) (561)

electron: a negatively charged particle that occurs in varying numbers in the electron clouds surrounding an atom's nucleus (26)

electron shell: the arrangement of electrons around an atom's nucleus according to the energy they contain; electrons with the least energy are in the shell closest to the nucleus, and those with more are in shells farther from the nucleus (26)

electron transport system: the process in which electrons transfer from one carrier molecule to another in photosynthesis and in cell respiration; results in storage of some of the energy in ATP molecules (133)

element: a substance composed of atoms that are chemically identical, alike in their numbers of protons (24)

ecosistema: comunidad biológica y su ambiente abiótico (56)

ectodermo: la capa externa de células en el estado de gástrula del embrión de un animal (266)

efector: músculo o glándula activado por impulsos nerviosos u hormonas (551)

eyaculación: expulsión de semen, fluido que contiene espermatozoides, desde la uretra (337)

potencial eléctrico: energía creada por una diferencia en la distribución de carga (por ejemplo, como en los dos lados de una membrana) (561)

electrón: partícula cargada negativamente que se presenta en cantidades variables en las nubes electrónicas que rodean al núcleo de un átomo (26)

capa electrónica: el arreglo de electrones alrededor del núcleo de un átomo de acuerdo con la energía que contengan; los electrones con menor energía están en la capa más cercana al núcleo y aquéllos con más están en las capas más alejadas del núcleo (26)

sistema de transporte de electrones: proceso en el cual los electrones se transfieren de una molécula transportadora a otra en la fotosíntesis y en la respiración celular; da como resultado el almacenamiento de la energía en las moléculas de ATP (133)

elemento: sustancia compuesta de átomos que son químicamente idénticos, iguales en el número de protones (24)

embryo: an organism in its earliest stages of development (262)

embryonic induction: the influence of one embryonic tissue on another, causing the second tissue to specialize (284)

endocytosis: the cellular uptake of materials in which the plasma membrane surrounds and engulfs extracellular materials (85)

endoderm: the innermost layer of the three germ layers, primary tissue layers, in an animal embryo (266)

endoplasmic reticulum (ER): an extensive membranous network in eukaryotic cells; composed of ribosome-studded (rough) and ribosome-free (smooth) regions (166)

endosperm: a nutrient-rich structure formed by the union of a sperm cell and a large cell having two nuclei; double fertilization is complete when a second sperm cell fertilizes the egg; the endosperm provides nourishment to the developing embryo in seeds of flowering plants (293)

endosymbiont: an organism having a mutually beneficial symbiotic relationship with a host organism while living in the host's body (456)

endosymbiont hypothesis: the hypothesis suggesting that eukaryotic cells evolved from prokaryotes living symbiotically inside other prokaryotes (457)

energy pyramid: a representation of the trophic structure of an ecosystem in which a decrease in the area at each level of the diagram illustrates the loss of energy available to those organisms living in that trophic level (638)

entropy: a measure of the degree of disorganization of a system, that is, how much energy in a system has become so dispersed that it is no longer available to do work (59)

enzyme: a protein or part-protein molecule made by an organism and used as a catalyst in a specific biochemical reaction (61)

epidermis: the outer covering of animals and plants (176)

epiglottis: flap of cartilaginous tissue at the base of the tongue in mammals; prevents food from entering the trachea, the airway to the lungs, during swallowing (70)

epistasis: condition in which one gene affects the expression of another, independently inherited gene (380)

erythrocyte: a red blood cell; a type of cell in vertebrates that contains hemoglobin and carries oxygen (204)

embrión: organismo en sus etapas iniciales de desarrollo (262)

inducción embriónica: influencia de un tejido embrionario en otro que causa la especialización del segundo tejido (284)

endocitosis: la ingestión celular de materiales en la cual la membrana plasmática rodea al núcleo y envuelve materiales extracelulares (85)

endodermo: capa más interna de las tres capas germinales, capas de tejido primario, en un embrión animal (266)

retículo endoplásmico (ER): red membranosa extensiva en células eucariotas; compuesta de regiones salpicadas de ribosomas (rugosas) y libres de ribosomas (lisas) (166)

endosperma: estructura rica en nutrientes formada por la unión de una célula de esperma y una gran célula que contiene dos núcleos, la doble fecundación se completa cuando un segunda célula espermática fecunda el óvulo; el endosperma provee de los nutrientes para el embrión en desarrollo en las semillas de las plantas con flores (293)

endosimbionte: organismo que tiene una relación simbiótica benéfica mutua con un organismo huésped mientras vive en el cuerpo del huésped (456)

hipótesis endosimbionte: hipótesis que sugiere que las células eucariotas evolucionaron de las procariotas que viven simbióticamente dentro de otros procariotas (457)

pirámide energética: una representación de la estructura trófica de un ecosistema en la cual un decremento en el área de cada nivel del diagrama ilustra la pérdida de energía disponible para aquéllos organismos que viven en ese nivel trófico (638)

entropía: medida del grado de desorganización de un sistema, es decir, la cantidad de energía en un sistema que se ha vuelto tan dispersa que ya no está disponible para realizar trabajo (59)

enzima: proteína o parte de una molécula proteica hecha por un organismo y usada como un catalizador en una reacción bioquímica específica (61)

epidermis: la cubierta exterior de animales y plantas (176)

epiglotis: lengüeta de tejido cartilaginoso en la base de la lengua de los mamíferos; evita que los alimentos entren a la tráquea, la vía respiratoria hacia los pulmones, cuando se ingieren (70)

epistasis: condición en la cual un gene afecta la expresión de otro gene heredado independientemente (380)

eritrocito: glóbulo rojo; un tipo de célula en vertebrados que contiene hemoglobina y transporta oxígeno (204)

estrogen: family of hormones that stimulate the development of female secondary sexual characteristics (334)

estrus: a limited period of sexual receptivity that occurs around ovulation in female mammals, except higher primates (524)

ethylene: a gaseous plant growth regulator that promotes fruit ripening while inhibiting plant growth in roots and stems (304)

Eubacteria: bacterial group including the cyanobacteria; sometimes called "true bacteria," they differ from archaebacteria in their ribosomal RNA and transfer RNA and in other ways (476)

eukaryote: an organism whose cells have a membrane-enclosed nucleus and organelles; a protist, a fungus, a plant, or an animal (160)

excision repair: a DNA-repair process where enzymes remove a damaged portion of DNA, synthesize a replacement section in place, and attach it to the neighboring DNA segments (221)

exocytosis: the release of macromolecules from a cell by the fusion of vesicles with the plasma membrane (85)

exon: a segment of DNA that is transcribed into RNA and translated into protein, specifying the amino acid sequence of a polypeptide (245)

exotic: describes a species that is not native to a given area but was either intentionally transplanted from another region or introduced accidentally (663)

exponential growth: rapid growth in which some value, such as population size, increases by a certain percentage of the total during a given period of time (645)

external fertilization: the joining of gametes outside the bodies of the two parents, as in many fishes and amphibians (330)

extracellular digestion: the breakdown of nutrient molecules outside of cells (67)

estrógeno: familia de hormonas que estimula el desarrollo de las características sexuales secundarias femeninas (334)

estro: periodo limitado de receptividad sexual que ocurre alrededor de la ovulación en mamíferos femeninos, excepto entre los primates más evolucionados (524)

etileno: regulador gaseoso del crecimiento de las plantas que promueve la maduración de las frutas mientras que inhibe el crecimiento de la planta en raíces y tallos (304)

eubacteria: grupo de bacterias que incluyen a las cianobacterias, algunas veces llamadas "bacterias verdaderas"; difieren de las arquebacterias en su RNA ribosomal y de transferencia y de otras maneras (476)

eucariota: cuyas células tienen un núcleo encerrado en una membrana y organelos; un protista un hongo, una planta, un animal (160)

reparación de escisión: proceso de reparación del DNA donde las enzimas eliminan una porción dañada del DNA, sintetizan una sección de reemplazo en su lugar y la unen a los segmentos vecinos de DNA (221)

exocitosis: liberación de macromoléculas de una célula por la fusión de vesículas con la membrana plasmática (85)

exón: segmento del DNA que se transcribe al RNA y se traduce en una proteína, especificando la secuencia de aminoácidos de un polipéptido (245)

exótico: describe una especie no oriunda de un área dada pero que fue transplantada intencionalmente de otra región o introducida accidentalmente (663)

crecimiento exponencial: crecimiento rápido en el cual algún valor, como el tamaño de la población, aumenta por cierto porcentaje del total durante un periodo dado de tiempo (645)

fecundación externa: la unión de gametos fuera de los cuerpos de los dos progenitores, como en muchos peces y anfibios (330)

digestión extracelular: la ruptura de moléculas de nutrientes fuera de las células (67)

F₁ (first filial) generation: the first generation of hybrid offspring in a genetic cross (352)

F₂ (second filial) generation: offspring resulting from interbreeding of the hybrid F₁ generation (352)

facilitated diffusion: the spontaneous passage of molecules and ions, bound to specific carrier proteins, across a biological membrane down their concentration gradients (84)

F₁ primera generación filial: la primera generación de progenie híbrido en un cruce genético (352)

F₂ segunda generación filial: progenie que resulta de un apareamiento cruzado del híbrido de la primera generación (352)

difusión facilitada: paso espontáneo de moléculas y iones, enlazadas a proteínas transportadoras específicas, a través de una membrana biológica por sus gradientes de concentración (84)

facultative aerobe: an organism that is normally anaerobic but can also grow in the presence of oxygen (140)

FAD: flavin adenine dinucleotide; combines with two hydrogen atoms during cell respiration to form FADH$_2$ (133)

familial hypercholesterolemia: a genetically dominant condition characterized by excess cholesterol in the bloodstream, which can lead to heart disease, hardening of the arteries, heart attacks, and strokes (203)

family: in taxonomy, a group of related genera (468)

feces: the waste material expelled from the digestive tract (70)

feedback regulation: the regulation of the activity of an enzyme by one of its products (96)

fermentation: the release of energy during the chemical breakdown of food, especially sugars, in the absence of oxygen (136)

fertilization: the union of the nucleus of an ovum, the unfertilized egg, and a sperm nucleus (262)

fetus: an older human embryo (273)

fibrin: an insoluble, fibrous protein that forms a network of fibers around which a blood clot develops (208)

first law of thermodynamics: the law derived from the principle of the conservation of energy stating that energy can be neither created nor destroyed, but it can be transferred or transformed (58)

fixed action pattern: a highly stereotyped behavior that is innate and species-specific (579)

flagella (singular: *flagellum*): long cellular appendages specialized for locomotion; in eukaryotes, they contain a core of nine outer doublet microtubules and two single inner microtubules; many protists and certain animal cells have flagella (162)

food web: the overlapping food chains of an ecosystem (56)

founder effect: colonization by a small population that differs in its genetic makeup from its source population (426)

frameshift mutation: the insertion or deletion of one or more nucleotides in a gene, causing disruption of the reading frame (the correct groupings of three nucleotide sequences, called the codons, which will produce the encoded gene product) (407)

free energy: energy that is available to do work (54)

aerobio facultativo: un organismo que es normalmente anaeróbico pero que puede crecer en presencia de oxígeno (140)

FAD: dinucleótido de flavina y adenina; se combina con dos átomos de hidrógeno durante la respiración celular para formar FADH$_2$ (133)

hipercolesterolemia familiar: condición genéticamente dominante caracterizada por un exceso de colesterol en el torrente sanguíneo, la cual conduce a enfermedades cardíacas, endurecimiento de arterias, ataques al corazón e infartos (203)

familia: en taxonomía, un grupo de géneros relacionados (468)

heces: el material de desecho expelido por el tracto digestivo (70)

regulación por retroalimentación: regulación de la actividad de una enzima por uno de sus productos (96)

fermentación: liberación de energía durante la descomposición química de proteínas, especialmente azúcares, en la ausencia de oxígeno (136)

fecundación: la unión del núcleo de un óvulo, el huevo sin fecundar y un núcleo de espermatozoides (262)

feto: un embrión humano más maduro (273)

fibrina: una proteína fibrosa insoluble que forma una red de fibras alrededor de la cual se desarrolla un coágulo de sangre (208)

primera ley de la termodinámica: la ley derivada del principio de conservación de la energía que establece que la energía no puede ser creada ni destruida, pero puede ser transferida o transformada (58)

patrón de acción fija: un comportamiento altamente estereotipado que es innato y específico de una especie (579)

flagelos: apéndices celulares largos especializados en la locomoción, en eucariotas, contienen un centro de nueve microtúbulos dobles externos y dos microtúbulos sencillos internos; muchos protistas y ciertas células animales tienen flagelos (162)

red alimenticia: cadenas alimenticias traslapadas de un ecosistema (56)

efecto fundador: colonización de una pequeña población que difieren en su composición genética de su población de procedencia (426)

mutación de desplazamiento de marco: la inserción o borrado de uno o más nucleótidos en un gene, trastocando el marco de lectura (los agrupamientos correctos de tres secuencias de nucleótidos, llamados codones, los cuales producirán el producto genético codificado) (407)

energía libre: energía disponible para hacer un trabajo (54)

free radical: a highly reactive molecular fragment that contains one or more unpaired electrons   (673)

Fungi: a kingdom of heterotrophic organisms that develop spores; fungi feed by absorbing, rather than ingesting, other organic matter; many are decomposers (478)

radical libre: un fragmento molecular altamente reactivo que contiene uno o más electrones no apareados   (673)

hongo: un reino de organismos heterótrofos que desarrolla esporas; los hongos se alimentan por absorción, más que por ingestión, de otro material orgánico; muchos son descomponedores   (478)

G_0 phase: a resting stage of the cell cycle in which DNA replication and cell division stop   (215)

G_1 phase: the first growth phase of the cell cycle starting just after offspring cells form   (215)

G_2 phase: the second growth phase of the cell cycle beginning after DNA synthesis   (215)

gamete: a haploid reproductive cell, either an ovum (the unfertilized egg) or a sperm, formed by meiosis   (262)

ganglia (singular: *ganglion*): clusters of nerve cell bodies (557)

gastrin: a digestive hormone secreted by the stomach lining; stimulates the secretion of fluid by gastric glands in the stomach   (71)

gastrula: the two-layered, cup-shaped embryonic stage (266)

gene: the fundamental physical unit of heredity, which transmits a set of specifications from one generation to the next; a segment of DNA that codes for a specific product   (44)

gene amplification: an increase in the number of copies of a particular segment of DNA   (409)

gene flow: the loss or gain of alleles in a population due to migration   (425)

gene pool: all the genes in a population at any one time (418)

gene therapy: the introduction of a gene into a cell to correct a hereditary disorder   (410)

genetic anticipation: the phenomenon in which the severity of symptoms associated with a genetic disorder increases (or the symptoms appear at an earlier age) in successive generations   (382)

genetic code: the "language" of the genes in which the nucleotide sequence of DNA (in triplets, or codons) specifies the amino acid sequences of proteins   (234)

genetic drift: changes due to chance in the gene pool of a small population   (426)

genome: the total genetic content of a haploid cell from any given species; the full set of genes in any individual   (350)

fase G_0: una fase de reposo en el ciclo celular en el cual la replicación del DNA y la división celular se detienen (215)

fase G_1: la primera fase de crecimiento del ciclo celular que empieza justo después de la formación de las células descendientes   (215)

fase G_2: la segunda fase de crecimiento del ciclo celular que empieza después de la síntesis del DNA   (215)

gameto: célula reproductora haploide, ya sea un óvulo (el huevo sin fecundar) o un espermatozoide, formado por meiosis   (262)

ganglios: grupos de cuerpos celulares nerviosos   (557)

gastrina: hormona digestiva secretada por las paredes del estómago; estimula la secreción de fluido por las glándulas gástricas en el estómago   (71)

gástrula: el estado embrionario en forma de copa y de dos capas   (266)

gene: la unidad física fundamental de la herencia, la cual transmite un conjunto de especificaciones de una generación a la siguiente; un segmento de DNA que codifica para un producto específico   (44)

amplificación genética: aumento en el número de copias de un segmento particular de DNA   (409)

flujo genético: pérdida o ganancia de alelos en una población debido a la migración   (425)

caudal de genes: todos los genes en una población en cualquier tiempo   (418)

terapia genética: introducción de un gene en una célula para corregir un trastorno hereditario   (410)

anticipación genética: fenómeno en el cual aumenta la gravedad de los síntomas asociados con un trastorno genético (o los síntomas aparecen a una edad más temprana) en generaciones sucesivas   (382)

código genético: el "lenguaje" de los genes en el cual la secuencia de nucleótidos del DNA (en tripletes o codones) especifica las secuencias de aminoácidos de las proteínas   (234)

deriva genética: cambios debidos al azar en el caudal de genes de una población pequeña   (426)

genoma: contenido genético total de una célula haploide de cualquier especie dada; el conjunto completo de genes en cualquier individuo   (350)

genomic imprinting: activation or inactivation of certain genes that depends on the gene's location on a chromosome and its parental origin   (377)

genotype: the genetic makeup of an organism   (352)

genus (plural: *genera*): in taxonomy, a group of related species   (468)

germination: the sprouting of a seed   (294)

germ-line therapy: the repair or replacement of a defective gene within the gamete-forming tissues, which produces an inheritable change in an organism's genetic constitution   (410)

gestation: pregnancy   (335)

gibberellins: a group of plant growth regulators that stimulate elongation of stems and leaves; trigger seed germination, and, with auxins, stimulate development of the fruit, which is the ripened ovary   (303)

glial cell: a nonconducting cell of the nervous system that supports, protects, or insulates neurons   (557)

glomerular capsule: the cup of a nephron, which filters blood; also called Bowman's capsule   (94)

glomerulus: a ball of capillaries surrounded by a glomerular capsule in the nephron; the site of filtration in the kidneys   (94)

glycolipid: a lipid covalently linked to a sugar or polysaccharide; an important part of animal cell membranes   (80)

glycolysis: the initial breakdown of a carbohydrate, usually glucose, into smaller molecules at the beginning of cell respiration or fermentation   (132)

glycoprotein: a protein linked to a sugar or polysaccharide; component of receptor molecules on the outer surface of cells   (80)

Golgi apparatus: an organelle in eukaryotic cells consisting of stacked membranes that modify and package materials in vesicles for export from the cell   (167)

gonad: reproductive organ in an animal; produces gametes and sometimes hormones; ovary in females and testis in males   (329)

gradualism: a model of evolution in which change takes place at a slow, steady rate, resulting in a steady increase in biological diversity   (513)

gravitropism: a positive or negative response of an organism to gravity; for example, the tendency of plant roots to grow downward in the direction of gravity and plant shoots to grow upward against gravity   (309)

impronta genética: activación o inactivación de ciertos genes que depende de la posición del gene en un cromosoma y de su procedencia   (377)

genotipo: constitución genética de un organismo   (352)

género: en taxonomía, un grupo de especies relacionadas   (468)

germinación: aparición de brotes de una semilla   (294)

terapia con células reproductoras: la reparación o reemplazo de un gene defectuoso dentro de los tejidos formadores de gametos, lo que produce un cambio heredable en la constitución genética de un organismo   (410)

gestación: embarazo   (335)

giberelinas: grupo de reguladores del crecimiento de plantas que estimulan el alargamiento de los tallos y las hojas, desencadenan la germinación de las semillas y, junto con las auxinas, estimulan el desarrollo de la fruta, que es el ovario madurado   (303)

célula glial: una célula no conductora del sistema nervioso que asiste, protege y aísla a las neuronas   (557)

cápsula glomerular: la copa de una nefrona, que filtra la sangre; también llamada cápsula de Bowman   (94)

glomérulo: masa redonda de capilares rodeada por una cápsula glomerular en la nefrona; el lugar donde se lleva a cabo la filtración en los riñones   (94)

glicolípido: lípido unido covalentemente a un azúcar o polisacárido; una parte importante de las membranas en las células animales   (80)

glicólisis: la descomposición inicial de un carbohidrato, usualmente glucosa, en moléculas más pequeñas al inicio de la respiración celular o fermentación   (132)

glicoproteína: proteína unida a un azúcar o polisacárido; componente de moléculas receptoras en la superficie externa de las células   (80)

aparato de Golgi: organelo en las células eucariotas que está formado por membranas apiladas que modifican y empacan materiales en las vesículas para exportarlos desde la célula   (167)

gónada: órgano reproductor en un animal; produce gametos y en ocasiones hormonas; ovarios en hembras y testículos en machos   (329)

gradualismo: modelo de evolución en el cual el cambio se lleva a cabo a un ritmo lento y constante, resultando en un incremento sostenido en la diversidad biológica   (513)

gravitropismo: respuesta positiva o negativa de un organismo a la gravedad; por ejemplo, la tendencia de las raíces de una planta a crecer hacia abajo en la dirección de la gravedad y de los retoños de la planta a crecer hacia arriba, contra la gravedad   (309)

ground tissue: tissue that makes up the bulk of a young plant, filling in the space between the dermal (outer surface tissue) and vascular tissue systems (296)

tejido fundamental: tejido que forma la mayor parte de una planta joven, llenando el espacio entre los tejidos dérmicos (tejido de la superficie externa) y los sistemas de tejidos vasculares (296)

habitat: type of place where an organism lives (56)

hábitat: tipo de lugar donde vive un organismo (56)

habituation: a simple type of learning that involves the loss of sensitivity to stimuli due to repeated exposure (580)

habituación: tipo de aprendizaje simple que involucra la pérdida de sensibilidad a un estímulo debido a una constante exposición (580)

haploid: a cell, a nucleus, or an organism containing only one set of unpaired chromosomes *(n)* (319)

haploide: una célula, un núcleo o un organismo que contiene solamente un conjunto de cromosomas no apareados *(n)* (319)

Hardy-Weinberg model: the model stating that, if certain conditions are met, the frequencies of different alleles of the same gene do not change from one generation to the next (421)

modelo de Hardy-Weinberg: modelo que establece que, si se cumplen ciertas condiciones, las frecuencias de alelos diferentes del mismo gene no cambian de una generación a la siguiente (421)

hemoglobin: the pigment of red blood cells; binds oxygen (204)

hemoglobina: el pigmento de los glóbulos rojos sanguíneos; se une con oxígeno (204)

herbivore: a plant-eating consumer; one of the class of consumers most closely associated with producers (637)

herbívoro: consumidor de plantas; uno de los tipos de consumidores más estrechamente relacionado con los productores (637)

heritability: an estimate of the importance of genetic factors in explaining how a trait or behavior varies within a specific population (587)

heredabilidad: cálculo de la importancia de los factores genéticos en la explicación de cómo una característica o comportamiento varía dentro de una población específica (587)

heterotroph: an organism that obtains carbon compounds from other organisms (54)

heterótrofo: organismo que obtiene compuestos de carbono de otros organismos (54)

heterotroph hypothesis: the hypothesis suggesting that the first life-forms used the supply of naturally occurring organic compounds for food (446)

hipótesis heterótrofa: hipótesis que sugiere que las primeras formas de vida usaron, como alimento, los compuestos orgánicos presentes de manera natural (446)

heterozygous: having two different alleles for a given trait (353)

heterocigoto: que tiene dos diferentes alelos para un rasgo dado (353)

HLA: human leukocyte antigens; the self proteins that T cells recognize (621)

HLA: antígenos de leucocitos humanos; las propias proteínas que reconocen las células T (621)

homeobox: a short DNA sequence that is virtually identical in certain homeotic genes; protein products of these genes regulate patterns of cell differentiation in a wide variety of organisms (270)

caja homeótica: secuencia corta de DNA que es virtualmente idéntica en ciertos genes homeóticos; productos proteicos de estos genes regulan los patrones de diferenciación celular en una amplia variedad de organismos (270)

homeostasis: the tendency for an organism to maintain a relatively stable internal environment by regulating its metabolism and adjusting to its environment (93)

homeostasis: la tendencia de un organismo por mantener un ambiente interno relativamente estable mediante la regulación de su metabolismo y ajustándose a su ambiente (93)

homeotic gene: a gene that determines which body parts are made at which locations on a developing organism; turns other genes on and off (270)

gene homeótico: gene qué determina qué partes del cuerpo y su ubicación se han de producir en un organismo en desarrollo; activa y desactiva a otros genes (270)

hominid: a primate of the family Hominidae, which includes modern humans, earlier members of the genus *Homo*, and australopithecines   (524)

homologous: of anatomical structures, sharing a common ancestor although the structure may look and function differently, such as a bird's wing and a reptile's forelimb; of chromosomes, carrying the same genes   (320)

homology: similarity of biological structures that results from evolution from a common ancestor   (466)

Homo sapiens: the human species   (527)

homozygous: having two identical alleles for a given trait (353)

hormone: a substance, secreted by cells or glands, that regulates the activities of cells and organs elsewhere in the body; a chemical messenger   (238)

Hox **genes:** a group of homeotic genes found in all animals; help establish the anterior-posterior axis (head to tail); named *Hox* genes for the *h*omeob*ox*es they contain   (271)

hydrogen bond: a weak attraction between hydrogen atoms and oxygen, nitrogen, or fluorine atoms; holds together the strands of DNA in their double helix (31)

hydrolysis: the splitting of a molecule by reaction with water   (146)

hydrophobicity: the tendency to repel water; substances that are hydrophobic are nonpolar and cannot hydrogen bond to water   (39)

hydrothermal: referring to hot water, such as that produced in the hydrothermal volcanic vents in the sea floor   (455)

hypothesis: a statement suggesting an explanation for an observation or an answer to a scientific problem   (8)

homínido: un primate de la familia Hominidae, que incluye a los humanos modernos, a los primeros miembros del género *Homo* y a los australopitecos   (524)

homólogo: en relación con estructuras anatómicas, que comparte un antecesor común, aunque la estructura pueda parecer y funcionar de manera diferente, tal como las alas de un ave y la extremidad delantera de los reptiles; en relación con cromosomas, aquéllos que poseen los mismos genes   (320)

homología: similaridad de estructuras biológicas que resulta de al evolución a partir de un antepasado común (466)

Homo sapiens: la especie humana   (527)

homocigoto: que tiene dos alelos idénticos para un rasgo dado   (353)

hormona: una sustancia, secretada por células o glándulas, que regula las actividades de células y órganos en cualquier otra parte del cuerpo; un mensajero químico   (238)

genes *Hox*: grupo de genes homeóticos que se hallan en todos los animales; ayuda a establecer el eje anterior-posterior (cabeza a cola); llamados genes *Hox* por el nombre en inglés de las cajas homeóticas (*h*omeob*ox*es) que contienen   (271)

puente de hidrógeno: atracción débil entre los átomos de hidrógeno y oxígeno, nitrógeno o átomos de flúor; mantiene unidas las cadenas de DNA en su doble hélice   (31)

hidrólisis: división de una molécula debido a una reacción con agua   (146)

hidrofobicidad: tendencia a repeler agua; las sustancias que son hidrofóbicas son no polares y no pueden unirse al agua por puentes de hidrógeno   (39)

hidrotérmico: se refiere al agua caliente, como aquélla producida por chimeneas volcánicas hidrotérmicas en el fondo del mar   (455)

hipótesis: declaración que sugiere la explicación de una observación o una respuesta a un problema científico (8)

identical twins: twins that develop from a single zygote (430)

immune memory: long-term resistance to previously encountered antigens   (607)

immune surveillance: process in which the immune system attacks small cancers before they can grow to a size that threatens health   (611)

gemelos idénticos: gemelos que se desarrollan de un solo cigoto   (430)

memoria inmunológica: resistencia a largo plazo contra antígenos reconocidos   (607)

vigilancia inmunológica: proceso en el cual el sistema inmunológico ataca pequeños cánceres antes de que puedan crecer hasta alcanzar un tamaño que amenace la salud   (611)

imprinting: a type of learning that can occur only during a specific period early in life (579)

inbreeding: sexual reproduction of closely related organisms (426)

inbreeding depression: reduction in vigor that results from extensive inbreeding (428)

incomplete dominance: pattern of gene expression in which the phenotype of a heterozygous individual is intermediate between those of the parents (358)

infection: an invasion of the body by pathogens (604)

inference: conclusion that follows logically from some form of evidence (26)

inflammation: a generalized response of vertebrates to tissue damage or foreign materials in which capillaries near an injury become more permeable, causing redness and swelling (605)

ingestion: the process of taking a substance from the environment, usually food, into the body (69)

innate: inborn; genetically inherited (578)

internal fertilization: the joining of gametes inside the female's body (330)

interneuron: a neuron that transmits nerve impulses from one neuron (such as a sensory neuron, which responds to stimuli) to another (such as a motor neuron, which carries information from the brain to muscles or glands) (557)

interphase: the parts of the eukaryotic cell cycle between cell divisions consisting of growth and DNA synthesis phases, G1, S, and G2 (214)

intertidal zone: the area of the seashore between the water levels at high tide and low tide (661)

intracellular digestion: the breakdown of nutrients within a cell (68)

intron: a segment of DNA that is transcribed into precursor messenger RNA; removed before the mRNA leaves the nucleus (245)

in vitro: literally means "in glass"; refers to laboratory procedures done in test tubes or in petri dishes (338)

involuntary: referring to a process or an activity that goes on without any conscious control (551)

ion: an atom or a molecule that has either gained or lost one or more electrons, giving it a positive or negative charge (30)

ionic bond: a chemical bond formed by the attraction between oppositely charged ions (30)

impronta: tipo de aprendizaje que sólo puede ocurrir durante un periodo específico a temprana edad (579)

endogamia: reproducción sexual de organismos estrechamente relacionados (426)

depresión endogámica: reducción en vigor que resulta de una endogamia excesiva (428)

dominancia incompleta: patrón de expresión genética en el cual el fenotipo de individuos heterocigotos es intermedio entre los de sus padres (358)

infección: invasión de agentes patógenos en un ser vivo (604)

inferencia: conclusión que proviene lógicamente de una forma de evidencia (26)

inflamación: respuesta generalizada de vertebrados al daño tisular o a materiales foráneos en la cual los capilares cercanos a una herida se vuelven más permeables, causando enrojecimiento e hinchazón (605)

ingestión: el proceso de introducir una sustancia de un ambiente a otro, usualmente comida, dentro del cuerpo (69)

innato: de nacimiento; heredado genéticamente (578)

fecundación interna: la unión de gametos dentro del cuerpo de la hembra (330)

interneurona: neurona que transmite impulsos nerviosos desde una neurona (como una neurona sensorial, la cual responde al estímulo) a otra (como una neurona motora, que transporta información del cerebro a los músculos o glándulas) (557)

interfase: las partes de un ciclo celular eucariota entre las divisiones celulares que consisten en las fases de crecimiento y de síntesis de DNA, G1, S y G2 (214)

zona intermareal: el área de la orilla del mar entre los niveles de agua en la marea alta y marea baja (661)

digestión intracelular: la descomposición de nutrientes dentro de una célula (68)

intrón: segmento del DNA que es transcrito dentro del RNA mensajero precursor; eliminado antes de que mRNA salga del núcleo (245)

in vitro: literalmente significa "en vidrio"; se refiere a los procedimientos de laboratorio hechos en tubos de ensayo o cajas petri (338)

involuntario: se refiere a un proceso o actividad que continúa sin ningún control consciente (551)

ion: átomo o molécula que ha ganado o perdido uno o más electrones, recibiendo una carga positiva o negativa (30)

enlace iónico: enlace químico formado por la atracción entre iones cargados positivamente (30)

isotope: one of multiple forms of an element having the same atomic number but a different atomic mass (differing in the number of neutrons) (27)

isótopo: una de las múltiples formas de un elemento que tiene el mismo número atómico pero una diferente masa atómica (difiere en el número de neutrones) (27)

karyotype: a display of the image of a set of chromosomes sorted by number (349)

cariotipo: despliegue de la imagen de un conjunto de cromosomas clasificados por número (349)

kidney: an organ that regulates water and salt levels, filters water and wastes from the blood, and excretes the end products (94)

riñón: órgano que regula el agua y los niveles salinos, filtra el agua y los desechos de la sangre y excreta los productos finales (94)

kinetochore: a disklike structure on the centromere; links chromosomes to the mitotic spindle (224)

cinetocoro: estructura en forma de disco en el centrómero, une cromosomas al huso mitótico (224)

kingdom: a taxonomic group composed of similar phyla or divisions; the system used in this book divides all living things into five kingdoms (470)

reino: grupo taxonómico compuesto de filos o divisiones similares; el sistema usado en este libro divide todos los seres vivos en cinco reinos (470)

Krebs cycle: the cycle in cell respiration that completes the breakdown of the intermediate products of glycolysis, releasing energy; also, a source of carbon skeletons for use in biosynthesis reactions (132)

ciclo de Krebs: ciclo en la respiración celular que completa la descomposición de productos intermedios de la glicólisis, liberando energía; también, fuente de cadenas carbonadas para usarse en reacciones de biosíntesis (132)

lactate: in mammals, to secrete milk; in chemistry, the ion of a 3-carbon acid (lactic acid) formed from pyruvic acid in fermentation (135)

lactar/lactato: en mamíferos, secretar leche; en química, el ión de un ácido de 3 carbonos (ácido láctico) formado de ácido pirúvico en fermentación (135)

lactic acid: substance that forms from pyruvic acid when glycolysis occurs in the absence of oxygen (136)

ácido láctico: sustancia que se forma del ácido pirúvico cuando ocurre la glicólisis en ausencia de oxígeno (136)

lactic-acid fermentation: an anaerobic pathway producing ATP when the conversion of pyruvate to lactate produces NAD+, which cycles back through glycolysis (136)

fermentación ácido–láctica: ruta anaerobia que produce ATP cuando la conversión de piruvato a lactato produce NAD+, el cual se cicla de regreso mediante la glicólisis (136)

lamina: a thin membrane or thin tissue (557)

lámina: membrana o tejido delgado (557)

larva: an immature stage of development in offspring of many types of animals, especially arthropods and some aquatic organisms (268)

larva: estado inmaduro de desarrollo de la progenie de muchos tipos de animales, especialmente artrópodos y algunos organismos acuáticos (268)

law of conservation of energy: the law stating that energy can be neither created nor destroyed, only changed from one form into another (58)

ley de conservación de la energía: ley que establece que la energía no se puede crear ni destruir; no se aplica al nivel subatómico (58)

law of conservation of matter: the law stating that matter can be neither created nor destroyed; does not hold true at the subatomic level (29)

ley de conservación de la materia: ley que establece que la materia no se puede crear ni destruir; no se aplica al nivel subatómico (29)

leukocyte: a white blood cell; any of the nucleated cells in the blood plasma that are the first line of defense against invading microorganisms; examples include lymphocytes (B cells, T cells) and macrophages (205)

leucocito: glóbulo blanco; cualquiera de las células nucleadas en el plasma sanguíneo que son la primera línea de defensa en contra de microorganismos invasores; ejemplos incluyen los linfocitos (células B, células T) y los macrófagos (205)

light reactions: the energy-capturing reactions in photosynthesis (106)

reacciones luminosas: reacciones de captación de energía en la fotosíntesis (106)

lignin: a hard organic compound that gives strength and rigidity to plants because it binds to and supports cellulose fibers in the cell walls   (187)

limiting factor: an environmental factor such as food, temperature, water, or sunlight that restricts growth, metabolism, or population size   (116)

lipase: a fat-digesting enzyme   (72)

lipid: a fat, an oil, a wax, or a fatlike compound that usually has fatty acids in its molecular structure; an important component of the plasma membrane   (36)

logistic growth: growth typically shown by populations in the natural world; the population size increases rapidly at first, then more slowly, finally leveling off at a value called the carrying capacity   (645)

lymphatic: referring to the vessels, nodes, and tissues of an organ system, which collects and returns fluid to the circulatory system and helps defend the body against infection   (206)

lymphocyte: a type of small leukocyte important in the immune response; examples include B cells and T cells   (606)

lymphoid organ: any organ where lymphocytes develop and collect   (609)

lysogenic: referring to a bacteriophage that infects a bacterium, inserts its DNA into the host's DNA, and replicates with the host   (255)

lysosome: a cell vesicle that contains digestive enzymes   (169)

lytic: referring to a virus that infects and destroys (lyses) susceptible cells   (255)

lignina: compuesto orgánico duro que da la fuerza y la rigidez a las plantas porque se une y sostiene a las fibras de celulosa en las paredes celulares   (187)

factor limitante: factor ambiental como la comida, temperatura, el agua o la luz solar que restringe el crecimiento, el metabolismo o el tamaño de una población   (116)

lipasa: enzima que ayuda en la digestión de las grasas   (72)

lípido: una grasa, un aceite, una cera o un compuesto adiposo que usualmente tiene ácidos grasos en su estructura molecular; un componente importante de la membrana plasmática   (36)

crecimiento logístico: crecimiento típicamente mostrado en poblaciones en el mundo natural; el tamaño de la población aumenta rápidamente al principio, después más lentamente y finalmente se nivela a un valor llamado capacidad de carga   (645)

linfático: se refiere a los vasos, ganglios y tejidos de un sistema de órganos, el cual recolecta y regresa fluidos al aparato circulatorio y ayuda a defender al cuerpo contra infecciones   (206)

linfocito: tipo de leucocito pequeño importante en la respuesta inmunológica; ejemplos incluyen a las células B y T   (606)

órgano linfoide: cualquier órgano donde los linfocitos se desarrollan y se colectan   (609)

lisogénico: se refiere a un bacteriófago que infecta a una bacteria, inserta su DNA en el DNA del huésped y se replica con el huésped   (255)

lisosoma: vesícula celular que contiene enzimas digestivas   (169)

lítico: se refiere a un virus que infecta y destruye (lisa) células susceptibles   (255)

M: mitosis; the phase of the cell cycle during which mitosis (nuclear division) occurs   (215)

macroevolution: evolutionary change on a large scale, including speciation, evolutionary trends, adaptive radiation, and mass extinction   (418)

macromolecule: a large, complex molecule   (34)

macrophage: a type of large, phagocytic white blood cell   (205)

major histocompatibility complex (MHC): the set of genes that code for cell-surface proteins important in antigen presentation to T cells   (613)

M: mitosis, la fase del ciclo celular durante la cual ocurre la mitosis (división nuclear)   (215)

macroevolución: cambio evolutivo a gran escala, incluyendo especiación, tendencias evolutivas, radiación adaptativa y extinción en masa   (418)

macromolécula: molécula grande y compleja   (34)

macrófago: tipo de glóbulo blanco grande y fagocítico   (205)

complejo principal de histocompatibilidad (MHC): conjunto de genes que codifican la formación de proteínas en la superficie celular, importantes en la presentación de antígenos a las células T   (613)

medulla: the inner portion of an organ, such as the kidney; also, the lower portion of the vertebrate brain, the hindbrain   (556)

meiosis: two successive cell divisions that produce gametes   (320)

memory B cell: a B lymphocyte produced in response to a primary immune response; remains dormant and can respond rapidly if the same antigen is encountered again   (612)

meninges: a group of three membranes that cover the brain and spinal cord   (554)

menstrual cycle: the female reproductive cycle that is characterized by regularly recurring changes in the uterine lining   (333)

mesoderm: in most animal embryos, a tissue layer between the ectoderm and endoderm   (267)

messenger RNA (mRNA): the RNA complementary to one strand of DNA; transcribed from genes and translated by ribosomes into protein   (234)

metabolism: the sum of all the chemical changes taking place in an organism   (62)

metamorphosis: in the life cycles of many animals, major changes in body form and function as the newly hatched young (larvae) mature into adults   (268)

metaphase: the stage in mitosis and meiosis in which chromosomes move to the center of the spindle, an array of microtubules, and become attached to it   (224)

metaphase plate: during nuclear division, an imaginary plane that is equidistant between the spindle's two poles   (224)

metastasis: the spread of cancer cells from their original site to other parts of the body, forming new tumors at distant sites   (228)

methanogens: archaebacteria that live in anaerobic environments and produce methane as a by-product of their metabolic processes   (455)

microevolution: a change in the gene pool of a population over generations   (418)

missense mutation: a change in DNA that results in a codon coding for a different amino acid than the original codon coded   (406)

mitochondria (singular: *mitochondrion*): the organelles in eukaryotic cells that carry on cell respiration; the site of ATP synthesis and of the Krebs cycle   (137)

mitochondrial DNA (mtDNA): DNA of the mitochondria; used to study the evolution of humans   (374)

mitosis: the process that distributes a copy of each chromosome to each new cell during eukaryotic cell division   (214)

médula: porción interna de un órgano, como el riñón; también la porción baja del cerebro vertebral, el cerebro posterior   (556)

meiosis: dos divisiones celulares sucesivas que producen gametos   (320)

célula B de memoria: linfocito B producido como respuesta a una respuesta inmunológica primaria; permanece latente y puede actuar rápidamente si el mismo antígeno es reconocido nuevamente   (612)

meninges: grupo de tres membranas que recubren el cerebro y la médula espinal   (554)

ciclo menstrual: el ciclo reproductor femenino que se caracteriza por cambios regulares recurrentes en las paredes del útero   (333)

mesodermo: en la mayoría de embriones, capa de tejido entre el ectodermo y el endodermo   (267)

RNA mensajero (mRNA): el RNA complementario a una cadena de DNA; se transcribe de los genes y los ribosomas las traducen en proteínas   (234)

metabolismo: la suma de todos los cambios químicos que suceden en un organismo   (62)

metamorfosis: en los ciclos vitales de muchos animales, cambios considerables en la forma y función corporales al convertirse los recién nacidos (larvas) en adultos   (268)

metafase: la etapa de la mitosis y meiosis en la cual los cromosomas se mueven hacia el centro del huso, un conjunto de microtúbulos y se pegan a él   (224)

plano de metafase: durante la división celular, un plano imaginario que está equidistante entre los dos polos del uso   (224)

metastasis: diseminación de células cancerosas de su sitio original hacia otras partes del cuerpo, formando nuevos tumores en sitios distantes   (228)

metanógenos: arqueobacterias que viven en ambientes anaerobios y producen metano como un producto derivado de sus procesos metabólicos   (455)

microevolución: cambio en el caudal de genes de una población a través de las generaciones   (418)

mutación sustitutiva: a un cambio en el DNA que resulta en un codón que codifica para un aminoácido diferente al que codificaba el codón original   (406)

mitocondria: los organelos de las células eucariotas que llevan a cabo la respiración celular: el sitio de la síntesis del ATP y del ciclo de Krebs   (137)

DNA mitocondrial (mtDNA): DNA de la mitocondria; empleado para estudiar la evolución de los humanos   (374)

mitosis: proceso que distribuye una copia de cada cromosoma a cada nueva célula durante la división celular de eucariotas   (214)

mitotic spindle: structure made up of microtubules and proteins, divides the chromatids during nuclear division (224)

molecule: the smallest unit of a compound; composed of atoms covalently bonded to one another (24)

Monera: a kingdom of bacteria and related one-celled microscopic organisms that have no clearly defined nucleus; prokaryotes (475)

monohybrid cross: a genetic cross between individuals differing in one trait (352)

monosaccharide: a simple sugar with three to seven carbon atoms in its carbon skeleton (34)

morphogenesis: the embryonic development of the structure of an organism (264)

motile: capable of movement from place to place; describes most animals (478)

motor system: the parts of the nervous system that control muscle movement (551)

multifactorial: referring to control of the expression of a trait by several genes and environmental factors (364)

multiple alleles: the existence of several alleles of a gene (359)

mutagenic: causing changes, or mutations, in DNA (221)

mutation: a structural change in a gene, a chromosome, or another genetic unit (221)

mutualism: symbiotic relationship that mutually benefits two species (639)

mycorrhizae (singular: *mycorrhiza*): root structures that result from mutualistic associations of plant roots and fungi (188)

myelin sheath: a fatty layer surrounding the long axons of motor neurons in the peripheral nervous system of vertebrates; composed of the membranes of Schwann cells (557)

myosin: a protein that, together with actin, is responsible for muscular contraction and relaxation (197)

NAD+: nicotinamide adenine dinucleotide; an electron and hydrogen carrier in cell respiration (132)

NADP+: nicotinamide adenine dinucleotide phosphate; a hydrogen carrier in photosynthesis (109)

natural selection: a mechanism of evolution whereby members of a population with the most successful adaptations to their environment are most likely to survive and reproduce (9)

huso mitótico: estructura formada por microtúbulos y proteínas, divide a las cromátidas durante la división nuclear (224)

molécula: la unidad más pequeña de un compuesto; formado por átomos unidos covalentemente entre sí (24)

Monera: reino de bacterias y organismos microscópicos unicelulares relacionados que no poseen un núcleo claramente definido; procariotas (475)

cruce monohíbrido: cruce genético entre individuos diferentes en un solo rasgo (352)

monosacárido: un azúcar simple con tres a siete átomos de carbono en su esqueleto carbonado (34)

morfogénesis: el desarrollo embriónico de la estructura de un organismo (264)

móvil: capaz de moverse de un lugar a otro; describe a la mayor parte de los animales (478)

sistema motor: las partes del sistema nervioso que controlan el movimiento muscular (551)

multifactorial: se refiere al control de la expresión de una característica por diversos genes y factores ambientales (364)

alelos múltiples: la existencia de varios alelos de un gene (359)

mutagénico: que causa cambios, o mutaciones, en el DNA (221)

mutación: cambio estructural en un gene, un cromosoma u otra unidad genética (221)

mutualismo: relación simbiótica que beneficia mutuamente a dos especies (639)

micorrizas: estructuras de raíz que resultan de asociaciones mutualistas de raíces de plantas y hongos (188)

vaina de mielina: capa adiposa que rodea a los axones largos de la neuronas motrices en el sistema nervioso periférico de los vertebrados; compuesta por las membranas de las células de Schwann (557)

miosina: proteína que, junto a la actina, es responsable de la contracción y relajación muscular (197)

NAD+: dinucleótido de nicotinamida y adenina; transportador de electrones e hidrógeno durante la respiración celular (132)

NADP+: fosfato dinucleótido de nicotinamida y adenina; transportador de hidrógeno en la fotosíntesis (109)

selección natural: mecanismo de evolución donde los miembros de una población con las adaptaciones más exitosas en su ambiente tienen una mayor probabilidad de sobrevivir y reproducirse (9)

nephron: the functional unit of a kidney consisting of a glomerulus, its associated capsule, and tubule, surrounded by capillaries   (94)

neritic zone: the shallow regions of the ocean overlying the continental shelf   (661)

nerve: a bundle of nerve fibers (axons)   (549)

nerve impulse: a wave of chemical and electrical changes that passes along a nerve fiber in response to a stimulus (562)

neural tube: the foundation of the nervous system that forms in an embryo at the gastrula stage   (268)

neuron: a nerve cell   (549)

neurotransmitter: a chemical messenger, often a hormone, that diffuses across a synapse and transmits an impulse from one neuron to another   (563)

neutron: a subatomic particle carrying no electrical charge   (26)

niche: the total of an organism's utilization of the biotic and abiotic resources of its environment   (638)

node: in plants, the point where leaves attach to the stem (295)

node of Ranvier: constriction in the myelin sheath of a neuron between Schwann cells, which wrap around the axon of a nerve cell like an athletic bandage around a knee   (557)

nondisjunction: failure of a pair of homologous chromosomes or chromatid pairs to separate during meiosis or mitosis   (363)

nonsense mutation: a change in DNA that changes a codon that specifies an amino acid to one that specifies a stop codon   (406)

notochord: a flexible, dorsal, rodlike structure that extends the length of the body of animals called chordates; in vertebrates, present only in the embryonic stages   (267)

nuclear division: the division of a cell's nucleus, as in mitosis and meiosis   (216)

nucleic acid: DNA or RNA; a polymer of nucleotides important in encoding instructions for cell processes (40)

nucleoid: a region in a prokaryotic cell consisting of a concentrated mass of DNA   (161)

nucleolus (plural: *nucleoli*): a structure in the nucleus that synthesizes ribosomal RNA   (163)

nucleosome: the basic package of chromatin in eukaryotes made up of DNA wound around a core of histone proteins   (220)

nucleotide: a subunit of DNA or RNA composed of a 5-carbon sugar, a nitrogen-containing base, and a phosphate group   (40)

nefrona: unidad funcional de un riñón que consiste en un glomérulo, su cápsula asociada y túbulo, rodeado por capilares   (94)

zona nerítica: las regiones poco profundas del océano que se acumulan en la plataforma continental   (661)

nervio: conjunto de fibras nerviosas (axones)   (549)

impulso nervioso: ola de cambios químicos y eléctricos que pasa a lo largo de la fibra nerviosa en respuesta a un estímulo   (562)

tubo neural: el precursor del sistema nervioso que se forma en un embrión en la etapa de gástrula   (268)

neurona: célula nerviosa   (549)

neurotransmisor: mensajero químico, con frecuencia una hormona, que se difunde en una sinapsis y transmite un impulso de una neurona a otra   (563)

neutrón: partícula subatómica que no posee carga eléctrica   (26)

nicho: el total de la utilización de un organismo de los recursos bióticos y abióticos de su ambiente   (638)

nudo: en las plantas, el punto donde las hojas se unen al peciolo   (295)

nódulo de Ranvier: constricción de la vaina de mielina de una neurona entre las células de Schwann, que envuelven al axón de una célula nerviosa como un vendaje atlético alrededor de una rodilla   (557)

no disyunción: el fracaso de un par de cromosomas homólogos o pares cromátides para separarse durante la meiosis o mitosis   (363)

mutación sin sentido: variación en el DNA que cambia un codón que especifica un aminoácido a uno que especifica un codón de terminación   (406)

notocordio: estructura dorsal flexible y cilíndrica, que extiende la longitud corporal de los animales llamados cordados; en los vertebrados, se presenta únicamente en etapas embrionarias   (267)

división nuclear: división de las células de un núcleo, como en la mitosis y la meiosis   (216)

ácido nucleico: DNA o RNA; un polímero de nucléotidos importante para la codificación de instrucciones para procesos celulares   (40)

nucleoide: región en una célula procariota que consiste en una masa concentrada de DNA   (161)

nucleolo: estructura en el núcleo que sintetiza el RNA ribosomal   (163)

nucleosoma: el nivel de organización básico de la cromatina en los eucariotas, compuesto por DNA enrollado a un centro de proteínas histonas   (220)

nucléotido: subunidad de DNA o RNA compuesto de un azúcar de 5 carbonos, una base nitrogenada y un grupo fosfato   (40)

nucleus (plural: *nuclei*): in atoms, the central core containing protons and neutrons; in eukaryotic cells, the membrane-bound organelle that houses the chromosomes   (161)

nutrient: a substance that supports the growth and maintenance of an organism   (54)

núcleo: en las átomos, la parte central que contiene protones y neutrones; en las células eucariotas, el organelo con membrana que aloja a los cromosomas   (161)

nutriente: sustancia que sostiene el crecimiento y mantenimiento de un organismo   (54)

obligate aerobe: a microorganism that requires oxygen to live   (140)

obligate anaerobe: a microorganism that lives without using oxygen (an anaerobe) and finds the presence of oxygen harmful   (140)

oceanic zone: the region of water lying over deep areas beyond the continental shelf   (662)

omnivorous: capable of eating both plant and animal material   (522)

open circulatory system: a system in which the blood moves in and out of open-ended blood vessels   (192)

operon: a genetic unit found in bacteria and bacteriophages made up of clusters of genes and their control sequences   (371)

opposable: of anthropoid thumbs, able to be positioned opposite the other fingers, which increases the precision of hand use   (520)

order: in taxonomy, a group of related families   (468)

organ: an organized group of tissues that carries on a specialized function in a multicellular organism   (177)

organelle: an organized structure within a cell with a specific function   (163)

organic: refers to compounds that are made up of carbon atoms and other elements; originally thought to be associated only with living things   (33)

organism: individual of a species; a single living thing   (23)

osmosis: the movement of water (or another solvent) through a selectively permeable membrane from a solution with a lower concentration of solutes to one with a higher concentration of solutes   (83)

ovaries (singular: *ovary*): the primary reproductive organs of a female; egg-cell-producing organs   (326)

oviduct: in vertebrates, a tube that carries eggs away from an ovary; in humans, a fallopian tube   (333)

aerobio obligado: microorganismo que requiere oxígeno para vivir   (140)

anaerobio obligado: microorganismo que vive sin usar oxígeno (un anaerobio) y que le perjudica la presencia de oxígeno   (140)

zona oceánica: la región de agua que se halla sobre las áreas profundas más allá de la plataforma continental   (662)

omnívoro: capaz de comer tanto material animal como vegetal   (522)

aparato circulatorio abierto: un sistema en el cual la sangre se mueve hacia adentro y hacia fuera de vasos sanguíneos abiertos   (192)

operón: unidad genética presente en bacterias y bacteriófagos compuesta por agrupaciones de genes y sus secuencias de control   (371)

oponible: en los pulgares de los antropoides, capaz de posicionarse de manera opuesta a los otros dedos, lo cual incrementa la precisión en el uso de la mano   (520)

orden: en taxonomía, un grupo de familias relacionadas   (468)

órgano: grupo especializado de tejidos que lleva a cabo una función especializada en un organismo multicelular   (177)

organelo: estructura organizada dentro de una célula con una función específica   (163)

orgánico: se refiere a los compuestos que están formados por átomos de carbono y otros elementos; originalmente se pensaba que estaban únicamente asociados con los seres vivos   (33)

organismo: individuo de una especie; un ser vivo individual   (23)

ósmosis: el movimiento del agua (u otro solvente) a través de membranas permeables selectivas, de una solución con concentración más baja de solutos a una con una mayor concentración de solutos   (83)

ovarios: los órganos reproductores primarios de una hembra; órganos productores de óvulos   (326)

oviducto: en los vertebrados, tubo que conduce a los óvulos fuera de los ovarios; en los humanos, la trompa de Falopio   (333)

ovulation: the release of an egg cell from a mature ovarian follicle (335)

ovule: a structure that develops in a plant ovary and contains the egg (326)

ovum (plural: *ova*): a mature egg cell; a female gamete, haploid *(n)* in chromosome number (320)

oxidation: the loss of electrons from a substance in a chemical reaction (64)

ozone hole: the reduction in the ozone layer over Antarctica; a layer of molecules of ozone (O_3) and oxygen (O_2) about 17–25 kilometers above Earth, which absorbs lethal wavelengths of ultraviolet light (673)

ovulación: la liberación de un óvulo de un folículo ovariano maduro (335)

óvulo: estructura que se desarrolla en el ovario de una planta y que contiene un huevo (326)

huevo: un óvulo maduro; un gameto femenino, haploide *(n)* en su número de cromosomas (320)

oxidación: la pérdida de electrones de una sustancia en una reacción química (64)

agujero de ozono: la reducción en la capa de ozono sobre la Antártida; una capa de moléculas de ozono (O_3) y oxígeno (O_2), aproximadamente a 17–25 kilómetros sobre la Tierra, la cual absorbe las longitudes de onda letales de la luz ultravioleta (673)

paleontology: the study of extinct organisms through fossils (501)

parasitism: a symbiotic relationship between two species in which one (the parasite) benefits at the expense of the other (the host) (639)

parasympathetic system: in vertebrates, one of two subdivisions of the autonomic nervous system; stimulates involuntary functions, such as digestion, and restores the body to normal after emergencies (553)

parental (P) generation: the parental organisms in a genetic cross (352)

passive transport: the diffusion of a substance through a biological membrane (84)

pathogen: an agent, such as a virus, bacterium, or fungus, that can cause disease (604)

pelagic zone: the area of the ocean past the continental shelf; open water often reaching to great depths (662)

pepsin: a protein-splitting enzyme secreted by the gastric glands of the stomach (72)

pepsinogen: the inactive form of pepsin (72)

peptide bond: a covalent chemical bond formed between two amino acids; bonds the amino group of each amino acid to the carboxyl group of the next (39)

perennial: a plant that lives for more than 2 years (654)

pericycle: in plants, a layer of cells around the vascular tissues from which branch roots grow (299)

peripheral nervous system (PNS): the sensory and motor neurons that connect to the vertebrate central nervous system (551)

peristalsis: the rhythmic waves of contraction of the smooth muscle that pushes food through the digestive tract (70)

paleontología: el estudio de organismo extintos a través de fósiles (501)

parasitismo: relación simbiótica entre dos especies en la cual una (el parásito) se beneficia a expensas del otro (el huésped) (639)

sistema parasimpático: en vertebrados, una de las dos subdivisiones del sistema nervioso autónomo; estimula las funciones involuntarias, como la digestión y lleva al cuerpo a la normalidad tras emergencias (553)

generación parental (P): los organismos padres en un cruce genético (352)

transporte pasivo: la difusión de una sustancia a través de una membrana biológica (84)

patógeno: organismo, como un virus, bacteria u hongo, que puede causar una enfermedad (604)

zona pelágica: el área de un océano más allá de la plataforma continental; mar abierto que frecuentemente alcanza grandes profundidades (662)

pepsina: enzima que inicia la digestión de las proteínas, secretada por las glándulas gástricas del estómago (72)

pepsinógeno: la forma inactiva de la pepsina (72)

enlace peptídico: enlace químico covalente entre dos aminoácidos; une el grupo amino de cada aminoácido con el grupo carboxilo del siguiente (39)

perenne: planta que vive por más de 2 años (654)

periciclo: en las plantas, una capa de células alrededor de los tejidos vasculares a partir de las cuales crecen las raíces secundarias (299)

sistema nervioso periférico (PNS): las neuronas motrices y sensoriales que se conectan al sistema nervioso central vertebrado (551)

peristalsis: oleadas rítmicas de contracciones del músculo liso que empuja la comida a través del tracto digestivo (70)

pH scale: a scale from 0 to 14 reflecting the concentration of hydrogen ions in solution; a number less than 7 denotes acidic conditions, and a number greater than 7 denotes basic conditions   (32)

phagocytic cell: specialized cell that ingests and destroys foreign particles or microorganisms   (605)

phenetic: a systematic method of classification that uses similarities based on phenotypic characteristics, giving equal importance to all characteristics   (472)

phenotype: the expression of a genotype in the appearance or function of an organism; an observed trait   (353)

pheromone: a chemical signal between members of the same species; acts much like hormones to influence physiology and behavior   (597)

phloem: a portion of the vascular system in plants consisting of living cells arranged into elongated tubes that transport sugar and other organic nutrients throughout the plant   (188)

photic zone: the shallow surface portion of the ocean where light penetrates sufficiently for photosynthesis   (660)

photoautotroph: an organism that derives energy from light and forms its own organic compounds (food) from abiotic carbon sources   (55)

photoheterotroph: an organism that derives energy from light but requires a source of organic compounds   (103)

photoinhibition: damage to the light-gathering process in photosynthesis; occurs when a chloroplast has absorbed too much light energy   (115)

photoperiodism: a biological response to the length of day or night, such as in flowering plants   (310)

photorespiration: a metabolic pathway in plants that consumes oxygen, produces carbon dioxide, generates no ATP, and reduces photosynthesis   (117)

photosynthesis: the process by which cells use light energy to make organic compounds from inorganic materials   (55)

phototropism: growth toward or away from a source of light   (309)

phylum: in taxonomy, a group of related classes; equivalent to a division for plants   (469)

phytochrome: a light-absorbing pigment involved in plant photoperiodism   (310)

phytoplankton: small, floating aquatic organisms, many microscopic, that carry on photosynthesis   (660)

pigment: any coloring matter or substance   (103)

escala de pH: escala del 0 al 14 que refleja la concentración de iones hidrógeno en solución; un número menor que 7 denota condiciones ácidas y un número mayor que 7 denota condiciones básicas   (32)

células fagocitas: células especializadas que ingieren y destruyen partículas foráneas o microorganismos foráncos   (605)

fenético: método sistemático de clasificación que usa similaridades basadas en características fenotípicas, dando igual importancia a todas las características   (472)

fenotipo: expresión de un genotipo en la apariencia o función de un organismo; rasgo observable   (353)

feromona: señal química entre miembros de la misma especie; funciona de manera similar a las hormonas, influyendo en la fisiología y el comportamiento   (597)

floema: porción del sistema vascular en las plantas, que consiste en células vivas ordenadas en tubos alargados los cuales transportan azúcar y otros nutrientes orgánicos a través de la planta   (188)

zona fótica: porción superficial de baja profundidad del océano, donde la luz penetra lo suficiente para que ocurra fotosíntesis   (660)

fotoautótrofo: organismo que obtiene energía de la luz y forma sus propios compuestos orgánicos (alimento) a partir de fuentes abióticas de carbono   (55)

fotoheterótrofo: organismo que obtiene energía de la luz, pero que requiere de fuentes de compuestos orgánicos   (103)

fotoinhibición: daño a los procesos recolectores de luz en la fotosíntesis; ocurre cuando un cloroplasto ha absorbido demasiada energía luminosa   (115)

fotoperiodismo: respuesta biológica a la duración del día o la noche, tal como en las plantes con flores   (310)

fotorrespiración: una ruta metabólica en plantas, la cual consume oxígeno, produce dióxido de carbono, no genera ATP y reduce la fotosíntesis   (117)

fotosíntesis: el proceso por el cual las células emplean energía luminosa para formar compuestos orgánicos a partir de materiales inorgánicos   (55)

fototropismo: crecimiento hacia o alejándose de una fuente de luz o en dirección contraria a ella   (309)

filo: en taxonomía, grupo de clases relacionadas; equivalente a una división en las plantas   (469)

fitocromo: pigmento fotoabsorbente involucrado en el fotoperiodismo de las plantas   (310)

fitoplancton: organismos acuáticos, pequeños y flotantes, muchos de ellos microscópicos, que llevan a cabo fotosíntesis   (660)

pigmento: cualquier materia o sustancia que colorea   (103)

placenta: a structure in the uterus for exchange of materials between a fetus and the mother's blood supply; formed from the uterine lining and embryonic membranes (273)

Plantae: the plant kingdom (476)

plant growth regulator (PGR): a substance that promotes, inhibits, or alters plant growth (301)

plasma: the liquid portion of the blood in which the blood cells are suspended (205)

plasma cell: antibody-producing cell that develops as a result of the proliferation of sensitized B lymphocytes (612)

plasma membrane: the membrane at the boundary of every cell, which serves as a selective barrier to the passage of ions and molecules (163)

plasmid: a small ring of DNA in bacteria that carries genes separate from those of the chromosome (161)

platelet: a small cell fragment found in blood that contributes to blood clotting at the site of a wound; releases substances that begin clot formation (207)

point mutation: a change in a single base pair of a DNA sequence in a gene (406)

polar body: small haploid cell produced in meiosis during egg-cell formation (324)

pollen: tiny grains that contain male gametes; released from the anthers of flowers (325)

pollination: the placement of pollen by wind or animal onto the stigma of a carpel; a prerequisite to fertilization (326)

polymerase chain reaction (PCR): a technique that uses double-stranded DNA and two primers as templates for numerous duplications; each round of duplication roughly doubles the amount of DNA (400)

polymorphic: having more than one form; for example, if different alleles of the same gene are found within a population, the population is considered to be polymorphic at that genetic locus (419)

polypeptide: a long chain of chemically bonded amino acids (39)

polyploidy: condition in which a cell, a nucleus, or an organism has more than two sets of chromosomes (510)

polysaccharide: a complex carbohydrate composed of many simple sugars (monosaccharides) chemically bonded in a chain; for example, starch and cellulose (35)

placenta: estructura en el útero para el intercambio de materiales entre el feto y el suministro sanguíneo de la madre; formada a partir de las paredes del útero y de las membranas del embrión (273)

Plantas: el reino vegetal (476)

regulador del crecimiento vegetal (PGR): sustancia que promueve, inhibe o altera el crecimiento vegetal (301)

plasma: la porción líquida de la sangre en la cual están suspendidas las células sanguíneas (205)

célula plasmática: célula productora de anticuerpos que se desarrolla como resultado de la proliferación de linfocitos B sensibilizados (612)

membrana plasmática: la membrana en los límites de cada célula, que funciona como una barrera selectiva al paso de iones o moléculas (163)

plásmido: anillo pequeño de DNA en las bacterias que conduce genes diferentes a aquéllos del cromosoma (161)

plaqueta: pequeño fragmento celular presente en la sangre, que contribuye al coagulado sanguíneo en el sitio de una herida; libera sustancias que comienzan la formación del coágulo (207)

mutación puntual: cambio en un par de bases sencillo de una secuencia de DNA en un gene (406)

cuerpo polar: pequeña célula haploide producida por la meiosis durante la formación de óvulos (324)

polen: pequeños granos que contienen gametos masculinos; liberados de las anteras de las flores (325)

polinización: la colocación del polen por el viento o animales en el estigma de un carpelo; un prerrequisito para la fecundación (326)

reacción en cadena de la polimerasa (PCR): una técnica que usa DNA de doble cadena y dos cebadores como plantillas para duplicaciones numerosas; cada ronda de duplicación aproximadamente dobla la cantidad de DNA (400)

polimórfico: que tiene más de una forma; por ejemplo, si alelos diferentes del mismo gene se encuentran dentro de una población, la población se considera polimórfica en ese sitio genético (419)

polipéptido: cadena larga de aminoácidos químicamente unidos (39)

poliploide: condición en la que una célula, un núcleo o un organismo tiene más de dos conjuntos de cromosomas (510)

polisacárido: un carbohidrato complejo compuesto de muchos azúcares simples (monosacáridos) químicamente unidos en una cadena; por ejemplo, almidón y celulosa (35)

population bottleneck: a decline in population size that causes the gene pool to become less diverse   (427)

population density: number of organisms per unit of habitat area   (645)

population genetics: the study of the genetics of groups of interbreeding individuals   (418)

postsynaptic: located or occurring after a synapse   (563)

predation: the killing and consumption of prey   (638)

predator-prey cycle: a pattern of oscillations in the population density of prey followed by a corresponding change in the population density of the prey's predator   (647)

presynaptic: located or occurring before a synapse   (563)

primary germ layers: in animals, the three cell groups— endoderm, ectoderm, and mesoderm— that give rise to all the tissues of the body   (266)

primary growth: growth in the length of a plant's roots and stems   (295)

primary immune response: the initial immune response to an antigen; appears after a lag of several days   (614)

primary structure: the first level of organization of a protein or nucleic acid; refers to the specific sequence of amino acids or nucleotides   (39)

primary succession: process of colonization and species replacement on a site that was previously uninhabited   (665)

principle of independent assortment: the principle stating that the inheritance of alleles residing on one pair of chromosomes does not affect the inheritance of alleles on a different pair of chromosomes   (355)

principle of segregation: the principle stating that homologous chromosomes, with the genes they carry, separate into different gametes during meiosis such that the two alleles for a given trait each appear in a different gamete   (352)

probability: the chance that an event will occur; the number of times an event is expected to occur divided by the total number of opportunities for the event to occur   (351)

probe: a molecule that is labeled in some way, such as with radioactive isotopes or with a fluorescent marker, and that selectively binds to a specific nucleotide sequence so that it can be isolated or identified   (404)

producer: an autotroph; any organism that produces its own food   (56)

reducción de diversidad (cuello de botella demográfica): declinación en el tamaño de la población que causa una disminución en la diversidad del caudal de genes   (427)

densidad demográfica: número de organismos por unidad de área de hábitat   (645)

genética de poblaciones: el estudio de la genética de grupos de individuos de intercruzamiento   (418)

postsináptico: que ocurre o se localiza después de una sinapsis   (563)

depredación: la matanza y consumo de presas   (638)

ciclo predador-presa: patrón de oscilaciones en la densidad demográfica de presas, seguido por un cambio correspondiente en la densidad demográfica de los predadores de la presa   (647)

presináptico: que se localiza u ocurre antes de la sinapsis   (563)

capas germinales primarias: en los animales, los tres grupos celulares—endodermo, ectodermo y mesodermo— que dan origen a todos los tejidos del cuerpo   (266)

crecimiento primario: crecimiento de la longitud de las raíces y tallos de una planta   (295)

respuesta inmunológica primaria: la respuesta inmunológica inicial a un antígeno; aparece tras un periodo de varios días   (614)

estructura primaria: el primer nivel de organización de una proteína o ácido nucleico; se refiere a la secuencia específica de aminoácidos o nucleótidos   (39)

sucesión primaria: proceso de colonización y reemplazo de especies en un lugar anteriormente deshabitado   (665)

principio de distribución independiente: principio que establece que la herencia de alelos que residen en un par de cromosomas no afecta la herencia de alelos en un par de cromosomas diferentes   (355)

principio de segregación: que establece que los cromosomas homólogos, con los genes que poseen, se separan en dos gametos diferentes durante la meiosis de tal manera que los dos alelos para un rasgo dado aparecen cada uno en un gameto diferente   (352)

probabilidad: la posibilidad de que un evento vaya a ocurrir; el número de veces que se espera que ocurra un evento dividido entre el número total de oportunidades para que ocurra el evento   (351)

sonda: molécula que se rotula de alguna manera, como con isótopos radioactivos o con un marcador fluorescente y que se une selectivamente a una secuencia de específica de nucleótidos para que ésta pueda aislarse o identificarse   (404)

productor: autótrofo; cualquier organismo que produce su propio alimento   (56)

productivity: the rate at which producers generate biomass (638)

progesterone: a female hormone, secreted by the corpus luteum and the placenta, that prepares the uterus for pregnancy and the mammary glands for lactation (334)

prokaryote: an organism whose cells do not have membrane-enclosed nuclei or organelles; a moneran (bacterium) (160)

promoter: a site on DNA to which RNA polymerase will bind and initiate transcription (371)

prophase: the stage in mitosis during which replicated strands of chromosomes condense, the nuclear envelope (the membrane around the nucleus) begins to disappear, and a spindle forms (223)

protein: an organic compound composed of one or more polypeptide chains of amino acids; most structural materials and enzymes in a cell are proteins (37)

Protista: a kingdom of mostly aquatic, mostly unicellular eukaryotes (476)

proton: a particle bearing a positive electrical charge found in the nuclei of all atoms (26)

proto-oncogene: gene that regulates cell division in eukaryotic cells; mutations in proto-oncogenes can produce oncogenes, genes having the potential to produce changes in cellular metabolism, leading to cancerous growth (228)

protozoan (plural: *protozoa*): a one-celled, mobile protist (476)

pseudogene: a nonfunctioning DNA segment that is similar in sequence to a functioning gene (506)

pseudoscience: false science; research that does not meet the criteria of science (16)

psychoactive: referring to a drug that affects the central nervous system (567)

puberty: the stage of development in which the reproductive organs become functional (337)

punctuated equilibrium: a model of evolution in which speciation occurs in spurts of relatively rapid change followed by long periods of stability (515)

pyruvic acid: 3-carbon compound that is the end product of glycolysis (134)

productividad: el ritmo al cual los productores generan biomasa (638)

progesterona: hormona femenina, secretada por el cuerpo lúteo y la placenta, que prepara al útero para el embarazo y a las glándulas mamarias para la lactancia (334)

procariota: organismo cuyas células no tienen núcleos u organelos rodeados por membranas; un miembro del reino monera (bacteria) (160)

promotor: sitio en el DNA al cual se unirá la polimerasa del RNA e iniciará la transcripción (371)

profase: la etapa en la mitosis durante la cual las hebras replicadas de cromosomas se condensan, la cubierta nuclear (la membrana que rodea al núcleo) comienza a desaparecer y se forma el huso (223)

proteína: compuesto orgánico formado por dos o más cadenas de polipeptídicas aminoácidos; la mayor parte de los materiales estructurales y las enzimas en una célula son proteínas (37)

Protista: reino conformado por eucariotas mayormente acuáticos y unicelulares (476)

protón: partícula que posee una carga eléctrica positiva hallada en los núcleos de todos los átomos (26)

protooncogen: gene que regula la división celular en las células eucariotas; mutaciones en los protooncogenes pueden producir oncogenes, genes que tienen el potencial de producir cambios en el metabolismo celular, conduciendo a un crecimiento canceroso (228)

protozoario: protista móvil unicelular (476)

pseudogen: segmento de DNA que no funciona, que es similar en secuencia a un gene funcional (506)

pseudociencia: ciencia falsa; investigación que no cumple con los criterios científicos (16)

psicoactiva: se refiere a una droga que afecta al sistema nervioso central (567)

pubertad: etapa del desarrollo en la cual los órganos reproductores se vuelven funcionales (337)

equilibrio puntualizado: un modelo evolutivo en el cual la especiación ocurre en rachas de cambios relativamente rápidos, seguidas de largos periodos de estabilidad (515)

ácido pirúvico: compuesto de 3 carbonos que es el producto terminal de la glicólisis (134)

QTL mapping: a set of procedures for finding the chromosomal locations of genes that cause variation in quantitative traits (433)

mapeo QTL: un conjunto de procedimientos para hallar las posiciones cromosómicas de genes que causan variaciones en los rasgos cuantitativos (433)

quantitative trait: any trait such as height, weight, or color that varies continuously over a range; a multifactorial trait   (429)

quantitative trait locus (QTL)-(plural: *loci*): a gene that affects quantitative traits   (429)

quaternary structure: the shape of a complex protein defined by the three-dimensional arrangement of its polypeptide subunits   (40)

rasgo cuantitativo: cualquier característica como estatura, peso o color, que varía continuamente a lo largo de un rango; una característica multifactorial (429)

locus de rasgos cuantitativos (QTL): un gene que afecta rasgos cuantitativos   (429)

estructura cuaternaria: la forma de una proteína compleja definida por el arreglo tridimensional de sus subunidades polipeptídicas   (40)

rate: the amount of change over a period of time   (113)

recessive: a trait whose expression is masked in a heterozygous organism   (352)

reduction: the gain of electrons by a substance in a chemical reaction   (64)

reflex: an involuntary reaction or response to a stimulus (566)

reflex arc: a nerve pathway consisting of a sensory neuron and a motor neuron, which may contain one or more interneurons; forms the structural and functional basis for a reflex   (566)

replication origin: specific sequence of DNA at which DNA synthesis begins   (218)

replisome: a complex of DNA polymerase and other enzymes that catalyzes the synthesis of DNA   (218)

response: action or movement of all or part of a cell or an organism as a result of a stimulus   (578)

response element: in eukaryotes, a regulatory DNA sequence to which a hormone-receptor complex binds, activating a nearby gene   (372)

resting potential: the electric potential across the membrane of a nerve cell or muscle cell when an action potential is not occurring   (561)

restriction enzyme: an enzyme that recognizes specific nucleotide sequences in DNA and breaks the DNA chain at those points; used in genetic engineering, which alters life-forms by recombining DNA   (398)

restriction point: a point of no return in the cell cycle; once this point passes, a cell is committed to a full round of the cell cycle   (215)

retina: the photosensitive layer of the vertebrate eye; contains several layers of neurons and light-receptor cells (rods and cones); receives the image formed by the lens and transmits it to the brain via the optic nerve (550)

tasa: la cantidad de cambio en un periodo de tiempo (113)

recesivo: rasgo cuya expresión está enmascarada en un organismo heterocigoto   (352)

reducción: la ganancia de electrones de una sustancia en una reacción química   (64)

reflejo: reacción o respuesta involuntaria a un estímulo (566)

arco reflejo: ruta nerviosa, que consiste en una neurona sensorial y una neurona motora, la cual puede contener una o varias interneuronas; constituye la base estructural y funcional para un reflejo   (566)

origen de replicación: secuencia específica de DNA en la cual se inicia la síntesis de DNA   (218)

replisoma: un complejo de polimerasa de DNA y otras enzimas que catalizan la síntesis de DNA   (218)

respuesta: acción o movimiento de toda o parte de la célula u organismo como resultado de un estímulo (578)

elemento de respuesta: en los eucariotas, una secuencia regulatoria de DNA a la cual se une un complejo hormona-receptor, activando un gene cercano   (372)

potencial de reposo: el potencial eléctrico a través de una membrana o una célula nerviosa o muscular cuando un potencial de acción no está ocurriendo   (561)

enzima de restricción: enzima que reconoce secuencias específicas de nucleótido en el DNA y rompe la cadena de DNA en esos puntos; empleada en ingeniería genética, la cual altera formas vivientes por recombinación de DNA   (398)

punto de restricción: un punto sin retorno en el ciclo celular; una vez pasado este punto, la célula se compromete a llevar a cabo el ciclo celular completo (215)

retina: la capa fotosensible del ojo de los vertebrados; contiene varias capas de neuronas y de células receptoras de luz (bastones y conos); recibe la imagen formada por el lente y la transmite al cerebro a través del nervio óptico   (550)

retrotransposon: a segment of DNA that randomly inserts itself in a different chromosome within the cell using reverse transcriptase, which makes DNA using RNA as the template   (387)

RFLP analysis: restriction fragment length polymorphism analysis; differences in DNA fragment sizes cut by specific restriction enzymes, which are used as markers on genetic maps, to help identify the source of DNA and to determine the presence of specific alleles or genes (403)

ribosomal RNA (rRNA): a class of RNA molecules that combine with certain proteins to form ribosomes   (234)

ribosome: an organelle consisting of two subunits and functioning as the site of protein synthesis   (166)

ribozyme: enzymatic RNA molecule, RNA that performs a catalytic function   (451)

RNA: ribonucleic acid; a nucleic acid similar to DNA but having the sugar ribose rather than deoxyribose and uracil rather than thymine as one of the bases   (41)

RNA polymerase: an enzyme that catalyzes the assembly of an RNA molecule   (240)

rod: a light-sensitive nerve cell in the vertebrate retina; sensitive to very dim light and responsible for night vision   (550)

root cap: a layer of protective cells that covers the growing tip of a plant root   (296)

retrotransposón: segmento de DNA que se inserta a sí mismo aleatoriamente en un cromosoma distinto dentro de la célula usando una transcriptasa inversa, la cual fabrica DNA usando RNA como plantilla   (387)

análisis RFLP: análisis del polimorfismo de longitud de fragmentos de restricción; diferencias en el tamaño de fragmentos de DNA cortados por enzimas de restricción específicas, que se usan como marcadores en mapas genéticos, para ayudar a identificar la fuente de DNA y para determinar la presencia de alelos o genes específicos   (403)

RNA ribosomal (rRNA): un tipo de moléculas de RNA que se combina con ciertas proteínas para formar ribosomas   (234)

ribosoma: un organelo que consiste en dos subunidades y que funciona como sitio de síntesis proteica   (166)

ribozima: molécula de RNA enzimático; RNA que lleva a cabo una función catalítica   (451)

RNA: ácido ribonucleico; un ácido nucleico similar al DNA, pero que tiene un azúcar ribosa en lugar de una desoxirribosa y uracilo en lugar de timina como una de las bases   (41)

RNA polimerasa: enzima que cataliza el armado de una molécula de RNA   (240)

bastón: célula nerviosa sensible a la luz localizada en la retina de los vertebrados; sensible a luz tenue y responsable de la visión nocturna   (550)

cubierta de raíz: una capa de células protectoras que recubre la punta creciente de una raíz de planta   (296)

S

S: synthesis of DNA; the phase of the cell cycle during which DNA, in the form of chromosomes, is duplicated (215)

saliva: liquid secreted in the mouth; begins mechanical and chemical digestion   (69)

salivary amylase: an enzyme in saliva that begins digestion of starch; converts starch to disaccharides   (70)

savanna: a tropical grassland biome with scattered individual trees and large herbivores; water is the major limiting factor   (653)

scientific name: the two-part Latin name of a species   (470)

scrotum: a pouch of skin that encloses the testes   (336)

second law of thermodynamics: the law stating that energy transfers and transformations increase the entropy of the universe   (59)

secondary growth: the growth in thickness or diameter of a plant stem or root   (298)

S: síntesis de DNA; la fase del ciclo celular durante la cual el DNA, en forma de cromosomas, se duplica   (215)

saliva: líquido secretado en la boca; comienza la digestión química y mecánica   (69)

amilasa salivaria: enzima en la saliva que comienza la digestión del almidón; convierte el almidón en disacáridos   (70)

sabana: bioma tropical de pastizales con árboles individuales espaciados y grandes herbívoros; el agua es el principal factor limitante   (653)

nombre científico: el nombre latino de dos partes de una especie   (470)

escroto: saco de piel que envuelve a los testículos   (336)

segunda ley de la termodinámica: ley que establece que la energía se transfiere y que las transformaciones incrementan la entropía del universo   (59)

crecimiento secundario: el crecimiento de grosor o diámetro del tallo o raíz de una planta   (298)

secondary immune response: immune response when an animal encounters an antigen a second time; more rapid, larger, and longer than the primary immune response (615)

secondary sex characteristic: characteristic of a male or female animal other than gamete production; in humans, includes mature genitalia, female breasts, body and facial hair, muscular development, and deep male voice; typically develops in response to sex hormones (337)

secondary structure: in proteins, the shape of a folded polypeptide chain; results from hydrogen bonds between adjacent parts of the molecule (39)

secondary succession: the sequential replacement of species after a major disruption in a community (665)

seed coat: the tough, protective outer covering of a seed (293)

segmentation: in animals, division into a series of similar parts, such as is found in annelid (earthworm) and arthropod (insect, crab, and spider) body plans (270)

selectively permeable: of membranes, allowing some substances to cross and preventing others from crossing (80)

sensory receptor: a specialized sensory structure or an organ that detects stimuli (548)

sieve tube: a column of phloem cells in a plant (190)

signal sequence: the directions for the transport of proteins to different parts of the cell; provided by the first few amino acids synthesized on the ribosomes (252)

single nucleotide polymorphisms (SNPs): genetic variations in which alleles differ by only one or a few scattered nucleotides; used to study inheritance patterns and mutations that cause disease (403)

sister chromatids: the replicated copies of a chromosome that are joined by a centromere and that separate during nuclear division (222)

somatic: referring to the body of a plant or an animal aside from the gametes; relating to the nonreproductive cells of a multicellular organism (320)

somatic nervous system: the part of the vertebrate nervous system that controls the skeletal muscles (552)

speciation: the origin of new species as a result of evolutionary processes (507)

respuesta inmunológica secundaria: respuesta inmunológica cuando un animal se encuentra con un antígeno por segunda vez; es más rápida, más grande y más duradera que la respuesta inmunológica primaria (615)

característica sexual secundaria: característica de un animal hembra o macho diferente a la producción de gametos; en los humanos incluye a los genitales maduros, los pechos maduros, el vello corporal o facial, el desarrollo de músculos y la voz grave masculina; típicamente se desarrolla en respuesta a las hormonas sexuales (337)

estructura secundaria: en proteínas, la forma de una cadena polipeptídica doblada; resulta de los puentes de hidrógeno entre partes adyacentes de la molécula (39)

sucesión secundaria: el reemplazo secuencial de especies tras una gran interrupción en una comunidad (665)

cotiledón: la dura cubierta externa protectora que cubre a una semilla (293)

segmentación: en los animales, la división en una serie de partes similares, como las encontradas en los planos corporales de anélidos (lombrices) y artrópodos (insectos, cangrejos y arañas) (270)

selectivamente permeable: en las membranas, factor que permite que algunas sustancias crucen e impide que otras lo hagan (80)

receptor sensorial: estructura sensorial especializada o un órgano que detecta estímulos (548)

tubo criboso: una columna de células de floema en una planta (190)

secuencia de señales: las instrucciones para el transporte de proteínas a diferentes partes de la célula; provista por los primeros aminoácidos sintetizados en los ribosomas (252)

polimorfismos en nucleótido único (SNP): variaciones genéticas en las cuales los alelos difieren en uno o unos cuantos nucleótidos aislados; empleados para estudiar los patrones de herencia y mutaciones que causan enfermedades (403)

cromátides hermanas: las copias replicadas de un cromosoma que están unidas por un centrómero y que se separan durante la división nuclear (222)

somático: se refiere al cuerpo de una planta o animal además de los gametos; relacionado con las células no reproductoras de una organismo multicelular (320)

sistema nervioso somático: la parte de un sistema nervioso de vertebrados que controla los músculos esqueléticos (552)

especiación: el origen de nuevas especies como resultado de procesos evolutivos (507)

species: all individuals and populations of a particular type of organism that can interbreed with one another   (9)

sperm: male gametes, each haploid *(n)* in chromosome number   (262)

spindle pole: one of two cellular regions at the tips of the mitotic spindle, where the daughter nuclei form during mitosis   (224)

splicing: the process that joins portions of the primary RNA transcript (exons) after the introns are removed; the joining of two pieces of DNA   (245)

spore: one-celled reproductive cell that is usually resistant to harsh environmental conditions and may remain dormant for long periods   (320)

stasis: a state of balance among opposing forces   (511)

stimulus: a change or signal in the internal or external environment that causes an adjustment or a reaction by an organism; anything that triggers a behavior   (548)

stomate: the opening between two guard cells in the epidermis of a plant leaf through which gases are exchanged with the air   (91)

stromatolites: rocks formed from layers of sediments made up of the fossilized remains of photosynthetic organisms, such as algae, that once lived in shallow seas   (454)

subatomic: referring to small particles that make up atoms, including electrons, protons, and neutrons   (26)

substrate: a molecule on which enzymes act   (61)

succession: orderly, natural changes, and species replacements that take place in communities of an ecosystem over time   (665)

succulent: a plant, such as a cactus, with thick, fleshy tissues that store water   (655)

symbiosis: an ecological relationship between organisms of two species that live together in direct contact   (456)

sympathetic system: a subdivision of the autonomic nervous system of vertebrates that functions in an alarm response; increases heart rate and dilates blood vessels; inhibits functions such as digestion; mobilizes the animal for rapid response to stress, danger, or excitement   (553)

synapse: a junction between neurons, across which a neurotransmitter transmits an electrical impluse   (563)

synthesis: the process of building chemical compounds from smaller components by means of chemical reactions   (62)

system: a group of organs that interact to perform a set of related functions   (177)

especies: todos los individuos y poblaciones de un tipo particular de organismo que pueden intercruzarse unos con otros   (9)

espermatozoides: gametos masculinos, cada uno haploide (n) en su número de cromosomas   (262)

polo del huso: una de las dos regiones celulares en las puntas del huso mitótico, donde los núcleos hijos se forman durante la mitosis   (224)

empalme: el proceso que junta las porciones de la transcripción del RNA primario (exones) después de extraer los intrones; la unión de dos piezas de DNA   (245)

espora: célula reproductora unicelular que usualmente es resistente a condiciones ambientales poco propicias y puede permanecer en estado de latencia por largos periodos   (320)

estasis: estado de balance entre fuerzas opuestas   (511)

estímulo: cambio o señal en el ambiente interno o externo que causa un ajuste o una reacción de un organismo; cualquier cosa que desencadena un comportamiento   (548)

estoma: abertura entre dos células guardianas en la epidermis de la hoja de una planta a través de la cual se intercambian los gases con el aire   (91)

estromatolitos: rocas formadas por capas de sedimentos hechas de restos fosilizados de organismos fotosintéticos, como las algas, que una vez vivieron en mares poco profundos   (454)

subatómico: se refiere a partículas pequeñas que forman a los átomos, incluyendo electrones, protones y neutrones   (26)

sustrato: molécula sobre la cual actúa la enzima   (61)

sucesión: Cambios naturales y ordenados y el reemplazo de especies que ocurren en las comunidades de un ecosistema a lo largo del tiempo   (665)

suculenta: una planta, como el cactus, con tejidos gruesos y carnosos que almacenan agua   (655)

simbiosis: una relación ecológica entre organismos de dos especies que viven juntas en contacto directo   (456)

sistema simpático: subdivisión del sistema nervioso autónomo de vertebrados, que funciona en una respuesta de alarma; incrementa el ritmo cardiaco y dilata los vasos sanguíneos; inhibe las funciones como la digestión; moviliza al animal para una respuesta rápida al estrés, peligro o excitación   (553)

sinapsis: unión entre neuronas, a través de las cuales un neurotransmisor transmite un impulso eléctrico   (563)

síntesis: el proceso de formación de compuestos químicos a partir de componentes más pequeños por medio de reacciones químicas   (62)

sistema o aparato: un grupo de órganos que interactúan para llevar a cabo un conjunto de funciones relacionadas   (177)

T cells: a class of lymphocytes that are produced in the bone marrow and mature in the thymus; involved in the cell-mediated immune response; examples include helper T cells, killer T cells, and suppressor T cells (606)

taiga: the coniferous or northern forest biome characterized by considerable snow, harsh winters, short summers, and evergreen trees   (657)

taxonomy: the theories and techniques of describing, grouping, and naming living things   (462)

telophase: the final stage in mitosis, meiosis I, and meiosis II characterized by two new nuclei forming at opposite ends of the cell; frequently followed by cytokinesis (cell division)   (224)

temperate deciduous forest: a biome characterized by enough precipitation to support large trees that shed their leaves in fall   (657)

temperate grassland: a biome similar to savanna; characterized by low precipitation and lack of trees, except along stream courses, such as the prairies of North America   (656)

10% rule: the rule stating that roughly 10% of the energy available at each trophic level is incorporated into the organisms at the next higher level   (638)

tertiary structure: the three-dimensional folded structure of a polypeptide or protein molecule   (39)

testes (singular: *testis*): the primary reproductive organs of a male; sperm-cell-producing organs   (329)

testosterone: a male sex hormone secreted by the testes (337)

theory: a well-tested hypothesis that organizes knowledge in a field, fits existing data, explains how events or processes are thought to occur, and successfully predicts future observations   (7)

threshold: the strength of stimulus below which no nerve impulse will be initiated   (563)

tissue: a group of cells with a common function and structure   (177)

tolerance: the ability to withstand or survive a particular environmental condition; in the immune system, the failure of an individual's T cells to react against HLA (621)

células T: tipo de linfocitos que se producen en la médula ósea y maduran en el timo; están involucradas en la respuesta inmunológica mediadas por células; ejemplos incluyen a las células T auxiliares, las células T asesinas y las células T supresoras   (606)

taiga: el bioma de coníferas o bosques nórdicos caracterizados por nevadas considerables, inviernos crudos, veranos cortos y árboles de hoja perenne (657)

taxonomía: las teorías y técnicas para describir, agrupar y nombrar seres vivos   (462)

telofase: la etapa final en la mitosis, la meiosis I y la meiosis II, caracterizada por la formación de dos nuevos núcleos en extremos opuestos de la célula; con frecuencia precede a la citocinesis (división celular) (224)

bosque templado caducifolio: un bioma caracterizado por precipitación suficiente para sostener grandes árboles que pierden sus hojas en el otoño   (657)

pastizal templado: un bioma similar a la sabana; caracterizado por escasas precipitaciones y carencia de árboles, excepto a lo largo de los cursos de las corrientes de agua; por ejemplo, las praderas en Norte América   (656)

regla del 10%: regla que establece que aproximadamente el 10% de la energía disponible en cada nivel trófico se incorpora a los organismos del siguiente nivel superior (638)

estructura terciaria: la estructura plegada tridimensional de un polipéptido o una proteína   (39)

testículos: los órganos reproductores primarios de un macho; órganos productores de espermatozoides (329)

testosterona: hormona sexual masculina secretada por los testículos   (337)

teoría: hipótesis comprobada que organiza el conocimiento en un campo, acopla datos existentes, explica cómo se cree que ocurren los eventos o los procesos y predice exitosamente observaciones futuras   (7)

umbral: la fuerza o estímulo debajo del cual no se iniciará ningún impulso nervioso   (563)

tejido: un grupo de células con una función y estructura común   (177)

tolerancia: la capacidad para soportar o sobrevivir una condición ambiental en particular; en el sistema inmunológico, el fracaso de las células T de un individuo de reaccionar contra los HLA   (621)

tracheid: a water-conducting and supportive element of xylem composed of long, thin cells with tapered ends and hardened walls   (188)

the tragedy of the commons: the decline and ruin of a common pool resource that results when everyone uses the resource in their own self-interest   (671)

transcription: the enzyme-catalyzed assembly of an RNA molecule complementary to a strand of DNA; the product may be messenger RNA, transfer RNA, or ribosomal RNA   (234)

transcription factor: any factor (usually a protein) that controls the process of transcription   (372)

transfer RNA (tRNA): a class of small RNA molecules with two functional sites—one for attachment of a specific amino acid, the other carrying the nucleotide triplet (anticodon) for that amino acid; each type of tRNA transfers a specific amino acid to a growing polypeptide chain   (234)

translation: the assembly of a protein on ribosomes, using messenger RNA to direct the order of amino acids   (234)

translocated: in genes, the relocation of a segment of DNA from one chromosomal location to another; in plants, the transport of soluble nutrients from one part of a plant to another   (408)

transpiration: the loss of water to the atmosphere by a plant through the stomates in its leaves   (92)

transport protein: a protein that plays a role in the active or passive movement of specific substances through cell membranes   (79)

transposable element: a mobile segment of DNA that serves as the agent of genetic change by randomly inserting itself into different chromosomal locations within the cell   (385)

transposon: a segment of DNA that randomly inserts itself along with other functional genes, such as those for antibiotic resistance, into different chromosomal locations within the cell   (386)

trial-and-error learning: a series of responses that are tested and eliminated in a problem situation until a solution is found   (580)

trinucleotide repeat expansion: a steady increase in the number of copies of a repeated DNA sequence of three nucleotide bases from generation to generation; responsible for genetic anticipation   (382)

trophic structure: the nutritional relationships among species of an ecosystem that determine the path of energy flow and pattern of chemical cycling   (637)

traqueida: un elemento de soporte del xilema que conduce agua, compuesto por células largas y delgadas con extremos afilados y paredes endurecidas   (188)

la tragedia de los bienes comunitarios: la declinación y ruina de un recurso de propiedad comunitaria que es el resultado de que todos usen el recurso para su propio beneficio   (671)

transcripción: el ensamble catalizado por enzimas de una molécula de RNA complementaria a una cadena de DNA; el producto puede ser el RNA mensajero, RNA de transferencia o RNA ribosómico   (234)

factor de transcripción: cualquier factor (usualmente una proteína) que controla el proceso de transcripción (372)

RNA de transferencia (tRNA): clase de moléculas de RNA pequeñas con dos sitios funcionales: uno de adhesión a un aminoácido específico y el otro que lleva el triplete nucleotido (anticodón) para ese aminoácido; cada tipo de tRNA transfiere un aminoácido específico a una cadena polipeptidídica en crecimiento   (234)

traducción: el ensamble de una proteína sobre los ribosomas, usando el RNA mensajero para dirigir el orden de los aminoácidos   (234)

translocalizado: en genes, la reubicación de un segmento de DNA de un sitio cromosómico a otro, en plantas, el transporte de nutrientes solubles de una parte de una planta a otra   (408)

transpiración: la pérdida de agua a la atmósfera que sufre una planta a través de los estomas de sus hojas   (92)

proteína de transporte: proteína que juega un papel muy importante en el movimiento activo o pasivo de sustancias específicas a través de las membranas celulares   (79)

elemento transponible: segmento móvil del DNA que sirve como el agente de cambio genético al insertarse al azar en diferentes sitios cromosómicos dentro de la célula   (385)

transposón: segmento del DNA que se inserta al azar junto con otros genes funcionales, tales como aquéllos para la resistencia antibiótica, en diferentes lugares cromosómicos dentro de la célula   (386)

aprendizaje de prueba y error: serie de respuestas que son probadas y eliminadas en una situación problemática hasta que se encuentra la solución   (580)

expansión de repeticiones del trinucleótido: aumento constante en el número de copias de una secuencia repetida de DNA de tres bases nucleótidas de generación en generación; responsable de la anticipación genética   (382)

estructura trófica: las relaciones nutricionales entre especies de un ecosistema que determinan la ruta del flujo de energía y el patrón del ciclaje químico   (637)

tropical rain forest: the most complex of all biomes located near the equator, where rainfall is abundant; harbors more species of plants and animals than any other biome; light is the major limiting factor   (653)

tropism: growth of a plant or part of a plant toward or away from light, gravity, or another environmental stimuli   (309)

trypsin: an enzyme in pancreatic juice that breaks down protein molecules   (72)

tumor suppressor: a gene that inhibits the growth of tumors; deletion or inactivation of such a gene can result in cancer   (228)

tundra: a biome at the northernmost limits of plant growth and at high altitudes where plant forms are limited to low, shrubby, or matlike vegetation   (658)

turgor: a cell's swelling against its cell wall caused by the pressure of the cell's contents   (83)

bosque pluvial tropical: el más complejo de todos los biomas situado cerca del ecuador, donde la lluvia es abundante, sustenta más especies de plantas y animales que cualquier otro bioma; la luz es el principal factor limitante   (653)

tropismo: el crecimiento de una planta o parte de una planta hacia la luz o en dirección contraria, gravedad o cualquier otro estímulo ambiental   (309)

tripsina: enzima en el jugo pancreático que descompone moléculas de proteínas   (72)

supresor de tumores: gene que inhibe el crecimiento de tumores; la eliminación o inactivación de tal gene puede resultar en cáncer   (228)

tundra: bioma en los límites del crecimiento de plantas y a altas altitudes donde las formas vegetales se limitan a vegetación baja de tipo mata   (658)

turgor: hinchazón celular en contra de su pared celular causada por la presión de los contenidos celulares (83)

uniformitarianism: the principle stating that the geological structure of Earth resulted from continuous and observable processes   (9)

ureter: a muscular tube that carries urine from the kidney to the urinary bladder   (94)

urethra: the tube through which urine travels from the bladder to the outside of the body   (94)

urinary bladder: an organ that stores urine before it is discharged from the body through the urethra   (94)

urinary system: a vertebrate organ system that regulates levels of water and dissolved substances in the body, excreting wastes as urine   (94)

urine: the solution of wastes excreted from the kidney   (94)

uterus: a hollow muscular organ, located in the female pelvis, in which a fetus develops   (333)

uniformitarianismo: el principio que establece que la estructura geológica de la Tierra resultó de procesos continuos y observables   (9)

uréter: conducto muscular que transporta orina del riñón a la vejiga urinaria   (94)

uretra: conducto a través del cual la orina viaja de la vejiga al exterior del cuerpo   (94)

vejiga urinaria: órgano que almacena orina antes de que sea descargada del cuerpo hacia la uretra   (94)

aparato urinario: un sistema de órganos de los vertebrados que regula los niveles de agua y sustancias disueltas en el cuerpo, excretando los desechos como orina (94)

orina: la disolución de desechos excretada desde el riñón (94)

útero: órgano muscular hueco, situado en la pelvis femenina, en el cual se desarrolla el feto   (333)

vaccination: process of administering a vaccine to prevent a specific disease   (607)

vaccine: a substance, often a suspension of a dead or weakened pathogen, that contains antigens and stimulates the production of antibodies without causing disease   (607)

vacuole: a membrane-enclosed structure in the cytoplasm of a cell   (169)

vacunación: proceso de administración de una vacuna para prevenir una enfermedad específica   (607)

vacuna: sustancia, a menudo una suspensión de un agente patógeno debilitado o muerto, que contiene antígenos y estimula la producción de anticuerpos sin causar enfermedad   (607)

vacuola: estructura encerrada en una membrana en el citoplasma de una célula   (169)

vagina: a tubular organ that leads from the uterus to the opening of the female reproductive tract (333)

variations: small differences among individuals within a population or species that provide the raw material for evolution (10)

vascular cambium: a lateral meristem that produces secondary growth (xylem on the inner surface, phloem on the outer), increasing the diameter of stems and roots (298)

vascular tissue: in plants, tissue specialized for the transport of food, water, and minerals; helps support the plant body (296)

vein: in animals, a blood vessel that carries blood toward the heart; in plants, a bundle of vascular tissue (194)

ventricle: a chamber of the heart that pumps blood out through an artery (194)

vertebrate: a chordate animal with a backbone; includes mammals, birds, reptiles, amphibians, and fishes (478)

vesicle: a small intracellular, membrane-enclosed sac that stores or transports substances (167)

vessel element: a short, wide cell in a plant; when arranged from end to end with other such cells, forms a tube that transports water and dissolved minerals (188)

villus (plural: *villi*): a fingerlike projection of the small intestine that increases surface area for absorption of digested food (72)

virus: a nonliving, infectious particle of nucleic acid, protein, and sometimes, lipid membrane that can replicate only inside a living cell (254)

voluntary: referring to a process or an activity that can be controlled consciously (551)

vagina: órgano tubular que va desde el útero hasta la abertura del tracto reproductor femenino (333)

variaciones: pequeñas diferencias entre individuos dentro de una población o especie que proveen el material en bruto para la evolución (10)

cámbium vascular: meristema lateral que produce crecimiento secundario (xilema en la superficie interna, floema en la externa) aumentando el diámetro de los tallos y raíces (298)

tejido vascular: en las plantas, tejido especializado para el transporte de comida, agua y minerales; ayuda a sostener el cuerpo de la planta (296)

vena: en los animales, un vaso sanguíneo que transporta sangre hacia el corazón; en las plantas, un manojo de tejido vascular (194)

ventrículo: cavidad del corazón que bombea sangre hacia una arteria (194)

vertebrado: animal cordado con un esqueleto; incluye mamíferos, aves reptiles, anfibios y peces (478)

vesícula: un saco encerrado en una membrana, pequeño e intracelular que almacena o transporta sustancias (167)

elemento o segmento de vaso: célula ancha y corta en una planta; cuando se ordena de extremo a otro extremo con otras células iguales, forma un conducto que transporta agua y minerales disueltos (188)

vellosidad: proyección similar a un dedo del intestino delgado, que aumenta el área superficial para la absorción de alimento digerido (72)

virus: partícula inerte de ácido nucleico infecciosa, proteína y a veces de membrana lipídica que puede replicarse sólo dentro de una célula viviente (254)

voluntario: se refiere a un proceso o una actividad que puede ser controlada conscientemente (551)

X-linked trait: a trait whose gene is carried on the X chromosome (the female sex chromosome, as compared to the male Y chromosome) (361)

xylem: the tube-shaped, nonliving tissue of the vascular system in plants; carries water and dissolved minerals from the roots to the rest of the plant (187)

rasgo ligado a X: un carácter cuyo gene es transportado en el cromosoma X (el cromosoma sexual femenino comparado con el cromosoma Y masculino) (361)

xilema: el tejido no viviente tubular del sistema vascular de las plantas; transporta agua y minerales disueltos desde las raíces hasta el resto de la planta (187)

yolk: mass of nutrients stored in an ovum (262)

yema: masa de nutrientes almacenada en el óvulo (262)

Z

Z-line: the line formed in muscle cells by a protein structure to which actin filaments are attached (198)

zooplankton: very small, floating or feebly swimming, heterotrophic aquatic organisms, including some protists and small animals (660)

zygote: the diploid product of the union of haploid gametes in conception; a fertilized egg (262)

línea Z: línea formada en las células musculares por una estructura proteínica a la cual se encuentran unidos los filamentos de actina (198)

zooplancton: organismos acuáticos heterótrofos muy pequeños que flotan o nadan débilmente, incluyendo a algunos protistas y pequeños animales (660)

cigoto: el producto diploide de la unión de gametos haploides en la concepción; un huevo fecundado (262)

INDEX

Note to the reader: Italicized page numbers refer to figures, tables, and illustrations.

◆ Photo & Illustration Credits ◆

COVER: Jennifer Waters Schuler/Photo Researchers
PROLOGUE: 0 James King-Holmes/Photo Researchers; **13** Getty Images

CHAPTER 1: 20 Darlyne A. Murawski/Getty Images; **22** Steve Allen/Getty Images; **24** (l)Hulton-Deutsch Collection/CORBIS, (r)Larry Mulvehill/Photo Researchers; **33** (l)Jim Steinberg/Photo Researchers, (r)Corel Corporations; **43** (l)A. Barrington Brown/Science Source/Photo Researchers, (c)Photo Researchers, (r)AP/Wide World Photos; **46** Driscoll, Youngquist & Baldeschwieler, Caltech/SPL/Photo Researchers.

CHAPTER 2: 50 Jack Dykinga/USDA; **52** (l)Alfred Pasieka/SPL/Photo Researchers, (r)Carolyn Noble; **53** NASA/ JPL/Cornell University; **54** SIU/Visuals Unlimited; **58** Joe McDonald/Visuals Unlimited; **59** (tl b)BSCS by Doug Sokell, (tr)Richard Hutchings/CORBIS; **68** Jeff Lepore/Photo Researchers; **73** G. Shih-R. Kessel/Visuals Unlimited.

CHAPTER 3: 76 Derek Berwin/Getty Images; **78** Ralph A. Slepecky/Visuals Unlimited; **81** (l c r)BSCS by Doug Sokell; **91** (t bl)Nada Pecnik/Visuals Unlimited, (br)J.H. Troughton and K.A. Card, Physics and Engineering Laboratory Dept. of Scientific and Industrial Research, Lower Hutt, New Zealand; **92** Joe McDonald/Tom Stack & Associates; **93** Stephen Frink/CORBIS.

CHAPTER 4: 100 Art Wolfe/Getty Images; **102** Wolfgang Kaehler/CORBIS; **104** Dr. Lewis K. Shumway, College of Eastern Utah; **105** BSCS by Carlye Calvin; **109** BSCS; **114** (t)M.I Walker/Photo Researchers, (b)Charles D. Winters/Photo Researchers; **120** BSCS by Doug Sokell; **125** (t)Simon Fraser/SPL/Photo Researchers, (b)Charles Fisher, National Undersea Research Program.

CHAPTER 5: 128 Neil McIntyre/Getty Images; **136** Phil Schermeister/CORBIS; **137** Don Fawcett/ Visuals Unlimited; **143** Warren Morgan/CORBIS; **147** (l)C.P. George/ Visuals Unlimited, (r)D. Caragnaro/Visuals Unlimited; **149** (t)Fred Hossler/Visuals Unlimited, (b)Courtesy of The Jackson Laboratory.

CHAPTER 6: 152 WG/SPL/Photo Researchers; **154** Roland Birke/Phototake; **156** (tl)David M. Phillips/Visuals Unlimited, (tr)Cabisco/Visuals Unlimited, (bl)Kevin and Betty Collins/Visuals Unlimited, (br)Bruce Iverson; **157** (tl)SIU/ Visuals Unlimited, (tr)CNRI/Science Photo Library/Photo Researchers, (bl)SIU/Visuals Unlimited, (br)NIBSC/ Science Photo Library/Photo Researchers; **159** Kevin Collins/Visuals Unlimited; **160** Dr. Tony Brian & David Parker/Science Photo Library/Photo Researchers; **162** (l c)David M. Phillips/Visuals Unlimited, (r)George J. Wilder/Visuals Unlimited; **166** (t)Biophoto Associates/Photo Researchers, (b)Don W. Fawcett/Visuals Unlimited; **167** (t)K.G. Murti/Visuals Unlimited, (b)R. Bolender-D.Fawcett/Visuals Unlimited; **168** David M. Phillips/Visuals Unlimited; **169** K.G. Murti/Visuals Unlimited; **170** David M. Phillips/Visuals Unlimited; **171** (l)Don Fawcett/Visuals Unlimited, (r)M. Abbey/Visuals Unlimited; **172** Fred Hossler/Visuals Unlimited; **173** Sherman Thomson/Visuals Unlimited; **174** John D. Cunnigham/Visuals Unlimited; **175** Stanley Flegler/Visuals Unlimited; **176** Bruce Iverson; **177** (t)John D. Cunnigham/Visuals Unlimited, (b)Omikron/Photo Researchers; **179** (l)C.Gerald Van Dyke/Visuals Unlimited, (r)David M. Phillips/Visuals Unlimited.

CHAPTER 7: 184 Vince Michaels/Getty Images; **188** Stanley Flegler/Visuals Unlimited; **194** (b)Professor P.M. Motta A. 'aggaiati & G. Macchiarelli/ SPS/Photo Reseachers; **197** (l)Robert Caughey/Visuals Unlimited, (r)Biophoto Associates/Photo Researchers; **199** Linda Hall Library, Kansas City,MO; **203** Cabisco/Visuals Unlimited; **204** (l)David M.

Phillips/Photo Researchers, (r)Bruce Iverson; **205** Fred Hossler/Visuals Unlimited; **207** Dr. Yorgos Nikas/SPL/Photo Researchers; **208** R&D Systems, Inc.

CHAPTER 8: 212 M. I. Walker/Photo Researchers; **214** David M. Phillips/Visuals Unlimited; **219** (l)G.F. Bahr, (r)American Society for Microbiology Journals Department; **220** Nature Magazine; **222** Science VU; **223** Dr. Andrew S. Bajer, Department of Biology, University of Oregon, Eugene; **229** Moredum Animal Health Ltd./SPL/Photo Researchers.

CHAPTER 9: 232 Andrew Syred/Photo Researchers; **238** (t)Baiyer River Sancturary, New Guinea/Tom McHugh/Photo Researchers, (b)Fred Hossler/Visuals Unlimited; **239** AP/Wide World Photos; **240** Michael Gabridge/Visuals Unlimited; **244** Courtesy of O.L. Miller, Jr/University of Virginia; **257** (t)Hans Gelderblom/ Visuals Unlimited, (b)CORBIS.

CHAPTER 10: 260 Paul Nicklen/Getty Images; **262** David M. Phillips/Visuals Unlimited; **263** Dr. Judith Venuti, Columbia University; **265** (l)Fred Hossler/Visuals Unlimited, (cl)Science VU, (cr)David M. Phillips/Visuals Unlimited, (r)Bruce Iverson; **271** F.R. Turner, Indiana University; **272** Carnegie Institution of Washington, Department of Embryology, Davis Division; **274** (tl)Carnegie Institution of Washington, Department of Embryology, Davis Division, (tr br)Lennart Nilsson, A Child is Born, Dell Publishing Co., (c bl)From Conception to Birth by Robert Hugh, er al, Harper and Row, Inc.; **275** Ellen B. Senisi/Photo Researchers; **276** (t)CNRI/SPL/Photo Researchers, (c)Getty Images, (b)Chris Carroll/CORBIS; **280** Design & Print, Roslin Institute, Roslin, Miidlothian; **283** (t)Ken Lucas/Visuals Unlimited; **283** E.R. Gavis, Princeton University; **286** Reproduced from The Journal of Cell Biology, 1995, Vol. 128, p. 575 by copyright permission of The Rockefeller University Press.

CHAPTER 11: 290 Premium Stock/CORBIS; **292** Ripon Microslides, Inc., Ripon, WI; **294** Adam Hart-Davis/Science Photo Library/Photo Researchers; **296** (l c)Bruce Iverson, (r)John D. Cunnigham/Visuals Unlimited; **297** (t)R.A. Gregory/Visuals Unlimited, (bl)Harry Rogers/Photo Researchers, (bc)John Kaprielian/Photo Researches, (br)Bruce Iverson; **298** Bruce Iverson; **299** Bob von Neorman, FUN Corp.; **302** BSCS by Doug Sokell; **303** Don Luvisi, University of California Cooperative Extension; **304** B.O. Phinney; **305** Dean Engler, Ph.D; **306** Bruce Iverson; **307** Courtesy Enrico Coen and Rosemary Carpenter; **308** BSCS by Carlye Calvin; **309** (t)David Newman/Visuals Unlimited, (b)Scott MacCleery and John Kiss; **311** Matheisl/Getty Images.

CHAPTER 12: 314 David Nardini/Getty Images; **316** George Grall/Getty Images; **318** (l)Bruce Iverson, (r)Biophoto Associates/Photo Researchers; **319** (l)The National Park Service, (r)James Richardson/Visuals Unlimited; **320** Rob & Ann Simpson/Visuals Unlimited; **325** Cabisco/Visuals Unlimited; **326** Milton Rand/Tom Stack & Associates; **328** D. Cavagnaro/Visuals Unlimited; **329** (l)Ralph A. Clevenger/CORBIS, (c)David Muench/ CORBIS, (r)Spencer Grant/Photo Researchers; **330** Gregory K. Scott/Photo Researchers; **331** (t)Robert Pickett/CORBIS, (b)David M. Phillips/The Population Council/Photo Researchers; **336** Ed Reschke; **337** Joe McDonald/Visuals Unlimited; **338** Hank Morgan/Science Source/Photo Researchers; **339** BSCS by Carlye Calvin.

CHAPTER 13: 342 Art Wolfe/Getty Images; **344** Getty Images; **345** Bettmann/CORBIS; **347** (t)M. Long/Visuals Unlimited, (b)Getty Images; **348** Stanley N. Cohen, Stanford University; **349** (tl)David C. Peakman, Reproductive Genetics Center, Denver, CO, (tr)Vysis Inc, Downers Grove, IL., (c)Dr. Ram S.

Verma, (b)Courtesy of Molecular Diagnostics, The Hospital for Sick Children, Toronto, Canada; **350** Courtesy Human Genome Sciences, Inc.; **351** BSCS by Doug Sokell; **356** John A. Moore, University of California (Riverside); **361** Robert Calentine/Visuals Unlimited; **362** (l)Robert Calentine/Visuals Unlimited, (r)Carolyn Trunca, Ph.D.; **364** (l)Margery W. Shaw, M.D., J.D., University of Texas Health Sciences Center at Houston, (r)March of Dimes Birth Defects Foundation.

CHAPTER 14: 368 Eastcott Momatiuk/Getty Images; **375** Annette Coleman, Brown University; **376** Courtesy UCSD Mitochondrial and Metabolic Disease Center; **381** Carolyn A. McKeone/Photo Researchers; **385** BSCS by Doug Sokell.

CHAPTER 15: 390 David Parker/Science Photo Library/Photo Researchers; **394** James King Holmes/Science Photo Library/ Photo Researchers; **395** Courtesy Human Genome Sciences, Inc.; **397** Francoise Sauze/Science Photo Library/Photo Researchers; **400** David Parker/Science Photo Library/Photo Researchers; **402** Dr. Gabriel Waksman, Washington University School of Medicine and The EMBO Journal; **405** Courtesy of David M. Iovannisci; **407** Susan Kuklin/Photo Researchers; **409** University of Utah Cytogenetics Labortory; **410** Geoff Tompkinson/Science Photo Library/Photo Researchers; **412** Hal Beral/Visuals Unlimited.

CHAPTER 16: 416 Mark Jones/Animals Animals; **424** (t)L.M. Cook, (b)Stanley Flegler/Visuals Unlimited; **425** Dr. James F. Crow, University of Wisconsin; **427** (tl)Maurice Walker/CORBIS, (tr)Buddy Mays/CORBIS, (b)Tom McHugh/ Photo Reasearchers; **428** Getty Images; **430** Courtesy The Jackson Laboratory; **431** Mark E. Gibson/Visuals Unlimited.

CHAPTER 17: 436 Gary Bell/Imagestate; **438** John Reader/ Science Photo Library/Photo Researchers; **440** (t)NASA, (b)Hubble Heritage Team/ AURA/STScI/NASA; **442** Corel; **444** CORBIS; **448** (t)NASA, (b)Dr. William Lillar/NASA's International Halley Watch; **449** (t)OAR/National Undersea Research Program (NURP), (b)The University of Wyoming Research Corporation; **453** (t)Sidney Fox/Visuals Unlimited, (bl)M. Wurtz/Biozentrum, University of Basel/Science Photo Library/Photo Researchers, (br)K.G. Murti/ Visuals Unlimited; **454** BPS/S.M. Awramik, University of California/ Tom Stack & Associates; **455** (tl)Tom & Therisa Stack & Associates, (tr)Dave Watts/Tom Stack & Associates, (b)J.G. Zeikus; **456** Dr. Gonzalo Vidal, Lund University; **457** (l)E. White/Visuals Unlimited, (r)Eric Grave/ Photo Researchers.

CHAPTER 18: 460 Jeff Hunter/Getty Images; **463** (tl)Getty Images, (tcl)Barbara Penoyar/Getty Images, (tcr tr)Getty Images, (bl)Lawrence Migdale/Photo Researchers, (br)USDA Natural Resources Conservation Service/Getty Images; **464** (t)Cheryl A. Ertelt/Visuals Unlimited, (c)Jeff Lepore/Photo Researchers, (bl)George Herben/Visuals Unlimited, (br)Dan Guravich/Photo Researchers; **465** (t)BSCS by Doug Sokell, (b)Kaylene Silverster, Walnut Creek, CA; **470** R.B. Forbes/Mammel Images Library of the American Society of Mammalogists; **471** (tl)BSCS by Doug Sokell, (tr)Tom & Therisa Stack & Associates, (bl)Barbara Gerlach/Tom Stack & Associates, (bcl)Tom & Pat Lesson/Photo Researchers, (bcr)Kennth H. Thomas/Photo Researchers, (br)John Gerlach/Visuals Unlimited; **473** Sidney Bahrt/Photo Researchers; **476** (l)Professor K.O. Stetter, (c)Ed Reschke, (r)J.G. Zeikus; **477** (tl cr b)BSCS by Doug Sokell, (tc)M. Abbey/Photo Researchers, (tr)John Shaw/Tom Stack & Associates, (cl)BSCS by Carlye Calvin; **478** (tl)Stuart Westmorland/CORBIS, (tr)Valorie Hodgson/Visuals Unlimited, (cl)Hans Beral/Visuals Unlimited, (cr)Getty Images, (bl)BSCS by Doug Sokell, (bc)John Shaw/Tom Stack & Associates, (br)BSCS by Richard Tolman.

CHAPTER 19: 498 James L. Amos/CORBIS; **500** Mark C. Burnett/Photo Researchers; **501** (l)Mark A. Schneider/Visuals Unlimited, (r)Layne Kennedy/CORBIS; **502** Getty Images; **503** O. Louis Mazzatenta/National Geographic Society Image Collection; **504** (l)Joe McDonald/Visuals Unlimited, (r)Denver Museum of Natural History; **507** Kwangshin Kim/Photo Researchers; **508** Dr. Howard Stutz, Brigham Young University; **509** (tl)Jeff Foott/Tom Stack & Associates, (tr)Sylvester Allred/ Visuals Unlimited, (c)Hal H. Harrison/Photo Researchers, (b)Rob Simpson/Visuals Unlimited; **510** John A. Moore; **511** John D. Cunnigham/Visuals Unlimited; **512** Lawrence Berkeley Laboratory; **513** Andrew J. Martinez/Photo Researchers; **514** (t)W.Faucett/Visuals Unlimited, (l c r) S. Brande 2000.

CHAPTER 20: 518 John Reader/Science Photo Library/Photo Researchers; **520** (l)Steven Kaufman/CORBIS, (c)Kjell B.Sandved/ Visuals Unlimited, (r)W. Perry Conway/Tom Stack & Associates; **521** (tl)Dave Watts/Tom Stack & Associates, (tc)Hanumantha Rao/Photo Researchers, (tr)Tom McHugh/Photo Researchers, (bl)P. Davey/Bruce Coleman, Inc., (br)Phillip James Corwin/ CORBIS; **522** Joe McDonald/ CORBIS; **526** BSCS by Doug Sokell; **528** D.W. Faucett/ Visuals Unlimited; **530** Ken Lucas/ Visuals Unlimited; **532** Robert Eckhardt, Professor of Developmental Genetics & Evolutionary Morphology(Dept. of Kinesiology, Penn State); **533** BSCS by Doug Sokell; **534** (l)Science VU, (r)John Reader/Photo Researchers; **535 536 537** BSCS by Doug Sokell; **539** Francis G. Mayer/CORBIS.

CHAPTER 21: 544 James Balog/Getty Images; **546** Mike Powell/Getty Images; **549** Hank Morgan/Science Source/Photo Researchers; **557** (l)Dr. John Zajicek/Science Photo Library/Photo Researchers, (r)C. Raines/ Visuals Unlimited; **562** (l)Richard Nowitz/Photo Researchers, (r)Roger Ressmeyer/CORBIS; **568** Peter N. Witt, M.D., Raleigh, NC; **573** (t)Andrew G. Wood/ Photo Researchers, (b)Daniel W. Gotshall/Visuals Unlimited.

CHAPTER 22: 576 Kevin Schafer/Getty Images; **578** J.H. Robinson/Photo Researchers; **579** (t)Renee Lynn/Photo Researchers, (bl)Gary W. Carter/Visuals Unlimited, (br)Science Photo Library/Photo Researchers; **582** Lee Boltin Picture Library; **584** (l)Getty Images, (c)M. Long/Visuals Unlimited, (r)Vanessa Vick/Photo Researchers; **587** Benelux Press B.V./Photo Researchers; **588** AP/Wide World Photos; **590** Getty Images; **592** Rob & Ann Simpson/Visuals Unlimited; ; **594** Barbara Gerlach/Visuals Unlimited; **595** Jeff Lepore/Photo Researchers; **596** Barbara Gerlach/Visuals Unlimited; **598** Susan Kuklin/Photo Researchers.

CHAPTER 23: 602 J. Berger/Photo Researchers; **605** Professors P.M. Motta, T. Fujita & M. Muto/Science Photo Library/Photo Researchers; **606** NIBSC/Science Photo Library/Photo Researchers; **607** CORBIS; **608** (t)Oliver Meckes/Photo Researchers, (b)Anthony Bannister, Gallo Images/CORBIS; **618** Allen Green/Photo Researchers; **619** Dr. Andrejs Liepins/ Science Photo Library/Photo Researchers; **625** Dr. Kari Lounatumaa/Science Photo Library/ Photo Researchers; **626** Sinclair Stammers/Science Photo Library/ Photo Researchers.

CHAPTER 24: 630 Terry Donnelly/Getty Images; **632** Wes Walker/Getty Images; **634** Glenn M. Oliver/Visuals Unlimited; **635** D. Cavagnaro/Visuals Unlimited; **636** J. Troughton and L. A. Donaldson, Physics and Engineering Laboratory, DSIR, New Zealand; **639** (l)Gil Lopez-Espina/Visuals Unlimited, (r) Harry Rogers/Photo Researchers; **639** Tony Hamblin/ CORBIS; **643** (l)Carl Purcell/Photo Researchers, (r)Jim Steinberg/Photo Researchers; **644** Scripps Institution of Oceanography/University of California, San Diego.